International Handbooks of Population

Volume 10

Series Editor
Dudley L. Poston, Jr.
Dept. Sociology
Texas A & M University
College Station, TX, USA

The *International Handbooks of Population* offer up-to-date scholarly summaries and sources of information on the major subject areas and issues of demography and population. Each handbook examines its particular subject area in depth, providing timely, accessible coverage of its full scale and scope, discusses substantive contributions for deeper understanding, and provides reliable guidance on the direction of future developments.

Volumes will explore topics of vital interest such as: Population Ageing, Mortality, Rural Demography, Poverty, Family Demography, Gender and Demography, Applied Demography, Migration, Demography of Sexuality, Race and Ethnic Demography. Each volume will provide a state-of-the-art treatment of its respective area. The series will quickly prove useful to a broad audience including demographers, practitioners and scholars across a range of disciplines.

More information about this series at http://www.springer.com/series/8111

Lori M. Hunter • Clark Gray
Jacques Véron
Editors

International Handbook of Population and Environment

Editors
Lori M. Hunter
Institute of Behavioral Science
University of Colorado Boulder
Boulder, CO, USA

Clark Gray
Department of Geography
University of North Carolina at Chapel Hill
Chapel Hill, NC, USA

Jacques Véron
Campus Condorcet
National Institute for Demographic Studies
Aubervilliers Cedex, France

ISSN 1877-9204   ISSN 2215-1877   (electronic)
International Handbooks of Population
ISBN 978-3-030-76432-6   ISBN 978-3-030-76433-3   (eBook)
https://doi.org/10.1007/978-3-030-76433-3

© Springer Nature Switzerland AG 2022
This work is subject to copyright. All rights are reserved by the Publisher, whether the whole or part of the material is concerned, specifically the rights of translation, reprinting, reuse of illustrations, recitation, broadcasting, reproduction on microfilms or in any other physical way, and transmission or information storage and retrieval, electronic adaptation, computer software, or by similar or dissimilar methodology now known or hereafter developed.
The use of general descriptive names, registered names, trademarks, service marks, etc. in this publication does not imply, even in the absence of a specific statement, that such names are exempt from the relevant protective laws and regulations and therefore free for general use.
The publisher, the authors, and the editors are safe to assume that the advice and information in this book are believed to be true and accurate at the date of publication. Neither the publisher nor the authors or the editors give a warranty, expressed or implied, with respect to the material contained herein or for any errors or omissions that may have been made. The publisher remains neutral with regard to jurisdictional claims in published maps and institutional affiliations.

This Springer imprint is published by the registered company Springer Nature Switzerland AG
The registered company address is: Gewerbestrasse 11, 6330 Cham, Switzerland

Acknowledgments

Development of the *Handbook on Population and Environment* was supported by center Grant #P2C HD06613 awarded to the University of Colorado Population Center (CUPC) by the Eunice Kennedy Shriver National Institute of Child Health and Human Development. The content is solely the responsibility of the authors and does not necessarily represent the official views of NIH or CUPC.

Chapter 6 was supported by center Grant #R24 HD047873 and training Grant #T32 HD07014 awarded to the Center for Demography and Ecology at the University of Wisconsin-Madison, and by funds to Curtis from the Wisconsin Agricultural Experimental Station and the Wisconsin Alumni Research Foundation. The authors would like to thank Caitlin Bourbeau at the Applied Population Laboratory at the University of Wisconsin-Madison for her invaluable contributions to our figures.

Chapter 8 benefitted from the excellent research assistance of Dylan Kirkeeng and was supported by the Eunice Kennedy Shriver National Institute of Child Health and Human Development grant R03HD098357.

Regarding Chap. 10, Elizabeth Fussell is grateful for the support of the Population Studies and Training Center (NICHD P2CHD041020) and the Institute at Brown on Environment and Society. Fussell's research is supported by NICHD grant R01HD093002. Brianna Castro thanks the National Science Foundation Graduate Research Fellowship program, the Weatherhead Center for International Affairs, and Harvard University for their generous support.

Chapter 11 benefited from support provided to the University of Colorado Population Center (CUPC, Project 2P2CHD066613-06) from the Eunice Kennedy Shriver Institute of Child Health Human and Human Development. We also acknowledge support from the National Science Foundation (Grants 1416960 and 1416860). Finally, we have also benefited from dialogue at the CUPC Conference on Climate Change, Migration and Health (NICHD project 5R13HD078101). The content is solely the responsibility of the authors and does not necessarily represent the official views of NIH, NSF, or CUPC.

Chapter 15 was supported by the Eunice Kennedy Shriver National Institute for Child and Human Development, National Institutes of Health (R03HD042003 VanLandingham PI; R21HD057609 VanLandingham PI; and P01HD082032 VanLandingham, Abramson, and Waters co-PIs). Helpful comments from Elizabeth Frankenberg and Duncan Thomas are acknowledged with gratitude.

Contents

1 Integrating the Environment into Population Research   1
Lori M. Hunter, Clark Gray, and Jacques Veron

Part I Theoretical Perspectives

**2 Population and Environment Interactions: Macro
Perspectives**   15
Jacques Véron

**3 A Micro Perspective: Elaborating Demographic
Contributions to the Livelihoods Framework**   37
Sara R. Curran

**4 Vulnerability to Climate Change and Adaptive Capacity
from a Demographic Perspective**   63
Raya Muttarak

Part II Data & Methods

5 Household-Scale Data and Analytical Approaches   89
Brian C. Thiede

6 Spatial Data and Analytical Approaches   111
Rachel A. Rosenfeld and Katherine J. Curtis

**7 Qualitative Data and Approaches to Population–Environment
Inquiry**   139
Sabine Henry, Sebastien Dujardin, Elisabeth Henriet,
and Sofia Costa Santos Baltazar

Part III Migration & Environment

**8 Building a Policy-Relevant Research Agenda
on Environmental Migration in Africa**   167
Valerie Mueller

viii

9 Water Stress and Migration in Asia 183
David J. Wrathall and Jamon Van Den Hoek

10 Environmentally Informed Migration in North America 205
Elizabeth Fussell and Brianna Castro

11 Environmental Migration in Latin America 225
Daniel H. Simon and Fernando Riosmena

Part IV Health and Mortality

12 Air Pollution, Health, and Mortality 243
Melissa LoPalo and Dean Spears

13 Population and Water Issues: Going Beyond Scarcity 263
Stéphanie Dos Santos, Bénédicte Gastineau, and Valérie Golaz

14 Heat, Mortality, and Health 283
Heather Randell

15 Land Use Change and Health 301
William K. Pan and Gabrielle Bonnet

16 Health and Mortality Consequences of Natural Disasters ... 331
Mark VanLandingham, Bonnie Bui, David Abramson, Sarah Friedman, and Rhae Cisneros

Part V The Influence of Demographic Dynamics on the Environment

17 Cities and Their Environments 349
Mark R. Montgomery Jessie Pinchoff, and Erica K. Chuang

18 Population and Agricultural Change 375
Richard E. Bilsborrow

19 Population and Energy Consumption/Carbon Emissions: What We Know, What We Should Focus on Next 421
Brantley Liddle and Gregory Casey

Part VI Other Arenas

20 Environment and Fertility 441
Sam Sellers

21 Gender, Population and the Environment 463
Jessica Marter-Kenyon, Sam Sellers, and Maia Call

22 Socio-demographic Inequalities in Environmental Exposures 485
James R. Elliott and Kevin T. Smiley

Part VII Conclusion & Reflections

23 Reflections on the Past, Present, and Future of Population-Environment Research 509
Barbara Entwisle

24 Environmental Migration Scholarship and Policy: Recent Progress, Future Challenges 515
Robert McLeman

Integrating the Environment into Population Research

1

Lori M. Hunter, Clark Gray, and Jacques Veron

For decades, demographers have danced around the issue of environmental correlates of population processes. In 1998, then PAA-President Anne Pebley reviewed this dance and put forward arguments as to why the avoidance of central engagement. In the end, she urged demographers to become more involved in research on environmental issues (Pebley, 1998). Today, demographers' engagement with environmental topics becomes all the more imperative as we move further into the Anthropocene – an era characterized by major human impacts on ecological systems.

This *Handbook* represents substantial progress by the demographic community in engaging with environmental dimensions of population processes. The intellectual community and research base related to population-environment inquiry has grown impressively in quantity, quality and methodological sophistication over the past two decades. As examples, recent theorizing has taken the community past early efforts to stretch classical theory to integrate environmental considerations (e.g. Hunter, 2005) and conceptual advancements have more centrally brought climate pressures, for instance, into frameworks of migration decision-making (e.g. Black et al. 2011a, b). Social science research on the association between human migration and environmental conditions and change has also grown rapidly (Chaps. 8, 9, 10 and 11), and demographers have made important contributions to understanding climate adaptation and climate futures (e.g. Lutz, 2017; Lutz & Muttarak, 2017) . Yet while empirical work linking population-environment has burgeoned, coverage has been uneven as related to particular population processes. Specifically, the environmental dimensions of fertility have received relatively sparse empirical coverage (for examples, see: Biddlecom et al., 2005; Shreffler & DoDoo, 2009), but these efforts are provocative and leave room for important progress related to reproductive health and decision-making (see Chap. 19). And while demographers have

L. M. Hunter (✉)
CU Population Center, Institute of Behavioral Science, Department of Sociology, University of Colorado Boulder, Boulder, CO, USA
e-mail: Lori.Hunter@colorado.edu

C. Gray
Department of Geography, University of North Carolina at Chapel Hill, Chapel Hill, NC, USA
e-mail: cgray@email.unc.edu

J. Veron
Institut national d'études démographiques (Ined), Campus Condorcet, Aubervilliers Cedex, France

© Springer Nature Switzerland AG 2022
L. M. Hunter et al. (eds.), *International Handbook of Population and Environment*, International Handbooks of Population 10, https://doi.org/10.1007/978-3-030-76433-3_1

left most of the research on environmental aspects of health to epidemiologists, more population scientists are engaging questions of environmental health (e.g. Currie, 2011; Currie et al., 2011; Chaps. 12, 13, 14, 15 and 16) as well as the social distribution of environmental risks (e.g. Crowder & Downey, 2010; Chap. 21).

One of the first demographic engagements with the environment was the ecological research in the 1950s by Hawley (1950), Schnore (1958), Duncan (1959) and others using the so-called POET ecological complex, that is, population, organization, environment and technology. This research was concerned with how populations adapt their fertility, mortality and especially their migration patterns to an ever-changing and mediating environment and technology. They defined the environment very broadly, perhaps too broadly, as including both a physical dimension pertaining to such features as climate, topography and natural resources, and an ecumenic or social dimension pertaining to the characteristics of other populations and organizations having a potential influence on the populations being studied (Hawley, 1986). Some have noted that one reason why ecological demography is no longer as prominent a research perspective as it was in the 1960s and 1970s was its overly broad conceptualization of the environment (Poston & Frisbie, 2019). The chapters in this *Handbook* have a narrower and perhaps more appropriate and engaging conceptualization of the environment.

Below, population-environment scholarship – termed "environmental demography" – is briefly reviewed with a focus on the past 20 years. Progress has been made through professional associations, workshops, online community dialogue, and the advancement of knowledge through increasing presence in high-impact publications. Mainstream demographic theory is critically engaged, with questions about the need to more centrally integrate environmental concerns particularly in this era of contemporary climate change. We also present the *Handbook's* chapters and briefly explore their primary arguments.

Motivated, But on the Margins

The Population-Environment Research Network (PERN) was launched in February 2001 with initial funding from the John D. and Catherine T. MacArthur foundation. Since 2003, the Network has been supported by a number of sponsors, primarily the IUSSP. Over 2000 international members strong, the network is entirely 'virtual' and offers updates on opportunities, an online bibliographic database, and regular "cyberseminars" engaging emerging topics in environmental demography. PERN has been highly influential in the creation, and maintenance of, an intellectual community focused on population-environment questions. The subdiscipline's evolution can also be retrospectively viewed by browsing through historical PAA programs with one session in 1998 (year of Pebley's address) entitled "Population growth and environmental change" to 2021's offering of eight core sessions with topics ranging from natural disasters to methods and measurement to environmental migration. An IGERT grant to UNC-Chapel Hill (awarded 2003) and occasional topical workshops at a variety of institutions, and a relaunch of the Springer journal *Population & Environment* (2004) with a renewed focus, have all helped bring like-minded scholars together through the years. Below, the resulting progress in research on the environmental dimensions of core demographic processes is briefly reviewed as background to the *Handbook's* 20 in-depth chapters.

Environmental Dimensions of Migration

"Climate Refugees!" The alarmist call rattled the public and policymakers and rallied the research community. The specter of climate change's potential to intensify population movements brought an unfamiliar public and policy spotlight to migration research. The research community responded.

Recent migration-environment publications include conceptual overviews and extensions (e.g. Borderon et al., 2019; Hoffman et al., 2020; Hunter et al., 2015; Piguet et al., 2018), methodological summaries (e.g., Bilsborrow & Henry, 2012; Fussell et al., 2014), regional and national case studies (e.g., Gray & Mueller, 2012a, b; Nawrotzki et al., 2013; Grace et al., 2018), as well as policy reviews (e.g., Warner, 2010; Thomas & Benjamin, 2018. In addition, several books, chapters, and edited collections (e.g. Laczko & Aghazarm, 2009; McLeman, 2013; Piguet et al., 2011; McLeman & Gemenne, 2018) and journal special issues have been devoted to the topic (e.g. Adamo & Izazola, 2010; Black et al., 2011b).

Research on the environmental dimensions of migration has thus far yielded several central findings (Chaps. 8, 9, 10 and 11). Environmental 'push' factors include both short-term and acute environmental disasters (e.g. hurricanes) and long-term, chronic environmental stressors (e.g. drought). In addition, migration is often used as a risk management strategy with households sending migrants in response to environmental shocks (e.g., Dillon et al., 2011; Henry et al., 2004). Finally, environmental pressures do not act in isolation but, instead, interact with economic, social and political factors to ultimately shape migration decision-making. Social capital in the form of migration networks has proven to be particularly key in impacting the likelihood of migration in the face of environmental stress (e.g., Doevenspeck, 2011; Morrissey, 2014; Riosmena et al., 2018).

A central challenge for demographers is whether migration-environment scholarship has sufficiently advanced such that migration theory should more explicitly integrate environmental considerations. A review of classic migration theory by Hunter (2005) identifies areas where environmental factors have been indirectly incorporated into classic theory although not taking center stage. Speare's residential mobility framework (1974), for example, notes residential satisfaction as a factor with physical amenities or disamenities representing "locational characteristics" shaping satisfaction. Environmental hazards feature in Wolpert's stress-threshold model (1966).

More recent migration theories such as the New Economics of Labor Migration (NELM) hold potential for explicit integration of environmental factors as households engage migration as a strategy to diversify livelihoods and minimize risk. Migration – as an adaptive strategy in the face of strain – can be seen as an "ex ante" behavior designed to reduce future risk through, for example remittances fueling investments in agriculture. Alternatively, migration may represent an "ex post" behavior as response to an environmental shock such as prolonged drought. In rural Mexico, for example, rainfall shortages are associated with an increase in the probability of a rural household sending a migrant to the US, but only from regions with historically strong international migration networks (e.g. Hunter et al., 2013; Riosmena et al., 2018). The framework by Black and colleagues (2011a, b) nicely integrates short- and long-term environmental pressures as they interact with macro forces (e.g. political, economic, social), which, in turn, interact with personal and household characteristics to shape migration decision-making. More tightly coupling this migration framework with contemporary theory would be a particularly useful endeavor and we hope that the reviews provided here may be useful toward that end.

Environmental Dimensions of Fertility

First to mind when considering frameworks of fertility decision-making are likely Caldwell's (1978) classic "wealth flow" arguments and Bongaart's (1978) framework of "proximate determinants." As in the case of migration, neither explicitly includes environmental factors although recent scholarship has suggested such factors may have relevance.

Caldwell's "wealth flows" argues that fertility decision-making is shaped by broader political economic factors that determine the relative cost-benefit of bearing children. Per Caldwell (1978), fertility tends to be relatively higher in times where children bring household gain through, for example, labor for natural resource collection.

Yet as the flow of wealth reverses – and the cost of raising children increases vis-à-vis their household benefit – fertility rates lower.

An environmentally-related shift in wealth flows has been identified in rural Kenya where the total fertility rate (TFR) underwent a dramatic decline. In Kenyan's Central Province, a leader in the decline, the 1978 TFR of 8.4 had decreased to 3.7 by 1998 – a 66% reduction in only 20 years (Shreffler & Dodoo, 2009). Using retrospective qualitative data, Shreffler and DoDoo (2009) illustrate an environmental dimension to this rapid decline. Specifically, land inheritance is a cultural norm in rural Kenya and scarcity combined with diminishing farm size played a role in preferences for smaller family size. Interestingly, households invested in education of their children in lieu of land inheritance – as a substitute investment in children in the face of environmental scarcity.

An earlier effort to link fertility and environmental conditions was undertaken in the Chitwan Valley of Nepal (Biddlecom et al., 2005). Biddlecom and colleagues explore the "vicious circle" argument that degradation of the local environment may increase the demand for children in rural, subsistence settings. As wild resources – used for fuel, food, and fodder – become increasingly scarce, children's labor is also increasingly valued, potentially resulting in higher desired family size. Key here is the low status of women and children and, therefore, the low value placed on their resource collection time. By linking household-level survey data with flora count data from 127 plots, Biddlecom et al. (2005) find that households in areas with low levels of available natural resources, and that rely on public lands, both men and women desire larger family sizes.

In 1971, Richard Easterlin published in the *American Economic Review* an essay entitled "Does human fertility adjust to the environment?" He begins by engaging Malthus and Ehrlich who would argue "... man, following his natural instincts, will breed without restraint and population will grow until environment limits force a halt through higher mortality" (Easterlin, 1971:399). With a focus on the U.S., Easterlin contrasts fertility trends in frontier, new urban and old urban places from 1800–1960. Here, the 'environment' represents, in part, the social environment as Easterlin argues tastes and opportunities differ across settings and preferences for children will, therefore, be different. In urban areas, with more educational opportunity, greater consumption options, and more knowledge and access to fertility limitation, Easterlin argues fertility will be lower. In rural regions, particularly areas with abundant agricultural opportunities, the value of children will be higher and fertility, too. Evidence is presented through trends in number of children under 5 for white women in urban and rural regions by geographic region and Easterlin concludes that the dramatic declines in US fertility during this period, combined with distinctions across urban-rural regions, demonstrates "the immense potential for adjustment." (1971:407)

The demographic research community largely did not take up Easterlin's provocative environmentally-oriented refutation to "the analogy commonly drawn between human population growth and that of fruit flies in a jar." (Easterlin, 1971:400) Nor has there been much research bringing an explicit environmental dimension to Bongaart's influential "proximate determinants" framework which argues that fertility outcomes are shaped by indirect factors such as exposure, fertility control and natural marital factors such as intercourse frequency and lactational infecundability. While these determinants are categorized as 'socioeconomic, cultural and environmental variables' (emphasis added), the environment is again particularly referencing the social context mainly as it shapes health and nutrition. Recently, however, there have been important summaries put forward (e.g. Grace, 2017), and a growing number of scholars have begun to address these issues (e.g., Sellers, 2017; Eissler et al., 2019; Sellers & Gray, 2019). These exciting advancements in demographic research are reviewed in detail in Chap. 19.

Environmental Dimensions of Population Health and Mortality

There is a more prominent presence of environmental factors in research on population health, likely due to engagement by public health researchers and social epidemiologists, fields for whom context has long been central. Even so, 'context' in this body of work has traditionally included primarily socio-structural factors such as social class, gender, race, networks and capital (Honjo, 2004).

Specifying associations between environmental conditions and individual and population health is challenging due to variation – and unknowns – with regard to exposure, toxicity, and length of incubation. Even so, intriguing efforts include relatively recent work using the innovative community-based University of Washington Twin Registry, with which researchers have found greater access to green space associated with less depression, although less evidence was apparent for effects on stress or anxiety (Cohen-Cline et al., 2015). Another innovative approach, based in Great Britain, adds nuance to "greenspace" with small area measures of land cover type (woodland, grassland), bird species richness, water quality and protected or designated status. Good health prevalence was higher for individuals living nearby nearly all greenspace indicators (Wheeler et al., 2015).

Clearly, negative health can also have environmental components. For example, spatial analyses of urban heat islands reveal local variation in heat stress such that census tracts with concentrations of racial/ethnic minorities and low income residents are exposed to higher land surface temperatures (Mitchell & Chakraborty, 2014). The disastrous 2004 Indonesian tsunami revealed important patterns of vulnerability, as children, females and older adults were least to survive (Frankenberg et al., 2011, 2020). Research along the U.S.-Mexico border finds negative mental health outcomes common among colonia residents facing numerous social and environmental challenges (Marquez-Velarde et al., 2015), while the Indonesian tsunami also resulted in substantial post-traumatic stress reactivity among residents of heavily damaged areas (Frankenberg et al. 2008). Recent research by population scientists also reveals widespread and persistent costs to child and adult health from population exposures to high temperatures and air pollution (Barreca et al., 2016; Bakhtsiyarava et al., 2018; Spears et al., 2019), emphasizing the importance of reducing fossil fuel use and agricultural burning as well as other forms of climate change mitigation.

This *Handbook* offers several chapters summarizing specific topical areas linking population, health, and environment. These include links between health and mortality to air quality, water stress, extreme temperatures, land use change, and natural disasters (Chaps. 12, 13, 14, 15 and 16).

Time for Better Integration?

As environmental demography continues to develop, specific topical areas – namely migration-environment – may be reaching a tipping point with regard to being able to argue for the necessity of inclusion of environmental factors in migration research, particularly in regions characterized by natural resource dependence. If environmental factors aren't included in these empirical models, do we risk omitted variable bias?

Quantitative demographers know well the consequences of a misspecified model. The danger: what if environmental variables are correlated with some of our more regularly used predictors (e.g. income) when examining outcomes such as fertility, migration or health. Research is emerging to suggest that, in many cases, this association may in fact ring true. Consider fertility. A vast amount of research rightly includes income as fertility correlate. Yet we know that poverty is associated with local environments through land use decisions in the Brazilian Amazon (Guedes et al., 2014). We also know that poor and racial minority households are more likely to live in polluted environments in the U.S. setting (e.g. Pais et al., 2013; Zwickl, 2019). As a result, within particular settings, should measures of the local environmental more regularly be included in fertility models?

In addition to misspecification, important policy levers are potentially missed when environmental factors are ignored. Take the AIDS pandemic in sub-Saharan Africa as an example. Over the past couple decades, a strong body of research has emerged of the determinants of the pandemic's spread – both at the level of behavioral decision-making and ecological patterns and trends. Yet an innovative qualitative study undertaken on the Kenyan shores of Lake Victoria reveals underlying environmental determinants of the devastating pandemic – determinants neglected within quantitative models.

Along Lake Victoria's Kenyan shores, AIDS prevalence among the dominant ethnic group is 23% for women and 17% for men and research reveals that the pandemic has been fueled by "sex-for-fish" relationships exacerbated by localized industrial lake pollution and resulting migration of both fish and fisherman (Mojola, 2011). Mojola argues that HIV interventions focused on individual behaviors have, in part, been ineffective due to gender inequalities, poverty and the culture of risk-taking among fishermen. Interventions grounded in a more holistic view of the eco-social system might aim to stabilize fish ecosystems through environmental regulations, Mojola argues, which might reduce extended migration and the potential for additional sexual partners.

The theoretical frameworks guiding demographic research have for years sidelined – or entirely ignored – environmental factors shaping fertility, migration and health patterns and trends. Yet contemporary environmental demographic research has provided examples in each domain of environmental correlates and thus raise the question of the need for more central integration of environmental variables within demographic theory and scholarship. Perhaps it's time more regularly put "people into place" as argued by Barbara Entwisle in her 2007 PAA address (Entwisle, 2007). Although Entwisle's address focused on the importance of neighborhood context, her message certainly resonates with Pebley's call nearly a decade prior.

The sociology of science tells us that the pursuit of science is influenced by societal values and interests (Merton, 1973). Climate change is increasingly of concern among publics and policymakers and, in this way, the socio-political landscape within which we do our research is changing. New emphasis on understanding the social, economic and political dimensions of climate change calls for new ways of doing demography.

In addition, the natural landscapes within which we explore fertility, mortality, health and migration are changing – climate change continues to alter local environments, as well as increasing and intensifying natural disasters, and the demographic research community should be on alert. As examples, shifting temperatures and rainfall will alter the viability of natural-resource based livelihoods, including agriculture, while natural disasters and sea level rise will shape patterns of both urban migration and population health.

Science is also social product, one that is influenced by, and has influence on, social structure, including established norms (Merton 1973). As with all scientific inquiry, population research has been guided by what has been defined as normal science within the discipline. Yet as postulated by Kuhn (1962), science shifts when anomalies emerge that push paradigmatic limits. Might climate change ultimate demand a shift in demography's "normal science"? (Hunter and Menken 2015). Specifically, isit time for demography as a science, guided by institutional histories, structures and norms, to evolve through more central integration of the natural environment?

Given that theoretical and empirical momentum has been built along these lines and our environment(s) are certainly changing – demographic researchers can, and should, broadly respond. It is our hope that the chapters provided in this *Handbook* help move the field in this direction. To this end, the *Handbook* begins with an overview of theoretical perspectives with a macro lens offered by Jacques Veron where he engages classical arguments by Malthus and more contemporary formulations such as "IPAT" (Chap. 2). In Chap. 3, Sara Curran moves to a micro perspective with emphasis on the utility of a "livelihoods framework" for population-environment inquiry. Emphasizing demographic perspectives as related to climate change, Raya Muttark explores adaptive capacity as related to population processes in Chap. 4.

To inform empirical investigation of population-environment connections, three chapters are then presented that detail particularly data sources analytical lenses including household scale approaches as discussed by Brian Thiede. He reviews examples of census, survey, surveillance, and other sources of demographic data, as well as exploring challenging analytical issues such as determining causality. In Chap. 7, Sabine Henry and her colleagues explore the potential of qualitative investigation within population-environment scholarship. While less often used, the incredible potential offered by qualitative investigation is illustrated by several examples from their own work engaging diverse methods such as interviews, participatory mapping, participant observation and observant participation, photo-language and even the use of a game. Following, in Chap. 6, Rachel Rosenfeld and Katherine Curtis offer an essential overview of spatial methodologies including general definitions as well as foundational theoretical and methodological concepts. They use compelling spatial work on migration, fertility, and mortality to argue for more regular incorporation of spatial perspectives within population-environment inquiry.

Moving to specific population processes, the *Handbook* presents several chapters on migration and the environment as a consequence of deeper investigation into this specific demographic dynamic over the years. To organize coverage, research is clustered geographically with Chap. 8, written by Valerie Mueller, offering an overview of migration-environment connections across the African continent. Mueller's synthesis explores migration as an adaptive strategy while also covering important barriers to movement. With an emphasis on sea level rise and other relevant environmental pressures across Asia, in Chap. 9, David Wrathall and Jamon Van Den Hoek interrogate population-environment interactions

in Asia with an emphasis on water stress. Through the review, they identify geographical gaps where water stress is likely to occur in the coming century but the migration-water connection has not yet been thoroughly examined. Next, Elizabeth Fussell and Brianna Castro bring us to North America in Chap. 10 with thorough coverage of three types of migration in North America that has been associated with environmental factors. These include amenity migration, natural hazards-related migration, and anthropogenic hazards-related migration. They emphasize the importance of understanding the vulnerability of both people and places within the context of climate change. And finally, Daniel Simon and Fernando Riosmena build on their own scholarship on migration-climate in Mexico to provide a summary of the literature more broadly encompassing Latin American settings. Their overview engages research on both internal and international migration and they conclude by reflecting on knowledge gaps that must be addressed to better inform migration policy.

Several dimensions of health and mortality as related to environmental context comprise the *Handbook's* next section. Here, Melissa LoPalo and Dean Spears examine air quality and health building on a recent *Lancet* Commission that reported 6.5 million people worldwide die each year from exposure to air pollution. They emphasize global inequality as related to this health-environment connection and argue that, in some contexts, enough evidence has emerged to inform policy. Even so, they claim that in other settings, far more research is needed. In Chap. 13, Stephanie dos Santos and colleagues explore population dynamics and water with an emphasis on complexities related to governance and inequities in access. With a focus on temperature, in Chap. 14 Heather Randell brings a demographic perspective to understanding the health impacts of heat stress in settings across the globe. She explores both direct and indirect effects and emphasizes this connection for particularly vulnerable groups. Land use change is the topic of Chap. 15, written by William Pan and Gabrielle Bonnet. They structure the overview with a conceptual framework that identifies typologies of land use/land cover change and population dynamics connected with human health. Using a wide array of research examples, they link these typologies to non-zoonotic communicable diseases, non-communicable diseases, and accidents and injuries. In the end, they call for increased inter- and trans-disciplinary collaboration to improve understanding of complex environmental health challenges. The section on health, mortality, and the environment concludes with Chap. 16 which emphasizes. Written by Mark VanLandingham and colleagues, the chapter offers a summary and critical evaluation of recent theoretical and empirical research focusing on the health and mortality consequences of disasters. They offer the insightful recommendation that research should shift from the current primary emphasis on health and mortality in the immediate aftermath of disasters, to greater emphasis on the medium- and long-term effects which would allow for better understanding of trajectories of recovery.

Of course, the association between population dynamics and environmental context is reciprocal and two chapters explore specific topics reflecting this complexity. In Chap. 17, Mark Montgomery and colleagues provide a synthesis of the literature on cities and their environments. They review empirical regularities in urbanization while also examining health threats as they manifest in urban spaces. Such threats include heat, drought, natural disasters, and flooding. But specific to reciprocality, they also review several environmental challenges resultant of urbanization such as intensified risks related to wildfire vector-borne disease. As another important example of the reciprocal association between population and the nvironment, Richard Bilsborrow offers an overview in Chap. 18 of population dynamics and agriculture land use.

There are several additional topics within population-environment that deserved coverage but did not explicitly align with the broader sections above. These are presented in the three final topical chapters including Sam Sellers' overview of the fertility-environment connection. In Chap. 20, Sellers examines environmental influences on a variety of fertility-related processes including union formation, reproductive health, and birth outcomes. In Chap. 21, Jessica Marter-Kenyon and colleagues more centrally review gender as it mediates a variety of population-environment interactions including perception of environmental change and risk; human impacts on the environment; adaptation to environmental change; and well-being outcomes following environmental stress. They also highlight key methodological considerations to aid the development of future research. And finally, in Chap. 22, James Elliott and Kevin Smiley review the wide variety of socio-demographic inequalities in environmental exposures. A running theme throughout their chapter is how the twin forces of urbanization and industrialization continue to change the natural environment while also joining ongoing forces of racial oppression and economic exploitation to yield unequal exposures the world over.

In addition to the 20 topical chapters, the Handbook presents two reflections on emergence and future of population-environment scholarship from senior researchers in the field. In Chap. 23, Barbara Entwisle argues that the past 25 years has seen a tremendous growth in population-environment inquiry, although important work yet remains. She outlines four such areas including increased use of comparative perspectives and more systematically linking the literatures on climate change and pollution and other hazards. Robert McLeman's reflection emphasizes migration-environment research and, in the concluding chapter, he contends that environmental migration processes are far more complex than previously thought and attention to labor flows, thresholds as related to environmental stress, and issues around gender and policy all require additional scholarly emphasis.

Moving Forward

In all, it is our hope that this first *Handbook of Population and Environment* offers a useful summary of the state of the field. But beyond a summary, we hope that these chapters provide a foundation for further innovation, creativity, and expansion in environmental demography. As the world grapples with a changing climate, the population perspective becomes all the more imperative. We trust that our colleagues in settings across the globe will rise to the occasion and, perhaps by doing so, the environmental demographic perspective will increasingly become mainstreamed within population scholarship.

Acknowledgements We owe a great debt of gratitude to Dr. Dudley Poston for coverage in this chapter's introductory section of ecological research in the 1950s. In addition, this work has benefited from support provided by the University of Colorado Population Center (Project 2P2CHD066613-06), funded by the Eunice Kennedy Shriver National Institute of Child Health and Human Development. The content is solely the responsibility of the author and does not necessarily represent the official views of the CUPC, NIH or CU Boulder.

References

Adamo, S. B., & Izazola, H. (2010). Human migration and the environment. *Population & Environment, 32*(2), 105–108.

Bakhtsiyarava, M., Grace, K., & Nawrotzki, R. J. (2018). Climate, birth weight, and agricultural livelihoods in Kenya and Mali. *American Journal of Public Health*, e1–e7. https://doi.org/10.2105/AJPH.2017.304128

Barreca, A., Clay, K., Deschenes, O., et al. (2016). Adapting to climate change: The remarkable decline in the US temperature-mortality relationship over the twentieth century. *Journal of Political Economy, 124*, 105–159. https://doi.org/10.1086/684582

Biddlecom, A. E., Axinn, W. G., & Barber, J. S. (2005). Environmental effects on family size preferences and subsequent reproductive behavior in Nepal. *Population and Environment, 26*(3), 583–621.

Bilsborrow, R. E., & Henry, S. J. (2012). The use of survey data to study migration–environment relationships in developing countries: Alternative approaches to data collection. *Population and Environment, 34*(1), 113–141.

Black, R., Adger, W. N., Arnell, N. W., Dercon, S., Geddes, A., & Thomas, D. (2011a). The effect of environmental change on human migration. *Global Environmental Change, 21*, S3–S11.

Black, R., Arnell, N., & Dercon, S. (2011b). Migration and global environmental change - Review of drivers of migration. *Global Environmenal Change, 21* (Supplement 1, December).

Bongaarts, J. (1978). A framework for analyzing the proximate determinants of fertility. *Population and Development Review*, 105–132.

Borderon, M., Sakdapolrak, P., Muttarak, R., Kebede, E., Pagogna, R., & Sporer, E. (2019). Migration influenced by environmental change in Africa. *Demographic Research, 41*, 491–544.

Caldwell, J. C. (1978). A theory of fertility: From high plateau to destabilization. *Population and Development Review*, 553–577.

Cohen-Cline, H., Turkheimer, E., & Duncan, G. E. (2015). Access to green space, physical activity and mental health: A twin study. *Journal of Epidemiology and Community Health*, jech-2014.

Crowder, K., & Downey, L. (2010). Inter-neighborhood migration, race, and environmental hazards: Modeling micro-level processes of environmental inequality. *American Journal of Sociology, 115*, 4,1110–4,1149.

Currie, J. (2011). Traffic Congestion and Infant Health: Evidence from E-ZPass. *American Economic Journals-Applied, 3*(1), 65–90.

Currie, J., Heep, S., & Neidell, M. (2011). Quasi- experimental approaches to evaluating the impact of air pollution on children's health. *Health Affairs, 30*(12), 2391–2399.

Dillon, A., Mueller, V., & Salau, S. (2011). Migratory responses to agricultural risk in northern Nigeria. *American Journal of Agricultural Economncs, 93*(4), 1048–1061.

Doevenspeck, M. (2011). The thin line between choice and flight: Environment and migration in rural Benin. *International Migration, 49*(s1), e50–e68.

Duncan, O. D. (1959). Human ecology and population studies. In P. M. Hauser & O. D. Duncan (Eds.), *The study of population* (pp. 678–716). University of Chicago Press.

Easterlin, R. A. (1971). Does human fertility adjust to the environment? *The American Economic Review*, 399–407.

Eissler, S., Thiede, B. C., & Strube, J. (2019). Climatic variability and changing reproductive goals in Sub-Saharan Africa. *Global Environmental Change, 57*, 101912.

Entwisle, B. (2007). Putting people into place. *Demography, 44*(4), 687–703.

Frankenberg, E., Gillespie, T., Preston, S., Sikoki, B., & Thomas, D. (2011). Mortality, the family and the Indian Ocean tsunami. *The Economic Journal, 121*(554), F162–F182.

Frankenberg, E., Friedman, J., Gillespie, T., Ingwersen, N., Pynoos, R., Rifai, I. U., et al. (2008). Mental health in Sumatra after the tsunami. *American Journal of Public Health, 98*(9), 1671.

Frankenberg, E., Sumantri, C., & Thomas, D. (2020). Effects of a natural disaster on mortality risks over the longer term. *Nature Sustainability, 3*(8), 614–619.

Fussell, E., Hunter, L. M., & Gray, C. L. (2014). Measuring the environmental dimensions of human migration: The demographer's toolkit. *Global Environmental Change, 28*, 182–191.

Grace, K. (2017). Considering climate in studies of fertility and reproductive health in poor countries. *Nature Climate Change, 7*(7), 479–485.

Grace, K., Hertrich, V., Singare, D., & Husak, G. (2018). Examining rural Sahelian out-migration in the context of climate change: An analysis of the linkages between rainfall and out-migration in two Malian villages from 1981 to 2009. *World Development, 109,* 187–196.

Gray, C. L., & Mueller, V. (2012a). Natural disasters and population mobility in Bangladesh. *Proceedings of the National Academy of Sciences, 109*(16), 6000–6005.

Gray, C., & Mueller, V. (2012b). Drought and population mobility in rural Ethiopia. *World Development, 40*(1), 134–145.

Guedes, G. R., VanWey, L. K., Hull, J. R., Antigo, M., & Barbieri, A. F. (2014). Poverty dynamics, ecological endowments, and land use among smallholders in the Brazilian Amazon. *Social Science Research, 43,* 74–91.

Hawley, A. H. (1950). *Human ecology: A theory of community structure.* Ronald Press.

Hawley, A. H. (1986). *Human ecology: A theoretical essay.* University of Chicago Press.

Henry, S., Schoumaker, B., & Beauchemin, C. (2004). The impact of rainfall on the first out-migration: A multi-level event-history analysis in Burkina Faso. *Population and Environment, 25*(5), 423–460.

Hoffmann, R., Dimitrova, A., Muttarak, R., Cuaresma, J. C., & Peisker, J. (2020). A meta-analysis of country-level studies on environmental change and migration. *Nature Climate Change, 10*(10), 904–912.

Honjo, K. (2004). Social epidemiology: Definition, history, and research examples. *Environmental Health and Preventive Medicine, 9*(5), 193–199.

Hunter, L. M. (2005). Migration and environmental hazards. *Population and Environment, 26*(4), 273–302.

Hunter, L. M., Luna, J., & Norton, R. (2015). The environmental dimensions of human migration. *Annual Review of Sociology.* In press.

Hunter, L. M., & Menken, J. (2015). Will climate change shift demography's' normal science'? *Vienna Yearbook of Population Research, 13,* 23–28.

Hunter, L. M., Murray, S., & Riosmena, F. (2013). Rainfall patterns and US migration from rural Mexico. *International Migration Review, 47*(4), 874–909.

Kuhn, T. (1962). *The structure of scientific revolutions.* The University of Chicago Press.

Laczko, F., & Aghazarm, C. (Eds.). (2009). *Migration, environment and climate change: Assessing the evidence.* International Organization for Migration.

Lutz, W. (2017). How population growth relates to climate change. *Proceedings of the National Academy of Sciences, 114*(46), 12103–12105.

Lutz, W., & Muttarak, R. (2017). Forecasting societies' adaptive capacities through a demographic metabolism model. *Nature Climate Change, 7*(3), 177–184.

Marquez-Velarde, G., Grineski, S., & Staudt, K. (2015). Mental health disparities among low-income US Hispanic residents of a US-Mexico border Colonia. *Journal of Racial and Ethnic Health Disparities,* 1–12.

Merton, R. K. (1973). *The sociology of science: Theoretical and empirical investigations.* University of Chicago Press.

McLeman, R. A. (2013). *Climate and human migration: Past experiences, future challenges.* Cambridge University Press.

McLeman, R., & Gemenne, F. (Eds.). (2018). *Routledge handbook of environmental displacement and migration.* Routledge.

Mitchell, B. C., & Chakraborty, J. (2014). Urban heat and climate justice: A landscape of thermal inequity in Pinellas County, Florida. *Geographical Review, 104,* 459–480. https://doi.org/10.1111/j.1931-0846.2014.12039.x

Mojola, S. A. (2011). Fishing in dangerous waters: Ecology, gender and economy in HIV risk. *Social Science & Medicine, 72*(2), 149–156.

Morrissey, J. (2014). Environmental change and human migration in Sub-Saharan Africa. In *People on the move in a changing climate* (pp. 81–109). Springer.

Nawrotzki, R. J., Riosmena, F., & Hunter, L. M. (2013). Do rainfall deficits predict US-bound migration from rural Mexico? Evidence from the Mexican census. *Population Research and Policy Review, 32*(1), 129–158.

Pais, J., Crowder, K., & Downey, L. (2013). Unequal trajectories: Racial and class differences in residential exposure to industrial Hazard. *Social Forces,* sot099.

Pebley, A. R. (1998). Demography and the environment. *Demography, 35*(4), 377–389.

Piguet, E., Kaenzig, R., & Guélat, J. (2018). The uneven geography of research on "environmental migration". *Population and Environment, 39*(4), 357–383.

Piguet, E., Pécoud, A., & De Guchteneire, P. (2011). Migration and climate change: An overview. *Refugee Survey Quarterly, 30*(3), 1–23.

Poston, D. L., Jr., & Frisbie, W. P. (2019). Chapter 26: Ecological demography. In D. L. Poston Jr. (Ed.), *Handbook of population* (2nd ed., pp. 697–712). Springer.

Riosmena, F., Nawrotzki, R., & Hunter, L. (2018). Climate migration at the height and end of the great Mexican emigration era. *Population and Development Review, 44*(3), 455.

Schnore, L. F. (1958). Social morphology and human ecology. *American Journal of Sociology, 63,* 620–634.

Shreffler, K. M., & Dodoo, F. N. A. (2009). The role of intergenerational transfers, land, and education in fertility transition in rural Kenya: The case of Nyeri district. *Population and Environment, 30*(3), 75–92.

Sellers, S. (2017). Family planning and deforestation: Evidence from the Ecuadorian Amazon. *Population and Environment, 38*(4), 424–447.

Sellers, S. & Gray, C. (2019). Climate shocks constrain human fertility in Indonesia. World development, 117, 357–369.

Speare, A. (1974). Residential satisfaction as an intervening variable in residential mobility. *Demography, 11*(2), 173–188.

Spears, D., Dey, S., Chowdhury, S., Scovronick, N., Vyas, S., & Apte, J. (2019). The association of early-life exposure to ambient PM2.5 and later-childhood height-for-age in India: An observational study. *Environmental Health, 18*(1), 62.

Thomas, A., & Benjamin, L. (2018). Policies and mechanisms to address climate-induced migration and displacement in Pacific and Caribbean small island developing states. *International Journal of Climate Change Strategies and Management.*

Warner, K. (2010). Global environmental change and migration: Governance challenges. *Global Environmental Change, 20*(3), 402–413.

Wheeler, B. W., Lovell, R., Higgins, S. L., White, M. P., Alcock, I., Osborne, N. J., et al. (2015). Beyond greenspace: An ecological study of population general health

and indicators of natural environment type and quality. *International Journal of Health Geographics, 14*(1), 17.

Wolpert, J. (1966). Migration as an adjustment to environmental stress. *Journal of Social Issues, 22*(4), 92–102.

Zwickl, K. (2019). The demographics of fracking: A spatial analysis for four US states. *Ecological Economics, 161*, 202–215.

Part I
Theoretical Perspectives

Population and Environment Interactions: Macro Perspectives

2

Jacques Véron

Abstract

If human activity has a major impact on the ecosystems and on the planet's climate, one can wonder about the role played by the "population" itself. And which dimension of the population variable really matters? Population size? Rate of growth? Population density? This chapter reviews different ways of considering the relationships between population and environment at a macro level. Starting with the Malthusian principle of population, we present different ways in which the role of population dynamics in environmental change has been analyzed: not only are Earth's carrying capacity, the IPAT equation (Impact = Population, Affluence, Technology) and the Ecological Footprint considered but also models built to integrate interactions and feedback loops. Some models are used for projections, others to specify the consequences of different policy options. A great advance in research, particularly with regard to climate change, is the bringing together of scientific communities

that had little exchange between them before, which has allowed a renewal and enrichment of approaches, such as the Shared Socioeconomic Pathways combined in the Representative Concentration Pathways.

Introduction

Twenty-five years after the United Nations' "Earth Summit" was launched in Rio de Janeiro in 1992, came the "World Scientist's Warning to Humanity: A Second Notice." More than 15,000 scientists who signed this warning voiced their concern about rapid population growth contributing to ecological and societal threats (Ripple et al., 2017). The signatories of the appeal advocate for a more rapid decline in fertility. The debate about the impact of population growth on development and the environment is set to continue for some time. A source of confusion lies in an insufficient separation between "population" and "human activities." If we are living in an Anthropocene era, if the human impact on the world ecosystem is critical, it does not mean that the "population factor" by

J. Véron (✉)
Institut national d'études démographiques (Ined),
Campus Condorcet 9 cours des Humanités, CS 50004,
93322, Aubervilliers Cedex, France
e-mail: veron@ined.fr

© Springer Nature Switzerland AG 2022
L. M. Hunter et al. (eds.), *International Handbook of Population and Environment*, International
Handbooks of Population 10, https://doi.org/10.1007/978-3-030-76433-3_2

itself is the main – or even the only – cause of environmental degradation or climate change.

Two decades ago, Anne R. Pebley (1998) reviewed the main perspectives of population dynamics and the environment and provided a picture of the highly fragmented research in this field. We focus here exclusively on the macro perspectives concerning population dynamics and the environment: even though the topic seems limited, we already see a wide range of approaches.

Defining the "Environment"

The history of demo-ecological thought reveals many recurring questions, but generally these fall within two themes: the impacts of population on the environment, and the limits to population growth. Although these topics may, on the surface, appear simple, they are multifaceted and highly complex. To begin examination of the population/environment, we first need to clarify terminology. What do we mean by "population"? It may refer to the number of people in a given territory (world, region, country, etc.), to the age or sex structure, or to the spatial distribution. Population may be considered through ratios, as the proportion of people under 15 years or aged 65+; growth rates may also be considered, applied to the world population, to the urban population, etc. The term population may also refer to demographic phenomena: fertility, mortality, and mobility. What do we mean by "environment"? In demographic research, references to the environment encompass a wide variety of concerns, including air and water quality, resource shortages, land use, climate change, and consumption and production patterns.

The relationships between population and environment are complex and dynamic. They are also reciprocal. Thus, if population growth leads to an overexploitation of soils and a decline in agricultural yields, a more difficult life for people dependent on agriculture may encourage migration, thereby reducing the demographic pressure.

Population and the Environment Historically

The history of the world population is the story of different forms of human adjustments to environmental conditions (Livi-Bacci, 1997). According to Biraben (1979, 2003), from less than one million up to 35,000 B.C.E., the world population could have then fluctuated between four and eight million inhabitants until 10,000 B.C.E. (Fig. 2.1). Owing to the development of agriculture, the Neolithic Revolution then marks a change of pace in world population growth, with the number of inhabitants on earth estimated to be 250 million at the beginning of the first century C.E..

Until the Industrial Revolution, humanity was largely in a situation of subordination to the natural environment, as the main sources of food, fuel, and energy were environmentally derived (Cipolla, 1962 quoted by Livi-Bacci, 1997). But, from the 1800s onward, the pressure on land and animals increased continuously, and dramatically, resulting in at least "one-third of the total space away from the natural environment" (Livi-Bacci, 2017, p. 25). Through the subsequent decades, air and water quality worsened, water resources declined, and land quality suffered; new and particularly complex ecological issues emerged over time, including climate change (Véron, 2013). Examination of these long-running population–environment interactions tends to focus on the dramatic increase in the population size beginning in the 1700–1800s (Livi-Bacci, 1997). Examples of such schools of thought include the work of Malthus, Boulding, and Boserup, each of which is reviewed in the next section.

Malthus, Boulding, and Boserup: The Population-Subsistence Debate

In the first edition of *An Essay on the Principle of Population* (1798), Malthus postulates that population and the means of subsistence evolve at very distinct paces. In terms of numbers, the human species would increase following a geometric

2 Population and Environment Interactions: Macro Perspectives

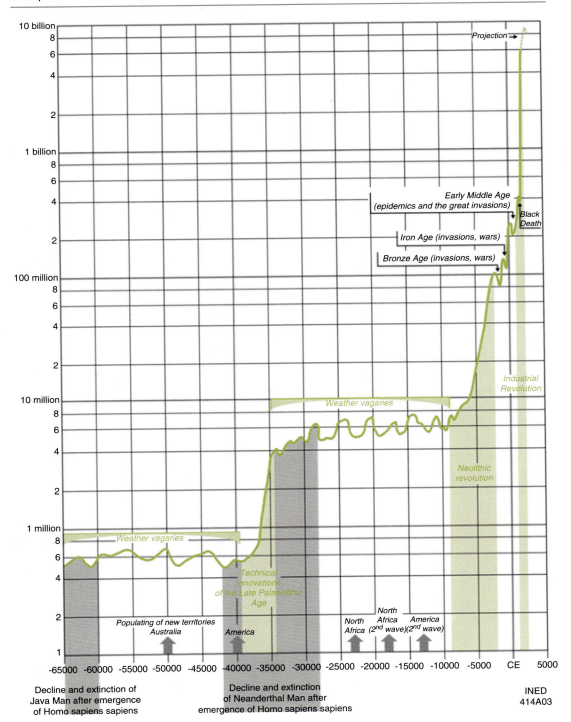

Fig. 2.1 Population growth of over 65,000 year. (Source: Biraben (2003))

progression (1, 2, 4, 8, 16, 32, 64, 128, 256, 512, etc.) and subsistence an arithmetic progression (1, 2, 3, 4, 5, 6, 7, 8, 9, 10, etc.). If the population doubles every 25 years "in two centuries and a quarter, the population would be to the means of subsistence as 512 to 10" (Malthus, 1798, p. 25 and 26). Population growth would inevitably hit the limit of subsistence: the "dismal theorem," or the "Malthusian trap," refers to the impossibility of economic progress in the face of unchecked population growth. Humanity is then condemned to be satisfied with the minimum of substance. As an example of this process, Malthus argues that cultivation of new land would yield a population increase that would ultimately outstrip the capacity to produce.

In response, Malthus suggests the presence of two "checks" to population growth. One is of a preventive nature: "a foresight of the difficulties attending the rearing of a family, acts as a preventive check" (Malthus, 1798, p. 62). People delay marriage or stay single. The whole society is concerned, regardless of the social rank of its members. The second check was named by Malthus "positive" because it "represses an increase which is already begun." It is expressed through increased mortality, especially in the form of child deaths from lack of food and attention. This check is "confined chiefly, though not perhaps solely, to the lowest orders of society" (Malthus, 1798, p. 71). Malthusian theory has generated major controversy. In particular, Malthus has been criticized for not having taken technological progress into account, although, when he wrote his *Essay,* England's industrial revolution had already started. This revolution was the moment when "technological progress and innovation became major factors in economic growth" and "progress took place in a wide range of industries and activities, not just in cotton and steam" (Meisenzahl & Mokyr, 2011). Increased standards of living and improvements in health have undergirded a global population increase from 1 billion in around 1800 to nearly 8 billion in 2021. And while technological progress facilitated this increase, it has also generated critically important environmental challenges.

An inspirational contribution to the debate was made by the economist Boulding (1955), considering "the Malthusian model as a general system." Overall, Boulding contends that there is a possibility of equilibrium between population and subsistence after introduction of a feedback effect. Boulding assumes that the "average standard of life" (i.e., the amount of subsistence per capita S) is a function of the population P: $S=F(P)$. In "the region of the population equilibrium the function is monotonically decreasing" (Boulding, 1955, p. 453) as a consequence of the law of diminishing returns: after reaching a maximum S declines as P increases. Further, the net reproduction ratio r, defined demographically as the number of girls who will reach reproductive age that a woman gives birth to (and more simply the ratio of the size of each generation to the size of the preceding generation), is assumed to be a function of the level of subsistence: $r = G(S)$. The population increases if $r > 1$, is stable if $r = 1$, and decreases if $r < 1$. Malthus postulates r increasing with S and thus, in the equilibrium, the standard of life equals the subsistence level, meaning that people have just enough to live on. In contrast to Malthus' "Dismal Theorem," Boulding states that this theorem "can be restated in a cheerful form" (Boulding, 1955, p. 455). To this end, it is sufficient to introduce a change in the r-curve, with a smaller net reproductive ratio associated with each level of S: as a consequence, the equilibrium standard of life rises, even if there is no progress at all. It is then possible to escape to the Malthusian trap and experience economic progress. The main argument put forward is that the subsistence level is "culturally determined" and is not exclusively of a physiological nature. Boulding also considers a dynamic system, but it is interesting to note that, with a slightly different hypothesis, economic progress is possible within the Malthusian framework.

If Boulding situates the Malthusian model in a larger framework, another scientist, Boserup, radically contests the Malthusian model. Boserup (1965) contended that population density fuels innovation and adaptation in the agricultural sector and, therefore, results in economic progress. As the adoption of new cultural techniques requires additional work, farmers will innovate only if demographic pressure (higher population density) gives them no choice. In *Population and Technology*, Boserup (1981) adopts a much larger definition of technology (including progress in sanitary methods or in administrative techniques) and she considers population and technology as reciprocally related (Boserup, 1981, p. ix). For their part, Pingali and Binswanger (1988) examined the leeway available to farmers to increase land and labor productivity. Re-examining the relationship between population density and intensity of land use, they conclude that in the case of Sub-Saharan Africa, for instance, the "farmer-generated technical change" is not sufficient to support "rapidly rising agricultural populations and/or rapidly rising non-agricultural demand for food" (Pingali & Binswanger, 1988, p. 52). They advocated for infrastructure investments (road network for instance) and for public sector intervention. If intensive use of farmer-generated inputs (land and labor) increases agricultural output, the "rate of growth in output is low in the absence of science-and industry based inputs" (Pingali & Binswanger, 1988, p. 67).

Malthus, Boulding, and Boserup all have lessons for humanity today, given that human activities have driven contemporary environmental challenges such as natural resource depletion, air and water pollution, and climate change (Véron, 2013). Issues of population related to subsistence remain highly relevant. In 2015, United Nations Member States adopted 17 Sustainable Development Goals (SDGs) as part of the 2030 Agenda for Sustainable Development. This agenda was a "call to action to end poverty, protect the planet and improve the lives and prospects of everyone, everywhere" and a "15-year plan to achieve the Goals."[1] "Zero Hunger" is the Sustainable Development Goal 2, just after "No Poverty." In this way, it is clear that the problem of subsistence, central in the Malthusian model, remains a core global concern even if our thinking on the population–environment connection has advanced beyond simplistic Malthusian understanding.

Developing Measures of the Earth's Carrying Capacity

Another way of considering the "problem of population" is to wonder "how many people can the Earth support" (Cohen, 1995)? There is no simple or consensus answer to this question, with divergent estimates proposed by the authors who have addressed this issue. Cohen recalls Ravenstein's pioneering calculations of the maximum world population to supply everybody with the necessary resources. In a paper presented at a meeting of the British Association for the Advancement of Science, Ravenstein (1891) aimed to determine the number of people on earth, "the area capable of being cultivated for the yield of food and other necessaries of life," and "the total number of people whom this land would be able to maintain" (quoted by Cohen, 1995, p. 162). Ravenstein gave estimates of "supportable people per square kilometre" in three different types of land: fertile region, steppe, and desert. The product in each case of the area (in square kilometers) by "supportable people per kilometer" gives the supportable population by type of land, and the sum of the three estimations the "total supportable population" (Table 2.1). Ravenstein estimated the number to be 5994 million inhabitants.

A first criticism regarding this estimate was made by the economist Alfred Marshall, who attended the meeting and considered "that a check to population in the future in temperate regions would come from the scarcity not of food but of fuel" (Cohen, 1995, p. 164). In more recent times, estimates of carrying capacity have been offered by Higgins et al. (1982) making use of an FAO framework. As per the framework, capacity is shaped by land productivity and potential productivity, which are in turn shaped by climatic conditions (e.g., temperature, sunshine, moisture), characteristics of the land and soil, types of crops grown, and farming practices (e.g., input levels and soil conservation). For these cal-

---

[1] https://www.un.org/sustainabledevelopment/development-agenda/

Table 2.1 Ravenstein's (1891) estimates of a "total[ly] supportable population"

	Fertile region	Steppe	Desert	Total
Area (km^2)	73,216,783	36,003,626	10,826,211	120,046,620
Supportable people per square kilometer	80	3.9	0.4	NA
Total supportable population (millions)	5851	139	4	5994

Source: from Cohen (1995, p. 163)

culations, a land resource inventory was necessary and Higgins et al. (1982) concluded that among 117 developing countries more than half would be unable to support their populations from land resources alone using production systems based on "low-input management level, with currently grown mixture of crops and local cultivars, no fertilizers, no chemical control of pest, disease or weed, no conservation measures, and manual labour with handtools" (Dudal et al., 1985, p. 7). In his review of the different values of human carrying capacity that have been proposed, Cohen (1995) shows how wide the range of estimates is, with the possibility for the earth to support 157 billion "with Japanese standards of food consumption and Asian standards of timber requirements" (Clark, 1967, 1977, quoted by Cohen, 1995, p. 409).

Within all such estimates, the role of technological progress in agriculture must be integrated. The Green Revolution starting in the middle of the 1960s changed the game and contradicted the Malthusian theory: over the past 50 years the population of the developing world has more than doubled, but in the same period cereal crops have tripled, with a limited increase in land area of only 30% (Pingali, 2012). The "First Green Revolution" (years 1966–1985 according to Pingali) was successful as a "combination of high rates of investment in crop research, infrastructure, and market development and appropriate policy" (Pingali, 2012, p. 12302). High-yielding varieties and an decrease in time to maturity allowed an increase in cropping intensity. Use of fertilizer and pesticides, as well as irrigation, contributed to the remarkable increase in agricultural output. But the Green Revolution also had negative effects on the environment: loss of biodiversity, and pollution and degradation of soils, groundwater, and rivers (Véron, 2013, p. 75).

To cope with expected population growth, knowing that climate change is an additional threat to food security, a major increase in food production is essential. As per Pingali (2012, p. 12307):

> Harnessing the best of scientific knowledge and technological breakthroughs is crucial for GR 2.0 [a new Green Revolution] as we attempt to reestablish agricultural innovation and production systems to meet today's complex challenges. New global public goods are needed that focus on shifting the yield frontier, increasing resistance to stress, and improving competitiveness and sustainability.

Today, the questions are less about maximum population and are more appropriately focused on the economic and ecological conditions necessary to support 10, 11, or 12 billion people. Of course, the question also arises as to the appropriate scale for such examinations. Context matters and analyses on a global scale neglect contextual variation. For instance, at the end of the 1990s, new varieties of crops covered 82% of total land planted in Asia compared with only 27% in Africa (Pingali, 2012, p. 12302). Another issue is production capacity compared with distribution in that a large proportion of the food produced is wasted (Guillou & Matheron, 2012). In developed countries, it is connected to the management of stocks of perishable foodstuffs. In developing countries there are many opportunities for post-harvest losses during transport, drying, or storage (torn bags, rodents, mold, etc.) (Guillou & Matheron, 2012, p. 121). Above all, the confrontation between human needs and the limits of Earth cannot ignore the fact that there is no "physical definition of needs" (Demeny, 1988). Underlying all approaches to estimation of Earth's carrying capacity is the sense of limits. In this way, such approaches undergird calls for population control, particularly in the context of rapid growth in the mid-twentieth century. It is

A Finite World: Arguments and Counterarguments on the Centrality of Population in Environmental Deterioration

In the 1960s, the biologist Paul R. Ehrlich published a best-selling book, *The Population Bomb* (1968), in which he argued that population numbers and growth were responsible for environmental degradation, writing: "The causal chain of deterioration is easily followed to its source. Too many cars, too many factories, too much detergent, too much pesticide, multiplying contrails, inadequate sewage treatment plants, too little water, too much carbon dioxide – all can be traced easily to *too many people*." (Ehrlich, 1968, p. 35–36). The same year, another biologist, Hardin (1968) stated in "The Tragedy of the Commons" that a constant population is inevitable: "A finite world can support only a finite population; therefore, population growth must eventually equal zero" (Hardin, 1968, p. 1243) Disputing the intercession of the "invisible hand" of Adam Smith (1776, quoted by Hardin, 1968, p. 1244), or the idea "that decisions reached individually will, in fact, be the best decisions for an entire society," Hardin defended the idea that "freedom in a commons brings ruin to all." (Hardin, 1968, p. 1244). Hardin's argument contrasts individual advantages and collective disadvantage, suggesting that self-interest will ultimately yield environmental degradation. He used an illustrative case of an open, collective pasture used by multiple herdsmen. Each herdsman has interest (positive utility) in adding an animal, but each additional animal heightens the pressure on the pasture resulting in a cost that is collectively borne by all herdsmen. Hardin concluded (1968, p. 1244): "each man is locked into a system that compels him to increase his herd without limit – in a world that is limited." These utility calculations may be applied to different situations such as pollution, for example: "the rational man finds that his share of the cost of the wastes he discharges into the commons is less than the cost of purifying his wastes before releasing them" (Hardin 1968, p. 1245). Hardin (1968) linked such pressure to human numbers, denouncing "the misery of overpopulation" (p. 1248) in that each individual acting in self-interest will ultimately threaten the whole.

The inevitability of the "tragedy" according to Hardin (1968) was countered by Elinor Ostrom (1990) through examples of successful collaboration between actors to address resource scarcity. For instance, Ostrom contended that voluntary organizations, such as the case a cooperative of 100 fishers of Alanya (Turkey), can locally solve the common-pool resources problem (Ostrom, 1990, pp. 18 and 19, quoting Berkes, 1986). In this example, there existed "an ingenious system for allotting fishing sites to local fishers" with fishers drawing lots for the attribution of "the named fishing locations," moving east one part of the year and then moving west to give "equal opportunities" to the fishers (Ostrom, 1990, p. 19). Ostrom (1990) put forward myriad examples of self-organized collective action, demonstrating that humanity was not necessarily on a path to self-destruction. Julian Simon (1981) also presented an argument that countered population pessimists. Simon considered the scarcity of resources to be relative, not absolute, as there are possibilities of substitution. For example, if copper scarcity is measured by its price relative to wages or its price relative to the consumer price index, for the period 1800–1980 there is a decreasing trend (Simon, 1981); scarcity is not increasing because of the substitutability between materials. To settle the question of absolute or relative scarcity, Simon and Ehrlich put forward a wager, with Simon asking Ehrlich to choose five minerals, to see how prices will evolve in the future. Indeed, Simon won the wager as the price of the minerals chosen by Ehrlich (i.e., copper, chromium, nickel, tin, tungsten) decreased in real terms between 1980 and 1990, supporting the perspective that there is no absolute scarcity.

Even earlier than Ostrom (1990) and Simon (1981), Ansley J. Coale (1970) made a related argument in his article "Man and His Environment" published in *Science*. Coale argued that the organization of the US economy and the use of harmful technological practices were ultimately responsible for environmental degradation. He further emphasized externality, defined "as a consequence (good or bad) that does not enter the calculation of gain or loss by the person who undertakes an economic activity" (Coale, 1970, p. 132). Coale critiques the exclusion of the externalities within the modern economy, arguing that environmental resources are not "free" and that their use comes at a cost. The way in which water is treated encourages the waste of this resource and discourages recycling. As related to population pressures, Coale contended that, as opposed to population growth per se, per capita production and consumption pressures are most impactful with regard to resources. In the USA, during the period 1940–1969, "the population has increased by 50 percent, but per capita use of electricity has been multiplied several times" (Coale, 1970, p. 135).

As pointed out by Keyfitz (1991a), "in the modern academy, knowledge comes packaged in disciplines" (Keyfitz, 1991a, p. 6). When economists conclude that population growth does not compromise development, they reason as if the world is infinite and they underestimate the impact of humans on the biosphere. The economic logic itself differs according to the timeframe considered. Keyfitz (1991b) gives an example of stress on the ecology of fish. If fish stocks decrease, short-term economic logic leads to fishing with more sophisticated equipment in deeper waters while medium- to long-term economic logic leads to restricting fishing to allow time for the resource to regenerate. "Economics makes people the exclusive object of terrestrial action; Biology takes them as one species among many in a web of life" (Keyfitz, 1993, p. 6). When economists consider biodiversity, it is in the interest of human beings: "Economics deals with a truncated part of the commodity circle, Ecology aims at the whole" (Keyfitz, 1993, p. 10).

Overall, the macro perspectives put forward by Ehrlich (1968), Hardin (1968), and others provide intriguing arguments with regard to population pressure and resource constraints. Still, they lack nuance, as put forward by Ostrom (1990) and Coale (1970) in that there are factors that intervene between population and environment that ultimately shape the population's impact. We now turn to a macro-scale equation put forward with the aim of capturing some of that complexity.

Complexity in a Simple Equation: The Development of IPAT

Before discussing the equation put forward by Ehrlich and Holdren reflecting the macro association between population and environment, it is important to present the five theorems they consider fundamentally true. They contend (Ehrlich & Holdren, 1971, p. 1212) that:

1. Population growth causes a *disproportionate* negative impact on the environment.
2. Problems of population size and growth, resource utilization and depletion, and environmental deterioration must be considered jointly on a global basis. [. . .]
3. Population density is a poor measure of population pressure, and redistributing population would be a dangerous pseudosolution to population progress.
4. "Environment" must be broadly construed to include such things as the physical environment of urban ghettos, and the epidemiological environment.
5. Theoretical solutions to our problems are often not operational and sometimes are not solutions.

Grounded in these assertions, Ehrlich and Holdren (1971) introduce a simple equation representing population–environment connections, where I represents environmental impact, P the population size, and F the impact per capita:

$$I = P \times F$$

Importantly, the impact of technology is included in the factor F; Ehrlich and Holdren (1971) argue that no technology can fully overcome the impact of consumption. A next step in this population–environment accounting is that the extension of per capita impact is itself being influenced by the factor of population:

$$I = P \times F \ (P)$$

"Consumption of energy and resources, and the associated impact on the environment, are themselves functions of the population size" (Ehrlich & Holdren, 1971, p. 1213). The shape of the function F depends on the relative importance of diminishing returns and of economies of scale. For Ehrlich and Holdren in the early 1970s, the diminishing returns were already dominating as a result of depletion of nonrenewable resources and exhaustion of water reserves: "in the case of partly renewable resources such as water (which is effectively non renewable when groundwater supplies are mined at rates far exceeding natural recharge) per capita costs and environmental impact escalate dramatically when the human population demands more than is locally available" (Ehrlich & Holdren, 1971, p. 1213). Furthermore, the possibility of a threshold effect has to be considered: when a biological system is overloaded because more and more people are polluting, for example, there comes a point when the system can no longer self-regulate.

In the final report of the Commission on Population Growth and the American Future, Ehrlich & Holdren (1972) proposed a slightly different equation more centrally integrating the interactions between population, environment, consumption, and technology:

$$I = P \ (I, F) \times F \ (P)$$

This "nonlinear" equation introduces more complexity. Here, environmental impact I is P times F with F depending on P; what is new is the fact that P depends on I and F. It is a "tangled relationship" (Ehrlich & Holdren, 1972, p. 371.) Let us consider P(I, F). If per capita consumption of energy is used to improve medical services, the death rate may decline and the population growth rate may increase. If the environment deteriorates with an effect on the health of the population, mortality can increase. F(P) is the classical relationship generally associated with the law of diminishing returns, a situation where "additional output resulting from each additional unit of input is becoming less and less" (Ehrlich & Holdren, p. 371). In the case of nonrenewable resources, population growth leading to depletion of resources forces "to use lower-grade ores, drill deeper, and extend supply networks" (Ehrlich & Holdren, p. 371). As a consequence, per capita use of energy increases, as does per capita impact on the environment.

In the same volume *Population, Resources and the Environment,* biologist Barry Commoner (1972) presents the equation of environmental impact as follows:

$$I = \text{Population} \times \frac{\text{Economic Good}}{\text{Population}} \times \frac{\text{Pollutant}}{\text{Economic Good}}$$

where Population is its size, Economic Good/Population the per capita production (or consumption) of any good (or "affluence") and Pollutant/Economic Good the amount of pollutant generated per unit of production or consumption.

Commoner showed, for instance, that since World War II, pollutant emissions had increased by 200–2000 percent (by the end of the 1960s) and the population only by 40% (Commoner, 1972, p. 346–347). Commoner gives figures for the pollution by nitrogen oxides due to increases in road traffic in the USA (Table 2.2). The impact on the environment I is 7.3; thus, the contribution of the increase in the emissions of nitrogen oxides by vehicle-miles is much more significant than the effect of population growth (2.58 versus 1.41).

A problem raised by Commoner is that it is possible to compute an Environmental Impact for each good, but this type of "analysis represents only small fragments of a complex whole. What is required is a full inventory of the various Environmental Impact Indices associated with the productive enterprise and the identification of the origins of theses impacts within the production process and of the ecosystems on which they intrude" (Commoner, 1972, p. 357).

Reacting to Commoner's publication, Ehrlich and Holdren rewrite the equation to make its

Table 2.2 Environmental impact of increases in road traffic in the USA (emissions of nitrogen oxide)

Year	Population (1000s)	Vehicle-miles/population	Nitrogen oxide[a]/vehicle-miles	Total index nitrogen oxide[a]
1946	140,686	1982	33.5	10.6
1967	197,849	3962	86.4	77.5
1967:1946	1.41	2.00	2.58	7.3
Percentage increase	41	100	158	630

Source: Commoner (1972, p. 359)

[a]Dimension: NO $\times$ (ppm) $\times$ gasoline consumption (gallons $\times$ 10^{-6})

components more compact. This effort resulted in the well-known equation:

$$I = P \times A \times T$$

A representing the factor "affluence" and T the factor "technology."

Kaya (1990), in "what is known in climate change literature as the Kaya identity" (O'Neill & Chen, 2002, p. 55), added a fourth term to the IPAT (Impact = Population, Affluence, Technology) equation to relate the CO_2 emissions to the main drivers:

$$C = P \times \frac{GDP}{P} \times \frac{E}{GDP} \times \frac{C}{E}$$

where C is carbon emissions per year, P the population size, GDP the gross domestic product, and E the total energy use. The CO_2 emissions are expressed as the product of population, per capita economic production, the amount of energy produced per unit of economic production (energy intensity), and the amount of carbon emitted per unit of energy produced (carbon intensity) (O'Neill & Chen, 2002, p. 55).

Building on IPAT, Dietz and Rosa (1994) reformulated the IPAT equation as STIRPAT ("Stochastic Impacts by Regression on Population, Affluence and Technology"). Their primary goal was to add nuance by allowing for differential impacts by each of the equation's components. To do so, they offered a stochastic reformulation of the IPAT equation:

$$I = aP^b \, A^c \, T^d \, e$$

where a, b, c, and d are parameters and e a residual term. After a logarithmic transformation, the equation becomes:

$$\ln I = \ln a + b \ln P + c \ln A + d \ln T + \ln e$$

In the equation above A is measured by the economic activity per person and T refers to the environmental impact per unit of economic activity and is "determined by the technology used for the production of goods and services and by the social organization and culture that determine how the technology is mobilized" (Dietz & Rosa, 1997, p. 175). If the coefficients b and c are constant, it means that "the effects of population and affluence are constant in proportions (linear in logarithms)"; it is possible to integrate thresholds replacing b and c with "more complex functions" (Dietz & Rosa, 1997, p. 176).

Applying their model to CO_2 emissions, using 1989 data for 111 countries, Dietz and Rosa (1997) observe diseconomies of scale for the population factor: the coefficient is not constant for the largest populations such as China and India. The effect of affluence reaches a maximum for an estimated value of per capita gross domestic product of around 10,000 dollars) and then declines for higher values. The T multiplier varies considerably from one country to another one depending on the type of energy used, climate, etc. They consider two scenarios depending on whether or not there is an increase in efficiency (smaller technology multiplier) to estimate the amount of CO_2 emissions by 2025.

Zagheni and Billari (2006) offered an alternative to STIRPAT, as presented by Dietz and Rosa (1997) to estimate the economic cost borne by a country to reduce its environmental impact. One of their intentions was to propose "a generalization of IPAT-based models" (Zagheni & Billari, 2006, p. 12).

Of course, the simplicity of IPAT has resulted in substantial critique (O'Neill & Chen, 2002). O'Neill and Chen (2002: p. 55) put it succinctly: "although as an identity it is always true by definition, when it is used as an explanatory model it implicitly assumes that there are only three relevant variables [...] all related in a simple linear fashion." They return to the distinction introduced by Shaw (1989) between "proximate factors" (i.e. direct causes) of environmental impact and "ultimate factors" (such as income distribution or attitudes and preferences). Among the ultimate causes are polluting technologies or affluence-related wastes, whereas population density or population growth and situation-specific factors are among the proximate causes.

A completely different perspective is that of the ecological footprint. As he said in an interview about "turning the carrying capacity ratio over," Rees asked himself the following question: "how much area is needed per organism per population, wherever it is?"[2] The concept of the ecological footprint was born.

The Ecological Footprint, Scientific or Political Concept?

As noted above, one of the limits to the IPAT approach is that each form of environmental impact may require a different estimation. As a consequence it is impossible to offer a global view of environmental change (or degradation). Another perspective is provided by the concept of the Ecological Footprint, which offers a synthetic measure of environmental load, in terms of productive land area necessary for the resource consumption and the waste assimilation requirements of a population (Wackernagel & Rees, 1996). The Ecological Footprint represents an inversion of

the carrying capacity ratio: instead of focusing on population by unit of land, the footprint indicator estimates the land required to support a typical individual (or on average) of a defined population (country, city, etc.). As all impacts are converted in terms of productive land areas, it is possible to aggregate the impacts and establish a global population–environment association. To obtain the per capita footprint of an average person (Wackernagel & Rees, 1996, p. 65–66), it is necessary to estimate the land area appropriated per capita aa_i to produce the consumption item i. The value aa_i is obtained by dividing average annual consumption of this item c_i (in kg per capita) by the average annual productivity or yield p_i (in kg/ha):

$$aa_i = c_i/p_i$$

The per capita footprint ef expressed in hectares is given by summation:

$$ef = \sum_1^n aa_i$$

The total Ecological Footprint Efp for a population p whose size is N is given by:

$$EFp = N.ef$$

Wackernagel and Rees give examples of ways of converting consumption into land areas. For instance, in 1991, the consumption of paper of each Canadian was estimated to 244 kg for the year; for each metric ton of paper 1.8 m^3 of wood is necessary; the wood productivity is assumed to be 2.3 m^3 per hectare for the year (ratio of land required to produce wood). With a correction of units to convert tons in kilos of paper (1000 in the denominator), the formula used is then (Wackernagel & Rees, 1996, p. 81):

$$\frac{244 \ (\text{kilos per capita and per year}) \times 1.8 \ (m^3 \ \text{per metric ton})}{1000 \ (\text{kilos per metric ton}) \times 2.3 \ (m^3 \ \text{per hectare per year})}$$

It gives 0.19 hectare per capita of productive land for the production of paper. The Ecological

---

[2]http://aurora.icaap.org/index.php/aurora/article/view/18/29

Table 2.3 The consumption–land use matrix for the average Canadian (1991 data)

Ecologically productive land in ha/capita	Energy (fossil energy consumed expressed in the land area necessary to sequester the corresponding CO_2)	Degraded land (or built-up environment)	Gardens (for vegetable and fruit production)	Crop land	Pastures (for dairy, meat, and wool production)	Forest (prime forest area)	Total
Food	0.33		0.02	0.60	0.33	0.02	1.30
Housing	0.41	0.08	0.00			0.40	0.89
Transportation	0.79	0.10					0.89
Consumer goods	0.52	0.01		0.06	0.13	0.17	0.89
Services	0.29	0.01					0.30
Total	2.34	0.20	0.02	0.66	0.46	0.59	4.27

Source: Wackernagel & Rees (1996, p. 82–83)

Footprint of the average Canadian, considering five types of consumption, was estimated to be 4.27 hectares in 1991 (Table 2.3).

Wackernagel and Rees (1996) present the Ecological Footprint not only as a scientific concept but also as a useful planning tool. That the estimated Ecological Footprints of more developed countries are very high compared with those for less developed countries fuels public debates. For instance, in 1991, the Ecological Footprint was estimated at 5.1 ha/person for the USA and only 0.4 for India (Wackernagel & Rees, 1996, p. 85.). Such a vast discrepancy renders visible the need for policies to promote sustainability.

Linked with this concept, are the notions of natural capital and natural income. The Ecological Footprint mays be expressed in "global hectares," that is to say, "adjusted hectares that represent the average yield of all bioproductive areas on Earth" (Wackernagel et al., 2006). If the Ecological Footprint exceeds the biological capacity, there is an ecological deficit. Thus, in 2002, he world's Ecological Footprint was estimated at 2.2 global hectares per capita, the biological capacity at 1.8, and therefore the ecological deficit at 0.4 (Wackernagel et al., 2006, p. 108). This means that the natural income was not sufficient and humankind made withdrawals from the natural capital, entering "overshoot." In the case of Canada, the Ecological Footprint was quite high in 2002

(7.5), but the biological capacity higher (14.3), which left an ecological reserve (Wackernagel et al., 2006, p. 108). Data on the Ecological Footprint by year and country are available on the website of the Global Footprint Network.[3] In 2017, the Ecological Footprint of Canada and the USA exceeded 8 ha/per person and is less than 1 in the poorest developing countries. According to Wackernagel and Lin (2019), the Ecological Footprint is useful for measuring "aggregate regeneration and demand" but is not "a full metric of sustainability," because it does not capture "human well-being or environmental quality." It is also not a forecasting instrument.

Overall, the IPAT formulation has been increasing in complexity over the past several decades. Even so, other macro perspectives have emerged, with the goal of better representing the complexities inherent in population–environment interactions.

Systems Modelling of Population-Environment Interactions

System dynamics, as the analysis of dynamic systems incorporating feedback, was developed

---

[3] See https://data.footprintnetwork.org/?_ga=2.1552182 99.1555743474.1613403998-1651254882.1613245314#/ compareCountries?type=EFCtot& cn=231&yr=2017

from the 1930s at the Massachusetts Institute of Technology (MIT) (Forrester, 1973, p. 13). From the end of the 1950s, system dynamics was applied to social systems under the leadership of Forrester. In the second edition of his *World Dynamics* Forrester (1973) intended to show "how the behaviour of the world system results from mutual interplay between its demographic, industrial, and agricultural subsystems" (Forrester, 1973, p. ix). Forrester integrated five components within the world system: population, capital investment, natural resources, fraction of capital devoted to agriculture, and pollution (Forrester, 1973, p. 19). Population growth increases the pressure on the world system, although a variety of mechanisms reciprocally influence population growth (birth rate and death rate are two loops affecting population). Such mechanisms include depletion of natural resources, pollution, crowding, insufficient food, and the feedback loops each process creates. Using this system, Forrester carried out an examination of "the transition from a world of growth to a world in equilibrium" (Forrester, 1973, p. 112) according to different hypotheses (pollution control, reduction of capital-investment rate, reduction of food productivity, etc.). If physical limits (a shortage of world resources, pollution, and inadequate food) play a major role in the model considered, Forrester insists on the need for social limits, which increase with population density and urbanization. In a context of growth, technology is used to relieve physical stress but it increases social stress (through the vulnerability of the social structure or stronger political control, for instance).

As an extension to Forrester's systems approach, 1 year later[4] Meadows et al. (1972) published the *The Limits to Growth*. Prepared by a MIT project team the report investigates the negative consequences of continuing population and economic growth as observed in the 1970s. In the same way as Forrester, the authors sought an understanding of the possibilities of "global equilibrium," with the population and capital stable. The world model is considered through five ele-

ments: population, food production, industrialization, pollution, and consumption of nonrenewable natural resources. It integrates interrelationships and feedback loops, as the elements are interdependent: "population cannot grow without food, food production is increased by growth of capital, more capital requires more resources, discarded resources become pollution, pollution interferes with the growth of both population and food" (Meadows et al., 1972, p. 97). A "standard" world model assumes exponential growth of population and industrial growth until natural resources start to run out, and leads to a decrease in industrial growth and then in the population. Next, the authors study the behavior of the world model: "how human society will respond to problems arising from the various limits to growth" (Meadows et al., 1972, p. 130). For instance, if resources are "unlimited" owing to technological solutions, the limit of arable land will decrease industrial output per capita and then the population. And how does the world system behave when "some policy to control growth deliberately" is included (Meadows et al., 1972, p. 164)? For instance, what happens if just population remains constant? The possibility of the depletion of nonrenewable resources constitutes a risk of collapse of the industrial system. What happens if technological policies, such as "resource recycling, pollution control devices, increased lifetime of all forms of capital, and methods to restore eroded and infertile soil" (Meadows et al., 1972, p. 168) complement growth-regulating policies? In that case, except for resources, all the values are constant and the "equilibrium value of industrial output per capita is three times the 1970 world average." If the same policies are instituted in 2000 and not in 1975, "the equilibrium state is no longer sustainable" owing to "food and resource shortages" (Meadows et al., 1972, p. 173).

Although intriguing, the "Meadows report" was criticized for neglecting the substitutability between resources (Simon, 1981) and also for insufficient attention to the property of emergence in the world system. Even today, the debate on the effects of technological innovation and substitution on the limits to growth remains open (Johnson, 2014).

---

[4]The first edition of *World Dynamics* was published in 1971.

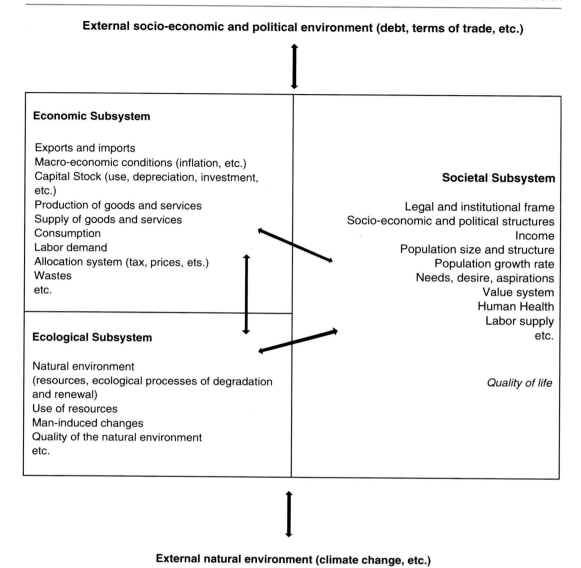

Fig. 2.2 The three subsystems of the sustainable development system. (Source: Shaw et al., 1992 (simplified diagram))

Whereas the *Bachue* models developed by the International Labour Organization (ILO) were essentially demo-economic models, with their three demographic, economic, and income distribution subsystems (Moreland 1984), the models developed at the International Institute for Applied Systems Analysis (IIASA) integrate environmental dimensions. The IIASA-based systems models were grounded in three subsystems: societal, ecological, and economic (Shaw et al., 1992; see Fig. 2.2). Population was represented within the societal subsystem through population size and structure, as well as the population growth rate. The economic system contained the core elements of production, consumption, capital, and labor demand. Finally, the ecological subsystem was considered in quantitative terms (natural capital stock) but also through the quality of natural environment, related to the degree of people's satisfaction.

Shaw et al. (1992) applied systems analysis to different case studies. One of these case studies concerned the water requirements in North Africa and the Middle East. They included relationships between the population of this region, the per capita agricultural water consumption, and the supply of renewable fresh water (precipitation minus evapotranspiration). They showed that the supply of renewable fresh water will not be sufficient to cope with the needs of the projected population in 2025. Climate change may affect the supply of fresh water: as there is no consensus on the predicted effect, the authors assumed a possible decrease of up to 20% or an increase of up to 20%. Shaw et al. (1992) concluded that "it is essential to account for linkages among sectors of the socio-ecological system, among geographical regions, and among generations if our resources are to be used in an efficient and equitable way" (Shaw et al., 1992, p. 19). In a changing world, with more restrictive ecological limits, systems analysis will help to have "a more imaginative and holistic" (Shaw et al., 1992, p. 27) management in the future.

In the 1990s, IIASA also developed and applied population development environment (PDE) models to several different countries and/or regions including Botswana, Cape Verde, Mauritius, Mozambique, Namibia, and the Yucatan peninsula (Lutz, 1994; Lutz et al., 2002b). The PDE models were "organized in three concentric circles, with population and development embedded in the environment" (Fig. 2.2).[5] The PDE models were context-specific in that HIV/AIDS was integrated for Namibia and water shortages were considered for Cape Verde. In the case of Mauritius, emphasis was placed on fertility, education, sugar, tourism, and textile exports. These context-specific models were used to simulate "alternative histories" in order to appreciate the relative importance of different system components. The model applied to Cape Verde allows the analysis of the effects of emigration, remittances and foreign aid on development (Wils, 1996). In the

different variants of an export boom scenario, "the gross national product trends are almost identical, regardless of the population size, or environmental investments. The only thing that makes some difference is the increase of domestic production, here, in the 'productive agriculture' variant" (Wils, 1996, p. 119).

The system dynamics modeling allows one to explore the consequences of policy options, through the play of interactions and feedback. IIASA model designers also considered that the PDE models might help to identify uncertainties and lack of knowledge. But combined with a high degree of data aggregation, those uncertainties limit the practical applications of the model (IIASA, 2013).

A decade or so later, extensions in systems modeling emerged in population, environment, development, and agriculture (PEDA) approaches (Lutz & Scherbov 2000; see Fig. 2.3). In contrast to PDE, the more comprehensive PEDA models adopted a more focused approach, electing to "make ceteris paribus assumptions on all factors that are not directly relevant to the hypothesis studied, even though such factors may be very significant for the future of a country under a more comprehensive approach" (Lutz & Scherbov, 2000, p. 311–312). Importantly, a "vicious circle" is central to this system, referring to the link between the high fertility, low education, and low status of women (Fig. 2.4). Such conditions lead to negative impacts of population growth on the environment resulting in declines in food production and, ultimately, heightened food insecurity. This model, adapted to the African context, has been applied in Mali, a setting characterized by serious land degradation, high fertility, and a high level of food insecurity (Lutz et al., 2002a). Using the PEDA systems model, the six scenarios considered indicate that between 30 and 65% of the Malian population would be food insecure in 2030. The value of 65% corresponds to the scenario with no change (all rates remain constant at their 1995 level, total fertility rate of 6.6) and the relatively optimistic value of 30% to a scenario of low fertility (total fertility rate of 2.2), strong educational effort, and an increase in fertilizer use (Lutz et al., 2002b, p. 208).

---

[5]https://iiasa.ac.at/web/home/research/researchPrograms/ WorldPopulation/Research/PDE-Concept.en.html

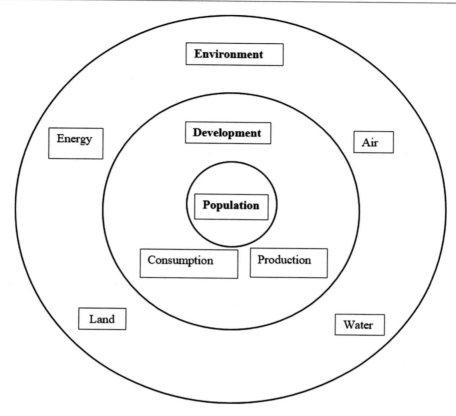

Fig. 2.3 Simplified structure of population development environment model. (Source: IIASA (2013))

In their state of the art summary concerning methods and analysis in the field of population (P) and environment (E) relationships, Lutz et al. (2002c) specified three criteria that P-E studies should, according to them, satisfy: "be explicit about both the P and the E dynamics" (Lutz et al., 2002c, p. 240); "be explicit about the specific mediating mechanisms between P and the aspect that you choose to address" (Lutz et al., 2002c, p. 242); "view the specific P-E question chosen in the broader context of all relevant interactions between a given population and its natural environment" (Lutz et al., 2002c, p. 243). One of the great difficulties that modelers encounter when they favor a global approach is to define the right level of complexity so as to best account for the phenomena and the P-E interactions while allowing the model to be operational (for the analysis of the consequences of a political option, for example).

Population as a Driver of Climate Change

The relationship between population and climate change is reciprocal, as climate change has important consequences for humans and vice versa. Yet a particularly important aspect of understanding future climate consequences is identifying the specific role of population numbers vis-à-vis human activities. This section reviews two approaches to the quantification of the influence of the population on the climate, a model developed within IIASA and the "Shared Socioeconomic Pathways" (SSPs) developed by the climate research community.

Between population and climate change there is a circle (Lutz, 2009) that represents human population (by age, sex, place of residence, and household structure), as it impacts greenhouse gas (GHG) emissions through consumption as

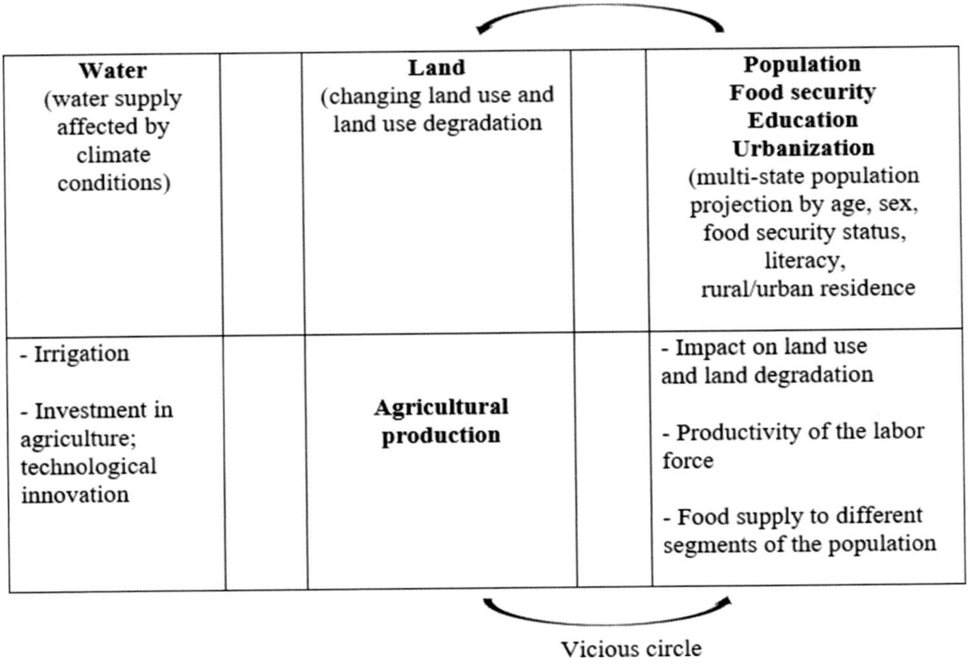

Water (water supply affected by climate conditions)		Land (changing land use and land use degradation		Population Food security Education Urbanization (multi-state population projection by age, sex, food security status, literacy, rural/urban residence
- Irrigation - Investment in agriculture; technological innovation		Agricultural production		- Impact on land use and land degradation - Productivity of the labor force - Food supply to different segments of the population

Vicious circle

Fig. 2.4 Basic structure of the population, environment, development and agriculture (PEDA) model linking population, food security, and the environment in Africa. (Source: Lutz et al., 2002b, p. 202. (simplified diagram))

well as innovation and technology (Fig. 2.5). In turn, GHG emissions shape global climate change with varying regional effects on temperature, humidity, extreme events, and sea-level rise. Estimation of human impacts involves modeling the differential vulnerability to climate change of various populations as a result of, for example, economic status or place of residence. Ultimately, environmental changes directly affect population health and mortality, as well as indirectly affecting livelihoods, which has further influence on health and mortality, as well as migration. The cycle continues as these shifts in population influence future GHG emissions. The PCC (Population and Climate Change) Program developed at IIASA was aimed in particular at a better understanding of the relationships between demographic change, energy use, and GHG emissions. O'Neill and his team produced "a comprehensive model which includes the effects of changing household size, age structure, and urbanization on energy use" (Lutz, 2009, p. 2). Furthermore, the population–environment relationships may be

considered under two different angles (O'Neill et al., 2001): adaptation (effect of climate change on food security or health) and mitigation (reduction of GHG emissions with lower population growth, more education, etc.)

Another approach to understanding population as a driver is the product of the research community working to standardize understanding of possible futures so that a more systematic understanding of future climate consequences might emerge (Hunter & O'Neill, 2014). According to the International Panel on Climate Change (IPCC, 2014), population growth is recognized as a key driving force in emissions in combination with economic growth. Initial climate scenarios developed by the IPCC integrated only total population size in an effort to understand climate futures (Lutz & Striessnig, 2015). To add nuance, several SSPs were subsequently developed that considered the population's age, sex, and educational composition, all key factors shaping future population growth (Lutz & Striessnig, 2015). The approach combining "representative

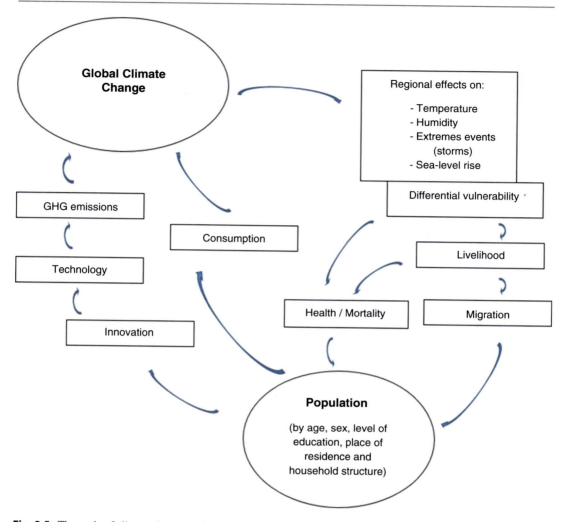

Fig. 2.5 The cycle of climate change and population relationships. (Source: Lutz, 2009)

concentration pathways" (RCPs), as the plausible future of atmospheric composition, and "socio-economic pathways" (SSPs), as the plausible evolution of social and natural systems, opens up a new avenue of research (Hunter & O'Neill, 2014).

The factors considered in shared socio-economic pathways fall under many areas: demographics, economic development, welfare, environmental and ecological factors, resources, institutions and governance, broader societal factors such as lifestyles (Hunter & O'Neill, 2014). The four RCPs (different levels of climate forcing) may be combined with the five SSPs ("Sustainability," "Middle of the Road," "Fragmentation," "Inequality," "Conventional Development") to build the "scenario matrix architecture" (van Vuuren et al., 2014, p. 377). Policy assumptions may then be considered, adding a new dimension. The SSP narratives may be translated in alternative demographic scenarios, as did Kc & Lutz (2017). On the other hand "the demographic research community can also provide insight into whether the SSPs neglect important aspects of regional or global demographic futures and/or population–climate interactions" and "context-specific research can also help in refinement of the pathways themselves" (Hunter & O'Neill, 2014, p. 238). One of the merits of this approach is to bring together research communities that have long tended to ignore each other.

Conclusion

In the 1960s and 1970s, the main question was whether humanity was experiencing a "population explosion" (Ehrlich, 1968) or a "civilization explosion" (Commoner, 1971). It was then a question of measuring the impact on the environment of three factors, population, affluence, and technology. A first renewal of the way of considering population–environment relationships was the introduction of systems analysis. Interactions and feedback loops were integrated into the models, with an architecture of a general system subdivided into subsystems (population, economy, environment). A second renewal was to bring together different scientific communities to make compatible the approach of the social sciences and those of the physical sciences, and in particular of the climate sciences. The progress yet to be made seems to be in the articulation between macro and micro perspectives. To what extent may the local context affect relationships observed at a macro level? And how to integrate the variety of potential situations in global models?

References

Biraben, J.-N. (1979). Essai sur l'évolution du nombre des hommes. *Population, 34*(1), 13–25.
Biraben, J.-N. (2003). The rising numbers of humankind. *Population & Sociétés, 394*. Ined.
Boserup, E. (1965). *The conditions of agricultural growth. The economics of agriculture under population pressure* (124 pp.). London and New York.
Boserup, E. (1981). *Population and technological change. A study of long-term trends* (245 pp.). University of Chicago Press and Blackwell.
Boulding, K. (1955, September). The Malthusian model as a general system. *Social and Economic Studies*, 453–463.
Cipolla, C. M. (1962). *The Economic History of World Population*. Penguin Books.
Clark C. (1967, 1977). *Population growth and land use*, Macmillan.

Coale, A. J. (1970). Man and his environment. *Science, 170*, 132–136.
Cohen, J. E. (1995). *How many people can the earth support?* Norton.
Commoner, B. (1971). The closing circle. In *Nature, Man, and Technology, Alfred a.* Knopf.
Commoner, B. (1972). The environmental cost of economic growth, population resources and the environment. In R. G. Ridker (Ed.), *Commission on population growth and the american future, research reports* (Vol. III, pp. 339–363).
Demeny, P. (1988). Demography and the limits to growth. In *Population and development review* (Supplement: Population and resources in western intellectual traditions) (Vol. 14, pp. 213–244).
Dietz, T., & Rosa, E. A. (1994). Rethinking the environmental impacts of population, affluence and technology. *Human Ecology Review, 1*(2), 277–300.
Dietz, T., & Rosa, E. A. (1997). Effects of population and affluence on CO_2 emissions. *Proceedings of the National Academy of Sciences, USA, 94*, 175–179.
Dudal, R., Higgins, G. M., & Kassam, A. H. (1985). Land, food and population in the developing world. In T. J. Davis (Ed.), *Population and food* (pp. 5–28). Proceedings of the Fifth Agriculture Sector Symposium, The World Bank.
Ehrlich, P. R. (1968). *The population bomb*. Ballantine Books.
Ehrlich, P. R., & Holdren, J. P. (1971). Impact of population growth. *Science, 171*, 1212–1217.
Ehrlich, P. R., & Holdren, J. P. (1972). Impact of population growth. In R. G. Ridker (Ed.), *Population resources and the environment* (Vol. of commission research reports) (pp. 365–377). U.S. Government Printing Office.
Forrester, J. W. (1973). *World dynamics, second edition (first edition in 1971)*. Wright-Allen Press.
Guillou, M., & Matheron, G. (2012). *The World's challenge: Feeding 9 billion people*. Editions Quae.
Hardin, G. (1968). The tragedy of commons. *Science, New Series, 162*(3859), 1243–1248.
Higgins, G. M., Kassam, A. H., & Naiken, L. (1982). *Potential population supporting capacities of lands in the developing world*. Land and Water Development Division, FAO.
Hunter, L. M., & O'Neill, B. C. (2014). Enhancing engagement between the population, environment, and climate research communities: The shared socio-economic pathway process. *Population and Environment, 35*(3), 231–242. Published online 2014 Feb 15. https://doi.org/10.1007/s11111-014-0202-7
IIASA. (2013). *PDE concept*. https://iiasa.ac.at/web/home/research/researchPrograms/WorldPopulation/Research/PDE-Concept.en.html
IPCC. (2014). *Climate change 2014: Synthesis report. Contribution of working groups I, II and III to the fifth assessment report of the Intergovernmental Panel on Climate Change*. IPCC.
Johnson, I. (2014). Towards a contemporary understanding of the limits to growth. In I. Goldin (Ed.), *Is the planet full?* (pp. 79–103). Oxford University Press.

Kaya, Y. (1990). *Impact of carbon dioxide emission control on GNP growth: Interpretation of proposed scenarios, IPCC response strategies working group memorandum 1989.* IPCC Energy and Industry Subgroup.

Kc, S., & Lutz, W. (2017). The human core of the Shared Socioeconomic pathways: Population scenarios by age, sex and level of education for all countries to 2100. *Global Environmental Change, 42,* 181–192. https://doi.org/10.1016/j.gloenvcha.2014.06.004

Keyfitz, N. (1991a). Population and development within the ecosphere: One view on the literature. *Population Index, 57*(I), 5–22.

Keyfitz, N. (1991b, June). Interdisciplinary analysis in four fields. In *Options.* IIASA.

Keyfitz, N. (1993). Are there ecological limits to population. *Proceedings of National Academy of Science USA, 90,* 6895–6899.

Livi-Bacci, M. (1997). *A Concise History of World Population* (Second ed.). Blackwell.

Livi-Bacci, M. (2017). *Our Shrinking Planet.* Polity Press.

Lutz, W. (Ed.). (1994). *Population-development-environment: Understanding their interactions in Mauritius.* Springer Verlag.

Lutz, W. (2009). *What can demographers contribute to understanding the link between Population and Climate Change* (Popnet, No 41). IIASA.

Lutz, W., & Scherbov, S. (2000). Quantifying vicious circle dynamics: The PEDA model for population, environment, development and agriculture in African countries. In E. J. Dockner, R. F. Hartl, M. Luptačik, & G. Sorger (Eds.), *Optimization, dynamics, and economic analysis, essays in honor of Gustav Feichtinger* (pp. 311–322). Springer. https://doi.org/10.1007/978-3-642-57684-3_26

Lutz, W., & Striessnig, E. (2015). Demographic aspects of climate change mitigation and adaptation. *Population Studies, 69*(Sup 1), S69–S76. https://doi.org/10.1080/00324728.2014.969929

Lutz, W., Prskawetz, A., & Sanderson, W. C. (2002a). Introduction. In *Population and environment: Methods of analysis* (Population and development review a supplement to volume 28, 2002) (pp. 1–21). Population Council.

Lutz, W., Scherbov, S., Prskawetz, A., Dworak, M., & Feichtinger, G. (2002b). Population, natural Resources, and food security: Lessons from comparing full and reduced-form models. In *Population and environment: Methods of analysis* (Population and development review a supplement to volume 28, 2002) (pp. 199–224). Population Council.

Lutz, W., Sanderson, W. C., & Wils, A. (2002c). Conclusion: Toward comprehensive P-E studies. In *Population and environment: Methods of analysis* (*Population and development review. a supplement to volume 28*(2002)) (pp. 225–250). Population Council.

Malthus, T. R. (1798). *An essay on the principle of population, as it affects the future improvement of society. With remarks on the speculations of Mr. Godwin, M. Condorcet and other writers.* London, printed for J. Johnston.

Meadows, D. H., Meadows, D. L., Randers, J., & Behrens, W. W., III. (1972). *The limits to growth.* A Potomac Associates Book.

Meisenzahl, R., & Mokyr, J. (2011). The rate and direction of invention in the British industrial revolution: Incentives and institutions, NBER working paper series, working paper 16993. http://www.nber.org/papers/w16993.

Moreland, R. S. (1984). *Population, development and income distribution, A modelling Approach.* Gower-St. Martin's Press for ILO.

O'Neill, B. C., & Chen, B. S. (2002). Demographic determinants of household energy use in the United States, in population and environment: Methods of analysis. *Population Council. Population and Development Review. A Supplement, 28,* 83–88.

O'Neil, B. C., MacKellar, F. L., & Lutz, W. (2001). *Population and climate change.* IIASA and Cambridge University Press.

Ostrom, E. (1990). *Governing the commons. The evolution of actions for collective action.* Cambridge University Press.

Pebley, A. R. (1998, November). Demography and the environment. *Demography, 35*(4), 377–389.

Pingali, P. L. (2012). Green revolution: Impacts, limits, and the path ahead. *PNAS, 109*(31), 12302–12308. www.pnas.org/cgi/doi/10.1073/pnas.0912953109.

Pingali, P. L., & Binswanger, H. P. (1988). Population density and farming systems. The changing locus of innovations and technical change. In R. Lee, W. B. Arthur, A. C. Kelley, G. Rodgers, & T. N. Srinivasan (Eds.), *Population, food and rural development* (pp. 51–76). Oxford Clarendon Press.

Ravenstein, E. G. (1891, Jan). Lands of the globe still available for European settlement. *Proceedings of the Royal Geographical Society and Monthly Record of Geography, 13*(1), 27–35.

Ripple, W. J., Wolf, C., Newsome, T. M., Galetti, M., Alamgir, M., Crist, E., Mahmoud, M. I., Laurance, W. F., 15, 364 scientist signatories from 184 countries. (2017, December 12). World scientists' warning to humanity: A second notice. *BioScience, 67,* 1026–1028. https://doi.org/10.1093/biosci/bix125.

Shaw, R. (1989). Rapid population growth and environmental degradation: Ultimate versus proximate factors. *Environmental Conservation, 16*(3), 199–208. https://doi.org/10.1017/S0376892900009279

Shaw, R., Gallopín, G., Weaver, P., & Öberg, S. (1992). Sustainable development. In *A systems approach, status report SR-92-6.* Laxenburg.

Simon, J. L. (1981). *The ultimate resource.* Princeton University Press.

Van Vuuren, D. P., Kriegler, E., O'Neill, B. C., Ebi, K. L., Riahi, K., Carter, T. R., Edmonds, J., Hallegatte, S., Kram, T., Mathur, R., & Winkler, H. (2014). A new scenario framework for climate change research: Scenario matrix architecture. *Climatic Change, 122,* 373–386. https://doi.org/10.1007/s10584-013-0906-1

Véron, J. (2013). *Démographie et écologie.* La Découverte.

Wackernagel, M., & Lin, D. (2019, August 5). *Ecological footprint accounting and its critics.* https://www.greenbiz.com/article/ecological-footprint-accounting-and-its-critics.

Wackernagel, M., & Rees, W. (1996). *Our ecological footprint. Reducing human impact on the earth.* New Society Publishers.

Wackernagel, M., Kitzes, J., Moran, D., Goldfinger, S., & Thomas, M. (2006). The ecological footprint of cities and regions: Comparingresource availability with resource demand. *Environment & Urbanization, 18*(1), 103–112. https://journals.sagepub.com/doi/pdf/10.1177/0956247806063978

Wils, A. (1996). *PDE-Cape Verde: A systems study of population, development, and environment* (WP 96–9). International Institute for Applied Systems Analysis.

Zagheni, E., & Billari, F. C. (2006). *An IPAT-type model of environmental impact based on stochastic differential equations, presented at Population Association of America 2006 Annual Meeting.* Los Angeles, CA. March 30 – April 1, 2006, 12 p.

A Micro Perspective: Elaborating Demographic Contributions to the Livelihoods Framework

3

Sara R. Curran

Abstract

This chapter offers a review of micro perspectives, or explanations, about the relationship between population and environment dynamics. The chapter reviews classic to contemporary micro perspectives in the demography and environment literature. Notably, as theoretical perspectives have evolved, there has been a forging of linkages to a livelihoods framework which has also come to the fore over the last decades in both applied and scholarly efforts around sustainable development. In the remainder of the chapter, using a livelihoods framework as an organizing lens, the chapter demonstrates the valuable contributions of demographic theory and evidence for a livelihoods framework. Specifically, demographic theorizing and evidence provide some conceptual insights on how to unpack and elaborate livelihood strategies – a key conceptual link within the framework. The conclusion of the chapter indicates how a revised and elaborated livelihood framework might serve to propel a future research agenda that can both address micro perspectives and account for macro linkages for the larger population and environment research agenda.

Introduction

This chapter offers a review of micro perspectives and explanations about the relationship between population and environment dynamics. A micro perspective focuses upon the social and behavioral dynamics and mechanisms located at the individual, family and household level (Cetina & Cicourel, 2014). This leaves macro perspectives to those explanations that emphasize mechanisms at higher orders of social and ecological organization. While these are not to be ignored in a micro perspective, because they frequently provide contextual or conditional effects for understanding micro level relationships, they are not the analytic focus (Turner, 2005). Instead a micro perspective focuses upon the patterns of relationships between individuals, and others in their immediate social sphere, with the environment (Cetina & Cicourel, 2014; Turner, 2005). This relationship might be two-way. That is, in one direction it is what individuals (and their social relations)

S. R. Curran (✉)
Center for Studies in Demography & Ecology, Raitt Hall, University of Washington, Seattle, WA, USA
e-mail: scurran@uw.edu

© Springer Nature Switzerland AG 2022
L. M. Hunter et al. (eds.), *International Handbook of Population and Environment*, International Handbooks of Population 10, https://doi.org/10.1007/978-3-030-76433-3_3

do to, take from, and perceive about, the natural environment. In the other direction it is how the natural environment and its ecological forces influence human perceptions and actions. This two-way interaction is difficult to disentangle, although empirically, scholars typically evaluate one or the other side of that two-way dynamic. One of the challenges across the scholarship in this field has been to articulate a unifying framework that can make sense of both the reciprocal relationship between population and environment and the multitude of possible pathways linking population behaviors with environmental conditions.

Increasingly, a livelihoods framework has come to the fore as a crucial way in which to understand both directions of the population and environment relationship and to link micro perspectives with macro perspectives (de Sherbinin et al., 2008; Ellis, 2000; Hunter et al., 2015). Additionally, a sustainable livelihoods framework now shapes much of the multilateral and practical development programs associated with a host of climate or environment and development projects (Serrat, 2017). There are two reasons a livelihoods framework has proven particularly fruitful for scientists and policy makers. First, it clearly integrates multiple, disciplinary lines of inquiry, providing a practical entrée and guidance for diverse teams to approach research and applications. Second, it is both durable and flexible in its relevance across localities around the globe, including for framing analyses in both more and less developed settings. In this chapter, I demonstrate how a livelihoods framework might provide a useful platform for integrating micro-level demographic theory and the new, burgeoning, and rich array of evidence and studies about the relationship between population and environmental conditions across the globe. Additionally, I suggest how theory and research in demography evaluating the dynamic relationship between population and environment offers important new conceptual and empirical elaborations for a livelihoods framework. By synthesizing insights from both

domains, it might be possible to propel a more comprehensive scientific research agenda linking micro and macro level factors and seeking to better understand adaptive and mitigative responses to global climate change.

Micro perspectives on the relationship between population and environment typically circle around how environment or ecological conditions might influence individual behavior that then yields demographic outcomes such as births, deaths, moving away from a current residence, or the formation and dissolution of families and households. Similarly, in the other direction, questions might be asked about how individual or family (household) behavior that leads to demographic outcomes such as births, deaths, moving away from or into a place, or forming or dissolving a family might, in turn, affect environmental conditions (quality or integrity of an ecological system). Ideas about how these pathways are understood and related to each other, at the micro level, stretch back to early Malthusian claims related to how individual decisions yield a tragedy of the commons (Hardin, 1968). Other intellectual antecedents include human ecological and archeological explanations for the emergence of settlements, or the more recent explanations from the 1960s onward that include arguments about individual and household behaviors that intensify resource use (e.g. Boserup, 1970). Demographers picked up these intellectual threads to propose a variety of explanations, such as: Davis' multiphasic model (1963, 1991), vicious cycle models (Dasgupta, 1995, 2000, Myrdal, 1957, O'Neill et al., 2001; Panayotou, 1994), or consumption and lifestyle explanations (Curran & De Sherbinin, 2004; Dauvergne, 2010; Newton & Meyer, 2013; Princen, 1999; Spaargaren, 2003). Integrating demographic insights into the livelihoods framework can elaborate and generate a more comprehensive framework for future research and practical policies and program designed for interventions in response to sustaining both humans and the environment in the face of climate change.

Classic to Contemporary Micro Perspectives

In this section, I outline five sets of micro level perspectives that capture the diverse array of demographic theories and evidence that characterize the population and environment literature. The emphases of these descriptions lie mostly with the theories or explanations and only relies on a few illustrative studies that have substantiated each of these explanations. The first set of explanations primarily focuses on early ideas about demographic behaviors and outcomes as a response to environmental endowments. The second set of explanations focus on general demographic impacts on environmental endowments and includes discussions about *intensification* and *extensification* in the *production* activities associated with family livelihoods. The third set of explanations focuses on studies that evaluated human migration impacts on the environment which have pointed to *extensification* explanations, as well as *diversification* behavior and activities. The fourth set of explanations focuses on the growing array of studies in recent decades and from around the world that examine demographic responses to environmental hazards or environmental shocks. These studies often employ a migration as *diversification* response to better manage shocks. However, an important contribution of this work has been the introduction of ideas about *vulnerabilities* and *resiliencies* at the individual and household level, as well as at the macro level. This perspective has usefully explained the heterogeneity of individual and household responses in demographic behaviors and outcomes. Finally, a fifth set of perspectives focuses on the role of *consumption* behavior and preferences as a conceptual domain mediating the relationship between population and environment. In this domain, the focus of the explanations at the micro level examines individual and household *conservation* behaviors, *lifestyle status* consumption and sustainable *investments* that might contribute to ecological footprints, regardless of household size or life cycle, or individual life course position. I have italicized key concepts organizing this literature and are useful ideas for further elaborating a sustainable livelihoods framework focused on micro perspectives.

Early Ideas About Environmental Endowment Impacts on Demographic Behavior

Environmental endowments are often used to explain the emergence of settlements and population growth (Scott, 2017; Southgate, 2019). These archeological accounts argue for the wealth of environmental endowments contributing to higher nutrition, greater sedentary behaviors, and higher fertility. On the flip side of this argument are explanations about how resource constraints lead to changes in demographic behaviors and outcomes. Kingsley Davis' (1963, 1991) multiphasic response theory, proposes that households in both traditional and modernizing societies may pursue a variety of demographic responses in the face of household strains. Household strains are generated by some combination of growing household size under resource limitations, including environmental endowments, or rising expectations in economic improvement, or both. For Davis, this might have been triggered by increasing rates of offspring survival and it may not necessarily lead to fertility control and might lead to out migration or changes in marital timing. In other words, demographic responses are a set of cascading ones across an array of demographic behaviors – hence the term 'multiphasic.' Households respond in real time and have variable abilities to influence their members and their own demographic behaviors. Thus, their responses might vary at different points of a household's life cycle or having learned from previous demographic behaviors. Easterlin's (1976) economic formulations added to this idea with an argument about how opportunity structures (land availability) might influence fertility, especially when considering future endowments for offspring. In other words, lowered resource strains might relieve constraints on family and household size, encouraging more fertility. These two sets of theoretical frameworks set the stage for a sizeable research agenda during the 1980s and 1990s examining how environmental endowments encourage population growth and vice versa (De Sherbinin et al. 2007).

One of the challenges with testing some of these ideas in earlier decades was the limited availability of data to link household mechanisms, individual demographic behavior and environmental conditions. Early work by Firebaugh (1982) in India, Feeney (1988) in Thailand, and Friedlander (1983) in Wales used historic records to link population density, settlements and land availability. Those studies suggest the plausibility of a resource constraint mechanism influencing population growth (and by implication family size) and higher fertility.

With richer data records in recent years, the role of environmental endowments for affecting fertility or related demographic behaviors has been consistently observed (Grace, 2017). For example, VanLandingham and Hirschman (2001) find evidence for land constraints leading to declines in fertility in Thailand. Yabiku (2006) finds that land availability is positively associated with marriage rates and earlier marriage timing in Nepal and does so employing the longitudinal Chitwan Valley Study. In addition, Carr et al. (2006) find that fertility declines with growing resource constraints in frontier regions of Ecuador when employing richly detailed and longitudinal household data. With a more nuanced, but similar argument, Shreffler and Dodoo (2009) elaborate upon Davis' multiphasic theory to add educational investments as substitutes for limited opportunities for land inheritance and both combine to explain lower family size preferences and lowered fertility through analyses of qualitative data from Kenya.

While some studies find that simple measures of land availability explain fertility, a competing set of arguments and evidence suggests a different mechanism mediating the relationship, namely land tenure and economic security (Sutherland et al., 2004). There are a number of studies that demonstrate how economic security, as measured through land tenure, reduces fertility and lengthens planning time horizons. The idea is that parents may make tradeoffs between the security of land versus the security of child survival in strate-

gizing about future fertility. With greater land security, there is less need for child labor. Early examples include cases from the Philippines (Hiday, 1978), Egypt (Schutjer et al., 1983), and Mexico (DeVaney & Sanchez, 1979). This line of research has been sustained until now, although not garnering as much attention as other strands of research at the nexus of human-environment studies. Interesting updates to this line of research include Sharif and Saha's finding of a downward and then upward pattern of the relationship between landholding and fertility (1993). Or, Grimm's (2019) economic history analyses about the American Southwest, that accounts for confounding explanations, shows how riskier environmental contexts yielded higher fertility and more children for farm households.

Overall Demographic Impacts on Environmental Endowments

Alternatively, a different set of studies suggests that poverty, natural resource dependencies, and the role of children as contributors to household livelihoods is associated with higher fertility, even in resource constrained environments. In large part, these arguments draw upon deeply-rooted Malthusian explanations (Malthus [1798]1960). Malthus is not typically associated with micro level perspectives, but the theory assumes individual level behavior is motivated by short term resource extraction to maximize livelihoods through an *extensification* mechanism, at the expense of long term ecological and population sustainability. Extensification is the broadening of resource extraction across the landscape. Individuals and households will extend or extensify their resource extraction, especially into eco-scapes that are public goods and not governed by any regulations. The threads of this idea have continued to appear in a number of realms, including the vicious cycle model (Myrdal, 1957). Aggarwal et al. (2001) make this argument in India, suggesting that for poor rural families having to search farther and farther for fuelwood is associated with larger family sizes, especially because children are engaged at an early age in providing labor in this endeavor. In other words, expanding family size is a way to respond to limited resources with no other prospects for sustaining livelihoods. Filmer and Pritchett (2002) also find evidence of an association with high fertility and resource degradation in high insecurity contexts and when children's labor is engaged at an early age. In a study from Nepal, Biddlecom et al. (2005) find higher fertility preferences associated with greater natural resource dependencies, which is also associated with higher levels of resource extraction and subsequent environmental degradation. This may not entirely contradict Yabiku's (2006) finding (noted in the previous sub-section), which also derived from Nepal and also used the same data, but each challenges the other to clarify the mechanisms and improved study design.

Vicious cycle models found a great deal of support throughout the 1980s and 1990s. Geist and Lambin (2002) summarize this literature, but forcibly argue for how there are proximate and distal processes at work linking population pressures and population density to deforestation. For example, Carr (2002) demonstrates how this may be the case at the macro level, but at the micro level interactions are far more complicated. One set of research sites that generated empirical support for vicious cycle models are those that examined population growth in frontier regions. The frontier regions often studied were forested and the concern was the rapid clearing of forests and large-scale conversion to pasture or farmland. Barbieri's (1997) review points to the coincident effects of household size and economic assets in explaining frontier land degradation through forest clearing in Latin America. These are substantiated by Pan et al. (2004) in the Ecuadorian Amazon, McCracken et al. (1999) in the Brazilian Amazon, and by Carr (2002) in Guatemala. Household size is a crucial variable associated with deforestation and explained by demands for food subsistence for consumption. However, complex explanations soon emerged as the evidence indicated a great deal of heterogeneity in behaviors and motives. For example, rather than solely focusing on household size or composition, scientists argued for a focus on

household life cycle status, arguing that households move through the life cycle, becoming older and perhaps dividing into new households that reside nearby or result in some members migrating away. Such dynamics better describe the likely macro demographic impacts on the environment. The household life cycle effect may shift the land demands in the older household to a new household, which may or may not deploy its members to extend its ecological footprint on the nearby landscape. Alternatively, a new household may return to urban settings and completely alleviate stress on environmental resources around the old household. Walker et al. (2002) offer a nice summary of these complicating, micro and macro level dynamics.

Contrary to vicious cycle expectations, Boserup (1965) also developed ideas about how household strain or population pressures might yield innovations in agricultural practices and productivity because more labor was available to manage less land. These ideas emerged coincidentally to Davis' postulations and continue to influence scholars studying population and environment dynamics. In her formulation, population growth might yield *intensification* or *diversification* of resource use through innovations in production practices. This happens through cultural evolution and, generally, the explanation occurs on a macro level or population level. However, it relies on assumptions about micro level relations, specifically those associated with livelihoods strategies that might yield population and ecological success. This somewhat optimistic account has been criticized mathematically (e.g. Richerson & Boyd, 1997), but also formally theorized (Chu & Tsai, 1998). At the micro level, the formulation assumes a learn-by-doing mechanism among individuals over a lifetime or across generations within a family or community. Marquette also argues that Boserup's ideas merely modified Malthus' predictions and allowed for more immediate behavioral modifications (1997). Similarly, Pender (1998) and Winfrey and Darity (1997) argue for the impact of population pressure on innovation and agricultural productivity.

And, Boserup's theoretical approach continues to receive attention during contemporary times, linking the theory to the emergence of innovations in sustainability (Turner & Fischer-Kowalski, 2010). The resurgence of Boserupian explanations aligns with arguments found in the forest transition research and new data with higher spatial resolution and longer temporal perspectives. For example, Kabubo-Mariara (2007) finds that population density at the community level and household land tenure security can yield land conservation practices that includes diversification of crops, not previously practiced in the Rift Valley of Kenya. Similarly, Rasmussen and Reenberg (2012) suggest that land use rationales, even under extreme environmental endowment constraints, will not lead to *extensification* in the Sahel of Burkina Faso, but a *diversification* of strategies that even yield conservation practices among some households. Diversification strategies are often described to include a variety of ways to invest, save, and extract resources with attention to a mix of goals that balance risk minimization or income maximization. The mechanisms that govern these strategies and balancing goals that appear to be at work are a mix of micro- and macro-economic, socio-cultural, and ecological ones. Similarly, Perz (2001) finds that out migration relieves immediate pressures on environmental endowments, while remittances support longer term investments in environmental endowments, such as perennial crops or reforestation (Perz, 2001).

Despite the long history of research in this field, there may still be the need for considerable work to elaborate the mechanisms undergirding micro perspectives related to demographic behaviors and outcomes and their impact on environmental endowments. Cumming and Petersen (2017) argue strongly for a unifying research framework. I suggest that a fully elaborated livelihoods framework might offer that paradigmatic solution. A mix of assets, alongside variable access to those assets, even in the same context and facing the same conditions might yield different livelihoods activities and subsequent livelihoods outcomes (Ellis, 2000). The key will be better explaining that heterogeneity. Incorporating demographic ideas about the demographic and environmental conditions that explain household intensification, diversification, and extensification and demonstrate how demographic behaviors and environmental practices are instantiations of intensification, diversification, and extensification will be crucial next steps in that research agenda.

Human Migration Impacts on Environmental Endowments

The impact of migration on environmental endowments explained with micro level arguments has often focused on how households seek to *extensify* or extend their access to resources by moving to places where resources are richer than they are in a current location. Households may extensify their extractive activities across the landscape, perhaps lengthening daily movements into forests or public lands, sending household members to distant frontiers, or relocating households entirely (Bilsborrow, 1992; Bilsborrow & Okoth-Ogendo, 2002) Employing an argument about population pressure yielding extensification, rather than intensification, Bilsborrow argues strongly for the role of livelihoods strategies as a mechanism employed by individuals and households to extend their spatial reach and settle frontier lands. The idea draws on Davis' multiphasic theory and Bilsborrow's articulations launched a wave of research in the 1990s that sought to better understand demographic dynamics in frontier areas.

Frontier areas are resource rich destinations that can serve to pull people from resource poor environments. In fact, lack of land and access to land in other places can push and pull people across the landscape as part of a livelihoods strategy (VanWey, 2003, 2005). Carr (2009) effectively summarizes the current state of knowledge about how rural migration matters for explaining deforestation in frontiers. When migrants move to frontier areas and settle in places with little land tenure security, usufruct claims often require settlement on a parcel and require clearing of that parcel. Whereas, if a migrant can be hired elsewhere, as a farm hand or day laborer, this may delay land clearing or provide the resources for investing in perennial crops. As migrant households grow in size in frontier areas their demands on land resources may also grow, including extending the impact on the environment. However, as they mature and evolve along a household life cycle (with changes in age, generational, and sex structures) these demands may change, as was mentioned in the earlier section (Walker et al., 2002). Household dynamics create heterogeneity in the impact of households on the landscape. Households may divide with one of the households including members at an earlier life cycle stage and the other household at an older stage. Each household will burden resources in different ways, making it difficult to generalize an effect of household size on environmental endowments. Similarly, households may pursue sequential moves across a frontier landscape after the initial move into a frontier area (Carr, 2009). The conditions that motivate households (and their members vary greatly) and each type of move has different effects on environmental outcomes. The micro level perspective is often difficult to generalize without a macro level perspective.

These varying behaviors fit well with multiphasic response theory, but they also fit with the New Economics of Labor Migration (NELM) (Stark, 1991; Stark & Bloom, 1985). With the integration of NELM into the multiphasic response theory, migration can become a *diversifying* income strategy for both short-and long-term purposes, serving to smooth incomes and hedge against shocks (in an origin household) (Paulson, 1995; Paxson, 1992; Rosenzweig & Oded, 1989; Yang & Choi, 2007). Or, migration can be a risk minimizing or income smoothing as a response following a shock (Bardsley & Hugo, 2010; Bylander, 2013). In fact, the multiphasic response theory is usefully elaborated by NELM and its proposed mechanisms for explaining risk diversification and income smoothing, instead of simpler economic explanations that argued for income or wealth maximization.

Numerous studies about frontier areas and population growth indicate these kinds of multiphasic dynamics. Extensification through the planting of upland crops occurs through both migrant remittances and available household labor (Entwisle et al., 2005). Or, in South Africa where Hunter et al. (2014) find a complex mix of assets and activities interact to inform livelihoods, including a number of different types of migratory behaviors. Similarly, in Cambodia (Nguyen et al., 2015) and in China (Qin, 2010), households undergo a number of activities to sustain livelihoods including rural-urban migration. In fact, numerous studies have shown how migration can be employed in numerous ways where households may stay but send some members away to work through short and long-distance moves, circular migration, and by sending remittances.

Alternatively, the forest transition literature (e.g. Rudel et al., 2002), which also draws upon Zelinsky's mobility transition (1971), argues that rural-to-urban migration can relieve pressure on both private and communal lands and lead to a return to forest cover, either by design or naturally. Baptista and Rudel (2006) find evidence of a rural-urban migration impact on forest recovery in the Atlantic Forest in Brazil. But, Izquierdo et al. (2008) after examining 30 years of land use change and demographic pressures in the Atlantic Forest of Argentina find mixed results and question the utility of migration as pressure release valve assumption. And, Perz (2007) in a summary of that literature suggests that more systematic attention is required to examine the more proximate household and individual mechanisms, such as a mix of behaviors like rural-urban migration, remittances, and other livelihoods activities related to forest recovery. Finally, Robson and Berkes (2011) suggest that while rural-urban migration may yield forest recovery, it does not necessarily improve biodiversity. In a very recent study in the Mixteca Alta region of Costa Rica, Lorenzen et al. (2020) suggest that it is not just rural-urban migration, but livelihoods diversification pathways operating at both the household and community levels that can lead to a forest transition. These include changes in technology and agricultural practices and the decline in the use of forest materials for housing and energy.

The variability of migrant impacts on the environment also depend on how migrants arrive and are incorporated into the local environments and how their departure changes institutional arrangements in origin (Robson & Nayak, 2010). One mechanism explaining how migrants can variably impact the environment is to assess how much access they have to migrant social capital (Curran 2002). For example, migrant social capital associated with communities in a destination may acculturate new migrants to sustainable environmental practices. For example, Cassels et al. (2005) finds that migrants arriving into coastal areas through marriage networks practice more sustainable fishing practices than do migrants arriving into similar areas but with few local ties connecting them to socio-cultural livelihoods practices and lifestyles. Alternatively, Kramer et al. (2009) find that in origin communities, while migration might relieve pressures on some environmental endowments, migrant networks to distant markets may also increase pressures on other environmental endowments shifting local resource extraction for the demands of those migrant-facilitated markets. Qin (2010) finds similarly mixed outcomes for rural communities in western China. Finally, Robson and Nayak (2010) in a comparison of Mexico and India illustrate the complex dynamics that might result from rural-urban migration, including the rearrangement of social institutions, organizations and relations in origin community. These findings increasingly point to ideational or cultural change not typically and systematically theorized or assessed in the preceding research.

While Robson and Nayak (2010) are relatively pessimistic about the impact of migration on environmental endowments because of the mediating forces surrounding migration and migrant networks, others are more optimistic. In some case studies, migrants are shown to bring not only financial remittances, but also social remittances (Levitt, 1999) that can change the relative preferences and value for environmental endowment conservation. Similar to relieving pressures on resource extractions, this preferential mechanism might also lead to investments in sustainable practices and conservation (Lira et al., 2016; Qin & Flint, 2012). Greiner and Sakdapolrak (2013) argue for a more nuanced assessment of household livelihoods strategies in light of households' location along a continuum of commodification and subsistence and in a context of trans-local networks which transmit people, identities, resources and information. Greiner and Sakdapolrak (2013) argue for the important transmission of social remittances and ideas as they may shift population-environment relations and values. The resources transmitted through trans-local networks can reshape behaviors and natural resource practices in origin communities. Trans-locality and migrant networks have yet to be fully elaborated in relation to the most proximate livelihoods practices related to natural resources, but they hold considerable promise in adjudicating significant heterogeneity of results currently found in the scientific literature. Especially fruitful lines of inquiry might be to study the transmission and diffusion of ideas that shape the valuation of environmental endowments and how that valuation relates to human-environment relations within households and among individuals. Such a research endeavor might involve collaborations among anthropologists, ecologists, demographers, and social or environmental psychologists. This agenda is slowly emerging in the new field of sustainability science (Waring et al., 2015).

Environmental Impacts on Migration and Health

The push-pull of environmental factors on migration decision making stretches back to early theorists and in relation to both more and less developed settings (Hunter, 2005). As Hunter (2005) describes, Wolpert (1966) posits that households may consider an array of environmental stressors (including pollution and crime) in their current residence as contributing to a strained livelihoods context and that may propel consideration of alternative residential locations with higher place utility. Similarly, Spear (1974) argues for the conceptual presence of thresholds of dissatisfaction with residential surroundings as a mediating explanation for a move to another location. Hunter (2005) further synthesizes the micro-economic arguments for how migration is influenced by current values about the origin places and expected values in the destination. An individual's or household's preferences in assessing these values might include non-pecuniary ones, such as environmental amenities or dis-amenities (p. 279). In an effective summary of the literature Hunter shows how negatively perceived and very degraded environmental conditions are a necessary but not sufficient factor pushing people out of a place. Furthermore, the link between perceptions, desires and movement is not unitary. Mobility requires resources and often those most poor and vulnerable are stuck in place, despite negative environmental amenities (Hunter, 2005).

The flip side of disamenities as push factors, is research on environmental amenities as pull factors. Most of this research has focused on regional planning and only addresses micro perspectives in a limited way. Environmental amenity-driven migration appears to be driven by psycho-social preferences for the quality and openness of a natural environments (Chen et al., 2009). These psycho-social mechanisms are only just beginning to be researched. Some hints about the value of environmental endowments for population health have recently emerged in the study of life expectancy (Poudyal et al., 2009), in a review of urban natural environment impacts on public health (van den Bosch & Sang, 2017), in an assessment of the impact of biodiverse environments (Lovell et al., 2014), and in a study of natural environment impacts on obesity (Michimi & Wimberly, 2012). These studies suggest ways in which morbidity and mortality might be influenced by environmental endowments, but a full elaboration of the pathways of micro level mechanisms have not yet been fully explored.

Environmental hazard impacts have received considerable attention, especially as they relate to health disparities. Environmental hazards disproportionally affect the most marginalized in most societies (Gee & Payne-Sturges, 2004; Kjellstrom et al., 2007; Marmot et al., 2012). In large part, this research points less towards micro level explanations and more towards macro perspectives, especially those related to structural inequalities of race, ethnicity, and class. To the extent that there are micro level explanations, explanations tend to emphasize stress burdens and psycho-social capacities at the individual level (Gee & Payne-Sturges, 2004). To the extent that these models focus on micro level demographic outcomes they tend to focus on the immobility, or limited mobility, of poor and marginal households. For example, in a longitudinal study of US metropolitan areas, single-mother and single-father families are far more likely to reside in highly polluted neighborhoods, but they are equally like to move out of those communities as two-parent families. However, single-parent families move to more polluted neighborhoods than do two-parent families (Downey et al., 2017).

Numerous research projects in the last three decades have also studied how environmental shocks may increase the likelihood of migration. In one of the earliest studies of this type, Findley (1994) shows how migration is a response to drought in Mali. Households with fewer resources are unable to move as far as those with more liquid resources to underwrite a move. Dillon et al.'s (2011) study in Nigeria indicates how heat and drought increase the probability of migration among some household members. Sometimes, minimizing natural resource dependencies and associated risks of drought, pests, or floods requires a diversification of household human capital to off-farm employment or migration. Increasingly, remittances from migration serve to smooth incomes and underwrite risky livelihoods in Mali (Findley, 1994), Ecuador (Gray, 2009), Ethiopia (Porter, 2012), Burkina Faso (Henry et al., 2003), or Brazil (VanWey et al., 2012).

As Hunter et al. (2015) describe, climate-related changes can be either slow-onset stresses or rapid-onset disasters, which may yield different strategies. For the former, households may adapt, stay in place, and change livelihoods practices. This might include short-term, or circular, migration to earn off-farm income or it may include migration to urban labor markets to earn remittances. The migration response is typically one of internal mobility, but not always. Rapid-onset disasters can also permanently and completely displace households and communities from the disaster-impacted location. There are numerous intervening or mediating factors that shape the distance of a move and the permanency of moves and these vary quite substantially across contexts.

Environmental stress may diminish livelihoods viabilities and lead to families sending one or two members to migrate and possibly send remittances to help underwrite the needs of the origin household. This was found by Findley (1994) in Mali, Bylander (2013) in Cambodia, and Dillon et al. in Nigeria (2011). Remittances might provide resources for new investments (VanWey et al., 2012) that also help with diversifying livelihoods in the origin community. And, Greiner and Sakdapolrak (2013), along with Schefren et al. (2012), demonstrate how social and financial remittances create new ways to organize life at home, adapt to climate change, and diversify livelihoods.

Rarely, but as sometimes observed, households may completely relocate to new destinations. Individual and household mobility, however, depends on the liquidity of resources with the very poor the least likely to move and likely the most vulnerable (Bardsley & Hugo, 2010). The distance traveled in response to environmental stress depends on household credit and wealth constraints, including larger amounts of natural capital (Curran & Meijer-Irons, 2014; Gray, 2009, 2010; Hunter et al., 2013). All-in-all, several summative reviews demonstrate how migration has been an adaptive response to ongoing environmental stress throughout histories in both global north and global south settings (Gutmann & Field, 2010; McLeman, 2014).

Vulnerabilities and *resiliencies* animate much of the literature concerning population responses to climate-related disasters (Adger, 2006; Cutter, 1996; Turner et al., 2003). For the purposes of this chapter, vulnerabilities and resiliencies are characteristics of systems and not individuals or households. In that sense, vulnerabilities and resiliencies should be understood to be observed at

the macro level. Vulnerabilities and resiliencies are a function of the mix of structured relationships within a system that make a system fragile or easily broken (in the case of vulnerabilities) or strong enough to absorb or recover (in the case of resiliencies) in reaction to a shock (Proag, 2014). The combination of individual and household assets that are structured by social institutions, organizations, and relations, however, can aggregate upwards to define a system's social vulnerabilities and resiliencies. When individuals and households are located within vulnerable systems, characterized by inequalities along one or more dimensions, they become vulnerable and less capable of recovery (Cutter, 1996). These concerns are central to both the livelihoods framework and the demography and environment literatures. Black et al.'s (2011) articulation brings these ideas together. And, Eakin (2005) provides an excellent theoretical synthesis and empirical example in an analysis of institutional change, climate risk and rural livelihoods vulnerability in Mexico. An important insight from the convergence of research around climate vulnerabilities and resiliencies with the demography and environment research is how marginalized and disempowered individuals, households, and communities end up forced to face the full onslaught of natural disasters resulting in higher levels of morbidity and mortality (Davis et al., 2010; Lichtveld, 2018). Importantly, in these formulations, vulnerabilities will manifest at the individual and household level along gender, age, race, ethnicity, or class characteristics and resulting from high amounts of societal level structural inequalities.

Growing evidence about these inequalities is emerging from across the globe, whether it is New Orleans (Rudowitz et al., 2006) or Nigeria (Ibrahim et al., 2019). One result of rapid-onset environmental disasters is that by destroying homes, businesses, and neighborhoods they can displace large numbers of individuals and households. Not all people return after displacement (Sastry & Gregory, 2014). An important element can be the perceived value of place or locational ties. In some cases, perceptions about, or identification with, the value of place can help with building social capital (Silver & Grek-Martin, 2015). However, in some cases, they may not have the resiliency or social capital assets to help them return and recover (Elliott et al., 2010). And, different people, perhaps with more advantages and resources, may move into their place (Aune et al., 2020; DeWaard et al., 2016). Explanations for these variable patterns rest with multi-level perspectives that link micro and macro perspectives.

The Demography of Household Consumption and Its Impacts on Environmental Sustainability

A vast majority of the research about the re-lationship between population and environment focuses on production activities at both the macro and micro levels and in both developed and less developed settings. The role of consumption and consumptive activities in contributing to environ-mental sustainability has received limited atten-tion (Princen, 1999). Curran and De Sherbinin (2004) argue for a focus on household behav-ioral studies focused on consumption in order to more completely understand the population and environment dynamic. In their argument they called for research that expands the understanding of households', families' and individuals' pref-erences and attitudes shaping consumptive be-havior. To the extent that consumption can be linked to demographic behavior it is in relation to the household life cycle and timing of household formation or division. Lifestyle preferences and behaviors, which may or may not be facilitated by technology, can have particularly important impacts, as Smil (2002) has demonstrated with systemic analyses linking the harnessing of ni-trogen, growth in agricultural production, and the consumption of meat.

Households as a focus for linking people to their environment through consumption behaviors can make considerable conceptual and empirical sense. Households are economically and socially integrated social units. Most households pool resources among members and act collectively with regards to consumption

decision-making. Growth in the number of households is understood to be better micro unit of analysis of consumption and its environmental impacts (O'Neill et al., 2001; O'Neill & Chen, 2002). Household decisions about construction, food and nutrition, and transportation and mobility explain more than 70% of an economy's material extraction and energy consumption, and more than 90% of its land use (Spangenberg & Lorek, 2002). In the U.S., Bin and Dowlatabadi (2005) employ a consumer lifestyle analysis (CLA) to show that after accounting for direct and indirect demands for energy by households, more than 80% of energy used and CO_2 emitted in the US is a consequence of consumer demand.

Moderating household behavior over the household life cycle is one focus in sustainability science studies (Gibson et al., 2011; Reid et al., 2010). Increasingly households are recog-nized as crucial places for social intervention (Collins, 2015). However, there continues to be some debate in the literature about whether consumption is shaped by needs or desires (Wilk, 2002). While needs might drive consumption in settings of poverties or vulnerabilities, desire-based consumption, or *status consumption,* might result from the social and cultural construction of habitus inveighed through marketing, politics, and in-group forces. If these assumptions about the deeply held nature of beliefs and practices are true, then the argument for interventions revolve around efforts to socialize or naturalize disruptions to household and individual practices through social, psychological and communicative interventions (Stern, 2000; Wilk, 2002).

Examples of how practices are deeply held with regard to household formation and consump-tion are found in the sustainability science liter-ature. The premise in this work is about how to shake loose persistent, well-ingrained habits and preferences that will not be change without large-scale promotion of environmental conservation norms and rules (Newton & Meyer, 2013). For ex-ample, when privacy preferences extend families across multiple households, Klocker et al. (2012) argue that resource consumption is unnecessarily redundant and increases the ecological footprint of families. Klocker et al. (2012) show how it is possible to achieve both privacy and greater *conservation* through shared resources rather than duplication. New lines of research are beginning to describe household preferences and practices, finding considerable variability in *conservation* and sustainability practices (Waitt et al., 2012). There is also some evidence that there is differentiation across household members doing the work of sustainability within households (Organo et al., 2013; Stanes et al., 2015). A social demography analysis might usefully explain these possibilities and their patterns across populations.

At the individual level, psycho-social models argue for a value-belief-norm theory of environmentalism (Stern, 2000). In this model, value formation around biospheric, altruistic, and egoistic realms ultimately shapes behavior when norms are activated (Stern et al., 1995). The agenda described by Stern and colleagues has yielded substantial research. Some important findings include efforts to understand when consumer behavior is activated. Thorgeson and Olander (2002) find in their review of the literature that environmentally friendly behavior is related to certain values and world views which can be traced to three basic causes: difference between generations, changing conditions over an individual's life course, and periodic influences such as major personal or society-wide events. Such explanations are remarkably demographic – namely cohort, age, and year effect. Demographers often seek to explain population level patterns or changes by attributing components of change to those contributed by a particular generation (or cohort), by the age and life cycle stage (older or younger) of a population, or by the uniform impact of a period or year that might have affected all generations and all ages in a population. Typically, some combination of cohort, period and age effects yields major population changes and social or environmental impacts.

As with health behavior research, there are identifiable gaps between knowledge, attitudes and behaviors related to consumption and sustainability (Hedlund-de Witt et al., 2014; Newton & Meyer, 2013). One is that the social, economic and geographic distancing of those consuming from those producing increases the gap between

values-beliefs-norms and behavior (Luna, 2008). Similarly, Dauvergne (2010) argues that the long shadow of ecological damage is frequently hidden from the consumer, blinding the activation of their value-beliefs-norms in the final decision of whether to buy beef or refrigerators, for example. Bringing these impacts out from the shadows has been the function of social marketing efforts around food miles or ecological certification schemes and is an explicit effort to trigger social psychological mechanisms (Curran et al., 2009; Dietsch & Philpott, 2009; Linton, 2009). There may also be generational differences in value systems that transform consumption and narrow the gap through cohort replacement, as well (Chhetri et al., 2014). More immediate approaches suggest that the human-nature linkage must be drawn more tightly and that those with 'closer' relationships to nature behave more sustainably (Braito et al., 2017). Nevertheless, Sanne (2002) has argued that values are largely 'locked in' because of structural conditions and that value changes are not a fundamental mechanism that might influence changes in environmental outcomes. Rather structural systems that shape values need to be changed before micro level processes can be influential (Sanne, 2002; Spaargen, 2011). In a case study evaluating Dutch public utilities policy, Spaargen argues for a structuration approach that equilibrates the influence of both macro (policy) and micro (individuals and household) perspectives in affecting environmental conservation. In this model, rather than one or the other (macro or micro perspectives) afforded theoretical pre-eminence, both are mutually constituitive (Giddens, 1984, 1991).

As suggested earlier in this section, exposure to the natural or environmental amenities improves life expectancy and public health. Similarly, a growing body of research related to sustainable development suggests that improving natural environments improves mental health (Schleicher et al., 2018). While, it could be that deeply ingrained psychological desires might be positively correlated with environmental endowments, there is also some recognition that such deep relationships are not sufficient. In fact, Kaiser et al. (2005) find that the planned behavior model was more highly related to conservation behaviors than the value-belief-norm model.

Despite these debates, there is a significant opening in research opportunities at the intersection of psychology, anthropology, and sociology that might productively investigate consumption and through the household life cycle and individual life course approaches. While research in sustainability science and environmental psychology is fairly well developed and emerging, its systematic linkages to demographic dynamics is still underdeveloped. Demographic mechanisms might help elaborate lifestyle preferences, attitudes, knowledge and behavior as either shared or differentiated within households and help explain how those might evolve over the life course of an individual or the life cycle of a household. Some of the most obvious mechanisms might be associated with family formation timing, size, and intergenerational co-residencies. Additionally, demographic methods and data might help to more completely describe preferences and sustainability practices across a population, with study designs that can evaluate plausible micro level explanations. Given the significant advances in measurement in environmental psychology field, there are several ongoing, representative and longitudinal panel studies of households and individuals in the U.S. that might be approached to add modules over several waves to link values, preferences, and behaviors around sustainability practices.

Demographic Contributions to the Livelihoods Framework: An Integrative Scientific Paradigm

While a livelihoods framework emerged from analyses of rural development in the global south (Chambers & Conway, 1992), the utility of the framework has extended its reach to include urban settings (Castells, 2002) in the global south, as well as along the rural-urban continuum in both global north and south settings (Elliott, 2012). The idea of sustainable livelihoods has its roots in the Institute for Development Studies, now based at the University of Sussex. At its heart

is an argument that has gained resonance, which is a pro-poor (Chambers, 2013) and capabilities-centric (Sen, 1991) agenda adopted in the early 2000s by the UK's Department for International Development (DFID). This rather radical agenda sought to look for agency and power in localities and in the hands of people most affected by exogenous economic, social, or environmental forces. Thus, the model focuses on individuals and households, or a micro perspective, at its core, while not ignoring contexts and conditions that may constrain or liberate individual and household decisions.

The most complete conceptual elaboration of the livelihoods framework is best traced to Ellis' *Rural livelihoods and diversity in developing countries* (2000). Livelihoods are defined as the assets (natural, physical, human, financial and social capital), the activities, and the access to these (mediated by institutions and social relations) that together determine the living gained by the individual or household (Ellis, 2000, 10). The assets and the activities are shaped by strategies, which, in turn, are shaped by social relations, institutions, and organizations. The point of the framework is to capture the complexity and diversity of life in multiple kinds of settings from the perspectives of individuals or the family or a household. Informing the relative importance of, and the nature or characterization of, framework elements depends on the particular contexts within which a research question is explored and must rely on a mix of preceding qualitative and quantitative evidence that describes those elements.

Figure 3.1 displays an illustration of the framework based on Ellis' summary (2000). The headings in white indicate the classic illustration of the framework. In it, the five types of *assets* are the starting point and *access* to them is determined by social relations, organizations and institutions which shape the constraints and opportunities to mobilize assets. Assets and access interact with each other, as illustrated by the two curved arrows that bleed into both pillars. In turn, there are an array of local and more distant, macro-level contexts that affect risks and opportunities mediated through and by social institutions, organizations and relations, for example natural and political

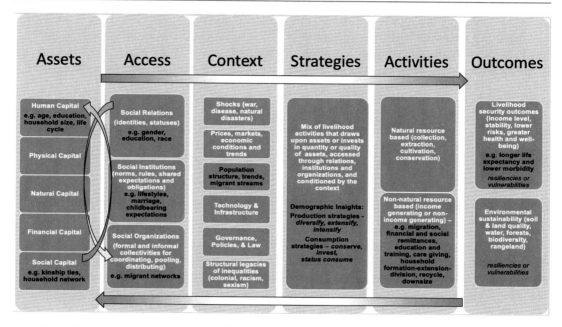

Fig. 3.1 An illustration of the sustainable livelihood framework (bold white lettering). (Adapted from Ellis (2000) and demographic insights from the population and environment literature (bold black lettering))

shocks, economic conditions, or technology or infrastructure. In total, combinations or mixes of assets, access, and context shape an individual's and family's strategies which in turn become instantiated in observed activities and subsequent outcomes in livelihoods security and environmental sustainability. This is illustrated by the linear arrow that points from left to right. There is also feedback arrow at the bottom of the figure, indicating how outcomes might, in turn, shift the array of assets and access attributes available for future livelihoods strategies. It should be noted that for most expressions of the livelihoods framework cultural and non-pecuniary attributes in relation to social capital or the three dimensions are located within the access pillar. For the most part, however, the framework is one about productivity and rarely does it emphasize preferences, values, ideas, or beliefs in a systematic conceptual or empirical way. The primary point of the original formulation was to link assets, capabilities, and options, empowering individuals and households in a productive sense (Sen, 1993).

The livelihoods framework reflected a major shift in understanding, evaluating, and intervening in economic development, especially in rural settings (Ellis, 2000). The paradigmatic shift was fairly profound. It was a move away from a solely sectoral foci for development interventions and policies after decades long critiques of failed sectoral efforts and years of research from critical development scholars and gender and development scholars that pointed towards the many ways poor, and rural, individuals or households pursued diverse strategies and strategies of diversification in order to smooth incomes and sustain themselves in conditions of under resourced or under developed social and fiscal institutions, as well as economic and ecological uncertainties and risks.

The livelihoods framework recognizes that individuals and households strategize to diversify resources and activities and that the mix of strategies is heterogeneous across individuals and households. Households are understood to extend beyond a local residence, where members may come and go or live far away but still contribute to household welfare (Bruce & Lloyd, 1992; Haddad et al., 1994). Heterogeneous communities and households are observed due to their differentiation in assets (land and education), income and social status and therefore do not share the same adversities and prospects. Their livelihoods, however, are shaped by local and distant institutions (customs, land tenure rule) and by social relations (gender, caste, kinship), as well as economic opportunities (Ellis, 2000). An important attribute in the livelihoods framework and highlighted by Ellis (2000) is the extent to which individuals and households have access to different capital types, opportunities, and services. Capital access is shaped by social relations, social organizations, and institutional norms and rules (Leach et al., 1997; North, 1990). For example, a household may have access to natural, physical, and human capital, but little access to financial or social capital. This mix of capitals is expected to yield a different set of strategies in relation to the natural environment.

A crucial epistemological aspect of a livelihoods framework is that it is built upon, and readily allows for, the integration of many disciplinary lines of social science inquiry from anthropology, demography, economics, geography, political science, and sociology (e.g. Cochrane, 2006; Odero, 2008; Reed et al., 2013). This is how demographers, especially population and environment researchers, have found value in the livelihoods framework. Demographers adopted the livelihoods framework in many contexts in large part because it fits well with a foundational theoretical perspective among demographers – multiphasic theory of demographic change – proposed and elaborated by Davis (1963, 1991).

As the preceding literature review in the second part of the chapter illustrates, livelihoods strategies are at the core for explaining micro level environment and population dynamics. Adger et al. (2002), De Sherbinin et al. (2008), and Hunter et al. (2015) make a strong case for a systematic linkage between the livelihoods framework and demographic dynamics to explain both population and ecological well-being. The relationship between migration and the environment easily adopted elements of the livelihoods framework (Adger et al., 2002; Hunter et al., 2015). The idea that the livelihoods framework offers a productive scientific paradigm for integrating micro level

perspectives around a household life cycle and environmental conditions is clearly described by De Sherbinin et al. (2008) in their account of the role of nuptiality, fertility, morbidity, mortality, and migration. These demographers' reviews, place population and environment research within a livelihoods framework offer significant elaborations around the assets and access pillars (for some examples, see: Barbieri et al., 2010, Bardsley & Hugo, 2010, Bremner et al., 2010, De Sherbinin et al., 2008, Hunter et al., 2013, Jankowska et al., 2012; Pascual & Barbier, 2006, Qin, 2010; Zenteno et al., 2013).

Building on these larger literature reviews from the past and this current chapter's updated review of the broad population and environment literature, Fig. 3.1 also elaborates the livelihoods framework with specific insights from the population and environment research. Those elaborations are bolded in black lettering. Whether it is in the micro perspective realm of the number of children, household size, members' educational attainment, relative ages of household members, or numbers of migrants or it is the macro perspective related to population composition, demographers and demographic models and concepts contribute valuable elaborations along every pillar of the livelihoods framework. The most obvious contributions are those related to human and social capitals – namely age, education, household size, and life cycle stage or kinship ties and household networks. Demographers have also usefully elaborated the contingent relationship between natural, social, and financial capital as it relates to human capital. In other words, these capital assets are not independent, but mutually constitutive.

Demographic research also provides insights on the access pillar and proximate matters related to social relations, institutions, and organizations – namely gender, education, race, lifestyles, family formation, childbearing norms, or migrant networks, to name but a few. Additionally, the relationships between the various types of capital and the domains within the access pillar are complex and reciprocal. And, as noted earlier demographers and especially those studying population, development and environment, have articulated the proximate forces shaping access to assets and presumed patterns of livelihood activities. As examples in section "Classic to Contemporary Micro Perspectives" of this chapter have illustrated, demographers studying micro level explanations are very sensitive to linkages with macro perspectives, have shown how social relations of gender, education, and race, or social institutions of marriage or childbearing expectations, or social organizations such as migrant networks, might variably explain how human, physical, natural, financial, and social capital are accessed and exploited for a livelihoods strategy. Again, the curved arrows indicate these complexities and suggest important avenues for continued scholarship and elaboration.

Overall population structures, trends and migration streams can shape access and assets in the livelihoods framework. This macro level component has not been specifically identified in the classic livelihoods framework literature, but the population and environment literature makes a strong case for singling out this factor as a crucial context. The age and sex structure of a population can create demographic deficits (e.g. Cooley & Henriksen, 2018), demographic dividends (e.g. Mason et al., 2016), or marriage market squeezes (e.g. Guilmoto, 2012), which in turn shift the value of assets or change the nature of access. Population and environment scholars have drawn on these demographic insights to elaborate the linkages between demographic dynamics and ecological outcomes through observational studies of livelihoods activities – such as migration, education, remittances, caregiving, household formation (or division) – and associated natural resource-based activities that involve collection, extraction, cultivation or conservation in either marine or forest ecologies. And, they have sought to understand the relationship to livelihoods outcomes and environmental sustainability, albeit not as systematically as might be suggested via the livelihoods framework.

Importantly, much of the work by demographers has theorized and induced several mechanisms associated with *production* strategies that could be added to that pillar of the livelihoods framework. As described in section "Classic to Contemporary Micro Perspectives" of this chapter, these include ideas about *extensification* and *intensification* of natural resource-based activities or diversification of non-natural resource-based activities and natural-resource based activities. At the micro level a preponderance of this research has focused on production and livelihoods in rural or global south settings. These insights from the population and environment literature, especially those examining migration and environment dynamics (Hunter et al., 2015), have offered invaluable conceptual elaboration about livelihoods strategies for the livelihoods framework.

Current articulations of the livelihoods framework offer either a mixed and undifferentiated set of livelihoods strategies that are highly contingent upon the mix of assets, access conditions, or contexts and with little guidance around expectations or hypotheses. Empirically, these are only discovered after observations of cases, making it difficult to fully evaluate interventions or assess predictions. If advocates of the livelihoods framework turned to the demographic literature (reviewed in section "Classic to Contemporary Micro Perspectives" of this chapter) they would discover how demographers have articulated a specific array of micro level strategies around production that have to do with *diversification, intensification,* and *extensification* (see sections "Early Ideas About Environmental Endowment Impacts on Demographic Behavior", "Overall Demographic Impacts on Environmental Endowments", "Human Migration Impacts on Environmental Endowments" and "Environmental Impacts on Migration and Health" of this chapter). This conceptual clarification effectively improves expectations about the conditions under which capital assets and access might influence livelihoods outcomes and environmental sustainability. Specifically tied to production.

A second, important contribution of the population and environment literature is a line of research inquiry focused on *consumption* and consumption strategies. Surprisingly, consumption is not explicitly articulated in the livelihoods framework. Nevertheless, while much of the literature about population and environment has predominantly focused upon production, there is no reason to exclude consumption behavior from within the livelihoods framework. In fact, there is a robust literature about consumption associated with strategies of status consumption, investments, or conservation and which is located at the micro level of individuals and households and related to individuals' and households' long term strategies to improve livelihoods and the environment (Bin & Dowlatabadi, 2005; Curran & De Sherbinin, 2004; Princen, 1999).

Furthermore, Princen (1999) argues that consumption, while distinct from production, importantly informs production. Without integrating consumption into conceptual models, sustainability is therefore poorly conceived and not achieved (Vergragt et al., 2014). The omission of consumptive aspirations from most applications of the livelihoods framework is a surprising omission, as Dorwand (2009) and Dorwand et al. note (2009). The challenge for the livelihoods framework is how best to account for shifting aspirations and the sustenance of well-being (Wright et al., 2016), while maintaining a pro-poor and capabilities standpoint. Nevertheless, Lorek (2016) suggests that the consumption domain must be included in any livelihoods framework, and better understood, in order to transition to sustainability. The growing research linking anthropology, social psychology, health, and demography is beginning to fill some of the scientific gaps at the micro level to identify the conceptual mechanisms within the consumption domain including status consumption, aspirations, and worldviews (Stern, 2000; Thorgerson & Olander, 2002) that may shift valuations and preferences with regards to capital assets and reorganize access with reformations in social institutions, organizations and relations. Demographers already have well-developed ideas about how these transformations might occur.

Along similar lines of thinking regarding the central role of the access pillar, Scoones' (2009) critical reformulation of the livelihoods framework suggests that earlier versions overlooked crucial assumptions about the meaning of good or bad livelihoods. Such critical valuation of livelihoods reveals how important it is to account for knowledge, scale, politics and dynamics, especially as it relates to both consumption and production strategies. For example, Holmes articulates such an example in a study of rural Australia (2008) as households diversify and shift landscape values from extraction to preservation and prioritize different forms of both production and consumption. Similarly, as noted earlier, social remittances via migrant networks can transmit knowledge and new forms of power that reshape values and social institutions and relations in origin communities and reprioritizing consumption and production strategies (Greiner & Sakdapolrak, 2013) and enhancing environmental sustainability in Kenya. Or, as Schneider and Niederle (2010) observe in the Brazilian context, farmers valuing autonomy and non-production imperatives can shift their livelihoods production and consumption strategies to enhance both livelihoods and the environment.

A review of the population and environment literature that evaluates micro level perspectives related to production, consumption, and demographic behavior can be valuable integrated to constructively elaborate the livelihoods framework. Furthermore, the explicit articulation of demographic behavior as it relates to the transmission of ideas and the nature of preferences links up well with recent critiques of the livelihoods framework. A summary of the consumption and sustainability literature reveals shortcomings in earlier versions of the framework, but those insights can be easily incorporated, and they provide an important elaboration. These elaborations address some of the shortcomings identified by critical poverty studies assessments (Scoones, 2009). Additional demography research, by scholars in the field of population and environment, elucidates the contextual role of population structures, the specific strategies of intensification, diversification, and extensification, the interdependent role of consumption and production strategies, and the complex mix of assets and access that interact to shape strategies. Finally, the population and environment literature is well-attuned to the micro- and macro- level linkages and the reciprocal relationships between population and environment conditions to help fully elaborate the livelihoods framework so as to inform a systemic understanding of the human-environment linkages informing sustainability, vulnerabilities and resiliencies in the short- and long-term.

Conclusion

This chapter's review of micro perspectives on the relationship between people and their environments began with a review of the longstanding linkages that demographers have theorized and evaluated between demographic behaviors and outcomes and environmental conditions. The review briefly presented Kingsley Davis' classic formulation, Boserup's theories about population and agricultural productivity through *intensification*, Easterlin's ideas about population growth and resource availability, and concludes with more recent, vicious cycle models that complicate the preceding formulations. This review of theory is complemented with an illustrative overview of the current state of knowledge substantiating the arguments.

The review continued along similar lines of inquiry and focused on micro perspectives in the migration and environment literature with a brief description of Bilsborrow's formulation to explain patterns of *extensification* (Bilsborrow, 1992) and which builds on Davis' multiphasic theory. The coincidental work on the New Economics of Labor Migration (NELM) (Stark, 1991) accelerated these formulations to include the concept of *diversification* in considerations about the role of migration in both responses to environmental conditions and as an impact on environmental outcomes. These ideas have found the most direct link to the sustainable livelihoods framework, which has also adopted migration into its own standard toolkit of strategies. With a few updates on

these formulations, I integrate evidence and ideas that provide more complexity about the role of migration in either conservation of resources, intensification resource use, or extensification of migrant impacts on environmental endowments. Importantly for a livelihoods framework, most of this work has focused on global south settings or rural settings in the global north and the research assumes that *production* processes rely on natural resources or environmental endowments and takes place within the habitus of the individual or household.

Micro perspective theorizing in recent decades has offered increasingly complicated understandings about the connections between population and environment. In particular, the scope of studies has widened to include cases from around the globe, both global north and south settings, with increasing recognition of the role of vulnerabilities and resilience as they are located at individual, household, communities, or landscape scales With the widening of the scope of settings, there is a growing recognition of commonalities regarding patterns of micro level responses. In the global north, a slightly different line of research has also examined how *consumption* is a mediating factor shaping the population and environment relationship at the micro level. Status lifestyle choices are often located at the household level and shape arguments about ecological footprints, no matter the size of households or other outcomes related to demographic behavior. These ideas have been extended to global south settings where urbanization and modern lifestyles may yield status consumption that has harmful effects on the environment.

This panoply of micro level approaches might be loosely characterized around the livelihoods framework. In the third section of the chapter, I summarize the contemporary livelihoods framework and demonstrate how it can serve as a valuable platform for integrating the wide array of micro level studies and explanations for population and environment relations that have burgeoned around the globe. Notably, the livelihoods framework can include demographic factors along most of the framework's conceptual pillars. Through an integration of micro level perspectives in demography and the livelihoods framework, I propose an elaboration of a more comprehensive research agenda that might usefully guide future scientific inquiries around the globe.

There are still important avenues for continued research related to production strategies, as noted through the review. There are also important avenues for systematically integrating consumption and production strategies into both understanding better the linked population-environment dynamic and practically intervening to limit systemic vulnerabilities and enhance systemic resiliencies. I would argue that the demography literature that focuses on population and environment has a lot to say about that linkage, especially as it relates to values, aspirations, and ideas. Demographers have frequently identified values, aspirations and ideas in relation to demographic behavior and this is no less important in the case of understanding population and environment dynamics.

Oftentimes, micro perspectives about demographic behavior and its impact indicate generational, period, or life course variables that profoundly influence numerous socioecological outcomes. Research in this field has provided valuable insights about livelihoods strategies as they relate to both production and consumption domains. With a comprehensive review of the literature from both directions of the population and environment relationship and with an eye towards elaborating both production and consumption strategies, this chapter's assessment suggests a reinvigoration of the livelihoods framework for addressing micro level relationships. Integrating micro level, demographic theorizing and variables into a livelihoods framework might significantly catalyze the kind of population and environment scientific agenda necessary to more completely understand the micro dimensions of linked human-nature systems. A more integrated framework could yield generalizable findings to inform programs and policies in relation to climate mitigation and adaptation at the individual and household level.

References

Adger, W. N. (2006). Vulnerability. *Global Environmental Change, 16*(3), 268–281.

Adger, W. N., Kelly, P. M., Winkels, A., Huy, L. Q., & Locke, C. (2002). Migration, remittances, livelihood trajectories, and social resilience. *Ambio, 31*(4), 358–366.

Aggarwal, R., Netanyahu, S., & Romano, C. (2001). Access to natural resources and the fertility decision of women: The case of South Africa. *Environment and Development Economics, 6*(2), 209–236.

Aune, K. T., Gesch, D., & Smith, G. S. (2020). A spatial analysis of climate gentrification in Orleans Parish, Louisiana post-Hurricane Katrina. *Environmental Research March, 12*, 109384.

Barbier, E. (1997). The economic determinants of land degradation in developing countries. *Philosophical Transactions of the Royal Society of London Series B: Biological Sciences., 352*(1356), 891–899.

Barbieri, A. F., Domingues, E., Queiroz, B. L., Ruiz, R. M., Rigotti, J. I., Carvalho, J. A., & Resende, M. F. (2010). Climate change and population migration in Brazil's Northeast: Scenarios for 2025–2050. *Population and Environment, 31*(5), 344–370.

Bardsley, D. K., & Hugo, G. J. (2010). Migration and climate change: Examining thresholds of change to guide effective adaptation decision-making. *Population and Environment, 32*(2–3), 238–262.

Baptista, S. R., & Rudel, T. K. (2006). A re-emerging Atlantic forest? Urbanization, industrialization and the forest transition in Santa Catarina, southern Brazil. *Environmental Conservation September, 1*, 195–202.

Biddlecom, A. E., Axinn, W., & Barber, J. S. (2005). Environmental effects on family size preferences and subsequent reproductive behavior in Nepal. *Population and Environment, 26*(3), 583–621.

Bilsborrow, R. E. (1992). Population growth, internal migration, and environmental degradation in rural areas of developing countries. *European Journal of Population/Revue Européenne de Démographie, 8*(2), 125–148.

Bilsborrow, R. E., & Okoth-Ogendo, H. W. O. (2002). Population-driven changes in land-use in developing countries. *Ambio, 21*(1), 37–45.

Bin, S., & Dowlatabadi, H. (2005). Consumer lifestyle approach to US energy use and the related CO2 emissions. *Energy Policy, 33*(2), 197–208.

Black, R., Adger, W. N., Arnell, N. W., Dercon, S., Geddes, A., & Thomas, D. (2011). The effect of environment change on human migration. *Global Environmental Change., 21*(S), S1–S3.

Boserup, E. (1965). *The conditions of agricultural growth.* Transactions.

Boserup, E. (1970). *Women's role in economic development.* George Allen & Unwin.

Braito, M. T., Böck, K., Flint, C., Muhar, A., Muhar, S., & Penker, M. (2017). Human-nature relationships and linkages to environmental behaviour. *Environmental Values, 26*(3), 365–389.

Bremner, J., López-Carr, D., Suter, L., & Davis, J. (2010). Population, poverty, environment, and climate dynamics in the developing world. *Interdisciplinary Environmental Review, 11*(2–3), 112–126.

Bruce, J., & Lloyd, C. B. (1992). *Finding the ties that bind: Beyond headship and household* (pp. 213–228). Intrahousehold Resource Allocation in Developing Countries. Population Council.

Bylander, M. (2013). Depending on the sky: Environmental distress, migration and coping in rural Cambodia. *International Migration, 53*(5), 135–147.

Carr, D. L. (2002). The role of population change in land use and land cover change in rural Latin America: Uncovering local processes concealed by macro-level data. In M. H. Y. Himiyama & T. Ichinose (Eds.), *Land use changes in comparative perspective* (pp. 133–148). Science Publishers.

Carr, D. L. (2009). Population and deforestation: Why rural migration matters. *Progress in Human Geography, 33*(3), 355–378.

Carr, D. L., Pan, W. K., & Bilsborrow, R. E. (2006). Declining fertility on the frontier: The Ecuadorian Amazon. *Population and Environment, 28*(1), 17.

Cassels, S., Curran, S. R., & Kramer, R. (2005). Do migrants degrade coastal environments? Migration, natural resource extraction and poverty in North Sulawesi, Indonesia. *Human Ecology, 33*(3), 329–363.

Castells, M. (2002). *Livable cities? Urban struggles for livelihood and sustainability*. University of California Press.

Cetina, K. K., & Cicourel, A. V. (Eds.). (2014). *Advances in social theory and methodology (RLE social theory): Toward an integration of micro-and macro-sociologies*. Routledge.

Chambers, R. (2013). *Ideas for development*. Routledge.

Chambers, R., & Conway, G. (1992). *Sustainable rural livelihoods: Practical concepts for the 21st century*. Institute of Development Studies.

Chen, Y., Irwin, E. G., & Jayaprakash, C. (2009). Dynamic modeling of environmental amenity-driven migration with ecological feedbacks. *Ecological Economics, 68*(10), 2498–2510.

Chhetri, P., Hossain, M. I., & Broom, A. (2014). Examining the generational differences in consumption patterns in South East Queensland. *City, Culture and Society, 5*(4), 1–9.

Chu, C. C., & Tsai, Y. C. (1998). Productivity, investment in infrastructure and population size: Formalizing the theory of Ester Boserup. In K. Arrow, Y. Ng, & X. Yang (Eds.), *Increasing returns and economic analysis* (pp. 90–107). Palgrave Macmillan.

Cochrane, P. (2006). Exploring cultural capital and its importance in sustainable development. *Ecological Economics, 57*, 318–330.

Collins, R. (2015). Keeping it in the family? Re-focusing household sustainability. *Geoforum, 60*, 22–32.

Cooley, T., & Henriksen, E. (2018). The demographic deficit. *Journal of Monetary Economics, 93*, 45–62.

Cumming, G. S., & Peterson, G. D. (2017). Unifying research on social–ecological resilience and collapse. *Trends in Ecology & Evolution, 32*(9), 695–713.

Curran, S. R. (2002). Migration, social capital, and the environment: Considering migrant selectivity and networks in relation to coastal ecosystems. *Population and Development Review, 28*, 89–125.

Curran, S. R., & De Sherbinin, A. (2004). Completing the picture: The challenges of bringing "consumption" into the population–environment equation. *Population and Environment, 26*(2), 107–131.

Curran, S. R., Linton, A., Cooke, A., & Schrank, A. (Eds.). (2009). *The global governance of food*. Routledge.

Curran, S. R., & Meijer-Irons, J. (2014). Climate variability, land ownership and migration: Evidence from Thailand about gender impacts. *Washington Journal of Environmental Law & Policy, 4*(1), 37–74.

Cutter, S. L. (1996). Vulnerability to environmental hazards. *Progress in Human Geography, 20*, 529–539.

Dasgupta, P. S. (1995). Population, poverty, and the local environment. *Scientific American., 272*, 40–46.

Dasgupta, P. (2000). Population and resources: An exploration of reproductive and environmental externalities. *Population and Development Review, 26*(4), 643–689.

Dauvergne, P. (2010). The problem of consumption. *Global Environmental Politics, 10*(2), 1–10.

Davis, J. R., Wilson, S., Brock-Martin, A., Glover, S., & Svendsen, E. R. (2010). The impact of disasters on populations with health and health care disparities. *Disaster Medicine and Public Health Preparedness, 4*(1), 30.

Davis, K. (1963). A theory of change and response in modern demographic history. *Population Index, 29*(4), 345–366.

Davis, K. (1991). Population and resources: Fact and interpretation. In D. Kingsley & M. S. Bernstanii (Eds.), *Resources, environment, and population—Present knowledge, future options* (pp. 1–24). Oxford University Press.

De Sherbinin, A., Carr, D., Cassels, S., & Jiang, L. (2007). Population and environment. *Annual Review of Environment and Resources, 32*, 345–373. https://doi.org/10.1146/annurev.energy.32.041306.100243

De Sherbinin, A., Van Wey, L. K., McSweeney, K., Aggarwal, R., Barbieri, A., Henry, S., Hunter, L., Twine, W., & Walker, R. (2008). Rural household demographics, livelihoods and the environment. *Global Environmental Change, 18*(1), 38–53.

DeVaney, A., & Sanchez, N. (1979). Land tenure structures and fertility in Mexico. *Review of Economics and Statistics., 61*, 67–72.

DeWaard, J., Curtis, K. J., & Fussell, E. (2016). Population recovery in New Orleans after Hurricane Katrina: Exploring the potential role of stage migration in migration systems. *Population and Environment, 37*(4), 449–463.

Dietsch, T., & Philpott, S. (2009). Linking consumers to sustainability: Incorporating science into eco-friendly certification. In S. R. Curran, A. Linton, A. Cooke, & A.

Schrank (Eds.), (pp. 247–258). The Global Governance of Food. Routledge.

Dillon, A., Mueller, V., & Salau, S. (2011). Migratory responses to agricultural risk in northern Nigeria. *American Journal of Agricultural Economics, 93*(4), 1048–1061.

Dorward, A. (2009). Integrating contested aspirations, processes and policy: Development as hanging in, stepping up and stepping out. *Development Policy Review, 27*(2), 131–146.

Dorward, A., Anderson, S., Bernal, Y. N., Vera, E. S., Rushton, J., Pattison, J., & Paz, R. (2009). Hanging in, stepping up and stepping out: Livelihood aspirations and strategies of the poor. *Development in Practice, 19*(2), 240–247.

Downey, L., Crowder, K., & Kemp, R. J. (2017). Family structure, residential mobility, and environmental inequality. *Journal of Marriage and Family, 79*(2), 535–555.

Eakin, H. (2005). Institutional change, climate risk, and rural vulnerability: Cases from Central Mexico. *World Development, 33*(11), 1923–1938.

Easterlin, R. A. (1976). Population change and farm settlement in the northern United States. *Journal of Economic History, 36*(1), 45–75.

Ellis, F. (2000). *Rural livelihoods and diversity in developing countries.* Oxford University Press.

Elliott, J. (2012). *An introduction to sustainable development.* Routledge.

Elliott, J. R., Haney, T. J., & Sams-Abiodun, P. (2010). Limits to social capital: Comparing network assistance in two New Orleans neighborhoods devastated by hurricane Katrina. *The Sociological Quarterly, 51*(4), 624–648.

Entwisle, B., Walsh, S. J., Rindfuss, R. R., & Van Wey, L. K. (2005). Population and upland crop production in Nang Rong, Thailand. *Population and Environment, 26*(6), 449–470.

Feeney, D. (1988). Agricultural expansion and forest depletion in Thailand, 1900-1975. In J. F. Richards & R. P. Tucker (Eds.), *World deforestation in the twentieth century* (pp. 425–454). Duke University Press.

Filmer, D., & Pritchett, L. H. (2002). Environmental degradation and the demand for children: Searching for the vicious cycle in Pakistan. *Environment and Development Economics, 7*, 123–146.

Findley, S. E. (1994). Does drought increase migration? A study of migration from rural Mali during the 1983–1985 drought. *International Migration Review, 28*, 539–553.

Firebaugh, G. (1982). Population density and fertility in 22 Indian villages. *Demography, 19*, 481–494.

Freidlander, D. (1983). Demographic responses and socioeconomic structure: Population processes in England and Wales in the Nineteenth Century. *Demography, 20*, 249–272.

Gee, G. C., & Payne-Sturges, D. C. (2004). Environmental health disparities: A framework integrating psychosocial and environmental concepts. *Environmental Health Perspectives, 112*(17), 1645–1653.

Geist, H. J., & Lambin, E. F. (2002). Proximate causes and underlying driving forces of tropical deforestation. *Bioscience, 52*(2), 143–150.

Gibson, C., Head, L., Gill, N., & Waitt, G. (2011). Climate change and household dynamics: Beyond consumption, unbounding sustainability. *Transactions of the Institute of British Geographers, 36*(1), 3–8.

Giddens, A. (1984). *The constitution of society.* Polity Press.

Giddens, A. (1991). *Modernity and self-identity.* Polity Press.

Grace, K. (2017). Considering climate in studies of fertility and reproductive health in poor countries. *Nature Climate Change, 7*(7), 479–485.

Gray, C. L. (2009). Rural out-migration and smallholder agriculture in the southern Ecuadorian Andes. *Population and Environment, 31*, 3–19.

Gray, C. L. (2010). Gender, natural capital, and migration in the southern Ecuadorian Andes. *Environment Planning A, 42*(3), 678–696.

Greiner, C., & Sakdapolrak, P. (2013). Rural–urban migration, agrarian change, and the environment in Kenya: A critical review of the literature. *Population and Environment, 34*(4), 524–553.

Grimm, M. (2019). Rainfall risk, fertility and development: Evidence from farm settlements during the American demographic transition. *Journal of Economic Geography*, 1–26.

Guilmoto, C. Z. (2012). Skewed sex ratios at birth and future marriage squeeze in China and India, 2005–2100. *Demography, 49*(1), 77–100.

Gutmann, M. P., & Field, V. (2010). Katrina in historical context: Environment and migration in the U.S. *Population and Environment, 31*(1–3), 3–19.

Haddad, L., Hoddinott, J., & Alderman, H. (1994). *Intrahousehold resource allocation: An overview.* The World Bank.

Hardin, G. (1968). The tragedy of the commons. In G. Hardin & J. Baden (Eds.), *Managing the commons.* WH Freeman.

Hedlund-de Witt, A., De Boer, J., & Boersema, J. J. (2014). Exploring inner and outer worlds: A quantitative study of worldviews, environmental attitudes, and sustainable lifestyles. *Journal of Environmental Psychology, 37*, 40–54.

Henry, S., Boyle, P., & Lambin, E. F. (2003). Modelling inter-provincial migration in Burkina Faso, West Africa: The role of socio-demographic and environmental factors. *Applied Geography, 23*(2), 115–136.

Hiday, V. A. (1978). Agricultural organization and fertility. *Social Biology, 25*, 69–79.

Holmes, J. (2008). Impulses towards a multifunctional transition in rural Australia: Interpreting regional dynamics in landscapes, lifestyles and livelihoods. *Landscape Research, 33*(2), 211–223.

Hunter, L. M. (2005). Migration and environmental hazards. *Population and Environment, 26*(4), 273–302.

Hunter, L. M., Luna, J. K., & Norton, R. M. (2015). Environmental dimensions of migration. *Annual Review of Sociology, 41*, 377–397.

Hunter, L. M., Murray, S., & Riosmena, F. (2013). Rainfall patterns and US migration from rural Mexico. *International Migration Review, 47*(4), 874–909.

Hunter, M., Nawrotski, R., Leyk, S., Maclurin, G. J., Twine, W., et al. (2014). Rural outmigration, natural capital, and livelihoods in South Africa. *Population, Space, and Place, 20*(5), 402–420.

Ibrahim, S. S., Ozdeser, H., & Cavusoglu, B. (2019). Vulnerability to recurrent shocks and disparities in gendered livelihood diversification in remote areas of Nigeria. *Environmental Science and Pollution Research, 26*(3), 2939–2949.

Izquierdo, A. E., De Angelo, C. D., & Aide, T. M. (2008). Thirty years of human demography and land-use change in the Atlantic Forest of Misiones, Argentina: An evaluation of the forest transition model. *Ecology and Society, 13*(2), 3–21.

Jankowska, M. M., Lopez-Carr, D., Funk, C., Husak, G. J., & Chafe, Z. A. (2012). Climate change and human health: Spatial modeling of water availability, malnutrition, and livelihoods in Mali, Africa. *Applied Geography, 33*, 4–15.

Kabubo-Mariara, J. (2007). Land conservation and tenure security in Kenya: Boserup's hypothesis revisited. *Ecological Economics, 64*(1), 25–35.

Kaiser, F. G., Hübner, G., & Bogner, F. X. (2005). Contrasting the theory of planned behavior with the value-belief-norm model in explaining conservation behavior. *Journal of Applied Social Psychology, 35*(10), 2150–2170.

Kjellstrom, T., Friel, S., Dixon, J., Corvalan, C., Rehfuess, E., Campbell-Lendrum, D., Gore, F., & Bartram, J. (2007). Urban environmental health hazards and health equity. *Journal of Urban Health, 84*(1), 86–97.

Klocker, N., Gibson, C., & Borger, E. (2012). Living together but apart: Material geographies of everyday sustainability in extended family households. *Environment and Planning A, 44*(9), 2240–2259.

Kramer, D. B., Urquhart, G., & Schmitt, K. (2009). Globalization and the connection of remote communities: A review of household effects and their biodiversity implications. *Ecological Economics, 68*(12), 2897–2909.

Leach, M., Mearns, R., & Scoones, I. (1997). Challenges to community-based sustainable development: Dynamics, entitlements, institutions. *IDS Bulletin, 28*(4), 4–14.

Levitt, P. (1999). Social remittances: A local-level, migration-driven form of cultural diffusion. *International Migration Review, 32*(124), 926–949.

Lichtveld, M. (2018). Disasters through the lens of disparities: Elevate community resilience as an essential public health service. *American Journal of Public Health, 108*(1), 28–30.

Linton, A. (2009). A niche for sustainability? Fair labor and environmentally sound practices in the specialty coffee industry. In S. R. Curran, A. Linton, A. Cooke, & A. Schrank (Eds.), *The global governance of food* (pp. 231–245). Routledge.

Lira, M. G., Robson, J. P., & Klooster, D. J. (2016). Can indigenous transborder migrants affect environmental governance in their communities of origin? Evidence from Mexico. *Population and Environment, 37*(4), 464–478.

Lorek, S. (2016). Sustainable consumption. In H. G. Brauch, U. O. Spring, J. Grin, & J. Scheffren (Eds.), *Handbook on sustainability transition and sustainable peace* (pp. 559–570). Springer.

Lorenzen, M., Orozco-Ramírez, Q., Ramírez-Santiago, R., & Garza, G. G. (2020). Migration, socioeconomic transformation, and land-use change in Mexico's Mixteca Alta: Lessons for forest transition theory. *Land Use Policy, 95*, 1–13. 104580. https://doi.org/95.104580

Lovell, R., Wheeler, B. W., Higgins, S. L., Irvine, K. N., & Depledge, M. H. (2014). A systematic review of the health and well-being benefits of biodiverse environments. *Journal of Toxicology and Environmental Health, Part B, 17*(1), 1–20.

Luna, M. (2008). Out of sight, out of mind: Distancing and the geographic relationship between electricity consumption and production in Massachusetts. *Social Science Quarterly, 89*(5), 1277–1292.

Malan, N. (2015). Urban farmers and urban agriculture in Johannesburg: Responding to the food resilience strategy. *Agrekon, 54*(2), 51–75.

Malthus, T. R. ([1798] 1960). An essay on the principle of population. In Himmelfarb, G (Eds.), On *Population: Thomas Robert Malthus*. Modern Library.

Marmot, M., Allen, J., Bell, R., Bloomer, E., & Goldblatt, P. (2012). WHO European review of social determinants of health and the health divide. *The Lancet, 380*(9846), 1011–1029.

Mason, A., Lee, R., & Jiang, J. X. (2016). Demographic dividends, human capital, and saving. *The Journal of the Economics of Ageing, 7*, 106–122.

McCracken, S., Brondizio, E., Nelson, D., Moran, E., Siqueira, A., & Rodriguez-Pedraza, C. (1999). Remote sensing and GIS at farm property level: Demography and deforestation in the Brazilian Amazon. *Photogrammetric Engineering and Remote Sensing, 65*(11), 1311–1320.

McLeman, R. A. (2014). *Climate and human migration: Past experiences, future challenges*. Cambridge University Press.

Michimi, A., & Wimberly, M. C. (2012). Natural environments, obesity, and physical activity in nonmetropolitan areas of the United States. *The Journal of Rural Health, 28*(4), 398–407.

Myrdal, G. (1957). *Rich lands and poor: The road to world prosperity*. Harper Brothers.

Newton, P., & Meyer, D. (2013). Exploring the attitudes-action gap in household resource consumption: Does "environmental lifestyle" segmentation align with consumer behaviour? *Sustainability, 5*(3), 1211–1233.

Nguyen, T. T., Do, T. L., Bühler, D., Hartje, R., & Grote, U. (2015). Rural livelihoods and environmental resource dependence in Cambodia. *Ecological Economics, 120*, 282–295.

North, D. (1990). *Institutions, institutional change and economic performance*. Cambridge University Press.

Odero, K. K. (2008). Information capital: 6th asset of sustainable livelihood framework. *Discovery and Innovation, 18*, 83–91.

O'Neill, B. C., MacKellar, F. L., & Lutz, W. (2001). *Population and climate change*. Cambridge University Press.

O'Neill, B. C., & Chen, B. S. (2002). Demographic determinants of household energy use in the United States. In W. Lutz, A. Prskawetz, & W. C. Sanderson (Eds.), *Population and environment: Methods of analysis, population and development review* (Vol. 28S, pp. 53–88).

Organo, V., Head, L., & Waitt, G. (2013). Who does the work in sustainable households? A time and gender analysis in New South Wales, Australia. *Gender, Place & Culture, 20*(5), 559–577.

Panayotou, T. (1994). The population, environment, and development nexus. In R. Cassen (Ed.), *Population and development: Old debates, new conclusions* (pp. 48–80). Overseas Dev. Council.

Pan, W. K., Walsh, S. J., Bilsborrow, R. E., Frizzelle, B., Erlien, C., & Baquero, F. (2004). Farm-level models of spatial patterns of land use and land cover dynamics in the Ecuadorian Amazon. *Agriculture, Ecosystems & Environment., 101*, 117–134.

Pascual, U., & Barbier, E. (2006). Deprived land-use intensification in shifting cultivation: The population pressure hypothesis revisited. *Agricultural Economics, 34*, 155–165.

Paulson, A. (1995). *Insurance motives for migration: Evidence from Thailand, draft*. Princeton University.

Paxson, C. (1992). Using weather variability to estimate the response of savings to transitory income in Thailand. *American Economic Review, 82*(1), 15–33.

Perz, S. G. (2001). Household demographic factors as life cycle determinants of land use in the Amazon. *Population Research and Policy Review, 20*(3), 159–186.

Perz, S. G. (2007). Grand theory and context-specificity in the study of forest dynamics: Forest transition theory and other directions. *The Professional Geographer, 59*(1), 105–114.

Pender, J. L. (1998). Population growth, agricultural intensification, induced innovation and natural resource sustainability: An application of neoclassical growth theory. *Agricultural Economics, 19*(1–2), 99–112.

Porter, C. (2012). Shocks, consumption and income diversification in Rural Ethiopia. *Journal of Development Studies, 48*(9), 1209–1222.

Poudyal, N. C., Hodges, D. G., Bowker, J. M., & Cordell, H. K. (2009). Evaluating natural resource amenities in a human life expectancy production function. *Forest Policy and Economics, 11*(4), 253–259.

Princen, T. (1999). Consumption and environment: Some conceptual issues. *Ecological Economics, 31*(3), 347–363.

Proag, V. (2014). The concept of vulnerability and resilience. *Procedia Economics and Finance, 18*, 369–376.

Qin, H. (2010). Rural-to-urban labor migration, household livelihoods, and the rural environment in Chongqing municipality, Southwest China. *Human Ecology, 38*(5), 675–690.

Qin, H., & Flint, C. (2012). The impacts of rural labor out-migration on community interaction and implications for rural community-based environmental conservation in Southwest China. *Human Organization, 71*(2), 135–148.

Rasmussen, L. V., & Reenberg, A. (2012). Land use rationales in desert fringe agriculture. *Applied Geography, 34*, 595–605.

Reed, M. S., Podesta, G., Fazey, I., Geeson, N., Hessel, R., Hubacek, K., Letson, D., Nainggolan, D., Prell, C., Rickenbach, M. G., Ritsema, C., Schwilch, G., Stringer, L. C., & Thomas, A. D. (2013). Combining analytical frameworks to assess livelihood vulnerability to climate change and analyze adaptation options. *Ecological Economics, 94*, 66–77. https://doi.org/10.1016/j.ecolecon.2013.07.007

Reid, L., Sutton, P., & Hunter, C. (2010). Theorizing the meso level: The household as a crucible of pro-environmental behaviour. *Progress in Human Geography, 34*(3), 309–327.

Richerson, P. J., & Boyd, R. (1997). Homage to Malthus, Ricardo, and Boserup: Toward a general theory of population, economic growth, environmental deterioration, wealth, and poverty. *Human Ecology Review. Dec, 1*, 85–90.

Robson, J. P., & Berkes, F. (2011). Exploring some of the myths of land use change: Can rural to urban migration drive declines in biodiversity? *Global Environmental Change, 21*(3), 844–854.

Robson, J. P., & Nayak, P. K. (2010). Rural out-migration and resource-dependent communities in Mexico and India. *Population and Environment, 32*(2–3), 263–284.

Rosenzweig, M., & Oded, S. (1989). Consumption smoothing, migration, and marriage: Evidence from rural India. *Journal of Political Economy, 97*(4), 905–926.

Rudel, T. K., Bates, D., & Machinguiashi, R. (2002). A tropical forest transition? Agricultural change, out-migration, and secondary forests in the Ecuadorian Amazon. *Annals of the Association of American Geographers, 92*(1), 87–102.

Rudowitz, R., Rowland, D., & Shartzer, A. (2006). Health care in New Orleans before and after Hurricane Katrina: The storm of 2005 exposed problems that had existed for years and made solutions more complex and difficult to obtain. *Health Affairs, 25*(Suppl 1), W393–W406.

Sanne, C. (2002). Willing consumers – Or locked in? Policies for a sustainable consumption. *Ecological Economics, 42*, 273–287.

Sastry, N., & Gregory, J. (2014). The location of displaced New Orleans residents in the year after hurricane Katrina. *Demography, 51*(3), 753–775.

Scheffren, J., Marmer, E., & Sow, P. (2012). Migration as a contribution to resilience and innovation in climate adaptation: Social networks and co-development in Northwest Africa. *Applied Geography, 33*, 119–127.

Schleicher, J., Schaafsma, M., Burgess, N. D., Sandbrook, C., Danks, F., Cowie, C., & Vira, B. (2018). Poorer without it? The neglected role of the natural environment in poverty and wellbeing. *Sustainable Development, 26*(1), 83–98.

Schneider, S., & Niederle, P. A. (2010). Resistance strategies and diversification of rural livelihoods: The construction of autonomy among Brazilian family farmers. *The Journal of Peasant Studies, 37*(2), 379–405.

Schutjer, W. A., Stokes, C. S., & Poindexter, R. (1983). Farm size, land ownership, and fertility in rural Egypt. *Land Economics, 59*, 393–403.

Scott, J. C. (2017). *Against the grain: A deep history of the earliest states*. Yale University Press.

Scoones, I. (2009). Livelihoods perspectives and rural development. *The Journal of Peasant Studies, 36*(1), 171–196.

Sen, A. (1991). Welfare, preference and freedom. *Journal of Econometrics, 50*(1-2), 15–29.

Sen, A. (1993). Capability and well-being. *Quality of Life, 30*, 270–293.

Serrat, O. (2017). The sustainable livelihoods approach. In O. Serrat (Ed.), *Knowledge solutions* (pp. 21–25). Springer. https://doi.org/10.1007/978-981-10-0983-9_5

Sharif, M., & Saha, R. K. (1993). The observed landholding-fertility relationship-is it monotonic? *The Journal of Development Studies, 29*(2), 319–341.

Shreffler, K. M., & Dodoo, F. N. A. (2009). The role of intergenerational transfers, land, and education in fertility transition in rural Kenya: The case of Nyeri district. *Population and Environment, 30*(3), 75–92.

Silver, A., & Grek-Martin, J. (2015). "Now we understand what community really means": Reconceptualizing the role of sense of place in the disaster recovery process. *Journal of Environmental Psychology, 42*, 32–41.

Smil, V. (2002). Nitrogen and food production: Proteins for human diets. *Ambio: A Journal of the Human Environment, 31*(2), 126–131.

Southgate, E. W. (2019). *People and the land through time: Linking ecology and history*. Yale University Press.

Spaargaren, G. (2003). Sustainable consumption: A theoretical and environmental policy perspective. *Society & Natural Resources, 16*(8), 687–701.

Spangenberg, J., & Lorek, S. (2002). Environmentally sustainable household consumption: From aggregate environmental pressures to priority fields of action. *Ecological Economics, 43*, 127–140.

Spear, A. (1974). Residential satisfaction as an intervening variable in residential mobility. *Demography, 11*(2), 173–188.

Stanes, E., Klocker, N., & Gibson, C. (2015). Young adult households and domestic sustainabilities. *Geoforum, 65*, 46–58.

Stark, O., & Bloom, D. E. (1985). The new economics of labor migration. *The American Economic Review, 75*(2), 173–178.

Stark, O. (1991). *The migration of labor*. Basil Blackwell.

Stern, P. C. (2000). New environmental theories: Toward a coherent theory of environmentally significant behavior. *Journal of Social Issues, 56*(3), 407–424.

Stern, P. C., Dietz, T., & Guagnano, G. A. (1995). The new ecological paradigm in social-psychological context. *Environment and Behavior, 27*(6), 723–743.

Sutherland, E. G., Carr, D. L., & Curtis, S. L. (2004). Fertility and the environment in a natural resource dependent economy: Evidence from Peten, Guatemala. *Poblacion y Salud en Mesoamerica, 2*(1), 1–14.

Thorgerson, J., & Olander, F. (2002). Human values and the emergence of a sustainable consumption pattern: A panel study. *Journal of Economic Psychology, 23*, 605–630.

Turner, B. L., & Fischer-Kowalski, M. (2010). Ester Boserup: An interdisciplinary visionary relevant for sustainability. *Proceedings of the National Academy of Sciences, 107*(51), 21963–21965.

Turner, B. L., Kasperson, R. E., Matson, P. A., McCarthy, J. J., Corell, R. W., Christensen, L., Eckley, N., Kasperson, J. X., Luers, A., Martello, M. L., & Polsky, C. (2003). A framework for vulnerability analysis in sustainability science. *Proceedings of the National Academy of Sciences, 100*(14), 8074–8079.

Turner, J. H. (2005). A new approach for theoretically integrating micro and macro analysis. In C. Calhoun, C. Rojek, & B. Turner (Eds.), *The sage handbook of sociology* (pp. 405–422). Sage.

van den Bosch, M., & Sang, Å. O. (2017). Urban natural environments as nature-based solutions for improved public health–a systematic review of reviews. *Environmental Research, 158*, 373–384.

Van Landingham, M., & Hirschman, C. (2001). Population pressure and fertility in pre-transition Thailand. *Population Studies, 55*(3), 233–248.

VanWey, L. K. (2003). Land ownership as a determinant of temporary migration in Nang Rong, Thailand. *European Journal of Population, 19*, 121–145.

VanWey, L. K. (2005). Land ownership as a determinant of international and internal migration in Mexico and internal migration in Thailand. *International Migration Review, 39*(1), 141–172.

VanWey, L. K., Guedes, G. R., & D'Antona, A. O. (2012). Out-migration and land-use change in agricultural frontiers: Insights from Altamira settlement project. *Population and Environment, 34*(1), 44–68.

Vergragt, P., Akenji, L., & Dewick, P. (2014). Sustainable production, consumption, and livelihoods: Global and regional research perspectives. *Journal of Cleaner Production, 63*, 1–12.

Waitt, G., Caputi, P., Gibson, C., Farbotko, C., Head, L., Gill, N., & Stanes, E. (2012). Sustainable household capability: Which households are doing the work of environmental sustainability? *Australian Geographer, 43*(1), 51–74.

Walker, R., Perz, S., Caldas, M., & Teixeira Silva, L. G. (2002). Land use and land cover change in forest frontiers: The role of household life cycles. *International Regional Science Review, 25*(2), 169–199.

Waring, T. M., Kline, A. M., Brooks, J. S., Goff, S. H., Gowdy, J., Janssen, M. A., & Jacquet, J. (2015). A multilevel evolutionary framework for sustainability analysis. *Ecology and Society, 20*(2), 34–49.

Wilk, R. (2002). Emulation, imitation and global consumerism. *Organization and Environment, 11*(3), 314–333.

Winfrey, W., & Darity, W., Jr. (1997). Increasing returns and intensification: A reprise on Ester Boserup's model of agricultural growth. *Metroeconomica, 48*(1), 60–80.

Wolpert, J. (1966). Migration as an adjustment to environmental stress. *Journal of Social Issues, 22*(4), 92–102.

Wright, J. H., Hill, N. A., Roe, D., Rowcliffe, J. M., Kümpel, N. F., Day, M., Booker, F., & Milner-Gulland, E. J. (2016). Reframing the concept of alternative livelihoods. *Conservation Biology, 30*(1), 7–13.

Yabiku, S. T. (2006). Land use and marriage timing in Nepal. *Population and Environment, 27*(5–6), 445–461.

Yang, D., & Choi, H. (2007). Are remittances insurance? Evidence from rainfall shocks in the Philippines. *The World Bank Economic Review, 21*(2), 219–248.

Zenteno, M., Zuidema, P. A., de Jong, W., & Boot, R. G. (2013). Livelihood strategies and forest dependence: New insights from Bolivian forest communities. *Forest Policy and Economics, 26*, 12–21.

Vulnerability to Climate Change and Adaptive Capacity from a Demographic Perspective

4

Raya Muttarak

Abstract

Human population is at the centre of the global climate system. The prominent contribution of demography in the field of climate change is in understanding how current and future population size, distribution and composition drive climate changing carbon emissions. Given evidence for rapid global warming, it is equally important to understand the impact of climate change on human population. Here demography can contribute to identifying vulnerable populations and their locations and provide forecasting of future societal vulnerability. This chapter focuses on vulnerability to climate change and adaptive capacity from a demographic perspective. Upon discussion on the relevance of demography in global climate change research, we present demographic concepts and tools that can be applied to the study of vulnerability and adaptive capacity. In particular, combining the empirical knowledge on differential impacts of climate change

on population subgroups with the theoretical concept of demographic metabolism allows for modelling of future social and demographic change. Here the demographic method of multidimensional cohort-component population projections is introduced as a fundamental tool for projections of population dynamics. Finally, based on the socioeconomic scenarios developed in the context of the Intergovernmental Panel on Climate change (IPCC), it is possible to forecast vulnerability and adaptive capacity of population in the future.

Introduction

Many effects of human-induced climate change can already be felt. Continuing increases in carbon dioxide emissions will eventually lead to irreversible adverse effects including atmospheric warming, precipitation changes and sea level rise (Solomon et al., 2009). Therefore, whilst mitigation efforts to curb global warming are fundamental, simultaneously adaptation is essential to reduce unavoidable effects of climate change. In order to design effective risk reduction and adaptation strategies, it is important to consider vulnera-

R. Muttarak (✉)
Wittgenstein Centre for Demography and Global Human Capital (IIASA, OeAW, University of Vienna), International Institute for Applied Systems Analysis, Laxenburg, Austria
e-mail: muttarak@iiasa.ac.at

© Springer Nature Switzerland AG 2022
L. M. Hunter et al. (eds.), *International Handbook of Population and Environment*, International Handbooks of Population 10, https://doi.org/10.1007/978-3-030-76433-3_4

bility and exposure and their relations with socio-economic processes (Core Writing Team et al., 2014). Likewise, successful adaptation requires knowledge of both the near-term and longer term climate risks as well as societies' capacity to adapt. Knowledge and methods in demography can be applied to improve our understanding of uncertainties because certain aspects of societal development can be quantified and forecasted (Lutz & Muttarak, 2017). This highlights the role social sciences can play in improving adaptation planning efforts through providing better understanding of social drivers of climate change vulnerability and unequal distribution of climate impacts (Thomas et al., 2019).

To that end, this chapter focuses on vulnerability and adaptive capacity from a demographic perspective. The chapter begins with the discussion on the relevance of demography in global climate change research. Subsequently, it presents demographic concepts and tools that can be applied to the study of vulnerability and adaptive capacity. This section first presents empirical studies on differential impacts of climate change on population subgroups and subsequently introduces the theoretical concept of demographic metabolism which provides a framework for modeling social change. Next, the demographic method of multidimensional cohort-component population projections is presented as a fundamental tool to provide projections for population dynamics in the future. The chapter then provides an example of empirical assessment and projection exercises with a focus on education as an illustration of a demographic characteristic relevant to vulnerability and adaptive capacity. The chapter concludes with suggestions on future applications of toolkits in demography in global climate change research.

Relevance of Demography to Global Climate Change Research

Population dynamics are closely linked to the global climate system as presented in Fig. 4.1. The lower left side of the figure shows that greenhouse gas (GHG) emissions that drive climate change are influenced by human activities such as energy consumption, transportation use and economic activities. The impact of human activities on the environment can be expressed in the equation: $I = PAT$ where environmental impact (I) is a product of population (P), affluence (A) and technology (T) (Ehrlich & Holdren, 1971). Conventionally, the effect of population on GHG emissions was assumed to be only through population size. However, recent studies have shown that population structure and its distribution including age structure, household size, rural/urban residence have significant effect on consumption and emissions (MacKellar et al., 1995; O'Neill et al., 2001, 2010; Cohen, 2010). Changes in both population size and composition thus pose influential consequences on the global climate system.

Population dynamics also matters greatly with respect to human vulnerability to the impacts of global climate change and the ability to cope with its consequences. Climate change-induced extreme weather events and temperature and rainfall fluctuations directly and indirectly affect human wellbeing and livelihood. This ranges from mortality risk to health, job and income loss and migration. Population size is linked with vulnerability since rapid population growth and high-density populations increase the number of people exposed to climate impacts as well as put pressure on the provision of basic services and infrastructure (Jiang & Hardee, 2011). However, the impacts are not distributed evenly across population subgroups and the ability to adapt and cope with global climate change varies substantially with population characteristics (Muttarak et al., 2016). Hence, understanding future compositional change of populations such as age, sex, education and urban and rural distribution is fundamental in coping with and adapting to global climate change.

The urgency of considering population dynamics and population heterogeneity in development and climate policy has already been put forward by international experts: first in preparation for the United Nations (UN) World Summit on Sustainable Development in 2002 (Lutz & Shah, 2002); and a decade later in preparation for the RIO+20 Earth Summit (Lutz et al., 2012). This

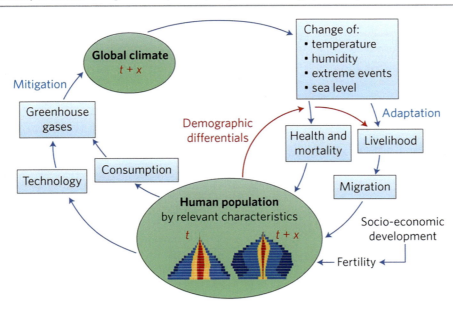

Fig. 4.1 Closing the full circle of mitigation (on the left) and vulnerability and adaptation (on the right) of population and climate change interactions. (Source: Lutz and Muttarak (2017), Figure 1)

highlights the major role demographers can play in providing relevant knowledge to worldwide climate change mitigation and adaption efforts.

Application of Demographic Concepts and Tools in Climate Change Research

Given the relevance of population dynamics in the global climate systems, concepts and tools readily available in demography can be applied to improve the understanding of vulnerability and adaptive capacity. The application of demographic toolkits has three components as presented in Fig. 4.2.

Lutz and Muttarak (2017) have proposed how the theory of demographic metabolism and the multi-dimensional population projection methods can be applied to forecast future societies adaptive capacity. In order to do so, this requires empirical evidence on how climate change affects population subgroups differentially, a theoretical concept with predictive power related to societal transformation, and a method that is able to provide projections for population dynamics in the future.

Demographically Differentiated Vulnerability

Policy design to reduce vulnerability or to enhance adaptive capacity to global climate change cannot be effective if demographic differentials are not taken into account. Not only are the impacts of climate change not distributed evenly, the capacity to adapt also depends on observable demographic characteristics such as age, gender, education, income and place of residence. In fact, the topic of differentials across population subgroups has long been a focus in demographic research. The estimation of the hazard function which originally focuses on estimating the risk of death by age and sex in demography can be applied to the study of differential vulnerability. In terms of policy, knowing *who* is vulnerable and to *what* allows for designing a policy response that appropriately addressing the needs of population subgroups.

Before looking closely into evidence on demographically differentiated vulnerability, it is useful to define what vulnerability is referred to in this context. The term vulnerability has been widely used in the context of climate change, but different scholars tend to use it in different ways,

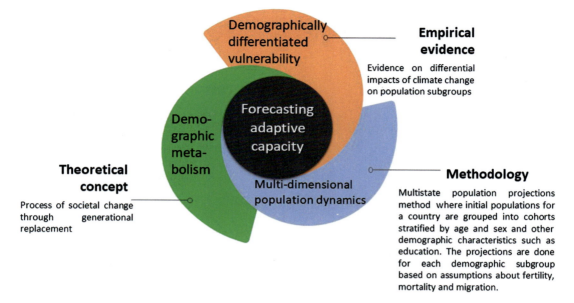

Fig. 4.2 Demographic concepts and tools relevant for understanding vulnerability and forecasting adaptive capacity

often without a clear definition. The Intergovernmental Panel on Climate Change (IPCC) in its recent Fifth Assessment Report defines vulnerability as "The propensity or predisposition to be adversely affected. Vulnerability encompasses a variety of concepts and elements including sensitivity or susceptibility to harm and lack of capacity to cope and adapt." (IPCC, 2014) In this context, vulnerability is clearly specified as a characteristic of a *living unit* that influences its susceptibility to be harmed by external shocks as well as capacity to anticipate, respond, cope with and recover from the impact of such shocks. The term vulnerability however has been confusedly used to refer to a physical dimension of a system or a place such as the potential climatic threats on the regions that are affected (IPCC, 1996). We argue that neither building structures nor specific regions can be vulnerable, but only the people and other species inhabiting in such locations are. Indeed, the original meaning of the Latin word *vulnerare* means "to wound, hurt or injure" suggesting that it is living beings who are vulnerable.

In accordance with the IPCC definition, we further contend that it is misleading to focus on vulnerability of the systems *per se*, but rather it is the adverse effect on people who are embedded in the systems that should be of primary concern. Natural or social systems can of course collapse but whether this is considered to be threatening or welcome – e.g., the collapse of a dictatorial regime – depends on its effect on human wellbeing. In fact, the concept of social vulnerability has been created to emphasize on inherent properties of human systems that make a person or a group susceptible to damage from external hazards as well as to characterize ability to respond to and recover from the impacts of hazards (Blaikie et al., 1994; Adger, 1999). While the term social vulnerability became commonly used to distinguish vulnerability of the human systems from non-human (nature) systems, it still does not explicitly focus on human wellbeing.

The concept of *demographically differentiated vulnerability* explicitly defines that a human being is the unit that is vulnerable (Muttarak et al., 2016). To understand why certain subgroups of population are more vulnerable to global climate change, it is useful to consider how risk factors are accumulated. Differential vulnerability results from differences in physiological susceptibility, hazard exposure and socioeconomic and psychosocial factors influencing risk perceptions and capacity to respond. While conventional demographic characteristics such as age, gender, race/ethnicity and income are considered as a key source of heterogeneity, here education is highlighted as an additionally important determinant of vulnerability. Both exposure and vulnerability to the impact of common exposures differ substantially by population subgroups as elaborated below.

Physiologic Susceptibility

Biological differences make different demographic groups more or less susceptible to extreme events and climatic shocks. Physiologically, the elderly have less tolerance and responsiveness to temperature extremes due to a declining ability for thermoregulation (Baccini et al., 2008; Blatteis, 2012; Wanka et al., 2014; Kenny et al., 2017). Older people are also susceptible to changes in body temperature due to a higher level of initial arterial diseases (Keatinge et al., 1984). Furthermore, some medications that older people take can reduce thirst sensation and consequently make them more prone to dehydration. Accordingly, the elderly (largely defined as a person aged 65 years or older) are commonly found to experience higher risk of morbidity and mortality during extreme heat events than other age groups. This pattern is found in many countries around the world including European countries, the USA, Australia, China and Korea (Robine et al., 2008; Schaffer et al., 2012; Bai et al., 2014; Kang et al., 2016; Sherbakov et al., 2018). Similarly, mortality risk and temperature-related illness and hospitalization are also found to be higher in older populations during cold spells, both in developed countries such as Sweden (Rocklöv et al., 2014) and England and Wales (Hajat et al., 2007) and in less developed countries such as India (Fu et al., 2018).

Differences in physiology and baseline metabolism also result in greater sensitivity to certain exposures among young children. A systematic literature review shows that the incidence of diarrheal diseases increases with rising ambient temperature as well as after heavy rainfall and flooding events (Levy et al., 2016). Dehydration caused by diarrheal diseases can be fatal for children because infants and young children are particularly susceptible to dehydration (Bennett & Friel, 2014). This is due to their higher body water content, higher metabolic rate and less capability to hydrate themselves or express their needs (Vega & Avva, 2019). Likewise, young children are also more susceptible to malaria, dengue and other vector-borne diseases since they have lower immunity. Expansion of the geographical range of conditions conducive to malaria, dengue and vector-borne diseases transmission due to increases in temperature thus can exacerbate morbidity and mortality from these diseases for young children (Loevinsohn, 1994; Martens et al., 1997; WHO, 2011).

Physiological differences by sex also contribute to differential susceptibility to climate change. Women have lower capacity to compensate to elevated temperatures since smaller body sizes allow rapid heat gain and a higher percentage of body fat reduces capacity to store heat (Burse, 1979). With lower sweat rates in heat stress, women also dissipate less heat (Kenney, 1985). As a consequence, most studies on heat-related mortality in Europe generally find that women have higher risk of dying in a heatwave than men (Kovats & Hajat, 2008). Similarly, gender differences in body size, body shape and composition also influence thermoregulatory response to cold. With lower muscle mass, women have greater heat loss-to-production ratios than men since maximum heat production is a function of total body mass (Burse, 1979). Larger body surface to mass ratio makes women lose heat faster due to the greater surface area for convective heat flux (Young et al., 1996). However, gender differences in cold-related illness and mortality are less well understood. Whilst some studies find that women are more vulnerable than men, as presented by higher mortality risk during cold spells in Australia and the UK (Wilkinson et al., 2004; Hajat et al., 2007; Yu et al., 2010), other studies report no significant gender differences or sometimes that men have higher risk than women

(Davídkovová et al., 2014; Han et al., 2017). Although there are physiological differences in response to heat and cold exposure by gender, these disparities are often moderated by socio-economic factors.

Whilst women seem to be more susceptible to extreme hot and cold temperatures than men, with respect to malnutrition, boys are generally more likely to be more exposed to the risk of undernutrition than girls (Kandala et al., 2011). Particularly in sub-Saharan Africa, the evidence that male children are more likely to be stunted than female children is consistent across countries (Wamani et al., 2007; Keino et al., 2014). Given that boys physiologically require more energy intake than girls while growing up, they are more susceptible to undernutrition, especially in poorer households vulnerable to food insecurity (Wamani et al., 2007). Note that whilst physiological susceptibility partially explains gender differences in the risk of malnutrition for children age under five, socio-cultural factors also play an important role in shaping malnutrition risk. Unlike in sub-Saharan Africa, in South Asian countries son preference over daughters results in differential feeding practices and consequently increases the risk of chronic malnutrition for girls (Klasen, 2008; Raj et al., 2015). Indeed, Muttarak and Dimitrova (2019) find that when households experience rainfall anomalies, the risk of stunting and wasting was higher for girls than for boys in southern India.

Moreover, physiology plays a key role in survivorship in certain hazard events. For instance, in the case of a tsunami, women, children aged <5 and the elderly aged >70 years had a clear mortality disadvantage since they have more limitations in mobility (Doocy et al., 2007). Differential vulnerability thus is a function of biological and physiologic differences between demographic groups, to a certain extent.

Differential Exposure

Apart from biological distinctions, differential social and behavioural patterns associated with demographic characteristics also influence hazard exposure. For instance, children have higher climate-sensitive exposures than adults because they are likely to spend more time outdoors and engage in activities such as playing on the ground and having hand-mouth contact. This increases their exposures to environmental pathogens and outdoor insect vectors (Pronczuk-Garbino, 2005). A study of dengue virus infection, for example, show that children aged 10–14 year olds had higher rates of infection than children aged 5–9 year olds and those <1 year of age (Sharp et al., 2013). This may be because children in the age group 10–14 year olds have higher exposure due to their outdoor activities than other age groups.

Similarly, differences in socially constructed gender roles, norms and social status also explain differential exposure to risks and capacity to cope with risks between men and women. The findings that men had higher mortality risks from floods and storms than women, for instance, are due to the fact that men are engaged in outdoor activities more frequently and more likely to work for emergency services (Jonkman & Kelman, 2005; Zagheni et al., 2016). This makes men generally have higher exposure to hydrological disasters than women. On the other hand, when mortality associated with natural disasters occurred at home or in the vicinity of a building such as in the case of landslides and avalanches, sex differentials in mortality risk become more balanced (Badoux et al., 2016). This is possibly because landslide fatalities are far more likely to occur indoors as compared to flood fatalities (Salvati et al., 2018). In this case, exposure to landslides are not necessarily higher in men than in women.

Exposure is also determined by expected behaviour associated with gender roles. For example, the mortality rates during the Indian Ocean tsunami in 2004 were higher for women than men partially due to the lack of ability to swim and/or restrictive clothing in certain societies like Sri Lanka (Neumayer & Plümper, 2007; James, 2016). Furthermore, women's family roles also contributed to higher female mortality in the Indian Ocean tsunami since women were likely to stay behind trying to save their children and other vulnerable family members (Yeh, 2010; Frankenberg et al., 2011). In this case, differential vulnerability is determined by socially constructed aspects of demographic categories and norms associated with these characteristics.

Besides socially constructed behavioural patterns, differential exposures to natural hazards are related to socioeconomic status. It is well-documented that socially and economically disadvantaged groups often live in poor housing and hazard-prone areas making them more exposed to natural hazards. A review of European literature has shown that less affluent population groups were more likely to live in low-quality housing (e.g. exposure to dampness, chemical contamination, temperature problems and poor sanitation) as well as poor residential location (e.g. close to hazardous waste sites, industrial facilities, proximity to pollution sites) (Braubach & Fairburn, 2010). Furthermore, low-income households disproportionately live in coastal or low-lying areas prone to storm surge and flooding such as the slum areas in Ho Chi Minh City, Vietnam (Bangalore et al., 2017), the two most deprived location deciles in the UK (Walker et al., 2006) and areas with higher social vulnerability in Mumbai, India (de Sherbinin & Bardy, 2016). Apart from underlying economic factors, members of racial or ethnic minority groups are also more likely to live in environmentally undesirable areas (Mohai & Saha, 2007; Mohai et al., 2009; Ard, 2015). Ueland & Warf, (2006) explained that discrimination at work and in housing markets coupled with prejudicial access to mortgage lending and exclusionary zoning contribute to unequal exposure of hazards among low-income and minority groups. Similarly, manual workers who engage in outdoor activities such as farming and construction naturally have greater exposure to heat and storm events (Balbus & Malina, 2009). These socioeconomic differentials in the probability of exposure partly explains the unequal distribution of climate risks across population subgroups.

Differential Adaptive Capacity

Other than influencing exposure, demographic and socioeconomic characteristics are principal drivers of populations' ability to prepare for, respond to, cope with and recover from natural disasters and impacts of climate change. The literature commonly identifies the elderly, children, women, the disabled, members of minority ethnic groups and individuals with low income as being more vulnerable to shocks (Cutter, 1995; Fothergill et al., 1999; Masozera et al., 2007). These subgroups of the population are generally less able to cope and respond to hazards or shocks due to their disadvantaged position: *socially* because of minority status (e.g. members of certain religious/ethnic groups), *economically* because they are poor, and *politically* because of lack of independence, decision making power and underrepresentation (e.g. women and children) (Gaillard, 2010).

That members of minority ethnic/racial groups have higher risks of heat-related morbidity and mortality is the case in point. Whilst some studies show that there are genetic differences in heat tolerance, the evidence remains inconclusive at best (Taylor, 2006; Yardley et al., 2011; Hansen et al., 2013). The vulnerability of members of minority ethnic/racial groups to heat extremes is rather due to their lower income, lower ownership of air conditioning, poor housing conditions and their living environment which have sparse vegetation and more heat-absorbing surfaces (Klein-Rosenthal et al., 2014). The lower economic status further prevents members of minority ethnic/racial groups from implementing adaptation strategies to temperature extremes.

The demographic and socioeconomic characteristics associated with vulnerability also intersect and are stratified based on social identities and positions of people and groups. Certain population subgroups uniformly lack access to economic, social and human resources or knowledge to cope with and respond to risks. Female-headed households in South Africa, for instance, are economically more vulnerable to climate variability than households headed by two adults, not only because of their lower level of education and greater economic disadvantages to start with but also due to gender differences in access to social networks and formal credit (Flatø et al., 2017). In this case, female-headed households are more vulnerable as a result of their gender as well as socioeconomic disadvantages associated with being a single-headed household.

Likewise, in the United States women and ethnic minorities consistently have higher risk perception than white males (the so-called "white male effect") because their reduced social and formal decision-making power make them more vulnerable to climate risks (Satterfield et al., 2004). Conversely, in Sweden where women have broadly the same opportunities and life chances as men, there is virtually no gender disparities in risk perception (Olofsson & Rashid, 2011). It is thus important to consider the intersectionality between different demographic and social categories in order to better understand differential vulnerability.

It is also important to note that population characteristics underlying vulnerability are not static and depend on the type of risk considered. Women, for instance, are not always more vulnerable to climate risks than men. There is evidence that lower risk perception and risk-taking behaviour (e.g. driving a car in flooded roads, crossing flooded bridges) make young and middle aged men more likely to die in floods (Ashley & Ashley, 2008; Doocy et al., 2013; Pereira et al., 2017). Furthermore, in certain situations, the shared sense of ethnicity serves as a basis for co-operative social relations and enables a minority ethnic group such as Vietnamese in New Orleans to return and recover from Hurricane Katrina at faster rate than the average population (Vu et al., 2009). Therefore, it is not possible to label one characteristic as always vulnerable since vulnerability is dynamic and multidimensional.

Education as a Key to Reducing Vulnerability and Enhancing Adaptive Capacity

It is worth highlighting here the role of education in reducing vulnerability and enhancing adaptive capacity in the context of global climate change. Not only is education a relevant source of heterogeneity in different life outcomes (Barakat & Blossfeld, 2010; Lutz & KC, 2010), it has been shown that education is an important

dimension in population projections (Lutz et al., 2014b). With education being strongly associated with fertility, mortality and migration intensities, changing educational distributions in the population affect the projected future population size. Population projections that do not account for educational difference in fertility behaviour and changes in future educational distribution can yield starkly different estimates as compared to the ones where education is explicitly included. The UN population projections which assume a constant fertility rate for Nigeria, for instance, project that the population in Nigeria will increase to 914 million in 2100 whilst the multidimensional population projections which include education as another source of heterogeneity expect the Nigerian total population will grow to 576 million under the medium scenario (Lutz et al., 2014a, b). These different projected population sizes and dynamics consequently pose a challenge to mitigation and adaptation planning (Jiang & Hardee, 2011).

Mechanisms Through Which Education Reduces Vulnerability

Education is critical for vulnerability reduction and adaptive capacity. At the individual level, barriers to the adoption of adaptation measures include a lack of awareness and understanding of climate change risk, doubt about efficacy of one's actions, lack of knowledge on how to change behaviour and lack of financial resources to implement changes. Accordingly, there are many sound reasons to assume that education can contribute to overcome these barriers both in direct and indirect manners.

First, directly formal schooling is a primary way individuals acquire knowledge, skills, and competencies that can influence their mitigation and adaptation efforts. Schooling provides a unique environment to engage in cognitive activities such as learning to read, write and use numbers. As students move to higher grades, cognitive skills required in school become progressively more demanding and involve meta-cognitive skills such as categorization, logical deduction and IQ (Ceci, 1991; Blair et al., 2005; Nisbett, 2009). This abstract cognitive exercise alters the way educated individuals think, reason and solve problems likewise (Baker et al., 2011). Indeed, experimental studies have shown that higher-order cognition improves risk assessment and decision making skills (Peters et al., 2006; Bruine de Bruin et al., 2007). These are relevant components of reasoning related to risk perception and making choices about adaptation actions.

Furthermore, education enhances the acquisition of knowledge, values and priorities, the capacity to plan for the future as well as efficiency in allocation of resources (Kenkel, 1991; Cutler & Lleras-Muney, 2010). Schooling can help individuals adopt, for instance, disaster preparedness measures by improving their knowledge of the relationship between preparedness and disaster risk reduction. Moreover, educated individuals may have better understanding of what measures to undertake and can make better choices with respect to safe construction practices and location decisions. Recent evidence also shows that education changes time preferences such that more educated people are more patient, more goal-oriented and thus make more investments (e.g., financial, health or education investments) for their future (Oreopoulos & Salvanes, 2011; Pérez-Arce, 2011; Chew et al., 2016). Such forward-looking attitudes can influence adoption of adaptation measures where benefits may only be expected by future generations.

Apart from the direct impacts, education may indirectly reduce vulnerability through many other means. Firstly, education improves socioeconomic status and generally increases earnings. This allows individuals to have command over resources such as purchasing costly disaster insurance, living in low risk areas and improving housing quality. Secondly, many empirical studies have shown that people with more years of formal education have access to more sources and types of information (Cotten & Gupta, 2004; Wen et al., 2011; Neuenschwander et al., 2012). The level of education is not only highly correlated with access to weather forecasts and warnings but the highly educated are also able to understand highly-complex environmental issues such as climate change better than less educated counterparts. Indeed, there is evidence that disaster risk perception and awareness of climate change is associated with education and the level of disaster preparedness is raised through these mechanisms (Hoffmann & Muttarak, 2017). Knowing where to get information on what adaptations to take allows individuals to change behaviour appropriately. On top of that, more educated individuals also have higher social capital (Lake & Huckfeldt, 1998; Huang et al., 2009). A perception of risk and motivation to take preventive action can be transferred via social networks such that individuals who participate regularly in social activities can benefit from an exchange of useful information and warnings (Witvorapong et al., 2015). Evidently, through increasing socio-economic resources, facilitating access to information and enhancing social capital, education can promote vulnerability reduction and foster adaptive capacity.

Evidence on the Role of Formal Education in Vulnerability Reduction

Indeed, recent empirical studies have demonstrated consistent evidence showing that countries, communities, households and individuals with higher average levels of education experience lower vulnerability to natural disasters – a proxy for risk associated with climate change in the near future (Butz et al., 2014). This applies to both developed and less developed countries as well as different dimensions of vulnerability including preparedness and response to disasters, mortality, morbidity, coping strategies, recovery from disasters and other relevant outcomes. The evidence discussed below is based on multivariate analysis of empirical data taking into account demographic and socioeconomic characteristics and, in some occasions, contextual factors determining disaster-related outcomes.

Many studies have established that higher educational attainment enhances disaster preparedness – measures taken to prepare for and reduce the impacts of disaster. This includes being prepared for earthquakes (Russell et al., 1995), hurricanes (Norris et al., 1999; Baker et al., 2011; Reininger et al., 2013), floods (Lave & Lave, 1991; Thieken et al., 2007), tsunamis (Muttarak & Pothisiri, 2013), terrorism (Eisenman et al., 2009; Lee & Lemyre, 2009; Bourque et al., 2012) and general emergencies (Al-Rousan et al., 2014; Smith & Notaro, 2009). Better knowledge about the disaster and higher capacity to perform effective emergency measures explain why individuals with higher levels of education have greater disaster preparedness (Lave & Lave, 1991; Thieken et al., 2007). There is also empirical evidence showing that the effect of education on disaster preparedness is mediated through social capital and risk perception. Hoffmann and Muttarak (2017) find that individuals with higher level of education were more risk averse, had better perception of disaster risk in a community and were more likely to engage in social participation – all pathways that contribute to disaster preparedness actions.

That schooled individuals are more able in performing abstract reasoning is evident in studies of disaster preparedness actions which include such– measures as stockpiling of food and water, having a family evacuation plan and buying disaster insurance. Figure 4.3 presents the probability of undertaking disaster preparedness by disaster experience and years of schooling. The proportion of people who reported undertaking disaster preparedness actions is notably higher in the Philippines (76%) as compared to only 32% in Thailand, possibly due to the fact that the Philippines has higher exposure to natural hazards. Although the level of disaster preparedness varies between the two countries, the effect of education on disaster preparedness is remarkably consistent in both countries. Indeed, education is found to substitute disaster experience in promoting disaster preparedness in both Thailand and the Philippines such that the highly educated carry out disaster preparedness measures without needing to first experience the harmful event and then learn later (Muttarak & Pothisiri, 2013; Hoffmann & Muttarak, 2017).

As a consequence, disaster preparedness among more educated individuals can provide protective effects when a disaster strikes. It has been found in Indonesia and India that individuals with higher educational attainment were more likely to survive from the 2004 Indian Ocean tsunami and had lower risk of injuries (Guha-Sapir et al., 2006; Frankenberg et al., 2013). Likewise, at the community level, communities with higher mean years of schooling were reported to experience lower losses in human lives due to floods and landslides in Nepal (KC, 2013). The evidence extends to the country level where countries with higher level of education, even after accounting for income per capita and other development indicators, experienced significantly lower mortality from climate-related disasters (Padli & Habibullah, 2009; Striessnig et al., 2013; Lutz et al., 2014c). With respect to morbidity associated with disasters, in general, there is not much evidence on the association between education and physical morbidity associated with natural hazards. The literature on mental health morbidity however has consistently shown that individuals with higher level of education had lower prevalence of distress, depression and post-traumatic stress disorder (PTSD) (Epstein et al., 1998; Frankenberg et al., 2013). With higher engagement in disaster preparedness and mitigation activities as well as better knowledge about where to obtain assistance after disasters, education and income facilitate faster recovery from disasters including psychosocial dimensions (Bourque et al., 2007). Indeed, the protective role of education on mental health is confirmed in the literatures reviewing quantitative studies on risk factors for psychological morbidity after natural disasters (Norris et al., 2002; Tang et al., 2014).

Apart from relatively lower disaster impacts on mortality and morbidity, damage to residential property and economic losses are found to be lower among communities and countries with higher mean years of schooling or higher literacy rates (Toya & Skidmore, 2007; Noy, 2009; KC, 2013). Education enhances awareness and knowledge of natural disasters, and educated citizens can make better choices related to disaster risk reduction measures such as construction practices and location decisions (Toya & Skidmore, 2007).

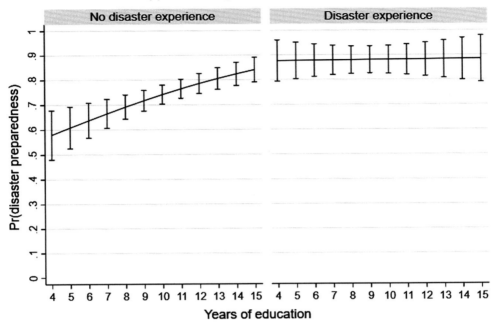

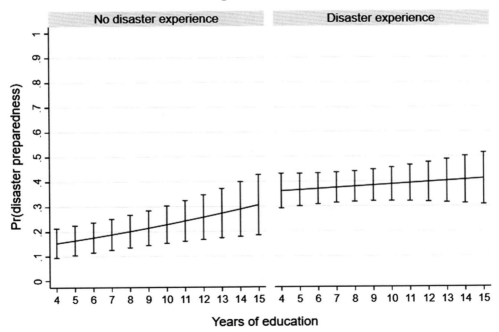

Fig. 4.3 Plots of marginal effects from the logistic regressions predicting the probability of undertaking disaster preparedness in the Philippines and Thailand by disaster experience controlling for demographic, social, economic and community characteristics. (Source: Hoffmann and Muttarak 2017, Figure 4)

This in turn mitigates disaster risk and reduces vulnerability.

Furthermore, education equips individuals and households with a variety of coping strategies following natural shocks. Natural disasters can disrupt livelihoods, destroy crops or damage homes and property. Consequently, households have to employ different mechanisms to smooth consumption i.e. maintain the same level of consumption when income is affected by transitory shocks. Households or communities with higher educational attainment are better able to maintain their welfare and level of consumption after disasters (Thomas et al., 2010; Garbero & Muttarak, 2013; Patnaik et al., 2016). There is evidence that households where household heads have higher levels of education have better access to loans and credit which facilitates stabilizing and increasing the consumption levels (Gitter & Barham, 2007). More educated heads of households have more success in staving off poverty and future poverty possibly because they have better information regarding aggregate risk and are able to make better decisions regarding this risk (Silbert & del Pilar Useche, 2012). With a wider coping strategies portfolio, highly educated households do not need to opt for coping options which involve disinvestment such as taking children out of school or reducing food consumption (Helgeson et al., 2013). Not only do children of literate mothers have better nutritional status in general, but even after experiencing floods, they are less likely to be stunted compared to their counterparts whose mothers are not literate (Muttarak & Dimitrova, 2019). There is also evidence that the protective effect of education is above and beyond economic status (Fuchs et al., 2010). It is found, for instance, that in India, children whose mother is literate but poor has the same probability of stunting with children whose mother is illiterate but rich when experiencing floods (Dimitrova & Muttarak, 2020).

With respect to adaptation to the changing climate, education is indeed highly relevant since individuals with higher level of education are also more likely to have better awareness of climate risk (Lee et al., 2015). Given that climate change is a relatively new form of risk, education facilitates the understanding of new ideas and concepts related to climate variability. Highly educated household heads are likely to have a better level of planning and access and understanding of early warning information which are relevant for climate variability adaptation (Opiyo et al., 2016). Education also enhances knowledge of what adaptation measures can be taken, and households with higher levels of education have a higher likelihood of carrying out adaptation actions such as changing crop types, planting and harvesting dates, or using improved type of seed (Deressa et al., 2009; Paul et al., 2016; Mugi-Ngenga et al., 2016). Likewise, education also increases options to diversify livelihoods. For instance, when facing climate pressure, farmers in rural Tanzania with higher levels of education were more able to switch to non-farm income earning activities (Van Aelst & Holvoet, 2016). Indeed, formal education is a fundamental tool to enhance comprehension of new messages and equip individuals with abilities to undertake behavioural change including capacity to adapt to the changing climate.

Demographic Metabolism Theory and Multi-dimensional Demography

As evident in the previous section, the level of vulnerability and adaptive capacity varies considerably across subgroups of population. Although the association between vulnerability or adaptive capacity and the demographic characteristics presented is mainly captured at the micro level, it is not unreasonable to assume that changes in population composition at the macro level will result in different trajectories of societal level vulnerability and adaptive capacity. Indeed, Billari (2015) has argued that a two-stage perspective where micro- and macro-level approaches are integrated can provide better understanding of demographic change and behaviour of human populations.

In particular, a macro-level model can be relevant for understanding the current state of society and how it will look like in the future. This is particularly important when designing development and climate policies. Along this line, Lutz (2013) develops the theory of demographic metabolism with its strength in having predictive power. Built upon Ryder's cohort approach (Ryder, 1965), Lutz (2013) holds that population change occur along cohort lines. Through the replacement of members with certain characteristics by members with different characteristics, this change in population composition stimulates societal change.

Lutz (2013) adds to the concept of "demographic metabolism" by adding a quantitative feature to the model. Given that individual characteristics such as education do not change for the entire population instantly, the transition between subcategories (e.g. primary to secondary level of education) can be defined though transition probabilities by age and sex (Keyfitz, 1985). For instance, education expansion results in 56.6% of young women aged 20–39 in Indonesia attained secondary level of education in 2015 compared to merely 29.4% of older women in the age group 40–64. As a result, the population with secondary education increases step-by-step as younger people move up the age pyramid in a predictable way and replace the older cohorts with much lower level of education. This dynamic can be captured and forecasted using the multidimensional cohort-component analysis.

The cohort-component model traditionally differentiates populations by age and gender. Population projections are then performed along cohort lines (for example, the cohort aged 20–24 in 1970 becomes 40–44 in 1990) adjusting for the three principal determinants of population dynamics: fertility, mortality and migration. Multi-dimensional population models, however, can further sub-divide populations by other measurable characteristics, particularly those that are relevant in influencing population change. Lutz et al. (1998) argue that educational attainment is the single most highly relevant source of observable population heterogeneity. It has consequently been shown that adding educational attainment to population analyses results in demographic outcomes and associated socioeconomic prospects which differ greatly from population projections that do not account for educational attainment (Lutz et al., 2014b, 2018).

Given that transitions from one educational level to another are centred around younger ages and only go in one direction (i.e. people only progress towards higher educational categories), projection can be done similarly along cohort lines as presented in Fig. 4.4. For example, the proportion of women with secondary level of education aged 20–24 in 1970 is a good predictor of the proportion of women aged 60–64

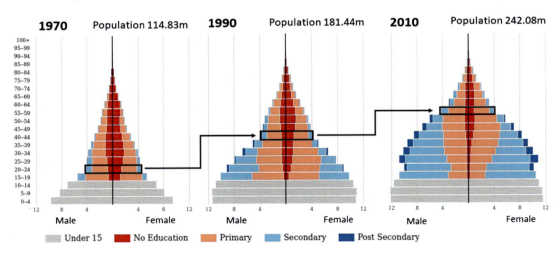

Fig. 4.4 Population of Indonesia by age, sex and educational attainment 1970–2010 in 20-year intervals. Colours indicate highest level of educational attainment. (Notes: Children aged 0–14 are marked in light grey. The dark grey line links the identical birth cohorts at different points in time when they are of different ages but maintain their highest level of education attained. Source: Wittgenstein Centre for Demography and Global Human Capital 2019)

with secondary education in 2010 after mortality and migration are adjusted for. This model of population change along cohort lines captures changing composition of human populations and this knowledge can subsequently be applied to forecast future societies' future adaptive capacity.

Forecasting Societies' Adaptive Capacity Based on Socioeconomic Development Scenarios

The empirical evidence on demographically differentiated vulnerability discussed earlier informs us about population heterogeneity that matters for vulnerability and adaptive capacity. The multidimensional cohort-component model considering highest educational attainment as a relevant population characteristic determining vulnerability and adaptive capacity provides projections of future population distribution and composition. This allows us to anticipate the future population dynamics through the process of demographic metabolism and assess future societies' capacity to adapt to climate change in the longer term.

However, as mentioned by Lutz et al. (2018) that "in the long-run demography is not destiny", how population may evolve in the long term depends on scenarios of socioeconomic development. Hunter and O'Neill (2014) put forward a new scenario framework – the Shared Socioeconomic Pathways (SSPs) – which can be applied to project demographic trends into the future. The SSPs provide narrative descriptions of five different paths of socioeconomic development over the next century at a macro, global scale. The SSPs included both a qualitative narrative and a quantitative component that includes numerical pathways for certain variables describing a broad range of future scenarios relevant to the main challenges associated with adaptation and mitigation (Riahi et al., 2017; O'Neill et al., 2017). Certain components have been quantified and projected along the SSPs storylines. This ranges from demographic dimensions such as population (KC & Lutz, 2014, 2017; Jones & O'Neill, 2016) and education (KC & Lutz, 2014, 2017) and migration (Abel, 2018), economic dimension such as in-

come (Leimbach et al., 2013; Crespo Cuaresma, 2017; Dellink et al., 2017) and inequality (Rao et al., 2018; Benveniste et al., 2021), and social and institutional dimensions such as human development (Crespo Cuaresma & Lutz, 2016), urbanization (Jiang & O'Neill, 2017), governance (Andrijevic et al., 2020) and civil conflict (Hegre et al., 2016). These dimensions are closely linked with vulnerability and adaptive capacity.

Human Core of the Shared Socioeconomic Pathways (SSPs)

KC and Lutz (2014, 2017) have translated the original narratives underlying the SSPs into the human core of the SSPs which essentially are demographic scenarios for 195 countries. The quantification of the SSPs for population was carried out based on the assessments of future fertility, mortality, migration and education trends for all parts of the world by 550 international population experts. The results were subsequently blended with statistical extrapolation models and translated into alternative demographic assumptions for the four components of population change for all countries of the world until 2060 (Lutz et al., 2014b). The specifications of these demographic scenarios follow the substantive narratives of the SSPs and are described in details in KC and Lutz (2014, 2017).

SSP1 is characterized by sustainability with small challenges in terms of mitigation and adaption thanks to a high rate of technological progress, which reduces resource intensity and fossil fuel dependency. With respect to the demographic consequences, rapid social development and rapid educational expansion corresponds to a rapid demographic transition. This implies low mortality and low levels of fertility in present-day high-fertility countries. Under this rapid development scenario, the overall population sizes are relatively low.

The medium scenario (SSP2) or the "middle of the road" scenario assumes the continuation of current trends typical of recent decades in terms of development and democratization. In SSP2, fertility, mortality and education expansion are assumed to follow a medium pathway resulting in slower decline in population growth as compared to SSP1.

SSP3 referred to as "fragmentation" or "stalled social development" scenario portrays a fragmented and polarized world. SSP3 assumes increasing global inequality with large fractions of the world population living in regions characterized by extreme poverty, pockets of moderate wealth and difficulty in maintain living standards for rapidly growing populations in many countries. The emphasis on security at the expense of international development results in a stall in educational expansion in developing countries which consequently leads to high levels of fertility, high mortality and high population growth rates in the currently high-fertility countries. In contrast, fertility is assumed to be low in the currently low-fertility countries resulting in low population growth. Migration is assumed to be low for all countries due to the emphasis on national security.

SSP4 represents the world of inequality with highly unequal investments in human capital and increasing economic and political inequalities both within and across countries. SSP4 assumes inequality in the distribution of education with a group of the very highly educated and large groups with low education. This means that a large number of disadvantaged people will have lower capacity to adapt to climate change. In SSP4 fertility is assumed to remain high in high fertility countries and low in low-fertility countries and high-income countries.

SSP5 corresponds to conventional development with an emphasis on technological progress and economic growth which as a consequence leads to high GHG emissions. The emphasis on economic growth results in rapid development of human capital which translates to high level of education throughout the world. As a consequence, mortality is assumed to be low across all countries. Fertility assumptions are highly differentiated with high fertility assumed for high income countries thanks to a very high standard of living but low fertility assumed for all other countries. The major difference with SSP1 is that emissions are higher in SSP5 posing high mitigation challenges but low adaptation challenges.

KC and Lutz (2017) present the time trends in total population of the world by educational attainment under five SSP scenarios. The world differs not only in terms of population size, but also educational distributions under different scenarios. In all scenarios, thanks to improvements in educational access, the number of people with secondary or tertiary education will increase in absolute terms. Even under SSP3, which assumes no further increase in school enrollment rates, the fact that younger age groups are better educated than the older ones creates momentum for educational improvement. In terms of population size and age and educational distribution, the rapid demographic transitions assumptions of SSP1 and SSP5 lead to a dramatic increase in the global population with higher education. These two scenarios assume that the global mean years of schooling of the total adult population will reach 12 years by 2050, which is similar to the current level in Europe (KC & Lutz, 2017). If the whole world were to achieve the level of education of Europe today in the next three decades, we can expect a variety of positive consequences typically found to associated with higher level of education such as good health, lower level of poverty and lower vulnerability. In contrast, SSP3 and SSP4 portray a rather pessimistic world future with stagnant school enrollment and large proportions of the population with low levels of education. These scenarios pose challenges to sustainable development since lower levels of education means the population is also less healthy and less wealthy, making them more vulnerable to global climate change.

Predicting Future Climate Vulnerability Under the SSPs

This section presents an example of how the demographic toolkit can be applied to understand the population-climate change connection in the future.

We draw on the study by Striessnig and Loichinger (2016) as an example. In the first step, a regression analysis is employed to estimate mortality from natural disasters as an indicator

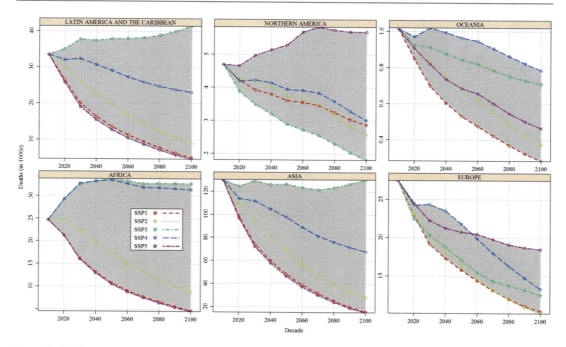

Fig. 4.5 Predicted number of decadal deaths in 1000s assuming constant natural hazard for six world regions, 2010–2100. (Source: Striessnig and Loichinger 2016, Figure 4)

of vulnerability from climate change. Based on the Emergency Events Data Base (EM-DAT), they employ negative binomial hurdle models to estimate deaths from natural disasters for 174 countries over 10-year intervals between 1970 and 2010. Apart from the population growth rate, which has a positive relationship with disaster mortality, it is found that the share of females aged 20–39 with at least secondary level of education is a strong predictor of national disaster mortality: a 1% increase in the proportion of females aged 20–39 with secondary education reduces the probability of disaster casualties by 55% ($\beta = -0.451$, $p < 0.01$) (See Table 1, Striessnig and Loichinger (2016)). In fact, contrary to the common assumption that income is a key predictor of vulnerability, it appears that income as measured by GDP per capita does not play a major role in determining disaster mortality (See Table A.2, Striessnig and Loichinger (2016)). The positive role of female education has similarly been found in other domains such as in promoting economic growth and the transition to democracy (Lutz et al., 2008, 2010, 2019).

Upon empirically establishing the relationship between education and vulnerability to climate change (as measured by disaster mortality), the next step is to project future disaster deaths. This can be done by applying the education coefficient estimates from the regression (Table 1, Striessnig and Loichinger (2016)) to the demographic scenarios underlying the SSPs.[1] In particular, this exercise aims to illustrate differential vulnerability in the future given different education distribution by the SSPs. Therefore, although in the future the impact of climate change on population vulnerability is likely to be more severe, in this analysis hazard levels are assumed to be unchanged throughout the century.

Figure 4.5 presents the trajectories of disaster mortality for six major world regions by five SSPs scenarios for the period 2010–2100. As presented in the regression results (Table 1, Striessnig and

---

[1] The data for global population projections for 201 countries by age, sex and education under five SSPs scenarios are made available by a team of researchers at the Wittgenstein Centre for Demography and Global Human Capital (IIASA, OeAW, University of Vienna) (WIC, 2019).

Loichinger (2016)), disaster mortality increases with population growth and decreases with an increase in educational attainment. Hence, Asia, Latin America and the Caribbean and Africa are more vulnerable to disaster mortality than other regions given their sheer size. However, what is more prominent is the differential levels of disaster mortality by the SSPs. Clearly, fatalities from natural disasters are much lower in SSP1 and SSP5 scenarios across the world regions given the high level of human capital investment and subsequently educational expansion in these scenarios. Disaster fatalities are projected to be the highest under SSP3 and SSP4 scenarios. However, the levels of projected mortality under SSP3 and SSP4 vary by world regions depending on fertility assumptions affecting population size and inequality in the expansion of education. Overall, this exercise presents an example of a procedure for how demographic concept and tools can be combined to forecast vulnerability to climate change in the next decades.

Discussion and Conclusion

This chapter argues that social science and demography can play a role in improving understanding of future societies and characteristics that are relevant to vulnerability and climate change adaptation. There is a need to incorporate socioeconomic components into assessment of climate change impacts precisely because the climate of tomorrow will not meet the society of today (Lutz & Muttarak, 2017). It is not meaningful to imagine future mitigation and adaptation challenges if societal change is not accounted for. As a scientific discipline that focuses on empirical evidence, demography can lend itself to assessing the differential impacts of global climate change on subgroups of population.

Furthermore, the study of future impacts of climate change requires methodological tools that allow for forecasting into the future. The method of multidimensional population projection is a good example of a methodological toolbox which already exists in demography to provide trajectories of human development. Coupled with demographic metabolism theory – a theory of social change – we can assume that cohort replacement will bring about societies of the future which differ from those of today. Hence, changing population composition such as by age, gender and education – a source of population heterogeneity relevant to vulnerability – implies different levels of vulnerability and adaptive capacity in the future. The readily available concept and tools in demography thus offer a meaningful way to quantitatively forecast societies' adaptive capacity to future climate change. Although there was historically a paucity of engagement of demographers in climate research, as noted by Anastasia J. Gage, the former President of the International Union for the Scientific Study of Population (2016), it is "the next best time for demographers to contribute to climate change research".

References

Abel, G. J. (2018). Non-zero trajectories for long-run net migration assumptions in global population projection models. *Demographic Research, 38*(54), 1635–1662. https://doi.org/10.4054/DemRes.2018.38.54

Adger, N. (1999). Social vulnerability to climate change and extremes in coastal Vietnam. *World Development, 27*(2), 249–269. https://doi.org/10.1016/S0305-750X(98)00136-3

Al-Rousan, T. M., Rubenstein, L. M., & Wallace, R. B. (2014). Preparedness for natural disasters among older US adults: A nationwide survey. *American Journal of Public Health, 104*(3), 506–511. https://doi.org/10.2105/AJPH.2013.301559

Andrijevic, M., Crespo Cuaresma, J., Muttarak, R., & Schleussner, C.-F. (2020). Governance in socioeconomic pathways and its role for future adaptive capacity. *Nature Sustainability, 3*(1), 35–41. https://doi.org/10.1038/s41893-019-0405-0

Ard, K. (2015). Trends in exposure to industrial air toxins for different racial and socioeconomic groups: A spatial and temporal examination of environmental inequality in the U.S. from 1995 to 2004. *Social Science Research, 53*, 375–390. https://doi.org/10.1016/j.ssresearch.2015.06.019

Ashley, S. T., & Ashley, W. S. (2008). Flood fatalities in the United States. *Journal of Applied Meteorology and Climatology, 47*(3), 805–818. https://doi.org/10.1175/2007JAMC1611.1

Baccini, M., Biggeri, A., Accetta, G., Kosatsky, T., Katsouyanni, K., Analitis, A., Anderson, H. R., Bisanti,

L., D'Ippoliti, D., Danova, J., Forsberg, B., Medina, S., Paldy, A., Rabczenko, D., Schindler, C., & Michelozzi, P. (2008). Heat effects on mortality in 15 European cities. *Epidemiology, 19*(5), 711–719. https://doi.org/10.1097/EDE.0b013e318176bfcd

Badoux, A., Andres, N., Techel, F., & Hegg, C. (2016). Natural hazard fatalities in Switzerland from 1946 to 2015. *Natural Hazards and Earth System Sciences, 16*(12), 2747–2768. https://doi.org/10.5194/nhess-16-2747-2016

Bai, L., Ding, G., Gu, S., Bi, P., Su, B., Qin, D., Xu, G., & Liu, Q. (2014). The effects of summer temperature and heat waves on heat-related illness in a coastal city of China, 2011–2013. *Environmental Research, 132*, 212–219. https://doi.org/10.1016/j.envres.2014.04.002

Baker, D. P., Leon, J., Smith Greenaway, E. G., Collins, J., & Movit, M. (2011). The education effect on population health: A reassessment. *Population and Development Review, 37*(2), 307–332. https://doi.org/10.1111/j.1728-4457.2011.00412.x

Balbus, J., & Malina, C. (2009). Identifying vulnerable subpopulations for climate change health effects in the United States. *Journal of Occupational and Environmental Medicine, 51*(1), 33–37. https://doi.org/10.1097/JOM.0b013e318193e12e

Bangalore, M., Smith, A., & Veldkamp, T. (2017). Exposure to floods, climate change, and poverty in Vietnam. *Natural Hazards and Earth System Sciences Discussions, 2017*, 1–28. https://doi.org/10.5194/nhess-2017-100

Barakat, B., & Blossfeld, H.-P. (2010). The search for a demography of education: Some thoughts. *Vienna Yearbook of Population Research, 8*(1), 1–8.

Bennett, C. M., & Friel, S. (2014). Impacts of climate change on inequities in child health. *Children, 1*(3), 461. https://doi.org/10.3390/children1030461

Benveniste, H., Cuaresma, J. C., Gidden, M., & Muttarak, R. (2021). Tracing international migration in projections of income and inequality across the Shared Socioeconomic Pathways. *Climatic Change, 166*(3), 39. https://doi.org/10.1007/s10584-021-03133-w

Billari, F. C. (2015). Integrating macro- and micro-level approaches in the explanation of population change. *Population Studies, 69*(sup1), S11–S20. https://doi.org/10.1080/00324728.2015.1009712

Blaikie, P. M., Cannon, T., Davis, I., & Wisner, B. (1994). *At risk: Natural hazards, people's vulnerability, and disasters*. Routledge.

Blair, C., Gamson, D., Thorne, S., & Baker, D. (2005). Rising mean IQ: Cognitive demand of mathematics education for young children, population exposure to formal schooling, and the neurobiology of the prefrontal cortex. *Intelligence, 33*(1), 93–106. https://doi.org/10.1016/j.intell.2004.07.008

Blatteis, C. M. (2012). Age-dependent changes in temperature regulation – A mini review. *GER, 58*(4), 289–295. https://doi.org/10.1159/000333148

Bourque, L. B., Siegel, J. M., Kano, M., & Wood, M. M. (2007). Morbidity and mortality associated with

disasters. In *Handbook of Disaster Research* (pp. 97–112). Springer.

Bourque, L. B., Mileti, D. S., Kano, M., & Wood, M. M. (2012). Who prepares for terrorism? *Environment and Behavior, 44*(3), 374–409. https://doi.org/10.1177/0013916510390318

Braubach, M., & Fairburn, J. (2010). Social inequities in environmental risks associated with housing and residential location – A review of evidence. *European Journal of Public Health, 20*(1), 36–42. https://doi.org/10.1093/eurpub/ckp221

Bruine de Bruin, W., Parker, A. M., & Fischhoff, B. (2007). Individual differences in adult decision-making competence. *Journal of Personality and Social Psychology, 92*(5), 938–956. https://doi.org/10.1037/0022-3514.92.5.938

Burse, R. L. (1979). Sex differences in human thermoregulatory response to heat and cold stress. *Human Factors, 21*(6), 687–699. https://doi.org/10.1177/001872087912210606

Butz, W. P., Lutz, W., & Sendzimir, J. (Eds.). (2014). *Education and differential vulnerability to natural disasters, special issue, ecology and society*. Resilience Alliance.

Ceci, S. J. (1991). How much does schooling influence general intelligence and its cognitive components? A reassessment of the evidence. *Developmental Psychology, 27*(5), 703–722. https://doi.org/10.1037/0012-1649.27.5.703

Chew, S. H., Yi, J., Zhang, J., & Zhong, S. (2016). Education and anomalies in decision making: Experimental evidence from Chinese adult twins. *Journal of Risk and Uncertainty, 53*(2), 163–200. https://doi.org/10.1007/s11166-016-9246-7

Cohen, J. E. (2010). Population and climate change. *Proceedings of the American Philosophical Society, 154*(2), 158–182.

Core Writing Team, Pachauri, R. K., & Meyer, L. (2014). *Climate change 2014: Synthesis report. Contribution of Working Groups I, II and III to the Fifth Assessment Report of the Intergovernmental Panel on Climate Change*. Intergovernmental Panel on Climate Change.

Cotten, S. R., & Gupta, S. S. (2004). Characteristics of online and offline health information seekers and factors that discriminate between them. *Social Science & Medicine, 59*(9), 1795–1806. https://doi.org/10.1016/j.socscimed.2004.02.020

Crespo Cuaresma, J. (2017). Income projections for climate change research: A framework based on human capital dynamics. *Global Environmental Change, 42*, 226–236. https://doi.org/10.1016/j.gloenvcha.2015.02.012

Crespo Cuaresma, J., & Lutz, W. (2016). The demography of human development and climate change vulnerability: A projection exercise. *Vienna Yearbook of Population Research, 13*(2015), 241–262. https://doi.org/10.1553/populationyearbook2015s241

Cutler, D. M., & Lleras-Muney, A. (2010). Understanding differences in health behaviors by education. *Jour-*

nal of Health Economics, 29(1), 1–28. https://doi.org/10.1016/j.jhealeco.2009.10.003

Cutter, S. L. (1995). The forgotten casualties: Women, children, and environmental change. *Global Environmental Change, 5*(3), 181–194. https://doi.org/10.1016/0959-3780(95)00046-Q

Davídkovová, H., Plavcová, E., Kynčl, J., & Kyselý, J. (2014). Impacts of hot and cold spells differ for acute and chronic ischaemic heart diseases. *BMC Public Health, 14*, 480. https://doi.org/10.1186/1471-2458-14-480

de Sherbinin, A., & Bardy, G. (2016). Social vulnerability to floods in two coastal megacities: New York City and Mumbai. *Vienna Yearbook of Population Research, 13*, 131–165. https://doi.org/10.1553/populationyearbook2015s131

Dellink, R., Chateau, J., Lanzi, E., & Magné, B. (2017). Long-term economic growth projections in the shared socioeconomic pathways. *Global Environmental Change, 42*, 200–214. https://doi.org/10.1016/j.gloenvcha.2015.06.004

Deressa, T. T., Hassan, R. M., Ringler, C., Alemu, T., & Yesuf, M. (2009). Determinants of farmers' choice of adaptation methods to climate change in the Nile Basin of Ethiopia. *Global Environmental Change, 19*(2), 248–255. https://doi.org/10.1016/j.gloenvcha.2009.01.002

Dimitrova, A., & Muttarak, R. (2020). After the floods: Differential impacts of rainfall anomalies on child stunting in India. *Global Environmental Change, 64*, 102130. https://doi.org/10.1016/j.gloenvcha.2020.102130

Doocy, S., Gorokhovich, Y., Burnham, G., Balk, D., & Robinson, C. (2007). Tsunami mortality estimates and vulnerability mapping in Aceh, Indonesia. *American Journal of Public Health, 97*(Supplement_1), S146–S151. https://doi.org/10.2105/AJPH.2006.095240

Doocy, S., Daniels, A., Murray, S., & Kirsch, T. D. (2013). The human impact of floods: A historical review of events 1980–2009 and systematic literature review. *PLoS Currents, 5*. https://doi.org/10.1371/currents.dis.f4deb457904936b07c09daa98ee8171a

Ehrlich, P. R., & Holdren, J. P. (1971). Impact of population growth. *Science, 171*(3977), 1212–1217. https://doi.org/10.1126/science.171.3977.1212

Eisenman, D. P., Zhou, Q., Ong, M., Asch, S., Glik, D., & Long, A. (2009). Variations in disaster preparedness by mental health, perceived general health, and disability status. *Disaster Medicine and Public Health Preparedness, 3*(1), 33–41. https://doi.org/10.1097/DMP.0b013e318193be89

Epstein, R. S., Fullerton, C. S., & Ursano, R. J. (1998). Posttraumatic stress disorder following an air disaster: A prospective study. *AJP, 155*(7), 934–938. https://doi.org/10.1176/ajp.155.7.934

Flatø, M., Muttarak, R., & Pelser, A. (2017). Women, weather, and woes: The triangular dynamics of female-headed households, economic vulnerability, and climate variability in South Africa. *World Development, 90*(Supplement C), 41–62. https://doi.org/10.1016/j.worlddev.2016.08.015

Fothergill, A., Maestas, E. G., & Darlington, J. D. (1999). Race, ethnicity and disasters in the United States: A review of the literature. *Disasters, 23*(2), 156–173.

Frankenberg, E., Gillespie, T., Preston, S., Sikoki, B., & Thomas, D. (2011). Mortality, the family and the Indian Ocean tsunami. *The Economic Journal, 121*(554), F162–F182. https://doi.org/10.1111/j.1468-0297.2011.02446.x

Frankenberg, E., Sikoki, B., Sumantri, C., Suriastini, W., & Thomas, D. (2013). Education, vulnerability, and resilience after a natural disaster. *Ecology and Society, 18*(2), 16. https://doi.org/10.5751/ES-05377-180216

Fu, S. H., Gasparrini, A., Rodriguez, P. S., & Jha, P. (2018). Mortality attributable to hot and cold ambient temperatures in India: A nationally representative case-crossover study. *PLoS Medicine, 15*(7), e1002619. https://doi.org/10.1371/journal.pmed.1002619

Fuchs, R., Pamuk, E., & Lutz, W. (2010). Education or wealth: Which matters more for reducing child mortality in developing countries? *Vienna Yearbook of Population Research, 8*, 175–199. https://doi.org/10.1553/populationyearbook2010s175

Gage, A. J. (2016). The next best time for demographers to contribute to climate change research. *Vienna Yearbook of Population Research, 13*, 19–22.

Gaillard, J. C. (2010). Vulnerability, capacity and resilience: Perspectives for climate and development policy. *Journal of International Development, 22*(2), 218–232. https://doi.org/10.1002/jid.1675

Garbero, A., & Muttarak, R. (2013). Impacts of the 2010 droughts and floods on community welfare in rural Thailand: Differential effects of village educational attainment. *Ecology and Society, 18*(4), 27. https://doi.org/10.5751/ES-05871-180427

Gitter, S. R., & Barham, B. L. (2007). Credit, natural disasters, coffee, and educational attainment in rural Honduras. *World Development, 35*(3), 498–511. https://doi.org/10.1016/j.worlddev.2006.03.007

Guha-Sapir, D., Parry, L., Degomme, O., Joshi, P., & Saulina, P. (2006). *Risk factors for mortality and injury: Post-tsunami epidemiological findings from Tamil Nadu. Catholic University of Louvain*. Centre for Research of the Epidemiology of Disasters (CRED).

Hajat, S., Kovats, R. S., & Lachowycz, K. (2007). Heat-related and cold-related deaths in England and Wales: Who is at risk? *Occupational and Environmental Medicine, 64*(2), 93–100. https://doi.org/10.1136/oem.2006.029017

Han, J., Liu, S., Zhang, J., Zhou, L., Fang, Q., Zhang, J., & Zhang, Y. (2017). The impact of temperature extremes on mortality: A time-series study in Jinan, China. *BMJ Open, 7*(4), e014741. https://doi.org/10.1136/bmjopen-2016-014741

Hansen, A., Bi, L., Saniotis, A., & Nitschke, M. (2013). Vulnerability to extreme heat and climate change: Is ethnicity a factor? *Global Health Action, 6*. https://doi.org/10.3402/gha.v6i0.21364

Hegre, H., Buhaug, H., Calvin, K. V., Nordkvelle, J., Waldhoff, S. T., & Gilmore, E. (2016). Forecasting civil conflict along the shared socioeconomic pathways.

Environmental Research Letters, 11(5), 054002. https://doi.org/10.1088/1748-9326/11/5/054002

Helgeson, J. F., Dietz, S., & Hochrainer-Stigler, S. (2013). Vulnerability to weather disasters: The choice of coping strategies in rural Uganda. *Ecology and Society, 18*(2), 2. https://doi.org/10.5751/ES-05390-180202

Hoffmann, R., & Muttarak, R. (2017). Learn from the past, prepare for the future: Impacts of education and experience on disaster preparedness in the Philippines and Thailand. *World Development, 96*, 32–51. https://doi.org/10.1016/j.worlddev.2017.02.016

Huang, J., Maassen van den Brink, H., & Groot, W. (2009). A meta-analysis of the effect of education on social capital. *Economics of Education Review, 28*(4), 454–464. https://doi.org/10.1016/j.econedurev.2008.03.004

Hunter, L. M., & O'Neill, B. C. (2014). Enhancing engagement between the population, environment, and climate research communities: The shared socio-economic pathway process. *Population and Environment, 35*(3), 231–242. https://doi.org/10.1007/s11111-014-0202-7

IPCC. (1996). *Climate change 1995: Impacts, adaptations and mitigation of climate change: Summary for policymakers*. Intergovernmental Panel on Climate Change and World Meteorological Organisation.

IPCC. (2014). Summary for policymakers. In C. B. Field, V. R. Barros, D. J. Dokken, K. J. Mach, M. D. Mastrandrea, T. E. Bilir, M. Chatterjee, K. L. Ebi, Y. O. Estrada, R. C. Genova, B. Girma, E. S. Kissel, A. N. Levy, S. MacCracken, P. R. Mastrandrea, & L. L. White (Eds.), *Climate change 2014: Impacts, adaptation, and vulnerability. Part A: Global and Sectoral Aspects. Contribution of Working Group II to the Fifth Assessment Report of the Intergovernmental Panel on Climate Change* (pp. 1–32). Cambridge University Press.

James, H. (2016). How do we re-make our lives? Gender and sustainability in the post-disaster context in Asia. In H. James & D. Paton (Eds.), *The consequences of disasters: Demographic, planning, and policy implications* (pp. 201–223). Charles C Thomas Publisher.

Jiang, L., & Hardee, K. (2011). How do recent population trends matter to climate change? *Population Research and Policy Review, 30*(2), 287–312. https://doi.org/10.1007/s11113-010-9189-7

Jiang, L., & O'Neill, B. C. (2017). Global urbanization projections for the shared socioeconomic pathways. *Global Environmental Change, 42*, 193–199. https://doi.org/10.1016/j.gloenvcha.2015.03.008

Jones, B., & O'Neill, B. C. (2016). Spatially explicit global population scenarios consistent with the shared socioeconomic pathways. *Environmental Research Letters, 11*(8), 084003. https://doi.org/10.1088/1748-9326/11/8/084003

Jonkman, S. N., & Kelman, I. (2005). An analysis of the causes and circumstances of flood disaster deaths. *Disasters, 29*(1), 75–97. https://doi.org/10.1111/j.0361-3666.2005.00275.x

Kandala, N.-B., Madungu, T. P., Emina, J. B., Nzita, K. P., & Cappuccio, F. P. (2011). Malnutrition among children under the age of five in the Democratic Republic of Congo (DRC): Does geographic location matter? *BMC Public Health, 11*, 261. https://doi.org/10.1186/1471-2458-11-261

Kang, S.-H., Oh, I.-Y., Heo, J., Lee, H., Kim, J., Lim, W.-H., Cho, Y., Choi, E.-K., Yi, S.-M., Sang, D. S., Kim, H., Youn, T.-J., Chae, I.-H., & Oh, S. (2016). Heat, heat waves, and out-of-hospital cardiac arrest. *International Journal of Cardiology, 221*, 232–237. https://doi.org/10.1016/j.ijcard.2016.07.071

KC, S. (2013). Community vulnerability to floods and landslides in Nepal. *Ecology and Society, 18*(1), 8. https://doi.org/10.5751/ES-05095-180108

KC, S., & Lutz, W. (2014). Demographic scenarios by age, sex and education corresponding to the SSP narratives. *Population and Environment, 35*(3), 243–260. https://doi.org/10.1007/s11111-014-0205-4

KC, S., & Lutz, W. (2017). The human core of the shared socioeconomic pathways: Population scenarios by age, sex and level of education for all countries to 2100. *Global Environmental Change, 42*, 181–192. https://doi.org/10.1016/j.gloenvcha.2014.06.004

Keatinge, W. R., Coleshaw, S. R., Cotter, F., Mattock, M., Murphy, M., & Chelliah, R. (1984). Increases in platelet and red cell counts, blood viscosity, and arterial pressure during mild surface cooling: Factors in mortality from coronary and cerebral thrombosis in winter. *British Medical Journal (Clinical Research Ed.), 289*(6456), 1405–1408. https://doi.org/10.1136/bmj.289.6456.1405

Keino, S., Plasqui, G., Ettyang, G., & van den Borne, B. (2014). Determinants of stunting and overweight among young children and adolescents in sub-Saharan Africa. *Food and Nutrition Bulletin, 35*(2), 167–178. https://doi.org/10.1177/156482651403500203

Kenkel, D. S. (1991). Health behavior, health knowledge, and schooling. *Journal of Political Economy, 99*(2), 287–305.

Kenney, W. L. (1985). A review of comparative responses of men and women to heat stress. *Environmental Research, 37*(1), 1–11. https://doi.org/10.1016/0013-9351(85)90044-1

Kenny, G. P., Poirier, M. P., Metsios, G. S., Boulay, P., Dervis, S., Friesen, B. J., Malcolm, J., Sigal, R. J., Seely, A. J. E., & Flouris, A. D. (2017). Hyperthermia and cardiovascular strain during an extreme heat exposure in young versus older adults. *Temperature, 4*(1), 79–88. https://doi.org/10.1080/23328940.2016.1230171

Keyfitz, N. (1985). *Applied mathematical demography* (2nd ed.). Springer.

Klasen, S. (2008). Poverty, undernutrition, and child mortality: Some inter-regional puzzles and their implicationsfor research and policy. *The Journal of Economic Inequality, 6*(1), 89–115. https://doi.org/10.1007/s10888-007-9056-x

Klein-Rosenthal, J., Kinney, P. L., & Metzger, K. B. (2014). Intra-urban vulnerability to heat-related mortality in New York City, 1997–2006. *Health & Place, 30*, 45–60. https://doi.org/10.1016/j.healthplace.2014.07.014

Kovats, R. S., & Hajat, S. (2008). Heat stress and public health: A critical review. *Annual Review of*

Public Health, 29(1), 41–55. https://doi.org/10.1146/annurev.publhealth.29.020907.090843

Lake, R. L. D., & Huckfeldt, R. (1998). Social capital, social networks, and political participation. *Political Psychology, 19*(3), 567–584.

Lave, T. R., & Lave, L. B. (1991). Public perception of the risks of floods: Implications for communication. *Risk Analysis: An Official Publication of the Society for Risk Analysis, 11*(2), 255–267.

Lee, J. E. C., & Lemyre, L. (2009). A social-cognitive perspective of terrorism risk perception and individual response in {Canada}. *Risk Analysis: An Official Publication of the Society for Risk Analysis, 29*(9), 1265–1280. https://doi.org/10.1111/j.1539-6924.2009.01264.x

Lee, T. M., Markowitz, E. M., Howe, P. D., Ko, C.-Y., & Leiserowitz, A. A. (2015). Predictors of public climate change awareness and risk perception around the world. *Nature Clim Change Advance Online Publication.* https://doi.org/10.1038/nclimate2728

Leimbach, M., Kriegler, E., Roming, N., & Schwanitz, J. (2013). Future growth patterns of world regions – A GDP scenario approach. *Global Environmental Change, 42*, 215–225. https://doi.org/10.1016/j.gloenvcha.2015.02.005

Levy, K., Woster, A. P., Goldstein, R. S., & Carlton, E. J. (2016). Untangling the impacts of climate change on waterborne diseases: A systematic review of relationships between diarrheal diseases and temperature, rainfall, flooding, and drought. *Environmental Science & Technology, 50*(10), 4905–4922. https://doi.org/10.1021/acs.est.5b06186

Loevinsohn, M. E. (1994). Climatic warming and increased malaria incidence in Rwanda. *The Lancet, 343*(8899), 714–718. https://doi.org/10.1016/S0140-6736(94)91586-5

Lutz, W. (2013). Demographic metabolism: A predictive theory of socioeconomic change. *Population and Development Review, 38*(Supplement), 283–301. https://doi.org/10.1111/j.1728-4457.2013.00564.x

Lutz, W., & KC, S. (2010). Dimensions of global population projections: What do we know about future population trends and structures? *Philosophical Transactions of the Royal Society B, 365*(1554), 2779–2791. https://doi.org/10.1098/rstb.2010.0133

Lutz, W., & Muttarak, R. (2017). Forecasting societies' adaptive capacities through a demographic metabolism model. *Nature Climate Change, 7*(3), 177–184. https://doi.org/10.1038/nclimate3222

Lutz, W., & Shah, M. (2002). Population should be on the Johannesburg agenda. *Nature, 418*(6893), 17–17. https://doi.org/10.1038/418017a

Lutz, W., Goujon, A., & Doblhammer-Reiter, G. (1998). Demographic dimensions in forecasting: Adding education to age and sex. *Population and Development Review, 24*(Supplementary Issue: Frontiers of Population Forecasting), 42–58. https://doi.org/10.2307/2808050

Lutz, W., Crespo Cuaresma, J., & Sanderson, W. C. (2008). The demography of educational attainment and economic growth. *Science, 319*(5866), 1047–1048. https://doi.org/10.1126/science.1151753

Lutz, W., Crespo Cuaresma, J., & Abbasi-Shavazi, M. J. (2010). Demography, education, and democracy: Global trends and the case of Iran. *Population and Development Review, 36*(2), 253–281. https://doi.org/10.1111/j.1728-4457.2010.00329.x

Lutz, W., Butz, W. P., Castro, M., Dasgupta, P., Demeny, P. G., Ehrlich, I., Giorguli, S., Habte, D., Haug, W., Hayes, A., Herrmann, M., Jiang, L., King, D., Kotte, D., Lees, M., Makinwa-Adebusoye, P. K., McGranahan, G., Mishra, V., Montgomery, M. R., Riahi, K., Scherbov, S., Peng, X., & Yeoh, B. (2012). Demography's role in sustainable development. *Science, 335*(6071), 918. https://doi.org/10.1126/science.335.6071.918-a

Lutz, W., Butz, W., KC, S., Sanderson, W. C., & Scherbov, S. (2014a). Population growth: Peak probability. *Science, 346*(6209), 561–561. https://doi.org/10.1126/science.346.6209.561-a

Lutz, W., Butz, W. P., & KC, S. (Eds.). (2014b). *World population and human capital in the twenty-first century.* Oxford University Press.

Lutz, W., Muttarak, R., & Striessnig, E. (2014c). Universal education is key to enhanced climate adaptation. *Science, 346*(6213), 1061–1062. https://doi.org/10.1126/science.1257975

Lutz, W., Goujon, A. V., Kc, S., Stonawski, M., & Stilianakis, N. (2018). *Demographic and human capital scenarios for the 21st century: 2018 assessment for 201 countries.* Publications Office of the European Union.

Lutz, W., Crespo Cuaresma, J., Kebede, E., Prskawetz, A., Sanderson, W. C., & Striessnig, E. (2019). Education rather than age structure brings demographic dividend. *PNAS, 201820362.* https://doi.org/10.1073/pnas.1820362116

MacKellar, F. L., Lutz, W., Prinz, C., & Goujon, A. (1995). Population, households, and CO2 emissions. *Population and Development Review, 21*(4), 849–865. https://doi.org/10.2307/2137777

Martens, W. J. M., Jetten, T. H., & Focks, D. A. (1997). Sensitivity of malaria, schistosomiasis and dengue to global warming. *Climatic Change, 35*(2), 145–156. https://doi.org/10.1023/A:1005365413932

Masozera, M., Bailey, M., & Kerchner, C. (2007). Distribution of impacts of natural disasters across income groups: A case study of New Orleans. *Ecological Economics, 63*(2–3), 299–306. https://doi.org/10.1016/j.ecolecon.2006.06.013

Mohai, P., & Saha, R. (2007). Racial inequality in the distribution of hazardous waste: A national-level reassessment. *Social Problems, 54*(3), 343–370. https://doi.org/10.1525/sp.2007.54.3.343

Mohai, P., Lantz, P. M., Morenoff, J., House, J. S., & Mero, R. P. (2009). Racial and socioeconomic disparities in residential proximity to polluting industrial facilities: Evidence from the Americans' Changing Lives Study. *American Journal of Public Health, 99*(Suppl 3), S649–S656. https://doi.org/10.2105/AJPH.2007.131383

Mugi-Ngenga, E. W., Mucheru-Muna, M. W., Mugwe, J. N., Ngetich, F. K., Mairura, F. S., & Mugendi, D. N. (2016). Household's socio-economic factors influencing the level of adaptation to climate variability in the dry zones of Eastern Kenya. *Journal of Rural Studies, 43*, 49–60. https://doi.org/10.1016/j.jrurstud.2015.11.004

Muttarak, R., & Dimitrova, A. (2019). Climate change and seasonal floods: Potential long-term nutritional consequences for children in Kerala, India. *BMJ Global Health, 4*(2), e001215. https://doi.org/10.1136/bmjgh-2018-001215

Muttarak, R., Lutz, W., & Jiang, L. (2016). What can demographers contribute to the study of vulnerability? *Vienna Yearbook of Population Research, 13*, 1–13. https://doi.org/10.1553/populationyearbook2015s1

Muttarak, R., & Pothisiri, W. (2013). The role of education on disaster preparedness: Case study of 2012 Indian Ocean earthquakes on Thailand's Andaman coast. *Ecology and Society, 18*(4), 51. https://doi.org/10.5751/ES-06101-180451

Neuenschwander, L. M., Abbott, A., & Mobley, A. R. (2012). Assessment of low-income adults' access to technology: Implications for nutrition education. *Journal of Nutrition Education and Behavior, 44*(1), 60–65. https://doi.org/10.1016/j.jneb.2011.01.004

Neumayer, E., & Plümper, T. (2007). The gendered nature of natural disasters: The impact of catastrophic events on the gender gap in life expectancy, 1981–2002. *Annals of the Association of American Geographers, 97*(3), 551–566. https://doi.org/10.1111/j.1467-8306.2007.00563.x

Nisbett, R. E. (2009). *Intelligence and how to get it: Why schools and cultures count.* WW Norton.

Norris, F. H., Smith, T., & Kaniasty, K. (1999). Revisiting the experience–behavior hypothesis: The effects of Hurricane Hugo on hazard preparedness and other self-protective acts. *Basic and Applied Social Psychology, 21*(1), 37–47. https://doi.org/10.1207/s15324834basp2101_4

Norris, F. H., Friedman, M. J., Watson, P. J., Byrne, C. M., Diaz, E., & Kaniasty, K. (2002). 60,000 disaster victims speak: Part I. An empirical review of the empirical literature, 1981–2001. *Psychiatry, 65*(3), 207–239.

Noy, I. (2009). The macroeconomic consequences of disasters. *Journal of Development Economics, 88*(2), 221–231. https://doi.org/10.1016/j.jdeveco.2008.02.005

O'Neill, B. C., MacKellar, F. L., & Lutz, W. (2001). *Population and climate change.* Cambridge University Press.

O'Neill, B. C., Dalton, M., Fuchs, R., Jiang, L., Pachauri, S., & Zigova, K. (2010). Global demographic trends and future carbon emissions. *PNAS, 107*(41), 17521–17526. https://doi.org/10.1073/pnas.1004581107

O'Neill, B. C., Kriegler, E., Ebi, K. L., Kemp-Benedict, E., Riahi, K., Rothman, D. S., van Ruijven, B. J., van Vuuren, D. P., Birkmann, J., Kok, K., Levy, M., & Solecki, W. (2017). The roads ahead: Narratives for shared socioeconomic pathways describing world futures in the 21st century. *Global Environmental Change, 42*, 169–180. https://doi.org/10.1016/j.gloenvcha.2015.01.004

Olofsson, A., & Rashid, S. (2011). The white (male) effect and risk perception: Can equality make a differ-

ence? *Risk Analysis, 31*(6), 1016–1032. https://doi.org/10.1111/j.1539-6924.2010.01566.x

Opiyo, F., Wasonga, O. V., Nyangito, M. M., Mureithi, S. M., Obando, J., & Munang, R. (2016). Determinants of perceptions of climate change and adaptation among Turkana pastoralists in northwestern Kenya. *Climate and Development, 8*(2), 179–189. https://doi.org/10.1080/17565529.2015.1034231

Oreopoulos, P., & Salvanes, K. G. (2011). Priceless: The nonpecuniary benefits of schooling. *Journal of Economic Perspectives, 25*(1), 159–184. https://doi.org/10.1257/jep.25.1.159

Padli, J., & Habibullah, M. S. (2009). Natural disaster death and socio-economic factors in selected Asian countries: A panel data analysis. *Asian Social Science, 5*(4), 65–71.

Patnaik, U., Das, P. K., & Bahinipati, C. S. (2016). Coping with climatic shocks: Empirical evidence from rural coastal Odisha, India. *Global Business Review, 17*(1), 161–175. https://doi.org/10.1177/0972150915610712

Paul, C. J., Weinthal, E. S., Bellemare, M. F., & Jeuland, M. A. (2016). Social capital, trust, and adaptation to climate change: Evidence from rural Ethiopia. *Global Environmental Change, 36*, 124–138. https://doi.org/10.1016/j.gloenvcha.2015.12.003

Pereira, S., Diakakis, M., Deligiannakis, G., & Zêzere, J. L. (2017). Comparing flood mortality in Portugal and Greece (Western and Eastern Mediterranean). *International Journal of Disaster Risk Reduction, 22*, 147–157. https://doi.org/10.1016/j.ijdrr.2017.03.007

Pérez-Arce, F. (2011). *The effect of education on time preferences.* Social Science Research Network.

Peters, E., Västfjäll, D., Slovic, P., Mertz, C. K., Mazzocco, K., & Dickert, S. (2006). Numeracy and decision making. *Psychological Science, 17*(5), 407–413. https://doi.org/10.1111/j.1467-9280.2006.01720.x

Pronczuk-Garbino, J. (Ed.). (2005). *Children's health and the environment: A global perspective: A resource manual for the health sector.* World Health Organization.

Raj, A., McDougal, L. P., & Silverman, J. G. (2015). Gendered effects of siblings on child malnutrition in South Asia: Cross-sectional analysis of demographic and health surveys from Bangladesh, India, and Nepal. *Maternal and Child Health Journal, 19*(1), 217–226. https://doi.org/10.1007/s10995-014-1513-0

Rao ND, Sauer P, Gidden M, Riahi K (2018) Income inequality projections for the shared socioeconomic pathways (SSPs). Futures 0(July):1–13. doi: https://doi.org/10.1016/j.futures.2018.07.001.

Reininger, B. M., Rahbar, M. H., Lee, M., Chen, Z., Alam, S. R., Pope, J., & Adams, B. (2013). Social capital and disaster preparedness among low income Mexican Americans in a disaster prone area. *Social Science & Medicine, 83*, 50–60. https://doi.org/10.1016/j.socscimed.2013.01.037

Riahi, K., van Vuuren, D. P., Kriegler, E., Edmonds, J., O'Neill, B. C., Fujimori, S., Bauer, N., Calvin, K., Dellink, R., Fricko, O., Lutz, W., Popp, A., Crespo Cuaresma, J., KC, S., Leimbach, M., Jiang, L., Kram, T., Rao, S., Emmerling, J., Ebi, K., Hasegawa, T.,

Havlik, P., Humpenöder, F., Da Silva, L. A., Smith, S., Stehfest, E., Bosetti, V., Eom, J., Gernaat, D., Masui, T., Rogelj, J., Strefler, J., Drouet, L., Krey, V., Luderer, G., Harmsen, M., Takahashi, K., Baumstark, L., Doelman, J. C., Kainuma, M., Klimont, Z., Marangoni, G., Lotze-Campen, H., Obersteiner, M., Tabeau, A., & Tavoni, M. (2017). The Shared Socioeconomic Pathways and their energy, land use, and greenhouse gas emissions implications: An overview. *Global Environmental Change, 42*, 153–168. https://doi.org/10.1016/j.gloenvcha.2016.05.009

Robine, J.-M., Cheung, S. L. K., Le Roy, S., Van Oyen, H., Griffiths, C., Michel, J.-P., & Herrmann, F. R. (2008). Death toll exceeded 70,000 in Europe during the summer of 2003. *Comptes Rendus Biologies, 331*(2), 171–178. https://doi.org/10.1016/j.crvi.2007.12.001

Rocklöv, J., Forsberg, B., Ebi, K., & Bellander, T. (2014). Susceptibility to mortality related to temperature and heat and cold wave duration in the population of Stockholm County, Sweden. *Glob Health Action, 7*. https://doi.org/10.3402/gha.v7.22737

Russell, L. A., Goltz, J. D., & Bourque, L. B. (1995). Preparedness and hazard mitigation actions before and after two earthquakes. *Environment and Behavior, 27*(6), 744–770. https://doi.org/10.1177/0013916595276002

Ryder, N. B. (1965). The cohort as a concept in the study of social change. *American Sociological Review, 30*(6), 843–861. https://doi.org/10.2307/2090964

Salvati, P., Petrucci, O., Rossi, M., Bianchi, C., Pasqua, A. A., & Guzzetti, F. (2018). Gender, age and circumstances analysis of flood and landslide fatalities in Italy. *Science of the Total Environment, 610–611*, 867–879. https://doi.org/10.1016/j.scitotenv.2017.08.064

Satterfield, T. A., Mertz, C. K., & Slovic, P. (2004). Discrimination, vulnerability, and justice in the face of risk. *Risk Analysis, 24*(1), 115–129. https://doi.org/10.1111/j.0272-4332.2004.00416.x

Schaffer, A., Muscatello, D., Broome, R., Corbett, S., & Smith, W. (2012). Emergency department visits, ambulance calls, and mortality associated with an exceptional heat wave in Sydney, Australia, 2011: A time-series analysis. *Environmental Health, 11*(1), 3. https://doi.org/10.1186/1476-069X-11-3

Sharp, T. M., Hunsperger, E., Santiago, G. A., Muñoz-Jordan, J. L., Santiago, L. M., Rivera, A., Rodríguez-Acosta, R. L., Feliciano, L. G., Margolis, H. S., & Tomashek, K. M. (2013). Virus-specific differences in rates of disease during the 2010 dengue epidemic in Puerto Rico. *PLoS Neglected Tropical Diseases, 7*(4), e2159. https://doi.org/10.1371/journal.pntd.0002159

Sherbakov, T., Malig, B., Guirguis, K., Gershunov, A., & Basu, R. (2018). Ambient temperature and added heat wave effects on hospitalizations in California from 1999 to 2009. *Environmental Research, 160*, 83–90. https://doi.org/10.1016/j.envres.2017.08.052

Silbert, M., & del Pilar Useche, M. (2012). *Repeated natural disasters and poverty in island nations: A decade of evidence from Indonesia*. University of Florida, Department of Economics.

Smith, D. L., & Notaro, S. J. (2009). Personal emergency preparedness for people with disabilities from the 2006–2007 behavioral risk factor surveillance system. *Disability and Health Journal, 2*(2), 86–94. https://doi.org/10.1016/j.dhjo.2009.01.001

Solomon, S., Plattner, G.-K., Knutti, R., & Friedlingstein, P. (2009). Irreversible climate change due to carbon dioxide emissions. *PNAS, 106*(6), 1704–1709. https://doi.org/10.1073/pnas.0812721106

Striessnig, E., & Loichinger, E. (2016). Future differential vulnerability to natural disasters by level of education. *Vienna Yearbook of Population Research, 13*(2015), 221–240. https://doi.org/10.1553/populationyearbook2015s221

Striessnig, E., Lutz, W., & Patt, A. G. (2013). Effects of educational attainment on climate risk vulnerability. *Ecology and Society, 18*(1), 16. https://doi.org/10.5751/ES-05252-180116

Tang, B., Liu, X., Liu, Y., Xue, C., & Zhang, L. (2014). A meta-analysis of risk factors for depression in adults and children after natural disasters. *BMC Public Health, 14*(1), 623. https://doi.org/10.1186/1471-2458-14-623

Taylor, N. A. S. (2006). Ethnic differences in thermoregulation: Genotypic versus phenotypic heat adaptation. *Journal of Thermal Biology, 31*(1–2), 90–104. https://doi.org/10.1016/j.jtherbio.2005.11.007

Thieken, A. H., Kriebich, H., Müller, M., & Merz, B. (2007). Coping with floods: Preparedness, response and recovery of flood-affected residents in Germany in 2002. *Hydrological Sciences Journal, 52*(5), 1016–1037. https://doi.org/10.1623/hysj.52.5.1016

Thomas, D. (2010). *Natural disasters and household welfare: Evidence from Vietnam*. The World Bank.

Thomas, K., Hardy, R. D., Lazrus, H., Mendez, M., Orlove, B., Rivera-Collazo, I., Roberts, J. T., Rockman, M., Warner, B. P., & Winthrop, R. (2019). Explaining differential vulnerability to climate change: A social science review. *Wiley Interdisciplinary Reviews: Climate Change, 10*(2), e565. https://doi.org/10.1002/wcc.565

Toya, H., & Skidmore, M. (2007). Economic development and the impacts of natural disasters. *Economics Letters, 94*(1), 20–25. https://doi.org/10.1016/j.econlet.2006.06.020

Ueland, J., & Warf, B. (2006). Racialized topographies: Altitude and race in southern cities. *Geographical Review, 96*(1), 50–78.

Van Aelst, K., & Holvoet, N. (2016). Intersections of gender and marital status in accessing climate change adaptation: Evidence from rural Tanzania. *World Development, 79*(Supplement C), 40–50. https://doi.org/10.1016/j.worlddev.2015.11.003

Vega, R. M., & Avva, U. (2019). *Pediatric dehydration*. StatPearls Publishing.

Vu, L., VanLandingham, M. J., Do, M., & Bankston, C. L. (2009). Evacuation and return of Vietnamese New Orleanians affected by Hurricane Katrina. *Organization and Environment, 22*(4), 422–436. https://doi.org/10.1177/1086026609347187

Walker, G., Burningham, K., Fielding, J., Smith, G., Thrush, D., & Fay, H. (2006). *Addressing environmental inequalities: Flood risk*. Environment Agency.

Wamani, H., Åstrøm, A. N., Peterson, S., Tumwine, J. K., & Tylleskär, T. (2007). Boys are more stunted than girls in Sub-Saharan Africa: A meta-analysis of 16 demographic and health surveys. *BMC Pediatrics, 7*(1), 17. https://doi.org/10.1186/1471-2431-7-17

Wanka, A., Arnberger, A., Allex, B., Eder, R., Hutter, H.-P., & Wallner, P. (2014). The challenges posed by climate change to successful ageing. *Zeitschrift für Gerontologie und Geriatrie, 47*(6), 468–474. https://doi.org/10.1007/s00391-014-0674-1

Wen, L. M., Rissel, C., Baur, L. A., Lee, E., & Simpson, J. M. (2011). Who is NOT likely to access the internet for health information? Findings from first-time mothers in Southwest Sydney, Australia. *International Journal of Medical Informatics, 80*(6), 406–411. https://doi.org/10.1016/j.ijmedinf.2011.03.001

WHO. (2011). *World malaria report 2011*. World Health Organization.

WIC. (2019). Wittgenstein Centre Data Explorer Version 2.0. www.wittgensteincentre.org/dataexplorer. Accessed 26 Mar 2019.

Wilkinson, P., Pattenden, S., Armstrong, B., Fletcher, A., Kovats, R. S., Mangtani, P., & McMichael, A. J. (2004). Vulnerability to winter mortality in elderly people in Britain: Population based study. *BMJ, 329*(7467), 647. https://doi.org/10.1136/bmj.38167.589907.55

Witvorapong, N., Muttarak, R., & Pothisiri, W. (2015). Social participation and disaster risk reduction behaviors in tsunami prone areas. *PLoS One, 10*(7), e0130862. https://doi.org/10.1371/journal.pone.0130862

Yardley, J., Sigal, R. J., & Kenny, G. P. (2011). Heat health planning: The importance of social and community factors. *Global Environmental Change, 21*(2), 670–679. https://doi.org/10.1016/j.gloenvcha.2010.11.010

Yeh, H. (2010). Gender and age factors in tsunami casualties. *Natural Hazards Review, 11*(1), 29–34. https://doi.org/10.1061/(ASCE)1527-6988(2010)11:1(29)

Young, A. J., Pandolf, K. B., & Sawka, M. N. (1996). Physiology of cold exposure. In B. Mariott & S. Carlson (Eds.), *Nutritional needs in cold and in high-altitude environments: Applications for military personnel in field operations*. National Academies Press.

Yu, W., Vaneckova, P., Mengersen, K., Pan, X., & Tong, S. (2010). Is the association between temperature and mortality modified by age, gender and socio-economic status? *Science of the Total Environment, 408*(17), 3513–3518. https://doi.org/10.1016/j.scitotenv.2010.04.058

Zagheni, E., Muttarak, R., & Striessnig, E. (2016). Differential mortality patterns from hydro-meteorological disasters: Evidence from cause-of-death data by age and sex. *Vienna Yearbook of Population Research, 13*(2015), 47–70. https://doi.org/10.1553/populationyearbook2015s47

Part II

Data & Methods

Household-Scale Data and Analytical Approaches

5

Brian C. Thiede

Abstract

Much of today's population-environment research centers on the interaction between households and individuals and their natural environment. The microdata used for such analyses are diverse with respect to their structure, substantive content, and geographic and temporal scope, raising both costs and benefits to users. The goal of this chapter is to put such issues in context by reviewing the major sources of household-level data and corresponding analytic methods used to examine the relationship between environmental change and demographic outcomes among individuals and households. In addition to reviewing examples of census, survey, surveillance, and administrative data, this chapter discusses key issues about the measurement of environmental exposures, adjustment for confounding variables, identification of causal pathways, and substantive interpretation of findings. It concludes by discussing opportunities for innovation, including in the

B. C. Thiede (✉)
Department of Agricultural Economics, Sociology, and Education, Pennsylvania State University, University Park, PA, USA
e-mail: bct11@psu.edu

manner that existing data are disseminated and in new data collection efforts. The recent, rapid expansion and improvement of data and methods provides population-environment researchers with many novel opportunities, but fundamental questions about conceptualization, measurement, and modeling remain salient.

Introduction

Demographers and other social scientists have given increased attention to the relationship between changes in the natural environment and human population dynamics over recent decades. Studies in this field address a wide range of substantive issues across many social and geographic contexts, spanning from projections of population displacements due to sea-level rise in the United States (Hauer, 2017), to estimates of the effects of temperature shocks on infant mortality in India (Banerjee & Maharaj, 2020), and to historical analyses of how climatic variability influenced marriage rates in the nineteenth and early-twentieth century Netherlands (Jennings & Gray, 2017). The diverse studies in this literature

© Springer Nature Switzerland AG 2022
L. M. Hunter et al. (eds.), *International Handbook of Population and Environment*, International Handbooks of Population 10, https://doi.org/10.1007/978-3-030-76433-3_5

are grounded by the common goal of understanding how environmental conditions, and changes therein, affect core demographic processes (i.e., fertility, mortality, and migration) and characteristics (i.e., population size, composition, and distribution). Doing so requires inter-disciplinary approaches that integrate concepts and data from the natural and social sciences, and in many cases also requires the use of spatial analytic methods to link data across sources and scales. Today's population scientists are fortunate to work at a time when high-quality demographic and environmental data are widely available (and proliferating) (Nawrotzki et al., 2016; Ruggles, 2014), as are the methodological tools to integrate and analyze these data (Auffhammer et al., 2013; Fussell et al., 2014b; Hsiang, 2016). However, researchers still face very real constraints and questions about what material and methods to use. These decisions are not always straightforward.

The goal of this chapter is to put such methodological issues in context by reviewing the major sources of household-level data and corresponding analytic methods used to examine the relationship between environmental change and demographic outcomes at the micro level, among individuals or households.[1] The discussion focuses on studies that assess the impact of environmental conditions on demographic outcomes. However, note that these methodological issues are also pertinent to much of the complementary literature on how demographic processes affect environmental outcomes (Crowder & Downey, 2010; De Sherbinin et al., 2008; Pais et al., 2013; Perz et al., 2006). Further, this chapter's focus on the micro level excludes ecological analyses, which examine the relationship between aggregate (e.g., province- or country-level) population characteristics and environmental conditions and which were employed in some of the foundational population-environment research (Henry et al., 2003; Hugo,

1996). Despite the value of such research, there is growing recognition that household-level data allow scholars to make stronger claims about population-environment interactions and to empirically test key hypotheses about how responses to, and consequences of, environmental change may vary within populations (Fussell et al., 2014b; Piguet, 2010).

The chapter proceeds as follows: the next section provides a general conceptual model that is common across the population-environment research addressed in this discussion. The third section describes the major categories of household-scale data and discusses their strengths and limitations, and the fourth section provides an overview of the analytic strategies that have been applied to these various data sources. The chapter's conclusion identifies common methodological challenges that researchers in this field must continue to deal with, as well as opportunities for innovation in micro-level population-environment research.

Conceptual Model

The literature on demographic responses to environmental change is substantively and geographically diverse. Studies within this field have focused on countries and regions across the development spectrum (Gray & Wise, 2016; Groen & Polivka, 2010; Thiede et al., 2016) and have examined many dimensions of the natural environment, including (but not limited to) temperature and precipitation (Mueller et al., 2014; Gray & Wise, 2016), earthquakes, floods, tsunamis and other natural disasters (Behrman & Weitzman, 2016; Bohra-Mishra et al., 2014; Nobles et al., 2015), and hazards such as pollution (Currie et al., 2017; Hill, 2018). The outcomes considered in this literature include the full range of demographic processes—fertility (Davis, 2017; Eissler et al., 2019), morbidity and mortality (Bandyopadhyay et al., 2012; Deschênes & Greenstone, 2011; Grace et al., 2012; Shi et al., 2015), and migration (Gray & Wise, 2016; Hunter et al., 2015)—as well as important dimensions of population composition, such as human capital attain-

---

[1] This chapter focuses on studies of demographic outcomes at the household or individual levels. It is also possible to aggregate household-scale data up to study ecological relationships (e.g., Abel et al., 2019; Mastrorillo et al., 2016). However, with one exception, such approaches are excluded from this discussion for brevity.

ment (Maccini & Yang, 2009; Randell & Gray, 2019).

A common conceptual model underlies these diverse studies (Fig. 5.1). In this framework, the demographic outcome of interest is a function of a given environmental variable(s), which measures individuals' exposure to the condition(s) of interest (e.g., precipitation, a natural disaster). Such environmental variables may represent the level of an environmental attribute (e.g. temperature) or its degree of change relative to a baseline level. For example, Grace et al. (2012) estimate the effects of total precipitation and average temperature levels on children's probability of chronic undernutrition (i.e., stunting). Gray and Wise (2016) model migration as a function of temperature and precipitation anomalies, which represent the standardized deviation from a baseline mean. As a final example, Nawrotzki and DeWaard's (2016) study of climate-induced migration in Mexico operationalized environmental exposures as the count of days with extremely high temperatures and precipitation. The demographic and environmental variables of interest are commonly measured over concurrent time periods or with a lag such that the measure represents prior years' environmental conditions. For example, Kumar et al. (2016) estimate the effects of *in utero* exposure to drought on the health of children ages 0–59 months. The assumption of such lagged effects is often plausible given delays in behavioral or biological responses or in the second-order impacts of environmental change that drive such responses.

Studies employing this common model also seek to control for other factors that may be correlated with both the demographic outcome and environmental predictor of interest. The suite of control variables that are appropriate and observable vary by context and study, but common controls include age, sex, socioeconomic status (e.g., education, income), and characteristics of individuals' place of residence (e.g., rural/urban community, region). For instance, Henry and Dos

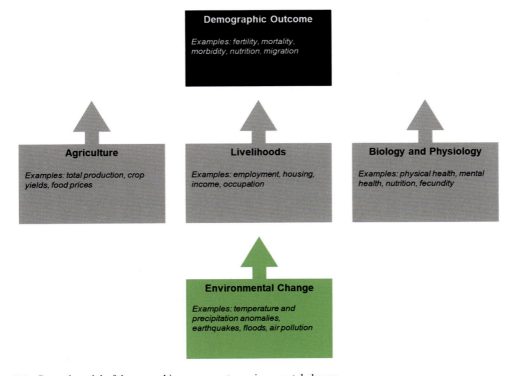

Fig. 5.1 General model of demographic responses to environmental change

Santos' (2013) statistical models of the relationship between rainfall and child mortality in the Sahel include controls for children's age, sex, birth order and season of birth, mother's education and age, the livelihood zone of children's residence, and time period. To take one of these variables as an example, mother's education may be positively associated with children's survival odds, and it may also be that populations in areas with high rainfall have lower levels of women's education than areas with average rainfall. In the absence of this control for education, Henry and Dos Santos (2013) may have erroneously concluded that rainfall is positively associated with child mortality when, in fact, their model shows the opposite. In these and other cases, it is important for researchers to draw on a clear conceptual framework to identify relevant control variables. The omission of controls and the inclusion of inappropriate (or "bad") controls can both bias estimates (Angrist & Pischke, 2008; Morgan & Winship, 2015).

In addition to controlling for these potentially confounding variables, many studies assess whether and how the effects of environmental change vary across sub-populations by estimating models with interaction terms or stratifying analyses by group. Such heterogeneous effects are often expected given differences in individuals' and households' exposure and ability to respond to environmental changes (Bohle et al., 1994). For example, Shively (2017) finds that access to transportation infrastructure—a proxy for access to markets and healthcare—moderates the effects of rainfall shocks on child malnutrition in Nepal and Uganda. In the U.S. context, Di et al.'s (2017) study of air pollution exposure and mortality includes multiple stratified models and post-hoc tests of between-group differences. Their results reveal, for example, that pollution exposure has a stronger effect on mortality among black and Hispanic adults than whites.

This general framework underlies most of the research examining environmental effects on micro-level demographic outcomes, but studies in this field vary with respect to how this conceptual model is operationalized in practice. Here, differences emerge with respect to at least four key issues. First, can environmental conditions and the outcome of interest be observed just once for each unit (i.e., individuals or households), or can they be tracked within units over time? Second, to what extent are potential confounders, moderators, or mediators observed in available data? Third, what is the spatial and temporal resolution of available demographic and environmental data, and to what extent are they compatible? Fourth and finally, do the data permit generalizable claims? Consideration of these issues will guide the discussion below of major micro-level data sources.

Data Sources

The objective of this section is to describe the data sources that population-environment researchers have commonly drawn on for micro-level analysis and identify major strengths and limitations of each. The discussion focuses on censuses, household surveys, demographic surveillance sites, and administrative data.

Censuses

The goal of a census is to enumerate an entire population—usually of a country or other political entity—on a given date, counting persons at either their place of usual or legal residence (*de jure*) or where they resided at the time of the census (*de facto*). These data have been implemented with growing frequency and quality throughout recent decades and are increasingly accessible due to a variety of dissemination efforts (Ruggles, 2014; UNFPA, 2019).[2] By definition, censuses aim to have complete geographic coverage and to capture accurate counts of an entire population. In many cases, censuses collect data on multiple social and demographic characteristics beyond simple population totals. These characteristics include but are not limited to age, education, marital status, race and (or) ethnicity, and dwelling characteristics. Many countries also collect informa-

---

[2]In many cases, a representative sample (e.g., 1%, 5%) of the census data is released to the public for analysis.

tion in censuses that can be used to study fertility (Hirschman & Guest, 1990), mortality (Leone, 2014), and migration (Nawrotzki et al., 2013) among populations. For example, numerous censuses across the world collect information on the number of children ever born to a woman and the number of such children who have survived (or died). It is also common to collect information on respondents' place of residence during a previous period (e.g., 5 years ago), which can be used to identify individuals who have migrated (Minnesota Population Center, 2019).

Most census data include detailed information on where individuals live since the governments that typically implement censuses are often interested in a population's geographic distribution. Censuses frequently classify individuals' place of residence in terms of administrative units, which may correspond to the different levels at which political power operates within countries. In some cases, however, the identifiers may simply correspond to one's location (e.g., residence on a given block). Geographic identifiers are often available at the second- or even third-order subnational level. For instance, U.S. census microdata include identifiers at the state and county levels,[3] and census microdata for Bangladesh include division-, district- *(zilla)*, and sub-district- *(upazila)* level identifiers. Censuses with identifiers for small, local administrative units are particularly useful for population-environment research since it permits precise measurement of environmental exposures. In the U.S. case, for example, knowing the average monthly temperature in an individual's county of residence is far more informative than the state- or country-wide means. However, the level of detail about individuals' location that is made available varies according to both data collection procedures and the implementing organizations' methods for, and regulations on, releasing census microdata to the public.

Censuses are typically implemented on a reoccurring basis, but often at a low frequency (e.g., every decade) compared with other sources, such as annual household surveys.[4] While individuals are likely be observed in multiple censuses over time, it is generally not possible to link observations across censuses given limitations on the identifiers that were collected, disclosure restrictions, or some combination of both. As a result, the population-environment literature has generally used census data for cross-sectional analysis, which examines relationships between variables at a single point in time. For example, Nawrotzki et al. (2013) use data from the 2000 Mexican census to examine the relationship between precipitation and international out-migration among rural households. In other cases, scholars have pooled multiple rounds of census data from the same country (or countries) and estimated pooled cross-sectional models, which analyze populations across multiple points in time but without tracking individuals from period to period. For example, Thiede, Gray, and Mueller (2016) analyze data from a total of 25 censuses implemented in eight South American countries to study the relationship between climatic variability and internal migration.

In only rare circumstances have population-environment researchers been able to use census data to construct longitudinal records, which track individuals over time. For instance, Bohra-Mishra et al. (2017) leveraged a unique set of migration data collected in the 2000 Philippines census to identify persons' locations at three time points—on census day and both 5 and 10 years prior—and study the effects of temperature, precipitation, and monsoon exposure on migration. However, the authors used these data to estimate and analyze province-level out-migration rates rather than treat the data as an individual-level panel. As such, for no known studies have sim-

---

[3]County identifiers are suppressed for some counties in the public-use samples to maintain confidentiality.

[4]While many population-environment researchers would prefer such data to be collected on a higher-frequency basis (Fussell et al., 2014b; Headey & Barrett 2015), such longer inter-censal intervals are likely preferred by implementing governments, citizens, and other stakeholders. Indeed, given resource constraints and the track record of censuses globally, a consistently implemented decadal census would represent a considerable improvement over the status quo.

ilarly creative approaches been implemented for micro-level analysis.

In general, censuses offer population-environment researchers nationally representative microdata that are collected on a regular, if infrequent basis. These data tend to include at least some information on the core demographic processes—fertility, mortality, and migration—and other social and demographic characteristics (e.g., age, sex, education). Census microdata also typically include information on individuals' place of residence within a country. However, the richness of the demographic and geographic data varies considerably across countries and over time and is a function of both data collection and data release procedures.

Household Surveys

Household surveys are commonly used by demographers and other social scientists to collect information about the socioeconomic and demographic characteristics of households and household members. Unlike censuses—which typically seek to count every individual in a country (or other population) for administrative purposes—surveys are often completed among a sample of the population of interest, and are implemented by a range of actors, including governments, researchers, and private-sector organizations. Unsurprisingly, then, population-environment research has drawn on a wide range of household surveys, which are diverse with respect to their geographic coverage, temporal scope, and original purpose. Broadly, one can distinguish between large-scale surveys that have been implemented across multiple countries in a standardized manner to facilitate cross-national comparison, large-scale surveys that have been implemented in a single country to collect data that are representative of all (or most) of that population, and smaller-scale surveys that are designed to answer some substantively- or geographically-limited questions. I discuss each of these survey types below and provide a list of examples in Table 5.1.

Large-Scale, Multi-National Surveys

The goal of large-scale, multi-national surveys, as defined here, is to collect comparable data across multiple countries and time periods. While these data permit the analysis of demographic dynamics within individual countries, they also facilitate geographically broad, multi-country studies. The latter are often of interest to population-environment researchers since they capture a high degree of diversity in environmental conditions, across which comparisons can be made.

Perhaps the most prominent example of a large-scale, multi-national survey used for population-environment research comes from the Demographic and Health Survey (DHS) Program. The DHS has collected demographic, health, and nutritional data by implementing over 400 surveys in at least 90 low- and middle-income countries across the world (DHS Program, 2019). DHS data have been used to study a diverse set of issues pertaining to population-environment interactions, with examples including the relationship between local land cover change and nutrition in Africa (Galway et al., 2018; Ickowitz et al., 2014), the effects of exposure to the 2010 earthquake in Haiti on fertility and related reproductive outcomes (Behrman & Weitzman, 2016), and the relationships between climatic variability and both nutrition and fertility in Africa and Asia (Bakhtsiyarava et al., 2018; Eissler et al., 2019; Grace et al., 2015; Tiwari et al., 2017).

In addition to collecting rich data on fertility, mortality, and health, the DHS has proven particularly useful for population-environment research for at least three reasons. First, the survey has been implemented multiple times within many countries and in a standardized manner, which facilitates comparisons over time and between countries. This approach allows researchers to reliably track changes in demographic outcomes before and after a given environmental change and (or) to compare affected and non-affected populations.

Second, the DHS sampling strategy involves collecting data from a large number of communities for each survey (e.g., 645 enumeration areas

5 Household-Scale Data and Analytical Approaches

Table 5.1 Examples of micro-level datasets used in population-environment research

Category	Dataset	Select uses of data
Large-scale, multi-national surveys	National population censuses	Bohra-Mishra et al. (2017), Nawrotzki et al. (2013) and Watmough et al. (2016)
	Demographic and Health Surveys	Behrman and Weitzman (2016), Galway et al. (2018) and Grace et al. (2015)
	Living Standards Measurement Study	Asfaw and Maggio (2018) and Mueller et al. (2020)
	Multiple Indicator Cluster Survey	Clifford et al. (2010) and Mason et al. (2005)
	Young Lives	Bahru et al. (2019)
	American Time Use Survey	Graff Zivin and Neidell (2014)
Large-scale, single-country surveys	China Health and Nutrition Survey	Mueller and Gray (2018)
	Chitwan Valley Family Study	Massey et al. (2010) and Williams and Gray (2020)
	Ethiopian Rural Household Survey	Gray and Mueller (2012) and Randell and Gray (2016)
	India Human Development Survey	Sedova and Kalkuhl (2020)
	Indonesian Family Life Survey	Bohra-Mishra et al. (2014), Sellers and Gray (2019) and Thiede and Gray (2017)
	Malawi Longitudinal Study of Families and Health	Anglewicz and Myroniuk (2018)
	Mexican Migration Project	Nawrotzki et al. (2015)
	Nang Rong Projects	Entwisle et al. (1998, 2016)
	Pakistan Panel Survey	Mueller et al. (2014)
	Panel Study of Income Dynamics	Elliott and Howell (2017)
	U.S. Current Population Survey	Cahoon (2006) and Groen and Polivka (2010)
Small-scale surveys	Bangladesh Environment and Migration Survey	Carrico and Donato (2019)
	Displaced New Orleans Residents Survey	Sastry (2009)
	Resilience in Survivors of Katrina	Waters (2016)
	Study of the Tsunami Aftermath and Recovery	Frankenberg et al. (2008), Gray et al. (2014) and Nobles et al. (2015)
Demographic surveillance systems	Agincourt Health and Demographic Surveillance System	Hunter et al. (2011, 2014) and Leyk et al. (2012)
	Matlab Health and Demographic Surveillance System	Call et al. (2017) and Islam et al. (2009)
Administrative records and vital statistics	Bangladesh Sample Vital Registration System	Chen and Mueller (2018)
	U.S. Internal Revenue Service tax returns	Deryugina et al. (2018)
	U.S. Centers for Medicare and Medicaid Services	Di et al. (2017)

in the 2016 Ethiopian DHS). In many cases, this sampling strategy helps researchers to capture sufficient geographic variation in environmental conditions to identify effects on demographic outcomes. Third, and relatedly, many DHS surveys provide publicly available geocoordinates (latitude and longitude) for each community included in the sample, which allows researchers to link DHS observations with environmental data with a high degree of precision.[5] As just one example, Galway et al. (2018) linked nutrition data from the DHS in 15 sub-Saharan African countries with tree cover estimates from the Global Forest Change dataset. They used DHS geocodes to measure the percentage of forest cover loss that occurred within a 5-km radius of each DHS community, and found that deforestation reduced dietary diversity (particularly in West Africa). In this and similar applications, researchers were able to construct more refined measures of environmental exposures than would be possible using administrative units or common alternatives.

The DHS is nonetheless characterized by important limitations. For one, the abovementioned geocoordinates are randomly displaced by a small distance before being released to the public, with the goal of reducing the likelihood that the places and people in the data can be identified.[6] This displacement introduces error into estimates of many contextual variables, such as tree cover or flood extent, within a given radius of the DHS cluster. More importantly, Elkies et al. (2015) demonstrate that such displacement can introduce systematic bias into estimates of the distance between DHS clusters and points of interest, such as the nearest water source. A second limitation pertains to the scope of the DHS. The primary

purpose of the survey is to track demographic and health outcomes, such as birth rates, contraceptive use, and child malnutrition, among others. As a result, the data for demographic and health outcomes are much richer than for socioeconomic characteristics that may function as confounders, moderators, or mediators of environmental effects on demographic outcomes. For instance, most rounds of the DHS include little to no information on income or recent in- and out-migration.

Third, the DHS is collected as a cross-sectional sample, and therefore largely precludes analysts from examining changes in environmental exposures and demographic outcomes within the same individual or household. The cross-sectional nature of the survey further limits the variables that can be included as controls since many of the socioeconomic characteristics that are observed are valid only at the time of the survey and may therefore have been affected by the environmental condition of interest. For example, theory may suggest the need to control for household wealth when modeling the relationship between recent flood exposure and child mortality. However, the economic disruptions associated with flood exposure may also affect household wealth, so if this variable is only measured at the time of the survey—and thus after the flood exposure of interest—it would represent a poor control variable.

With that said, it is worth noting that the DHS has collected birth records for women in the sample, which can be leveraged to construct longitudinal datasets of fertility and child mortality. Reproductive-aged female respondents were asked to provide information on the timing and survival of all births prior to the time of the survey. These records can then be used to produce individual-level panel data, such as those indicating whether a woman gave birth to a child during a given year or month or whether a child died during a given year or month.[7] This approach has the advantage of allowing analysts to study changes in the same individuals or households

---

[5]Even among the surveys that lack these geocodes, geographic identifiers and corresponding GIS data for first-order sub-national units are widely available.

[6]To reduce disclosure risk, the location of urban clusters (i.e., communities) are displaced up to two kilometers, and the location of rural clusters is displaced by up to five kilometers, with an additional (randomly selected) 1% of rural clusters displaced by up to 10 km. This displacement does have non-trivial consequences for estimating the effects of some contextual variables, particularly those measuring distance to a given point (Elkies et al., 2015). See Grace et al. (2019) for further discussion.

[7]Evidence of systematic misreporting of birth month (Larsen et al., 2019) suggests caution is needed when constructing such datasets at a monthly resolution.

over time, such as before and after a natural disaster. On the other hand, such birth records may suffer from recall bias and do not provide detailed information on the previous residence of women who have migrated.[8] Nonetheless—and similar to the example described in the discussion of censuses above—this case exemplifies the opportunities for creative use of existing data.

Other examples of large-scale, multi-national household surveys that have been used for population-environment research include the Living Standards Measurement Study (LSMS) and the Migration and Remittance Surveys (MRS). The MRS was implemented by the African Development Bank and the World Bank to collect standardized data on migration and remittances across sub-Saharan Africa. It was used by Gray and Wise (2016) to estimate and compare the effects of climate variability on migration across multiple African countries. The World Bank's LSMS program has supported the collection of household survey data since the 1980s in more than three dozen lower- and middle-income countries across the world. These surveys collect data on many dimensions of human welfare at multiple levels (e.g., individual, household, community), are typically georeferenced, and often follow individuals and households longitudinally. LSMS data are therefore well-suited for population-environment research. For example, Mueller et al. (2020) combine LSMS data from Ethiopia, Malawi, Tanzania, and Uganda to study the effects of temperature and precipitation variability on temporary migration across East Africa. They find that high temperatures and low precipitation reduce migration in urban areas. They then leverage the detailed employment data in the LSMS to demonstrate that climate variability also has negative effects on urban employment

opportunities (see Loebach, 2016; Ocello et al., 2015 for other examples of LSMS data used to study environment-induced migration). LSMS data have also been used to study the effects of weather shocks on food consumption. For example, Asfaw and Maggio's (2018) analysis of LSMS data from Tanzania finds that temperature shocks reduce food consumption and caloric intake and shows that these effects are strongest among female-headed households.

The Multiple Indicator Cluster Surveys (MICS), the Performance Monitoring and Accountability 2020 (PMA) data, and the Young Lives study are all examples of other multi-national datasets that could be used for population-environment research. However, these data have seemingly been under-utilized for this purpose to date. Limited exceptions include the use of MICS data to study the links between AIDS, drought, and nutrition in southern Africa (Mason et al., 2005) and the effects of drought on fertility and nuptiality in Tajikistan (Clifford et al., 2010); and the use of Young Lives data to study the effects of drought on child height in Ethiopia (Bahru et al., 2019). Beyond these examples, there are few (if any) large-scale, cross-national household surveys that have been used for micro-level population-environment research. Notably as well, the implementation of these surveys tends to be skewed toward lower- and middle-income countries, where limited research capacity has motivated donors to support such data collection efforts.

Large-Scale, Single-Country Surveys

Large-scale, single-country surveys aim to collect a broad set of demographic information that is representative of all (or most) of a given country. These surveys are often implemented multiple times, either as repeated cross-sections or as panels. Given the number and variety of such datasets, it is unsurprising that they have been widely used in the population-environment literature to date.

Many of these large-scale, single-country surveys have collected longitudinal data, which tracks individuals and households over time. Examples of panel studies used for population-

---

[8]The most common data on migration in these surveys includes (a) years in current residence; and (b) type of place of previous residence (e.g., rural vs. urban). However, even these two basic variables have not been collected in every round of the DHS. Notably, a number of recent DHS surveys have included information on place of previous residence (e.g., region of previous residence in the 2016 Ethiopian DHS).

environment research include, but are not limited to: the China Health and Nutrition Survey (CHNS), Chitwan Valley Family Study (CVFS), Ethiopian Rural Household Survey (ERHS), India Human Development Survey (IDHS), Indonesian Family Life Survey (IFLS), Mexican Family Life Study (MxFLS), Mexican Migration Project (MMP), Nang Rong Projects, Pakistan Panel Survey, and, in the U.S. context, the Panel Study of Income Dynamics (PSID).

These surveys have been used to study a wide range of substantive questions about population-environment interactions. For example, Mueller and Gray's (2018) analysis of CHNS data showed that exposure to heat increases the likelihood that Chinese adults (particularly the elderly) experienced underweight; while Elliott and Howell's (2017) study of natural hazard exposure and residential instability using the PSID found that county-level hazard damage is associated with increased residential instability. Such data have also been used to study the effects of drought on human migration and consumption in Ethiopia (Dercon et al., 2005; Gray & Mueller, 2012), the effects of climate variability on migration and human capital formation in Indonesia (Bohra-Mishra et al., 2014; Maccini & Yang, 2009; Thiede & Gray, 2017), the effects of climate on migration in Pakistan and India (Mueller et al., 2014; Sedova & Kalkuhl, 2020), the respective relationships between temperature and precipitation and migration in Mexico (Nawrotzki et al., 2015), the effects of environmental degradation and weather shocks on migration in Nepal (Massey et al., 2010; Williams & Gray, 2020), and the relationship between climate shocks, land cover change, and migration in northeast Thailand (Entwisle et al., 1998, 2016).

The longitudinal survey data used in these and other analyses have the advantage of tracking changes in environmental and demographic outcomes over time among the same individuals and households rather than comparing two different populations. Such data allow researchers to measure, and control for, confounding variables prior to the environmental change in question and to therefore avoid the time-order concerns that are common to cross-sectional data. More broadly, many of these datasets include more detailed and context-specific variables than large multi-national surveys (e.g., the DHS) that are designed with cross-national comparisons in mind. Given their context-specificity, these datasets often offer relatively refined measures of the demographic outcomes of interest. For example, the IFLS collects detailed information on geographic mobility in Indonesia that accounts for unique features of the country's migration system, such as the importance of circular migration (Hugo, 1982). These surveys also tend to include rich data on potential confounders, moderators, and mechanisms. For instance, the CHNS data allowed Mueller and Gray (2018) to discover that the effects of heat on adult underweight are driven by changes in illness rather than shifts in diet, income, or purchasing power.

Of course, there are drawbacks to studies focused on a single country. For one, a number of these datasets draw samples from a limited number of locations within the countries of interest or focus on geographically small or homogenous countries. In such cases, analysts will need to rely much more upon temporal than spatial variability in environmental conditions to identify the effects of interest. More broadly, the collection of context-specific data may, in some cases, preclude cross-national comparisons over time that are needed to assess the generalizability of findings.

In addition to these longitudinal studies, a number of large-scale cross-sectional surveys have been used in population-environment research. In many cases, these surveys have been implemented by national governments to track broad sets of demographic and economic indicators. For example, the American Time Use Survey (ATUS) is implemented by the U.S. Bureau of Labor Statistics with the goal of understanding how individuals use their time. Graff Zivin and Neidell (2014) used the ATUS to study the effects of temperature on time allocation in the United States, and found that high temperatures reduced the number of hours worked by individuals working in climate-sensitive industries. Likewise, Jacoby et al.

(2014) studied the effects of rising temperatures on agricultural productivity using employment data from the Indian government's National Sample Survey (NSS), which has regularly collected socioeconomic data across the country since 1950. Using these data, they estimate that future decades of warming will adversely affect the agricultural sector in rural India, where such climate shocks have been shown to influence migration patterns (Sedova & Kalkuhl, 2020).

Notably, in the aftermath of some natural disasters, large-scale surveys may be modified to explicitly assess their impact. For example, the U.S. Current Population Survey (CPS)—which is a monthly survey of approximately 60,000 households and the primary source of U.S. labor statistics—was modified in the aftermath of Hurricane Katrina. Procedures were changed, and a module added, to allow researchers to track socioeconomic and demographic outcomes among those affected (Cahoon, 2006; Groen & Polivka, 2010). This modification yielded a number of important findings, including evidence that the likelihood of returning to Katrina-affected areas in the year after the storm varied systematically by evacuees' age and income, and that the region experienced shifts in ethno-racial, age, and income composition after the storm (Groen & Polivka, 2010). Of course, large-scale cross-sectional surveys can be of value to population-environment researchers even in the absence of such modifications to the survey instrument or sampling strategy.

Small-Scale Surveys

Many population-environment studies have also used smaller-scale surveys. As defined here, these studies have focused on individual environmental events or interventions with environmental impacts, or have been designed to study specific subpopulations or regions within countries. A number of such studies have been fielded in the aftermath of natural disasters. For instance, multiple studies were fielded in the aftermath of Hurricane Katrina, including the Displaced New Orleans Residents Survey (DNORS, Sastry, 2009), the Hurricane Katrina Community Advisory Group (Kessler et al., 2008), the Gulf Coast Child and Family Health Study (Abramson et al., 2008), and the Resilience in Survivors of Katrina (RISK) Project (Waters, 2016). As one illustrative case, the RISK Project studied outcomes among a sample of 1019 young community college students—mainly black women with children—from New Orleans. This project leveraged a study (Opening Doors) that had collected data from these students during the year prior to the storm, providing a baseline. The RISK Project located 86% of the original sample for at least one post-storm interview, and collected a range of data to understand the trauma they experienced and how relocation affected their long-term wellbeing and recovery (Waters, 2016). Data from RISK and the other studies of Katrina have collectively been used to understand many demographic outcomes among the population affected by the storm, ranging from evacuation behavior and return migration to physical and mental health (Fussell et al., 2010; Graif, 2016; Kessler et al., 2006; Thiede & Brown, 2013).

Likewise, the Study of the Tsunami Aftermath and Recovery (STAR) was implemented in the wake of the December 2004 Indian Ocean tsunami. Like the RISK Project, STAR was designed to use another survey—in this case one regularly fielded by the Indonesian government (National Socioeconomic Survey, SUSENAS)—as a baseline (Gray et al., 2014). STAR conducted follow-up interviews with respondents from the 2004 SUSENAS who were living in coastal districts (*kabupaten*) of the provinces of Aceh and North Sumatra prior to the tsunami (n = 22,390), producing a dataset that allowed for pre- and post-tsunami comparisons. These innovative data have been used to study health and reproductive outcomes (Frankenberg et al., 2008; Nobles et al., 2015), among others.

Many other small-scale studies have been implemented outside the context of natural disasters. For example, López-Carr (2012) collected original household survey data in rural Guatemala to study the relationship between environmental change and out-migration. Since this study was focused on the environmental determinants of out-migration to agricultural frontiers, they drew on prior findings to design and implement the study in a set of rural Guatemalan *municipos* that were the primary origin of migrants to frontier settlements within the Maya Biosphere Reserve. As another example, a team of researchers collected micro-level data among five indigenous groups in the Ecuadorian Amazon in 2001 and 2012 to study demographic dynamics in an ecologically-sensitive environment (Bremner et al., 2009; Davis et al., 2015).

Many of these studies are characterized by limited geographic and (or) temporal scope. Findings may therefore be limited in terms of generalizability, but this is rarely the goal of such research. However, in many cases this limited scope makes it difficult to compare outcomes before and after a given environmental change or to identify comparison groups who were not "exposed" to an event, thus limiting causal claims. Yet, these datasets have the distinctive advantage of being designed explicitly for research on population-environment interactions or cognate issues and tailored to the local context. The value of such characteristics should not be understated.

Demographic Surveillance Systems

Demographic surveillance systems (DSSs) have been developed by public health researchers across the world to monitor vital events (births, deaths, and migration), typically in locations that would otherwise lack health information infrastructure.[9] The International Network for the continuous Demographic Evaluation of Populations and their Health (INDEPTH) is perhaps the most prominent network of DSSs. At the time of writing, this network includes 49 sites in 19 countries, collectively observing the vital events of more than 3 million people. DSSs provide longitudinal, detailed records of vital events that are of interest to many demographers and are intended to be more reliable sources of

data than what might otherwise be available in these contexts.

Such data have been used to conduct a number of innovative population-environment studies, including Hunter et al.'s (2014) study of natural capital and out-migration among rural South Africans and Call et al.'s (2017) study of environmental change and temporary migration in Bangladesh. Hunter et al. (2014) analyzed migration data from the Agincourt Health and Socio-Demographic Surveillance Site in South Africa and used remote-sensing data to measure vegetation cover and biomass production (as proxied by the Normalized Difference Vegetation Index, NDVI) around each village during the 3 years prior to (and through) the year over which migration was measured. They find that village-level natural resource availability is positively associated with temporary out-migration.

Despite such strengths, DSSs are characterized by at least two important limitations pertinent to population-environment research. First, many of these systems are often set up in limited geographic areas, and the data from a given system may lack spatial variation in environmental conditions. For example, the Agincourt Study Site is approximately 400 km^2 (Tollman et al., 1999), which may be characterized by spatial heterogeneity in variables such as vegetation and rainfall (Hunter et al., 2014) but that is unlikely to vary with respect to temperature. Researchers must therefore often use temporal rather than spatial variation to identify climate effects. Second, since DSSs are oriented toward the collection of vital records they may not collect detailed social and economic data needed to control for confounders. For example, Call et al.'s (2017) study of how flooding, precipitation, and temperatures affect migration in Bangladesh noted that data on household assets, housing characteristics, and land ownership were not available in the Matlab DSS but instead had to be drawn from census data.

Administrative Records and Vital Statistics

Population-environment researchers have also made use of other sources of microdata that

---

[9]These systems are sometimes also referred to as Health and Demographic Surveillance Sites (HDSSs).

might be broadly characterized as administrative records. These include, but are not limited to, government tax records and vital registration systems. For example, Chen and Mueller (2018) used migration records collected in Bangladesh's Sample Vital Registration System (SVRS) to examine the relationship between soil salinity and migration. They link data on household migration and sub-district of residence from the SVRS with measures of soil salinity derived from soil samples across the country's coastal regions, and estimate statistical models that show salinity is positively associated with internal migration and inversely associated with international moves.

In another example, Deryugina et al. (2018) use tax records to estimate the effects of Hurricane Katrina on household income, including assessing whether these effects vary between migrants and non-migrants. They find that individuals who left New Orleans around the storm and did not return fared worse than those who never left in the short run, but experienced greater long-run gains in labor income.[10] As a final illustration, Di et al. (2017) draw on a dataset from the U.S. Centers for Medicare and Medicaid Services that includes all Medicare beneficiaries in the United States from 2000 through 2012. Using beneficiaries' ZIP code of residence as a geographic identifier, they link these population data with pollution records from the U.S. Environmental Protection Agency and estimate statistical models that show exposure to $PM_{2.5}$ and ozone were associated with increased all-cause mortality. The diversity of such sources precludes a general discussion of strengths and weaknesses. However, it is important to acknowledge that researchers are conducting innovative new population-environment research by leveraging fundamentally new types of data.

Analytic Approaches

Quantitative analysis of the relationship between environmental conditions and micro-level demographic outcomes requires major methodological decisions in at least four domains: the measurement of environmental exposures, the treatment of confounding variables, the identification of causal mechanisms, and the substantive interpretation of estimates in the context of expected environmental changes. The following sections address each of these issues in turn.

Measuring Environmental Exposures

A first and fundamental step in most population-environment research is to measure individuals' exposures to the environmental conditions of interest as precisely as possible in relation to the demographic outcome of interest. At least three approaches have been taken, the choice between which may be a function of data availability and the substantive issue in question.

First, some studies have measured environmental exposures using self-reported retrospective data, which are typically derived from individuals' or households' responses to a question(s) about recent or current environmental conditions. For example, the ERHS asked respondents about historical exposure to droughts (Gray & Mueller, 2012), and the CVFS asked respondents about their perceptions of agricultural productivity, the time needed to collect firewood, and related questions about natural resources and environmental conditions. In some other cases, household surveys are complemented by the collection of community-level data—typically derived from structured interviews with community leaders—which may include information on environmental conditions and changes that are common across the community. For example, the STAR project collected information on tsunami damage in surveyed communities by asking local leaders for their own assessments and collecting data from the survey supervisors working in the

---

[10]A number of other studies have used U.S. Internal Revenue Service tax return data to study the effects of environmental change on migration. These important studies—including Curtis et al. (2015), Fussell et al. (2014a), and Hauer (2017)—aggregate these data (e.g., to the county level) and conduct ecological analyses, and are therefore not featured in the text above.

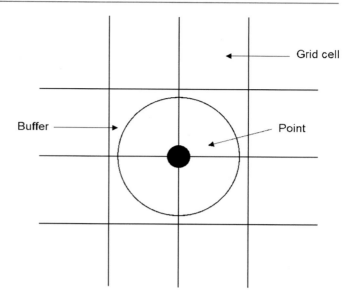

Fig. 5.2 Illustration of relationship between grid and point data

field (in addition to the use of satellite imagery) (Gray et al., 2014).

Second—and perhaps most commonly—researchers may combine demographic records with environmental data from other sources, such as those derived from weather stations, climate models, or earth-observation satellites. For example, many studies of the demographic impacts of climate variability have used the University of East Anglia Climate Research Unit's Time Series (CRUTS) (Call et al., 2019; Eissler et al., 2019; Gray & Wise, 2016), the Climate Hazards Group Infrared Precipitation with Stations (CHIRPS) dataset (Davenport et al., 2017; Grace et al., 2015), and (or) the Modern-Era Retrospective Analysis for Research and Applications (MERRA, Sellers & Gray, 2019; Thiede & Gray, 2017). Likewise, the Global Forest Change database has been used to study the relationship between deforestation and demographic dynamics (Galway et al., 2018), and data on earthquake intensity from the U.S. Geological Survey were used to study the demographic effects of earthquake exposure in Haiti (Behrman & Weitzman, 2016).

The precise manner that demographic and environmental data are linked varies according to both substantive concerns and the nature of the data being considered (Fig. 5.2). For example, in some cases demographic data include geospatial identifiers expressed as points (i.e., latitude and longitude) and environmental data are gridded, with values assigned for each n-by-n (e.g., 1 km-by-1 km) areal unit. Analysts may link the demographic records to the environmental data for the grid cell that the point of enumeration falls into (Eissler et al., 2019) or they may calculate the mean of all cells within a given buffer of that point (e.g., 13 km, Grace et al., 2012). In other instances, researchers may identify the administrative unit (e.g., province) that an individual or household is observed in—either using administrative identifiers in the data or by matching the point location with administrative units—and then calculate the mean of the environmental variable(s) of interest for all cells that fall within that unit (Thiede et al., 2016).[11] And in a final example, environmental data may only be available for particular points (e.g., weather stations) and then linked to demographic data by finding the nearest match (Maccini & Yang, 2009) or using various spatial interpolation techniques (Auffhammer et al., 2013; Nawrotzki et al., 2015).

---

[11] It is common to use the spatial mean, but conceptually it is possible to develop other measures of exposure such as population-weighted means.

Third and finally, some studies combine information about the time and (or) location of a given demographic observation with knowledge about the timing and (or) location of an environmental change to measure exposure without explicitly measuring environmental conditions. For example, Hoddinott and Kinsey (2001) studied the effects of drought exposure on the height of Zimbabwean children aged 12–24 months by comparing the growth rate of the cohort who passed through these ages during a drought year with that of similarly-aged children during the preceding and subsequent non-drought years. Children in the exposed cohort experienced reductions in growth rates of approximately 15–20%. A similar approach is taken by Nandi et al., (2018), who draw on information about the geographic distribution of earthquake intensity in India to identify a treatment group without explicitly integrating environmental data. Using this approach, they found that earthquake exposure was associated with decreased birth spacing, increased birth rates, and, in some cases, an increased male:female sex ratio.

Measurement decisions are often constrained by the data that are available. For example, many population-environment researchers would prefer to know the precise latitude and longitude of an individual's residence but work with data that is limited to far coarser geographic identifiers (e.g., an individual's province or district of residence). Beyond these constraints, however, researchers must also make measurement decisions—and weigh the pros and cons of sub-optimal choices—in the context of a clear conceptual framework. They must identify the process through which environmental conditions are expected to influence the demographic outcome of interest, and in doing so also define the appropriate spatial and temporal scales over which the environmental condition of interest may influence that outcome. For example, studies of how climatic variability affects child malnutrition have focused on both stunting (low height-for-age) and wasting (low weight-for-height) as outcome variables (Davenport et al., 2017; Tiwari et al., 2017). In such studies, it is important to account for the fact that stunting tends to be driven by chronic insults *in utero* and over the first years of life, while wasting is more likely to reflect acute illness. The implication is that the timing and duration of environmental exposure windows should differ for each outcome: stunting might be best modeled as a function of lifetime exposures while wasting may be modeled as a function of exposures during the prior 12 months, for example (Randell et al., 2020; Thiede & Gray, 2020).

In this and other cases, clearly defined frameworks are needed to alleviate concerns that the temporal and spatial scope of environmental exposure measures are arbitrary. Similarly, causal claims can be strengthened by evidence that environmental conditions during a theoretically-defined critical period have a strong effect on a demographic outcome of interest, but conditions during alternative, less-salient periods have weaker (or null) effects (i.e., a "placebo test"). For example, Beuermann and Pecha (2020) study the relationship between weather shocks and early childhood development in Jamaica by analyzing the impact of tropical storm exposure on children's weight, height, and birthweight. To test the robustness of their findings, they estimate models that predict child health outcomes as a function of tropical storms that occurred after the observation and thus could not plausibly affect the outcomes of interest. These estimates are not statistically significant, which increases the confidence that the main estimates of storm impacts (with proper time order) are capturing true effects.

Controlling for Confounding Variables

Statistical analyses of the relationship between environmental conditions and micro-level demographic outcomes have typically employed regression approaches to control for confounders. Here, at least two issues are salient. First is the degree to which available datasets observe such variables. Many of the datasets that population-environment researchers use—but particularly census and large-scale, multi-country surveys such as the DHS—include a limited set of relevant control variables. The choice of appropriate controls is also often limited by time-order concerns: variables that are themselves potentially affected by environmental conditions (the independent variable of interest in these studies) should not be included as controls unless they can be measured prior to the period of environmental exposure. This issue represents a particular challenge for analysis of cross-sectional data. For example, it may be theoretically appropriate to control for household wealth when estimating the effects of precipitation and temperature anomalies on children's birth weight (Grace et al., 2015). However, in cross-sectional data, wealth is typically measured at the time of the survey and therefore concurrent to the outcome variable and after the environmental exposure of interest.

A second and related issue is the extent to which the structure of the data allows researchers to account for variables that are not explicitly observed but could be adjusted for through the inclusion of fixed effects, which control for factors that are invariant within the unit of interest. Population-environment researchers have taken such approaches when analyzing pooled cross-sectional data and panel data. For example, in some cases scholars have combined multiple rounds of census data from the same country (or countries) and estimated pooled cross-sectional models with spatial and temporal fixed effects (e.g., Thiede et al., 2016). While one still cannot track individuals or households over time, such approaches allow researchers to control for time invariant contextual factors (e.g., province characteristics) and to control for temporal trends that are common across the population (or specific sub-populations). The analysis of longitudinal micro-level data allows researchers to track individuals or households over time. With such data, researchers are most likely to be able to include control variables that do not violate time order assumptions and can account for unobservable time-invariant

characteristics at high resolution, such as the household (e.g., Mueller et al., 2014). These approaches allow for more conservative hypothesis testing but are not panaceas since estimates may still be affected by time-varying or spatially heterogeneous confounders. Careful thinking about causality is needed regardless of data structure (Morgan & Winship, 2015).

Identifying Causal Mechanisms

Micro-level analysis of population-environment interactions are further strengthened by evidence of the processes that explain a given statistical association between environmental and demographic variables. Population-environment researchers have addressed this issue in at least three ways. First and as noted above, the careful selection of, and justification for, one's measures of environmental exposure vis-à-vis theory can provide indirect support for a given causal mechanism. For instance, if one hypothesizes that climate shocks influence child malnutrition rates by reducing agricultural production (Grace et al., 2012), then they might assess whether precipitation deficits during the agricultural growing season have a stronger effect on nutrition than during other months. Likewise, if one hypothesizes that environmental conditions influence educational attainment through health and other developmental pathways rather than by influencing households' ability to pay for school fees, they may compare the effects of *in utero* and later-childhood exposures on educational attainment.

Second, if a hypothesized causal mechanism can be expected to operate more strongly for some sub-populations than others, researchers can empirically test those mechanisms by comparing the strength of environmental effects between groups. For example, if climatic variability is expected to influence human migration by disrupting agricultural livelihoods, researchers can assess this expectation by comparing climate effects among agriculturally dependent populations versus others (Nawrotzki & Bakhtsiyarava, 2017;

Thiede & Gray, 2017). For example, Nawrotzki and Bakhtsiyarava (2017) assess the hypothesis that climate variability influences migration by disrupting agriculture using the cases of rural Burkina Faso and Senegal. They explicitly test this hypothesis by interacting measures of climate exposure with measures of the area within each respondent's province that is devoted to harvesting key crops (cotton in Burkina Faso and groundnuts in Senegal), and find that in Senegal, droughts decrease the likelihood of international moves in agriculturally dependent regions.

Third, researchers have compared the effects of environmental change on a demographic outcome of interest with the estimated effects of that same environmental change on hypothesized mechanisms. The consistency of the effects across these parallel models determines the degree of support (or lack thereof) for a given mechanism. For example, Mueller et al. (2014) examine the effects of heat stress on farm and non-farm income to understand the potential drivers of observed temperature effects on migration in Pakistan. Their results suggest that heat-related declines in both sources of income may drive increased migration among men. Likewise, Behrman and Weitzman (2016) complement their analysis of how the 2010 Haitian earthquake affected fertility by conducting parallel analyses of earthquake effects on potential pathways, including access to contraceptives, sexual activity, and fertility preferences. They showed that earthquake-related increases in pregnancies corresponded with increases in unmet need for contraceptives, declining access to condoms, and diminished negotiating power about condom use with partners.

Substantive Interpretation of Findings

In many cases, population-environment researchers estimate the demographic impacts of previous environmental changes in order to develop expectations for how individuals and households will respond to future changes. For example, the finding that permanent whole-household migration in Indonesia increases in response to spells of high temperatures could suggest that migration will increase as climate changes push long-term average temperatures above this threshold (Bohra-Mishra et al., 2014: 9784). In many such instances, researchers have attempted to make these inferences explicit, using their findings to estimate how a demographic outcome of interest will change under future scenarios of environmental change. For example, Chen and Mueller (2018) develop projections of change in within-district migration under plausible scenarios of future soil salinity. In other cases, researchers have not made claims about demographic outcomes under future scenarios of environmental change but have still made efforts to clarify the substantive interpretation of findings. Examples include reporting predicted probabilities across a range of plausible environmental conditions or, in the case of interaction models, illustrating how the effect of a given variable changes according to the value of the moderating variable (Eissler et al., 2019; Mueller et al., 2014).

Opportunities for Innovation

Population-environment research has made considerable advances over recent decades, in large part due to innovations in the collection, dissemination, and analysis of micro-level data. There is reason to believe these improvements will continue into the future, with at least four areas of development particularly encouraging. First is the continued proliferation of microdata. For example, throughout the past two decades the IPUMS project (e.g., IPUMS-International, IPUMS-Terra) has harmonized and made public data from hundreds of censuses. It now includes data for approximately 702 million persons, derived from over 300 censuses implemented from 1960 to the present. This project has also provided an interface for researchers to rapidly and seamlessly link population microdata to environmental variables, first through IPUMS-

Terra and increasingly through other IPUMS products (e.g., IPUMS-DHS). The increased volume of microdata will allow researchers to study population-environment interactions in new geographic and historical contexts, and the dissemination of already-integrated population-environment data has and will continue to rapidly accelerate research in this field. Increased access to new census and survey data may also open opportunities for researchers to exploit unique "quirks" in particular datasets—such as the multiple migration questions asked in the 2000 Philippines census described above—that allow researchers to test unique hypotheses or apply more powerful statistical techniques.

Second, recent efforts to develop methods for linking environmental and household data at high spatial resolutions but without violating confidentiality restrictions may lead to improved measurement of environmental exposures (Grace et al., 2019; Hunter et al., 2019). Increased precision in these measurements is especially useful in contexts with a high degree of spatial heterogeneity in environmental conditions, such as in Ethiopia, and where underlying environmental data can be acquired at similarly high resolutions.

Third, there has been a recent increase in what might be considered "non-traditional" microdata, such as cell phone records, social media data, and crowdsourced information (e.g., Lu et al., 2016). While there remain a number of important questions to address with respect to these data—from privacy issues to questions about representativeness—many such efforts are underway. The full potential of these data for population-environment research remains uncertain, but additional work in this area is merited.

Fourth and finally, there is a fundamental need for investments in large-scale, high-frequency data collection efforts. A similar argument has been made by economists interested in food security and resilience (Headey & Barrett, 2015), suggesting there is widespread demand for such data from across multiple disciplines. This type of data would be particularly helpful for population-environment researchers because they would allow for temporally precise measures of environmental exposures and behavioral responses across a geographically diverse population. As Headey and Barrett (2015) point out, such ambitious efforts have been launched to address data limitations in the past, such as the U.S. National Science Foundation's Long-Term Ecological Research Network. Given the pressing challenge associated with global environmental change, this and other innovations to improve knowledge and inform interventions are much needed.

Conclusion

This chapter has outlined the types of micro-level data that population-environment researchers currently employ, as well as the related conceptual and methodological issues that they encounter. Today's population-environment researchers are fortunate to work at a time when microdata are widely available, research methods are improving, and computational advances promise continued innovation into the future. They nonetheless continue to face fundamental challenges to properly measuring environmental exposures and demographic outcomes, appropriately controlling for confounding variables, and clearly communicating the substantive meaning and importance of their results. To sustain the major improvements in population-environment research over the past decades, scholars in the field must simultaneously embrace innovative new data and methods and double-down on the fundamentals of conceptualization, measurement, and modeling.

References

Abel, G. J., Brottrager, M., Cuaresma, J. C., & Muttarak, R. (2019). Climate, conflict and forced migration. *Global Environmental Change, 54*, 239–249.

Abramson, D., Stehling-Ariza, T., Garfield, R., & Redlener, I. (2008). Prevalence and predictors of mental health distress post-Katrina: Findings from the Gulf Coast Child and Family Health Study. *Disaster Medicine and Public Health Preparedness, 2*(2), 77–86.

Anglewicz, P., & Myroniuk, T. W. (2018). Shocks and migration in Malawi. *Demographic Research, 38*, 321–334.

Angrist, J. D., & Pischke, J. S. (2008). *Mostly harmless econometrics: An empiricist's companion.* Princeton University Press.

Asfaw, S., & Maggio, G. (2018). Gender, weather shocks and welfare: Evidence from Malawi. *The Journal of Development Studies, 54*(2), 271–291.

Auffhammer, M., Hsiang, S. M., Schlenker, W., & Sobel, A. (2013). Using weather data and climate model output in economic analyses of climate change. *Review of Environmental Economics and Policy, 7*(2), 181–198.

Bahru, B. A., Bosch, C., Birner, R., & Zeller, M. (2019). Drought and child undernutrition in Ethiopia: A longitudinal path analysis. *PLoS One, 14*(6), e0217821.

Bakhtsiyarava, M., Grace, K., & Nawrotzki, R. J. (2018). Climate, birth weight, and agricultural livelihoods in Kenya and Mali. *American Journal of Public Health, 108*(S2), S144–S150.

Bandyopadhyay, S., Kanji, S., & Wang, L. (2012). The impact of rainfall and temperature variation on diarrheal prevalence in Sub-Saharan Africa. *Applied Geography, 33*, 63–72.

Banerjee, R., & Maharaj, R. (2020). Heat, infant mortality, and adaptation: Evidence from India. *Journal of Development Economics, 143*, 102378.

Behrman, J. A., & Weitzman, A. (2016). Effects of the 2010 Haiti earthquake on women's reproductive health. *Studies in Family Planning, 47*(1), 3–17.

Beuermann, D. W., & Pecha, C. (2020). The effects of weather shocks on early childhood development: Evidence from 2 years of tropical storms. *Economics & Human Biology, 100851.*

Bohle, H. G., Downing, T. E., & Watts, M. J. (1994). Climate change and social vulnerability: Toward a sociology and geography of food insecurity. *Global Environmental Change, 4*(1), 37–48.

Bohra-Mishra, P., Oppenheimer, M., & Hsiang, S. M. (2014). Nonlinear permanent migration response to climatic variations but minimal response to disasters. *Proceedings of the National Academy of Sciences, 111*(27), 9780–9785.

Bohra-Mishra, P., Oppenheimer, M., Cai, R., Feng, S., & Licker, R. (2017). Climate variability and migration in the Philippines. *Population and Environment, 38*(3), 286–308.

Bremner, J., Bilsborrow, R., Feldacker, C., & Holt, F. L. (2009). Fertility beyond the frontier: Indigenous women, fertility, and reproductive practices in the Ecuadorian Amazon. *Population and Environment, 30*(3), 93–113.

Cahoon, L. S. (2006). The current population survey response to Hurricane Katrina. *Monthly Labor Review, 129*, 40–51.

Call, M., Gray, C., & Jagger, P. (2019). Smallholder responses to climate anomalies in rural Uganda. *World Development, 115*, 132–144.

Call, M. A., Gray, C., Yunus, M., & Emch, M. (2017). Disruption, not displacement: Environmental variability and temporary migration in Bangladesh. *Global Environmental Change, 46*, 157–165.

Carrico, A. R., & Donato, K. (2019). Extreme weather and migration: Evidence from Bangladesh. *Population and Environment, 41*(1), 1–31.

Chen, J., & Mueller, V. (2018). Coastal climate change, soil salinity and human migration in Bangladesh. *Nature Climate Change, 8*(11), 981.

Clifford, D., Falkingham, J., & Hinde, A. (2010). Through civil war, food crisis and drought: Trends in fertility and nuptiality in post-Soviet Tajikistan. *European Journal of Population, 26*(3), 325–350.

Crowder, K., & Downey, L. (2010). Interneighborhood migration, race, and environmental hazards: Modeling microlevel processes of environmental inequality. *American Journal of Sociology, 115*(4), 1110–1149.

Currie, J., Greenstone, M., & Meckel, K. (2017). Hydraulic fracturing and infant health: New evidence from Pennsylvania. *Science Advances, 3*(12), e1603021.

Curtis, K. J., Fussell, E., & DeWaard, J. (2015). Recovery migration after Hurricanes Katrina and Rita: Spatial concentration and intensification in the migration system. *Demography, 52*(4), 1269–1293.

Davenport, F., Grace, K., Funk, C., & Shukla, S. (2017). Child health outcomes in sub-Saharan Africa: A comparison of changes in climate and socio-economic factors. *Global Environmental Change, 46*, 72–87.

Davis, J. (2017). Fertility after natural disaster: Hurricane Mitch in Nicaragua. *Population and Environment, 38*(4), 448–464.

Davis, J., Gray, C., & Bilsborrow, R. (2015). Delayed fertility transition among indigenous women: A case study in the Ecuadoran Amazon. *International Perspectives on Sexual and Reproductive Health, 41*(1), 1–10.

Dercon, S., Hoddinott, J., & Woldehanna, T. (2005). Shocks and consumption in 15 Ethiopian villages, 1999–2004. *Journal of African Economies, 14*(4), 559.

Deryugina, T., Kawano, L., & Levitt, S. (2018). The economic impact of Hurricane Katrina on its victims: Evidence from individual tax returns. *American Economic Journal: Applied Economics, 10*(2), 202–233.

De Sherbinin, A., VanWey, L. K., McSweeney, K., Aggarwal, R., Barbieri, A., Henry, S., . . . Walker, R. (2008). Rural household demographics, livelihoods and the environment. *Global Environmental Change, 18*(1), 38–53.

Deschênes, O., & Greenstone, M. (2011). Climate change, mortality, and adaptation: Evidence from annual fluctuations in weather in the US. *American Economic Journal: Applied Economics, 3*(4), 152–185.

DHS Program. (2019). Available at: https://www.dhsprogram.com/

Di, Q., Wang, Y., Zanobetti, A., Wang, Y., Koutrakis, P., Choirat, C., . . . Schwartz, J. D. (2017). Air pollution and mortality in the Medicare population. *New England Journal of Medicine, 376*(26), 2513–2522.

Eissler, S., Thiede, B. C., & Strube, J. (2019). Climatic variability and changing reproductive goals in Sub-Saharan Africa. *Global Environmental Change, 57*, 101912.

Elkies, N., Fink, G., & Bärnighausen, T. (2015). "Scrambling" geo-referenced data to protect privacy induces bias in distance estimation. *Population and Environment, 37*(1), 83–98.

Entwisle, B., Walsh, S. J., Rindfuss, R. R., & Chamratrithirong, A. (1998). Land-use/land-cover and population dynamics, Nang Rong, Thailand. People and pixels: Linking remote sensing and social science, 121–144.

Elliott, J. R., & Howell, J. (2017). Beyond disasters: A longitudinal analysis of natural hazards' unequal impacts on residential instability. *Social Forces, 95*(3), 1181–1207.

Entwisle, B., Williams, N. E., Verdery, A. M., Rindfuss, R. R., Walsh, S. J., Malanson, G. P., . . . Heumann, B. W. (2016). Climate shocks and migration: An agent-based modeling approach. *Population and Environment, 38*(1), 47–71.

Frankenberg, E., Friedman, J., Gillespie, T., Ingwersen, N., Pynoos, R., Rifai, I. U., . . . Thomas, D. (2008). Mental health in Sumatra after the tsunami. *American Journal of Public Health, 98*(9), 1671–1677.

Fussell, E., Curtis, K. J., & DeWaard, J. (2014a). Recovery migration to the City of New Orleans after Hurricane Katrina: A migration systems approach. *Population and Environment, 35*(3), 305–322.

Fussell, E., Hunter, L. M., & Gray, C. L. (2014b). Measuring the environmental dimensions of human migration: The demographer's toolkit. *Global Environmental Change, 28*, 182–191.

Fussell, E., Sastry, N., & VanLandingham, M. (2010). Race, socioeconomic status, and return migration to New Orleans after Hurricane Katrina. *Population and Environment, 31*(1–3), 20–42.

Galway, L. P., Acharya, Y., & Jones, A. D. (2018). Deforestation and child diet diversity: A geospatial analysis of 15 Sub-Saharan African countries. *Health & Place, 51*, 78–88.

Grace, K., Davenport, F., Funk, C., & Lerner, A. M. (2012). Child malnutrition and climate in Sub-Saharan Africa: An analysis of recent trends in Kenya. *Applied Geography, 35*(1–2), 405–413.

Grace, K., Davenport, F., Hanson, H., Funk, C., & Shukla, S. (2015). Linking climate change and health outcomes: Examining the relationship between temperature, precipitation and birth weight in Africa. *Global Environmental Change, 35*, 125–137.

Grace, K., Nagle, N. N., Burgert-Brucker, C. R., Rutzick, S., Van Riper, D. C., Dontamsetti, T., & Croft, T. (2019). Integrating environmental context into DHS analysis while protecting participant confidentiality: A new remote sensing method. *Population and Development Review, 45*(1), 197–218.

Graff Zivin, J., & Neidell, M. (2014). Temperature and the allocation of time: Implications for climate change. *Journal of Labor Economics, 32*(1), 1–26.

Graif, C. (2016). (Un) natural disaster: Vulnerability, long-distance displacement, and the extended geography of neighborhood distress and attainment after Katrina. *Population and Environment, 37*(3), 288–318.

Gray, C., & Mueller, V. (2012). Drought and population mobility in rural Ethiopia. *World Development, 40*(1), 134–145.

Gray, C., Frankenberg, E., Gillespie, T., Sumantri, C., & Thomas, D. (2014). Studying displacement after a disaster using large-scale survey methods: Sumatra after the 2004 tsunami. *Annals of the Association of American Geographers, 104*(3), 594–612.

Gray, C., & Wise, E. (2016). Country-specific effects of climate variability on human migration. *Climatic Change, 135*(3–4), 555–568.

Groen, J. A., & Polivka, A. E. (2010). Going home after Hurricane Katrina: Determinants of return migration and changes in affected areas. *Demography, 47*(4), 821–844.

Hauer, M. E. (2017). Migration induced by sea-level rise could reshape the US population landscape. *Nature Climate Change, 7*(5), 321–325.

Headey, D., & Barrett, C. B. (2015). Opinion: Measuring development resilience in the world's poorest countries. *Proceedings of the National Academy of Sciences, 112*(37), 11423–11425.

Henry, S. J., & Dos Santos, S. (2013). Rainfall variations and child mortality in the Sahel: Results from a comparative event history analysis in Burkina Faso and Mali. *Population and Environment, 34*(4), 431–459.

Henry, S., Boyle, P., & Lambin, E. F. (2003). Modelling inter-provincial migration in Burkina Faso, West Africa: The role of socio-demographic and environmental factors. *Applied Geography, 23*(2–3), 115–136.

Hill, E. L. (2018). Shale gas development and infant health: Evidence from Pennsylvania. *Journal of Health Economics, 61*, 134–150.

Hirschman, C., & Guest, P. (1990). Multilevel models of fertility determination in four Southeast Asian countries: 1970 and 1980. *Demography, 27*(3), 369–396.

Hoddinott, J., & Kinsey, B. (2001). Child growth in the time of drought. *Oxford Bulletin of Economics and Statistics, 63*(4), 409–436.

Hsiang, S. (2016). Climate econometrics. *Annual Review of Resource Economics, 8*, 43–75.

Hugo, G. J. (1982). Circular migration in Indonesia. *Population and Development Review*, 59–83.

Hugo, G. (1996). Environmental concerns and international migration. *International Migration Review, 30*(1), 105–131.

Hunter, L. M., Twine, W., & Johnson, A. (2011). Adult mortality and natural resource use in rural South Africa: Evidence from the Agincourt health and demographic surveillance site. *Society and Natural Resources, 24*(3), 256–275.

Hunter, L. M., Nawrotzki, R., Leyk, S., Maclaurin, G. J., Twine, W., Collinson, M., & Erasmus, B. (2014). Rural outmigration, natural capital, and livelihoods in South Africa. *Population, Space and Place, 20*(5), 402–420.

Hunter, L. M., Luna, J. K., & Norton, R. M. (2015). Environmental dimensions of migration. *Annual Review of Sociology, 41*, 377–397.

Hunter, L. M., Twine, W., & Talbot, C. (2019). *Working toward effective anonymization for surveillance data: innovation at South Africa's Agincourt health and demographic surveillance site*. Paper presented at the

2019 annual meeting of the Population Association of America, Austin, TX.

Ickowitz, A., Powell, B., Salim, M. A., & Sunderland, T. C. (2014). Dietary quality and tree cover in Africa. *Global Environmental Change, 24*, 287–294.

Islam, M. S., Sharker, M. A. Y., Rheman, S., Hossain, S., Mahmud, Z. H., Islam, M. S., ... Rector, I. (2009). Effects of local climate variability on transmission dynamics of cholera in Matlab, Bangladesh. *Transactions of the Royal Society of Tropical Medicine and Hygiene, 103*(11), 1165–1170.

Jacoby, H. G., Rabassa, M., & Skoufias, E. (2014). Distributional implications of climate change in rural India: A general equilibrium approach. *American Journal of Agricultural Economics, 97*(4), 1135–1156.

Jennings, J. A., & Gray, C. L. (2017). Climate and marriage in the Netherlands, 1871–1937. *Population and Environment, 38*(3), 242–260.

Kessler, R. C., Galea, S., Jones, R. T., & Parker, H. A. (2006). Mental illness and suicidality after Hurricane Katrina. *Bulletin of the World Health Organization, 84*, 930–939.

Kessler, R. C., Galea, S., Gruber, M. J., Sampson, N. A., Ursano, R. J., & Wessely, S. (2008). Trends in mental illness and suicidality after Hurricane Katrina. *Molecular Psychiatry, 13*(4), 374.

Kumar, S., Molitor, R., & Vollmer, S. (2016). Drought and early child health in rural India. Population and Development Review, 53–68.

Larsen, A. F., Headey, D., & Masters, W. A. (2019). Misreporting month of birth: Diagnosis and implications for research on nutrition and early childhood in developing countries. *Demography, 56*(2), 707–728.

Leone, T. (2014). Measuring differential maternal mortality using census data in developing countries. *Population, Space and Place, 20*(7), 581–591.

Leyk, S., Maclaurin, G. J., Hunter, L. M., Nawrotzki, R., Twine, W., Collinson, M., & Erasmus, B. (2012). Spatially and temporally varying associations between temporary outmigration and natural resource availability in resource-dependent rural communities in South Africa: A modeling framework. *Applied Geography, 34*, 559–568.

Loebach, P. (2016). Household migration as a livelihood adaptation in response to a natural disaster: Nicaragua and Hurricane Mitch. *Population and Environment, 38*(2), 185–206.

López-Carr, D. (2012). Agro-ecological drivers of rural out-migration to the Maya Biosphere Reserve, Guatemala. *Environmental Research Letters, 7*(4), 045603.

Lu, X., Wrathall, D. J., Sundsøy, P. R., Nadiruzzaman, M., Wetter, E., Iqbal, A., ... Bengtsson, L. (2016). Unveiling hidden migration and mobility patterns in climate stressed regions: A longitudinal study of six million anonymous mobile phone users in Bangladesh. *Global Environmental Change, 38*, 1–7.

Maccini, S., & Yang, D. (2009). Under the weather: Health, schooling, and economic consequences of early-life rainfall. *American Economic Review, 99*(3), 1006–1026.

Mason, J. B., Bailes, A., Mason, K. E., Yambi, O., Jonsson, U., Hudspeth, C., ... Martel, P. (2005). AIDS, drought, and child malnutrition in southern Africa. *Public Health Nutrition, 8*(6), 551–563.

Massey, D. S., Axinn, W. G., & Ghimire, D. J. (2010). Environmental change and out-migration: Evidence from Nepal. *Population and Environment, 32*(2–3), 109–136.

Mastrorillo, M., Licker, R., Bohra-Mishra, P., Fagiolo, G., Estes, L. D., & Oppenheimer, M. (2016). The influence of climate variability on internal migration flows in South Africa. *Global Environmental Change, 39*, 155–169.

Minnesota Population Center. (2019). Integrated public use microdata series, International: Version 7.2 [dataset]. IPUMS.

Morgan, S. L., & Winship, C. (2015). *Counterfactuals and causal inference*. Cambridge University Press.

Mueller, V., & Gray, C. (2018). Heat and adult health in China. *Population and Environment, 40*(1), 1–26.

Mueller, V., Gray, C., & Kosec, K. (2014). Heat stress increases long-term human migration in rural Pakistan. *Nature Climate Change, 4*(3), 182.

Mueller, V., Sheriff, G., Dou, X., & Gray, C. (2020). Temporary migration and climate variation in eastern Africa. *World Development, 126*, 104704.

Nandi, A., Mazumdar, S., & Behrman, J. R. (2018). The effect of natural disaster on fertility, birth spacing, and child sex ratio: Evidence from a major earthquake in India. *Journal of Population Economics, 31*(1), 267–293.

Nawrotzki, R. J., & Bakhtsiyarava, M. (2017). International climate migration: Evidence for the climate inhibitor mechanism and the agricultural pathway. *Population, Space and Place, 23*(4), e2033.

Nawrotzki, R. J., & DeWaard, J. (2016). Climate shocks and the timing of migration from Mexico. *Population and Environment, 38*(1), 72–100.

Nawrotzki, R. J., Riosmena, F., & Hunter, L. M. (2013). Do rainfall deficits predict US-bound migration from rural Mexico? Evidence from the Mexican census. *Population Research and Policy Review, 32*(1), 129–158.

Nawrotzki, R. J., Hunter, L. M., Runfola, D. M., & Riosmena, F. (2015). Climate change as a migration driver from rural and urban Mexico. *Environmental Research Letters, 10*(11), 114023.

Nawrotzki, R. J., Schlak, A. M., & Kugler, T. A. (2016). Climate, migration, and the local food security context: Introducing Terra Populus. *Population and Environment, 38*(2), 164–184.

Nobles, J., Frankenberg, E., & Thomas, D. (2015). The effects of mortality on fertility: Population dynamics after a natural disaster. *Demography, 52*(1), 15–38.

Ocello, C., Petrucci, A., Testa, M. R., & Vignoli, D. (2015). Environmental aspects of internal migration in Tanzania. *Population and Environment, 37*(1), 99–108.

Pais, J., Crowder, K., & Downey, L. (2013). Unequal trajectories: Racial and class differences in residential exposure to industrial hazard. *Social Forces, 92*(3), 1189–1215.

Perz, S. G., Walker, R. T., & Caldas, M. M. (2006). Beyond population and environment: Household demographic life cycles and land use allocation among small farms in the Amazon. *Human Ecology, 34*(6), 829–849.

Piguet, E. (2010). Linking climate change, environmental degradation, and migration: A methodological overview. *Wiley Interdisciplinary Reviews: Climate Change, 1*(4), 517–524.

Randell, H., & Gray, C. (2016). Climate variability and educational attainment: Evidence from rural Ethiopia. *Global Environmental Change, 41*, 111–123.

Randell, H., & Gray, C. (2019). Climate change and educational attainment in the global tropics. *Proceedings of the National Academy of Sciences, 116*(18), 8840–8845.

Randell, H., Gray, C., & Grace, K. (2020). Stunted from the start: Early life weather conditions and child undernutrition in Ethiopia. *Social Science & Medicine, 261*, 113234.

Ruggles, S. (2014). Big microdata for population research. *Demography, 51*(1), 287–297.

Sastry, N. (2009). Tracing the effects of Hurricane Katrina on the population of New Orleans: The displaced New Orleans residents pilot study. *Sociological Methods & Research, 38*(1), 171–196.

Sedova, B., & Kalkuhl, M. (2020). Who are the climate migrants and where do they go? Evidence from rural India. *World Development, 129*, 104848.

Sellers, S., & Gray, C. (2019). Climate shocks constrain human fertility in Indonesia. *World Development, 117*, 357–369.

Shi, L., Kloog, I., Zanobetti, A., Liu, P., & Schwartz, J. D. (2015). Impacts of temperature and its variability on mortality in New England. *Nature Climate Change, 5*(11), 988.

Shively, G. E. (2017). Infrastructure mitigates the sensitivity of child growth to local agriculture and rainfall in Nepal and Uganda. *Proceedings of the National Academy of Sciences, 114*(5), 903–908.

Thiede, B. C., & Brown, D. L. (2013). Hurricane Katrina: Who stayed and why? *Population Research and Policy Review, 32*(6), 803–824.

Thiede, B. C., & Gray, C. L. (2017). Heterogeneous climate effects on human migration in Indonesia. *Population and Environment, 39*(2), 147–172.

Thiede, B. C., & Gray, C. (2020). Climate exposures and child undernutrition: Evidence from Indonesia. *Social Science & Medicine, 265*, 113298.

Thiede, B., Gray, C., & Mueller, V. (2016). Climate variability and inter-provincial migration in South America, 1970–2011. *Global Environmental Change, 41*, 228–240.

Tiwari, S., Jacoby, H., & Skoufias, E. (2017). Monsoon babies: Rainfall shocks and child nutrition in Nepal. *Economic Development and Cultural Change, 65*(2), 167–188.

Tollman, S. M., Herbst, K., Garenne, M., Gear, J. S., & Kahn, K. (1999). The Agincourt demographic and health study-site description, baseline findings and implications. *South African Medical Journal, 89*(8).

UNFPA. (2019). *UNFPA strategy for the 2020 round of population & housing censuses (2015–2024): Because everyone counts*. United Nations.

Waters, M. C. (2016). Life after Hurricane Katrina: The resilience in survivors of Katrina (RISK) project. *Sociological Forum, 31*(S1), 750–769.

Watmough, G. R., Atkinson, P. M., Saikia, A., & Hutton, C. W. (2016). Understanding the evidence base for poverty–environment relationships using remotely sensed satellite data: An example from Assam, India. *World Development, 78*, 188–203.

Williams, N. E., & Gray, C. (2020). Spatial and temporal dimensions of weather shocks and migration in Nepal. *Population and Environment, 41*(3), 286–305.

Spatial Data and Analytical Approaches

6

Rachel A. Rosenfeld and Katherine J. Curtis

Abstract

Given increasingly accessible spatially-referenced environmental data, environmental demographers have the unique opportunity to use spatial data and spatial analytical techniques to unpack the complex relationships between demographic processes and the biophysical environment. A distinguishing feature of spatial studies, demographic or otherwise, is the explicit consideration of one location relative to others. This chapter provides an overview of how spatial data and analyses are used in research at the population-environment nexus and identifies how spatial data and analytic strategies might advance our theoretical knowledge of today's most significant environmental issues. In Part 1, we provide general definitions, and foundational theoretical and methodological concepts. In Part 2, we highlight compelling spatial work occurring in the population and environment literature organized by demography's traditional themes: migration, fertility, and mortality. We conclude with suggestions

for future directions researchers might take to incorporate spatial thinking from both methodological and theoretical standpoints. By underscoring spatial differentiation within population-environment research via reflexive theory and appropriate modeling strategies, environmental demographers are uniquely positioned to draw connections between demographic processes, social inequality, and environmental hazards.

Introduction

In sociologist Anne Pebley's 1998 Presidential Address for the Population Association of America, she calls for increased attention to environmental concerns in demographic research. Pebley highlights a series of waves of concern about the environment that subsequently inform the types of questions researchers ask (Pebley, 1998). The first wave of the 1940s and 1950s focused on limited natural resources, such as inadequate food production and the exhaustion of nonrenewable resources. The second wave of the 1960s and 1970s centered on by-products of production and consumption, including pesticide

R. A. Rosenfeld (✉) · K. J. Curtis
Department of Community and Environmental Sociology, University of Wisconsin–Madison, Madison, WI, USA
e-mail: rrosenfeld@wisc.edu

© Springer Nature Switzerland AG 2022
L. M. Hunter et al. (eds.), *International Handbook of Population and Environment*, International Handbooks of Population 10, https://doi.org/10.1007/978-3-030-76433-3_6

and fertilizer use, waste disposal, noise pollution, air and water pollution, and radioactive and chemical contamination. And the third wave of the 1980s and 1990s emphasized global environmental change, including acid rain, ozone depletion, and climate change. Subsequent academic and policy research over the past two decades has demonstrated an increased interest in the population-environment nexus far beyond the earlier demographic-environmental research in the context of population growth and economic development. The central objective of this chapter is to highlight the spatial focus adopted in recent scholarship advancing the relationships between demographic and environmental patterns and processes.

Given increasingly accessible spatially-referenced environmental data (i.e., data that can be linked to a specific geographic entity through geographic coordinates), environmental demographers have the unique opportunity to use spatial data and spatial analytical techniques to unpack the complex relationships between demographic processes and the biophysical environment. The goals of this chapter are twofold: (1) to provide an overview of how spatial data and analyses are used in research at the population-environment nexus; and (2) to identify how spatial data and analytic strategies might advance our theoretical knowledge of today's most significant environmental issues. In Part 1, we provide general definitions, and foundational theoretical and methodological concepts. In Part 2, we highlight compelling spatial work occurring in the population and environment literature organized by the field's traditional themes: migration, fertility, and mortality. We conclude with suggestions for future directions researchers might take to incorporate spatial thinking from both methodological and theoretical standpoints.

Laying the Groundwork: Key Spatial Definitions and Concepts

Defining Spatial

Spatial data are widely used in demography, geography, sociology, economics, regional science, and environmental studies. One of the most famous and earliest examples of spatial analysis dates back to British Dr. John Snow and the Broad Street Pump cholera outbreak. In mid-nineteenth century London, Snow generated a map linking the location of communal water pumps and cholera deaths. Through careful visual analysis of the map, Snow concluded that a pump on the corner of Broad Street and Cambridge Street was the source of cholera deaths (de Castro, 2007). Snow's spatial analysis was elementary compared to many of the tools and techniques used today, yet Snow's approach demonstrates the importance of considering spatial location to understand the environmental conditions affecting population health and well-being.

A distinguishing feature of spatial studies, demographic or otherwise, is the explicit consideration of one location relative to others. In theorizing space, geographer Massey (1994) contrasts space with place by stressing that place is reduced to an enclosed identity (e.g., nationalist, regionalist, localist), whereas space transcends this limited, enclosed, static definition through connections that extend beyond any one place. Unlike a discrete entity, spatial entities exist because of their relational quality to each other at a given point in time and over time (Massey, 1994). Geographer Weeks (2004) underscores this distinction by asserting that incorporating information on a physical location and place-based characteristics does not automatically qualify as a spatial analysis. Place-based data reflect the characteristics of a place, neighborhood, or region. Without explicit geospatial reference, such data can be considered aspatial because the consideration of the spatial relationships between places is not of primary analytical concern.

Until the mid-twentieth century, demography was inherently spatial in terms of both conceptual awareness and analytical techniques (Weeks, 2004). Prior to the extensive availability of public use microdata files, most demography in the United States analyzed population trends and changes within geographic entities (e.g., counties, states, metropolitan/nonmetropolitan aggregates). Administrative areas were the primary units of analysis for empirical demographers in the United States (Voss, 2007). Yet later in the twentieth century, a large part of demographic research shifted focus to the individual and away from spatial units of analysis (Voss, 2007). This shift happened primarily because of three factors: (1) widespread accessibility of microdata at the individual and household levels (microdata is data on the characteristics of units of a population, such as individuals or households, collected by a census or survey); (2) a desire to avoid ecological fallacy or aggregation bias (i.e., inaccurate use of aggregate data to draw inferences about individuals); and (3) a tactical approach to avoid statistical and computational challenges that commonly arise when studying the aggregate level (e.g., boundaries may change over time) (Voss, 2007). Nevertheless, not all demographers, and certainly not all environmental demographers, shifted focus away from aggregate social data. There has always been a core subset of demographers interested in identifying the spatial patterning of demographic outcomes. For example, using data from the 1911 census for England and Wales, historical demographers Garrett and Reid show that infant and child mortality were connected to where a child lived (Garrett & Reid, 1995).

To further illustrate, the Chicago School of urban and community development of the 1920s and 1930s incorporated many ecological elements to studying sociological questions. In order to fully understand both ecological and sociological processes, Chicago School scholars highlighted the importance of studying individuals in the context of the structure, change, and behavior of aggregate groups. Sociological human ecology explores how humans are impacted by the environment in both space and time (McKenzie, 1924). This perspective informs the ways sociologists consider the spatial distribution of population, and specifically for the Chicago School spatial differentiation in urban systems (Hawley, 1950). For example, Park et al. (1925) developed a precursor to the concentric zonal hypothesis, which aimed to delineate how the city of Chicago has a central core with rings spreading outward. The rings denote zones defined and labeled by their attributes. The original hypothesis included five zones from center circle moving outward in a series of rings: central business district, transition zone, workingman zone, residential zone, and commuter zone. By understanding the patterns of the urban growth, scholars at the Chicago School could better understand the social impacts of these boundaries (Park et al., 1925; see also, Porter & Howell, 2012b).

Earlier, and arguably an inspiration to the Chicago School approach, rural sociologist Charles Galpin pioneered the use of an isotropic map as a tool for understanding social organization in 1915, and mid-twentieth century urban and rural demographers, especially sociologists, paid attention to the ways that social heterogeneity mapped itself onto ecological processes and patterns (Porter & Howell, 2012b; Voss, 2007). Predating these efforts is W.E.B. Du Bois's urban and community studies, which strongly featured statistical analysis of different local areas to capture diversity and heterogeneity across communities (Itzigsohn & Brown, 2020). A key distinguishing feature of Du Bois's approach from that adopted by the Chicago School is the explicit consideration of the historical and social forces that shape local context and give rise to heterogeneity, as opposed to homogeneity, among communities.

Although spatial thinking has been a part of demography and the history of the discipline and related fields to varying degrees, the reemergence of spatial thinking in demography is partly owed to computational advances and improved data availability, and partly due to an increased desire to understand patterns at various levels of aggregation (Voss, 2007). Sociologists Porter and Howell (2012a) discuss geographical sociology and the ways in which researchers have increased their attention to studying the geographic contexts and ecological settings in which social processes occur and, at the same time, have developed better tools to operationalize and test the core spatial concepts of containment, proximity, and adjacency. Indeed, despite the Chicago School's emphasis on human-environment interactions, its analysis largely was limited to high-income regions and was rarely embraced in lower-income settings (Balk & Montgomery, 2015). Whether a product of data availability, computational capacity, or limited theoretical reflexivity, today's scholarship is not bounded in the same ways or to the same degree as were earlier spatial studies.

Types of Spatial Data

Spatial data, also known as geospatial data and geographic information, refer to measurements of specific characteristics at specified locations. Within the social sciences, spatial data tend to correspond with areal aggregates – of or pertaining to social science attributes in the context of an area or space – and include demographic, social, economic, and environmental attributes aggregated to a level within a geographic hierarchy (Voss, 2007). Spatial data are "spatial" because they refer to the relationships the data have to other points or features on a map. Data are spatial, as opposed to aspatial place-, neighborhood-, or region-based, when the analyst is interested in the spatial arrangement or relational position of the data on a map, which may or may not be visualized on a map. Generally, spatial data represent the location, size, or shape of a place or feature of the earth, and these locational data are relational data since spatial data reference relative positions of locations by identifying exact locations (e.g., by use of geographic coordinates) or natural or constructed boundaries (Haining, 1990).

Spatial data can be discrete or continuous (de Castro, 2007; Haining, 1990), and examples of discrete data are point data (e.g., geographic coordinates) and lattice data (e.g., data aggregated at an administrative level such as a county, district, or province), whereas continuous data do not have well-defined breaks between the values or features (e.g., precipitation, temperatures, or topographical features). The essential element for environmental demographers is most accurately bringing environmental and social data together, since environmental data often are continuous whereas social data tend to be socially constructed discrete measurements.

Of course, as with data in general, spatial data can be transformed between continuous and discrete data structures as the analyst sees suitable (for a computational primer on data transformations, see Clarke, 1995: 226–248; see also, for new directions on spatial data use and manupulation in R, Pebesma, 2017). Further, as Fig. 6.1 presents, when using geographic information system (GIS) processing, data come in two general processing formats: raster and vector. Raster data are a matrix of cells (or pixels) which are arranged into rows and columns (i.e., on a grid). Each cell on the grid holds a value that represents information, such as temperature or rain (e.g., environmental data). Vector data, on the other hand, are vertices and paths, representing the features of the world by points, lines, and polygons via XY coordinates (latitude and longitude lines). Vector data do not necessarily fill entire grid cells, and tend to be socially constructed, such as country borders, roads, and land parcels (however, a river can also be represented in vector data). Vectors typically are considered discrete, whereas raster data may be continuous or discrete.

The central point is that spatial data come in varying forms, and the analyst must reconcile the varying forms with the research objective, which can be especially challenging when combining environmental and social data of different formats. For example, analysts using the Normalized Difference Vegetation Index (NDVI) can use the data in a continuous form, with values ranging from -1 to 1 and varying by pixel (raster), or convert the data to a discrete form to reflect the type

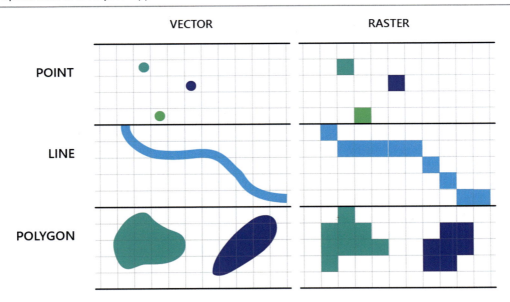

Fig. 6.1 Vector versus raster processing for spatial data, including examples of a point, line, and polygon within each

of land cover (e.g., woodland, urban) by polygon (vector). In other instances, analysts work in the opposite direction, transforming discrete data into continuous data. For example, analysts use interpolation methods to convert rainfall measured at weather stations to rainfall estimates that span a continuous spatial region.

Traditional sources of environmental spatial data include archival sources, such as maps, census materials, areal photographs, field observation from survey data and from direct observation, and simulation work in a laboratory environment. For example, systems ecologist Tian et al. (2014) use historical archives along with contemporary satellite data to follow the history of land use in India from 1880–2010 as they explore high deforestation rates under British rule, extensive cropland expansion between 1950–1980, and Government of India forest promotion activities in the 1980s. The archives include extensive information on changes in croplands, forests, grasslands, and the built environment over the study period. Computationally, Tian et al. calibrate land use and land cover records from the historical archives with contemporary Resourcesat-1 datasets to generate land use and land cover data at 5 arc minute resolution for the 130 year study period. The area in time i is compared to the area in time j using a calibration coefficient, so they are able to look at various changes over time as a ratio and subsequently reconstruct maps over the study period.

In another example, to investigate the linkages between household demographic change and land use and land cover change in the Brazilian Amazon, D'Antona et al. (2008) present an innovative approach to collecting data on the spatial organization of land use within a rural property. D'Antona et al. generate sketch maps from the point of view of their survey respondents about the spatial organization and infrastructure of their properties. These sketch maps often elicit more detailed responses from the respondent and later help clarify verbal survey responses as well. These sketches are then used in combination with remotely sensed imagery to contribute to the understanding of various types of land use. With this approach the researchers facilitate better interviewer-interviewee interaction, by enhancing the interviewers' understanding of land use on the interviewees' properties. This extra interaction promoted by generating sketch maps increases confidence in the land use and land cover information recorded in the questionnaires. Furthermore, the sketch maps are compared with satellite imagery to improve land use classifications, including considering local categories of land use and land cover from the interviewee's perspective such as capoeira, *juquira*, and new cocoa plantations (D'Antona et al., 2008).

As technology has developed, more modern sources have incorporated remotely sensed satellite imagery and newer government and commercial spatially-referenced databases (e.g., Haining, 1990). As early as the 1970s, satellite photos of land cover in West Africa indicated substantial deforestation (Vittek et al., 2014), and satellite imagery showed overgrazing in the Western Sahel countries (Schmidt, 2001). Yet, it was not until the 1990s that social scientists embraced the possibilities of remotely sensed data, partly motivated by books such as *People and Pixels Linking Remote Sensing and Social Science* (Liverman et al., 1998), and partly because these data became increasingly accessible. For example, created in 1985 but experiencing greater research usage in the 1990s, the Famine and Early Warning Systems (FEWS) became an important tool for researchers exploring vulnerability to famines (Hutchinson, 1998). Today, FEWS incorporates a wide variety of remotely sensed data from the United States Geological Survey, the United States National Oceanic and Atmospheric Administration, and the Climate Hazards Group along with food security classifications, livelihoods zones, and food pricing and trade. It remains a rich source for social scientists focusing on food security (FEWS, 2020).

As introduced above, one modern and widely used data source is the Normalized Difference Vegetation Index (NDVI), which through remotely sensed measurements estimates biophysical variables of temperate vegetation (Foody et al., 2001; Hunter et al., 2014). This index, for example, can be used as an indicator for drought by identifying barren areas. For rainfall data, the Climate Hazards Group InfraRed Precipitation with Station dataset (CHIRPS) provides more than 35-years of global precipitation data incorporating 0.05°

resolution satellite imagery with in-situ station data to generate gridded rainfall time series for monitoring seasonal droughts as well as periods of excess rain (Funk et al., 2015).

From a public-access standpoint, CHIRPS data plus environmental data from other sources are joined with demographic variables and disseminated through IPUMS Terra's integrated population and environmental database. This database decreases the amount of time researchers need to collect, process, and integrate data from many sources (IPUMS Terra, 2019). In addition to matching population variables and environmental variables, IPUMS Terra makes temporal and spatial comparisons of these variables easier for researchers by utilizing consistent geographic codes over time. Data from IPUMS Terra can be used for a range of questions such as exploring health outcomes dependent on variation in the biophysical environment, investigating the ways deforestation and agricultural trends are related to community composition, and many others. Additionally, the Climatic Research Unit (CRU) at the University of East Anglia releases high resolution (0.5° or finer) monthly precipitation and temperature data. Using multiple climate data sources in tandem is increasingly popular for researchers. For example, geographers Gray and Wise (2016) explore the country-specific effects of climate variability on international migration in eastern and western African countries by comparing findings based on CRU temperature and precipitation data with analyses using monthly mean land surface temperature and total surface precipitation from the NASA Modern Era-Retrospective Analysis for Research and Applications (MERRA) (for more information on MERRA data, see Rienecker et al., 2011). This enabled the researchers to explore the ways in which both temperature and precipitation are related to migration outcomes in Kenya, Uganda, Nigeria, Burkina Faso, and Senegal. They find increased migration is associated with temperature anomalies in Uganda, whereas decreased migration is associated with temperature anomalies in Kenya and Burkina Faso, and they do not identify a consistent relationship between temperature anomalies in Nigeria and Senegal.

An important use of environmental spatial data in the social sciences is for understanding the spatial and temporal changes in social vulnerability (and social resilience) to environmental hazards. Geographers and disaster risk researchers Cutter and Finch (2008) examine the spatial and temporal patterns in social vulnerability in the United States from the 1960s through the early 2000s. The notion of social vulnerability identifies sensitive populations in place that may be less able to respond to, tackle, and recover from an environmental disaster. They create a Social Vulnerability Index (SoVI), incorporating 42 independent variables on the characteristics that influence social vulnerability to natural hazards, and they recalculate the index for each decade investigated. SoVI is created and mapped at the U.S. county level, and spatial clusters are identified using the Global Moran's I testing for spatial dependence and the Local Indicator of Spatial Association (LISA) for exploring spatial heterogeneity (additional discussion of spatial dependence and spatial heterogeneity in 1.3 Spatial Analytical Approaches). In short, they find that the spatial patterning of high social vulnerability to natural hazards has changed over time – initially concentrated in the Deep South, Southwest, and in Florida, but in the 2000s is concentrated in the lower Mississippi Valley, the South Texas borderlands, California's Central Valley, and the upper Great Plains. Social vulnerability is multidimensional and dynamic, changing over time and across space as the demographic composition of the United States changes and exposure to natural hazards likely increases throughout many parts of the country. Conceptualizing and modeling the spatial patterning of social vulnerability is key to understanding how people perceive and react to environmental hazards (see Entwisle, 2007 to elucidate different vulnerabilities, risks, and exposures).

Spatial Analytical Approaches

The availability and quality of spatial data are continually changing, both as a result of technological advances that enable the capture, processing, and analysis of new data, and of shifting priorities among agencies that fund data collection, compilation, and dissemination. Yet, data alone do not create a spatial analysis – whether a study is spatial depends upon the chosen method of analysis (e.g., Weeks, 2004). As discussed above in most social science research, space is not defined as a point in isolation, but rather as locations, regions, or areas in relation to one another (e.g., lattice data) (see Voss, 2007). An initial step in conducting a spatial analysis is identifying spatial patterns or spatial concentration of an observed pattern of a phenomenon or behavior (Haining, 1990).

With this orientation in mind, spatial thinking or awareness about *spatial differentiation* across a study region is crucial, whether the analyst is observing differences in an outcome or in relationships between key variables among areas within the study site (e.g., province, county). For example, in rural sociological studies, scholars have fundamentally factored spatial continuity and heterogeneity into studies of differentiation across the rural-urban continuum to better estimate the full range of lived experiences across the full spectrum of human settlement patterns. A researcher may be interested in exploring how both population size and adjacency to a metropolitan area impact poverty outcomes for nonmetropolitan areas. The researcher can approach this question by using the United States Department of Agriculture Rural-Urban Continuum Codes that divide the Office of Management and Budget's single nonmetropolitan category into six categories (additional recent examples using a rural-urban continuum include Curtis et al., 2019a; Lee & Sharp, 2017; Ward & Shackleton, 2016; Winkler & Johnson, 2016). Improved spatial understanding about heterogeneous population trends and livelihoods across the rural-urban gradient will help researchers interpret the different population responses to environmental events among different types of affected areas.

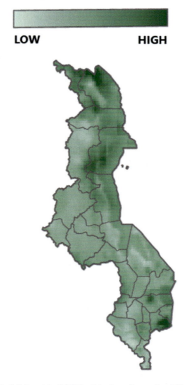

Fig. 6.2 Left panel: Locations where the Demographic and Health Surveys interviews occurred in Malawi in 2015–2016. Right panel: Variation in cumulative precipitation in Malawi in 2015 with data from the Climate Hazards Group InfraRed Precipitation with Station (CHIRPS)

A frequent and major problem that occurs when studying population-environment interactions concerns *boundary alignment*. Attributes, especially environmental attributes such as precipitation or soil quality, in a specific spatial region are generally subject to continuous change and not discrete change. Using current work in Malawi by Grant and the authors (2020), Fig. 6.2 highlights the differences between discrete (in this case, social) and continuous (in this case, environmental) data. The left panel of the figure represents the clusters where Demographic and Health Survey respondents were interviewed, whereas the right panel illustrates cumulative precipitation in 2015. In short in our research, we have converted the continuous rainfall measures to discrete measures that harmonize with the location of the social data. Sharp, socially-constructed regional boundaries are typically features of political decisions, and rarely does an environmental attribute assigned to these regions uniformly align with the designated social boundary. Biophysical attributes and environmental variables, such as air pollution, rarely, if ever, perfectly match political and social sphere boundaries. Consequently, researchers must determine how to best align social and environmental dimensions (Haining, 1990; Liverman et al., 1998; Weeks, 2004).

Posing another and longstanding analytical challenge, intrinsically, spatial data have a certain amount of *spatial autocorrelation*, which yields many standard statistical methods invalid (Haining, 1990; Voss et al., 2006). In defining positive spatial autocorrelation, Tobler's First Law establishes that neighboring values are more similar to one another than values among

6 Spatial Data and Analytical Approaches

those farther apart (Tobler, 1970). While positive spatial autocorrelation is more common, negative spatial autocorrelation may also arise (i.e., neighboring values systematically differ from one another). The central takeaway is that when there is a nonrandom spatial distribution of values across a study site, analysts must deal with the observed spatial pattern(s) in the data. In testing for spatial autocorrelation, the researcher tests a hypothesis that the data are autocorrelated, positively or negatively, in space relative to the null hypothesis of randomly distributed values across spatial regions (Haining, 1990; Ward & Gleditsch, 2008). The strength and statistical significance of spatial autocorrelation can be measured with global measures for identifying spatial dependence across the study area and with local statistics for revealing local spatial heterogeneity (Chi & Zhu, 2020; for details on frequently used statistics to detect spatial autocorrelation see p. 43–49). Upon identifying spatial autocorrelation, analysts use diagnostics to inform the types of analyses best suited to investigate the phenomenon or behavior of concern. In the analysis, the researcher may control for spatial autocorrelation to isolate the core association of interest, or in other cases, spatial autocorrelation may be a primary focus.

Depending on the question at hand and the structure of the spatial data, *exploratory spatial data analysis (ESDA)* can inform more complex statistical analyses or can be used for the full analysis. Data analysis using statistical methods begins with exploratory data analysis that helps the researcher familiarize herself with the data by using graphics and summary statistics prior to fitting models. ESDA, in particular, helps summarize the spatial patterns in the data by visualizing the spatial patterns in the data, identifying spatial clusters and spatial outliers, and diagnosing misspecifications of spatial aspects of statistical models that might be used (Chi & Zhu, 2020). Rigorous ESDA is crucial for the identification of outliers, the interpretation of the distribution (e.g., location, skew, spread, and tail properties), the identification of spatial properties (e.g., spatial outliers and spatial relationships), and the identification of spatial relationships between observa-

tional units (Haining, 1990; Wooldridge, 2013). ESDA may also offer direction for alternative approaches by identifying a location to study where spatial differences require additional, qualitative exploration.

Building from ESDA, researchers can advance to more spatially-explicit approaches, including spatial regression techniques. Best practice directs analysts to fit a standard (linear) regression model with the purpose of diagnosing possible spatial dependence and spatial heterogeneity before considering which spatial regression models to use, or whether to use one at all. If the data are spatially autocorrelated, whether from spatial dependence and/or spatial heterogeneity, their structure will violate many of the core assumptions of ordinary least squares (OLS) regression (e.g., identically and independently distributed errors). From a spatial perspective, researchers focus on various residual plots to assess patterns that depart from OLS assumptions including nonlinear relationships, unequal variances, and/or nonnormality of errors. Upon identifying spatial autocorrelation for a given dataset, a researcher determines whether the spatial autocorrelation is adequately explained by including another variable in the analysis (ESDA can be especially helpful for this decision process) or through data transformations (i.e., smoothing or detrending). In cases where these approaches insufficiently address spatial autocorrelation, the scholar can consider which spatial regression model will be most appropriate to address the spatial structure – spatial dependence and/or spatial heterogeneity – at play in the particular dataset.

Common spatial regression models include spatial lag regression, spatial error regression, spatial regime standard linear regression, and more advanced spatial regression models, including spatio-temporal regression and spatial regression forecasting models. For most spatial regression approaches, researchers must specify a spatial weights matrix (the scientific community lacks consensus on how to optimally specify a spatial weights matrix, and for more discussion on this concern, see Chi & Zhu, 2020: 68 and 78). Further, in deciding between spatial regression models, a "data-driven approach" offers some insight (Chi & Zhu, 2020: 79; see also Voss & Chi, 2006), but theory-based approaches to spatial model selection can generate clearer and sometimes more defensible results (Doreian, 1980; see also O'Loughlin et al., 1994). Ultimately, conceptually distinguishing between spatial heterogeneity and spatial dependence and between spatial autocorrelation and temporal (serial) autocorrelation underlies the researcher's decision between modeling strategies. An example of spatial dependence is spatial diffusion, such as smoke from a wildfire. Smoke from the wildfire starts at a clear point of origin and spreads to nearby places over time. By comparison, an example of spatial heterogeneity is an exogenous force that affects a spatial region, such as windfall from a hurricane.

With the advent of numerous options for spatial panel data, which refer to data containing time series observations of spatial units (e.g., county, province, subdistrict), scholars often aim to consider both spatial and temporal dependence in their research. In their new textbook, sociologist Chi and statistician Zhu (2020) highlight two general approaches for spatio-temporal regression modeling. In the first, researchers can fit spatial regression models separately for each point in time and subsequently compare the findings. For example, Fussell and colleagues examine spatial and temporal variability of hurricanes and tropical storms and the effects on population growth in the United States from 1980–2012. They highlight different population growth patterns for counties experiencing a current year hurricane versus cumulative hurricanes, and find that current year hurricanes and associated losses tend to suppress population growth, whereas cumulative hurricanes and associated losses tend to increase population growth (Fussell et al., 2017). However, the temporal findings generated through this approach are limited by comparing the temporal difference of model parameters rather than parameterizing the temporal dependence itself. In the second and preferred approach, researchers can employ a spatio-temporal regression model to formally and simultaneously parameterize spatial and temporal

dependence (for more details about spatio-temporal specifications, see Chi & Zhu, 2020: 157–182; see also Elhorst, 2001). For example, Curtis et al. (2019b) test systematic variation in the relationship between poverty and industry in the Upper Midwest between 1970–2010 using models that account for both spatial and temporal autocorrelation as well as heterogeneity. They employ a temporal regime autoregressive model and a spatial regime autoregressive model and find that industry is more protective against poverty in certain places and in certain time periods than in others. In another example, Reyes et al. (2012) employ a variety of spatio-temporal models and a simulation to examine the impact of temperature, precipitation, and elevation on a mountain pine beetle outbreak in British Columbia, Canada. Spatio-temporal approaches are underutilized and ripe for application in environmental demography research.

Key Challenges and Limitations

All data analyses confront certain empirical problems that might lead to incorrect theoretical conclusions, especially when statistical and modeling assumptions are not met. These issues apply to spatial data. Yet, spatial data also bring their own unique concerns. In this section, we focus on the (1) modifiable areal unit problem (MAUP); (2) limitations of dichotomous rural-urban classifications; (3) challenges of alternative spatial scales and possibilities of spatially-harmonized datasets, and (4) confidentiality concerns.

First, the modifiable areal unit problem (MAUP) is a common analytical challenge when using spatial data. In longitudinal study contexts, while the data may answer the same question over a long period of time, the boundaries of a particular area might change because of administrative zoning modifications or because different agencies collect data on similar questions (de Castro, 2007; Haining, 1990). When MAUP arises, researchers may try aggregating up to a larger, more encompassing geographic unit to avoid certain boundary changes, but using this cruder level might mask the relationships observed at the more local level. In addition to changing boundaries, researchers identify different relationships between variables when looking at the same variables at different scales or levels of aggregation. For example, analysts tend to uncover different findings when focusing on attributes at the U.S. census tract level rather than at the county level. Typically, the former biases the individual and urban rather than the spatial attributes because there are many more census tracts in urban areas. Tracts are created based on population size whereas counties are based on geographic, political, administrative, and/or historical boundaries. Census data generally are constructed to facilitate data gathering for administrative purposes. For example, zip codes can be used as a level of analysis, but these codes were originally developed to aid in postal service delivery. Sadler (2019) demonstrates the flaws of using zip codes as a geographic unit of analysis for public health scholarship and policy. Specifically, he refers to the water crisis in Flint, Michigan because Flint zip codes do not align well with the city of Flint and its water system. Further, Sadler proposes using a finer unit of analysis (e.g., census block groups, wards, planning districts, municipal designations for neighborhoods) that is more geographically appropriate for assessing exposure to lead prevalence in water and human exposure to it.

There are many different types of census units from small areas (e.g., census block) to relatively large areas (e.g., state), but nonetheless, most groupings tend to bias urban areas as opposed to more sparsely populated, geographically isolated areas (Irwin, 2007). Figure 6.3 juxtaposes two counties in Wisconsin – one of the most urban counties in the state, Milwaukee County, and one of the most rural counties in the state, Iron County. This figure demarcates the county boundaries, highlights the watersheds that run through the counties, and delineates the census tract boundaries within each county. We overlay population density by census tract. Milwaukee County consists of 297 census tracts, whereas Iron County only includes 3 sparsely populated tracts. Therefore, if one is conducting research at the census tract level in Wisconsin, one faces the risk of biasing urban Milwaukee County relative to more rural Iron County.

Second, dichotomous classifications, often used in traditional statistical analyses at the individual level (e.g., urban versus rural, metropolitan versus nonmetropolitan), are often presumed applicable to place attributes, yet these groupings only function appropriately when they match the spatial cohesion of the social relationships across the space investigated. Far too often a dichotomous urban-rural definition fails to adequately characterize the full range of human settlement patterns around the world. Dorélien et al. (2013) highlight the absence of a common framework at the global level for understanding the spatial boundaries of urban areas, and crucially this lack of a common conceptual framework makes cross-country comparisons and aggregations challenging. Definitions for urban areas throughout the world may be based on differing population size thresholds, varying classifications of economic activities, nationally defined administrative functions, or combinations. One nation's urban measures are not directly comparable across all countries, and assuming that they are can generate potentially biased findings.

Third, in the ecological context, alternative spatial scales (including those not considered administrative boundaries, e.g., watershed) can reveal more about a place's characteristics than scales that reflect administrative boundaries (e.g., municipality, province) (Irwin, 2007). The spatial analytical scale often depends on the availability of data which are frequently produced first and foremost for institutional or agency purposes (Fussell et al., 2014b; Voss, 2007). The origins of most spatial data reflect the perspectives and goals of the agencies creating them. For example, the U.S. Geological Survey focuses on data collection and complication in a manner best suited to capture biophysical environments, whereas the U.S. Census Bureau gathers and synthesizes data in a manner intended to best understand human geographies. Of further example, researchers often use U.S. Census data when considering spatial and demographic questions, and thus analysts

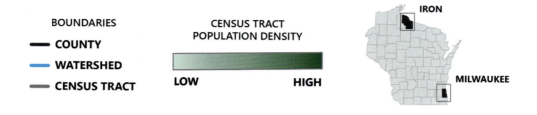

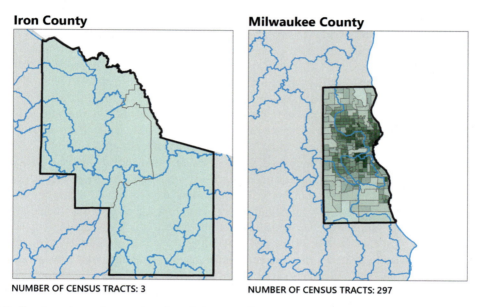

Fig. 6.3 County, water, and census tract boundaries, as well as census tract population density, for Iron County and Milwaukee County, Wisconsin

may observe a mismatch between the census geographic units and the demographic concepts they aim to examine. For example, the demographic data from the Census may correspond to a county or a census tract, whereas the Geological Survey data may correspond to daily stream flows or a watershed. Further, the misalignment becomes more frequent and apparent when incorporating various biophysical environment variables into analyses. As Fig. 6.3 illustrates, the watershed boundaries do not neatly conform to census tract or county boundaries. Without reflection and analytical adaptation, differences in data structures and various levels of aggregation can hinder the ability to understand the interactions between population and the biophysical environment.

In recent years, scholars have made great strides in developing spatially-harmonized measures (i.e., spatially consistent measures over time that accommodate administrative boundary changes and difference). For example, IPUMS-Terra is useful for within-country and cross-country comparisons because it synthesizes microdata with area-level data for geographic units including spatially-explicit environmental information such as climate and land use (e.g., data derived from remote sensing, vector-based geographic information system data, and weather station data) (IPUMS Terra, 2019). The Global Rural-Urban Mapping Project (GRUMP) also offers researchers georeferenced data that can contribute to their understanding of the full range of human settlement patterns from densely populated to sparsely populated and isolated areas through a variety of measurements including population counts, population density, geographic unit area grids, among other data (GRUMP, 2018). GRUMP incorporates standardized, aggregate measures for international geographic comparisons that enable scholars to improve the understanding of population-environment interactions along the rural-urban continuum globally.

Finally, confidentiality is a concern when using data that includes humans, and this issue is becoming increasingly relevant for spatial data as researchers collect data rich in fine-scale geographic identifiers. Data can be reported at a fine-scale (e.g., city block, exact geographic coordinates) or reported at a coarser scale (e.g., province, county, metropolitan statistical area). The central issue is that the confidentiality of an individual or household could be violated if a dataset reveals a location for the individual or household (de Castro, 2007; VanWey et al., 2005). Many funding and data sources that publish spatial data, such as the National Science Foundation, the National Institutes of Health, the Social Security Administration, and the Internal Revenue Service, follow rigorous protocols to ensure confidentiality of the respondents in their data (de Castro, 2007; VanWey et al., 2005). While privacy is essential, preserving confidentiality may make it difficult to link social data to fine-scale environmental data; the fine-scale might be lost by the need to aggregate up to the cruder scale of the social data; or the "swapping" of respondents between spatial units to preserve confidentiality might create specious associations between social and environmental attributes (Shlomo et al., 2010). Additionally, newer spatially-explicit data sources, such as open-source, volunteered geographic information (i.e., geospatial content generated by the public using mapping systems that are freely available on the Internet to identify crises, such as hurricanes, earthquakes, or famines), pose a challenge to preserving confidentiality (Goodchild, 2007). Presently, concerns around privacy are changing the way in which the US Decennial Census 2020 data will be processed (post-enumeration) and released for analysis. The US Census 2020 will be the first large-scale application of differential privacy, a new set of methods to capture the patterns of groups while withholding information about individuals. Ruggles et al. (2019) contend that differential privacy conceals individual characteristics rather than respondent identities, and the application of differential privacy in the Census could have grave implications on the quality of research conducted with publicly available data. Abowd and Schmutte (2019) argue that there always is a trade-off between privacy and data usability, a fact that environmental demographers continually confront.

Spatially-Explicit Work in the Population and Environment Literature

Environmental demographers focus heavily on improving our understanding of social and environmental vulnerabilities and interactions between the two dimensions. Indeed, environmental demographers are uniquely positioned to draw connections between social inequality (as well as social vulnerability or resilience) and environmental hazards (as well as environmental vulnerability or resilience). Identifying these intersections, thus, informs how or whether society moves closer toward protection from environmental degradation, prevention of adverse health impacts from worsening environmental situations, and restoration of the environment in collaboration with human rights principles for all people (Cutter, 1995). The spatial distribution of environmental degradation, such as acid rain, ozone depletion, and transboundary water pollution, transcends national borders and adds to the challenge of conducting spatial research and subsequently designing and implementing policies. As researchers thinking about spatial differentiation, we need to consider carefully the variables we select and our assumptions attached to our measurement, models, and theoretical frameworks. While few environmental demographers might self-identify as activist-scholars, the policy implications of research at the population-environment nexus are obvious and undeniable.

As with any analysis, a key factor for environmental demographers is whether we are most interested in a demographic process, an ecological process as an outcome of population-environment interactions, or modelling the reciprocal relationship within a system. Historically, most of the environmental demography work identified demographic processes as the outcome rather than environmental processes, but this is changing. For example, using survey data from agricultural households in Uganda combined with high-resolution satellite data on forest cover change, geographer Call et al. (2017b) examine the socio-environmental drivers of deforestation and reforestation in rural Uganda. Their findings highlight that social drivers of forest change vary greatly across spatial scales (i.e., the parcel scale and the community scale). Using a multi-scale approach, they demonstrate that tree planting is more common on parcels with secure land tenure, whereas land-rich, agrarian communities tend to display higher amounts of deforestation. Furthermore, scholars are not only researching the role of environmental degradation in triggering population displacement (McLeman & Hunter, 2010) and in suppressing population mobility (Gray & Mueller, 2012), but also the positive and negative impacts of migration processes on the biophysical environment (e.g., coastal ecosystems or the natural environment where a refugee camp is located) (Curran, 2002; Kelly et al., 2019).

Moving into the realm of projections, earth system scientist O'Neill and team (2018) created the Benefits of Reduced Anthropogenic Climate changE (BRACE) project. Using a range of simulations produced with the Community Earth System Model (CESM), O'Neill and colleagues investigate both the biophysical and socioeconomic impacts of climate change across different climate scenarios (i.e., a higher emissions future with a global average temperature increasing about 3.7 degrees Celsius above pre-industrial levels by 2100 and a moderate emissions future with a global average temperature increasing about 2.5 degrees Celsius). Through these simulations the BRACE study quantifies avoided impacts on biophysical and societal systems via measures of extreme events, health outcomes, agricultural productivity, and tropical cyclone frequency. Understanding the future of climate change is complex, both in terms of human contributions to climate change and the ways humans are affected, and utilizing simulation techniques is one of the important tools for improving our understanding of potential climate change impacts. In the following sections, we highlight spatially-explicit population-environment scholarship that has advanced research within the traditional demographic themes: migration, fertility, and mortality.

Migration and the Environment

There is a rich history of social science theory describing human migration. In the late 1880s, English-German geographer and cartographer Ernst Georg Ravenstein proposed the first well-known laws of migration, and was one of the first to advocate for using an aggregate approach as opposed to an individual approach to studying migration (Rowland, 2003). In mid-twentieth century contemporary demography, Lee (1966) expanded upon Ravenstein's laws of migration and highlighted that there are push and pull factors that cause people to move between the origin and destination locations. In order to bring the biophysical environment into our understanding of migration laws, environmental health scholar Findley (1994) assessed the connections between drought and migration in rural Mali between 1983–1985, and geographer Hugo (1996) emphasized that environmental migration is more likely to result in internal migration due to the political and socioeconomic costs of cross-border moves. Geographer Findlay (2011) stressed the importance of not only identifying environmental change as a potential driver of mobility (or immobility) in the origin but also distinguishing which destinations will be most impacted by environmental migration, especially for new destinations concerned with climate adaptation planning. Today, both domestic and international scholars think about the connections between places and changing environmental conditions, and migration is a large component of this story. To wit, demographer Wachter indicates that "'migration' is the usual label for the spatial subfield of demography, and movement is its preoccupation" (Wachter, 2005: 15299). Furthermore, sociologist Voss underscores that "migration is an area of research often located within a broader field of spatial data analysis dealing specifically with 'spatial interaction data'" (Voss, 2007: 458).

Sociologist Fussell and colleagues (2014b) offer examples of incorporating spatial awareness through the environmental dimensions of human migration. In several cases they explore how drought and rainfall cause variability in local natural resources in developing contexts, such as wood for fuel, building materials, and medicinal plants, and how this variability affects the likelihood of out-migration for resource-dependent households. Emphasizing both spatial and temporal dimensions of migration, the scholars argue natural resource unpredictability may lessen a family's agricultural productivity or ability to produce and sell goods created from natural resources (e.g., wicker baskets). Subsequently, as agricultural productivity declines, families may send a family member to migrate as a household livelihood diversification strategy, causing the family to rely more heavily on remittances (i.e., the new economics of labor migration [Massey et al., 1998]). Further, another challenge for the family may be that the natural resources available in their origin community may drastically differ from their new destination. Correspondingly, sociologist Hunter et al. (2014) empirically explore changes in spatial and temporal associations between temporary out-migration and natural resource availability in Agincourt, South Africa (See also Leyk et al., 2012). Hunter and colleagues investigate migration and environment associations using demographic survey data along with satellite imagery depicting natural resource availability (i.e., NDVI measuring vegetation levels). Their findings highlight that natural capital may offer households economic stability, and because of

this economic security, a household member may be able to migrate away temporarily. Yet, their findings suggest that natural capital is not associated with permanent out-migration. By linking these data, they are able to explore the associations between out-migration (typically for income-generating activities) and environmental variables (e.g., resource availability measured via NDVI) that vary significantly across the study space and, thus, highlight the importance of considering spatial differentiation (i.e., relationship heterogeneity).

While natural resource availability and changes in environmental conditions may induce out-migration, recent research has demonstrated that disadvantaged households may be restricted in their migration options because of upfront economic and social migration costs (Fussell et al., 2014b; Gray & Mueller, 2012). Using the Mexican Migration Project, Hunter et al. (2013) link rainfall patterns to commonly understood migration predictors, such as employment opportunities, political instability, and family ties. This work emphasizes the temporal and social dimensions that give rise to spatial heterogeneity in human-environment interactions by showing that rural Mexican households experiencing drought at least 2 years prior and possessing relatively deeper migration networks and histories are more likely to send a family member to the United States compared to households experiencing more recent droughts and possessing weaker migration networks. Recent work takes this a step further to consider not only observed migration responses to climate shocks, but also ex ante migration in conjunction with local labor responses to rural climate shocks in Mexico (Quiñones, 2018).

Migration researchers study many types of migration streams, and while researchers have identified that climate change likely increases population mobility in some cases and suppresses it in others, there is minimal research on temporary migration and changing environmental conditions. To fill this gap, Call et al. (2017a) investigate the extent to which precipitation, flooding, and temperature can predict temporary migration in Matlab, Bangladesh. In short, they find that these environmental changes do not consistently increase temporary migration, but rather environmental variability disrupts household livelihood strategies, including suppressing temporary migration and diminishing agriculture production.

While many of the examples we cite focus on rural contexts, in recent years a growing number of scholars have considered rural transformations and increased urbanization in the context of environment and migration. Using event history analysis to explore changing human settlement patterns in the southern Ecuadorian Andes, geographer Gray (2009) explores the effects of land-ownership and environmental conditions (e.g., land quality and harvest conditions) on out-migration to local, internal, and international destinations. Gray finds that, while the determinants of out-migration differ by type of migration stream, environmental degradation and landlessness do not always increase out-migration.

Until recently, although most population growth occurs in urban centers, scholars do not have a consistent, well-established measurement for urban land and populations. Demographer Balk and team (2018) use a time-series of the Global Human Settlement Layer for 1990–2010 to improve our understanding of urban structure and changes in the United States. Their overarching goal is to work toward generating globally and temporally consistent proxies to guide the modeling of urban change. In considering the spatial changes occurring in conjunction with both migration shifts and changing environmental conditions, geo-information scientist and geographer de Sherbinin et al. (2007) discuss the connections between climate change vulnerability and large global coastal cities. Using extensive data from the Centre for Research on the Epidemiology of Disasters, de Sherbinin et al. focus on regional increases in the severity and periodicity of hazard events in combination with the large and growing proportion of the world's population residing in coastal cities. Using case study examples of Mumbai, Rio de Janeiro, and Shanghai, they project changes in temperature (i.e., emissions related), consider a range of changing precipitation events and sea-level rise, increased urbanization, and changes in several socioeconomic characteristics. The authors provide a range of spatially and temporally measured characteristics and the relationships between these attributes to highlight the vulnerability of large coastal cities – vulnerabilities that may have been missed or disguised in a more rudimentary vulnerability calculation (See also Fussell et al., 2014a for an example about New Orleans and Hurricane Katrina).

Fertility and the Environment

While today much population growth, especially in more urban, developed contexts, occurs because of migration, spatial differentiation in fertility is substantial. Fertility researchers have long-explored the spatial patterning of fertility decline starting with the European Demographic Transition (Coale & Watkins, 1986; Goldstein & Klüsener, 2014). In her 1980 "A Suggested Framework for Analysis of Urban-Rural Fertility Differentials with an Illustration of the Tanzanian Case," Findley describes her project as a framework for analyzing how individual characteristics and urban or rural "place" factors interact to influence fertility choices (Findley, 1980: 237). To elaborate, rural women in Tanzania tend to have more children than urban women, yet rural fertility is not uniform across place or across age groups for a given tribal group. Findley explores the fertility patterns among the Chagga who live in the Mount Kilimanjaro area. While fertility is high for Chagga women, young Chagga women (younger than 25) have fewer children than other young women elsewhere in rural Tanzania. Findley demonstrates that several place-based variables, unique to slopes of Kilimanjaro and the nearest municipality, Moshi, influence childbearing patterns for young Chagga women. These place-based variables include education rates, job access including non-agricultural sources of income, marriage patterns, health care access, and perceptions of desired family size. While her article does not describe the complex relationships between all places along the rural-urban continuum, Findley presents a distinct spatial awareness of the range of different types of human settlement patterns throughout Tanzania, and within a small region (the Mount Kilimanjaro area) that is considered rural yet has very distinct, local characteristics, and how these various place characteristics may affect women's fertility decisions.

About three decades ago, fertility experts began incorporating remotely sensed data and GIS technologies in combination with census data and historical demography to improve our understanding of spatial and non-spatial attributes influencing fertility. Using a decomposition analysis, Weeks et al. (2000) explore the spatial variability in fertility in rural Menuofia, Egypt in 1976 and again in 1986. Descriptively, they find that by 1986, lower fertility is clustered in the northern portions of the study area and higher fertility in the southern portions. In 1976, the researchers find that 39% of the fertility variation is due to the spatial component including diffusion of ideas (e.g., female illiteracy and proportion married because these variables display high spatial autocorrelation and spatial clustering). Whereas in 1986, 51% of the variation in fertility is explained by spatial components, and the remainder of variation is a combination of demographic non-spatial characteristics in both periods. Results suggest that the spatial influences of fertility patterns increased over time. By combining spatial data from maps and Global Positioning System coordinates with sociodemographic data and administrative data, Entwisle et al. (1997) use spatial network analysis in Nang Rong, Thailand to better understand contraceptive choice and usage.[1] This innovative spatial network analysis is one of the studies that has paved the way for more spatial analysis in fertility research, and it reveals the importance of considering the history of accessibility in a

location for reaching family planning clinics, road composition and travel time, and options for alternative sources for contraceptives. In certain parts of the world, birth seasonality is a large factor in understanding fertility patterns. Linking fertility data from 1992–1993 to atmospheric temperature data between 1987–1991 in West Bengal, India, Chatterjee and Acharya (2000) find that conceptions are greater in the first quarter of a calendar year and the distribution of conceptions over calendar months is negatively associated with the average monthly temperature – a distinct birth seasonality. Warmer temperatures may affect the seasonality of conception because of reduced semen quality, reduced coital frequency (also linked to the husband's occupation especially during intensive periods for agricultural laborers), marriage timing (traditions of marrying during the cooler months in West Bengal), and reduced fetal viability.

In thinking about recent spatial work at the fertility-environment nexus, scientists are considering precipitation and temperature change influences on fertility rates and timing of births, the effects of deforestation on family planning decisions, and exposure to particulate matter and fertility rates to name a few (Dorélien, 2016; Sellers, 2017; Sellers & Gray, 2019; Simon, 2017; Xue & Zhang, 2018). Considering birth seasonality, demographer Dorélien (2016) produced one of the most comprehensive articles on birth seasonality in sub-Saharan Africa. She looks at amplitude and periodicity in 31 countries, and explores birth amplitudes between 1980–2000, with aggregated data by country (also sub-national levels) and ecological zone. Overall, in most countries, she finds that births are seasonal at the national level. Regionally, she highlights that births in Western and Central African ecological zones are highly seasonal, while many Eastern and Southern African ecological zones are not seasonal or are not seasonal to the same degree as in West and Central Africa. The Eastern and Southern ecological zones are larger and their environments and demographics are more heterogeneous than other zones, which might mask more local birth seasonality patterns. Additionally, Dorélien uses the Food and Agriculture Organization classifi-

---

[1] Spatial network analysis is a way through which to understand the routing and allocating of resource flows through a set of linear features. For example, this case looks at distance optimization decisions, directionality of resource flows, and various pathways traveled by people and resources as villagers seek family planning services at specific locations.

cations for ecological zones, and she suggests that other ecological factors and classifications for zones may produce different seasonality patterns. Shifting to research in Mexico, sociologist Simon (2017) finds that the fertility-environment connection differs greatly by climate zones and is context-specific, but notably recent fertility timing is associated with recent changes in rainfall patterns for households living in drier, rural areas. In addition to rainfall and temperature changes, deforestation and land availability may shape decisions regarding family size, especially in areas with high natural resource dependence.

While there is a rich literature base investigating the relationships between population growth and forest cover change, researchers have to a lesser degree explored the relationship between fertility decisions and contraceptive usage on forest cover change. Demographer and ecologist Sellers (2017) aims to explore fertility decisions in the Northern Ecuadorian Amazon and how these decisions impact deforestation and reforestation. Sellers finds that contraceptive usage is not associated with changes in forest cover, but this discovery also implies that the use of family planning is unlikely to alter deforestation trends. Finally, there is emerging environmental demography work in the environmental pollution and spatial epidemiology literatures. To enrich the epidemiological examinations on the infertility effects of air pollutants, earth scientists Xue and Zhang (2018) investigate long-term concentrations of ambient fine particles with a spatial resolution of 0.1 degrees (2009–2010) and county-level fertility rates in 2010, and find declines in fertility rates as outdoor particulate matter increases.

Mortality, Health, and the Environment

There is extensive literature on spatial analysis in mortality studies, especially in the realm of spatial epidemiology, wherein scientists model death relative to spatial clustering of various environmental risk factors and spatial patterns of disease transmission (de Castro, 2007). As briefly noted earlier, using the 1911 census for England and Wales, Garrett and Reid examine spatial and social differences in infant and child mortality in England and Wales going back to 1895–1911, and find that infant and child mortality were associated with a father's education and other socioeconomic characteristics but notably also with where the child lived (Garrett & Reid, 1995). They show that the social class relationship with infant mortality is related to the "location (where a child lives) within spatially structured disease environments" because social classes tend to concentrate in certain places (Ibid., 85).

In mid-twentieth century Chicago, Illinois, sociologists Kitagawa and Hauser (1973) explore geographical mortality differentials through indices of excess mortality by location by linking death certificates with Census information on educational attainment and income levels. They demonstrate that lower socioeconomic status is associated with higher risks of dying and lower life expectancy, and importantly they emphasize differences in mortality between nonmetropolitan and metropolitan residences as well as within metropolitan areas (Myers, 1974). As with reseach citing *The American Occupational Structure* by Blau and Duncan (1967), future research citing this seminal piece by Kitagawa and Hauser tends to focus on the socioeconomic dimension ignoring geographic variation. In a rousing call to both researchers and policymakers about the increased opportunity of spatially harmonized variables in the Demographic and Health Surveys (DHS), Balk et al. (2003) employ a range of spatial variables to explore childhood mortality in West Africa. The selected spatial variables capture climatic parameters (e.g., rainfall, aridity, farming systems, length of growing season, and stability of malaria transmission) and the rural-urban continuum (e.g., proximity to urban areas and population density). They find that incorporating the spatial attributes within their models has a rousing albeit modest effect on both infant and childhood mortality, and that these spatial characteristics explain much of the country-specific variations in childhood mortality.

Focusing on health, geographer López-Carr et al. (2016) use DHS data to consider climate-related impacts on undernutrition in East Africa's Lake Victoria Basin. Unique to this project is the use of DHS data as well as the Climate Hazards Group Infrared Precipitation with Stations (CHIRPS) and NDVI. Overall, the researchers find that areas experiencing increased rainfall and a negative change in NDVI are likely to be more vulnerable to climate-related child stunting. This study is the first application of time series NDVI data to climate-health vulnerability research, as well as the first application of vulnerability for those under 5 years of age at the 5-kilometer resolution for precipitation measures. As computing power increases and access to remotely sensed data continues to expand, we can expect novel combinations of spatial data and other socioeconomic and demographic data to emerge.

In a critical call to action, Barbara Entwisle's Presidential Address to the Population Association of America (2007) challenged demographers to focus attention on the intersection of human agency, neighborhoods, and health to advance our research on inequality. Fundamental to her argument, Entwisle emphasizes that individuals operate within specific social, spatial, and biophysical environments, and individuals are constrained by these environments – a large concern since the vast majority of literature focused on the individual level rather than the neighborhood level. Furthermore, the neighborhood effects scholarship at the time did not sufficiently reflect the influence of spatial and environmental contexts on human health (e.g., the spheres of space and the environment – health services, built environment, and toxins). Partly addressing Entwisle's call, demographer Jones and sociologist Pebley (2014) explore the measurement of the spatial context by suggesting that researchers treat the activity space system (the areas within which people move or travel or work during the course of their daily activities) as the object of inquiry. They highlight the potential of probing human activities within multiple spatial units, perhaps with shared and overlapping boundaries, and by identifying which populations systematically follow which activity space pathways (i.e., treating the activity space system as the object of inquiry). Activity spaces display more heterogeneous social characteristics

relative to residential neighborhoods, and these heterogeneous social characteristics are essential for identifying how people experience segregation and environmental exposure, not only where they live, but across a variety of spaces frequently used. Sociologist Sharp et al. (2015) also investigate how the structures of health inequalities are shaped not only by residential neighborhoods but also by additional activity spaces. By expanding their spatial lens to consider additional spaces, they find evidence of spatial effects on health outcomes characteristic of a relative deprivation process. Chief among their findings is that adults living in the most disadvantaged neighborhoods are more likely to report worse health when they spend more time working and/or commuting in more advantaged neighborhoods, and adults who live in more advantaged neighborhoods report worse health when they spend more time in disadvantaged neighborhoods. The authors show that spatial effects extend beyond the traditionally measured residential neighborhood and include other spaces with(in) which people interact, and that the nature of the extra-local spatial effect is conditional on how the spaces rank relative to one another.

Future Directions

A key element to understanding spatial relationships between environmental and demographic patterns and processes is probing various scales of analysis to uncover different relationships in terms of strength, direction, and type. Linking back to our discussion of certain administrative unit boundaries and classifications in Part 1, it is important to consider our unit of analysis in terms of the spatial unit. When studying the United States, researchers frequently use census tracts because of their analytical convenience, but scholars often wrongfully assume that a census tract represents a homogenous neighborhood. In the context of exposure to environmental risks, population density and distribution matter, and by looking at census tracts without understanding population density and settlement patterns we miss part of the story. Similarly, we may miss certain relationships, especially in terms of health and exposure to environmental hazards, if we do not account for activity spaces (Jones & Pebley, 2014; Sharp et al., 2015).

Shifting to the international horizon, global comparisons in social science research often rely on the nation-state and/or the national economy as the primary unit of analysis. This is problematic for at least two reasons. First, spatially, environmental degradation does not conform to administrative boundary units. Second, from a conceptual perspective, resources that are consumed, especially in developed countries, tend to come from other less developed countries (York & Rosa, 2003). Such cross-country dependence makes social, economic, environmental, and spatially differentiated vulnerabilities difficult to measure. Given global concerns about population-environment interactions, correctly specifying relationships between space, demographic processes, and the environment is crucial for generating valid research that ultimately informs policy suggestions.

Spatial data collection methods will hopefully continue to improve in the coming decades. Today, researchers have widespread access to often freely available data to purse spatial analysis of population-environment interactions. At its foundations, spatial analysis is rooted in an objective to theorize relationships between places (i.e., spatial differentiation and, less often, spatial homogeneity). Over time, improved data availability and advanced computing power have allowed us to test many existing theoretical assumptions and develop new frameworks. Indeed, innovative technologies that capture spatial data and the increase in spatially harmonized datasets have positioned today's environmental demographers to advance population-environment theory. We propose three theoretical directions to begin rethinking space and theory at the population-environment nexus: (1) embracing a multi-scalar approach; (2) considering differences in spatial patterns and processes in developing and developed country contexts; and (3) enriching our spatial understanding of social-environmental vulnerability. Our hope is that advancement in these intertwined, theoretical directions will frame population and environment research in a lens that moves us closer toward addressing the many cogent needs (e.g., social, environmental, and spatial inequalities) confronting populations across the globe.

Multi-scalar Approach

Porter and Howell (2012a) consider geographical sociology (geo-sociology) a "synergy between the ecologically centered macro-theory and the application of spatially-centered research methods" to sociological questions (1). We extend this position to questions asked in the population sciences. While in some ways incorporating a multi-scalar approach can be methodological in nature (e.g., running a multilevel regression), the approach is fundamentally rooted in theoretical assumptions and frameworks with clear statements about the spatial associations we seek to understand. Social spaces are multi-scalar (i.e., individual and household level; neighborhood and activity spaces; village, town, and urban center; county or province; country sub-region; country; global sub-region) (Chowdhury et al., 2011), and, as environmental philosopher Norton (1996) stresses, environmental problems are scalar problems. The researcher's choices for perspective and scale are an essential element in every description of the biophysical environment. Physical scientists consider vast ecological scales down to the microscopic and all the way up to the macro (i.e., universe). Yet for the purposes of population and environment research, Norton proposes that humans experience their world on roughly three spatio-temporal scales: the individual scale (and short-term); the community scale (and intergenerational); and a global scale (Norton, 1996). Figure 6.4, adapted from ecologist Martín-López et al. (2017), demonstrates a conceptual framework for a multi-scalar approach that incorporates both social and ecological components. Further, the figure highlights the spatial scale from local to global as well as the temporal scale from slow rates to fast rates of change. Improved conceptualizations about how and why multi-scalar

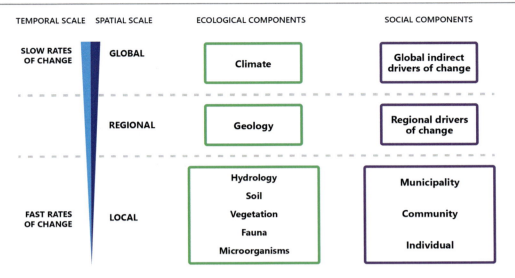

Fig. 6.4 Conceptual framework for a multi-scalar approach incorporating both social and ecological components [adapted from Martín-López et al., 2017]

processes occur and interact in the social and environmental contexts will aid researchers in deepening theories of demographic-ecological relationships and, subsequently, inform analytical methods and improve policy.

Spatial Differentiation Across the Globe

The population-environment literature offers a more global perspective (i.e., less U.S.-European centric) relative to other subareas of demography, largely due to numerous studies centered on livelihoods and natural resources across a variety of contexts. Yet, while theoretical variants that account for different global contexts are sometimes implied, more work remains to improve theory on population-environment interactions in both developing and developed country contexts (Gendreau et al., 1996). Sociologist and economic-historian Wallerstein developed World-Systems Theory (WST) in the 1970s as a critique in part against Eurocentric modernization theory. While this theory was largely developed to understand economic and political factors, the theory highlights that "core nations are the predominant global producers and consumers, but they extract the basic resources they need for production from, and export (often hazardous) waste to, peripheral nations" (York et al., 2003: 287).

In considering environmental degradation within WST, do types and amounts of environmental degradation vary by a nation's position within the world system? Certainly, as environmental demographers, we argue yes. But the question is more complex when focusing on differences between wealthier, weaker nations versus powerful, larger, poorer ones (Roberts & Grimes, 2002). Global development inequalities and sustainable natural resource livelihoods are one of the root causes of vulnerability to environmental hazards (Mercer et al., 2007). Within WST, scholars wonder to what degree the "periphery" has agency and power in this severely state-centered and hierarchical framework. Returning to spatial heterogeneity, within the WST framework, environmental demography can be utilized to highlight potential differences in environmental risk management both across countries and within countries. For example, availability and access to various forms of insurance are not spatially uniform. Namely, access to crop insurance can be especially important in agriculturally dependent areas, potentially reducing the impact of climate change and protecting against food insecurity, but in many areas is not easily accessible or is viewed undesirably (The Economist, 2018). Explicit spatial analysis has the potential to identify which areas and populations are most vulnerable to environmental change, through which mechanisms, and the subsequent consequences.

Social-Environmental Vulnerability

Another dimension for theory development focuses on spatial measures of social and environmental vulnerability globally. Definitions of vulnerability vary across disciplines, and we highlight a few and offer our own working definition. Human geographer Adger (2006) defines vulnerability as "the state of susceptibility to harm from exposure to stresses associated with environmental and social change and from the absence of capacity to adapt" (268). In their research on spatial and temporal changes to social vulnerability in the United States, Cutter and Finch (2008) outline several angles for exploring the notion of vulnerability. First, human environmental vulnerability research focuses on largescale global environmental processes, especially climate change ranging from local to global effects, and second, hazard vulnerability and mitigation research addresses vulnerability and resilience to natural hazards and disasters. Exposure, sensitivity, and response (i.e., carrying capacity or resilience) broadly create the factors of vulnerability, measured by both environmental and social systems. Social vulnerability is a multidimensional concept that is difficult to capture with only one variable. Race/ethnicity, nationality, socioeconomic status, gender, age (both children and elderly), exposure to natural hazards, and land tenure are a few of the characteristics often used to capture social vulnerability.

When exploring vulnerability in spatially-informed environmental demography literature, vulnerability stems from the interactions between environmental and social processes and varies across time and space. Additionally, vulnerability

occurs at various levels, ranging from individual to household to community to regional. In spatial analysis, exposure to hazards and adaptive capacity of people and places are compared across locations. Notably, one of the key challenges for researchers addressing vulnerability is to develop robust and credible measures that incorporate individual and societal perceptions of risk and exposure, but also account for institutional (e.g., government) mechanisms that can mediate vulnerability and promote adaptive action and resilience across space and over time (Adger, 2006). The Social Vulnerability Index (SoVI) is a step in this direction. The index measures the social vulnerability of U.S. counties to environmental hazards, and one of its innovative features is providing a spatial measure that enables comparison across U.S. counties (For more information, see SoVI, 2019).

While researchers do not have a robust, harmonized index measuring spatial and temporal dimensions of social-environmental vulnerability on the international scale, researchers generate internationally applicable frameworks and propose methods for investigating vulnerability throughout the world. For example, using a wide range of physical, climate, and socioeconomic characteristics, de Sherbinin et al. (2007) explore human-environment interactions in Mumbai, India; Rio de Janeiro, Brazil; and Shanghai, China to better understand the vulnerabilities emerging from climate hazards. The goal of the SoVI data source and de Sherbinin et al.'s research is to help determine which locations may need additional attention immediately following and during longer term recovery after a natural hazard event because certain populations appear vulnerable and have lower capacity to respond. Vulnerability is often considered in the context of shorter natural disasters, but it is just as crucial to understanding the mitigation and adaptation processes various populations will undertake in the coming years as slower onset environmental changes are felt. Theoretically, environmental demographers have the opportunity to encourage researchers and policymakers alike to fully interrogate the mechanisms that generate different vulnerabilities (and resilience) in systematic and predictable ways between spatial regions, especially if we are to move toward an environmentally just planet.

This chapter provides an overview of how spatial awareness, spatial data, and spatial analysis are used in population-environment scholarship to aid future research exploring the multifaceted relationships between demographic processes and the natural environment. We begin by defining spatial data and describing tools for using spatial data and analysis. Subsequently, we offer research examples from the population and environment literature base, and finally, we propose several future directions for the incorporation of spatial awareness into methodological and theoretical perspectives. Using a theoretical toolkit that includes multi-scalar investigation, recognizing spatial heterogeneity globally, and developing a deeper spatial understanding of social-environmental vulnerability, environmental demographers are empowered to forge a reflexive approach toward understanding the population-environment nexus.

References

Abowd, J. M., & Schmutte, I. M. (2019). An economic analysis of privacy protection and statistical accuracy as social choices. *American Economic Review, 109*(1), 171–202. https://doi.org/10.1257/aer.20170627

Adger, W. N. (2006). Vulnerability. *Global Environmental Change, 16*(3), 268–281. https://doi.org/10.1016/j.gloenvcha.2006.02.006

Balk, D. L., Leyk, S., Jones, B., Montgomery, M. R., & Clark, A. (2018). Understanding urbanization: A study of census and satellite-derived urban classes in the United States, 1990–2010. *PLoS One, 13*(12), e0208487. https://doi.org/10.1371/journal.pone.0208487

Balk, D. L., & Montgomery, M. R. (2015). Guest Editorial: "Spatializing demography for the urban future". *Spatial Demography, 3*(2), 59–62. https://doi.org/10.1007/s40980-015-0017-x

Balk, D. L., Pullum, T., Storeygard, A., Greenwell, F., & Neuman, M. (2003). *Spatial analysis of childhood mortality in West Africa*. ORC Macro and Center for International Earth Science Information Network (CIESIN), Columbia University.

Blau, P. M., & Duncan, O. D. (1967). *The American occupational structure*. Wiley.

Call, M. A., Gray, C. L., Yunus, M., & Emch, M. (2017a). Disruption, not displacement: Environmental variability and temporary migration in Bangladesh. *Global Environmental Change, 46*, 157–165. https://doi.org/10.1016/j.gloenvcha.2017.08.008

Call, M. A., Mayer, T., Sellers, S., Ebanks, D., Bertalan, M., Nebie, E., & Gray, C. (2017b). Socioenvironmental drivers of forest change in rural Uganda. *Land Use Policy, 62*, 49–58. https://doi.org/10.1016/j.landusepol.2016.12.012

Chatterjee, U., & Acharya, R. (2000). Seasonal variation of births in rural West Bengal: Magnitude, direction and correlates. *Journal of Biosocial Science, 32*(4), 443–458. https://doi.org/10.1017/S0021932000004430

Chi, G., & Zhu, J. (2020). *Spatial regression models for the social sciences*. Sage Publications.

Chowdhury, R. R., Larson, K., Grove, M., Polsky, C., Cook, E., Onsted, J., & Ogden, L. (2011). A multiscalar approach to theorizing socio-ecological dynamics of urban residential landscapes. *Cities and the Environment, 4*(1), 1–21. https://doi.org/10.15365/cate.4162011

Clarke, K. C. (1995). Data structure transformations. In *Analytical and computer cartography* (2nd ed., pp. 226–248). Prentice Hall.

Coale, A., & Watkins, S. C. (Eds.). (1986). *The decline of fertility in Europe*. Princeton University Press.

Curran, S. (2002). Migration, social capital, and the environment: Considering migrant selectivity and networks in relation to coastal ecosystems. *Population and Development Review, 28*, 89–125. Retrieved from http://jstor.org/stable/3115269

Curtis, K. J., DeWaard, J., Fussell, E., & Rosenfeld, R. A. (2019a). Differential recovery migration across the rural-urban gradient: Minimal and short-term population gains for rural disaster-affected Gulf Coast counties. *Rural Sociology*.

Curtis, K. J., Lee, J., O'Connell, H. A., & Zhu, J. (2019b). The spatial distribution of poverty and the long reach of the industrial makeup of places: New evidence on spatial and temporal regimes. *Rural Sociology, 84*(1), 28–65. https://doi.org/10.1111/ruso.12216

Cutter, S. L. (1995). Race, class and environmental justice. *Progress in Human Geography, 19*(1), 111–122. https://doi.org/10.1177/030913259501900111

Cutter, S. L., & Finch, C. (2008). Temporal and spatial changes in social vulnerability to natural hazards. *Proceedings of the National Academy of Sciences of the United States of America, 105*(7), 2301–2306. https://doi.org/10.1073/pnas.0710375105

D'Antona, Á. O., Cak, A. D., & VanWey, L. K. (2008). Collecting sketch maps to understand property land use and land cover in large surveys. *Field Methods, 20*(1), 66–84. https://doi.org/10.1177/1525822X07309354

de Castro, M. C. (2007). Spatial demography: An opportunity to improve policy making at diverse decision levels. *Population Research and Policy Review, 26*(5–6), 477–509. https://doi.org/10.1007/s11113-007-9041-x

de Sherbinin, A., Schiller, A., & Pulsipher, A. (2007). The vulnerability of global cities to climate hazards. *Environment and Urbanization, 19*(1), 39–64. https://doi.org/10.1177/0956247807076725

Doreian, P. (1980). Linear models with spatially distributed data: Spatial disturbances or spatial effects? *Sociological Methods & Research, 9*(1), 29–60. https://doi.org/10.1177/004912418000900102

Dorélien, A. (2016). Birth seasonality in Sub-Saharan Africa. *Demographic Research, 34*(27), 761–796. https://doi.org/10.4054/DemRes.2016.34.27

Dorélien, A., Balk, D., & Todd, M. (2013). What is urban? Comparing a satellite view with the demographic and health surveys. *Population and Development Review, 39*(3), 413–439. https://doi.org/10.1111/j.1728-4457.2013.00610.x

Elhorst, J. P. (2001). Dynamic models in space and time. *Geographical Analysis, 33*, 119–140. https://doi.org/10.1111/j.1538-4632.2001.tb00440.x

Entwisle, B. (2007). Putting people into place. *Demography, 44*(4), 687–703. https://doi.org/10.1353/dem.2007.0045

Entwisle, B., Rindfuss, R. R., Walsh, S. J., Evans, T. P., & Curran, S. R. (1997). Geographic information systems, spatial network analysis, and contraceptive choice. *Demography, 34*(2), 171–187. https://doi.org/10.2307/2061697

Famine Early Warning Systems Network. (2020). Famine Early Warning Systems Network. https://fews.net. Accessed 5 May 2020.

Findlay, A. M. (2011). Migrant destinations in an era of environmental change. *Global Environmental Change, 21*, S50–S58. https://doi.org/10.1016/j.gloenvcha.2011.09.004

Findley, S. E. (1980). A suggested framework for analysis of urban-rural fertility differentials with an illustration of the Tanzanian case. *Population and Environment, 3*(3/4), 237–261. https://doi.org/10.1007/BF01255341

Findley, S. E. (1994). Does drought increase migration? A study of migration from rural Mali during the 1983–1985 drought. *The International Migration Review, 28*(3), 539–553. https://doi.org/10.2307/2546820

Foody, G. M., Cutler, M. E., McMorrow, J., Pelz, D., Tangki, H., Boyd, D. S., & Douglas, I. (2001). Mapping the biomass of Bornean tropical rain forest from remotely sensed data. *Global Ecology and Biogeography, 10*(4), 379–387. https://doi.org/10.1046/j.1466-822X.2001.00248.x

Funk, C., Peterson, P., Landsfeld, M., Pedreros, D., Verdin, J., Shukla, S., et al. (2015). The climate hazards infrared precipitation with stations—A new environmental record for monitoring extremes. *Scientific Data, 2*, 1–21. https://doi.org/10.1038/sdata.2015.66

Fussell, E., Curran, S. R., Dunbar, M. D., Babb, M. A., Thompson, L., & Meijer-Irons, J. (2017). Weather related hazards and population change: A study of hurricanes and tropical storms in the United States, 1980–2012. *The Annals of the American Academy of Political and Social Science, 669*(1), 146–167. https://doi.org/10.1177/0002716216682942

Fussell, E., Curtis, K. J., & DeWaard, J. (2014a). Recovery migration to the city of New Orleans after Hurricane Katrina: A migration systems approach. *Population and Environment, 35*(3), 305–322. https://doi.org/10.1007/s11111-014-0204-5

Fussell, E., Hunter, L. M., & Gray, C. L. (2014b). Measuring the environmental dimensions of human migration: The demographer's toolkit. *Global Environmental Change, 28*, 182–191. https://doi.org/10.1016/j.gloenvcha.2014.07.001

Garrett, E., & Reid, A. (1995). Thinking of England and taking care: Family building strategies and infant mortality in England and Wales, 1891–1911. *International Journal of Population Geography, 1*(1), 69–102. https://doi.org/10.1002/ijpg.6060010106

Gendreau, F., Gubry, P., & Véron, J. (Eds.). (1996). *Populations et environnement dans les pays du Sud* [People and environment in developing countries]. Karthala, CEPED.

Global Rural-Urban Mapping Project (GRUMP), v1. (2018). Socioeconomic Data and Applications Center (SEDAC). http://sedac.ciesin.columbia.edu/data/collection/grump-v1. Accessed 17 Jan 2018.

Goldstein, J. R., & Klüsener, S. (2014). Spatial analysis of the causes of fertility decline in Prussia. *Population & Development Review, 40*(3), 497–525. https://doi.org/10.1111/j.1728-4457.2014.00695.x

Goodchild, M. F. (2007). *Citizens as sensors: The world of volunteered geography.* Resource document. National Center for Geographic Information and Analysis, UC Santa Barbara. https://doi.org/ncgia.ucsb.edu/projects/vgi/docs/position/Goodchild_VGI2007.pdf. Accessed 20 Feb 2018.

Grant, M. J., Curtis, K. J., & Rosenfeld, R. A. (2020). *Fertility decline and rainfall variation in Malawi* [Center for Demography and Ecology Working Paper Series].

Gray, C. L. (2009). Environment, land, and rural outmigration in the southern Ecuadorian Andes. *World Development, 37*(2), 457–468. https://doi.org/10.1016/j.worlddev.2008.05.004

Gray, C. L., & Mueller, V. (2012). Natural disasters and population mobility in Bangladesh. *Proceedings of the National Academy of Sciences of the United States of America, 109*(16), 6000–6005. https://doi.org/10.1073/pnas.1115944109

Gray, C. L., & Wise, E. (2016). Country-specific effects of climate variability on human migration. *Climatic Change, 135*(3), 555–568. https://doi.org/10.1007/s10584-015-1592-y

Haining, R. (1990). *Spatial data analysis in the social and environmental sciences.* Cambridge University Press.

Hawley, A. H. (1950). *Human ecology; A theory of community structure.* Ronald Press.

Hugo, G. (1996). Environmental concerns and international migration. *The International Migration Review, 30*(1), 105–131. https://doi.org/10.2307/2547462

Hunter, L. M., Murray, S., & Riosmena, F. (2013). Rainfall patterns and U.S. migration from rural Mexico. *International Migration Review, 47*(4), 874–909. https://doi.org/10.1111/imre.12051

Hunter, L. M., Nawrotzki, R., Leyk, S., Maclaurin, G. J., Twine, W., Collinson, M., & Erasmus, B. (2014). Rural outmigration, natural capital, and livelihoods in South

Africa. *Population, Space and Place, 20*(5), 402–420. https://doi.org/10.1002/psp.1776

Hutchinson, C. F. (1998). Social science and remote sensing in famine early warning. In D. Liverman, E. F. Moran, R. R. Rindfuss, & P. C. Stern (Eds.), *People and pixels: Linking remote sensing and social science* (pp. 189–196). National Academies Press.

In Africa, agricultural insurance often falls on stony ground. (2018, December 15). *The Economist*. Retrieved from www.economist.com/finance-and-economics/2018/12/15/in-africa-agricultural-insurance-often-falls-on-stony-ground

IPUMS Terra: Integrated population and environmental data. (2019). Minnesota Population Center. https://terra.ipums.org. Accessed 12 Sept 2019.

Irwin, M. D. (2007). Territories of inequality: An essay on the measurement and analysis of inequality in grounded place settings. In L. M. Lobao, G. Hooks, & A. R. Tickamyer (Eds.), *The sociology of spatial inequality* (pp. 85–109). State University of New York Press.

Itzigsohn, J., & Brown, K. L. (2020). *The sociology of W. E. B. Du Bois: Racialized modernity and the global color line*. New York University Press.

Jones, M., & Pebley, A. R. (2014). Redefining neighborhoods using common destinations: Social characteristics of activity spaces and home census tracts compared. *Demography, 51*(3), 727–752. https://doi.org/10.1007/s13524-014-0283-z

Kelly, J., Burl, T., & VanRooyen, M. (2019, April). *Links between resilience and political conflict in northern Congo*. Presented at the Population Association of America Annual Meetings, Austin, Texas.

Kitagawa, E. M., & Hauser, P. M. (1973). *Differential mortality in the United States: A study in socioeconomic epidemiology*. Harvard University Press.

Lee, B. A., & Sharp, G. (2017). Ethnoracial diversity across the rural-urban continuum. *The Annals of the American Academy of Political and Social Science, 672*(1), 26–45. https://doi.org/10.1177/0002716217708560

Lee, E. (1966). A theory of migration. *Demography, 3*(1), 47–57. https://doi.org/10.2307/2060063

Leyk, S., Maclaurin, G. J., Hunter, L. M., Nawrotzki, R., Twine, W., Collinson, M., & Erasmus, B. (2012). Spatially and temporally varying associations between temporary outmigration and natural resource availability in resource-dependent rural communities in South Africa: A modeling framework. *Applied Geography, 34*, 559–568. https://doi.org/10.1016/j.apgeog.2012.02.009

Liverman, D., Moran, E. F., Rindfuss, R. R., & Stern, P. C. (Eds.). (1998). *People and pixels: Linking remote sensing and social science*. National Academy Press.

López-Carr, D., Mwenda, K. M., Pricope, N. G., Kyriakidis, P. C., Jankowska, M. M., Weeks, J., et al. (2016). Climate-related child undernutrition in the Lake Victoria Basin: An integrated spatial analysis of health surveys, NDVI, and precipitation data. *IEEE Journal of Selected Topics in Applied Earth Observations and Remote Sensing, 9*(6), 2830–2835. https://doi.org/10.1109/JSTARS.2016.2569411

Martín-López, B., Palomo, I., García-Llorente, M., Iniesta-Arandia, I., Castro, A. J., García Del Amo, D., Gómez-Baggethun, E., & Montes, C. (2017). Delineating boundaries of social-ecological systems for landscape planning: A comprehensive spatial approach. *Land Use Policy, 66*, 90–104. https://doi.org/10.1016/j.landusepol.2017.04.040

Massey, D. (1994). *Space, place, and gender*. University of Minnesota Press.

Massey, D. S., Arango, J., Hugo, G., Kouaouci, A., Pellegrino, A., & Taylor, J. E. (1998). *Worlds in motion: Understanding international migration at the end of the millenium*. Oxford University Press.

Mercer, J., Dominey-Howes, D., Kelman, I., & Lloyd, K. (2007). The potential for combining indigenous and western knowledge in reducing vulnerability to environmental hazards in small island developing states. *Environmental Hazards, 7*(4), 245–256. https://doi.org/10.1016/j.envhaz.2006.11.001

McKenzie, R. D. (1924). The ecological approach to the study of the human community. *American Journal of Sociology, 30*(3), 287–301.

McLeman, R. A., & Hunter, L. M. (2010). Migration in the context of vulnerability and adaptation to climate change: Insights from analogues. *Wiley Interdisciplinary Reviews Climate Change, 1*(3), 450–461. https://doi.org/10.1002/wcc.51

Myers, G. C. (1974). Review of differential mortality in the United States: A study in socioeconomic epidemiology by E. M. Kitagawa & P. M. Hauser. *American Journal of Sociology, 80*(2), 532–534.

Norton, B. (1996). A scalar approach to ecological constraints. In P. C. Schulze (Ed.), *Engineering within ecological constraints* (pp. 45–64). National Academy Press.

O'Loughlin, J., Flint, C., & Anselin, L. (1994). The geography of the Nazi vote: Context, confession, and class in the Reichstag election of 1930. *Annals of the Association of American Geographers, 84*(3), 351–380.

O'Neill, B. C., Done, J. M., Gettelman, A., Lawrence, P., Lehner, F., Lamarque, J.-F., et al. (2018). The benefits of reduced anthropogenic climate change (BRACE): A synthesis. *Climatic Change, 146*(3), 287–301. https://doi.org/10.1007/s10584-017-2009-x

Park, R. E., Burgess, E. W., & McKenzie, R. D. (1925). *The city*. The University of Chicago Press.

Pebesma, E. (2017, July 4). Spatial data in R: New directions. https://edzer.github.io/UseR2017/#manipulating-geometries. Accessed 12 Sept 2019.

Pebley, A. R. (1998). Demography and the environment. *Demography, 35*(4), 377–389. https://doi.org/10.2307/3004008

Porter, J. R., & Howell, F. M. (2012a). *Geographical sociology: Theoretical foundations and methodological applications in the sociology of location*. Springer.

Porter, J. R., & Howell, F. M. (2012b). Roots of space in sociology: Community sociology and the Wisconsin and Chicago schools. In *Geographical sociology: Theoretical foundations and methodological applications in the sociology of location* (pp. 25–33). Springer.

Quiñones, E. J. (2018, December). Anticipatory migration and local labor responses to rural climate shocks. Presented at the Demography Training Seminar, University of Wisconsin-Madison.

Reyes, P. E., Zhu, J., & Aukema, B. H. (2012). Selection of spatial-temporal lattice models: Assessing the impact of climate conditions on a mountain pine beetle outbreak. *Journal of Agricultural, Biological, and Environmental Statistics, 17*(3), 508–525.

Rienecker, M. M., Suarez, M. J., Gelaro, R., Todling, R., Bacmeister, J., Liu, E., et al. (2011). MERRA: NASA's modern-era retrospective analysis for research and applications. *Journal of Climate, 24*(14), 3624–3648. https://doi.org/10.1175/JCLI-D-11-00015.1

Roberts, J. T., & Grimes, P. E. (2002). World-systems theory and the environment: Toward a new synthesis. In R. E. Dunlap, F. H. Buttel, P. Dickens, & A. Gijswijt (Eds.), *Sociological theory and the environment: Classical foundations, contemporary insights* (pp. 167–194). Rowman & Littlefield Publishers.

Rowland, D. T. (2003). *Demographic methods and concepts.* Oxford University Press.

Ruggles, S., Fitch, C., Magnuson, D., & Schroeder, J. (2019). Differential privacy and census data: Implications for social and economic research. *AEA Papers and Proceedings, 109*, 403–408. https://doi.org/10.1257/pandp.20191107

Schmidt, L. J. (2001). From the dust bowl to the Sahel. NASA. https://earthobservatory.nasa.gov/features/DustBowl. Accessed 26 Feb 2018.

Sadler, R. C. (2019). Misalignment between ZIP codes and municipal boundaries: A problem for public health. *City, 21*(3), 335–340. https://doi.org/10.2307/26820661

Sellers, S. (2017). Family planning and deforestation: Evidence from the Ecuadorian Amazon. *Population and Environment; New York, 38*(4), 424–447. https://doi.org/10.1007/s11111-017-0275-1

Sellers, S., & Gray, C. (2019). Climate shocks constrain human fertility in Indonesia. *World Development, 117*, 357–369. https://doi.org/10.1016/j.worlddev.2019.02.003

Sharp, G., Denney, J. T., & Kimbro, R. T. (2015). Multiple contexts of exposure: Activity spaces, residential neighborhoods, and self-rated health. *Social Science & Medicine, 146*, 204–213. https://doi.org/10.1016/j.socscimed.2015.10.040

Shlomo, N., Tudor, C., & Groom, P. (2010). Data swapping for protecting census tables. *Privacy in Statistical Databases*, 41–51. https://doi.org/10.1007/978-3-642-15838-4_4

Simon, D. H. (2017). Exploring the influence of precipitation on fertility timing in rural Mexico. *Population and Environment; New York, 38*(4), 407–423. https://doi.org/10.1007/s11111-017-0281-3

SoVI. (2019). Hazards & Vulnerability Research Institute, University of South Carolina. http://artsandsciences.sc.edu/geog/hvri/sovi%C2%AE-0

Tian, H., Banger, K., Bo, T., & Dadhwal, V. K. (2014). History of land use in India during 1880-2010: Large-scale land transformations reconstructed from satellite data and historical archives. *Global and Planetary Change, 121*, 78–88. https://doi.org/10.1016/j.gloplacha.2014.07.005

Tobler, W. (1970). A computer movie simulating urban growth in the Detroit region. *Economic Geography, 26*, 234–240. https://doi.org/10.2307/143141

VanWey, L. K., Rindfuss, R. R., Gutmann, M. P., Entwisle, B., & Balk, D. L. (2005). Confidentiality and spatially explicit data: Concerns and challenges. *Proceedings of the National Academy of Sciences, 102*(43), 15337–15342. https://doi.org/10.1073/pnas.0507804102

Vittek, M., Brink, A., Donnay, F., Simonetti, D., & Desclée, B. (2014). Land cover change monitoring using Landsat MSS/TM satellite image data over West Africa between 1975 and 1990. *Remote Sensing, 6*(1), 658–676. https://doi.org/10.3390/rs6010658

Voss, P. R. (2007). Demography as a spatial social science. *Population Research and Policy Review, 26*(5), 457–476. https://doi.org/10.1007/s11113-007-9047-4

Voss, P. R., & Chi, G. (2006). Highways and population change. *Rural Sociology, 71*(1), 33–58. https://doi.org/10.1526/003601106777789837

Voss, P. R., Curtis White, K. J., & Hammer, R. B. (2006). Explorations in spatial demography. In W. A. Kandel & D. L. Brown (Eds.), *Population change and rural society* (pp. 407–429). https://doi.org/10.1007/1-4020-3902-6_19.

Wachter, K. W. (2005). Spatial demography. *Proceedings of the National Academy of Sciences, 102*(43), 15299–15300. https://doi.org/10.1073/pnas.0508155102

Ward, C. D., & Shackleton, C. M. (2016). Natural resource use, incomes, and poverty along the rural–urban continuum of two medium-sized, South African towns. *World Development, 78*, 80–93. https://doi.org/10.1016/j.worlddev.2015.10.025

Ward, M. D., & Gleditsch, K. S. (2008). *Spatial regression models.* Sage Publications.

Weeks, J. R. (2004). Role of spatial analysis in demographic research. In M. F. Goodchild & D. G. Janelle (Eds.), *Spatially integrated social science* (pp. 381–399). Oxford University Press.

Weeks, J. R., Gadalla, M. S., Rashed, T., Stanforth, J., & Hill, A. G. (2000). Spatial variability in fertility in Menoufia, Egypt, assessed through the application of remote-sensing and GIS technologies. *Environment and Planning A: Economy and Space, 32*(4), 695–714. https://doi.org/10.1068/a3286

Winkler, R. L., & Johnson, K. M. (2016). Moving toward integration? Effects of migration on ethnoracial segregation across the rural-urban continuum. *Demography, 53*(4), 1027–1049. https://doi.org/10.1007/s13524-016-0479-5

Wooldridge, J. M. (2013). *Introductory econometrics: A modern approach* (5th ed.). South-Western Cengage Learning.

Xue, T., & Zhang, Q. (2018). Associating ambient exposure to fine particles and human fertility rates in China. *Environmental Pollution, 235*, 497–504. https://doi.org/10.1016/j.envpol.2018.01.009

York, R., & Rosa, E. A. (2003). Key challenges to ecological modernization theory. *Organization & Environment, 16*(3), 273–288. https://doi.org/10.1177/1086026603256299

York, R., Rosa, E. A., & Dietz, T. (2003). Footprints on the earth: The environmental consequences of modernity. *American Sociological Review, 68*(2), 279–300. https://doi.org/10.2307/1519769

Qualitative Data and Approaches to Population–Environment Inquiry

7

Sabine Henry, Sebastien Dujardin, Elisabeth Henriet, and Sofia Costa Santos Baltazar

Abstract

This chapter highlights the contribution of qualitative methods for understanding the population–environment nexus. A brief overview is offered of a variety of qualitative methods: in-depth interview, focus groups, participant observation, ethnography, and drawing. Then, the chapter presents three concrete examples of population–environment research aimed at illustrating the selection and application of appropriate tools to build qualitative knowledge. These research examples engage various time periods and diverse settings, and they combine to illustrate core dimensions of qualitative investigation including diverse methodologies such as interviews, participatory mapping, participant observation, and observant participation, photo-language, and the use of a game. The first example addresses the intersection between scientific and local knowledge, the second explores individuals' experiences within their environment, and the third illustrates the participatory action research process. Within the description of each example, the decision-making processes around data collection and analysis are detailed. In each case, the process of interpretation is also described.

Introduction

Qualitative research represents approaches allowing for detailed examination of experiences, attitudes, and behaviors. The qualitative approach can include in-depth interviews, focus group discussions, observation, content analysis, and the collection of personal or institutional histories (Hennink et al., 2020). In general, qualitative researchers seek to understand the meanings and interpretations that individuals attribute to their behavior, as well as to objects and events (Taylor et al., 2015; Hennink et al., 2020). Qualitative data include observations of what people say, draw, and/or write, in addition to how they act, interact, and express themselves. Addressing

S. Henry (✉) · S. Dujardin · S. C. S. Baltazar
Geography Department and Institute of
Life-Earth-Environment, University of Namur, Namur,
Belgium
e-mail: sabine.henry@unamur.be

E. Henriet
Geography Department and Transitions Institute,
University of Namur, Namur, Belgium

© Springer Nature Switzerland AG 2022
L. M. Hunter et al. (eds.), *International Handbook of Population and Environment*, International
Handbooks of Population 10, https://doi.org/10.1007/978-3-030-76433-3_7

the population and environment nexus qualitatively involves uncovering clues in the form of expressions that inform and reflect individuals' environmental experience. That reality is often more nuanced and complex than the most detailed questionnaire can capture.

In this chapter, we review qualitative research in the context of population–environment inquiry. We first provide a brief overview of a variety of qualitative methods and then present three concrete examples of applied research aimed at illustrating the selection and application of appropriate tools to build qualitative knowledge. As with any research methodology, of core concern is the need for a critical approach to the definition of research objectives, data collection methods, analyses and interpretation.

Overview of Qualitative Methods and Concerns

Several approaches can be used to capture the meanings individuals give to elements of their lives. As mentioned by Taylor et al., "in qualitative methodology, the researcher looks at settings and people holistically; people, settings, or groups are not reduced to variables, but are viewed as a whole" (Taylor et al., 2015, p. 9). Qualitative studies seek to extract insights from a wide variety of sources using a diverse array of techniques, but with the common aim of uncovering new knowledge, or revealing novel perspectives on existing knowledge (Kennedy & Thornberg, 2018). Often, qualitative methods engage an **inductive approach**, which seeks to develop understanding from empirical data – observations and experiences – in order to identify patterns and relationships, and generate meaning. Starting from empirical observation rather than from pre-existing theory, the research process then entails moving back and forth between data and theory: examining how data marry with existing theory, but also paying attention to how modifications in existing understanding are required after analyzing data. Scholars have identified a wide range of research techniques that fall within the broad

ambit of qualitative inquiry. These are explored in depth in numerous texts, both at foundation and advanced levels (see for example DeLyser et al., 2010; Kitchin & Tate, 2013; Taylor et al., 2015; Flick, 2018); thus, we limit our attention here to providing just a brief review identifying key elements of five principal techniques employed in the three case studies that follow in section "Qualitative research examples". These are: in-depth interviews, focus groups, participant observation, ethnography, and drawing.

In-depth interviews involve dialogue as it emerges from carefully designed open-ended questions, thereby giving interviewees the opportunity to elaborate their responses. Such interviews are aimed at eliciting detailed information on experiences, feelings, and opinions (Roulston & Choi, 2018). Interviews also allow for information to be gleaned from the social interaction between interviewer and respondent, and the context in which the interview takes place. Successful interviewers have good conversational skills and can readily guide a conversation, while also allowing scope for the respondent to explore and elaborate.

Generally, in-depth interviews follow a systematic schedule, asking questions in the same order across interviews. Yet, these can also take a wide variety of forms, including: specific questions (e.g., to a migrant: if you have experienced difficulties in the visa procedure, please can you describe them?); a list of topics to be covered during the interview (e.g., the visa procedure, costs, and timelines), an approach that offers greater flexibility in sequencing and wording; or using an entirely open agenda in which questions are asked in the context of a conversation (e.g., about the visa process), allowing the respondent to raise issues that may not have been anticipated by the researcher (Kitchin & Tate, 2013).

A successful interview strategy calls for the interviewer to develop a trusting relationship with the respondent while maintaining a neutral position on the topics being discussed – a challenging issue (Kitchin & Tate, 2013). Interviewer characteristics (e.g., gender, color, age), clothes (e.g., formal or casual), and attitude (e.g., attentive, distracted, bored) shape respondent engagement and their willingness to contribute meaningful insights.

Participant recruitment is another key issue for in-depth interviews. Recruitment can be challenging in that interviews require a significant time commitment from respondents, as such interviews are generally of lengthy duration (Kitchin & Tate, 2013; Roulston & Choi, 2018). By their very nature, in-depth interviews are also likely to involve sensitive topics that respondents may be reluctant to discuss. Careful planning is therefore needed to convince people to participate and increase the chance of successful outcomes. Conventional sampling strategies (Kitchin & Tate, 2013), such as commonly used in quantitative surveys, may therefore be more difficult to apply, but for specific target groups snowballing strategies[1] are often used, with respondents recruited via word of mouth, and reassured by other community members (see Schreier, 2018, for additional detail about sampling). Despite their manifest challenges, in-depth interviews have the potential to elicit a much greater range and depth of insights and experiences than those obtained from conventional quantitative surveys, which are often confined to eliciting factual or simple attitudinal information using closed questions. As demonstrated by our examples in section "Qualitative research examples", this technique has particular merit for eliciting local, grounded understandings of environmental processes.

Focus groups represent another qualitative tool designed to elicit information, observations, experiences, and reactions based on responses and interactions among a small, selected group of individuals (Morgan & Hoffman, 2018). Implementations vary, but such groups commonly involve 5–10 or so participants, meeting for perhaps 90 min. More than being just a quick way of collecting data from multiple people, focus groups provide for a free flow of conversation guided by the moderator, while observers keep a record of the interaction between participants, thereby allowing for a deeper exploration of their knowledge and experiences. Commonly, the moderator starts with a broad question and each participant then shares their point of view. Participants are invited to talk to others, commenting on their responses, adding their own points of view, and even asking questions to other participants.

Focus groups reveal collective perceptions, highlighting views that are widely shared among participants, while also providing an opportunity for individuals to share experiences and contribute unique perspectives, and enabling the moderator to observe reactions from other members of the group (Kitchin & Tate, 2013). For example, in the context of population–environment relations, a moderator might invite key people in the village (e.g., a priest, the oldest person, a primary school teacher, the mayor, returning migrants, etc.) and start the conversation with a question such as "Do you think that out-migrants are a good thing for your village?"

Selection of participants is a key issue for this method (Morgan & Hoffman, 2018). Although all participants have some knowledge of the topic, the relationship between participants is also important in shaping the dynamics of the discussion. Each participant should feel secure in giving their opinion and the moderator plays a crucial role in avoiding conflict, uneven power relations, and feelings of discomfort. A participant may avoid speaking if they are concerned about how others will react. Homogeneous groups (in terms of gender, age, sexual orientation, power, etc.) facilitate the sharing of opinion, but heterogeneous groups are likely to draw out a wider range of views and result in discussion that may reveal perspectives and issues that would otherwise remain hidden. Indeed, the interaction between participants can be a key source of data, providing insights that would likely not come to light using tools such as one-on-one interviews. Again, as long as the moderator is able to maintain a comfortable, balanced ambience for the meeting, focus groups offer particular benefits for research

---

[1] This term, used in the social sciences but also in human geography, describes "a technique used by researchers whereby one contact, or participant, is used to help recruit another, which in turn puts the researcher in contact with another" (Clifford et al., 2010, p. 535.). The number of participants increases rapidly or forms a "snowball."

on topics such as environmental processes and management, where individual interests are often strongly contested.

In practice, focus groups ideally require two researchers: a moderator who manages the conversation by posing questions, providing the time and opportunity for each participant to contribute, and re-orienting discussion when needed, and an assistant whose role is to take notes, record contributions and observe the interactions between participants (Morgan & Hoffman, 2018).

Participant observation represents another core element in the qualitative methods toolbox. By observing and/or participating in activities, the researcher is able to witness and experience life as it unfolds in a natural social setting (Kawulich, 2005). This method is particularly useful when the researcher wants to set aside their own perspective and "stand in the shoes of others," as cogently expressed in the title of the paper by Savage (2000). It implies a strong commitment from the researcher because it involves living temporarily in another setting with unknown people and participating in activities that might not be those of choice (Kitchin & Tate, 2013). The process essentially consists of the researcher observing the context, behavior, activities and interactions of those with whom they are living, or spending time, and recording their impressions, as these relate to the aims of the study. Researchers can also record nonverbal expressions of feelings and watch how participants communicate, including the often important subtleties of word choice (Schmuck, 1997). For example, in a study of environmentally led migration, an approach based on participant observation might typically involve the researcher living, for a period, in a village with a high rate of out-migration. By participating in the daily, weekly or seasonal round of village activities, a researcher is directly exposed to the mix of factors that trigger migration decisions, is able to record movements to and from the village in detail, and their reasons (Chapman, 1975) and observe the impacts of out-migration first-hand.

Ethnography involves long-term, systematic observation of a community or society in which the researcher shares the daily lives of people,

learning about their way of life through observation and interaction, often using a variety of qualitative tools including discussion and formal or informal interviews (Buscatto, 2018). Overtly or covertly, the researcher follows and observes people in one or more of their varied settings. Through direct observation, they aim to derive a comprehensive picture of societal practices and how they are socially produced. The approach therefore contrasts with survey questionnaires in which respondents are required to describe their own practices; at the same time, ethnography generally involves a broader ambit of activity, and wider focus of interest, than is involved in simple participant observation. This very time-consuming method requires a significant personal commitment by the ethnographer in the field, both intellectually and personally. Being in situ, the researcher has access to the foundations underlying societal practices and can observe these over time, including those that guide attitudes and responses to environmental issues (Kitchin & Tate, 2013). Environment itself, however, is rarely the sole or even primary focus of ethnographic studies.

Drawing and other visual tools such as films, photographs, maps, and painting may also be used in qualitative analysis, and have considerable application in environmental research (Eberle, 2018). Visual aids assist people to share their opinions and ideas (Henwood et al., 2018). For some, it is easier to draw than to talk. For example, asking a respondent to draw a map of their environment reveals how they perceive space and their surroundings, and can serve to enhance interaction between participants in a focus group or similar setting (Safra de Campos et al., 2017). Maps or drawings may help respondents to represent the layout of a local neighborhood, identify key features of their environment, and draw attention to perceived environmental risks. The sketch they provide may be explicit and detailed or barely formed in their consciousness, but in either case the features that they choose to represent, the order in which they are drawn, and their juxtaposition provide valuable insights into the way they perceive and experience the world (Richardson, 1981).

Table 7.1 summarizes the main characteristics of the methods presented here. Although the overview is necessarily brief, additional detail can be found in *SAGE Handbook of Qualitative Data Collection* (Flick, 2018).

Irrespective of the technique used to gather the data, the subsequent task for the researcher is to identify patterns, commonalities, and exceptions from the text or other information that has been assembled. Transcription may be needed and some form of **coding** system is generally implemented to help to meaningfully categorize information. With qualitative data, this often takes the form of an inductive process, starting with a small number of codes representing broad classes of information, then adding categories to proceed from the general to the specific. The ultimate aim is to identify evidence, connections, associations or links between ideas and opinions and hence progressively refine understanding of the available data (Kitchin & Tate, 2013). Computer software may prove helpful in analysis, but the real value of qualitative data often lies in the richness of detail that comes, not from coded data, but from respondent descriptions, impressions, and perceptions.

Content analysis is useful not only for analyzing in-depth interviews and focus group conversations, but also for examining secondary data sources, such as diaries, letters, (auto)biographies, or official documents (Kitchin & Tate, 2013). Coding is generally applied to textual information, but it can also be used with drawings. For example, key elements such as roads, river, mountains, and coasts can be coded using the same principles as with textual elements, noting their prominence, frequency, or juxtaposition, for example. The researcher interprets the drawing in the same way as for a sentence, a text, a book, or a face-to-face interview. Step by step, the aim is to build a conceptual framework that helps to answer a research question.

When addressing the population and environment nexus qualitatively, the researcher faces two major concerns: (i) how to deal with subjectivity and (ii) how to capture and represent the perceptions of the population that has been studied.

Subjectivity is central to any qualitative approach and for scientists from a positivist tradition, it is seen as a bias that should be controlled or eliminated (Maxwell, 2018). Qualitative data collection explicitly acknowledges the relationship between the researcher and the participants. A researcher's personal beliefs and motives may thus influence their conclusions and hence their validity (Maxwell, 2018). This is especially significant in the case of population–environment research because studies often involve communities that are culturally, socio-economically, and politically distant from the researchers. The realities faced by these communities are remote from the problematic that is of interest to the researchers. Members of such communities have their own priorities and ways of seeing the population–environment questions under study. For example, people affected by Typhoon Haiyan, which buffeted the Philippines in 2013, do not necessarily relate the typhoon to climate change (Henriet et al., 2021). Some consider the event as belonging to the past and are not concerned with the risk it might foreshadow in terms of future typhoon occurrences and strengths.

These differences in perception and focus can lead to a 'dialogue of the deaf' and the collection of inadequate qualitative data (Cloke et al., 2004). Awareness of diverging ideological backgrounds, interests, and ways of seeing the world are needed to guide selection of the most appropriate methods of qualitative data collection, and associated questions, with particular attention to the concepts used (e.g., 'climate change'). Reflective notes made during the research process can be helpful in assessing how the researcher's own identity (background, assumptions, feelings, and values) may have shaped interpretations and impacted research findings (Maxwell, 2018).

After controlling for subjectivity, the researcher's task is to understand the views of the respondent population by examining the various forms of evidence that have been assembled, including their discourses. According to the socio-nature thesis, the material existence of entities such as trees, birds, and rivers cannot be separated from the knowledge we have of them (Castree, 2014). This creates a challenge

Table 7.1 Main characteristics of the most common qualitative research methods

Method	Sensitive topics	Time-consuming	Interactions between respondents	Confidentiality	Researcher involvement	Type of information collected
In-depth interview	Very suitable	Medium	None	Very high	Low	In-depth information
Focus groups	Difficult	Low	Strong	Difficult/imposs	Low	Interactions between respondents
Participant observation	Possible	High	Strong	Possible	High	Practices and interaction between people
Ethnography	Possible	High	Limited	Possible	High	Practices and interaction between people
Drawing	Difficult	Low	Strong	Difficult/imposs	Medium	Representation and interaction between people

for the researcher in trying to interpret how the subject population sees their environment. For example, in his analysis of the lived reality of hazardous places, Mustafa (2005) demonstrates the importance of discourses in understanding how a community manage river floods. The hazardous location under study was the Lai Nullah Watershed in the Rawalpindi-Islamabad conurbation in Pakistan. The watershed is hazardous because its main river, Lai Nullah, is managed without consideration for the overall riverine regime and local perceptions of the river. In the local language, "Lai" is the generic term used to refer to rivers, whereas 'Nullah' is associated with a sewerage drain. A semantic shift occurred as 'Nullah' progressively replaced the original name 'Lai Nullah,' which resulted in locals treating the river like a sewerage drain. The material existence of the river is thus closely interrelated to the discourse, which in turn influences the lived reality of hazardous places.

In the same way, the term 'drought,' understood as a 'period of negative rainfall anomalies in respect to the mean,' differed from perceptions of drought among farmers and pastoralists in Northern Ethiopia, as observed by Meze-Hausken (2004) in a study contrasting meteoro-logical records with local perceptions. Indeed, although 2002 was seen by local people as the worst drought year in human memory, meteorological statistics would not classify it as a dry year. It was the combination of a heavy loss of animals and low agricultural yield in 2002, that underpinned the use of the term 'drought' by local people: a mix of social and physical impacts, rather than an assessment of rainfall.

As seen in Meze-Hausken's (2004) findings and as demonstrated by Mustafa (2005), studying discourses that describe people's local realities is an important source of knowledge in people–environment research. Yet, scholars who have adopted a qualitative approach and notions of discourse analysis insist that there is no extra-discursive immediate access to reality; all analysis needs an encounter with the discursive, given the impossibility of apprehending reality beyond language and representation (Howarth et al., 2000). By defining concepts and formulating interpretations, researchers themselves shape their own discourses that in turn may have important consequences for reality. What counts, then, for those who engage in this kind of approach is not necessarily the degree of correspondence to reality, but by whom and

7 Qualitative Data and Approaches to Population–Environment Inquiry

how discourse about people and the environment is produced, how it works, and what it does (Mels, 2009). Scientists, business, institutions, nongovernment organizations, the media and many others produce discourses that become received 'truths' because of social processes and power relations. These discourses (and counter-discourses) also need to be questioned and confronted in the study of human–environment systems and the ways they are connected.

Qualitative Research Examples

This section presents three examples of population–environment research, conducted by the authors at various times and in diverse settings, which illustrate core facets of qualitative investigation. The first example addresses the intersection between scientific and local knowledge, the second explores individuals' experiences within their environment, and the third illustrates the participatory action research process. All three detail how the researchers selected and used qualitative methodologies to answer their research questions. Diverse methodologies are represented, including interviews, participatory mapping, participant observation, and observant participation, as well as photo-language and the use of a game. In each case, we also describe the process of interpretation.

Example 1. Scientific Versus Local Knowledge

Delineating the Research Question Climate change, viewed from the perspective of the Intergovernmental Panel on Climate Change (IPCC), is understood almost exclusively by way of positivist, quantitative science. As a result, it is often represented in terms that are wholly disconnected from cultural understandings of climate (Hulme, 2008). The multidimensional nature of climate, and climate change, therefore, require constant reinterpretation when applied in different contexts. Although recent scientific

literature has tended to support a binary division between scientific and local knowledge (Raymond et al., 2010; Gaillard & Mercer, 2013; Kettle et al., 2014), we consider experiential knowledge as a hybrid notion involving both the local and scientific domains. Derived from individual life experiences, experiential knowledge is defined here as the ways in which 'a variety of publics make sense of climate change, as witnessed and responded to in ordinary, everyday-life scenarios' (Geoghegan & Leyson, 2012, p. 289).

Against this background, this case study draws attention to the way in which social actors engage with the idea of climate change and adaptation (i.e., the process of adjustment to actual or expected climate and its effects – de Coninck et al., 2018). The research is set in coastal municipalities in the province of Bohol, Philippines, areas that are particularly vulnerable to the adverse impacts of climate change (Capili et al., 2005; Rincón & Virtucio, 2008). This vulnerability poses serious challenges for sustainable land use and management and, as such, this case study focuses on planners. Although national strategies that mainstream adaptation into local development planning are often implemented by scientifically derived, top–down approaches, the integration of adaptation into local practices requires reinterpretation by planners in ways that align with local cultures. To inform this work, this case study addressed the following research questions:

- How do planners engage with the idea of climate change?
- What types of knowledge are involved in these accounts and what is their spatial ordering?
- How is experiential knowledge applied to action within planners' practices?

Who and How? The Philippines coastal municipalities offer an ideal setting in which to examine the experiences of local government planners, especially municipal planning and development coordinators (MPDC). Their knowledge of the interaction between the biophysical and social contexts was important in understanding local processes, including efforts to balance competing priorities and values in policymaking and resource management (Berkes & Folke, 2002; Picketts et al., 2012). In addition, their position within the planning system potentially provides a bridge between top–down and bottom–up actions. In this sense, planners were considered to have a key role in bringing together local and scientific understandings of climate change.

To investigate planners' discourses and experiences with climate change, we drew on a number of resources, including in-depth face-to-face interviews conducted between June and September 2013 with 29 municipal planning and development coordinators from the 30 coastal municipalities of Bohol (one informant declined the interview invitation). Most (21) were male, with experience as head of the office ranging from 2 months to 32 years (median of 20 years). All interviews took place onsite at Municipality Halls and lasted between 30 and 75 min. Interviews focused on perceptions of local climate change and associated risks, as well as the role of local development planning in addressing climate adaptation. A list of key questions allowed for in-depth exploration of multiple ways of understanding and engaging with climate change. For instance, respondents were asked questions such as: do you perceive that climate change is happening in your municipality?; What can your office do to address climate change?; Do you have specific plans or programs that address climate change?; What role do such plans and programs play in addressing climate change?

Interviews also made use of a simplified map of the Municipality (Box 7.1) displaying the names and boundaries of *barangays* (the smallest political administrative divisions), river and road networks, and topographical contours. Interviewees were asked to pinpoint the "critical places" that were discussed as they explained the impacts of climate change observed within their municipality. The main benefit of these mental maps was the provision of place-specific descriptions of climate-related risks and the characterization of planners' territorial representations of both current and desirable adaptive measures. Finally, a set of optional sub-questions addressed key recurrent themes such as flood management, relocation, and land use conflicts. The 29 interviews proceeded until "thematic saturation" was reached, that is, when no new information or themes were forthcoming from interviewee responses (Baxter & Eyles, 1997).

Box 7.1: Conducting Face-to-Face Interviews with the Support of Mental Maps

Mainly used in behavioral geography for the study of environmental perceptions (Lynch, 1960; Gould & White, 2012), mental maps (or sketch maps) have been widely used in the study of "spatial cognitive representation" (Freundschuh, 1992, p. 288). Such maps allow for respondent representation and social construction of places emphasizing social factors such as gender or age. As Signorino and Beck (2014) described, mental maps can be collected using various methods, including:

1. **White sheet.** Respondents are asked to draw their environment (such as the district in which they live) on a blank sheet of paper. They are free to choose the scale, number of elements they represent, and so on.
2. **Simplified or detailed map.** Respondents are presented with a map and asked to locate a specific area that relates to the interview topic.
3. **Spatial reconstruction game.** In this method, respondents are asked to build, with wooden pieces and strings for instance, a three-dimensional map of their environment (Ramadier & Bronner, 2006).

Irrespective of the approach, researchers can then connect the mental maps with interview transcripts in order to take into account the ways in which space influences environmental perceptions.

Analysis: The Artisan Researcher In order to analyze interview transcripts, we embraced what Cloke et al. (2004, p. 312) describe as the 'artisan researcher' approach. Through a detailed and close reading of interview transcripts, this strategy provides a means of understanding the planner's worldview, as revealed in their everyday experiences, encounters, and utterances. It offers a means of standing in the shoes of the informant and seeing the world through their eyes. Although various methods can be used to gain empathetic understanding, the artisanal approach draws upon the best known and most widely applied technique (not least in human geography) of 'grounded theory', which seeks to develop concepts ('theory') that have their roots – and thus are 'grounded' – in the concepts voiced by informants (Charmaz, 2006). The artisanal approach further embraces a set of procedures that offer clear guidelines for practice, without being overly rigid or prescriptive. We implemented the procedure through a coding scheme in two steps.

Step 1 The first step involved selecting and grouping interview extracts of a couple of sentences into qualitative themes (thematic coding). Throughout the analysis, both manifest (content) and latent (discourses) data were considered. Manifest data included specific terms and themes named by the respondent such as erratic weather, typhoon, or Municipal Planning and Development plan. Latent data referred to the underlying meaning in respondent discourses such as the fear of flood events or distrust of national politics. As Miller and Fox (1997) suggest, a combination of 'top–down' and 'bottom–up' approaches is used to build analytical bridges that can be mutually informative. With the assistance of qualitative data analysis software, three overarching thematic codes ('Observations of climate change'; 'Perceived impacts of climate change'; 'Municipal responses to climate change') were derived directly from the research questions (top–down). At the same time, sub-themes emerged as the analysis progressed (bottom–up). Within the first main theme, for instance, categories such as 'Stronger *habagats* [monsoon winds],' 'Heavy rains,' 'Erratic weather,' were created. These sub-themes were then grouped under new codes 'Erratic weather pattern,' 'Sea level rise,' and 'Unexpected extreme events' as a means of more accurately classifying the initial sub-themes.

Step 2 The second step involved selective coding. By selecting relevant themes created during the first step, we built interpretive, analytical constructs such as 'Types of experiential knowledge,' 'Lived experience of climate change,' and 'Spatial ordering of climate risk knowledge.' Thematic and theoretical constructs therefore evolved in parallel, instead of being pre-determined and represented as 'factual' structures. Thus, codes were formulated as "signposts that support the identification of relevant text passages and help to make them available for further interpretation and analysis" (Seidel, 1995, p. 484). In this way, our understanding of planning officers' discourses evolved during the analytical process. In essence, the textual coding was not mechanical, but instead required creative input from the researcher.

Results: Planners' Experiential Ways of Knowing About Climate Change Planning officers considered climate change simultaneously as (1) a reality they observe on a daily basis, (2) a problem they should address through their planning activities, and (3) an agenda affecting their tools and practices (Table 7.2). This diversity of answers betrays what Brace and Geoghegan (2010, p. 2) describe as the 'definitional ambiguity' of climate change, but also shows that planners relate to climate change by mobilizing hybrid forms of knowledge involving various sets of experiences. The experiential way of knowing about climate change provides them with a grounded sense of climate's material effects on their responsibilities. They also understand climate change as an agenda for integrating adaptation into local development plans. In practice, MPDCs rely heavily on the integration of local knowledge with geo-hazards maps developed by national government agencies. These maps contain a form of knowledge generated by technical experts at the national level and are utilized by local planning officers at the municipal level for determining hazard-prone areas. The maps

Table 7.2 Summary of planners' experiential ways of knowing about climate change within the coastal municipalities of Bohol (Dujardin et al., 2018)

Experiential way of knowing about climate change	Main evidence provided by planning officers
Reality	The weather has become erratic (unseasonal rains, high-heat days)
	Municipalities experience abnormal tides (coastal and estuarine areas)
	Unexpected extreme events are increasing (heavier rainfalls, stronger *habagat* [monsoon] winds, bigger waves or storm surges)
Problem	Greater disaster risks resulting from flood, landslides, sea level rise, droughts, coastal erosion, storms, and typhoons
	Negative impacts on livelihoods (e.g., lower yields and catches for farmers and fishers respectively)
	Land use change (e.g., storm surges and sea salt intrusions lead to the conversion of rice fields into mangrove plantations)
Agenda	Climate change has become part of planners' mandate and duties since the *2009 Climate Change Act*
	Recent attendance to training programs led by the Housing and Land Use Regulatory Board in Cebu
	Current planning activities focused on the integration of CCA into the CLUP via the DRRM plan and geo-hazard maps

CLUP comprehensive land use plan, *DRRM* disaster risk reduction and management, *CCA* climate change adaptation

therefore carry climate-related knowledge that flows smoothly between the central production of scientific knowledge and the periphery where it is combined with experiential knowledge to guide plan implementation and adaptation.

Applying Experiential Knowledge to Action
The three main experiential ways in which planners know about climate change challenge the idea that 'scientific facts build the appropriate foundation for knowing how to act in the world' (Sarewitz, 2004, p. 385). It follows that scientific and local forms of knowledge should not be viewed as mutually exclusive for addressing climate-related risks. Instead, they serve complementary roles. To illustrate, one respondent recalled his approach to alerting people in his community to impending disaster:

> During signal number 2, about 36 hours before the typhoon arrives, I will start contacting all the barangay officials through our communication systems and instruct them to cut all the leaves of the coconut palms within the vicinity of houses [. . .] [As] the tree can fall down because of the wind, if you cut those palms you lessen the stress on the branches. [People] have been doing that as early as . . . A long time ago. It's been tested already. So it's just our way of informing them. This mechanism system is a way of mitigating. Simplest way. (MPDC officer, male, 20 years of work experience, 2013)

This narrative involves both scientific and local ways of recognizing and responding to storm hazard. The planning officer responds to the early warning information (signal number 2, which specifies potential light to moderate damage from the wind) coming from the national scientific weather agency (PAGASA) by invoking a vernacular disaster risk reduction practice. In this case, the planner's experiential knowledge shows that scientific and nonscientific information are both valid and function in a complementary manner in reducing exposure to weather variability, preventing typhoon damage, and avoiding casualties.

The Merits of a Constructivist Qualitative Approach This example demonstrates how qualitative research can highlight the ways in which individuals' social context shapes their reality. It helped to reveal important links between climate and society. The range of interacting factors identified by planning officers revealed that understandings of climate risks are complex and multi-dimensional. As Yager (2015, p. 146) argued: "insights into the causality and consequences of local and global environmental change are complicated by the fact that climate change is being accompanied by rapid societal changes." Indeed, during face-to-face interviews, the extent

7 Qualitative Data and Approaches to Population–Environment Inquiry

to which climatic or social changes were entangled became abundantly clear. Our interpretive discourse analysis enabled a better understanding of the many ways in which individuals perceive socio-climatic change. Although planners' observations of weather variability and sea level change may differ from those derived from climate science, planning officers saw these observations as a way of both conveying and acting upon climate change. This is not to say that climate change should not be measured quantitatively, but rather to emphasize that climate change must also be examined qualitatively to reveal its multiple understandings across both the scientific and local domains. Such an understanding is essential for the development of locally informed adaptation policies and programs.

Example 2. Individual Experiences in Their Environment

Delineating the Research Question Because environmental change adversely affects living conditions (IOM, 2008), it influences the migration decisions of households and individuals (Black et al., 2011). Environmental strain can be sudden or progressive. Sudden-onset events include natural disasters such as earthquakes, tsunamis, floods, volcanoes, storms, and hurricanes. Slow-onset events include longer-term shifts in environmental and climatic conditions such as progressive increases in the incidence or intensity of drought (Roncoli et al., 2001).

In their conception of research designs for establishing the link between migration and environmental strain, some scholars have tried to capture the changing environment by using rainfall variability indicators measured at synoptic stations (Henry et al., 2003; De Longueville et al., 2019). Others have preferred to take into account households' perceptions of these environmental modifications (De Longueville et al., 2020).

The research project reviewed here adopts a distinctive perspective by considering the emotions related to the living environment. It is based on the residential choice literature where emo-

tions are known to shape the evaluative phase of the decision-making process, alongside cognitive aspects, representations, and perceptions (Griffond-Boitier et al., 2016). Some research examining environmental strain and migration evokes the possible role played by emotions. Naik (2009) suggests that a possible reason for people *not* to move following a natural hazard is that fear and terror eventually diminish, increasing the psychological ability to stay, whereas migrant dissatisfaction with their new place may also be reflected in their emotional state (IOM, 2009). Similarly, a majority of respondents in migration research conducted in Funafuti, the capital of Tuvalu, gave "lifestyle" reasons, including the low-stress working environment and their enjoyment of the natural environment for their unwillingness to move (Mortreux & Barnett, 2009). Yet, the island is threatened by rising sea levels due to climate change, according to the country's prime minister (Mortimer, 2015).

The research reported here was conducted in a part of the Philippines that was devastated by typhoon Haiyan in 2013, and assesses the emotional experience of residents who decided to stay. The aim was to improve understanding of the role of the environment in decision-making by people who decided *not* to move, in the face of a natural disaster.

Who and How? We first conducted exploratory interviews in 2015 asking residents who did not relocate about their environmental experiences and perceptions. These few interviews revealed three main challenges to our qualitative approach, each with lessons for the prospective qualitative scholar. First, we had expected respondents to reveal information about their subjective experiences, especially emotions. In practice, responses were very pragmatic and rational, and did not meet our expectations of revealing deeper meaning. For example, one respondent said: *"It would be really nice if the hospital will be clean and nice if it's to be transferred back there. The current hospital is only for 5 years."* It seemed necessary to design a data collection setting that would help respondents to elaborate their answers, especially because the link between emotions and

Fig. 7.1 A game session: the household head selects emotions to associate with a picture of the household's environment (Philippines, July 2016, by E. Henriet)

the environment is not straightforward. A second difficulty was that the researchers, as white westerners, were seen as being affiliated to a nongovernmental organization that arrived after the typhoon to deliver material assistance. Most respondents seized the opportunity to highlight their needs and difficulties, which led to answers that strayed from the primary topics of interest. Moreover, people with whom we interacted seemed to be intimidated by the presence of a "white," "foreigner," and English speaker. We could feel a "power imbalance" (Cefaï, 2010) as people expected assistance from us and showed signs of high respect. For example, respondents would offer the best chair while themselves sitting on the ground. The interactional structure of a classical interview was therefore challenging and we felt the need to change the participation status of the various stakeholders (respondents, interviewers, and translator). Translation represented a third challenge. We had hired a local teacher to interpret during the interviews, although we had the sense that the translation was sometimes problematic, especially because we were trying to elicit nuanced responses conveying people's emotions.

Given the specificities of the aimed for data and of the research context, we returned a year later with another data collection strategy. This involved a game in which respondents associated images of their household's environment with particular emotions (Fig. 7.1). Collecting data with a tool as the game helped to foster the immersion of participants and improve the retrospective evaluations of emotional experiences (Box 7.2). That activity guided the respondent to provide answers indirectly instead of being asked direct questions, lowering obstacles to translation and better capturing the abstract nature of the study object. Data quality was also enhanced by improving connections between the participants. Indeed, the particular interactional structure of the game was expected to lower cultural barriers as well as the biases due to expectation or power imbalance. The game was also expected to create dialogue, encourage explanation, and therefore allow us to collect quantitative and qualitative data.

7 Qualitative Data and Approaches to Population–Environment Inquiry

Box 7.2: Developing a Game to Collect the Data

Games are usually employed in research in the form of role-playing whereby players are asked to perform specific actions with respect to real-world issues (Costanza et al., 2014; Hertzog et al., 2014). An example might be acting out the management of irrigated and cultivated land in the context of available water resources (Hertzog et al., 2014), where stakeholders could foresee the possible consequences of various scenarios; define a strategy for water management; and increase their awareness of each other's positions and strategies.

The objective with the use of the game described here was to enhance interaction between household members by reassigning leadership within the game and facilitate the emergence of data, rather than to simulate a specific issue. The emoticons were discussed between a husband/wife and daughter/son with the researcher in a less powerful position.

To connect the game to our research question, we integrated photo elicitation, a well-established technique (Oldrup & Carstensen, 2012; Glaw et al., 2017). Researchers in environmental psychology unanimously confirm that color photos are a valid proxy for on-site responses (Moser & Weiss, 2003).

Another advantage of using games for data collection is that it improves motivation (Henriet et al., 2021). It also facilitates the elaboration of answers, with consequent enhancement in data quality.

The game proceeded in a series of steps. First, respondents chose meaningful pictures from a series of images that represented local environments, knowing that they would have to associate them with emotions. In the set of photos offered to the players, some clearly represented areas destroyed or affected by the typhoon. The respondent was then asked to associate the images with emoticons (e.g., fear, happiness, anger, etc.), more than one of a particular type of emoticon indicating a stronger association. The third step of the game consisted of a partner of the respondent and the researchers trying to guess the picture by looking at the emotions the respondent associated with it. The third step was timed: the partner and the researchers had to hurry to choose the picture they thought was the right one. The "guessing" led the respondents to associate emotions with pictures in a way that has to be understood not only by themselves but also by their partner, leading them to make associations concerning their own lived environment. The method of distributing points in the game encouraged respondents to avoid emotion–picture associations that were too general (they gained fewer points if the researchers were able to guess), and encouraged them instead to make more personal associations. In some cases, unexpected responses triggered discussions that revealed useful qualitative insights; at other times, respondents were asked to explain the associations (Henriet et al., 2021).

Although the game did not entirely meet expectations in terms of revealing interactions between the various stakeholders, it did provide a rich dataset. The use of pictures and tokens that allow players to accumulate points to win the game, as well as being a fun activity framed by rules, motivating and thus immersive, encouraged respondents to engage. This type of game can be effective at launching discussion because it puts respondents at their ease (Safra de Campos et al., 2017). Many aspects of the respondent's experience with their environment were reflected in the associations between images and emotions including: for example, happiness linked to the beauty of a place, happiness from picking bananas or going to church, annoyance with traffic, fear of ghosts, and disgust related to dirtiness. In addition, the game also created quantitative data, with the number of emoticons indicating the intensity of emotions.

Interpretation Strategy The material collected through the game provided valuable insights into how a disaster shaped participants' judgements

Fig. 7.2 Data collection example: a picture of coconut trees was associated with three 'happy' emoticons (happy to be with the family) and one 'afraid' emoticon (afraid that the weather conditions are not good) (Philippines, July 2016, by E. Henriet)

about various places, and their emotions with respect to the activities commonly occurring in those places. Although the number of emoticons indicated the intensity of participants' emotions,[2] the explanations they provided added meaning and interpretative depth (Fig. 7.2).

Figure 7.3 illustrates how the quantitative data were combined with the qualitative material to reveal participants' emotions with respect to the environment, and to understand how the typhoon interfered with these emotions. Data related to particular emotions and a particular picture could be combined to reveal respondent reactions to their environment. A picture of the road, for example, was occasionally associated with excitement because it represented returning home, but some linked it with anger that the roads were poorly maintained. A picture of the school elicited happiness in relation to the pleasure of being at school, but also to the fact that students would be able to resume school once it is rebuilt.

First, we computed the mean number of emoticons by emotion for all the emotions–pictures associations that people related to the typhoon and compared these with the emotions–pictures associations that they did not relate to the typhoon. This provided a global picture of how the typhoon influenced people's affective judgements about their environment in general. Surprisingly, both positive and negative emotions were related to the typhoon. The explanations provided meaning to these counterintuitive emotions: on the one hand, respondents expressed sadness because buildings and local landscapes had been destroyed; on the other hand, they expressed happiness at the recovery. However, a comparison of emotion levels showed that the positive emotions related to the typhoon were less intense than those unrelated to the typhoon, whereas the converse was true for negative emotions.

Second, we compared emotion intensities between pictures. Although we expected the sea to yield negative emotions as the typhoon provoked a devastating storm surge, it was actually connected to positive emotions: fishermen were happy when they were fishing again and most respondents were excited to visit the beach during their leisure time. If fear was related to the sea it was mainly because of the everyday risk of

---

[2]There is a long tradition of measuring the intensity of emotions related to environmental setting in environmental psychology (Russell & Lanius, 1984). More recently, emoticons have been used next to lexical-anchored scales and prove to be valid indicators (Phan et al., 2019).

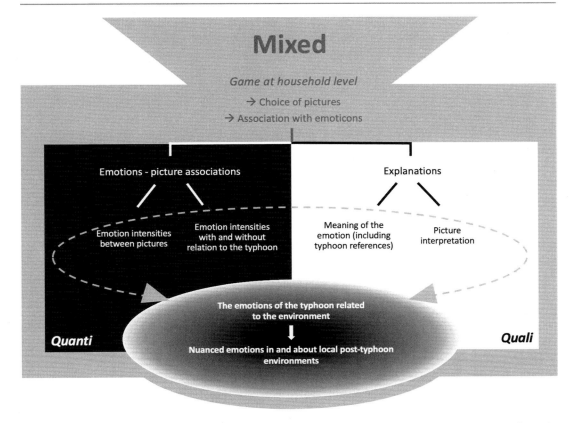

Fig. 7.3 Methodological scheme: the data collection setting is a game through which emoticons are associated with pictures. Explanations for the associations provide information about the meaning of the emotions (including typhoon references) and the interpretation of the picture by the respondent. The mean number of emoticons by emotion are computed following two categorizations: typhoon versus not typhoon and by picture. The emotions elicited by the typhoon, the meaning of the emotion and the picture interpretation reveal emotions with respect to the environment

drowning, or because of the creatures encountered while swimming. On the other hand, the picture of the historical church that had been destroyed by the typhoon elicited very high levels of fear, surprise, and sadness: the church that dated back centuries had been destroyed in a single night by the typhoon.

Mixed methods have proven to be an effective alternative to purely qualitative or quantitative research (see, for example, Onwuegbuzie & Leech, 2004, 2005; Tashakkori & Teddlie, 2009). In this exploratory research, we used quantification to generalize and to indicate the strength of qualitative observations. Conversely, qualitative information was used to reveal the meaning of patterns evident in the quantitative data. The mixed method approach, therefore, delivered a nuanced picture of emotions toward the environment for those affected, but also revealed how the typhoon shaped those emotions differently with respect to particular features of the environment.

Example 3. Participatory Action Research into Local Practices and Regional Dynamics of Seed Management

Delineating the Research Question Participatory approaches are recognized as an asset for fostering innovation in farming practices (MacMillan & Benton, 2014; Stirling, 2014; Hazard et al.,

2016; Biekart, 2017; Ortolani et al., 2017). Within this approach, there are various types of "participation" (Pimbert, 2011; Cuéllar-Padilla & Calle-Collado, 2011). In its more passive forms, the actors concerned are simply informed or consulted, whereas the researcher observes social processes unfold. In more active forms of participation, the researcher engages actively with the actors and encourages them to be a co-producer of scientific knowledge that serves to develop their farming practices.

Participatory action research (PAR) is aimed at empowering those observed as they engage in activities with the researcher. Moreover, the term "action research" reflects a desire on the part of researchers to consider experience or action as a source of knowledge and to assume a stance of engagement throughout participation (Bradbury-Huang, 2010; Morvan, 2013). In this case, the researcher has a support role based on the difficulties presented by the actors concerned as a starting point for action research (Morvan, 2013).

In this section, we present a research project anchored in political agroecology and adopting a participatory action research (PAR) approach. Guided by the principle of "science with people," agro-ecological approaches are based on the participation of farmers and other actors in the food system in transforming it toward sustainability. The main objective of the research was to understand how to breed seed varieties adapted to local socio-agro-ecosystems. As a case study, we focused on varieties of bread cereals in Andalucía (Spain) and Wallonia (Belgium).

Bread has always been an emblematic staple food. Yet, bread wheat (*Triticum aestivum L.*) cropping today is part of a large-scale and globalized bread-baking industry (Delcour et al., 2014) and relies exclusively on commercial varieties grown for high-input farming and mechanized processing. These technological advances have enabled substantial yield gains but have also led to a decrease in mineral content and thus nutritional value (Fan et al., 2008). Pure lines were gradually substituted in farmers' seed selection practices, resulting in a genetic and cultural erosion (Vara-Sánchez & Cuéllar-Padilla, 2013; Bonnin et al., 2014). This diversity loss reduces options for adapting to changing conditions and thus threatens the resilience of farming systems (Bueren & Myers, 2012). However, the cereal system is such that it prevents the development of varietal innovations, even when they exist (Vanloqueren & Baret, 2008). However, alternative pathways are emerging in Europe to develop more resilient and locally adapted cereal production systems. The research reported here focused on exploring the emergence of such alternative pathways. Following a PAR approach, it sought to describe and understand local practices and regional dynamics of in situ seed management, as well as to support these as more sustainable approaches to agriculture.

Who and How? The flowchart of the participatory action research (PAR) process varies among researchers, but typically involves three main steps:

1. *Reflection*: question the problem(s) and consider possible solutions to be explored.
2. *Research*: investigate said problem(s).
3. *Action*: develop and support the implementation of actions.

These steps are repeated, which, combined with reflexive analysis, creates an iterative process: each new cycle builds on the lessons of the previous cycle to improve the process and outcomes of PAR. In addition, within the same PAR, several iterations can take place simultaneously (McIntyre, 2008). Finally, the reflection stage can engage only researchers or research participants as well.[3]

As this process requires intention and facilitation, several authors draw attention to the importance of the exploratory phase of PAR (Faure et al., 2010; Ganuza et al., 2010; Guzmán-Casado

---

[3]This conceptualization of a PAR process should not suggest that PAR must necessarily correspond to a succession of predefined steps. PAR can be more or less formalized and seen as an emerging process (McIntyre, 2008), that is, "not something you always say at the beginning, but rather a progression that can be achieved with good intentions and committed actors" (Méndez et al., 2017).

et al., 2013; Méndez et al., 2017). This initial step affords a better understanding of the context being examined, which is essential for building trust, setting expectations, and refining research questions (Méndez et al., 2017). The qualitative approach is particularly relevant in this exploratory phase as it allows for the development of a detailed foundational understanding of the issue.

Building upon this exploratory phase, our research was conducted as follows:

1. A systemic **diagnosis**: exploring local practices and regional dynamics based on representations by stakeholders using alternative cereal systems that differed from commercial approaches.
2. The **co-construction of a collective of stakeholders in order to support local development**: support and analysis of the first cereal seed network in Wallonia.
3. The **exploration of an emergent question** from the PAR process in Wallonia: studying the interaction between on-farm seed management and wheat-cropping systems based on a reduction in seeding rates.

To serve these diverse purposes, we used a series of methodological tools: semi-structured interviews, observant participation and participant observation, bibliographic research, and agronomic trials on farms. Only the latter research component produced quantitative data, and is not discussed further here. Most of the other data produced by these tools is qualitative in nature, and these provide the focus below.

Semi-structured Interviews We conducted and analyzed 36 interviews, most of them as part of the systemic diagnosis represented in Step 1 above. In Andalusia, we interviewed 26 key stakeholders: 11 farmers, 4 bakers, 4 millers, 3 researchers, 3 representatives of associations, and 1 private breeder. We initially targeted interviewees according to the research themes[4] and used a stratified sampling strategy based on three categories: stakeholder types, seed system, and provinces. Local partners provided the first contacts and we then used the snowball approach[2] to expand the sample. We identified other stakeholders through a review of literature and of secondary sources, as well as through networking and interviews. Space and time constraints, difficulties in contacting some individuals, and unexpected opportunities influenced the final sample.

In view of our research objectives, the semi-structured interview was deemed the most appropriate approach. In this type of data collection, the researcher develops an interview guide with a series of themes and open-ended questions to explore a particular topic. The guide is used more as a frame of reference than as a questionnaire, leaving room for flexibility in order for themes to emerge spontaneously. For this reason, the researcher must demonstrate good listening skills (Harding, 2013). We developed a specific interview guide for each type of actor. All interviews were conducted face-to-face, except in Andalusia where three interviews were conducted by telephone given logistical constraints. Interviews ranged from 15 min to 3 h and were recorded. Details of the interview context were also described in a field notebook. Such details include the process of organizing the interview, details of the setting, interactions, and general impressions.

Participant Observation and Observant Participation In this research we used both participant observation (PO) and observant participation (OP). This distinction reflects a concern for methodological clarity as to the stance of the researcher and level of involvement (participation) in the field. The term "OP" is used preferentially when observation has taken precedence over participation, such as at general public conferences and scientific meetings, for example. This was

---

[4]Mainly, personal trajectory, information on cultivated varieties and seed management, perception of cereal and seed systems and their evolution, and priorities in action and research.

especially the case for the fieldwork in Andalusia, where the relatively short lengths of stay did not allow for full participation with the individuals or groups under examination. The term "PO" is used when involvement as a participant is at least as important as that of an observer, for example, when assisting with farm work in Belgium and Spain, or attending meetings and activities of the seed network.[5] In both cases, observations and impressions were systematically collected in a field notebook and, in some cases (e.g., conferences), the interactions were recorded, always with the agreement of those present.

Bibliographic Research and Document Analysis In addition to fieldwork, we conducted a review of the scientific literature to contextualize the research. We also used internal or public documents (the "gray literature") and audiovisual material (websites, e-mails, etc.) as secondary data sources. These tools can be useful in multiple ways. In our case, such material aided in the preparation of the interviews, providing background information on activities that were the subject of POs or OPs, or cross-referencing, confirming, or even invalidating the results produced by the other research tools.

Interpretation Strategy For the analysis of qualitative data, we relied mainly on the inductive thematic analysis method. Commonly used in qualitative research, this approach can involve data from interviews but also from OPs or secondary documents (Guest et al., 2012). Details are provided in Box 7.3, based on guidelines from Harding (2013). These steps, although presented in a linear fashion, are actually iterative in that they should be undertaken repeatedly throughout the research process. In our case, we repeated Steps 1–3 successively. In addition, the organization of the codes and the analysis (Steps 3 and 4) were sometimes carried out simultaneously, with reflections arising as the process unfolded.

Box 7.3: Main Steps of the Inductive Thematic Analysis (Based on Harding, 2013)

1. Transcribing records, which entailed converting recordings and/or observations to text.
2. Reading secondary documents (gray literature, etc.) and writing summaries.
3. Coding:
   (a) Identifying the initial categories, which correspond to the main research themes[6] covered (e.g., titles of the interview guide).
   (b) Code: i.e., assign one or more codes (words or sentences) to a portion of text.
4. Code organization:
   (a) Revise the list of codes based on certain criteria (e.g., redundancies between codes, number of sources).
   (b) Revise the list of categories and create sub-categories, based on the codes.
   (c) In each sub-category, reorganize the codes by grouping some, deleting some.
5. Analysis of themes emerging in each category.
   (a) Read the coded text carefully.
   (b) Examining similarities, differences and links across themes, as well as relationships with other categories.
   (c) After comparing the coding results with the original research questions and the literature, determine which themes should be fully analyzed for the purposes of this specific project.

The analysis can be performed by hand or by using **computer-assisted qualitative data analysis software**. The process includes storing, organizing, and locating data from different sources (text, audio, and video), transcribing audio material, facilitating coding and linking, creating

---

[5]Bringing together a plurality of actors, this network facilitates the exchange of seeds and knowledge (Baltazar, 2019).

[6] For this project, see footnote 4.

memos, and generating data representations. In the example provided here, we used the software NVivo, because this software was designed primarily for **inductive**[7] approaches and had been used by the team previously. Importantly, software does not replace the researcher's interpretative process but instead provides a set of tools that can increase efficiency. It is therefore essential to use these tools in a reflective way. For example, throughout our analytical process, we recorded different sets of codes, but also noted underlying detail on the choices made and their justifications. Such detail is essential and should be recorded in methodological notes or "**memos**."

Main Results The objectives and methodology of this PAR emerged from issues expressed by the cereal seed networks with whom we were collaborating. Despite its exploratory nature, two primary results emerged from this work. First, the systemic diagnosis shed light on the diversity of practices and motivations of stakeholders, as well as allowing for identification of the obstacles they face in the development of varietal alternatives. To overcome these obstacles, networking has been viewed as the key during this PAR process (Baltazar, 2019). Second, this research accompanied, through observant participation, the creation of the first cereal seed network in Wallonia, the Li Mestère network (Baltazar et al., 2018). Bringing together a plurality of stakeholders, this network facilitates the exchange of seeds and knowledge and supports the development of an artisanal cereal system in Wallonia. As researchers, our reflexive approach to examining this process highlighted the potential and challenges of such a network. Finally, we explored the interaction between on-farm management of seed- and wheat-cropping systems based on a reduction in seeding rates, an issue that emerged during the participatory action research process. A first qualitative characterization of these different systems, as experienced by farmers in Northern Europe, was provided by qualitative interviews. Agronomic trials confirmed the value of combining genetically diversified varieties and low seeding rates to facilitate participatory breeding, while ensuring correct yields under variable conditions. They also highlighted the challenges of on-farm experimentation for both farmers and researchers.

Participation and Qualitative Approaches for Transdisciplinary Research The research presented above was characterized by its focus on action and participation. We aimed not only to understand the socio-ecological process, but also to test possible solutions such as collaborating in the creation of a cereal seed network. In accordance with the iterative nature of PAR, the succession of several cycles of reflection, research, and action made it possible to gradually deepen our understanding of seed management and wheat-cropping systems. Unlike linear research approaches, the recursive links between theoretical background and field research allowed for iterative testing of research questions, interpretations, and applications.

In this case, the qualitative approach proved particularly appropriate for tackling unexplored issues such as the interaction between on-farm management of seed and wheat cropping systems based on a reduction in seeding rates. Besides, we unraveled the bundle of relationships linking the biotechnical and social dimensions of agricultural and environmental issues. In our opinion, this requires an interdisciplinary or even transdisciplinary approach, in which the qualitative perspective plays an important role. **Transdisciplinarity**, that is, the valorization and integration of different forms of knowledge, plays an essential role in the democratization of research processes and supports local action and social transformation (Sevilla Guzmán, 2011), two components of the change in the food production systems toward greater sustainability. In the PAR process presented above, we practiced transdisciplinarity in two ways. First, we combined theoretical and methodological frameworks from different scientific disciplines in both the human and natural sciences (agronomy, ecology, sociology, geography) to address different dimen-

---

[7] See section "Overview of qualitative methods and concerns".

sions of the issue. Second, nonscientific knowledge was considered on an equal footing with scientific knowledge. To understand how to select varieties of bread cereals adapted to local socio-agroecosystems, we found it necessary to start from the perceptions and knowledge of the stakeholders involved; thus, we chose interviews and participant observation as our main tools. At the same time, the opening of a safe learning space such as the Li Mestère network allowed for the circulation of different forms of knowledge, which nourished the research, but also facilitated individual and collective action.

Despite its manifest advantages, transdisciplinarity poses practical challenges, including the difficulty of integrating insights from the social and natural sciences given their different epistemologies, and the time discrepancy between field (i.e., the farming season) and academic (i.e., the PhD duration) research. (Darnhofer et al., 2012). Within this approach, the researcher/facilitator must be flexible and comfortable with a diversity of ideas, surprises, and the unusual (Pimbert, 2011). Nevertheless, such collaborative approaches offer great potential as new insights and novel forms of knowledge are directly produced (and adopted) by the actors involved.

Examining the Validity of Qualitative Methods

As explained in the three examples, qualitative approaches explicitly engage the subjectivity of the researcher. In this context, there is no consensus on methods to measure validity and even the use of validity as a criterion is subject to debate (Creswell, 2009). Nevertheless, some authors recommend a set of techniques aimed at improving the validity of qualitative research. These include (Harding, 2013; Vanwindekens, 2014):

- The sustained commitment of the researcher in the field before, during, and after initial

data collection in order to acquire an acute understanding of the process under study.
- The triangulation of data whereby the researcher collects information from various sources and perspectives in order to determine whether results from the analysis converge, differ, or combine.
- Returning to the original data, for example, by re-reading transcription notes and comparing these against the results.
- Crafting a full, detailed, and accurate description of the study context and results in order to evaluate their transferability to other settings.[8]
- Connecting with participants to check the accuracy of data and results.
- Discussing methods and results with peers at workshops, conferences, and in other professional venues.

Researchers can also improve the validity and utility of qualitative methods by acknowledging and applying the concept of reflexivity in their research practices. Reflexivity requires researchers to undertake self-examination regarding their role in data collection and the production of results (Harding, 2013). For this purpose, notes documenting all decisions taken during the observation, participation, and/or analysis processes are crucial. Putting reflexivity into practice thus helps to improve transparency in the research process. More often used in social sciences and humanities, reflexivity is not widely practiced in the natural sciences. Yet, it becomes especially important when using qualitative methods, especially participatory research, where a change in the stance

---

[8]Note that qualitative approaches are often considered as "case specific." The researcher's responsibility is then to discuss the extent to which any place-specific finding is transferable to other places in the world. The main advantage of qualitative approaches lies in their ability to provide an in depth understanding of a particular issue, highlighting the multiplicity of interplaying factors. The risks lie in over-generalizing findings, which are better placed within a larger framework defined by quantitative approaches.

adopted by the researcher is required. As Pimbert (2011) states, "this form of co-operative inquiry and participatory knowledge creation implies a significant reversal from the dominant roles, locations and ways of knowing."

Conclusions

The benefits of qualitative approaches for population and environment research include (i) the ability to capture the complexity and multiplicity of dynamics between population and environment as they unfold in individuals' perceptions, emotions, and realities; (ii) explicit consideration of subjectivity; (iii) the ability to confront different perspectives and record the product of the interaction between participants; and (iv) the ability to address complex issues in depth, such as during one-to-one interviews.

Using three research examples, this chapter has highlighted the contribution of qualitative methods to understanding the population–environment nexus. Although the subdiscipline has been largely dominated by quantitative approaches (Henry et al., 2003; Gray & Mueller, 2012; Gray & Bilsborrow, 2013; De Longueville et al., 2019), interesting and valuable insights can be garnered from qualitative inquiry. First, they can help to prepare quantitative surveys by developing unusual hypotheses suggested by local people, by using a classification relevant for and/or suggested by local communities to characterize the variables used as explanatory factors, and by focusing attention on historical and contextual details in order to complete missing information. Participatory inquiry can help to capture the attitudes, beliefs, and perceptions among the target population. Qualitative tools may also shed light on statistical associations by exploring with actors the processes that underlie the apparent connections, thus aiding understanding, or at least reducing misinterpretation of the results. In the case of sensitive issues, a qualitative approach is often better suited than direct questions. By providing an in-depth understanding of the population–environment nexus, qualitative approaches may also be preferred.

The main characteristics of the qualitative approach are summarized in Box 7.4. Data collected using qualitative methods are, of course, not necessarily representative of the total population, as can be obtained with a statistical sample. The number of respondents is often small, as the approach is time-consuming: in-depth questioning simply requires more interviewer time. The role of the interviewer also takes on added importance; thus, their quality and experience are key to the success of the approach.

In practice, coupling quantitative and qualitative approaches potentially offer the most powerful method of data collection and analysis. Qualitative approaches show great promise in addressing the limitations of quantitative approaches, and vice versa. We therefore argue in favor of addressing a research issue through multiple approaches, both as a means of enhancing understanding, advancing theory, and enabling practice. Greater attention to the use of qualitative approaches takes on particular importance in the context of population–environment relations, because such approaches are especially suited to understanding the dynamics, perceptions, and issues faced by individuals, families, and communities at the local level. Qualitative methods, in their various forms, tease out nuances in the population–environment nexus that are overlooked, or even obscured, by regional, national, or global studies. The very act of carrying out such studies also serves a critical role, in and of itself, by engaging the very people and organizations at a grassroots level who are most directly affected by environmental change, and are ultimately the self-same people and organizations that have to confront environmental issues on the ground, in their daily lives.

> **Box 7.4: Main Characteristics of the Qualitative Approach (Harding, 2013; Creswell, 2013)**
>
> - Focus on participants' perceptions.
> - A holistic approach, aimed at taking into account the multiple perspectives and complexity of the situation as a whole.
> - An emerging design: research questions, data collection, and samples can evolve with field work.
> - Multiple data sources, collected primarily in the context of the participants, to which the researcher must be sensitive.
> - Particularity rather than generalization.
> - Reflexivity.

References

Baltazar, S. (2019). *Pratiques locales et dynamiques régionales de gestion in situ de la diversité cultivée des céréales panifiables: une recherche-action participative en Wallonie et en Andalousie*. Presses universitaires de Namur.

Baltazar, S., Visser, M., & Dendoncker, N. (2018). *Au-delà des idées reçues: L'exemple de Li Mestère, réseau de semences wallon*. Etudes rurales 2018/2.

Baxter, J., & Eyles, J. (1997). Evaluating qualitative research in social geography: Establishing 'rigour' in interview analysis. *Transactions of the Institute of British Geographers, 22*(4), 505–525.

Berkes, F., & Folke, C. (2002). Back to the future: Ecosystem dynamics and local knowledge. In *Panarchy: Understanding transformations in human and natural systems* (pp. 121–146). Island Press.

Biekart, K. (2017). Contributing to civic innovation through participatory action research. *European Public & Social Innovation Review, 2*, 34–44.

Black, R., Bennett, S. R. G., Thomas, S. M., & Beddington, J. R. (2011). Migration as adaptation. *Nature, 478*, 447–449. https://doi.org/10.1038/478477a

Bonnin, I., Bonneuil, C., Goffaux, R., et al. (2014). Explaining the decrease in the genetic diversity of wheat in France over the 20th century. *Agriculture, Ecosystems & Environment, 195*, 183–192. https://doi.org/10.1016/j.agee.2014.06.003

Brace, C., & Geoghegan, H. (2010). Human geographies of climate change: Landscape, temporality, and lay knowledges. *Progress in Human Geography, 35*, 284–302.

Bradbury-Huang, H. (2010). What is good action research?: Why the resurgent interest? *Action Research, 8*, 93–109. https://doi.org/10.1177/1476750310362435

Buscatto, M. (2018). Doing ethnography: Ways and reasons. In U. Flick (Ed.), *The SAGE handbook of qualitative data collection* (pp. 327–343). SAGE Reference.

Capili, E. B., Ibay, A. C. S., & Villarin, J. R. T. (2005). Climate change impacts and adaptation on Philippine coasts. In *Proceedings of the international oceans 2005 conference*. Washington, DC, pp. 1–8.

Castree, N. (2014). *Socializing nature: Theory, practice, and politics*. Blackwell Publishers.

Cefaï, D. (2010). *L'engagement ethnographique*. EHESS.

Chapman, M. (1975). Mobility in a non-literate society: Method and analysis for two Guadalcanal communities. In L. A. Kosiński & R. M. Prothero (Eds.), *People on the move: Studies on internal migration* (pp. 129–145). Methuen; distributed by Harper & Row, Barnes and Noble Import Division.

Charmaz, K. (2006). *Constructing grounded theory: A practical guide through qualitative research*. SAGE.

Clifford, N., French, S., & Valentine, G. (2010). *Key methods in geography*. SAGE.

Cloke, P., Cook, I., Crang, P., et al. (2004). *Practising human geography*. SAGE.

Costanza, R., Chichakly, K., Dale, V., et al. (2014). Simulation games that integrate research, entertainment, and learning around ecosystem services. *Ecosystem Services, 10*, 195–201. https://doi.org/10.1016/j.ecoser.2014.10.001

Creswell, J. W. (2009). *Research design: Qualitative, quantitative, and mixed methods approaches* (3rd ed.). Sage publications.

Creswell, J. W. (2013). *Research design: Qualitative, quantitative, and mixed methods approaches*. SAGE.

Cuéllar-Padilla, M., & Calle-Collado, Á. (2011). Can we find solutions with people? Participatory action research with small organic producers in Andalusia. *Journal of Rural Studies, 27*, 372–383. https://doi.org/10.1016/j.jrurstud.2011.08.004

Darnhofer, I., Gibbon, D., & Dedieu, B. (2012). Farming systems research: An approach to inquiry. In I. Darnhofer, D. Gibbon, & B. Dedieu (Eds.), *Farming systems research into the 21st century: The new dynamic* (pp. 3–31). Springer Netherlands.

De Coninck, H., Revi R, Babiker M, et al. (2018). Strengthening and implementing the global response. In V. Masson-Delmotte, P. Zhai, & Pörtner H.-O. (Eds.), *Global warming of 1.5°C. IPCC – The Intergovernmental Panel on Climate Change* (pp. 313–443). Cambridge University Press.

De Longueville, F., Zhu, Y., & Henry, S. (2019). Direct and indirect impacts of environmental factors on migration in Burkina Faso: Application of structural equation modelling. *Population and Environment, 40*, 456–479. https://doi.org/10.1007/s11111-019-00320-x

De Longueville, F., Ozer, P., Gemenne, F., et al. (2020). Comparing climate change perceptions and meteorological data in rural West Africa to improve the understanding of household decisions to migrate. *Cli-*

matic Change, 160, 123–141. https://doi.org/10.1007/s10584-020-02704-7

Delcour, A., Stappen, F. V., Gheysens, S., et al. (2014). Etat des lieux des flux céréaliers en Wallonie selon différentes filières d'utilisation. *Biotechnologie, Agronomie, Société et Environnement, 18*, 181–192.

DeLyser, D., Herbert, S., Aitken, S., et al. (2010). *The SAGE handbook of qualitative geography*. SAGE.

Dujardin, S., Hermesse, J., & Dendoncker, N. (2018). Making space for experiential knowledge in climate change adaptation? Insights from planning officers in Bohol, Philippines. *Journal of Disaster Risk Studies: Jàmbá, 10*, 1–10. https://doi.org/10.4102/jamba.v10i1.433

Eberle, T. S. (2018). Collecting images as data. In U. Flick (Ed.), *The SAGE handbook of qualitative data collection* (pp. 392–411). SAGE Reference.

Fan, M.-S., Zhao, F.-J., Fairweather-Tait, S. J., et al. (2008). Evidence of decreasing mineral density in wheat grain over the last 160 years. *Journal of Trace Elements in Medicine and Biology, 22*, 315–324. https://doi.org/10.1016/j.jtemb.2008.07.002

Faure, G., Gasselin, P., Triomphe, B., et al. (2010). *Innover avec les acteurs du monde rural: La recherche-action en partenariat*. Editions Quae.

Flick, U. (Ed.). (2018). *The SAGE handbook of qualitative data collection*. SAGE Reference.

Freundschuh, S. (1992). Is there a relationship between spatial cognition and environmental patterns? In A. U. Frank, I. Campari, & U. Formentini (Eds.), *Theories and methods of spatio-temporal reasoning in geographic space* (pp. 288–304). Springer.

Gaillard, J.-C., & Mercer, J. (2013). From knowledge to action bridging gaps in disaster risk reduction. *Progress in Human Geography, 37*, 93–114.

Ganuza E, Olivari L, Paño P, et al (2010) La Democracia en Acción: una visión desde las metodologías participativas.

Geoghegan, H., & Leyson, C. (2012). On climate change and cultural geography: Farming on the Lizard Peninsula, Cornwall, UK. *Climatic Change, 113*, 55–66.

Glaw, X., Inder, K., Kable, A., & Hazelton, M. (2017). Visual methodologies in qualitative research: Autophotography and photo elicitation applied to mental health research. *International Journal of Qualitative Methods, 16*, 1–8. https://doi.org/10.1177/1609406917748215

Gould, P., & White, R. (2012). *Mental maps*. Routledge.

Gray, C., & Bilsborrow, R. (2013). Environmental influences on human migration in rural Ecuador. *Demography, 50*, 1217–1241. https://doi.org/10.1007/s13524-012-0192-y

Gray, C. L., & Mueller, V. (2012). Natural disasters and population mobility in Bangladesh. *Proceedings of the National Academy of Sciences of the United States of America, 109*, 6000–6005. https://doi.org/10.1073/pnas.1115944109

Griffond-Boitier, A., Mariani-Rousset, S., Frankhauser, P., et al. (2016). The wheres and hows of residential choice. In P. Frankhauser & D. Ansel (Eds.), *Deciding where to live* (p. 332). Springer.

Guest, G., Namey, E. E., & Mitchell, M. L. (2012). *Collecting qualitative data: A field manual for applied research*. SAGE.

Guzmán-Casado, G. I., López, D., Román, L., & Alonso, A. M. (2013). Participatory action research in agroecology: Building local organic food networks in Spain. *Agroecology and Sustainable Food Systems, 37*, 127–146. https://doi.org/10.1080/10440046.2012.718997

Harding, J. (2013). *Qualitative data analysis from start to finish*. SAGE.

Hazard, L., Gauffreteau, A., & Borg, J. (2016). *L'innovation à l'épreuve d'un climat et d'un monde changeant rapidement: intérêt de la co-conception dans le domaine des semences*. Fourrages.

Hennink, M., Hutter, I., & Bailey, A. (2020). *Qualitative research methods*. SAGE.

Henriet, E., Burnay, N., Dalimier, J., et al. (2021). Challenges and opportunities of field-based data collection with a game. Analysis of the development and use of a game to collect data on people's emotional experience in their environment. *Bulletin of Sociological Methodology/Bulletin de Méthodologie Sociologique, 149*, 7–29. https://doi.org/10.1177/0759106320960885

Henry, S., Schoumaker, B., & Beauchemin, C. (2003). The impact of rainfall on the first out-migration: A multi-level event-history analysis in Burkina Faso. *Population and Environment, 25*, 423–460. https://doi.org/10.1023/B:POEN.0000036928.17696.e8

Henwood, K., Shirani, F., & Groves, C. (2018). Using photographs in interviews: When we lack the words to say what practice means. In U. Flick (Ed.), *The SAGE handbook of qualitative data collection* (pp. 599–614). SAGE Reference.

Hertzog, T., Poussin, J. C., Tangara, B., et al. (2014). A role playing game to address future water management issues in a large irrigated system: Experience from Mali. *Agricultural Water Management, 137*, 1–14. https://doi.org/10.1016/j.agwat.2014.02.003

Howarth, D. R., Howarth, D. R., Howarth, D. J., et al. (2000). *Discourse theory and political analysis: Identities, hegemonies and social change*. Manchester University Press.

Hulme, M. (2008). Geographical work at the boundaries of climate change. *Transactions of the Institute of British Geographers, 33*, 5–11.

IOM. (2008). *Expert seminar: Migration and the environment*. International Organization for Migration.

IOM. (2009). *Environment and climate change: Assessing the evidence*. International Organization for Migration.

Kawulich, B. B. (2005). Participant observation as a data collection method. *Forum Qualitative Sozialforschung/Forum: Qualitative Social Research, 6*, 43.

Kennedy, B. L., & Thornberg, R. (2018). Deduction, induction, and abduction. In U. Flick (Ed.), *The SAGE handbook of qualitative data collection* (pp. 49–64). SAGE Reference.

Kettle, N. P., Dow, K., Tuler, S., et al. (2014). Integrating scientific and local knowledge to inform risk-based management approaches for climate adaptation.

Climate Risk Management, 5, 17–31. https://doi.org/10.1016/j.crm.2014.07.001

Kitchin, R., & Tate, N. (2013). *Conducting research in human geography: Theory, methodology and practice.* Routledge.

Lynch, K. (1960). *The image of the city.* MIT Press.

MacMillan, T., & Benton, T. G. (2014). Agriculture: Engage farmers in research. *Nature, 509*, 25–27. https://doi.org/10.1038/509025a

Maxwell, J. A. (2018). Collecting qualitative data: A realist approach. In U. Flick (Ed.), *The SAGE handbook of qualitative data collection* (pp. 19–32). SAGE Reference.

McIntyre, A. (2008). *Participatory action research.* SAGE.

Mels, T. (2009). Analysing environmental discourses and representations. In *A companion to environmental geography.* Wiley Blackwell.

Méndez, V. E., Caswell, M., Gliessman, S., et al. (2017). Integrating agroecology and participatory action research (PAR): Lessons from Central America. *Sustainability, 9*, 705. https://doi.org/10.3390/su9050705

Meze-Hausken, E. (2004). Contrasting climate variability and meteorological drought with perceived drought and climate change in northern Ethiopia. *Climate Research, 27*, 19–31. https://doi.org/10.3354/cr027019

Miller, G., & Fox, K. J. (1997). Building bridges: The possibility of analytic dialogue between ethnography, conversation analysis and Foucault. In D. Silverman (Ed.), *Qualitative research: Theory, method and practice.* SAGE.

Morgan, D. L., & Hoffman, K. (2018). Focus groups. In U. Flick (Ed.), *The SAGE handbook of qualitative data collection* (pp. 250–263). SAGE Reference.

Mortimer, C. (2015). Tuvalu prime minister begs for help to stop his country disappearing off the face of the Earth. *The Independent.*

Mortreux, C., & Barnett, J. (2009). Climate change, migration and adaptation in Funafuti, Tuvalu. *Global Environmental Change, 19*, 105–112. https://doi.org/10.1016/j.gloenvcha.2008.09.006

Morvan, A. (2013). Recherche-action. In I. Casillo, R. Barbier, L. Blondiaux, et al. (Eds.), *Dictionnaire critique et interdisciplinaire de la participation.* GIS Démocratie et Participation.

Moser, G., & Weiss, K. (2003). *Espaces de vie.* Armand Col.

Mustafa, D. (2005). The production of an urban hazardscape in Pakistan: Modernity. *vulnerability, and the range of choice, 95*, 566–586.

Naik, A. (2009). Migration and natural disasters. In F. Lacsko & C. Aghazarm (Eds.), *Migration, environment and climate change: Assessing the evidence* (p. 441). International Organization for Migration.

Oldrup, H. H., & Carstensen, T. A. (2012). Producing geographical knowledge through visual methods. *Geografiska Annaler, Series B: Human Geography, 94*, 223–237.

Onwuegbuzie, A. J., & Leech, N. L. (2004). Enhancing the interpretation of "significant" findings: The role of mixed methods research. *The Qualitative Report, 9*, 770–792.

Onwuegbuzie, A., & Leech, N. (2005). On becoming a pragmatic researcher: The importance of combining quantitative and qualitative research methodologies. *International Journal of Social Research Methodology: Theory and Practice, 8*, 375–387. https://doi.org/10.1080/13645570500402447

Ortolani, L., Bocci, R., & Bàrberi, P. (2017). Changes in knowledge management strategies can support emerging innovative actors in organic agriculture: The case of participatory plant breeding in Europe. *Organic Farming, 3*, 20–33. https://doi.org/10.12924/of2017.03010020

Phan, W. M. J., Amrhein, R., Rounds, J., & Lewis, P. (2019). Contextualizing interest scales with emojis: Implications for measurement and validity. *Journal of Career Assessment, 27*, 114–133. https://doi.org/10.1177/1069072717748647

Picketts, I. M., Curry, J., & Rapaport, E. (2012). Community adaptation to climate change: Environmental planners' knowledge and experiences in British Columbia, Canada. *Journal of Environmental Policy & Planning, 14*(2), 119–137.

Pimbert, M. P. (2011). *Participatory research and on-farm management of agricultural biodiversity in Europe.* IIED.

Ramadier, T., & Bronner, A.-C. (2006). Knowledge of the environment and spatial cognition: JRS as a technique for improving comparisons between social groups. *Environment and Planning. B, Planning & Design, 33*(2), 285–299.

Raymond, C. M., Fazey, I., Reed, M. S., et al. (2010). Integrating local and scientific knowledge for environmental management. *Journal of Environmental Management, 91*. https://doi.org/10.1016/j.jenvman.2010.03.023

Richardson, M. (1981). Commentary on the "superorganic" in American cultural geography. *Annals of the Association of American Geographers, 71*, 284–287.

Rincón, M. F. G., & Virtucio, F. K. (2008). Climate change in the Philippines: A contribution to the country environmental analysis. In *Proceedings of the country environmental analysis consultative workshops.* Manila, Philippines.

Roncoli, C., Ingram, K., & Kirshen, P. (2001). The costs and risks of coping with drought: Livelihood impacts and farmers responses in Burkina Faso. *Climate Research, 19*, 119–132.

Roulston, K., & Choi, M. (2018). Qualitative interviews. In U. Flick (Ed.), *The SAGE handbook of qualitative data collection* (pp. 233–249). SAGE Reference.

Russell, J. A., & Lanius, U. F. (1984). Adaptation level and the affective appraisal of environments. *Journal of Environmental Psychology, 4*, 119–135. https://doi.org/10.1016/S0272-4944(84)80029-8

Safra de Campos, R., Bell, M., & Charles-Edwards, E. (2017). Collecting and analysing data on climate-related local mobility: The MISTIC toolkit: Collecting data on climate-related mobility: The MISTIC toolkit.

Population, Space and Place, 23, e2037. https://doi.org/10.1002/psp.2037

Sarewitz, D. (2004). How science makes environmental controversies worse. *Environmental Science & Policy, 7*, 385–403.

Savage, J. (2000). Participative observation: Standing in the shoes of others? *Qualitative Health Research, 10*, 324–339. https://doi.org/10.1177/104973200129118471

Schmuck, R. A. (1997). *Practical action research for change*. IRI/Skylight Training and Pub.

Schreier, M. (2018). Sampling and generalization. In U. Flick (Ed.), *The SAGE handbook of qualitative data collection* (pp. 84–98). SAGE Reference.

Seidel, J. (1995). Different functions of coding in the analysis of textual data' in U. In Kelle (Ed.), *Computer-aided qualitative data analysis: Theory, methods and practice*. SAGE.

Sevilla Guzmán, E. (2011). *Sobre los orígenes de la agroecología en el pensamiento marxista y libertario*. AGRUCO: Plural Editores: CDE, Centre for Development and Environment: JACS-Sud America, La Paz.

Signorino, G., & Beck, E. (2014). Risk perception survey in two high-risk areas. In M. Pierpaolo, T. Benedetto, & M. Marco (Eds.), *Human health in areas with industrial contamination* (pp. 232–245). World Health Organization – Europe.

Stirling, A. (2014). *Towards innovation democracy? Participation, responsibility and precaution in innovation governance*. Social Science Research Network.

Tashakkori, A., & Teddlie, C. (2009). *Foundations of mixed methods research: Integrating quantitative and qualitative approaches in the social and behavioral sciences*. SAGE.

Taylor, S. J., Bogdan, R., & DeVault, M. (2015). *Introduction to qualitative research methods: A guidebook and resource*. Wiley.

Van Bueren, E. T. L., & Myers, J. R. (2012). *Organic crop breeding*. Wiley.

Vanloqueren, G., & Baret, P. V. (2008). Why are ecological, low-input, multi-resistant wheat cultivars slow to develop commercially? A Belgian agricultural 'lock-in' case study. *Ecological Economics, 66*, 436–446. https://doi.org/10.1016/j.ecolecon.2007.10.007

Vanwindekens, F. (2014). *Les pratiques dans la gestion des systèmes socio-écologiques : développements méthodologiques & application à la gestion des prairies en région herbagère belge*. UCL – Université Catholique de Louvain.

Vara-Sánchez, I., & Cuéllar-Padilla, M. (2013). Biodiversidad cultivada: una cuestión de coevolución y transdisciplinariedad. *Revista Ecosistemas, 22*, 5–9.

Yager, K. (2015). Satellite imagery and community perceptions of climate change impacts and landscape change. In J. Barnes & M. Dove (Eds.), *Climate cultures: Anthropological perspectives on climate change* (p. 146). Yale University Press.

Part III

Migration & Environment

Building a Policy-Relevant Research Agenda on Environmental Migration in Africa

8

Valerie Mueller

Abstract

Periodic drought or heat stress during the main growing season, torrential flooding during harvesting periods, and soil erosion are just a few factors that have historically jeopardized household income. These sources of income variability incline households to adapt through informal channels, such as engaging in seasonal or permanent migration. Scarcity of wage employment opportunities, poor transportation infrastructure, and weak endowments in financial capital pose strong barriers to use of migration as an adaptation strategy in Africa. Within this unique context, I synthesize the main findings of the literature on environmental migration in Africa. Inconsistencies in migration definitions, the absence of evaluations of slow onset events, and limited evidence to support the mechanisms underlying migratory responses to environmental change are identified as major knowledge gaps. I discuss how new methods and data may be used to address

the above gaps, while informing the types of polices and interventions being considered by development agencies and governments.

Introduction

The aim of this chapter is to reveal the steps we, as scholars in the field of population and the environment, may take towards building a foundation for policy-relevant research on environmental migration in Africa. Migration in Africa is often motivated by household interest in accessing income auxiliary to subsistence farming (Lucas & Stark, 1985; Hoddinott, 1994), marital arrangements (Kudo, 2015; Hoddinott & Mekasha, 2020), and the desire to diversify household labor activities away from agriculture in order to minimize income risk (Grimard, 1997; Mazzucato, 2009; Dillon et al., 2011). Periodic drought or heat stress during the main growing season, torrential flooding during harvesting periods, and soil erosion are just a few factors that have historically jeopardized household income, inclining households to adapt through informal channels, such as engaging in seasonal or permanent migration. After synthesizing the main messages from

V. Mueller (✉)
School of Politics and Global Studies, Arizona State University and International Food Policy Research Institute, Tempe, AZ, USA
e-mail: vmuelle1@asu.edu

© Springer Nature Switzerland AG 2022
L. M. Hunter et al. (eds.), *International Handbook of Population and Environment*, International Handbooks of Population 10, https://doi.org/10.1007/978-3-030-76433-3_8

the limited research on Africa, I provide an in-depth discussion of how to expand the evidence base, focusing on ways to overcome existing challenges in data collection.

The composition of African labor markets render use of migration as a risk management strategy more challenging than in other contexts. First, opportunities for labor diversification are scarce. Studies corroborating labor diversification as a coping strategy focus on Indian rural villages, where agricultural and non-agricultural wage labor markets may be more likely to accommodate the excess labor supply (Kochar, 1999; Rose, 2001; Jayachandran, 2006). Arguably, there is considerable uncertainty as to the expected return of migration in response to risk in the African context given the potential lower demand for wage labor (Poelhekke, 2011; Mueller et al., 2020b). Moreover, strong sectoral linkages (e.g., food production and food trading) render all labor markets vulnerable to environmental risks further exacerbating the ability for would-be-migrants to secure jobs and cope (Mueller et al., 2020b).

Second, there are strong barriers to movement across locales in Africa, which can raise the costs of migration (Jedwab & Moradi, 2016; Jedwab et al., 2017). A qualitative survey of migrants in Tanzania illustrates the common obstacles faced by workers interested in searching for employment outside of their home location (Ingelaere et al., 2018). The price of a bus ticket, for example, can preclude migrants from traveling to distant cities where the probability of securing a rewarding job may be higher. High transportation costs may also prevent migrants from returning home when the move is considered unsuccessful. These constraints can make individuals, who might otherwise move as an adaptation strategy, reluctant to take the risk (Bryan et al., 2014).

The unique features of the labor market and direct and indirect mobility costs may explain more broadly the relatively low seasonal or permanent migration rates in Africa relative to the rest of the world (Lagakos et al., 2020; Mueller et al., 2020b). They further call for a revision in our expectations of how the environment will influence migration patterns, especially in Africa.

While some households or individuals may move as an adaptation strategy, others may aspire to move but are trapped in place (Black et al., 2011; Black & Collyer, 2014). Thus, previous pleas to precisely measure the absolute number of climate migrants may be less instrumental for guiding policy design in Africa (Gemmene, 2011). Rather, understanding the mechanisms underlying extant environmental migration patterns and identifying which subpopulations are capable of utilizing migration as a risk management strategy will be crucial to inform donors and governing agencies. Such analyses can provide insights on the types of interventions necessary to promote household resilience to specific environmental shocks, as well as improve the targeting of policies aimed at fostering the economic integration of environmental migrants in receiving areas.

Which Environmental Risks Are Pervasive in Africa?

In this section, I take stock of which natural hazards are most pervasive and in which locations of Africa. Such descriptive summaries can serve to highlight the discrepancies between the environmental stressors and geographic areas emphasized by policymakers versus researchers. After reviewing which events frequently occur in Africa, I provide a brief discussion of how these natural hazards may affect livelihoods sufficiently to encourage people to move.

While changes in temperature clearly affect livelihoods reliant on agriculture, extreme temperature events, for example, are rarely reported in Sub-Saharan Africa. According to EM-DAT, 18 have been documented since 1980 and occur mainly in the north Africa region: Algeria (2), Egypt (4), Liberia (1), Morocco (5), Nigeria (2), South Africa (3), and the Sudan (1) (EM-DAT, 2019). The aforementioned information in no way suggests that policymakers need not address the vulnerability posed by gradual changes in temperature. In our review of recent work, it will become clear that researchers have favored studying the implications of rapid onset (rather than slow onset) changes in temperature.

8 Building a Policy-Relevant Research Agenda on Environmental Migration in Africa

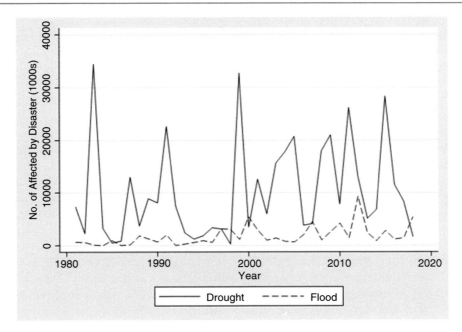

Fig. 8.1 Prevalence and exposure to hydrological disasters in Africa. (Source: EM-DAT, 2019)

The increasing frequency and intensity of hydrological disasters, such as droughts and floods, have been the main causes for concern among policymakers in recent years (UNEP, 2005, 2011; Rigaud et al., 2018). Figure 8.1 exhibits the prevalence of droughts and floods in the Africa region in the last 40 years. Droughts clearly affect a greater number of people in the region and their duration can extend beyond a year. Floods are also prevalent but typically affect fewer people in the short-term as they tend to be highly localized. Since 2010, a significant share of the people affected by droughts were in Ethiopia, Kenya, Malawi, and Somalia (Table 8.1). Although almost half of the people affected by floods since 2010 were in Nigeria,[1] the occurrence of at least one flooding event appears widespread across the region (Table 8.1).

Droughts continue to endanger the food security of the broader African population (Dercon, 2004; Loewenberg, 2011; UNEP, 2011; Rigaud et al., 2018). Future climate change projections

Table 8.1 Share of people affected by hydrological disaster by African Country since 2010

Country	Drought %	Flood %
Angola	0.03	
Burkina Faso	0.07	
Chad	0.04	0.03
Ethiopia	0.16	0.03
Ghana		0.05
Kenya	0.13	0.04
Malawi	0.09	0.04
Mali	0.04	
Mauritania	0.05	
Mozambique		0.03
Namibia		0.02
Niger	0.07	0.06
Nigeria		0.46
Somalia	0.08	0.08
South Africa	0.03	
South Sudan	0.04	0.03
Sudan	0.04	0.05
Zimbabwe	0.04	

Source: EM-DAT (2019)

---

[1] Nigeria experienced a major flooding in 2012 which affected 7 million people in the south-central part of the country.

suggest that higher temperatures and droughts will threaten the ability of farmers, herders, and fisherman on the continent to sustain crop yields (Schlenker & Lobell, 2010; Thornton et al., 2011; Serdeczny et al., 2017), livestock (Thornton et al., 2011) and small-scale fishing enterprises (UNEP, 2011; Rigaud et al., 2018). Rural households have begun to adapt to climate variability by diversifying their labor away from the agricultural sector (Porter, 2012). However, rural employment opportunities during climatic shocks can be finite due to a weak demand for hired labor (Mueller et al., 2020b). Moreover, self-employment, such as small-scale trading and retail, may only be an option for workers with sufficient physical capital (Cunguara et al., 2011; Nagler & Naude, 2017). Migration is often the only available strategy for rural households to diversify income (Dillon et al., 2011; Gray & Mueller, 2012a). In countries with high transportation costs, family members are often relegated to migrate rather than commute to obtain auxiliary sources of income for the house through employment in nearby towns or cities (Ingelaere et al., 2018).

Households living in areas prone to flooding can utilize similar risk management strategies (Silva & Matyas, 2014). A unique feature of a flooding event is its ramifications on infrastructure, i.e. housing, water systems, electricity, and roads (UNEP, 2011; Chinowsky & Arndt, 2012; Munzhedzi, 2017), and health, in terms of increasing incidence of vector-borne diseases (e.g., malaria and dengue) (Duo et al., 2010) and water-borne diseases (e.g., cholera and typhoid) (Kondo et al., 2002; Ahern et al., 2005). The recent occurrences of cyclones in Mozambique highlight how such expansive and frequent flooding events can bear quite devastating impacts on national well-being and displacement. When Cyclone Idai hit on March 14, 2019, 500,000 hectares of cropland during the main harvesting cycle were destroyed (UNDP, 2019). Cyclone Kenneth struck northern Mozambique 6 weeks later. Together, both cyclones were responsible for the displacement of 170,000 people (UNICEF, 2020). Given the substantive decline in cereal production, international donor agencies initiated humanitarian aid programs to support farmers in the region subjected to income losses as well as food insecurity in the region (WFP, 2019).

While both disaster types may affect voluntary mobility, involuntary mobility, or aspirations to move, the factors that prompt individuals to relocate are poorly understood. Idiosyncratic factors, such as reliance on the agricultural sector and attachment to place, influence the decision to migrate. Furthermore, pull factors, such as rural-urban wage differences and remoteness, affect the relative appeal of using migration as an adaptation strategy. The push and pull factors in which drive migration decisions will likely vary by the context under investigation. The role disasters have on migration will depend on their magnitude, type, and location, particularly as the event itself can attract varying degrees of aid for reconstruction which can expedite recovery and potentially return migration (Strömberg, 2007; Gillespie et al., 2014). In what follows, I synthesize the lessons ascertained from the current climate migration literature and identify the knowledge gaps that are particularly relevant for the research community as well as African policymakers today.

Building a Consensus on Observed Climate Migration Patterns in Africa

In this section, I review the body of literature that measures the migration consequences of climate using country or household level data. The majority of studies focus on the role of precipitation variability on migration, paying scant attention to other environmental factors that affect livelihoods such as gradual changes in temperature, flooding, and soil degradation. We then touch on the few studies that, in addition to quantifying migration effects, relate changes in the environment to measures of well-being. This secondary piece of empirical evidence is crucial for informing policy response, as the findings raise awareness of the extent specific livelihoods are responsible for current migration practices and which barriers preclude populations from migrating to aid in income diversification.

Thus far, the climate-migration literature in Africa has centered on the effects of precipitation on mobility. In their global analysis of 78 countries from 1960 to 1990, Barrios et al. (2006) first demonstrate the importance of rainfall deficits on urbanization (a proxy for rural-urban migration) in sub-Saharan African countries relative to other countries. Poelhekke (2011) corroborates these findings when extending the spatial coverage to include 163 countries and when focusing on a more recent 30-year period (1970–2000). In essence, both studies provide empirical evidence that push factors are primarily responsible for the migration patterns witnessed in Africa. In other words, the inability to sustain livelihoods in the short term when exposed to a negative climatic shock drives households to migrate to cities.

While these seminal pieces are crucial for informing future investments in African countries, there are three main criticisms of this work for the purpose of understanding climate migration trends. In focusing on urbanization, the studies ignore the dominant migration pattern in Africa which is from rural to other rural areas (Wouterse & Taylor, 2008; Beegle et al., 2011; Grace et al., 2018). The second limitation is their use of urbanization as a measure of migration. Clearly, inferring migratory responses from urbanization relationships may be misguided as fertility (Eissler et al., 2019) and mortality (Deschenes & Moretti, 2009; Deschenes & Greenstone, 2011; Barreca et al., 2016; Kudamatsu et al., 2016; Carleton, 2017; Geruso & Spears, 2018) vary with climate. Moreover, urbanization also captures reclassification of rural to urban areas, which varies over time. The studies also suffer from omitted variable bias due to the exclusion of other features of climate change (such as temperature) that influence migration and are correlated with the covariate of interest (in this case, rainfall) (Auffhammer et al., 2013).

Marchiori et al. (2012) address the above concerns when measuring the effect of temperature and rainfall anomalies on both international migration and urbanization rates in 39 sub-Saharan countries from 1960 to 2000. Temperature appears to be an integral driver of net-migration rates, especially among countries reliant on the agricultural sector for income. Extending the spatial coverage to 115 countries and focusing on international migration, Cattaneo and Peri (2016) show use of emigration as an adaptation strategy to temperature variability is limited for poor countries, where their classification of poor countries mostly comprises nations in Africa.

A parallel set of studies complements the aforementioned literature through examination of households' capacity to adapt. Henry et al. (2004) confirm that rainfall shortages in Burkina Faso increase the tendency for men to move to rural areas. However, similar events decrease the tendency for men to move abroad and for women to move to urban areas. The authors ascribe the decline in long-distance migration to the inability to finance the transport costs required to move. Similar constraints to the use of migration to cope with rainfall shocks have been established in Malawi (Lewin et al., 2012) and Mali (Grace et al., 2018).

In an attempt to connect migration more directly with the income risks faced by farming households, Dillon et al. (2011) investigate whether migration is responsive to variations in temperature degree days over the main growing season. Nigerian men are more likely to move during hot episodes, while women are more likely to be retained. They attribute the gender-differentiated migration effects to two mechanisms. First, the returns to migration may be higher for men than women during a hot spell given gender segmentation in the labor market. For example, the demand for male labor may increase if the heat spell requires more labor-intensive tasks on farms to mediate the damages of heat on agricultural production. The demand for female labor, in turn, may decline if heat reduces the disposable income of households to afford domestic and other services typically provided by migrant women. Second, bride prices may fall with temperature, reducing the benefits of marriage migration to the household.

It is important to note that the narrow geographic scope of the previous household studies may affect the ability to detect relationships between migration and climate (Henry et al., 2004; Dillon et al., 2011; Lewin et al., 2012; Grace et al., 2018). Increasing the spatial scale can augment the likelihood that the study captures a distribution of meteorological conditions in the sample. Gray and Wise (2016) use nationally representative household data in 5 African countries (Kenya, Uganda, Nigeria, Burkina Faso, and Senegal) to evaluate the effects of temperature and precipitation anomalies on internal and international migration. They find temperature anomalies continue to have a pervasive effect on migration patterns but the direction of the relationship is context- specific. Migration increases (decreases) with temperature in Uganda (Kenya and Burkina Faso[2]). Migration patterns remain unaffected by climate variability in Nigeria and Senegal. Interestingly, the authors show that there are no robust relationships between rainfall and migration in these quite different countries, using several model specifications and climate data products to quantify exposure to rainfall.[3]

Expanding the spatial representation of households can also be beneficial from the perspective of increasing the visibility of a range of subpopulations and economic activities in the analysis. Previous work has leveraged heterogeneity in demographic characteristics, physical asset endowments, and population density to provoke discussion on the implicit barriers of engaging in migration as an adaptation strategy. Gray and Mueller (2012a) find that women and children are less inclined to move when a drought occurs,

while the movement of men coming from land poor households is enhanced. Mastrorillo et al. (2016) confirm droughts (high temperatures and rainfall shortages) contribute to internal migration in South Africa, particularly for those who identify as black and low-income. In contrast to the previous two articles, Nawrotzki and De-Waard (2018) demonstrate that only the wealthy are able to use migration to cope with heat and drought exposure in Zambia, while the poor overcome constraints to moving through connections with their migrant networks. These studies offer quite distinct outlooks for policymakers. In Ethiopia and South Africa, disadvantaged populations may require interventions to improve crop resilience and protect the vulnerable from being displaced. In Zambia, the evidence suggests that there are benefits from migration in the wake of drought. Thus, an intervention that reduces moving costs or facilitates matches between employees and employers (the operative function of an informal network) may facilitate the adaptation of underprivileged households.

Finally, few studies in this literature empirically identify why migration rises in response to a climatic event (Lilleør & Van de Broeck, 2011; Marchiori et al., 2017). From a policy-maker's perspective, knowing the root cause of the threat to household livelihoods could provide him/her with leads as to which programs can be implemented to proactively prevent households from engaging in migration as an adaptation strategy in areas lacking strong labor markets to support this coping mechanism. Marchiori et al. (2012) quantify the climatic damages to the gross domestic product per capita and triangulate that income variability is the driving force behind the observed migration effects. In the same vein as Marchiori et al. (2012), Hirvonen (2016) attributes temperature-induced migration to welfare losses, by providing reduced-form estimates of the consumption losses attributable to temperature to complement his predictions of climate migration. Kubik and Maurel (2016) use an instrumental variables model to directly link drought-induced crop losses to variations in migration outcomes. One of the main reasons why researchers may be reluctant to adopt the instrumental variables

---

[2]Nawrotzki and Bakhtsiyarava (2017) confirm the relationship between heat and international migration in Burkina Faso with an alternative data source.

[3]Nawrotzki and Bakhtsiyarava (2017) indicate that excessive precipitation—the percentage of times in which rainfall was above one standard deviation above the long-term average over the 6-year period leading up to the interview year—induces individuals to migrate abroad from Senegal. The cross-sectional analysis may be underlying the inconsistency across studies. In this case, the authors are unable to generalize their results beyond the year in which the census was collected due to their inability to control for the cyclical trends that coincide with climate anomalies as well as affect mobility within the country.

technique is because it is very unlikely that the exclusion restriction will be satisfied in all contexts (Maystadt et al., 2016). In this case, for the instrumental variables model to be valid, the exclusion criterion requires that the only channel in which climate affects migration is through losses to crop income. Other climate impact assessments have not only given reason for researchers to question the credibility of this assumption, but also challenge whether the observed migration patterns are motivated solely by household's potential losses to agricultural yields (Pascual et al., 2006; Zhou et al., 2008; Paaijmans et al., 2010; Graff Zivin & Neidell, 2014; Burke et al., 2015; Graff Zivin et al., 2018; Zhang et al., 2018).

Future Directions in Environmental Migration Research in Africa

In this section, I describe three shortcomings in the literature. First, in taking advantage of available data, existing studies apply different definitions of migration, and, therefore, there is a lack of consensus with respect to whether environmental migration is temporary or permanent in this region. Moreover, the geographical representation of research in this area is inhibited by where data is being collected, which may not reflect countries most susceptible to environmental migration. Second, the majority of studies focus on a few environmental stressors, such as precipitation and temperature anomalies, when there are a variety of environmental risks that jeopardize livelihoods in Africa. Third, we know little about the mechanisms underlying the migratory responses observed empirically, providing minimal guidance to policymakers on what solutions may be necessary to mitigate patterns of displacement. After discussion each of these shortcomings, I try to provide suggestions of how we might overcome these issues in future research.

Measurement of Migration

Perhaps the biggest obstacle to performing research on this topic in Africa is having access to surveys that frequently collect information on individual migration behavior, include metrics that allow researchers to distinguish between permanent and temporary migration, as well as provide geo-referenced information for the origin and destination of migrants at a sufficient scale to represent highly localized climatic events. The studies discussed in the previous section all classify migration in quite distinct ways often due to the variety of data available to measure migration. Each form of data presents advantages and disadvantages in migration analyses (see Carletto et al., 2014, for more details).

The census often asks the respondent to recall his previous residence a year ago. On this basis, researchers may make assumptions that the move reflected in the census is permanent, when the interview timing might have coincided with a temporary migration episode. One main advantage, however, of using a census relative to other forms of data is that the patterns observed are nationally representative. One way to establish whether the migratory responses may, in fact, be permanent is to relax the recall period from 1 to 5 years. A subset of censuses in Africa, for example, include questions over a recall period of 1 and 5 years, and others also include the year in which the respondent moved. Researchers can refine their analysis to check how sensitive the results are to definitions of migration based on the multiple sources of migration information in the census. In the future, national statistical agencies should consider expanding the recall period of the migration period to 5 years, as well as including the year of move.

Another common source of migration information is from household surveys. Some household surveys include migration history modules, which document whether each member of the household moved, the departure and return dates, as well as their final destinations. This information can be used to investigate topics related to environmental migration and distinguish between temporary and permanent movements. Yet, the main challenge with using such information is that the migration outcomes themselves may be subject to measurement error. The information is likely less precise as you extend the recall reference period and when using a proxy respondent, such as the head of household, to answer the questions.

The final source of migration information is from longitudinal surveys and demographic surveillance systems (DSS), which through the diligence of the survey implementers track households over time. There are few examples in this region where careful attention has been paid to finding individuals that leave the household over time and re-surveying them (Beegle et al., 2011). The advantage of tracking migrants is that you can ask him/her about the duration of the migration episode and verify the destination. The challenge in streamlining the collection of this type of data is that recovering information from individuals who enter and exit the household can amplify the survey budget substantively.

There is also the issue of using the aforementioned standard instruments in politically unstable countries. For example, Table 8.1 shows that Somalia has the fourth greatest population afflicted by droughts in recent years, but it is left underrepresented in the literature due to data constraints. While there are several programs for data collection led by the World Bank (e.g., the Living Standards Measurement Study—Integrated Surveys on Agriculture), the International Food Policy Research Institute, and other organizations with mandates to provide data to the public (IPUMS International), these rich data sources are rare in locations where climate change may serve as a risk multiplier for migration.

Given the challenges of the current sources of migration information, the use of call detail records (CDR) data may be a promising direction going forward. Lai et al. (2019) provide one of the first proof of concepts of the benefits of CDR data for measuring migration behavior over a longer timeframe in the African context. In particular, they show the migration rates constructed from CDR in Namibia correlate closely to the migration rates produced from censuses. They also demonstrate that the census data may potentially overestimate the number of migrants in regions experiencing a flood. They find a greater number of respondents in Zambezi identified themselves as migrants in the 2011 census relative to the

number of migrants detected in the CDR data. The authors owe the discrepancy to recall bias in the census data. Zambezi experienced a flood in 2010, a few months prior to the 12-month reference period for migration on the census questionnaire. Thus, Lai et al. (2019) claim that the overestimate in migratory flows generated from the census data collected in Zambezi likely stems from respondents claiming to have migrated within the 12-month window when their migration episode might have occurred before the reference period and their move was actually temporary.

Lai et al. (2019) also attempt to bound the selection bias in CDR data, since mobile phone ownership is far from universal, using auxiliary data collected in Namibia. Using the Demographic Health Survey (DHS), they show what characteristics differentiate mobile phone users from non-users. Additional data sources, such as the DHS, can be used to complement the analysis from CDR to construct sampling weights that try to correct for bias driven by analyzing data from a select sample of mobile phone users in migration models (Fitzgerald et al., 1998; Thomas et al., 2012). Lai et al. (2019) offer a promising avenue for environmental migration research in African contexts as telecommunications companies become open to more research collaborations.

A second modality for collecting migration information in African countries is the use of mobile phone-based surveys. Experimental evidence in South Africa (Garlick et al., 2019) and Tanzania (Arthi et al., 2018) validate the accuracy of the information collected from high-frequency phone-based surveys for measuring micro-enterprise behavior and participation in agricultural activities, respectively, throughout the year. While the sampling frame may be limited to mobile phone owners, as when analyzing CDR data, phone-based surveys offer the added benefit of being able to collect individual migration histories in addition to the SIM card users' demographic and employment information over time. Phone-based surveys also can be used to verify that the SIM card reflects one identity, since they can include questions about the number of owners and users of the phone over frequent time steps. As long as countries in the region have a reliable signal from the cellphone towers, mobile phone surveys potentially offer another viable alternative to collecting high-frequency migration data for the purpose of understanding which sub-populations are susceptible to moving when exposed to a change in the environment, as well as the duration of the moves and the popular destinations.

Exposure to Environmental Stressors

The intent of this section is to focus on how we can increase the number of influential studies for African policymakers, by accounting for two environmental stressors that rarely are addressed in the migration literature: flooding and soil quality. In the previous section, I show that the dominant approach used in the environmental migration literature is to quantify migratory responses to climatic events through the inclusion of temperature and precipitation variables in empirical models.[4] Recent research has expanded the purview to consider the role of droughts in driving migration, using soil moisture indices (Kubik & Maurel, 2016; Henderson et al., 2017), self-reported information on drought exposure (e.g., Gray and Mueller (2012a) in Ethiopia; Ocello et al. (2015) in Tanzania, and Koubi et al. (2016) in Uganda),[5] and the Normalized Difference Vegetation Index (NDVI) (Leyk et al., 2012; Hunter et al., 2014), as explanatory variables. The advantage of these type of metrics is they inherently capture the aspect of drought that influences yield losses directly. Insights from models that integrate readily-available indices, like the NDVI, can be useful for policy responses to natural disasters for the targeting and design of humanitarian programming (IFAD, 2017; Okpara et al., 2017).

In contrast, impact of flooding on displacement in Africa is rarely quantified. While precipitation-based metrics are typically integrated in migration models, risk of flooding is a function of proximity to riverine or coastal water sources, topography, and other factors independent of rainfall. For this reason, the inclusion of hazard measures, based on validated remote sensing techniques, yield quite different migration predictions than precipitation-based measures (Hunter et al., 2014; Call et al., 2017; Chen & Mueller, 2018). An emerging body of work demonstrates the utility of coupling flooding metrics based on satellite data with empirical models of migration to quantify vulnerability to flooding (Call et al., 2017; Chen & Mueller, 2018). Using data from NASA's MODIS satellite, Chen and Mueller (2018) construct the Modified Normalized Difference Water Index (MNDWI) for each pixel over an 8-day period (Xu, 2006). Each pixel is classified as water based on whether the index exceeds a threshold. A pixel is determined to be inundated if it is considered to be water in any of the 8-day periods within the season. This information is then used to construct a measure of inundation at the sub-district level, which is the difference between the percentages of pixels in the sub-district classified as water in the monsoon and dry seasons. The authors offer a promising direction for combining remotely sensed imagery with high-frequency migration data (such as those obtained through CDR) to measure the temporary and/or permanent displacement effects from flooding events, such as Hurricane Idai in Mozambique.

An additional knowledge gap is related to how land degradation and poor soil quality affect migration (UNEP, 2011). Unsustainable farming practices, such as the burning of crop residues, and deforestation, caused by resource extractive activities, jeopardize yields and livestock production. Yet, there are few studies that examine the extent the slow deterioration of land displaces agrarian households. Gray (2011) takes advantage of existing longitudinal

---

[4]There are a number of articles that discuss the advantages of various types of *in situ* and satellite data available globally to measure climate variability (Auffhammer et al., 2013; Dell et al., 2014; Donaldson & Storeygard, 2016).

[5]The use of self-reported information on drought exposure comes with the additional concern that its inclusion in an empirical model of migration may be subject to parameter bias. This is because household perceptions of risk are correlated with other characteristics unobserved by the researcher, such as level of risk aversion, farming ability of the household, and so on.

studies in Kenya and Uganda to understand the extent soil quality affects migration.[6] The baseline survey collected a breadth of information regarding the properties of the soil: the percent carbon, the percent nitrogen, calcium, potassium, phosphorus, percent clay, and the pH. Gray (2011) consolidates the information gathered for each factor into an index that he uses to capture the natural capital of the farm in terms of its productive potential.[7] He shows that the relationship between soil quality and migration is context-specific. Temporary labor migration is negatively associated with good soil quality in Kenya and permanent non-labor migration is positively associated with good soil quality in Uganda. Innovations in soil testing may facilitate collecting this information in the future (Stirzaker et al., 2017). Combining these cost-effective technologies for monitoring changes in soil quality with longitudinal surveys going forward will open the door to a new set of studies which can measure how natural capital will affect population mobility.

Mechanisms

The literature on environmental migration typically presumes the migration decision is driven by households' attempts to mitigate their income risk. Few studies empirically validate this assumption (Gray & Mueller, 2012b; Mueller et al., 2014; Cai et al., 2016). One of the main reasons why this practice is not widespread is due to the paucity of socioeconomic data in African countries. For example, data sources for migra-

tion information (e.g., census) often exclude information on agricultural production or income. Newly available remote sensing data products can be used to complement existing migration data for the purpose of evaluating whether changes in agrarian livelihoods may be responsible for the observed changes in migration (Weiss et al., 2014; Zhang, 2015). As African economies experience structural transformation, it is likely that the threat to income extends beyond implications environmental factors have on the agricultural sector (Burke et al., 2015; Mueller et al., 2020b). To investigate the economic impact of environmental changes more broadly, one might also consider incorporating night lights data into the analysis as a proxy for economic growth (Donaldson & Storeygard, 2016).

Migration decisions are multi-faceted, and, thus, it will be important to consider the role of other factors in influencing migration as an adaptation strategy in future research. First, not only may health risks increase with changes in the environment, but those risks may be unevenly distributed across family members. For example, if climate change affects the health of a household breadwinner, households may suffer from a loss of income unrelated to agricultural yields (Carleton & Hsiang, 2016; Carleton, 2017). Where Demographic Health Surveys are available, it could be useful to explore how the deterioration of family members' physical and/or mental health may be positively associated with climate migration. Use of CDR data can serve dual purposes in terms of giving researchers the ability to jointly measure population mobility and exposure to health risks under different environmental scenarios (Wesolowski et al., 2012; Bengtsson et al., 2015; Peak et al., 2018).

Second, migration patterns may change because of the associated infrastructural damages caused by environmental factors, such as inundation. Workers' commutes may be disrupted by the destruction of roads due to torrential floods (Strobl & Valfort, 2013). Loss of housing can render certain locations inhabitable. Displacement patterns could reverse if the magnitude of the event attracts resources for reconstruction (Del Ninno et al., 2003; Frankenberg et al., 2019).

---

[6] Although individual migration data is available over multiple rounds of the study, soil samples were only collected at baseline. The implements needed to collect soil samples at the household level, as well as the technical expertise required to analyze the data render performing the procedure cost-prohibitive for all survey rounds.

[7] Gray (2011) notes one of the main limitations of the index is that soil properties depend on the agricultural activities underlying the household rendering the explanatory variable endogenous. To address this concern, he performs a robustness check replacing the index with a measure of soil carbon which is less reliant on the farming techniques applied on the parcel.

Innovative techniques, such as machine learning, could be used to document how changes in infrastructure affect mobility over time (e.g., Tusting, 2019).

Finally, the above analyses ignore how behavioral factors influence tendencies to move. Household immobility during a natural disaster may stem from perceptions of employment or attachment to place. For example, migration may not appear as a feasible adaptation strategy to the decision-maker, as long as s/he regards securing a job or earning greater wages at a destination as highly uncertain. Similarly, moorings may drive households to withstand frequent environmental stressors, despite the ramifications on member welfare. From a policy point of view, identifying these reservations is crucial, as they suggest the need for different interventions. To introduce certainty over employment, policymakers may consider tools that effectively disseminate information about jobs at multiple destinations as a way of promoting adaptation to environmental stressors (Bryan et al., 2014). Alternatively, if communal ties keep households trapped in poor climate conditions, then policy interventions along the lines of voluntary communal resettlement may be more appropriate (Mendola & Franklin, 2015; Randell, 2016).

Recent work applies agent-based models (ABMs) to better understand the degree in which behavioral parameters drive migration decisions (Kniveton et al., 2011; Entwisle et al., 2016; Bell et al., 2019). The advantage of using an ABM over a parametric empirical model is that it can provide insights on the relative importance of behavioral dimensions in affecting location decisions. The main challenge with the ABMs' predictions is giving them legitimacy through validation of the underlying model's assumptions. For example, model predictions draw heavily on restrictions imposed on the functional form of household preferences. Survey experiments can be conducted to estimate the parameters required to retrieve an empirical preference function, when additional resources for primary data collection are available.

Going forward, use of ABM and other modelling approaches will be crucial to inform a range of interventions to promote resilience and adaptation among households. These approaches will become particularly important for determining how out-of-sample changes in environmental stressors may affect behavior in the future, such as what is required for understanding adaptation to sea level change (Wrathall et al., 2019). Global initiatives, like the Global Compact of Migration, which require the coordination of resources to both facilitate adaptation processes and cushion the potential effects of resource and job scarcity in receiving areas may benefit from preliminary predictions of where people may move and under what behavioral contexts.

Conclusion

The literature has made significant strides in the last few years to provide case studies of environmentally-induced human migration in Africa. There is consistent evidence that temperature and precipitation anomalies affect migration patterns broadly, but the magnitude of the effects and the direction of the relationships varies by context. Yield losses and food insecurity during climatic events often coincide with increased migratory responses. This suggests that climate shocks serve as a push factor. At the same time, studies indicate that climatic shocks reduce the financial capital available to support migration investments and affect opportunities for migrant work. These latter two phenomena tend to immobilize individuals who might typically migrate during normal climate conditions. Thus far, conclusions have been based on responses to rapid onset events, leaving how households will adapt to slow onset events, such as gradual changes in temperature, soil erosion, and changes in sea level an open question.

Identifying the factors that push or pull migrants to relocate during a climatic event is important for designing policies that aim to reduce involuntary migration and the associated negative consequences in receiving areas (Ruiz & Vargas-Silva, 2017; Maystadt et al., 2019).

Developing a broader understanding of the limitations of adaptation and the extent psychosociological aspects contribute to immobility should not be underestimated. As African governments phase in social protection programs, researchers can begin to study their role in promoting household adaptation–whether it be through facilitating seasonal migration for the liquidity-constrained or as a means of reducing involuntary migration (Hoddinott & Mekasha, 2020; Mueller et al., 2020a). Such studies would not only increase the desirability of these programs through demonstration of auxiliary benefits, but would also exhibit the extent populations are indeed trapped and whether financing migration serves as the main barrier.

References

Ahern, M., Kovats, R., Wilkinson, P., Few, R., & Matthies, F. (2005). Global health impacts of floods: Epidemiologic evidence. *Epidemiologic Reviews, 27*(1), 36–46.

Arthi, V., Beegle, K., De Weerdt, J., & Palacios-Lopez, A. (2018). Not your average job: Measuring farm labor in Tanzania. *Journal of Development Economics, 130*, 160–172.

Auffhammer, M., Hsiang, S., Schlenker, W., & Sobel, A. (2013). Using weather data and climate model output in economic analyses of climate change. *Review of Environmental Economics and Policy, 7*(2), 181–198.

Barreca, A., Clay, K., Deschenes, O., Greenstone, M., & Shapiro, J. (2016). Adapting to climate change: The remarkable decline in the US temperature-mortality relationship over the twentieth century. *Journal of Political Economy, 124*(1), 105–159.

Barrios, S., Bertinelli, L., & Strobl, E. (2006). Climatic change and rural-urban migration: The case of sub-Saharan Africa. *Journal of Urban Economics, 60*, 357–371.

Beegle, K., De Weerdt, J., & Dercon, S. (2011). Migration and economic mobility in Tanzania: evidence from a tracking survey. *Review of Economics and Statistics, 93*(3), 1010–1033.

Bell, A., Calvo-Hernandez, C., & Oppenheimer, M. (2019). Migration, intensification, and diversification as adapative strategies. *Socio-Environmental Systems Modelling, 1*, 16102. https://iopscience.iop.org/article/10.1088/1748-9326/abdc5b

Bengtsson, L., Gaudart, J., Lu, X., Moore, S., Wetter, E., Sallah, K., Rebaudet, S., & Piarroux, R. (2015). Using mobile phone data to predict the spatial spread of cholera. *Scientific Reports, 5*, 8923.

Black, R., & Collyer, M. (2014). "Trapped" populations: limits on mobility at times of crisis. In S. Martin, S. Weerasinghe, & A. Taylor (Eds.), *Humanitarian crises and migration: Causes, consequences and responses.* Routledge.

Black, R., Bennett, S., Thomas, S., & Beddington, J. (2011). Migration as adaptation. *Nature, 478*, 447–449.

Bryan, G., Chowdhury, S., & Mobarak, A. M. (2014). Under-investment in a profitable technology: The case of seasonal migration in Bangladesh. *Econometrica, 82*(5), 1671–1748.

Burke, M., Hsiang, S., & Miguel, E. (2015). Global non-linear effect of temperature on economic output. *Nature, 527*, 235–239.

Cai, R., Feng, S., Oppenheimer, M., & Pytlikova, M. (2016). Climate variability and international migration: The importance of the agricultural linkage. *Journal of Environmental Economics and Management, 79*, 135–151.

Call, M., Gray, C., Yunus, M., & Emch, M. (2017). Disruption, not displacement: Environmental variability and temporary migration in Bangladesh. *Global Environmental Change, 46*, 157–165.

Carleton, T. (2017). Crop-damaging temperatures increase suicide rates in India. *Proceedings of the National Academy of Sciences, 114*(33), 8746–8751.

Carleton, T., & Hsiang, S. (2016). Social and economic impacts of climate. *Science, 353*(6304), aad9837.

Carletto, C., Larrison, J., & Ozden, C. (2014). Chapter: Informing migration policies: A data primer. In R. E. B. Lucas (Ed.), *International handbook on migration and economic development.* Edward Elgar.

Cattaneo, C., & Peri, G. (2016). The migration response to increasing temperatures. *Journal of Development Economics, 122*, 127–146.

Chen, J., & Mueller, V. (2018). Coastal climate change, soil salinity and human migration in Bangladesh. *Nature Climate Change, 8*, 981–985.

Chinowsky, P., & Arndt, C. (2012). Climate change and roads: A dynamic stressor-response model. *Review of Development Economics, 16*(3), 448–462.

Cunguara, B., Langyintuo, A., & Darnhofer, I. (2011). The role of nonfarm income in coping with the effects of drought in southern Mozambique. *Agricultural Economics, 42*(6), 701–713.

Del Ninno, C., Dorosh, P., & Smith, L. (2003). Public policy, markets and household coping strategies in Bangladesh: Avoiding a food security crisis following the 1998 floods. *World Development, 31*(7), 1221–1238.

Dell, M., Jones, B., & Olken, B. (2014). The new climate—Economy literature. *Journal of Economic Literature, 52*(3), 740–798.

Dercon, S. (2004). Growth and shocks: Evidence from rural Ethiopia. *Journal of Development Economics, 74*(2), 309–329.

Deschenes, O., & Greenstone, M. (2011). Climate change, mortality, and adaptation: Evidence from annual fluctuations in weather in the U.S. *American Economic Journal: Applied Economics, 3*(4), 152–185.

Deschenes, O., & Moretti, E. (2009). Extreme weather events, mortality, and migration. *Review of Economics and Statistics, 91*(4), 659–681.

Dillon, A., Mueller, V., & Sheu, S. (2011). Migratory responses to agricultural risk in northern Nigeria. *American Journal of Agricultural Economics, 93*(4), 1048–1061.

Donaldson, D., & Storeygard, A. (2016). The view from above: Applications of satellite data in economics. *Journal of Economic Perspectives, 30*(4), 171–198.

Du, W., Fitzgerald, G., Clark, M., & Hou, X. (2010). Health impacts of floods. *Prehospital and Disaster Medicine, 25*(3), 265–272.

Eissler, S., Thiede, B., & Strube, J. (2019). Climatic variability and changing reproductive goals in sub-Saharan Africa. *Global Environmental Change, 57*, 101912.

EM-DAT. (2019). *EM-DAT: The emergency events database—Universite Catholique de Louvain (UCL)—CRED. D. Guha-Sapir-www.emdat.be,* Brussels, Belgium.

Entwisle, B., Williams, N., Verdery, A., Rindfuss, R., Walsh, S., Malanson, G., Mucha, P., Frizzelle, B., McDaniel, P., Yao, X., Heumann, B., Prasartkul, P., Sawangdee, Y., & Jampaklay, A. (2016). Climate shocks and migration: An agent-based modeling approach. *Population and Environment, 38*(1), 47–71.

Fitzgerald, J., Gottschalk, P., & Moffitt, R. (1998). An analysis of sample attrition in panel data: The Michigan Panel Study of Income Dynamics. *Journal of Human Resources, 33*(3), 251–299.

Frankenberg, E., K. Karrigan, P. Katz, E. Peshkin, C. Sumantri, and D. Thomas (2019). Using high resolution imagery and neural networks to measure destruction and reconstruction after a disaster. Paper presented at the Population Association of America Conference, Austin, Texas.

Garlick, R., Orkin, K., & Quinn, S. (2019). Call me maybe: Experimental evidence in using mobile phones to survey microenterprises. *World Bank Economic Review, 32*(3), 656–675.

Gemmene, F. (2011). Why the numbers don't add up: A review of estimates and the predictions of people displaced by environmental changes. *Global Environmental Change, 21*(1), S41–S49.

Geruso, M., & Spears, D. (2018). *Heat, humidity, and infant mortality in the developing world* (National Bureau of Economic Research Working Paper No. 24870).

Gillespie, T., Frankenberg, E., Chum, K. F., & Thomas, D. (2014). Night-time lights time series of tsunami damage, recovery, and economic metrics in Sumatra, Indonesia. *Remote Sensing Letters, 5*(3), 286–294.

Grace, K., Hertrich, V., Singare, D., & Husak, G. (2018). Examining rural Sahelian out-migration in the context of climate change: An analysis of the linkages between rainfall and out-migration in two Malian villages from 1981-2009. *World Development, 109*, 187–196.

Graff Zivin, J., & Neidell, M. (2014). Temperature and the allocation of time: Implications for climate change. *Journal of Labor Economics, 32*, 1–26.

Graff Zivin, J., Hsiang, S., & Neidell, M. (2018). Temperature and human capital in the short and long run. *Journal of the Association of Environmental and Resource Economists, 5*(1), 77–105.

Gray, C., & Mueller, V. (2012a). Drought and population mobility in rural Ethiopia. *World Development, 40*(1), 134–145.

Gray, C., & Mueller, V. (2012b). Natural disasters and population mobility in Bangladesh. *Proceedings of the National Academy of Sciences, 109*(16), 6000–6005.

Gray, C., & Wise, E. (2016). Country-specific effects of climate variability on human migration. *Climatic Change, 135*(3), 555–568.

Grimard, F. (1997). Household consumption smoothing through ethnic ties: Evidence from Cote d'Ivoire. *Journal of Development Economics, 53*, 391–422.

Henderson, V., Storeygard, A., & Deichmann, U. (2017). Has climate change driven urbanization in Africa? *Journal of Development Economics, 124*, 60–82.

Henry, S., Schoumaker, B., & Beauchemin, C. (2004). The impact of rainfall on the first out-migration: A multi-level event-history analysis in Burkina Faso. *Population and Environment, 25*(5), 423–460.

Hoddinott, J. (1994). A model of migration and remittances applied to western Kenya. *Oxford Economic Papers, 46*(3), 459–476.

Hoddinott, J., & Mekasha, T. (2020). Social protection, household size, and its determinants: Evidence from Ethiopia. *Journal of Development Studies.* Available at: https://www.tandfonline.com/doi/full/10.1080/00220388.2020.1736283

Hunter, L., Nawrotzki, R., Leyk, S., Maclaurin, G., Twine, W., Collinson, M., & Erasmus, B. (2014). Rural outmigration, natural capital, and livelihoods in South Africa. *Population, Space and Place, 20*(5), 402–420.

Ingelaere, B., Christiaensen, L., De Weerdt, J., & Kanbur, R. (2018). Why secondary towns can be important for poverty reduction—A migrant perspective. *World Development, 105*, 273–282.

International Fund of Agricultural Development (IFAD). (2017). *Remote sensing for index insurance: Findings and lessons learned for smallholder agriculture.* IFAD.

Jayachandran, S. (2006). Selling labor low: Wage responses to productivity shocks in developing countries. *Journal of Political Economy, 114*(3), 538–575.

Jedwab, R., & Moradi, A. (2016). The permanent effects of transportation revolution in poor countries: Evidence from Africa. *Review of Economics and Statistics, 98*(20), 268–284.

Jedwab, R., Kerby, E., & Moradi, A. (2017). History, path dependence and development: Evidence from colonial railways, settlers and cities in Kenya. *Economic Journal, 127*(603), 1467–1494.

Kniveton, D., Smith, C., & Wood, S. (2011). Agent-based model simulations of future changes in migration flows for Burkina Faso. *Global Environmental Change, 21*(S1), S34–S40.

Kochar, A. (1999). Smoothing consumption by smoothing income: Hours-of-work responses to idiosyncratic agricultural shocks in rural India. *Review of Economics and Statistics, 81*(1), 50–61.

Kondo, H., Seo, N., Yasuda, T., Hasizume, M., Koido, Y., Ninomiya, N., & Yamamoto, Y. (2002). Post-flood—

Infectious diseases in Mozambique. *Prehospital and Disaster Medicine, 17*(3), 126–133.

Koubi, V., Spilker, G., Schaffer, L., & Böhmelt, T. (2016). The role of environmental perceptions in migration decision-making: Evidence from both migrants and non-migrants in five developing countries. *Population and Environment, 38*(2), 124–163.

Kubik, Z., & Maurel, M. (2016). Weather shocks, agricultural production and migration: Evidence from Tanzania. *Journal of Development Studies, 52*(5), 665–680.

Kudamatsu, M., Persson, T., & Strömberg, D. (2016). *Weather and infant Mortality in Africa.* Unpublished. Accessed online on 3 Apr 2019 at: https://docs.google.com/viewer?a=v&pid=sites&srcid=ZGVmYXVsdGRvbWFpbnxta3VkYW1hdHN1fGd4OjRkMTZkYzJlMmIxYWFiMmQ

Kudo, Y. (2015). Female migration for marriage: Implications from the land reform in rural Tanzania. *World Development, 65*, 41–61.

Lagakos, D., Marshall, S., Mobarak, A. M., Vernot, C., & Waugh, M. (2020). Migration costs and observational returns to rural-urban migration in the developing world. *Journal of Monetary Economics.* Available online at: https://doi.org/10.1016/j.jmoneco.2020.03.013

Lai, S., Zu Erbach-Schoenberg, E., Pezzulo, C., Ruktanonchai, N., Sorichetta, A., Steele, J., Li, T., Dooley, C., & Tatem, A. (2019). Exploring the use of mobile phone data for national migration statistics. *Palgrave Communications, 5*(34), 1–10.

Lewin, P., Fisher, M., & Weber, B. (2012). Do rainfall conditions push or pull rural migrations: Evidence from Malawi? *Agricultural Economics, 43*, 191–204.

Leyk, S., Maclaurin, G., Hunter, L., Nawrotzki, R., Twine, W., Collinson, M., & Erasmus, B. (2012). Spatially and temporally varying associations between temporary outmigration and natural resource availability in resource-dependent rural communities in South Africa: A modeling framework. *Applied Geography, 34*, 559–568.

Lilleør, H. B., & Van de Broeck, K. (2011). Economic drivers of migration and climate change in LDCs. *Global Environmental Change, 21*(1), S70–S81.

Loewenberg, S. (2011). Global food crisis takes heavy toll on East Africa. *The Lancet, 378*(9785), 17–18.

Lucas, R. E. B., & Stark, O. (1985). Motivations to remit: Evidence from Botswana. *Journal of Political Economy, 93*(5), 901–918.

Marchiori, L., Maystadt, J.-F., & Schumacher, I. (2012). The impact of weather anomalies on migration in sub-Saharan Africa. *Journal of Environmental Economics and Management, 63*, 355–374.

Marchiori, L., Maystadt, J.-F., & Schumacher, I. (2017). Is environmentally induced income variability a driver of human migration? *Migration and Development, 6*(1), 33–59.

Mastrorillo, M., Licker, R., Bohra-Mishra, P., Fagiolo, G., Estes, L., & Oppenheimer, M. (2016). The influence of climate variability on internal migration flows in South Africa. *Global Environmental Change, 39*, 155–169.

Maystadt, J.-F., Mueller, V., & Sebastian, A. (2016). Environmental migration and labor markets in Nepal. *Journal of the Association of Environmental and Resource Economists, 3*(2), 417–452.

Maystadt, J.-F., Hirvonen, K., Mabiso, A., & Vandercasteelen, J. (2019). Impacts of hosting forced migrants in poor countries. *Annual Review of Resource Economics, 11*, 439–459.

Mazzucato, V. (2009). Informal insurance arrangements in Ghanaian migrants transnational networks: The role of reverse remittances and geographic proximity. *World Development, 37*(6), 1105–1115.

Mendola, M., & Simtowe, F. (2015). The welfare impact of land redistribution: Evidence from a quasi-experimental initiative in Malawi. *World Development, 72*(C), 53–69.

Mueller, V., Gray, C., & Kosec, K. (2014). Heat stress increases long-term human migration in rural Pakistan. *Nature Climate Change, 4*, 182–185.

Mueller, V., Gray, C., Handa, S., & Seidenfeld, D. (2020a). Do social protection programs foster short-term and long-term migration adaptation strategies? *Environment and Development Economics, 25*, 135–158.

Mueller, V., Sheriff, G., Dou, X., & Gray, C. (2020b). Temporary migration and climate variation in East Africa. *World Development, 126*, 104704.

Munzhedzi, S. (2017). *Perspective for SADC climate change adaptation Southern African Development Community (SADC).* Accessed online on 14 Mar 2017: https://www.sanbi.org/sites/default/files/documents/documents/ltas-factsheet-1.pdf

Nagler, P., & Naude, W. (2017). Non-farm entrepreneurship in rural sub-Saharan Africa: New empirical evidence. *Food Policy, 67*, 175–191.

Nawrotzki, R., & Bakhtsiyarava, M. (2017). International climate migration: Evidence for the climate inhibitor mechanism and the agricultural pathway. *Population, Space and Place, 23*(4), e2033.

Nawrotzki, R., & DeWaard, J. (2018). Putting trapped populations into place: Climate Change and inter-district migration flows in Zambia. *Regional Environmental Change, 18*(2), 533–546.

Ocello, C., Petrucci, A., Testa, M., & Vignoli, D. (2015). Environmental aspects of internal migration in Tanzania. *Population and Environment, 37*(1), 99–108.

Okpara, J., Afiesimama, E., Anuforom, A., Owino, A., & Ogunjobi, K. (2017). The applicability of standardized precipitation index: Drought characterization for early warning system and weather index insurance in West Africa. *Natural Hazards, 89*(2), 555–583.

Paaijmans, K., Blanford, S., Bell, A., Blanford, J., Read, A., & Thomas, M. (2010). Influence of climate on malaria transmission depends on daily temperature variation. *Proceedings of the National Academy of Sciences, 103*(15), 5829–5834.

Pascual, M., Ahumada, J., Chaves, L., Rodo, X., & Bouma, M. (2006). Malaria resurgence in the East African highlands: Temperature trends revisited. *Proceedings of the National Academy of Sciences, 103*(15), 5829–5834.

Peak, C., Wesolowski, A., Erbach-Schoenberg, E. Z., Tatem, A., Wetter, E., Lu, X., Power, D., Weidman-Grunewald, E., Ramons, S., Moritz, S., Buckee, C. O., & Bengtsson, L. (2018). Population mobility reductions associated with travel restrictions during the Ebola epidemic in Sierra Leone: Use of mobile phone data. *International Journal of Epidemiology, 47*(5), 1562–1570.

Poelhekke, S. (2011). Urban growth and uninsured rural risk: Booming towns in bust times. *Journal of Development Economics, 96*(2), 461–475.

Porter, C. (2012). Shocks, consumption and income diversification in rural Ethiopia. *Journal of Development Studies, 48*(9), 1209–1222.

Randell, H. (2016). The short-term impacts of development-induced displacement on wealth and subjective well-being in the Brazilian Amazon. *World Development, 87*, 385–400.

Rigaud, K. K., de Sherbinin, A., Jones, B., Bermann, J., Clement, V., Ober, K., Schewe, J., Adamo, S., McCusker, B., Heuser, S., & Midgley, A. (2018). *Groundswell: Preparing for internal climate migration.* World Bank.

Rose, E. (2001). Ex ante and ex post labor supply response to risk in a low-income areas. *Journal of Development Economics, 64*, 371–388.

Ruiz, I., & Vargas-Silva, C. (2017). The consequences of forced migration for host communities in Africa. *Revue D'Economie du Developpement, 25*(3), 135–154.

Schlenker, W., & Lobell, D. (2010). Robust negative impacts of climate change on African agriculture. *Environmental Research Letters, 5*(1), 014010.

Serdeczny, O., Adams, S., Baarsch, F., Coumou, D., Robinson, A., Hare, W., Schaeffer, M., Perrette, M., & Reinhardt, J. (2017). Climate change impacts in sub-Saharan Africa: From physical changes to their social repercussions. *Regional Environmental Change, 17*(6), 1585–1600.

Silva, J., & Matyas, C. (2014). Relating rainfall patterns to agricultural income: Implications for rural development in Mozambique. *Weather, Climate, and Society, 6*(2), 218–237.

Stirzaker, R., Mbakwe, I., & Mziray, N. R. (2017). A soil water and solute learning system for small-scale irrigators in Africa. *International Journal of Water Resources Development, 33*(5), 788–803.

Strobl, E., & Valfort, M.-A. (2013). The effect of weather-induced internal migration on local labor markets: Evidence from Uganda. *World Bank Economic Review, 29*(2), 385–412.

Strömberg, D. (2007). Natural disasters, economic development, and humanitarian aid. *Journal of Economic Perspectives, 21*, 199–222.

Thomas, D., Witoelar, F., Frankenberg, E., Sikoki, B., Strauss, J., Sumantri, C., & Suriastini, W. (2012). Cutting the costs of attrition: Results from the Indonesia Family Life Survey. *Journal of Development Economics, 98*(1), 108–123.

Thornton, P., Jones, P., Ericksen, P., & Challinor, A. (2011). Agriculture and food systems in sub-Saharan Africa in a 4°C+ world. *Philosophical Transactions of the Royal Society A, 369*, 117–136.

Tusting, L. (2019). Mapping changes in housing in sub-Saharan Africa from 2000 to 2015. *Nature, 568*, 391–394.

United Nations Development Programme (UNDP). (2019). *Mozambique: UNDP mobilizes 1 million in emergency funding, as early recovery phase begins.* Posted on March 27, 2019. Accessed on 23 Apr 2019 online at: http://www.africa.undp.org/content/rba/en/home/presscenter/pressreleases/2019/mozambique%2D%2Dundp-mobilizes-1-million-in-emergency-funding%2D%2Das-ea.html

United Nations Environment Programme (UNEP). (2005). *Africa environment tracking: Issues and development.* United Nations Environment Programme.

United Nations Environment Programme (UNEP). (2011). *Livelihood security: Climate change, migration and conflict in the Sahel.* United Nations Environment Programme.

United Nations International Children's Emergency Fund (UNICEF). (2020). *Cyclone Idai and Kenneth: For the first time in recorded history two strong tropical cyclones have hit Mozambique in the same season.* Posted on March 27, 2019. Accessed online on 9 Sept 2020 at: https://www.unicef.org/mozambique/en/cyclone-idai-and-kenneth

Weiss, D., Atkinson, P., Bhatt, S., Mappin, B., Hay, S., & Gething, P. (2014). An effective approach for gap-filling continental scale remotely sensed time-series. *ISPRS Journal of Photogrammetry and Remote Sensing, 98*, 106–118.

Wesolowski, A., Eagle, N., Tatem, A., Smith, D., Noor, A., Snow, R., & Buckee, C. (2012). Quantifying the impact of human mobility on malaria. *Science, 338*(6104), 267–270.

World Food Program (WFP). (2019, September). *World Food Program: Mozambique country brief.* Accessed on 13 Apr 2020 at: https://docs.wfp.org/api/documents/WFP-0000109792/download/?_ga=2.104649280.1776866508.1586816488-1619767801.1586816488

Wouterse, F., & Taylor, J. E. (2008). Migration and income diversification: Evidence from Burkina Faso. *World Development, 36*(4), 625–640.

Wrathall, D., et al. (2019). Meeting the looming policy challenge of sea-level change and human migration. *Nature Climate Change, 9*, 898–901.

Xu, H. (2006). Modification of normalized difference water index (NDWI) to enhance open water features in remotely sensed imagery. *International Journal of Remote Sensing, 27*, 3025–3033.

Zhang, X. (2015). Reconstruction of a complete global time series of daily vegetation index trajectory from long-term AVHRR data. *Remote Sensing of Environment, 156*, 457–472.

Zhang, P., Deschenes, O., Meng, K., & Zhang, J. (2018). Temperature effects on productivity and factor reallocation: Evidence from a half million Chinese manufacturing plants. *Journal of Environmental Economics and Management, 88,* 1–17.

Zhou, X.-N., Yang, G.-J., Yang, K., Wang, X.-H., Hong, Q.-B., Sun, L.-P., Malone, J., Kristensen, T., Bergquist, N., & Utzinger, J. (2008). Potential impact of climate change on schistosomiasis transmission in China. *American Journal of Tropical Medicine and Hygiene, 78*(2), 188–194.

Water Stress and Migration in Asia

David J. Wrathall and Jamon Van Den Hoek

Abstract

This chapter provides a geospatial review of the empirical research on the link between migration and water stress in Asia. Despite Asia's large landmass, populous countries and diverse cultures and environments, water stress has widely contributed to migration across the region. What generalities can be made? One of the most compelling findings emerging from recent research recognizes the link between surface temperature variation, physical water scarcity, agricultural change, and migration. Temperature appears to be a contributing factor to migration in settings affected not only by drought, but also rainfall extremes. In diverse cases, the mechanism by which changes in temperature influence migration patterns appears to be water scarcity's negative impacts on seasonal agricultural production. While much is known about migration and water stress in Asia, this paper identifies the geographical gaps where migration has not been studied, but

where water stress is likely to occur in the coming century, as well as thematic research gaps in the region. This review concludes with a summary of the key implications for policymakers emerging from research.

Introduction

A stable water cycle provides a basis for human livelihoods, and therefore over the long-term, sustainable development (Smit & Wandel, 2006). As global temperatures rise, water stress, or the circumstances in which the demand for water is not met due to changes in availability, quality, or stability, may undermine societies' place-specific livelihood systems (Adams et al., 2009; Smith et al., 2009), and in turn produce new patterns of human migration as people adjust their livelihood strategies (Black et al., 2011). Water stress is a human-centered problem, which results from climatic drivers of water scarcity (i.e. drought, dry spells, changing seasonality of rainfall) and weather extremes (i.e. flooding, severe weather), but also from human decisions about water distribution and provisioning. This chapter aims to

---

D. J. Wrathall (✉) · J. Van Den Hoek
College of Earth, Ocean and Atmospheric Sciences, Oregon State University, Corvallis, OR, USA
e-mail: david.wrathall@oregonstate.edu;
wrathald@oregonstate.edu

© Springer Nature Switzerland AG 2022
L. M. Hunter et al. (eds.), *International Handbook of Population and Environment*, International Handbooks of Population 10, https://doi.org/10.1007/978-3-030-76433-3_9

evaluate the relationship between migration and water stress in Asia.

In water-stressed areas of Asia, migration is at once a strategy for adapting to changes in water availability and, at the same time, evidence of deficiencies in people's capacities to adapt to change. Migration is a universal strategy that households use to buffer livelihoods from risk. This view holds that individual migrants' capacity to absorb the costs and extract the benefits of migration is highly uneven across societies (see Todaro, 1969), an insight supported here. This review lends support to the "migration as adaptation" thesis originally laid out in McLeman and Smit (2006), which holds that when people experience detrimental environmental change, they can meet their individual needs by reallocating their labor to wage-earning opportunities elsewhere. This perspective explains migration as a mechanism for rural adaptation to environmental stress through individual labor effort and remitted income, in which migrants personally bear the cost of adaptation (Barnett & Webber, 2010). This perspective acknowledges that many social, political and demographic factors influence people's livelihoods and expose them to risk, altering the necessity or desirability of migration (Black et al., 2011).

Migration is used differentially across social categories in response to water stress (Curran, 2002; Henry et al., 2004). For example, men and women may migrate differently as may elderly and young people, and the relatively wealthy and poor. The capacity to migrate of the most vulnerable populations may be eroded gradually over time (see Black et al., 2011). Those unwilling or unable to engage in labor migration despite a necessity for adaptation must also be considered (see Adams, 2016).

To examine the ways in which water stress interacts with population and migration, we identified four categories of water stress. These are not mutually exclusive:

(i) **Physical water scarcity:** This generally refers to stress originating from changes in water quantity, i.e. rainfall variability, increasing temperature, changing sea-sonality of rainfall, and/or drought. This category also applies to instances where water demand increases above supply from economic or demographic drivers. Finally, this category also refers to changes in water quality, e.g. salinity, which reduces water availability. Physical water scarcity results in a supply that fails to meet a population's water demands, specifically for drinking and agriculture.

(ii) **Economic Water Scarcity:** This category refers to chronic water infrastructure problems, when there is physically enough water, but scarcity arises due to pricing, poor infrastructure, or poor management. Likewise, water management, such as allocation of water to agriculture instead of domestic consumption may result in stress that mirrors climate-induced stress (i.e. drought, and changes to growing period).

(iii) **Political Water Scarcity and Conflict:** This category refers to social or political conflict over access to, or use of, or quality of water resources. Intergroup conflict can affect water access or quality.

(iv) **Riverine and Coastal Flooding:** The final category of water stress is related to excesses of water in rivers and along coastlines due to weather extremes or slow changes such as sea level rise. Flooding may also be related human-driven environmental changes in watersheds, such as deforestation or farming.

Under these criteria, water stress may result from climate change, demographic change, water management practices, or all of the above.

The aim of this review is to synthesize empirical knowledge on the nexus of water stress and human migration and to uncover the key dynamics in Asia. This paper geospatially evaluates 184 studies that examine the link between water stress and human migration. We examined observed and projected changes in precipitation and surface temperature of the respective river basins in which those studies occurred under modeled climate change scenarios. Based on these criteria, we surveyed the literature on environmentally-induced migration and identified studies that considered at least one of these four categories of water stress. A review of literature shows that temperature (as a component of water-stress) will be amplified in multiple sub-regions of Asia, where this link has not been clearly evaluated (notable exceptions include Mueller et al., 2014). Temperature thus appears to be one factor linking migration to anthropogenic climate change effects and *predicting* the future impacts of climate change on rates of migration. While this study does not account for differences in water policy, this exercise illuminates the potential for water stress and migration, while also identifying the geographic gaps in our knowledge of water-migration relationships. Basic research should prioritize a better fundamental understanding of how current and future temperature increases interact with other more well-established drivers of migration, such as urbanization, shifting labor conditions and violent conflict. This exercise concludes that, in Asia, there are hotspots of water stress that may be driving current migration and will likely drive future migration. This analysis serves a call to direct future research efforts to these places.

Identifying Research on Water Stress in Asia

This review inherits much from previous reviews of scholarly literature on migration-environment linkages. Due to the interdisciplinary nature of the research community, with contributions from demography, anthropology, geography, and economics, it has been necessary to frequently synthesize and recapitulate the most common methods and findings. Methodological reviews (such as Piguet, 2010, Piguet et al., 2011, McLeman, 2011, and Fussell et al., 2014) have served an essential role in the maturation of this interdisciplinary research community by identifying the diversity of empirical strategies and evaluating their quality and contribution. This paper also takes important cues from Obokata et al. (2014), which provided the first comprehensive summary of empirical research on environmentally-induced international migration. Other reviews have grouped case studies by theme, such as McLeman (2011), which gathered analogues for the most severe adverse consequences of climate change: settlement abandonment. This chapter follows this example, and narrows its scope to migration and water stress, and emphasizes the geography of studies in Asia.

To synthesize the state of empirical research examining the relationship between water stress and migration in Asia, we collected a global sample of 184 peer reviewed publications that explicitly examined linkages between water stress and migration. These included original empirical research reports and reviews of empirical

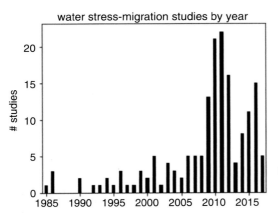 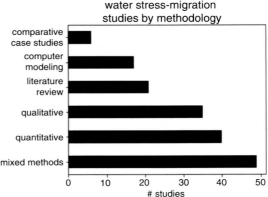

Fig. 9.1 Summary of Studies by year and research method. (**a**) Annual distribution of reviewed publications. (**b**) Frequency of methodological strategies used in reviewed publications

research. To maintain a standard for quality, we selected only papers that had undergone a process of external review.

We followed four sampling strategies. First, we sampled key scientific journals that publish on population-environment issues, such as *Population & Environment, Climatic Change, Global Environmental Change,* and *Nature Climate Change.* Second, upon identifying key publications, we implemented a snowball sampling approach to identify additional publications through the references cited. Third, we examined the publication histories of specific authors with a record of research on water-migration linkages. Fourth, we restricted the search to English language scientific publications; while this restriction biases the total number of publications included in the sample, it does not likely bias the results of such synthesis reporting (Morrison et al. 2012). Nevertheless, we are cognizant that relevant, high quality non-English language scientific publications address key gaps in the literature. Papers reviewed were by and large published in the last ten years, reflecting the recent emergence of the topic (Fig. 9.1).

We then coded papers by: (1) the methodological strategy, (2) study claims and conclusions (Sect. 9.3), (3) study location or extent (Sect. 9.4), and (4) manner of engagement with climate change projections. On this last point, 45% of reviewed studies were explicitly designed to examine migration resulting from water stress introduced by climate change (see McLeman & Hunter, 2010).

The sample includes research from anthropological research, computer modeling, environmental field research methods, literature reviews, historical reviews, qualitative research, quantitative research (making use of census data and nationally representative household survey data), mixed research methods, geospatial methods, and purpose-fit survey research. For the purposes of this review, we narrowed our focus to a set of mutually exclusive categories: (1) mixed methods, (2) quantitative, (3) qualitative, (4) literature review, (5) computer modeling, and (6) comparative case studies (Fig. 9.1b). The largest group of research reports employed a mix

of research methods, which is unsurprising given that population-environment research requires attention to both environmental and social metrics. A majority of papers from this category included methods that were also employed in quantitative research reports (i.e. survey research, census data, or household questionnaires).

We geocoded study location or extent from 116 articles to examine the geographic distribution of reviewed literature. We reviewed each article for evidence of an explicit study location (i.e. a place name or latitude-longitude) or region, which may include several sending and receiving areas. Thirty-four studies were included in the sample that examined water stress and migration at a global scale, or examined multiple global regions. Sixty-eight studies could not be geocoded given the vague nature of the study site description (e.g. rural areas of a country), association with a natural feature rather than a specific location (e.g. a river of substantial length), or inaccurate transliteration or outdated naming conventions. Overall, we found that Asia was far less studied than North America and Africa, and that studies of South Asian countries were more abundant than those in the Near East and East Asia: Bangladesh was studied in 13 articles, India in 11, and a relatively smaller number of studies came from Thailand, Vietnam and Pakistan. (Fig. 9.2).

Mixed methods research dominates the empirical study of migration and water stress (Fig. 9.2). We speculate that knowledge gaps have formed in specific research contexts where states do not implement censuses due to institutional fragility (e.g. Afghanistan); are not supportive of data transparency (e.g. Gulf States, Myanmar); or do not make census data readily available for research (e.g. China). Notable too are the lacunae in countries that have experienced civil war or political instability, such as where armed violence may hinder data collection, the ability of researchers to engage in field research, or both (e.g. Iraq, Syria, Afghanistan). In this respect, little work has been done in drought-prone Afghanistan, and research in and around Syria on water stress and migration has come into the fore only in recent years due to the outbreak of the Syrian Conflict in 2011 (Fig. 9.3).

9 Water Stress and Migration in Asia

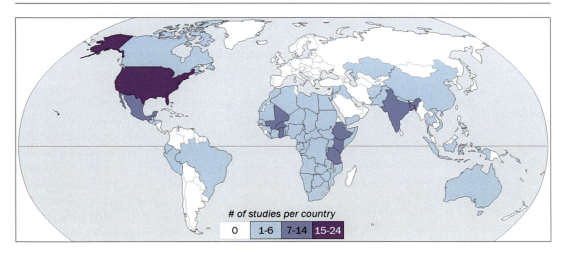

Fig. 9.2 Country-level frequency of 116 geocoded reviewed publications. The distribution of 34 global scale studies is not represented

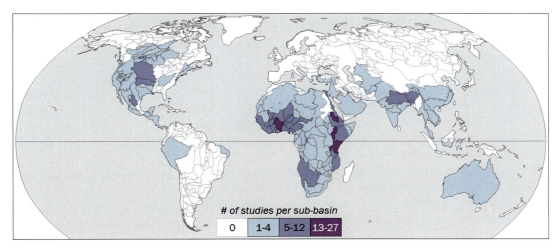

Fig. 9.3 The distribution of 116 geocoded reviewed publications across global sub-basins derived from HYDRO1k dataset. (USGS, 2014)

To complement the national perspective, we measured the global distribution of studies within sub-basins derived from the HYDRO1k drainage basin dataset (USGS, 2014). In each sub-basin, we counted the total number of reviewed articles and identified geographic gaps in coverage (Fig 9.3). Finally, we examined the global distribution of studies with regard to changes in near surface temperature and annual cumulative precipitation between historic (1951–1980) and projected (2041–2060) conditions using multi-model ensemble (MME) products developed for Coupled Model Intercomparison Project, Phase 5 (CMIP5) (Fig. 9.5). Under this framework for comparison, we see that the Ganges-Brahmaputra River Basin is the geographic focus of studies in Asia, followed by the Mekong Delta. Riverine and coastal water stress is the emphasis of research, whereas water stress related to precipitation is less frequent. This exercise reveals the sub-basins that are critically understudied given projected climate change hazards (*viz.* changing rainfall and temperature) most likely to amplify water stress.

Findings

Our geospatial survey of the literature shows that globally the greatest research lacuna on migration and water stress remains in Asia (see Figs. 9.4 and 9.5). Some of the major breakthroughs in research linking water stress to migration (namely the determining influence of temperature, a key ingredient of water stress) have been established in research from Asia: in Pakistan (Mueller et al., 2014), the Philippines (Bohra-Mishra et al., 2017), Bangladesh (Gray & Mueller, 2012), Vietnam (Koubi et al., 2016), and Indonesia (Bohra-Mishra et al., 2014). Virtually every study included in this review concluded that migration within specific life stages is common, and to a large degree, migration patterns can be explained by changes in the national and regional economies. However, the literature from Asia, in agreement with global research, finds that water stress interrupts livelihood activities, destroys assets, and alters rates at which people change residences on a temporary, seasonal, cyclical, or permanent basis. Nevertheless, migrant networks and corridors are not new, and when water stress drives new patterns of migration, they usually fold into existing spatial, temporal, seasonal, and economic patterns of migration. Migration is not a pathological state that arises only under circumstances of environmental stress, but a universal strategy for protecting against risk – and environmental stress can be one driver among many. As for research gaps in Asia, we cannot know why research has not been conducted, but various global studies pointed to a general lack of publicly available data across the region. Nevertheless, to bridge the gap, research in Asia has exploited promising new data sources and techniques.

Declining Agricultural Production Is a Commonly Cited Mechanism for Translating Water Stress into Migration

Recent consensus has emerged around the finding that rainfall extremes and catastrophic flooding impacts on homes or assets are a much weaker predictor of permanent migration than chronic water stress. Studies from a variety of contexts showed that flooding and disaster have modest or only short-term effects on migration (Gray & Mueller, 2012; Lu et al., 2016) but can increase the likelihood of impoverishment. Instead, water stress that results in declined smallholder and subsistence agricultural production is more likely to drive new forms of migration (e.g. Cai et al., 2016). Factors contributing to water stress, such as drought, dry spells, and changing rainfall variability, undermine agricultural production and accelerate or amplify migration patterns as a livelihood adjustment. This has been shown in various settings, including Pakistan (Mueller et al., 2014), the Philippines (Bohra-Mishra et al., 2017), Bangladesh (Gray & Mueller, 2012), Vietnam (Koubi et al., 2016), Indonesia (Bohra-Mishra et al., 2014). Nevertheless, responses to water stress are context dependent and may vary region to region.

Recent consensus has also emerged around the finding that rising temperatures are a stronger predictor of migration than other components of water stress, i.e. drought, rainfall variability and extremes, or flooding. Heat amplifies the potential for migration across a range of water stresses. This has been shown to be the case in country-level studies in Pakistan (Mueller et al., 2014), Indonesia (Bohra-Mishra et al., 2014), the Philippines (Bohra-Mishra et al., 2017), as well as global comparative studies (see Cai et al., 2016). Significantly, migration responses to heat are non-linear, and changes in migration are observed beyond specific temperature thresholds

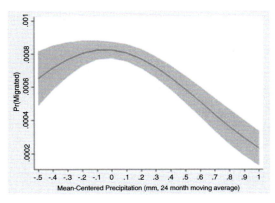
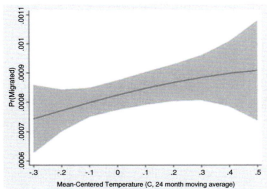

Fig. 9.4 The predicted probabilities of migration in Bangladesh as a function of precipitation and temperature (from Call et al., 2017). The counter-intuitive finding is that while changing rainfall does not produce additional migration (left panel), temperature does (right panel). The principal finding of Cai et al. (2016) is that migration follows a similar response curve around temperature extremes in agriculture-dependent countries

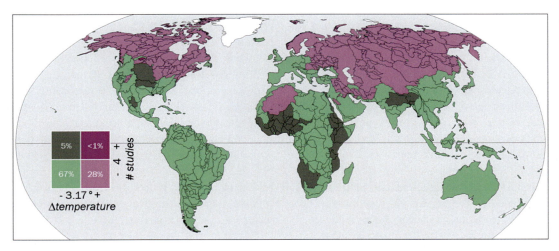

Fig. 9.5 The average change in near surface air temperature from 1951–1980 to 2041–2060 (based on CMIP5 multi-model ensemble outputs) measured within global sub-basins and compared to the number of review studies. Sub-basins are colored relative to whether projected changes in temperature exceed or fall below the sub-basin average increase of 3.17 °C, as well as the number of review studies

in the form of extreme temperatures (see Cai et al., 2016) as well as long-lasting temperatures (see Nawrotzki et al., 2017).

One potential explanation is that extreme or sustained heat increases evaporation and decreases the water content of the topsoil to the detriment of shallow root crops commonly associated with subsistence agricultural production. This phenomenon would explain migration as a response to water shortages, in which water access, storage, or sufficient irrigation capacity does not meet agricultural water demands. As Cai et al. (2016) recently reported in a study of 163 countries, there is a positive and statistically significant relationship between temperature and international out-migration in the most agriculture-dependent countries, where "an additional 100 [hours] of exposure to 30 °C+ weather during the growing season would raise the out-migration rate by 5 percentage points"

(p. 143). Further, the temperature–migration relationship observed in various publications (such as Call et al., 2017 and Cai et al. 2016, see Fig. 9.4) resembled a nonlinear temperature–yield curve for many crops. This may indicate that the decision to migrate by agriculturally dependent people does not immediately follow environmental stress but rather after people have exhausted other adaptation options. This reinforces the notion that changes in agricultural production and migration are tightly coupled.

The causal interaction of temperature, water stress, and migration is still not clearly understood. However, recent papers showing the temperature effect of migration do not only find that it occurs in circumstances of drought (i.e. when water storage and supply diminish) but also in association with rainfall extremes (i.e. when water storage continues). Curiously, Bohra-Mishra (Bohra-Mishra et al., 2014, 2017) found that in Indonesia and the Philippines anomalously high temperatures amplified the migration effect in both instances of drought and extreme precipitation. Future work also requires more systematic study of water demand with sensitivity to regional variations in agricultural land uses, livelihood opportunities, and socioeconomic conditions. In addition, key variables to include in future research are related to the organismal responses of crop varieties to changes in water and temperature.

While our review outlines the ways that water stress drives migration, we found little evidence of migration as a driver of water stress. However, literature from Asia outlined three pathways by which migration can adversely affect water. First, rapid urbanization and growth of informal settlements, in particular, may increase water demand in settings without adequate water supply capacity to support continued in-migration, and could outpace water supply or infrastructure capacity to deliver supply, or treat wastewater (Chatterjee, 2010). At the same time, long-term urban migration can require forms of land use that are water intensive or result in degradation of land, forests or other ecosystems that sustain local water cycles (Xiao et al., 2013). Finally, migration may disrupt land tenure systems that incentivize water con-

servation, and introduce competition for scarce water resources (Bhavanani & Lacina 2015).

Study Geography in Asia Is Broadly Out of Step with Greatest Projected Changes in Temperature and Precipitation

Considering the distribution across global sub-basins of studies linking water stress and human migration, we present the geography of existing research (Fig. 9.5). Yet despite some of the planet's largest international migration corridors running through central Asia and the Middle East (see Abel & Sander, 2014), most watersheds in these regions are conspicuously absent of research. In fact, Asia north of the Himalayan Mountain Rim is virtually unrepresented, and Southeast Asia was relatively understudied given its long coastlines, low-lying river deltas, heavy exposure to tropical flooding and extreme vulnerability. East Asia is likewise understudied despite the historically unprecedented migrations underway in contemporary China, toward burgeoning industrialized coastal cities like Shanghai and Guangzhou. Thematically, for a region replete with rapidly growing cities and peri-urban regions, there are relatively few studies linking water stress to urbanization (Jim et al., 2010).

Globally, the greatest increases in near surface air temperature by mid-century is expected in sub-basins in Central Asia, and much of the Near East; however these settings are virtually unstudied. The best studied watershed in Asia, the Indo-Gangetic Plain, is actually expected to see *below* average (3.17 °C) increases in near surface air temperature (Fig. 9.5). Likewise, there is a paucity of research on migration in areas with projected decreasing annual precipitation –areas where water stress and migration may become a problem— including the entirety of Central Asia, the Persian Gulf and Northwest India (Fig. 9.6). On the other hand, existing research on water stress and migration in the Indo-Gangetic Plain (including India and Bangladesh) may offer a foundation for future research linkages between water stress and migration in regions where some

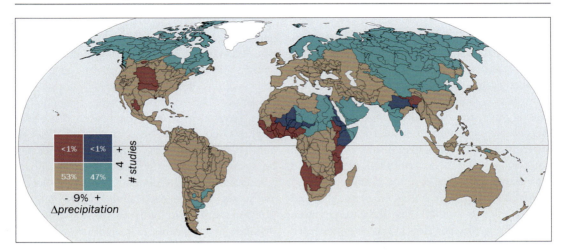

Fig. 9.6 The average relative change in cumulative annual precipitation from 1951–1980 to 2041–2060 (based on CMIP5 multi-model ensemble outputs) measured within global sub-basins and compared to the number of review studies. Sub-basins are colored relative to whether projected relative changes in precipitation exceed or fall below the sub-basin average relative increase of 9%, as well as by the number of review studies

of the greatest increases in precipitation is expected, including South Asia and South East Asia. While modeled changes in precipitation and temperature are minimal in Southeast Asia and Oceania, it is still possible these densely populated regions may experience water stress related to decreasing water quality and/or poor water management.

Aggregating studies at the sub-basin level not only illuminates sub-national study concentrations, it also recognizes the physiographic conditions that influence water stress and asserts the often transboundary character of water stress and migration, as the case of India and Bangladesh so clearly represents. Increasing the number of studies grounded at the sub-basin level, or basin country unit for international transboundary basins, would be a promising advancement for future research on water-migration relationships. Similarly, identifying where significant groundwater resources exist or are being utilized at the sub-basin scale could refine the understanding of water stress and migration.

Overcoming Data Gaps

Nearly as many conceptual papers have addressed water-stress and migration as there are empirical research papers. This is true globally, but particularly the case in Asia. One hypothesis for the abundance of conceptualizations is that the topic is important but difficult to study. Representative household survey data have provided the basis for our knowledge of causal mechanisms driving climate induced migration (see Henry et al., 2004). However, data limitations are a key challenge in some of the most water-stressed regions, including the Middle East and Central Asia, where representative socio-economic data are notoriously scarce.

Another obvious gap emerged around existing migration corridors in water stressed regions, where migrants typically enter neighboring countries without authorization. Well studied contexts include Bangladesh and India. However, the most dominant migration corridors on the planet, in Central Asia and the Near East, remain relatively unstudied.

While household surveys will likely remain the cornerstone of empirical research on this topic, as Lu et al. (2016) point out, they have limited ability to resolve some fundamental knowledge gaps. In particular, migration resulting from water stress is complex, dynamic and evolving over time and across vast spaces. Such migration may include a sequence of moves of differing durations and spatial scales, which are not easily captured in traditional survey-based information. Surveys, too, are rarely conducted with reference to the events –the droughts and tropical storms— that may induce migration and mobility. Studies from Asia have pioneered methodologies that overcome these hurdles. Lu et al. (2016) relied on usage of data from mobile phone networks to reveal fundamental features of human mobility in the context of water stress. A key research problem that new data sources can help resolve is better characterizing the relationship between short-term forms of mobility (evacuation and displacement) and long-term migration. Similarly, remote sensing enabled assessments of human settlements have improved population estimates where censuses do exist (Jean et al., 2016) and hold promise for revealing hidden populations.

Institutions and Policies Have an Important Intervening Role in Asia

In Asia, the literature emphasizes the strong mediating role of institutions on the link between water stress and migration. Speculations that the origins of the Syrian Conflict included water stress factors such as drought, agricultural decline and urban migration (Kelley et al., 2015; Gleick, 2014) have received vociferous push-back (see Selby et al., 2017). In Syria, the origins of civil conflict are not related to climate change, nor migrants, but rather stem from a region-wide reform movement, the Arab Spring, as well as a set of ill-conceived agricultural policies implemented by the Assad regime (Richani, 2016). A handful of articles have shown that conflicts can emerge between host and migrant communities when migrants' needs are ignored or exacerbated by policy. For example, Bhavanani and Lacina (2015) investigated the Sons of the Soil movement in India, in which migrants (moving in part due to drought) have been implicated in destructive riots. They find that politically well-connected host communities have resources available for appeasing disquieted migrants and can thus resolve large-scale social conflict more efficiently than poorly-connected host communities. Without these political resources, hosts and migrants resort to violence. While this is one case study, it highlights a political mechanism for the conversion of natural disasters and, possibly, climate change impacts into violence. It signals a widely applicable strategy for mitigating the risk of migration on violence: ensuring resource access and use.

Discussion

Given the likelihood of severe water stress in combination with high heat stress in the twenty-first century, the analysis points to a geographic gap in research in Asia, particularly Central Asia and the Near East. This exercise makes a clear case for more research on the influence of changing temperature regimes on migration in these priority areas. Research from Asia has included breakthroughs, including the pioneering use of novel data sources such as user-generated cell-phone data (Lu et al., 2016), nevertheless, this review reveals three research themes that should be prioritized in future research:

First, there is a need to understand migration thresholds related to temperature and precipitation (see Bohra-Mishra et al., 2014; Nawrotzki et al., 2017), as well as sea-level rise related hazards such as coastal inundation (Hauer, 2017) and soil salinity (Chen & Mueller, 2018). Evidence from Pakistan reveals that after an initial exposure to water stress, rates of migration may not change, but rather households are able to adapt in place; however multiple, repeating shocks increased the likelihood of migration (Mueller et al., 2014). In other regions, evidence also suggests a delayed migration response to the experience of water stress, but as households cycle through various *potential* adaptation strategies the likelihood of migration increases (Nawrotzki & DeWaard, 2016). More such studies would reveal the strategies that enable people to stay in place, and the moments at which migration becomes necessary.

Secondly, water stress may increase potential for the formation of mobility traps in Asia, in which people are unable or unwilling to migrate despite increasing vulnerability. In some settings (e.g. West Africa), an empirical relationship has been observed between heat, drought and *decreasing* rates of migration; however as of yet, this relationship has not been documented with great nuance in Asia. In Ethiopia, for example, migration of new brides was shown to decrease during the drought periods because less money was available for dowry (Gray & Mueller, 2012). Such findings reveal the impoverishing consequences of water stress, but also how water stress affects the social relations that mediate migration. Because of the mass of Asian populations are in coastal cities which will be chronically exposed to sea-level rise and extreme precipitation over the coming century, the issue of mobility traps is a great concern (Bell et al., 2021).

Finally, in recent years Asia has experienced two migration "crises": refugees fleeing the Syrian civil war and the influx of Rohingya refugee population into Bangladesh. These are case studies of sudden, large-scale, destabilizing migration flows occurring around conflict and water stress. While it is important to underscore that climate change cannot be considered a decisive, determining cause of migration in either case, water stress is an important underlying factor (see Selby et al., 2017). The broader research community could benefit from studies that provide insight and draw implications from these cases of sudden, large-scale migration.

The Policy Implications of Research

This review exposes pressing public problems facing Asia that may demand policy responses, including: rural poverty; rapid urbanization and the expansion of informal settlements in urban peripheries (such as the slum dwellers surrounding mega-cities in India and Bangladesh); urban poverty; the condition of irregular labor migrants who live and work outside of their countries of citizenship (such as migrants from South Asian countries working abroad in Gulf countries); as well as the plight of refugees (such as Syrian and Rohingya refugees). To help global leaders evaluate the potential benefits of pursuing climate mitigation and adaptation policies through a *migration* lens, *there is a need to know* key driving and mitigating factors that influence water-migration relationships.

Five key implications for policymakers are outlined here. First, studies in this review highlighted the link between agricultural problems and migration flows both within countries and across countries. For example, Kartiki (2011) found that climate-related international migration from Bangladesh to India was predominantly irregular. This suggests that one approach to regularizing migration policy is regional-scale investments in adaptation to climate change. Dollar for dollar, investing in programs to adapt to heat and water stress may be more effective in reducing irregular border crossings, and may be more conducive of cooperative international relations than border securitization and militarization.

Secondly, policymakers might consider the long-term development trade-offs of investing in rural adaptation to water stress *versus* urban development. Without societal-level investments and strategies for adapting to water stress, individuals and households bear the costs of adapting to water stress, which they generally accomplish through their migration and wage earning, usually in urban labor markets (Blaikie et al., 2014). Investments in rural agricultural adaptation techniques and technologies and in rural livelihood diversification can diminish the economic imperatives that drive vulnerable people's reliance on migration as a means of coping with water stress. Studies in multiple settings imply that public investments in rural agricultural adaptation may attenuate rapid urbanization where this is a desirable policy goal. A key consideration for Asian countries struggling internally with the challenges of planning rapid urban growth and the expansion of slums and informal settlements at urban peripheries is whether to invest in rural communities' agricultural adaptation and livelihood diversification. Strategies for adapting to rural water stress might be included among urban policy and planning priorities.

Third, policymakers need to envision pathways for development *despite* environmental problems, so that *when* livelihoods are negatively impacted or rendered inoperable (not *if*), migration decisions are chosen and not forced by water stress. Public investments in adaptation may reduce the most deleterious dimensions of human mobility, such as forced displacement and mobility traps. However, when water stress makes migration or relocation necessary, policies can result in desirable outcomes when they prioritize human rights, guarantee entitlements, and expand choices available to migrants (Tanner et al., 2015). Examples include:

- sustained investment in local livelihood diversification
- robust programs of vocational training
- social protections aimed at people whose livelihoods are especially vulnerable
- safety nets for migrants, including portable health and education entitlements
- universal guaranteed minimum wages

Fourth, international water governance can also influence how people experience water stress. In this way, water management agencies play a *de facto* role as migration policymakers. This is especially the case where water stress originates in transboundary watersheds shared among nations, such as the Ganges-Brahmaputra, the Indus and the Mekong Rivers, where migrants are crossing international boundaries (McAdam 2012). Migration should be a consideration in water governance decisions.

Finally, international diplomacy may be required to address two problems that will likely arise in a future where climate change and regional rivalry continues unabated. First, it is not certain how to treat citizens of small island developing states, including Asian countries such as the Maldives, whose entire landmasses are vulnerable to sea level rise (Stojanov et al., 2017). While at present no legal category affording special protections for cross-border climate migrants exists (i.e. "climate refugees"), policy norms have been advanced that ask states to mutually recognize the rights and entitlements of non-citizens migrating across borders due to climate impacts (McAdam, 2016). Secondly, the international community is not well prepared for international migration where conflict and climate change vulnerability are co-occurring. Examples include the Syrian civil war and subsequent refugee "crisis," neither of which can be attributed to climate change (Selby et al., 2017), and the Rohingya refugee population in Bangladesh. Policymakers should understand the destabilizing and deleterious effect of anti-immigrant and nativist rhetoric.

Appendix 1: Additional Articles Considered in the Geospatial Review

Abu, M., Codjoe, S. N. A., & Sward, J. (2014). Climate change and internal migration intentions in the forest-savannah transition zone of Ghana. *Population and Environment*, *35*(4), 341–364.

Adamo, S. B. 2010. "Environmental Migration and Cities in the Context of Global Environmental Change." *Current Opinion in Environmental Sustainability* 2 (3): 161–65.

Adams, V., Van Hattum, T., and English, D. "Chronic Disaster Syndrome: Displacement, Disaster Capitalism, and the Eviction of the Poor from New Orleans." American Ethnologist 36.4 (2009): 615–636.

Adano, R et al. "Climate Change, Violent Conflict and Local Institutions in Kenya's Drylands." Journal of Peace Research 49.1 (2012): 65–80.

Afifi, T. and Warner, K. "The Impact of Environmental Degradation on Migration Flows across Countries." (2008): Accessed from: collections.unu.edu.

Afifi, T., Milan, A., Etzold, B., Schraven, B., Rademacher-Schulz, C., Sakdapolrak, P., Reif, A., van der Geest, K., and Warner, K. (2016). "Human Mobility in Response to Rainfall Variability: Opportunities for Migration as a Successful Adaptation Strategy in Eight Case Studies." *Migration and Development* 5 (2): 254–274.

Afifi, T., Liwenga, E., and Kwezi, L. (2014) "Rainfall-Induced Crop Failure, Food Insecurity and out-Migration in Same-Kilimanjaro, Tanzania." *Climate and Development* 6 (1): 53–60.

Afifi, T. (2011). Economic or Environmental Migration? The Push Factors in Niger. *International Migration* 49: e95–e124.

Afolayan, A. A., and I. O. Adelekan. (1999). "The Role of Climatic Variations on Migration and Human Health in Africa." Environmentalist 18 (4): 213–18.

Akokpari, J. K. (1998). The State, Refugees and Migration in Sub-Saharan Africa. *International Migration* 36 (2): 211.

Albert, S. et al. (2016). Interactions between Sea-Level Rise and Wave Exposure on Reef Island Dynamics in the Solomon Islands. *Environmental Research Letters* 11.5: 054011.

Angus, S. D., B. W. Parris, and B. Hassani-M. (2009). Climate Change Impacts and Adaptation in Bangladesh: An Agent-Based Approach. *Proceedings of the 18th World IMACS Congress and MODSIM09 International Congress on Modelling and Simulation*: Interfacing Modelling and Simulation with Mathematical and Computational Sciences, January, 2720–26.

Anthoff, D. et al. (2006). "Global and Regional Exposure to Large Rises in Sea-Level: A Sensitivity Analysis." Monograph.

Arenstam, G, Sheila J., and Nicholls, R.J. Island Abandonment and Sea-Level Rise: An Historical Analog from the Chesapeake Bay, USA. *Global Environmental Change* 16.1 (2006): 40–47.

Ballard, C. (1986). Drought and Economic Distress: South Africa in the 1800s. *The Journal of Interdisciplinary History* 17 (2): 359–78.

Barbieri, A. F., Domingues, E., Queiroz, B. L., Ruiz, R. M., Rigotti, J. I., Carvalho, J. A., & Resende, M. F. (2010). Climate change and population migration in Brazil's Northeast: scenarios for 2025–2050. *Population and Environment*, 31(5), 344–370.

Barbieri, Alisson F., Edson Domingues, Bernardo L. Queiroz, Ricardo M. Ruiz, José I. Rigotti, José A. M. Carvalho, and Marco F. Resende. 2010. "Climate Change and Population Migration in Brazil's Northeast: Scenarios for 2025–2050." *Population and Environment* 31 (5): 344–70.

Bardsley, D. K., and Hugo, G.J. (2010). Migration and Climate Change: Examining Thresholds of Change to Guide Effective Adaptation Decision-Making. *Population and Environment* 32 (2–3): 238–62.

Barnett, J.R.., and Webber, M. (2010). "Accommodating Migration to Promote Adaptation to Climate Change." SSRN Scholarly Paper ID 1589284. Rochester, NY: Social Science Research Network.

Barnett, J. (2003). Security and Climate Change. *Global Environmental Change* 13 (1): 7–17.

Barrios, S., Bertinelli, L., and Strobl, E.. (2006). Climatic Change and Rural–urban Migration: The Case of Sub-Saharan Africa. *Journal of Urban Economics* 60.3: 357–371.

Bassett, T.J., and Turner, M.D. (2007). Sudden Shift or Migratory Drift? FulBe Herd Movements to the Sudano-Guinean Region of West Africa. *Human Ecology* 35 (1): 33–49.

Biermann, F., and Ingrid B. (2010). Preparing for a Warmer World: Towards a Global Governance System to Protect Climate Refugees. *Global Environmental Politics* 10 (1): 60–88.

Bilsborrow, R.E., and DeLargy, P.F.. (1990). Land Use, Migration, and Natural Resource Deterioration: The Experience of Guatemala and the Sudan. *Population and Development Review* 16: 125–47.

Biswas, A. K., & Quiroz, C. T. (1996). Environmental impacts of refugees: a case study. *Impact Assessment,* 14(1), 21–39.

Black, R. (1994). Forced migration and environmental change: the impact of refugees on host environments. *Journal of Environmental Management,* 42(3), 261–277.

Black, R., Adger, W. N., Arnell, N. W., Dercon, S., Geddes, A., & Thomas, D. (2011). The effect of environmental change on human migration. *Global Environmental Change,* 21, S3-S11.

Bogardi, J., and Warner, K. (2009). "Here Comes the Flood." *Nature Climate Change,* no. 0901 (January): 9–11.

Bohra-Mishra, P., & Massey, D. S. (2011). Environmental degradation and out-migration: New evidence from Nepal. *Migration and climate change,* 74–101.

Bohra-Mishra, P., Oppenheimer, M., & Hsiang, S. M. (2014). Nonlinear permanent migration response to climatic variations but minimal response to disasters. *Proceedings of the National Academy of Sciences,* 111(27), 9780–9785.

Bunce, M., Rosendo, S., & Brown, K. (2010). Perceptions of climate change, multiple stressors and livelihoods on marginal African coasts. *Environment, Development and Sustainability,* 12(3), 407–440.

Burrows, K., & Kinney, P. L. (2016). Exploring the climate change, migration and conflict nexus. *International journal of environmental research and public health,* 13(4), 443.

Byravan, S., & Rajan, S. C. (2010). The Ethical Implications of Sea-Level Rise Due to Climate Change. *Ethics & International Affairs,* 24(3), 239–260.

Byravan, S., & Rajan, S. C. (2015). Sea level rise and climate change exiles: A possible solution. *Bulletin of the Atomic Scientists,* 71(2), 21–28.

Cai, R., Feng, S., Oppenheimer, M., & Pytlikova, M. (2016). Climate variability and international migration: The importance of the agricultural linkage. *Journal of Environmental Economics and Management,* 79, 135–151.

Campbell, K. M., Gulledge, J., McNeill, J. R., Podesta, J., Ogden, P., Fuerth, L., ... & Weitz, R. (2007). *The age of consequences: The foreign policy and national security implications of global climate change.* Center for Strategic and International Studies, Washington DC.

Curtis, K. J., & Schneider, A. (2011). Understanding the demographic implications of climate change: estimates of localized population predictions under future scenarios of sea-level rise. *Population and Environment,* 33(1), 28–54.

Curtis, K. J., Fussell, E., & DeWaard, J. (2015). Recovery migration after Hurricanes Katrina and Rita: Spatial concentration and intensification in the migration system. *Demography,* 52(4), 1269–1293.

Curtis, K. J., & Schneider, A. (2011). Understanding the demographic implications of climate change: estimates of localized population predictions under future scenarios of sea-level rise. *Population and Environment,* 33(1), 28–54.

de Sherbinin, A., Levy, M., Adamo, S., MacManus, K., Yetman, G., Mara, V., ... & Pistolesi, L. (2012). Migration and risk: net migration in marginal ecosystems and hazardous areas. *Environmental Research Letters,* 7(4), 045602.

Deshingkar, P. (2012). Environmental risk, resilience and migration: implications for natural resource management and agriculture. *Environmental Research Letters,* 7(1), 015603.

DeWaard, J., Curtis, K. J., & Fussell, E. (2016). Population recovery in New Orleans after Hurricane Katrina: Exploring the potential role of stage migration in migration systems. *Population and environment,* 37(4), 449–463.

De Weijer, F. (2007). Afghanistan's kuchi pastoralists: Change and adaptation. *Nomadic Peoples,* 11(1), 9–37.

Dietz, T. (1986). Migration to and from dry areas in Kenya. *Tijdschrift voor economische en sociale geografie,* 77(1), 18–26.

Dillon, A., Mueller, V., & Salau, S. (2011). Migratory responses to agricultural risk in northern Nigeria. *American Journal of Agricultural Economics,* 93(4), 1048–1061.

Dun, O. (2011). Migration and Displacement Triggered by Floods in the Mekong Delta. *International Migration,* 49 (2011): e200–e223.

Eckstein, G. E. (2009). Water scarcity, conflict, and security in a climate change world: challenges and opportunities for international law and policy. *Wis. Int'l LJ, 27,* 409.

Ezra, M. (2000). Leaving-home of young adults under conditions of ecological stress in the drought prone communities of northern Ethiopia. *Genus,* 121–144.

Ezra, M., & Kiros, G. E. (2001). Rural Out-migration in the Drought Prone Areas of Ethiopia: A Multilevel Analysis 1. *International Migration Review, 35*(3), 749–771.

Ezra, M. (2001). Demographic responses to environmental stress in the drought-and famine-prone areas of northern Ethiopia. *International Journal of Population Geography, 7*(4), 259–279.

Farbotko, C., and Lazrus, H. (2012) The First Climate Refugees? Contesting Global Narratives of Climate Change in Tuvalu. *Global Environmental Change* 22.2: 382–390.

Feng, S., Krueger, A. B., & Oppenheimer, M. (2010). Linkages among climate change, crop yields and Mexico–US cross-border migration. *Proceedings of the national academy of sciences, 107*(32), 14,257–14,262.

Findley, S. E. (1994). Does drought increase migration? A study of migration from rural Mali during the 1983–1985 drought. *International Migration Review, 28*(3), 539–553.

Fishman, R., Jain, M., & Kishore, A. (2013). Patterns of migration, water scarcity and caste in rural northern Gujarat. *International Growth Centre, London, UK.*

Fratkin, E., and Roth, E.. (1990). Drought and Economic Differentiation among Ariaal Pastoralists of Kenya. *Human Ecology* 18 (4): 385–402.

Fussell, E. (2009). Post-Katrina New Orleans as a new migrant destination. *Organization & Environment, 22*(4), 458–469.

Fussell, E., Curran, S. R., Dunbar, M. D., Babb, M. A., Thompson, L., & Meijer-Irons, J. (2017). Weather-Related Hazards and Population Change: A Study of Hurricanes and Tropical Storms in the United States, 1980–2012. *The Annals of the American Academy of Political and Social Science,* 669(1), 146–167.

Fussell, E., Curtis, K. J., & DeWaard, J. (2014). Recovery migration to the City of New Orleans after Hurricane Katrina: a migration systems approach. *Population and environment, 35*(3), 305–322.

Fussell, E., Sastry, N., & VanLandingham, M. (2010). Race, socioeconomic status, and return migration to New Orleans after Hurricane Katrina. *Population and environment, 31*(1–3), 20–42.

Gila, O. A., Zaratiegui, A. U., & De Maturana Diéguez, V. L. (2011). Western Sahara: Migration, exile and environment. *International Migration, 49*(s1).

Gilbert, G., & McLeman, R. (2010). Household access to capital and its effects on drought adaptation and migration: a case study of rural Alberta in the 1930s. *Population and Environment, 32*(1), 3–26.

Goff, James R., and Bruce G. McFadgen. 2003. "Large Earthquakes and the Abandonment of Prehistoric Coastal Settlements in 15th Century New Zealand." Geoarchaeology 18 (6): 609–23.

Gray, C., & Bilsborrow, R. (2013). Environmental influences on human migration in rural Ecuador. *Demography, 50*(4), 1217–1241.

Gray, C., and Mueller, V. (2012). Drought and Population Mobility in Rural Ethiopia. *World Development* 40 (1): 134–45.

Gutmann, M. P., & Field, V. (2010). Katrina in historical context: environment and migration in the US. *Population and environment, 31*(1–3), 3–19.

Gutmann, M. P., & Field, V. (2010). Katrina in historical context: environment and migration in the US. *Population and environment, 31*(1–3), 3–19.

Haque, C. E. (1985). Impacts of river bank erosion on population displacement in the lower Brahmaputra (Jamuna) floodplain. *Population geography: a journal of the Association of Population Geographers of India, 8*(1–2), 1–16.

Hartmann, B. (2010). Rethinking climate refugees and climate conflict: Rhetoric, reality and the politics of policy discourse. *Journal of International Development: The Journal of the Development Studies Association,* 22(2), 233–246.

Hassani-Mahmooei, B., & Parris, B. W. (2012). Climate change and internal migration patterns in Bangladesh: an agent-based model. *Environment and Development Economics,* 17(6), 763–780.

Haug, R. (2002). Forced migration, processes of return and livelihood construction among pastoralists in northern Sudan. *Disasters,* 26(1), 70–84.

Henry, S., Boyle, P., & Lambin, E. F. (2003). Modelling inter-provincial migration in Burkina Faso, West Africa: the role of socio-demographic and environmental factors. *Applied Geography,* 23(2–3), 115–136.

Henry, S., Piché, V., Ouédraogo, D., & Lambin, E. F. (2004). Descriptive analysis of the individual migratory pathways according to environmental typologies. *Population and Environment,* 25(5), 397–422.

Hinkel, J., Nicholls, R. J., Vafeidis, A. T., Tol, R. S., & Avagianou, T. (2010). Assessing risk of and adaptation to sea-level rise in the European Union: an application of DIVA. *Mitigation and adaptation strategies for global change,* 15(7), 703–719.

Hoffman, Charles S. (1938). Drought and depression migration into Oregon, 1930 to 1936. *Monthly Labor Review* 46 (1): 27–35.

Hugo, G. (1996). Environmental Concerns and International Migration. *The International Migration Review* 30 (1): 105.

Huho, J. M., Ngaira, J. K., & Ogindo, H. O. (2011). Living with drought: the case of the Maasai pastoralists of northern Kenya. *Educational Research,* 2(1), 779–789.

Hunter, B., & Biddle, N. (2011). Migration, labour demand, housing markets and the drought in regional Australia. *Research paper,* 49.

Hunter, L. M. (2005). Migration and environmental hazards. *Population and environment,* 26(4), 273–302.

Jacobsen, K. (1997). Refugees' environmental impact: the effect of patterns of settlement. *Journal of Refugee Studies, 10*(1), 19–36.

Jónsson, G. (2010). "The Environmental Factor in Migration Dynamics-a Review of African Case Studies." Accessed from: https://ora.ox.ac.uk/objects/uuid:cece31bd-0118-4481-acc2-e9ca05f9a763.

Joseph, G., & Wodon, Q. (2013). Is Internal Migration in Yemen Driven by Climate or Socioeconomic Factors?. *Review of International Economics,* 21(2), 295–310.

Jülich, S. (2011). Drought Triggered Temporary Migration in an East Indian Village. *International Migration* 49 (s1).

Kartiki, K. (2011). Climate change and migration: a case study from rural Bangladesh. *Gender & Development,* 19(1), 23–38.

Kayastha, S. L., & Yadava, R. P. (1985). Flood induced population migration in India: a case study of Ghaghara Zone. In *Population redistribution and development in south Asia* (pp. 79–88). Springer, Dordrecht.

Kniveton, D., Smith, C., & Wood, S. (2011). Agent-based model simulations of future changes in migration flows for Burkina Faso. *Global Environmental Change,* 21, S34-S40.

Lerer, L. B., & Scudder, T. (1999). Health impacts of large dams. *Environmental Impact Assessment Review,* 19(2), 113–123.

Leyk, S., Runfola, D., Nawrotzki, R. J., Hunter, L. M., & Riosmena, F. (2017). Internal and international mobility as adaptation to climatic variability in contemporary Mexico: Evidence from the integration of census and satellite data. *Population, space and place,* 23(6), e2047.

Lilleør, H. B., & Van den Broeck, K. (2011). Economic drivers of migration and climate change in LDCs. *Global Environmental Change,* 21, S70-S81.

Lindtjørn, B., Alemu, T., & Bjorvatn, B. (1993). Population growth, fertility, mortality and migration in drought prone areas in Ethiopia. *Transactions of the royal society of tropical medicine and hygiene,* 87(1), 24–28.

Linke, A. M., O'Loughlin, J., McCabe, J. T., Tir, J., & Witmer, F. D. (2015). Rainfall variability and violence in rural Kenya: Investigating the effects of drought and the role of local institutions with survey data. *Global Environmental Change, 34*, 35–47.

Locke, J. T. (2009). Climate change-induced migration in the Pacific Region: sudden crisis and long-term developments 1. *Geographical Journal, 175*(3), 171–180.

Lu, X., Wrathall, D. J., Sundsøy, P. R., Nadiruzzaman, M., Wetter, E., Iqbal, A., ... & Bengtsson, L. (2016a). Detecting climate adaptation with mobile network data in Bangladesh: anomalies in communication, mobility and consumption patterns during cyclone Mahasen. *Climatic Change, 138*(3–4), 505–519.

Marchildon, G. P., Kulshreshtha, S., Wheaton, E., & Sauchyn, D. (2008). Drought and institutional adaptation in the Great Plains of Alberta and Saskatchewan, 1914–1939. *Natural Hazards, 45*(3), 391–411.

Marino, E., & Lazrus, H. (2015). Migration or forced displacement? The complex choices of climate change and disaster migrants in Shishmaref, Alaska and Nanumea, Tuvalu. *Human Organization, 74*(4), 341–350.

Martin, M., Billah, M., Siddiqui, T., Abrar, C., Black, R., & Kniveton, D. (2014). Climate-related migration in rural Bangladesh: a behavioural model. *Population and environment, 36*(1), 85–110.

Mastrorillo, M., Licker, R., Bohra-Mishra, P., Fagiolo, G., Estes, L. D., & Oppenheimer, M. (2016). The influence of climate variability on internal migration flows in South Africa. *Global Environmental Change, 39*, 155–169.

Mbonile, M. (2005). Migration and intensification of water conflicts in the Pangani Basin, Tanzania. *Habitat International, 29*(1), 41–67.

McGranahan, G., Balk, D., & Anderson, B. (2007). The rising tide: assessing the risks of climate change and human settlements in low elevation coastal zones. *Environment and urbanization, 19*(1), 17–37.

McLeman, R., & Smit, B. (2006). Migration as an adaptation to climate change. *Climatic change, 76*(1), 31–53.

McLeman, R. (2006). Migration out of 1930s rural eastern Oklahoma: insights for climate change research. *Great Plains Quarterly, 26*(1), 27–40.

McLeman, R. A. (2009). On the origins of environmental migration. *Fordham Envtl. L. Rev., 20*, 403.

McLeman, R. A., & Hunter, L. M. (2010). Migration in the context of vulnerability and adaptation to climate change: insights from analogues. *Wiley Interdisciplinary Reviews: Climate Change, 1*(3), 450–461.

McLeman, R. A., & Ploeger, S. K. (2012). Soil and its influence on rural drought migration: insights from Depression-era Southwestern Saskatchewan, Canada. *Population and Environment, 33*(4), 304–332.

McLeman, R., Mayo, D., Strebeck, E., & Smit, B. (2008). Drought adaptation in rural eastern Oklahoma in the 1930s: lessons for climate change adaptation research. *Mitigation and Adaptation Strategies for Global Change, 13*(4), 379–400.

McLeman, R., Herold, S., Reljic, Z., Sawada, M., & McKenney, D. (2010). GIS-based modeling of drought and historical population change on the Canadian Prairies. *Journal of Historical Geography, 36*(1), 43–56.

McMichael, C., Barnett, J., & McMichael, A. J. (2012). An ill wind? Climate change, migration, and health. *Environmental health perspectives, 120*(5), 646.

McMichael, C. (2015). Climate change-related migration and infectious disease. *Virulence, 6*(6), 548–553.

McNamara, K. E., & Gibson, C. (2009). We do not want to leave our land': Pacific ambassadors at the United Nations resist the category of 'climate refugees. *Geoforum, 40*(3), 475–483.

Metulini, R., Tamea, S., Laio, F., & Riccaboni, M. (2016). The Water Suitcase of Migrants: Assessing Virtual Water Fluxes Associated to Human Migration. *PLoS ONE, 11*(4).

Meze-Hausken, E. (2000). Migration caused by climate change: how vulnerable are people inn dryland areas?. *Mitigation and Adaptation Strategies for Global Change, 5*(4), 379–406.

Milan, A., & Ruano, S. (2014). Rainfall variability, food insecurity and migration in Cabricán, Guatemala. *Climate and Development, 6*(1), 61–68.

Milman, A., & Arsano, Y. (2014). Climate adaptation and development: Contradictions for human security in Gambella, Ethiopia. *Global environmental change*, 29, 349–359.

Mortreux, C., & Barnett, J. (2009). Climate change, migration and adaptation in Funafuti, Tuvalu. *Global Environmental Change*, 19(1), 105–112.

Müller, M. F., Yoon, J., Gorelick, S. M., Avisse, N., & Tilmant, A. (2016). Impact of the Syrian refugee crisis on land use and transboundary freshwater resources. *Proceedings of the national academy of sciences*, 113(52), 14,932–14,937.

Munshi, K. (2003). Networks in the modern economy: Mexican migrants in the US labor market. *The Quarterly Journal of Economics*, 118(2), 549–599.

Naudé, W. (2008). Conflict, Disasters and No Jobs: Reasons for International Migration from Sub-Saharan Africa. Research paper / UNU-WIDER, 2008. Accessed from: www.econstor.eu.

Nawrotzki, R. J., & Bakhtsiyarava, M. (2017). International climate migration: Evidence for the climate inhibitor mechanism and the agricultural pathway. *Population, Space and Place*, 23(4).

Nawrotzki, R. J., Hunter, L. M., Runfola, D. M., & Riosmena, F. (2015). Climate change as a migration driver from rural and urban Mexico. *Environmental Research Letters*, 10(11), 114,023.

Nawrotzki, R. J., Riosmena, F., Hunter, L. M., & Runfola, D. M. (2015). Amplification or suppression: Social networks and the climate change—migration association in rural Mexico. *Global Environmental Change*, 35, 463–474.

Nawrotzki, R. J., Riosmena, F., Hunter, L. M., & Runfola, D. M. (2015). Undocumented migration in response to climate change. *International journal of population studies*, 1(1), 60.

Nawrotzki, R. J., Runfola, D. M., Hunter, L. M., & Riosmena, F. (2016). Domestic and International Climate Migration from Rural Mexico. *Human Ecology*, 44(6), 687–699.

Nawrotzki, R. J., Schlak, A. M., & Kugler, T. A. (2016). Climate, migration, and the local food security context: introducing Terra Populus. *Population and environment*, 38(2), 164–184.

Nawrotzki, R. J., Riosmena, F., & Hunter, L. M. (2013). Do rainfall deficits predict US-bound migration from rural Mexico? Evidence from the Mexican census. *Population research and policy review*, 32(1), 129–158.

Nicholls, R. J., Hoozemans, F. M., & Marchand, M. (1999). Increasing flood risk and wetland losses due to global sea-level rise: regional and global analyses. *Global Environmental Change*, 9, S69-S87.

Nkonya, E., Gerber, N., Baumgartner, P., von Braun, J., De Pinto, A., Graw, V., ... & Walter, T. (2011). The economics of desertification, land degradation, and drought toward an integrated global assessment. *ZEF-Discussion Papers on Development Policy*, (150).

Obioha, Emeka E. (2008). Climate change, population drift and violent conflict over land resources in northeastern Nigeria. *J. Hum. Ecol*, 23(4), 311–324.

Ocello, C., Petrucci, A., Testa, M. R., & Vignoli, D. (2015). Environmental aspects of internal migration in Tanzania. *Population and Environment*, 37(1), 99–108.

Pedersen, J. (1995). Drought, migration and population growth in the Sahel: the case of the Malian Gourma: 1900–1991. *Population Studies*, 49(1), 111–126.

Perch-Nielsen, S. L., Bättig, M. B., & Imboden, D. (2008). Exploring the link between climate change and migration. *Climatic change*, 91(3), 375–393.

Radel, C., Schmook, B., & McCandless, S. (2010). Environment, transnational labor migration, and gender: case studies from southern Yucatan, Mexico and Vermont, USA. *Population and Environment*, 32(2–3), 177–197.

Rain, D., Engstrom, R., Ludlow, C., & Antos, S. (2011). Accra Ghana: A city vulnerable to flooding and drought-induced migration. *Case study prepared for cities and climate Change: Global Report on Human Settlements, 2011*, 1–21.

Roy, C., & Guha, I. (2013). Climate change induced migration: A case study from Indian Sundarbanas. *ACUMEN-Marian Journal of Social Work, 5*(1), 72–93.

Runfola, D. M., Romero-Lankao, P., Jiang, L., Hunter, L. M., Nawrotzki, R., & Sanchez, L. (2016). The influence of internal migration on exposure to extreme weather events in Mexico. *Society & natural resources, 29*(6), 750–754.

Salauddin, M., & Ashikuzzaman, M. (2012). Nature and extent of population displacement due to climate change triggered disasters in southwestern coastal region of Bangladesh. *International Journal of Climate Change Strategies and Management, 4*(1), 54–65.

Saldaña-Zorrilla, S. O., & Sandberg, K. (2009). Impact of climate-related disasters on human migration in Mexico: a spatial model. *Climatic change, 96*(1), 97–118.

Scheffran, J., Marmer, E., & Sow, P. (2012). Migration as a contribution to resilience and innovation in climate adaptation: Social networks and co-development in Northwest Africa. *Applied geography, 33*, 119–127.

Schultz, J., & Elliott, J. R. (2013). Natural disasters and local demographic change in the United States. *Population and Environment, 34*(3), 293–312.

Schwabach, A. (2003). Ecocide and Genocide in Iraq: International Law, the Marsh Arabs and Environmental Damage in Non-international Conflicts. *TJSL Public Law Research Paper*, (03–08).

Scoones, I. (1992). Coping with drought: responses of herders and livestock in contrasting savanna environments in southern Zimbabwe. *Human ecology, 20*(3), 293–314.

Selby, J., & Hoffmann, C. (2012). Water scarcity, conflict, and migration: a comparative analysis and reappraisal. *Environment and planning C: government and policy, 30*(6), 997–1014.

Shah, A. (2001). Water scarcity induced migration: can watershed projects help?. *Economic and Political Weekly*, 3405–3410.

Shen, S., & Gemenne, F. (2011). Contrasted views on environmental change and migration: the case of Tuvaluan migration to New Zealand. *International Migration, 49*, e224-e242.

Shen, S., & Binns, T. (2012). Pathways, motivations and challenges: contemporary Tuvaluan migration to New Zealand. *GeoJournal, 77*(1), 63–82.

Small, I., Van der Meer, J., & Upshur, R. E. (2001). Acting on an environmental health disaster: the case of the Aral Sea. *Environmental Health Perspectives, 109*(6), 547–549.

Smith, P. J. (2007). Climate change, mass migration and the military response. *Orbis, 51*(4), 617–633.

Swain, A. (1996). Environmental migration and conflict dynamics: focus on developing regions. *Third World Quarterly, 17*(5), 959–974.

Tacoli, C. (2009). Crisis or adaptation? Migration and climate change in a context of high mobility. *Environment and urbanization, 21*(2), 513–525.

Talozi, S., Al Sakaji, Y., & Altz-Stamm, A. (2015). Towards a water–energy–food nexus policy: realizing the blue and green virtual water of agriculture in Jordan. *International Journal of Water Resources Development, 31*(3), 461–482.

Toufique, K. A., & Islam, A. (2014). Assessing risks from climate variability and change for disaster-prone zones in Bangladesh. *International Journal of Disaster Risk Reduction, 10*, 236–249.

Uscher-Pines, L. (2009). Health effects of relocation following disaster: a systematic review of the literature. *Disasters, 33*(1), 1–22.

Van der Geest, K. (2011). North-South migration in Ghana: what role for the environment?. *International Migration, 49*, e69-e94.

Vidyattama, Y., Cassells, R., Li, J., & Abello, A. (2016). Assessing the significance of internal migration in drought affected areas: A case study of the Murray-Darling Basin. *Australasian Journal of Regional Studies, The, 22*(2), 307.

Warner, K., and Afifi, T. (2014). Where the Rain Falls: Evidence from 8 Countries on How Vulnerable Households Use Migration to Manage the Risk of Rainfall Variability and Food Insecurity. *Climate and Development, 6* (1): 1–17.

Warner, K., Ehrhart, C., de Sherbinin, A., Adamo, S., & Chai-Onn, T. (2009). In search of shelter: Mapping the effects of climate change on human migration and displacement. In *In search of shelter: Mapping the effects of climate change on human migration and displacement* (pp. 26–26).

Warner, K., Hamza, M., Oliver-Smith, A., Renaud, F., & Julca, A. (2010). Climate change, environmental degradation and migration. *Natural Hazards, 55*(3), 689–715.

Wiesmann, U., Gichuki, F. N., Kiteme, B. P., & Liniger, H. (2000). Mitigating conflicts over scarce water resources in the highland-lowland system of Mount Kenya. *Mountain Research and Development, 20*(1), 10–15.

Wrathall, D. J. (2012). Migration Amidst Social-Ecological Regime Shift: The Search for Stability in Garífuna Villages of Northern Honduras. *Human Ecology,* 40.4: 583–596.

Yamamoto, L., & Esteban, M. (2010). Vanishing island states and sovereignty. *Ocean & Coastal Management, 53*(1), 1–9.

Yan, T., & Qian, W. Y. (2004). Environmental migration and sustainable development in the upper reaches of the Yangtze River. *Population and Environment, 25*(6), 613–636.

References

Abel, G. J., & Sander, N. (2014). Quantifying global international migration flows. *Science, 343*(6178), 1520–1522.

Adams, H. (2016). Why populations persist: Mobility, place attachment and climate change. *Population and Environment, 37*(4), 429–448.

Adams, H. D., Guardiola-Claramonte, M., Barron-Gafford, G. A., Villegas, J. C., Breshears, D. D., Zou, C. B., … Huxman, T. E. (2009). Temperature sensitivity of drought-induced tree mortality portends increased regional die-off under global-change-type drought. *Proceedings of the National Academy of Sciences, 106*(17), 7063–7066.

Barnett, J. R., & Webber, M. (2010). Accommodating migration to promote adaptation to climate change. *World bank policy research working paper,* (5270).

Bell, A. R., Wrathall, D. J., Mueller, V., Chen, J., Oppenheimer, M., Hauer, M., … Slangen, A. B. A. (2021). Migration towards Bangladesh coastlines projected to increase with sea-level rise through 2100. *Environmental Research Letters, 16*(2), 024045.

Bhavnani, R. R., & Lacina, B. (2015). The effects of weather-induced migration on sons of the soil riots in India. *World Politics, 67*(4), 760–794.

Black, R., Bennett, S. R., Thomas, S. M., & Beddington, J. R. (2011). Climate change: Migration as adaptation. *Nature, 478*(7370), 447–449.

Blaikie, P., Cannon, T., Davis, I., & Wisner, B. (2014). *At risk: Natural hazards, people's vulnerability and disasters.* Routledge.

Bohra-Mishra, P., Oppenheimer, M., & Hsiang, S. M. (2014). Nonlinear permanent migration response to climatic variations but minimal response to disasters. *Proceedings of the National Academy of Sciences, 111*(27), 9780–9785.

Bohra-Mishra, P., Oppenheimer, M., Cai, R., Feng, S., & Licker, R. (2017). Climate variability and migration in the Philippines. *Population and Environment, 38*(3), 286–308.

Cai, R., Feng, S., Oppenheimer, M., & Pytlikova, M. (2016). Climate variability and international migration: The importance of the agricultural linkage. *Journal of Environmental Economics and Management, 79,* 135–151.

Call, M. A., Gray, C., Yunus, M., & Emch, M. (2017). Disruption, not displacement: Environmental variability and temporary migration in Bangladesh. *Global Environmental Change, 46,* 157–165.

Chatterjee, M. (2010). Slum dwellers response to flooding events in the megacities of India. *Mitigation and Adaptation Strategies for Global Change, 15*(4), 337–353.

Chen, J., & Mueller, V. (2018). Coastal climate change, soil salinity and human migration in Bangladesh. *Nature Climate Change, 8*(11), 981–985.

Curran, S. (2002). Migration, social capital, and the environment: Considering migrant selectivity and networks in relation to coastal ecosystems. *Population and Development Review, 28,* 89–125.

Fussell, E., Hunter, L. M., & Gray, C. L. (2014). Measuring the environmental dimensions of human migration: The demographer's toolkit. *Global Environmental Change, 28,* 182–191.

Gleick, P. (2014). Water, drought, climate change, and conflict in Syria. *Weather, Climate, and Society, 6*(3), 331–340.

Gray, C., & Mueller, V. (2012). Natural disasters and population mobility in Bangladesh. *Proceedings of the National Academy of Sciences, 109*(16), 6000–6005.

Hauer, M. E. (2017). Migration induced by sea-level rise could reshape the US population landscape. *Nature Climate Change, 7*(5), 321–325.

Henry, S., Schoumaker, B., & Beauchemin, C. (2004). The impact of rainfall on the first out-migration: A multi-level event-history analysis in Burkina Faso. *Population & Environment, 25*(5), 423–460.

Jean, N., Burke, M., Xie, M., Davis, W. M., Lobell, D. B., & Ermon, S. (2016). Combining satellite imagery and machine learning to predict poverty. *Science, 353*(6301), 790–794.

Jim, C. Y., Yang, F. Y., & Wang, L. (2010). Social-ecological impacts of concurrent reservoir inundation and reforestation in the three gorges region of China. *Annals of the Association of American Geographers, 100*(2), 243–268.

Kartiki, K. (2011). Climate change and migration: A case study from rural Bangladesh. *Gender and Development, 19*(1), 23–38.

Kelley, C. P., Mohtadi, S., Cane, M. A., Seager, R., & Kushnir, Y. (2015). Climate change in the Fertile Crescent and implications of the recent Syrian drought. *Proceedings of the National Academy of Sciences, 112*(11), 3241–3246.

Koubi, V., Spilker, G., Schaffer, L., & Bernauer, T. (2016). Environmental stressors and migration: Evidence from Vietnam. *World Development, 79*(2016), 197–210.

Lu, X., Wrathall, D. J., Sundsøy, P. R., Nadiruzzaman, M., Wetter, E., Iqbal, A., . . . Bengtsson, L. (2016). Unveiling hidden migration and mobility patterns in climate stressed regions: A longitudinal study of six million anonymous mobile phone users in Bangladesh. *Global Environmental Change, 38*, 1–7.

McAdam, J. (2012). *Climate change, forced migration, and international law*. Oxford University Press.

McAdam, J. (2016). From the Nansen initiative to the platform on disaster displacement: Shaping international approaches to climate change, disasters and displacement. *UNSWLJ, 39*, 1518.

McLeman, R. A. (2011). Settlement abandonment in the context of global environmental change. *Global Environmental Change, 21*, S108–S120.

McLeman, R. A., & Hunter, L. M. (2010). Migration in the context of vulnerability and adaptation to climate change: insights from analogues. *Wiley Interdisciplinary Reviews: Climate Change, 1*(3), 450–461.

McLeman, R., & Smit, B. (2006). Migration as an adaptation to climate change. *Climatic Change, 76*(1), 31–53.

Morrison, A., Polisena, J., Husereau, D., Moulton, K., Clark, M., Fiander, M., . . . & Rabb, D. (2012). The effect of English-language restriction on systematic review-based meta-analyses: A systematic review of empirical studies. *International Journal of Technology Assessment in Health Care, 28*(2), 138–144.

Mueller, V., Gray, C., & Kosec, K. (2014). Heat stress increases long-term human migration in rural Pakistan. *Nature Climate Change, 4*(3), 182–185.

Nawrotzki, R. J., & DeWaard, J. (2016). Climate shocks and the timing of migration from Mexico. *Population and Environment, 38*(1), 72–100.

Nawrotzki, R. J., DeWaard, J., Bakhtsiyarava, M., & Ha, J. T. (2017). Climate shocks and rural-urban migration in Mexico: Exploring nonlinearities and thresholds. *Climatic Change, 140*(2), 243–258.

Obokata, R., Veronis, L., & McLeman, R. (2014). Empirical research on international environmental migration: A systematic review. *Population and Environment, 36*(1), 111–135.

Piguet, E. (2010). Linking climate change, environmental degradation, and migration: A methodological overview. *Wiley Interdisciplinary Reviews: Climate Change, 1*(4), 517–524.

Piguet, E., Pécoud, A., & De Guchteneire, P. (2011). Migration and climate change: An overview. *Refugee Survey Quarterly, 30*(3), 1–23.

Richani, N. (2016). The political economy and complex interdependency of the war system in Syria. *Civil Wars, 18*(1), 45–68.

Selby, J., Dahi, O., Fröhlich, C., & Hulme, M. (2017). Climate change and the Syrian civil war revisited: A rejoinder. *Political Geography, 60*, 253–255.

Smit, B., & Wandel, J. (2006). Adaptation, adaptive capacity and vulnerability. *Global Environmental Change, 16*(3), 282–292.

Smith, J. B., Schneider, S. H., Oppenheimer, M., Yohe, G. W., Hare, W., Mastrandrea, M. D., . . . Füssel, H. M. (2009). Assessing dangerous climate change through an update of the Intergovernmental Panel on Climate Change (IPCC)"reasons for concern". *Proceedings of the National Academy of Sciences, 106*(11), 4133–4137.

Stojanov, R., Duží, B., Kelman, I., Němec, D., & Procházka, D. (2017). Local perceptions of climate change impacts and migration patterns in Malé, Maldives. *The Geographical Journal, 183*(4), 370–385.

Tanner, T., Lewis, D., Wrathall, D., Bronen, R., Cradock-Henry, N., Huq, S., . . . Alaniz, R. (2015). Livelihood resilience in the face of climate change. *Nature Climate Change, 5*(1), 23.

Todaro, M. P. (1969). A model of labor migration and urban unemployment in less developed countries. *The American Economic Review, 59*(1), 138–148.

USGS, U. (2014). HYDRO1K documentation. *Survey, USG, ed, 2014.*

Xiao, L. B., Fang, X. Q., & Ye, Y. (2013). Reclamation and revolt: Social responses in eastern Inner Mongolia to flood/drought-induced refugees from the north China plain 1644–1911. *Journal of Arid Environments, 88*, 9–16.

Environmentally Informed Migration in North America

10

Elizabeth Fussell and Brianna Castro

Abstract

The natural environment shapes North American settlement patterns by pulling people toward natural resources and amenities and, to a lesser extent, pushing them away from hazards. Environmental management and disaster recovery policies have minimized concerns with settlements in hazardous places. However, in the Anthropocene era, more frequent and intense environmental hazards threaten to overwhelm society's capacity to protect against hazard-related death and destruction. This chapter reviews research on three types of environmental migration in North America: amenity migration, natural hazards-related migration, and anthropogenic hazards-related migration. Spatial and social forces are found to strongly influence all three types of environmental migration. The chapter concludes by focusing on research that identifies the places and people that are most vulnerable to climate change-related migration, and the policies and practices that influence residential location decisions, particularly those influencing residential mobility after disasters. As the climate crisis unfolds, such policies are more likely to inform environmental migration behavior in North America.

Introduction

Despite growing scientific and public concern about the effect of the environment on population, and vice versa, research on the environment as a driver of human migration remains a narrow specialty area in population studies (Hunter & Menken, 2015; McDonald, 2015; Pebley, 1998; Ruttan, 1993). North America is a prime site for research on environmental migration, especially as it relates to climate change, due to its geographic variability in coupled human-natural systems, a wealth of spatial and population data, and the large scholarly community. To date, there are over 60 case studies based in the US and

E. Fussell (✉)
Population Studies and Training Center, Institute at Brown on Environment and Society, Brown University, Providence, RI, USA
e-mail: elizabeth_fussell@brown.edu

B. Castro
Sociology Department, Harvard University, Cambridge, MA, USA
e-mail: briannacastro@g.harvard.edu

© Springer Nature Switzerland AG 2022
L. M. Hunter et al. (eds.), *International Handbook of Population and Environment*, International Handbooks of Population 10, https://doi.org/10.1007/978-3-030-76433-3_10

10 in Canada, making it one of the most well researched areas of the world on this topic (Piguet et al., 2018). The aim of this chapter is to assess multidisciplinary research on environmental migration in North America,[1] emphasizing population scientists' contributions, and to highlight future research directions.

Population scientists have long argued that the environment is both a push and a pull factor for migration (Lee, 1966). Migration theories mainly conceptualize the environment as an indirect influence on migration, operating through unemployment or wage differentials that exist between origin and destination communities (Greenwood, 1997) or political, social, and demographic drivers (Black et al., 2011). The environment is largely invisible in much of the contemporary migration literature, not because it is irrelevant, but because it is a relatively stable feature that has been of less interest to social scientists than the economic and social forces that redistribute population across space and among places (Pebley, 1998). It is mainly when environmental conditions change or a sudden event occurs that the environment becomes part of the migration research agenda.

We take this demographic insight – that the environment pushes people from an origin area and pulls them toward a destination – to organize the literature on environment and migration in North America. In the next section, we review the tremendous investment in environmental management in North America during the 19th and 20th centuries that facilitated the redistribution of the population across the variegated North American geography. By managing natural hazards, these investments pulled European settlers to places inhabited by indigenous people and allowed the formation of large cities. The main section brings us to the more recent past and focuses on three types of environmental pushes and pulls on migrants: amenity migration, natural hazards-related migration, and anthropogenic hazards-related migration. Next, a future-facing section considers the literature on place and social

vulnerability to climate hazards and policies influencing environmental migration. We conclude with guidance for future research.

Development, Environmental Management, and Population Redistribution in North America

The environment is central in North American human migration dynamics, although its role has evolved over time. The focus of this chapter is on the Anthropocene, a period that begins with the industrial era (1800–1945) when human activities began to "overwhelm the forces of nature" (Crutzen, 2002; Steffen et al., 2007). In the industrial era, human consumption of fossil fuel for energy began to raise concentrations of atmospheric pollutants and created a greenhouse effect that raised average global temperatures. After World War II, the growth rate of atmospheric pollutants sped up sharply, producing observable increases in sea levels, glacial melting, and extreme weather (Steffen et al., 2007). Another way in which humanity has overwhelmed the forces of nature is through environmental management, that is, changes to the natural environment that allow humans to live safely and comfortably in a wide range of climates and terrains. The history of environmental management provides lessons for how humans may or may not be able to adapt to the environmental changes we are currently experiencing and that will become more frequent, intense, and destructive in the future.

Environmental Management and Human Settlement in North America

Census records from the US and Canadian governments document population movement across the continent, making evident the strong association between settlement patterns and environmental conditions.[2] The 100th meridian

---

[1] In this chapter, North America is defined as the United States and Canada. Other chapters of the handbook cover Latin America and the Caribbean.

[2] The first US decennial census was taken in 1790, counting all free whites and enslaved blacks. Censuses were taken every ten years since and, beginning in 1860, included American Indians living off reservations and those living

approximates a climatic boundary that divides North America into an arid region to the west, beginning in the Great Plains, and a wet region to the east (Leifert, 2018). Until about 1840, when the US frontier was still east of the 100th meridian, settlement followed an "Eastern pattern" of contiguous, densely populated areas (Lang et al., 1995). The fur trade and the Gold Rush of 1848 enticed pioneers further west, but government programs were crucial for encouraging settlement in the arid region with its daunting and vast grass prairies, deserts, and mountains. The US Homestead Act (1862) and the Canadian Dominion Lands Act (1872) granted public land to settlers who agreed to establish farms, thus building up larger settlements. The government-subsidized construction of the transcontinental railroads spurred migration further west, with the Pacific Railroad connecting San Francisco to the eastern railroads network in 1869 and the Canadian Pacific Railway spanning British Columbia to eastern Canada in 1885 (White, 2011). Settlements west of the 100th meridian took a "Western pattern" of cities surrounded by large, low-density populated land that persists to the present (Lang et al., 1995). These settlement patterns reflect the ways in which technology and markets overcame environmental barriers to settlement and allowed for migration and the growth of cities (Cronon, 1991).

Water access and management shaped both the Eastern and Western settlement patterns (Fang & Jawitz, 2019; Rappaport & Sachs, 2003). Private landowners' efforts were insufficient to manage the streams, rivers, and lakes that provided fresh water for rural and urban settlements. In the early eighteenth century, Congress charged the US Army Corps of Engineers (US ACE) with building and maintaining the nation's infrastructure (Barry, 2007). A central part of its mission was to build the dams, levees, canals, and reservoirs necessary to maintain waterways for transportation, create and protect water supplies for human consumption and agricultural use, and prevent and control flood hazards (Shallat, 1994; US ACE, n.d.). Motivated by the catastrophic Mississippi River flood of 1927 and unemployment during the Great Depression, Congress passed the 1936 Flood Control Act, which made flood control a federal activity involving multiple agencies and many public works projects (Barry, 2007). In Canada, private, provincial, and territorial resources directed and funded dam construction and management (Canadian Dam Association, n.d.). Dam construction and other fortifications against hazards prevent flood damage, but they also concentrate population in flood vulnerable areas (Lindsay, 2014).

The Red River flood of 1997 is an example of how the failure of water management created a disaster and displaced the residents who had settled in an environmentally vulnerable area. That spring an early thaw released a tremendous amount of water stored in the frozen Red River, overwhelming the holding capacity of flood control infrastructure in the Canadian Province of Southern Manitoba and the US states of North Dakota and Minnesota (International Joint Commission, 2000). In the greater Grand Forks area of North Dakota and Minnesota floodwater stretched as far as three miles beyond the river's banks, affecting over 100,000 people and damaging over US$5 billion worth of crops and property. After the event, a joint task force recommended structural and nonstructural changes to prevent catastrophic impacts from future flooding, including demolishing entire neighborhoods in the floodplain and relocating residents to higher ground in Grand Forks (IJC 2000). Relocating residents after an extreme event is unusual, although managed retreat from hazards is an emergent topic in environmental migration that we discuss in Sect. 10.4.

The relocation of Grand Forks' residents is an exception to the governmental approach of restoring people and places to their pre-disaster conditions. The US ACE began providing formal disaster relief after the Mississippi River Flood of 1882. Since then, Congress continued

---

on reservations in 1900 (US Census, n.d.-a). The Canadian decennial census began in 1871 and included all residents, including aboriginals (Statistics Canada, n.d.).

to greenlight both water-related infrastructure projects and social protections for communities affected by natural and anthropogenic disasters. In the US, federally subsidized social protections include the Federal Crop Insurance Program (voluntary 1938; mandatory 1994–present), the National Flood Insurance Policy (1968–present), and the Federal Emergency Management Agency (formed 1979) to assist communities and individuals in recovery from all types of disasters (Schroeder et al., 2001). Since 1959, the Canadian government has provided federally subsidized crop and livestock insurance against losses due to flood, hail, and drought (Lindsay, 2014; Public Safety Canada, n.d.). However, it was not until 2014 that the Canadian government introduced the National Disaster Mitigation Program focused on reducing flood risk and costs and facilitating private residential insurance for overland flooding (Public Safety Canada, n.d.). Programs insuring against losses from environmental hazards allow residents to remain in places they might otherwise deem too risky, thereby producing a sense of security, and even complacency, among residents (Di Baldassarre et al., 2018; Burby, 2001).

Urbanization and the Environment as an Amenity

The environmental management policies described in the previous section created a more secure and stable environment-migration regime in North America, in which human settlements are engineered to withstand the effects of natural hazards and, when disasters occur, most residents are able to recover in place. Research on the migratory response to the Dustbowl of the 1930s and subsequent internal migration provides evidence of the success of this regime (Gutmann et al., 2005; McLeman et al., 2013; Parton et al., 2007). A multiyear drought and poor land management practices throughout the Great Plains – an area extending from the Canadian Provinces of Alberta, Saskatchewan, and Manitoba southward to Oklahoma and Texas in the US – resulted in soil degradation and dust storms. These environmental conditions combined with low commodity prices and multi-sectoral unemployment to increase out-migration from and lower in-migration to the region, especially the hardest hit areas in the Southern Great Plains (Gregory, 1989; Long & Siu, 2016; McLeman & Smit, 2006). Gutmann et al. (2005) take a long view of the relationship between environment and population change in the US Great Plains between 1930 and 1990 to show that, all else equal, the association between drought and net out-migration has diminished since the 1930s. In later decades, Great Plains counties with environmental amenities, such as higher altitudes and water features, even experienced net in-migration relative to those without such amenities. Since the 1950s, natural amenities exert greater influence on environmental migration than harsh environmental conditions since hazard mitigation infrastructure, improved agricultural technology, governmental policies, and public and private insurance are able to manage risks associated with extreme weather events and flooding (Gutmann et al., 2005). These environmental management methods reduce environmental limits on where people are able to settle.

As industrialization drew Americans into cities, more Americans, especially urbanites, came to appreciate the natural environment as an amenity rather than a basis for their livelihood (Cronon, 1991, 1996). Urbanization trends illustrate the scope of this transformation of North Americans' lives: between 1850 and 1960, the percentage of the US population living in urban areas grew from 15% to 70% (US Census Bureau, 2012) and from 13% to 76% in Canada (Statistics Canada, 2011). Rates of urbanization slowed in subsequent years, and new patterns of suburbanization, regional redistribution of population from the East to the South and West, and increased urban-to-rural migration emerged (Frey & Speare, 1988; Johnson & Beale, 1994; Manson & Groop, 2000). Today, North America is largely urban, with just over 80% of the population in both countries living in metropolitan areas (Statistics Canada, 2018; US Census, n.d.-b).

The processes underlying the spatial distribution of population are complex and more than can be summarized here (Fossett (2006) reviews this literature). To date, social scientists and demographers have only incorporated the natural environment and environmental management into their explanations of spatial population distribution to a very limited degree, if at all. It is undeniable, however, that the hazard mitigation and response technologies described above, alongside improved transportation and communications technologies, the abundance of low cost fossil fuel-based energy, and more efficient home heating and air conditioning have allowed residents to live safely and comfortably in a wider variety of natural and built environments (Biddle, 2008; Rappaport, 2007). From this accounting, it is apparent that many North Americans have adapted in place rather than migrated away from hazards (Dronkers et al., 1990; Gutmann & Field, 2010). These long-run secular trends provide the relatively stable context in which social scientists examine short and medium-term changes in migration patterns. However, as climate change produces more variability in weather and increased flooding from sea level rise, migration may become a more attractive adaptation option. To orient demographers and social scientists to these anticipated changes, we review the kinds of environmental migration that have occurred in the North American context.

Three Types of Environmental Migration

Building on the insight that the environment is both a push and pull factor in migration dynamics, we consider three types of environmental migration: amenity migration, natural hazard migration, and pollution-related migration. All three types of environmental migration involve the environment as either a push or pull factor. Research on environmental migration asks two distinct questions: Do people move in response to the environment? If they do move, are some types of households or people more likely to

move? Typically, data to study migration and environment relationships in North America is available at two levels of analysis: an aggregated geography – such as residential blocks, census tracts, or counties – and an individual household or person level (Bilsborrow & Henry, 2012; Fussell et al., 2014; Hsiang, 2016). Research at the county-level focuses on changes in population or net migration rates between counties. Counts of movers and non-movers can be broken down by a limited set of social and demographic characteristics. However, analyses at the aggregate level are subject to the ecological inference fallacy; this occurs when inferences about individuals are made, wrongly, from group-level statistics (Diez-Roux, 1998). Research at the individual or household level overcomes this problem and allows for more insight into factors that inform migration decision-making. However, individual or household level data often lack the geographic coverage and specificity available in aggregate level data. Therefore, research at both levels of analysis contributes to knowledge of the environmental dimensions of migration. This section relies heavily on US-based research, reflecting the geographic focus of existing scholarship.

Amenity Migration

In the second half of the twentieth century, migration, both internal and international, has driven the growth of cities and their surrounding suburbs (Johnson et al., 2005; Plane et al., 2005). Throughout this time, people have sorted themselves into different types of places according to their stage in the life course (Fuguitt & Beale, 1993; Johnson et al., 2005; Plane et al., 2005; Plane & Jurjevich, 2009). Central cities gained young adults in their twenties. The suburbs of metropolitan areas, where population density is lower but jobs are still within commuting range, tended to gain slightly older working age adults who were more likely to have children. Rural areas, where amenities were higher and cost-of-living was lower, added older adults who were nearing retirement or at retirement age. These life-stage patterns of migration were clearly driven by place-based amenities, whether they are economic, social, or environmental.

Regional science provides evidence that environmental amenities have become an important factor in migration (Graves, 1980; Rappaport, 2007; Partridge, 2010). In the US, Rappaport (2007) shows that since 1970 people have been moving to places with warm winters, as well as to places with cooler summers. Since these are not always the same places, this pattern suggests that while not everyone agrees on what a pleasing climate is, they agree that such a climate is worth moving for. Many of these pleasing places are on the Atlantic, Gulf, and Pacific coasts, especially the shoreline counties, where population density is increasing in these already densely populated counties (US NOAA, 2013). Environmental features also draw migrants seeking recreation to service-based, amenity rich counties in the Mountain West (Shumway & Otterstrom, 2001). In Canada, the population is already concentrated in warmer, coastal regions, suggesting that climate amenities play a role in migration decisions there (Brown & Scott, 2012).

While both jobs and place-based amenities contribute to this regional redistribution, a small literature has emerged examining which factor is more important (e.g. Brown & Scott, 2012; Chen & Rosenthal, 2008; Partridge, 2010). Partridge (2010) argues from the empirical record for the US that place-based amenities, including environmental and social amenities, have become more important than jobs. However, when environmental amenities are capitalized in the cost-of-living of more highly endowed places, such as coastal or mountain areas developed for tourism, only wealthier migrants can afford to relocate to those places (Keenan et al., 2018; McNamara & Keeler, 2013). Employers competing for workers with specialized skills and high levels of human capital obtain an advantage in the labor market by locating employment sites in places with natural and other types of amenities (Sutton & Day, 2004). In this way, environmental amenities contribute to residential class segregation by sorting more advantaged groups into high amenity locations and less advantaged groups into places with fewer amenities or even disamenities. Freed from labor

market constraints, retirees often migrate for environmental amenities (Fuguitt & Heaton, 1995; Johnson et al., 2005; Walker, 2016). Florida and the Southwestern states of Arizona and Nevada attract this population (Johnson et al., 2005). This is not so for internal migrants in Canada, although part-year "snow birds" often travel to warmer climates in the US and elsewhere (Northcott & Petruik, 2013). What is clear from research on amenity migration is that climate and other environmental features, such as coasts, lakes, and mountains, play an increasingly important role in migration as North Americans have come to value the environment as a consumption amenity.

Natural Hazard-Related Migration

The natural environment is viewed as an amenity by many movers, but it is also a hazard. Natural hazards are meteorological or geological events, such as hurricanes, earthquakes, or even heavy precipitation. A disaster occurs when a natural hazard overwhelms the capacity of the built and social environment to withstand the event, resulting in property damages, injuries and fatalities (Wisner et al., 2004). The low-probability that a disaster will occur in a given time or place means that few people incorporate these into their residential decisions, although they may decide to relocate after a disaster event. Since a natural hazard and the built and social environment all contribute to a disaster event, social scientists argue that there is no such thing as a "natural" disaster, and even less so in the Anthropocene era, when human activity has increased the destructive power of natural hazards (Cutter et al., 2003; Wisner et al., 2004).

Sea level rise is a particular concern for population scientists, since a disproportionate number of the world's human population lives in the low elevation coastal zone (McGranahan et al., 2007; Hauer et al., 2020). As oceans warm and sea level rises, hurricanes, tropical storms and associated storm surges have become more intense and destructive, and flood events due to king tides are more frequent (Field et al., 2012). The intersection of these coastal hazards with coastal populations, especially urban coastal populations, has already increased property and crop losses (Pielke Jr. et al., 2008; Weinkle et al., 2018) and is expected to cause future human population displacement (Curtis & Schneider, 2011; Hauer et al., 2016; Preston, 2013; Weinkle et al., 2018). However, to date, research provides no evidence of a reversal of the pattern of amenity migration toward coastal counties described in the previous section (US NOAA, 2013; Hauer et al., 2019).

Hazards research provides insight into why migration is not a more common response to hazard exposure. Post-disaster mobility takes many forms including emergency sheltering, temporary evacuation, short-term displacement followed by return, and, only rarely, long-term relocation (Peacock et al., 2018; Smith & McCarty, 1996; Smith & McCarty, 2009). Only the last of these is migration, which demographers define as a change of permanent residence.[3] As already noted, in the US and Canada, disaster recovery focuses on rebuilding homes, businesses, and communities. Therefore, the challenge is to explain why some groups are not able to recover in place. In general, homeowners have more financial resources than renters, including income, savings, and private and public insurance, to rebuild their housing and replace damaged property. Furthermore, barring governmental decisions that preclude

---

[3]Conceptually, migration involves a change of usual residence (Lee, 1966). However, migration has spatial and temporal dimensions, and may sometimes reference migrants' agency or volition (Fussell, 2012; McLeman, 2014). The spatial dimension refers to the type of boundary crossed, most commonly distinguishing between international and internal boundaries, or the characteristics of origin and destination, such as rural-to-urban or rural-to-rural moves. Most environmental migrants choose nearby destinations within national boundaries (Findlay, 2011). The temporal dimension of migration involves duration as well as frequency of moves, and may be described as seasonal, temporary, recurrent, continuous, indefinite, or permanent (McLeman, 2014). Most measures of migration impose temporal or spatial criteria, making it difficult to compare across studies and data sets. Designations of migrations as voluntary or involuntary are even more challenging since migrants rarely report their reason for moving. This dimension is viewed typically as a continuum (Hugo, 1996; Fussell, 2012).

rebuilding, homeowners have legal authority to rebuild their homes. Renters whose homes have been damaged have limited legal claim on their residence (Peacock, Dash, Zhang, et al. 2018). Consequently, housing tenure is a key factor in post-disaster residential mobility (Fussell & Harris, 2014; Elliott & Howell, 2017). In addition, groups that tend to rent or to have few financial resources are also more likely to move after a disaster. These groups include racial and ethnic minorities, low-income and poor households, members of female-headed households, and people with disabilities (Cutter et al., 2003; Morrow-Jones & Morrow-Jones, 1991; Fothergill et al., 1999; Fothergill & Peek, 2004). An emerging body of research examines demographic and social differentials in disaster-related residential mobility.

Using US county-level data to establish a relationship between hazard exposure and population change or net migration, there is little evidence that such exposure causes population loss. For example, four metropolitan areas affected by three "billion dollar storms" of the early 1990s experienced post-event population growth (Pais & Elliott, 2008). Considering losses from all hazards in all US counties, a county experiencing one million dollars in total property damage from hazards during the 1990s grew by an average of 3.2 percentage points more than a county experiencing no disaster-related property damage (Schultz & Elliott, 2013). In contrast, between 1970 and 2005 Gulf Coast counties with higher levels of hurricane-related damage experienced reduced growth, but not necessarily population loss, for up to three years after the hurricane event (Logan et al., 2016). Data on disasters affecting US counties from 1930 to 2010 shows that only counties affected by severe disasters – defined as a disaster causing 25 or more deaths – lost population through net out-migration, while less severe disasters did not have an effect (Boustan et al., 2020). Data on all hurricane affected US counties from 1980 to 2012 shows that only high-density counties with growing populations registered any change in population growth after a hurricane; in those counties, current year hurricanes and related losses suppressed future growth, and greater cumulative hurricane-related losses in the previous decade elevate future growth, likely because federal disaster recovery programs and private insurance spurred economic growth (Fussell et al., 2017). Altogether, this research indicates that the relationship between hazards and population change is hardly deterministic; instead it suggests that complex mechanisms underlie this association.

Several of these studies investigate relationships between hazard exposure and population change among subgroups of the county population. For example, counties experiencing greater cumulative hazard-related property losses during the 1990s experienced increased income inequality in the form of higher median family income but no change in the number of residents living in poverty (Schultz & Elliott, 2013). Similarly, hazard-related property losses increase residential mobility overall, especially among racial and ethnic minorities (Elliott, 2015). However, among hurricane-affected counties in the Gulf Coast between 1970 and 2005, the proportion of whites and younger residents declined in hurricane-affected counties, while the relative size of blacks and elderly residents increased (Logan et al., 2016). Authors interpret these apparently contradictory findings as being consistent with the idea that advantaged groups fare better than disadvantaged groups after disasters, although sometimes this divergence occurs through recovery in place and other times through out-migration. While these findings are suggestive, associations made at the aggregate level do not necessarily reflect relationships at the individual level.

Studies that use survey data to examine the association between household or individual level characteristics, environmental exposures, and migration avoid the ecological fallacy and provide greater insight into the mechanisms that produce different migration responses. An early study supporting the social vulnerability thesis found that homeowners, female householders, African Americans, older householders, and householders with lower incomes and lower levels of education are more likely to move because of a disaster than to move for other reasons (Morrow-Jones & Morrow-Jones, 1991). More recent research following residents' residential trajectories shows that residents of counties that experienced higher levels of hazard-related property damage also experienced greater residential instability, net of other factors. Furthermore, it demonstrates that Black and Latina women have especially high levels of residential instability because of differences in personal and social resources, such as being a renter, that make them more vulnerable to the effects of hazard damages (Elliott & Howell, 2017). These individual and household level studies support the social vulnerability thesis and suggest that race, gender, and age may interact with housing tenure, income, and human capital to shape post-disaster outcomes.

Survey research is challenging in a post-disaster context, limiting research on disasters and residential mobility. However, when Hurricane Katrina struck four Gulf Coast states, including the City of New Orleans on August 29, 2005, displacing over a million residents, the Census Bureau took extra steps to ensure and enhance data collection of the American Community Survey (ACS) and the Current Population Survey (CPS) (Cahoon et al., 2006). Analyses show sociodemographic differences in the likelihood that displaced residents returned to their pre-disaster home among all Gulf Coast residents: on average, displaced residents who were less than age 40, non-white, female, had low incomes, or low levels of education were less likely to have returned to their pre-disaster homes within a year after the disaster (CPS: Groen & Polivka, 2010). Among New Orleans residents exposed to Hurricane Katrina, Black residents regardless of education and non-Black residents with low levels of education were more likely to be living elsewhere in Louisiana, in Texas, or elsewhere in the South in the year after Hurricane Katrina; non-Black residents with higher levels of education were more likely to be living in the New Orleans metropolitan area (ACS: Sastry & Gregory, 2014). Similarly, among New Orleans residents exposed to Hurricane Katrina, Black residents and those with less than a college education returned more slowly than non-Black residents and college degree holders, although these differentials were diminished by differences in housing damage (Fussell et al., 2010). As a group, these findings support the thesis that some sociodemographic groups are more vulnerable to displacement from disasters than others, although all of the findings above are specific to Hurricane Katrina.

Anthropogenic Hazards and Migration

In contrast to natural hazards, anthropogenic hazards originate in human activity and, once emplaced, they become relatively permanent features of the built environment, such as industrial or power plants that produce toxic waste, landfills, brown fields, and hazardous waste treatment, storage, and disposal facilities. Such sites leave behind relict waste even after they are closed, remediated, or repurposed (Frickel & Elliott, 2018). To the extent that these hazardous sites are public knowledge, residents may avoid moving into surrounding neighborhoods. Similarly, once a hazard is sited in a neighborhood or becomes evident, residents may move out. These disamenities are capitalized in housing costs, leading to lower property values and housing that is more affordable to low-income households. The cumulative effect of migration patterns around hazardous locations is to concentrate residents least able to avoid exposure to the hazard (Bullard, 1994).

Since the 1970s, a large multidisciplinary literature has established spatial correlations between the locations of hazardous waste sites and neighborhoods with disproportionately high numbers of African American residents (see reviews by Brulle & Pellow, 2006; Mohai et al., 2009). The strength of this relationship has formed the basis for the concept of environmental racism, which refers to policies and practices that intentionally or unintentionally differentially affect individuals, groups, or communities based on race or ethnicity. Environmental justice researchers have sought to determine the role of migration as a mechanism that increases or decreases pollution exposure, usually referring to this process as residential segregation rather than migration. Although early studies had inconsistent findings, improvements in data and methods converge on the finding that areas with higher proportions of black and poor residents are more likely to be close to hazardous sites (Anderton et al., 1994; Hunter et al., 2003; Mohai & Saha, 2007, 2015; Mohai et al., 2009).

Two key migration-related topics reveal how environmental racism unfolds over time and space. First, researchers ask whether polluting industries were more likely to select neighborhoods with higher proportions of black residents to locate their facilities, or once the facilities were established, were black residents more likely to move in and white residents more likely to move out? Evidence from regional historical case studies is consistent with the former rather than the latter: neighborhoods with larger proportions of black residents were preferentially selected as locations for hazardous facilities (e.g. Pastor et al., 2001; Saha & Mohai, 2005; Mohai & Saha, 2015).

The second related topic is whether racial residential segregation meant that black and other non-white racial groups Americans and other non-white groups were more exposed to hazardous sites over their lifetimes. A few studies using individual level survey data combined with geolocated data on hazardous facilities showed that, even after controlling for socioeconomic status, there are large racial and ethnic differences in residential proximity to industrial pollution with African Americans being more exposed than other racial and ethnic groups (Crowder & Downey, 2010; Mohai et al., 2009). Furthermore, residential mobility does little to reduce this disparity over a 16-year period; immobile and mobile African Americans alike experienced increased exposure to industrial hazards as a function of industrial siting and constrained neighborhood mobility choices, respectively (Pais et al., 2014). Similarly, single mother and single father households experience greater levels of pollution, compared to two-parent households, and single parent households are more likely to move into neighborhoods with higher pollution levels than are mobile two-parent families, even when racial differentials are statistically controlled (Downey et al., 2017). Not only are these findings supportive of the environmental racism thesis, they are also consistent with the social vulnerability thesis, by showing that some groups are better able and more willing than others to use migration as a strategy for moving out of harms' way.

The Future of North American Environmental Migration

Theories and frameworks specify multiple links in the causal chain connecting the environment and human migration. So far, we show that, first, economic development and environmental management have made the environment less of an obstacle to human settlement. Second, we find that, on balance, environmental amenities have pulled people toward areas prone to natural hazards rather than pushed them away. In short, the environment does not operate exclusively as a push or pull factor for migration, but in the Anthropocene era it has been more of a pull than a push factor in North America. Evidence of this comes mainly from spatial associations, with much less evidence to show how households and individuals perceive and respond to environmental signals.

As North Americans confront the realities of climate change, there is a need for more evidence of how environmental change affects human migration decisions, a need that requires more research at the individual or household level. With an eye toward the future, we review two emergent literatures that contribute to a better understanding of human migration in response to the environment: first, research that identifies the places and people in Canada and the US that are most vulnerable to climate change-related migration, and second, policies and practices that influence residential location decisions, particularly those influencing housing recovery after disasters.

Place and Social Vulnerability to Climate Change

Vulnerability generally refers to susceptibility to harm and the lack of ability to cope with change; it can describe the physical and built environment as well as the residential population (Adger et al., 2005; Wisner et al., 2004). One popular approach to measure vulnerability in a data-rich context like North America is to create indices from available metrics of the land, assets, or population exposed to a hazard. Many vulnerability indices focus on physical hazards, for example sea level rise (Boruff et al., 2005; Gornitz, 1991; Thieler & Hammar-Klose, 1999) or flooding vulnerability (Qiang, 2019). Social vulnerability has been operationalized with county-level measures of population characteristics, housing, economic, governance, and environmental measures (e.g., Borden et al., 2007; Cutter et al., 2003; Cutter & Finch, 2008; Flanagan et al., 2011; Oulahen et al., 2015).

Social vulnerability indices have been adapted for particular risks, such as heat (Reid et al., 2009), hurricane displacement (Esnard et al., 2011), wildfires (Wigtil et al., 2016), flooding (Qiang, 2019), and climate-sensitive hazards (Emrich & Cutter, 2011). Validation of indices with data from hazard events increases their value for hazard response planners, urban planners, and academics alike (e.g. Bakkensen et al., 2017; Finch et al., 2010; Fekete, 2009). Sometimes these indices are used to identify "hot spots", places where specific climate change related hazards are particularly pronounced and where populations are especially vulnerable (de Sherbinin, 2014; de Sherbinin et al., 2019). However, most hot spot identification occurs at the subnational level and focuses on specific hazards unique to that region (Lemmen & Warren, 2016; US NOAA, 2013; Reidmiller et al., 2018).

Conceptually, hot spots are places that concentrate risks and are thought to be close to a tipping point or threshold that will make that place uninhabitable (Bardsley & Hugo, 2010; Kopp et al., 2016; McLeman, 2018). A threshold is the point at which *in situ* adaptations to a hazard are exhausted and out-migration becomes the only option (Adger et al., 2009). In North America, environmental management has raised the habitability threshold in most places, so populations typically persist and even grow after environmental changes or events (Pais & Elliott, 2008; Fussell et al., 2017). To date, the historical record provides very few examples of settlement abandonment (McLeman, 2011; Vale & Campanella, 2005). The sparse historical record suggests that after a series of closely timed environmental events or changes, often of different types of hazards, some places may be abandoned or experience a significant population decline (McLeman, 2011). In North America, environmental migration due to hazards is rare. In the next section, we turn our attention to examples of this type of environmental migration in the US and Canada.

Policies Influencing Environmental Migration

A systematic approach to hazard mitigation and disaster resiliency is embodied in the United Nation's Sendai Framework on Disaster Risk Reduction (United Nations, 2015) and other initiatives, such as the Rockefeller Foundations' 100 Resilient Cities program (now concluded) (Martín & McTarnaghan, 2019). These approaches typically aim to "build back better" and maintain existing settlement footprints and residential populations. For example, Vancouver, Canada, plans *in situ* changes to deal with hotter, drier summers, wetter springs, and warmer winters along with one meter of sea level rise by 2100. These plans include upgrading buildings' cooling capacity, increasing tree canopy, and building new coastline protections (Vancouver Climate Adaptation Strategy, 2012). New Orleans, Louisiana, a city that already experiences regular flood events, continues to invest in levees to protect itself from coastal waters and pumps to remove stormwater, while at

the same time restoring wetlands, incorporating green spaces and technologies, and incentivizing residents to raise existing buildings (City of New Orleans, 2015). Although public debate over the wisdom of rebuilding occurs after every catastrophic disaster event, especially hurricanes, to date these debates have not lead to coordinated action plans involving relocation of residents.

However, a public discussion regarding managed retreat is gaining traction as hazards become more frequent and intense. Managed retreat is the "strategic relocation of structures or abandonment of land to manage natural hazard risk" (Hino et al., 2017: 364). From a social science perspective, managed retreat means relocating neighborhoods and residents away from hazardous areas, including riverine, estuarine, and coastal lands subject to periodic flooding. Other areas, such as the wildland-urban interface, are also candidates for managed retreat (Radeloff et al., 2018). Resistance to the idea of managed retreat is based on multiple sources, including the institutional structures that create a moral hazard by fostering residents' confidence in their ability to withstand environmental hazards (Burby, 2006; Di Baldassarre et al., 2018; Ludy & Kondolf, 2012). This complacency is augmented by Americans' sense that climate change is both a temporally and spatially distant threat (Covi & Kain, 2016; Harvatt et al., 2011). Political leadership is lacking as well; if residents do not lead the conversation on managed retreat, local governments and political leaders lack incentives to rally around the idea (Burby, 2006; Gibbs, 2016; Hino et al., 2017). Consequently, in the North American context, these conversations mostly arise in a post disaster context, and involve federal programs to buyout privately owned homes or to relocate communities (Greer & Binder, 2017; Koslov, 2016; Lynn, 2017).

Federally funded home buyout programs have been available to states and localities as a tool for disaster recovery and hazard mitigation since the late 1970s, but they became a more important tool after the historic 1993 Midwest Flood of the Mississippi and Missouri Rivers (National Wildlife Federation, 1998; Siders, 2017). The federal government promoted voluntary property buyouts to address the escalating federal costs of disaster preparedness, relief, and recovery, as well as costs for private insurers and relief agencies. These escalating costs were the logical but unintended consequence of decades of economic development in high-risk flood plains, made possible by successful environmental management and hazard insurance programs (National Wildlife Federation, 1998). In a home buyout program the government purchases disaster-affected housing at pre-disaster property values and demolishes the structures to create open space. Between 1993 and 2017, the US federal government implemented a property buy-out program that acquired about 38,500 properties in 44 states with the aim of creating open spaces (Siders, 2017). The buy-out programs were initiated, designed, and implemented by a state or local agency with the approval of FEMA, and supported by federal funds allocated through the Housing and Urban Development (HUD) agency's Community Development Block Grants (CDBG) (Greer and Binder 2017). Given the variability in the program designs, it is challenging to assess the success of these programs (Greer and Binder 2017; Siders, 2019).

For our interest in environmental migration, the key question is whether home buyout programs are successful in moving residents from hazardous areas to safer areas. However, most of the literature focuses instead on the program design and participants' satisfaction with the program and their new neighborhoods and homes (Barile et al., 2019; Binder & Greer, 2016; Koslov, 2016; Lynn, 2017). A recent review of the empirical literature on home buyouts found that federal buyout programs have design flaws, such as lack of transparency, vulnerability to political influence, and biased cost-benefit analyses, that make them prone to exacerbate existing social inequalities (Siders, 2019). A statistical analysis of buyout programs provides some support for this contention: buyout programs tend to occur in wealthier municipalities, but the purchased homes have lower than average property values within those municipalities, suggesting that buyout programs may be an instrument for eliminating less desirable homes or neighborhoods (Kraan et al., 2019). It is rare that buyout programs record the new address of homeowners who accepted a buyout, allowing for an assessment of whether residents are presumably safer after the move. When movers are tracked, they tend to move to neighborhoods with a similar or "whiter" racial composition, adding to evidence that home buyout programs increase neighborhood race and class segregation (Loughran & Elliott, 2019).

Typically, when homeowners are given the option to rebuild or relocate, they tend to either rebuild in place or relocate over short distances (Loughran & Elliott, 2019; Gotham, 2014). For example, after Hurricane Katrina, Louisiana designed the Road Home Program to support housing recovery of residents with heavily damaged homes. This massive rebuilding program offered homeowners a one-time grant with the option to rebuild their damaged home or sell it to a state corporation in a buyout program, with the grant value discounted for those who purchased a new home outside of Louisiana. The overwhelming majority of beneficiaries opted to rebuild their homes rather than sell and relocate even though the grants typically fell short of rebuilding costs, especially in neighborhoods with lower property values (Green & Olshansky, 2012). The imbalance in homeowners' funds-to-costs of rebuilding in low-income compared to wealthier neighborhoods reinforced and intensified existing inequalities in the city and generated lower rebuilding rates in low-income, historically black neighborhoods compared to wealthier, majority white neighborhoods in New Orleans (Gotham, 2014).

In contrast, community-wide property buyout programs that convert bought out areas to open space or flood prevention infrastructure effectively remove properties with repeated hazard-related loss and damage claims. Indeed, cities, towns, and neighborhoods that initiate community-wide buyouts often insist on repurposing the land rather than allowing redevelopment (Koslov, 2016). For example, after the catastrophic Red River Flood in 1997, Grand Forks, North Dakota successfully orchestrated a buyout of 802 homes riverside of new dike construction.

Residents initially resisted the program, feeling that political leaders were leveraging substantially damaged properties to force participation in the program (Rakow et al., 2003; Siders, 2019). The program garnered broader support when politicians messaged that bought out land would be turned into public parks (Fraser et al., 2003; Tate et al., 2016). More recently, residents of Oakwood Beach, Staten Island, whose homes were heavily damaged by Hurricane Sandy, insisted that bought out land remain undeveloped, explicitly engaging in the language of managed retreat (Koslov, 2016).

However, the North American experience with the permanent, planned relocation of entire communities as a result of environmental change is still quite limited (Ferris, 2014). In the US and Canada, several native American or first nations communities have requested government assistance in relocating from tribal lands threatened by coastal hazards, although such requests are fraught with the troubled history of government relocation of tribes (Maldonado et al., 2013). For example, in Kashechewan (Northern Ontario, Canada), Shishmaref, Kivalina, and Newtok (Alaska, US) and Isle de Jean Charles (Louisiana, US), tribal communities face displacement and loss of livelihoods from seasonal flooding, erosion, and damage from extreme weather events (Bronen & Chapin, 2013; Maldonado et al., 2013; Marino, 2012). To preserve their communities and cultures, these tribes have requested governmental assistance to relocate to areas where they can continue to live together in an appropriate new location and to preserve tribal lands. However, tribal leaders and residents advocating for relocation must bring different government agencies together and reach agreements within their own communities to develop a plan and identify an appropriate destination (Maldonado et al., 2013). To date, these communities have not relocated. More broadly, research finds that in Arctic Alaska there is little evidence of population loss through "climigration" away from the region (Hamilton et al., 2016; Maldonado & Peterson, 2018). Consequently, these highly vulnerable residents remain in climate change hotspots.

Conclusion

This chapter reviews historical evidence of environmental influences on North American migration and shows that adaptations to the environment have allowed people to inhabit a wider range of climates, often increasing exposure to hazard risk. It is uncertain how much longer these adaptations will effectively manage hazards and prevent disasters in an ever more rapidly changing climate. In the US, damages and losses due to increases in temperature are expected to diminish gross domestic product overall, and to have greater impacts in the South and Northeast (Hsiang et al., 2017; Reidmiller et al., 2018). In Canada, which is warming twice as fast as the world overall, changes will occur sooner than in other world regions (Bush et al., 2019; Council of Canadian Academies, 2019). Along coasts, sea level rise is expected to cause more and more disruption to human systems and livelihoods (Reidmiller et al., 2018). To date, there is little research to guide us on the limits of technical solutions to mitigate hazards and likely habitability thresholds, making it difficult to predict when and where we can expect to see hazard-related environmental migration in the North American context. Instead, amenity migration will likely continue.

Still, the anticipated effects of climate change on the habitability of North American settlements drives scholarly and public interest in environmental migration. Social scientists have a critical role to play in addressing the demand for new knowledge in this area. Demographers and human geographers are mapping places where environmental migration away from hazards is likely to occur due to repeated coastal events and sea level rise (Hauer et al., 2019). However, a broader range of social scientists are needed to investigate the role of social and psychological factors such as place attachment, risk perception, loss aversion, and social networks that inform residential decision-making in relation to environmental hazards (e.g., Adams & Kay, 2019; Adger et al., 2012; Dandy et al., 2019; Hunter, 2005; Hunter et al., 2015; Huntington et al., 2018; Simms, 2017). While social and psychological factors are important for gauging willingness to move, such ideas only translate into action when resources are available to facilitate residential mobility. In this way, social and economic inequalities interact with social and psychological concerns to create inequality in environmentally-informed migration (Elliott & Howell, 2017). A more comprehensive understanding of the links that connect environment to migration in the North American context is key to developing policies that can prevent or mitigate disaster-related losses and damage of material, cultural, and human resources, and thereby create more climate change resilient people and places.

References

Adams, H., & Kay, S. (2019). Migration as a human affair: Integrating individual stress thresholds into quantitative models of climate migration. *Environmental Science and Policy, 93*, 129–138.

Adger, W. N., Arnell, N. W., & Tompkins, E. L. (2005). Successful adaptation to climate change across scales. *Global Environmental Change, 15*(2), 77–86.

Adger, W. N., Lorenzoni, I., & O'Brien, K. L. (2009). Adaptation now. Adapting to climate change: Thresholds, values, governance. In W. N. Adger, I. Lorenzoni, & K. L. O'Brien (Eds.), (pp. 1–22). Cambridge University Press.

Adger, W. N., Barnett, J., Brown, K., Marshall, N., & O'Brien, K. (2012). Cultural dimensions of climate change impacts and adaptation. *Nature Climate Change, 3*, 112–117.

Anderton, D. L., Anderson, A. B., Oakes, J. M., & Frasier, M. (1994). Environmental equity: The demographics of dumping. *Demography, 31*(2), 21–40.

Bakkensen, L. A., Fox-Lent, C., Read, L. K., & Linkov, I. (2017). Validating resilience and vulnerability indices in the context of natural disasters. *Risk Analysis, 37*(5), 982–1004.

Baldassarre, D., Giuliano, H. K., Vorogushyn, S., Aerts, J., Arnbjerg-Nielsen, K., Barendrecht, M., Bates, P., Borga, M., Botzen, W., Bubeck, P., De Marchi, B., Llasat, C., Mazzoleni, M., Molinari, D., Mondino, E., Mard, J., Petrucci, O., Scolobig, A., Biglione, A., & Philip J. Ward. (2018). Hess opinions: An interdisciplinary research agenda to explore the unintended consequences of structural flood protection. *Hydrology and Earth System Sciences, 22*, 5629–5637.

Bardsley, D. K., & Hugo, G. J. (2010). Migration and climate change: Examining thresholds of change to guide effective adaptation decision-making. *Population and Environment, 32*(2–3), 238–262.

Barile, J. P., Binder, S. B., & Baker, C. K. (2019). Recovering after a natural disaster: Differences in quality of life across three communities after Hurricane Sandy. *Applied Research in Quality of Life*, Online first: April 18, 2019.

Barry, J. M. (2007). *Rising tide: The great Mississippi flood of 1927 and how it changed America*. Simon and Schuster.

Biddle, J. (2008). Explaining the spread of residential air conditioning, 1955–1980. *Explorations in Economic History, 45*, 402–423.

Bilsborrow, R. E., & Henry, S. J. F. (2012). The use of survey data to study migration-environment relationships in developing countries: Alternative approaches to data collection. *Population and Environment, 34*(1), 113–141.

Binder, S. B., & Greer, A. (2016). The devil is in the details: Linking home buyout policy, practice, and experience after Hurricane Sandy. *Politics and Governance, 4*(4), 97–106.

Black, R., Neil Adger, W., Arnell, N. W., Dercon, S., Geddes, A., & Thomas, D. S. G. (2011). The effect of environmental change on human migration. *Global Environmental Change, 21S*, S3–S11.

Borden, K. A., Chmidtlein, M. C. S., Emrich, C. T., Piegorsch, W. W., & Cutter, S. L. (2007). Vulnerability of U.S. cities to environmental hazards. *Journal of Homeland Security and Emergency Management, 4*(2), Article 5.

Boruff, B. J., Emrich, C., & Cutter, S. L. (2005). Erosion vulnerability of US coastal counties. *Journal of Coastal Research, 21*(4), 932–942.

Boustan, L. P., Kahn, M. E., Rhode, P. W., & Yanguas, M. L. (2020). *The effect of natural disasters on economic activity in US counties: A century of data*. National Bureau of Economic Research Working Paper No. 23410, Accessed on September 30, 2020 at: https://www.nber.org/papers/w23410

Bronen, R., & Chapin, F. S. (2013). Adaptive governance and institutional strategies for climate-induced community relcoations in Alaska. *Proceedings of the National Academy of Science, 110*(23), 9320–9325.

Brown, W. M., & Scott, D. M. (2012). Human capital location choice: Accounting for amenities and thick labor markets. *Journal of Regional Science, 52*(5), 787–808.

Brulle, R. J., & Pellow, D. N. (2006). Environmental justice: Human health and environmental inequalities. *Annual Review of Public Health, 27*, 103–124.

Bullard, R. D. (1994). *Dumping in dixie: Race, class and environmental quality*. Westview Press.

Burby, R. J. (2001). Flood insurance and floodplain management: The US experience. *Environmental Hazards, 3*, 111–122.

Burby, R. J. (2006). Hurricane Katrina and the paradoxes of government disaster policy: Bringing about wise governmental decisions for hazardous areas. *Annals of the American Academy of Political and Social Science, 604*, 171–191.

Bush, E., Gillett, N., Bonsal, B., Cohen, S., Derksen, C., Flato, G., Greenan, B., Shepherd, M., & Zhang, X. (2019). *Canada's changing climate report*. Government of Canada. https://changingclimate.ca/

Cahoon, L. S., Herz, D. E., Ning, R. C., Polivka, A. E., Reed, M. E., Robison, E. L., & Weyland, G. D. (2006). The current population survey response to Hurricane Katrina. *Monthly Labor Review*, August 2006: 40–51.

Canadian Dam Association. (n.d.). Regulation of Dams in Canda. Website accessed July 14, 2019: https://www.cda.ca/EN/Dams_in_Canada/Regulation/EN/Dams_In_Canada_Pages/Regulation.aspx

Chen, Y., & Rosenthal, S. S. (2008). Local amenities and life-cycle migration: Do people move for jobs or fun? *Journal of Urban Economics, 65*, 519–537.

City of New Orleans. (2015). *Resilient New Orleans: Strategic actions to shape our future city*. Accessed 15 July 2019 at: http://resilientnola.org

Council of Canadian Academies. (2019). *Canada's top climate change risks*. The Expert Panel on Climate Change Risks and Adaptation Potential, Council of Canadian Academies.

Covi, M. P., & Kain, D. J. (2016). Sea-level rise risk communication: Public understanding, risk perception, and attitudes about information. *Environmental Communication, 10*, 612–633.

Cronon, W. (1991). *Nature's Metropolis: Chicago and the Great West*. WW Norton and Company.

Cronon, W. (1996). The trouble with wilderness: Or, getting back to the wrong nature. *Environmental History, 1*(1), 7–28.

Crowder, K., & Downey, L. (2010). Interneighborhood migration, race, and environmental hazards: Modeling microlevel processes of environmental inequality. *American Journal of Sociology, 115*(4), 1110–1149.

Crutzen, P. J. (2002). Geology of mankind, *Nature*: 415, 3 January 2002.

Curtis, K. J., & Schneider, A. (2011). Understanding the demographic implications of climate change: Estimates of localized population predictions under future scenarios of sea-level rise. *Population and Environment, 33*, 28–54.

Cutter, S. L., & Finch, C. (2008). Temporal and spatial changes in social vulnerability to natural hazards. *Proceedings of the National Academy of Sciences, 105*(7), 2301–2306.

Cutter, S. L., Boruff, B. J., & Shirley, W. L. (2003). Social vulnerability to environmental hazards. *Social Science Quarterly, 84*(2), 242–261.

Dandy, J., Horwitz, P., Campbell, R., Drake, D., & Leviston, Z. (2019). Leaving home: Place attachment and decisions to move in the face of environmental change. *Regional Environmental Change, 19*, 615–620.

de Sherbinin, A. (2014). Climate change hotspots mapping: What have we learned? *Climatic Change, 123*, 23–37.

de Sherbinin, A., Bukvic, A., Rohat, G., et al. (2019). Climate vulnerability mapping: A systematic review and future prospects. *WIREs Climate Change, 2019*, 10:e600.

Diez-Roux, A. V. (1998). Bringing context back into epidemiology: Variables and fallacies in multilevel analysis. *American Journal of Public Health, 88*(2), 216–222.

Downey, L., Crowder, K., & Kemp, R. J. (2017). Family structure, residential mobility, and environmental inequality. *Journal of Marriage and Family, 79*, 535–555.

Dronkers, J., Gilbert, J. T. E., Butler, L. W., Carey, J. J., Campbell, J., James, E., McKenzie, C., Misdorp, R., Quin, N., Ries, K. L., Schroder, P. C., Spradley, J. R., Titus, J. G., Vallianos, L., & von Dadelszen, J. (1990). *Coastal zone management. Climate change: The intergovernmental panel on climate change response strategies*. Intergovernmental Panel on Climate Change.

Elliott, J. R. (2015). Natural hazards and residential mobility: General patterns and racially unequal outcomes in the United States. *Social Forces, 93*(4), 1723–1747.

Elliott, J. R., & Howell, J. (2017). Beyond disasters: A longitudinal analysis of natural hazards' unequal impacts on residential instability. *Social Forces, 95*(3), 1181–1207.

Emrich, C. T., & Cutter, S. L. (2011). Social vulnerability to climate-sensitive hazards in the Southern United States. *Weather, Climate, and Society, 3*, 193–208.

Esnard, A.-M., Sapat, A., & Mitsova, D. (2011). An index of relative displacement risk to hurricanes. *Natural Hazards, 59*, 833–859.

Fang, Y., & Jawitz, J. W. (2019). The evolution of human population distance to water in the USA from 1790 to 2010. *Nature Communications, 10*, 430.

Fekete, A. (2009). Validation of a social vulnerability index in context to river-floods in Germany. *Natural Hazards and Earth System Sciences, 9*, 393–403.

Ferris, E. (2014). Planned relocations, disasters, and climate change: Consolidating good practices and preparing for the future. Background paper, Sanremo Consultation, Sanremo, Italy, March 12–14. Washington, DC: Brookings Institution. www.brookings.edu//media/research/files/papers/2014/03/14-planned-relocations-climate-change/planned-relocations-background-paper-march-2014.pdf

Field, C. B., Barros, V., Stocker, T. F., Qin, D., Dokken, D. J., Ebi, K. L., Mastrandrea, M. D., Mach, K. J., Plattner, G.-K., Allen, S. K., Tignor, M., & Midgley, P. M. (Eds.). (2012). *Managing the risks of extreme events and disasters to advance climate change adaptation*. Cambridge University Press.

Finch, C., Emrich, C., & Cutter, S. L. (2010). Disaster disparities and differential recovery in New Orleans. *Population and Environment, 31*(4), 179–202.

Findlay, A. M. (2011). Migrant destinations in an era of environmental change. *Global Environmental Change, 21S*, S50–S58.

Flanagan, B. E., Gregory, E. W., Hallisey, E. J., Heitgerd, J. L., & Lewis, B. (2011). A social vulnerability index for disaster management. *Journal of Homeland Security and Emergency Management, 8*(1), Article 3.

Fossett, M. (2006). Urban and spatial demography (Chapter 16). In D. L. Poston & M. Micklin (Eds.), *Handbook of population* (pp. 479–524). New York, Springer.

Fothergill, A., & Peek, L. A. (2004). Poverty and disasters in the United States: A review of recent sociological findings. *Natural Hazards, 32*(1), 89–110.

Fothergill, A., Maestas, E. G. M., & Darlington, J. A. D. R. (1999). Race, ethnicity, and disasters in the United States: A review of the literature. *Disasters, 23*(2), 156–173.

Fraser, J., Elmore, R., Godschalk, D., & Rohe, W., (2003). *Implementing floodplain land acquisition programs in urban localities*. The Center for Urban & regional studies, University of North Carolina at Chapel Hill FEMA.

Frey, W. H., & Speare, A. (1988). *Regional and metropolitan growth and decline in the United States*. Russell Sage Foundation.

Frickel, S., & Elliott, J. R. (2018). *Sites unseen: Uncovering hidden hazards in American cities*. Russell Sage Foundation.

Fuguitt, G. V., & Beale, C. L. (1993). The changing concentration of the older nonmetropolitan population, 1960–90. *Journal of Gerontology: Social Sciences, 48*(6), 278–288.

Fuguitt, G. V., & Heaton, T. B. (1995). The impact of migration on the nonmetropolitan population age structure, 1960–1990. *Population Research and Policy Review, 14*, 215–232.

Fussell, E. (2012). Space, time, and volition: Dimensions of migration theory. In M. R. Rosenblum & D. J. Tichenor (Eds.), *The Oxford handbook of the politics of international migration*. Oxford University Press.

Fussell, E., & Harris, E. (2014). Homeownership and housing displacement after Hurricane Katrina among low-income African-American mothers in New Orleans. *Social Science Quarterly, 95*(4), 1086–1100.

Fussell, E., Sastry, N., & VanLandingham, M. (2010). Race, socioeconomic status, and return migration to New Orleans after Hurricane Katrina. *Population and Environment, 31*, 20–42.

Fussell, E., Hunter, L. M., & Gray, C. L. (2014). Measuring the environmental dimensions of human migration: The demographers' toolkit. *Global Environmental Change, 28*(1), 182–191.

Fussell, E., Curran, S. R., & Dunbar, M. D. (2017). Weather-related hazards and population change: A study of hurricanes and tropical storms in the United States, 1980–2012. *Annals of the American Academy of Political and Social Science, 669*(1), 146–167.

Gibbs, M. T. (2016). Why is coastal retreat so hard to implement? Understanding the political risk of coastal adaptation pathways. *Ocean and Coastal Management, 130*, 107–114.

Gornitz, V. (1991). Global coastal hazards from future sea level rise. *Palaeogeography, Palaeoclimatology Palaeoecology (Global Planetary Change Section), 89*, 379–398.

Gotham, K. F. (2014). Racialization and rescaling: Post-Katrina rebuilding and the Louisiana Road Home Program. *International Journal of Urban and Regional Research, 38*(3), 773–790.

Graves, P. E. (1980). Migration and climate. *Journal of Regional Science, 20*, 227–238.

Green, T. F., & Olshansky, R. B. (2012). Rebuilding housing in New Orleans: The Road Home Program after the Hurricane Katrina disaster. *Housing Policy Debate, 22*(1), 75–99.

Greenwood, M. J. (1997). Internal migration in developed countries (Chapter 12). *Handbook of population and family economics*, volume 1, Part B: 647–720.

Greer, A., & Binder, S. B. (2017). A historical assessment of home buyout policy: Are we learning or just failing? *Housing Policy Debate, 27*(3), 372–392.

Gregory, J. N. (1989). *American exodus: The dust bowl migration and Okie culture in California*. Oxford University Press.

Groen, J. A., & Polivka, A. E. (2010). Going home after hurricane Katrina: Determinants of return migration and changes in affected areas. *Demography, 47*(4), 821–844.

Gutmann, M. P., & Field, V. (2010). Katrina in historical perspective: Environment and migration in the U.S. *Population and Environment, 31*, 3–19.

Gutmann, M. P., Deane, G. D., Lauster, N., & Peri, A. (2005). Two population-environment regimes in the Great Plains of the United States, 1930–1990. *Population and Environment, 27*(2), 191–225.

Hamilton, L. C., Saito, K., Loring, P. A., Lammers, R. B., & Huntington, H. P. (2016). Climigration? Population and climate change in Arctic Alaska. *Population and Environment, 38*, 115–133.

Harvatt, J., Petts, J., & Chilvers, J. (2011). Understanding householder responses to natural hazards: Flooding and sea-level rise comparisons. *Journal of Risk Research, 14*, 63–83.

Hauer, M. E., Evans, J. M., & Mishra, D. R. (2016). Millions projected to be at risk from sea-level rise in the continental United States. *Nature Climate Change, 6*, 691–695.

Hauer, M. E., Hardy, R. D., Mishra, D. R., & Pippin, J. S. (2019). No landward movement: Examining 80 years of population migration and shoreline change in Louisiana. *Population and Environment, 40*(4), 369–387.

Hauer, M. E., Fussell, E., Mueller, V., Burkett, M., Call, M., Abel, K., McLeman, R., & Wrathall, D. (2020). Sea-level rise and human migration. *Nature Reviews Earth & Environment, 1*, 28–39.

Hino, M., Field, C. B., & Mach, K. J. (2017). Managed retreat as a response to natural hazard risk. *Nature Climate Change, 7*, 364–370.

Hsiang, S. (2016). Climate econometrics. *Annual Review of Resource Economics, 8*, 43–75.

Hsiang, S., Kopp, R., Jina, A., Rising, J., Delgado, M., Shashank, M., Rasmussen, D. J., Muir-Wood, R., Wilson, P., Oppenheimer, M., Larsen, K., & Houser, T. (2017). Estimating economic damage from climate change in the United States. *Science, 356*, 1362–1369.

Hugo, G. (1996). Environmental concerns and international migration. *International Migration Review., 30*(1), 105–131.

Hunter, L. M. (2005). Migration and environmental hazards. *Population and Environment, 26*(4), 273–302.

Hunter, L. M., & Menken, J. (2015). Will climate change shift demography's 'normal science'? *Vienna Yearbook of Population Research, 13*, 23–28.

Hunter, L. M., White, M. J., Little, J. S., & Sutton, J. (2003). Environmental hazards, migration, and race. *Population and Environment, 25*(1), 23–39.

Hunter, L. M., Luna, J. K., & Norton, R. M. (2015). Environmental dimensions of migration. *Annual Review of Sociology, 41*, 377–397.

Huntington, H. P., Loring, P. A., Gannon, G., Gearheard, S. F., Craig Gerlach, S., & Hamilton, L. C. (2018). Staying in place during times of change in Arctic Alaska: The implications of attachment, alternatives, and buffering. *Regional Environmental Change, 18*, 489–499.

International Joint Commission. (2000). Living with the Red: A Report to the Governments of Canada & the U.S. on Reducing Flood Impacts in the Red River Basin. Website accessed 14 July 2019: https://www.ijc.org/en/living-red-report-governments-canada-us-reducing-flood-impacts-red-river-basinx000d

Johnson, K. M., & Beale, C. L. (1994). The recent revival of widespread population growth in nonmetropolitan areas of the United States. *Rural Sociology, 59*, 655–667.

Johnson, K. M., Voss, P. R., Hammer, R. B., Fuguitt, G. V., & McNiven, S. (2005). Temporal and spatial variation in age-specific net migration in the United States. *Demography, 42*(4), 791–812.

Keenan, J. M., Hill, T., & Gumber, A. (2018). Climate gentrification: From theory to empiricism in Miami-Dade County, Florida. *Environmental Research Letters, 13*(5), Article 054001.

Kopp, R. E., Shwom, R. L., Wagner, G., & Yuan, J. (2016). Tipping elements and climate-economic shocks: Pathways toward integrated assessment. *Earth's Future, 4*, 346–372.

Koslov, L. (2016). The case for retreat. *Public Culture, 28*(2(79)), 359–387.

Kraan, C., Mach, K. J., Miyuji H., Siders, A.R., Johnston, E. M., Tsai, Y.-L., & Field, C. B.. (2019). The landscape of voluntary property buyouts to manage flood risk in the United States. Presentation from National Adaptation Forum and Managed Retreat Conference. https://www.nationaladaptationforum.org/program/wednesday/concurrent-sessions-4/relocation-and-innovative-solutions-managing-future-flood-risk

Lang, R. E., Popper, D. E., & Popper, F. J. (1995). "Progress of the nation": The settlement history of the enduring American frontier. *Western Historical Quarterly, 26*, 289–307.

Lee, E. S. (1966). A theory of migration. *Demography, 3*(1), 47–57.

Leifert, H. (2018). Dividing line: The past, present and future of the 100th Meridian. *Earth*. January 2018. Accessed 6 July 2019. https://www.earthmagazine.org/article/dividing-line-past-present-and-future-100th-meridian

Lemmen, D. S., & Warren, F. J. (2016). Synthesis: Canada's marine coasts in a changing climate. In D. S. Lemmen, F. J. Warren, T. S. James, & C. S. L. M. Clarke (Eds.), (pp. 17–26). Government of Canada. Accessed 15 July 2019 at: https://www.nrcan.gc.ca/sites/www.nrcan.gc.ca/files/earthsciences/files/pdf/NRCAN_fullBook%20%20accessible.pdf

Lindsay, J. (2014). The power to react: Review and discussion of Canada's emergency measures legislation. *The International Journal of Human Rights, 18*(2), 159–177.

Logan, J. R., Issar, S., & Zengwang, X. (2016). Trapped in place?: Segmented resilience to hurricanes in the Gulf Coast, 1970–2005. *Demography, 53*(5), 1511–1534.

Long, J., & Siu, H. E. (2016). *Refugees from dust and shrinking land: Tracking the Dust Bowl migrants*. National Bureau of Economic Research, Working Paper 22108. Accessed: http://www.nber.org/papers/w22108

Loughran, K., & Elliott, J. R. (2019). Residential buyouts as environmental mobility: Examining where homeowners move to illuminate social inequities in climate adaptation. *Population and Environment, 41*, 52–70.

Ludy, J., & Kondolf, G. M. (2012). Flood risk perception in lands "protected" by 100-year levees. *Natural Hazards, 61*, 829–842.

Lynn, K. A. (2017). Who defines 'whole': An urban political ecology of flood control and community relocation in Houston, Texas. *Journal of Political Ecology, 24*, 951–967.

Maldonado, J. K., & Peterson, K. (2018). A community-based model for resettlement: Lesson from coastal Louisiana. In R. McLeman & F. Gemmenne (Eds.), *Routledge handbook of environmental displacement and migration* (pp. 289–299). Routledge.

Maldonado, J. K., Shearer, C., Bronen, R., Peterson, K., & Lazrus, H. (2013). The impact of climate change on tribal communities in the US: Displacement, relocation, and human rights. *Climatic Change, 120*, 601–614.

Manson, G. A., & Groop, R. E. (2000). U.S. Intercounty migration in the 1990: People and income move down the urban hierarchy. *The Professional Geographer, 52*(3), 493–504.

Marino, E. (2012). The long history of environmental migration: Assessing vulnerability construction and obstacles to successful relocation in Shishmaref, Alaska. *Global Environmental Change, 22*, 374–381.

Martín, C., & McTarnaghan, S. (2019). *Institutionalizing urban resilience*. A midterm monitoring and evaluation report of 100 resilient cities. Accessed 16 July 2019 at: http://www.100resilientcities.org/wp-content/uploads/2019/03/100RC-2018-Urban-Institute-Midterm-Report.pdf

McDonald, P. (2015). Engagement of demographers in environmental issues from a historical perspective. *Vienna Yearbook of Population Research, 13*, 15–17.

McGranahan, G., Balk, D., & Anderson, B. (2007). The rising tide: Assessing the risks of climate change and human settlements in low elevation coastal zones. *Environment & Urbanization, 19*(1), 17–37.

McLeman, R. A. (2011). Settlement abandonment in the context of global environmental change. *Global Environmental Change, 21S*, S108–S120.

McLeman, R. (2014). *Climate and human migration: Past experiences, future challenges.* Cambridge University Press.

McLeman, R. (2018). Thresholds in climate migration. *Population and Environment, 39*(4), 319–338.

McLeman, R. A., & Smit, B. (2006). Migration as adaptation to climate change. *Climatic Change, 76*(1–2), 31–53.

McLeman, R. A., Dupre, J., Ford, L. B., Ford, J., Gajewski, K., & Marchildon, G. (2013). What we learned from the dust bowl: Lessons in science, policy, and adaptation. *Population & Environment, 25*, 417–440.

McNamara, D. E., & Keeler, A. (2013). A coupled physical and economic model of the response of coastal real estate to climate risk. *Nature Climate Change, 3*, 559–562.

Mohai, P., & Saha, R. (2007). Racial inequality in the distribution of hazardous waste: A national-level reassessment. *Social Problems, Volume, 54*(3), 343–370.

Mohai, P., & Saha, R. (2015). Which came first, people or pollution? Assessing the disparate siting and post-siting demographic change hypothesis of environmental justice. *Environmental Research Letters, 10*, Article115008.

Mohai, P., Pellow, D., & Roberts, J. T. (2009). Environmental justice. *Annual Review of Environment and Resources, 34*, 405–430.

Morrow-Jones, H. A., & Morrow-Jones, C. R. (1991). Mobility due to natural disaster: Theoretical considerations and preliminary analyses. *Disasters, 15*, 126–132.

National Wildlife Federation. (1998). *Higher ground: A report on voluntary property buyouts in the nation's floodplains.* Washington, DC, http://www.nwf.org

Northcott, H. C., & Petruik, C. R. (2013). Trends in the residential mobility of seniors in Canada, 1961–2006. *The Canadian Geographer, 57*(1), 43–55.

Oulahen, G., Mortsch, L., Tang, K., & Harford, D. (2015). Unequal vulnerability to flood hazards: "Ground truthing" a social vulnerability index of five municipalities in Metro Vancouver, Canada. *Annals of the Association of American Geographers, 105*(3), 473–495.

Pais, J. F., & Elliott, J. R. (2008). Places as recovery machines: Vulnerability and neighborhood change after major hurricanes. *Social Forces, 86*(4), 1415–1453.

Pais, J. F., Crowder, K., & Downey, L. (2014). Unequal trajectories: Racial and class differences in residential exposure to industrial hazard. *Social Forces, 92*(3), 1189–1215.

Parton, W. J., Gutmann, M. P., & Ojima, D. (2007). Long-term trends in population, farm income, and crop production in the Great Plains. *Bioscience, 57*(9), 737–747.

Partridge, M. D. (2010). The dueling models: NEG vs amenity migration in explaining US engines of growth. *Papers in Regional Science, 89*(3), 513–536.

Pastor, M., Sadd, J., & Hipp, J. (2001). Which came first? Toxic facilities, minority move-in and environmental justice. *Journal of Urban Affairs, 23*, 1–21.

Peacock, W. G., Dash, N., Zhang, Y., & Van Zandt, S. (2018). Post-disaster sheltering, temporary housing, and permanent housing recovery (Chapter 27). In H. Rodríguez, E. L. Quarantelli, & R. R. Dynes (Eds.), *Handbook of disaster research* (pp. 569–594). Springer.

Pebley, A. R. (1998). Demography and the environment. *Demography, 35*(4), 377–389.

Pielke, R. A., Jr., Gratz, J., Landsea, C. W., Collins, D., Saunders, M. A., & Musulin, R. (2008). Normalized hurricane damage in the United States: 1900–2005. *Natural Hazards Review, 9*, 29–42.

Piguet, E., Kaenzig, R., & Guélat, J. (2018). The uneven geography of research on "environmental migration.". *Population and Environment, 39*, 357–383.

Plane, D. A., & Jurjevich, J. R. (2009). Ties that no longer bind? The patterns and repercussions of age-articulated migration. *The Professional Geographer, 61*(1), 4–20.

Plane, D. A., Henrie, C. J., & Perry, M. J. (2005). Migration up and down the urban hierarchy and across the life course. *Proceedings of the National Academy of Sciences of the United States of America., 102*(43), 15313–15318.

Preston, B. (2013). Local path dependence of US socioeconomic exposure to climate extremes and the vulnerability commitment. *Global Environmental Change, 23*(4), 719–732.

Public Safety Canada. (n.d.). Disaster Assistance Programs. Website accessed August 29, 2019: https://www.publicsafety.gc.ca/cnt/mrgnc-mngmnt/rcvr-dsstrs/dsstr-ssstnc-prgrms/index-en.aspx.

Qiang, Y. (2019). Disparities of population exposed to flood hazards in the United States. *Journal of Environmental Management, 232*, 295–304.

Radeloff, V. C., Helmers, D. P., Kramer, H. A., Mockrin, M. H., Alexandre, P. M., Bar-Massada, A., Butsic, V., Hawbaker, T. J., Martinuzzi, S., Syphard, A. D., & Stewart, S. I. (2018). Rapid growth of the US wildland-urban interface raises wildfire risk. *Proceedings of the National Academy of Sciences, 115*(13), 3314–3319.

Rakow, L. F., Belter, B., Dyrstad, H., Hallsten, J., Johnson, J., & Indvik, K. (2003). The talk of movers and shakers: Class conflict in the making of a community disaster. *Southern Journal of Communication, 69*(1), 37–50.

Rappaport, J. (2007). Moving to nice weather. *Regional Science and Urban Economics, 37*, 375–398.

Rappaport, J., & Sachs, J. D. (2003). The United States as a coastal nation. *Journal of Economic Growth, 8*, 5–46.

Reid, C. E., O'Neill, M. S., Gronlund, C. J., Brines, S. J., Brown, D. G., Diez-Roux, A. V., & Schwartz, J. (2009). Mapping community determinants of heat vulnerability. *Environmental Health Perspectives, 117*(11), 1730–1736.

Reidmiller, D. R., Avery, C. W., Easterling, D. R., Kunkel, K. E., Lewis, K. L. M., Maycock, T. K., & Stewart, B. C. (2018). Fourth national climate assessment. In *Impacts, risks, and adaptation in the United States* (Vol. II). US Global Change Research Program.

Ruttan, V. W. (1993). Population growth, environmental change, and innovation: Implications for sustainable growth in agriculture. In C. L. Jolly & B. B. Torrey (Eds.), *Population and land use in developing countries* (pp. 124–156). National Academy Press.

Saha, R., & Mohai, P. (2005). Historical context and hazardous waste facility siting: Understanding temporal patterns in Michigan. *Social Problems, 52*(4), 618–648.

Sastry, N., & Gregory, J. (2014). The location of displaced New Orleans residents in the year after Hurricane Katrina. *Demography, 51*, 753–775.

Schroeder, A., Wamsley, G., & Ward, R. (2001). The evolution of emergency management in America: From a painful past to a promising but uncertain future. In A. Farazmand (Ed.), *Handbook of crisis and emergency management* (pp. 357–418). Marcel Dekker, Inc.

Schultz, J., & Elliott, J. R. (2013). Natural disasters and local demographic change in the United States. *Population and Environment, 34*(3), 293–312.

Shallat, T. (1994). *Structures in the stream: Water, science, and the rise of the U.S. Army Corps of Engineers.* University of Texas Press.

Shumway, J. M., & Otterstrom, S. M. (2001). Spatial patterns of migration and income change in the mountain west: The dominance of service-based, amenity-rich counties. *The Professional Geographer, 53*(4), 492–502.

Siders, A. R. (2017). Past US floods give lessons in retreat. *Nature, 548*, 281.

Siders, A. R. (2019). Social justice implications of US managed retreat buyout programs. *Climatic Change, 152*, 239–257.

Simms, J. R. Z. (2017). "Why would I live anyplace else?": Resilience, sense of place, and possibilities of migration in coastal Louisiana. *Journal of Coastal Research, 33*(2), 408–420.

Smith, S. K., & McCarty, C. (1996). Demographic effects of natural disasters: A case study of Hurricane Andrew. *Demography, 33*(2), 265–275.

Smith, S. K., & McCarty, C. (2009). Fleeing the storm(s): An examination of evacuation behavior during Florida's 2004 hurricane season. *Demography, 46*(1), 127–145.

Statistics Canada. (2011). Canada's rural population since 1851. Website. Accessed 8 July 2019: https://www12.statcan.gc.ca/census-recensement/2011/as-sa/98-310-x/98-310-x2011003_2-eng.cfm

Statistics Canada. (2018). *Canada goes urban.* Website. Accessed 8 July 2019: https://www150.statcan.gc.ca/n1/pub/11-630-x/11-630-x2015004-eng.htm

Statistics Canada. (n.d.). History of the Census of Canada. Website. accessed July 6, 2019: https://www12.statcan.gc.ca/english/census01/Info/history.cfm#1871

Steffen, W., Crutzen, P. J., & McNeill, J. R. (2007). The Anthropocene: Are humans now overwhelming the great forces of nature? *Ambio, 36*(8), 614–621.

Sutton, P. D., & Day, F. A. (2004). Types of rapidly growing counties of the US, 1970–1990. *The Social Science Journal, 41*, 251–265.

Tate, E., Strong, A., Kraus, T., & Xiong, H. (2016). Flood recovery and property acquisition in Cedar Rapids, Iowa. *Natural Hazards, 80*(3), 2055–2079.

Thieler, R. E., & Hammar-Klose, E. S. (1999). National assessment of coastal vulnerability to sea-level rise: Preliminary results for the US Atlantic Coast. US Geological Survey Open-File Report 99-593. Website accessed July 31, 2019: https://pubs.usgs.gov/of/1999/of99-593/

U.S. Army Corps of Engineers. (n.d.). The U.S. Army Corps of Engineers: A brief history. Website accessed July 6, 2019: https://www.usace.army.mil/About/History/Brief-History-of-the-Corps/

U.S. Census Bureau. (2012). United States summary: 2010: Population and housing unit counties, 2010 Census of population and housing. CPH-2-1. U.S> Department of Commerce, economics and statistics Administration. Accessed 12 July 2019: https://www.census.gov/prod/cen2010/cph-2-1.pdf

U.S. Census Bureau. (n.d.-a). *Censuses of American Indians.* Website Accessed 6 July 2019: https://www.census.gov/history/www/genealogy/decennial_census_records/censuses_of_american_indians.html

U.S. Census Bureau. (n.d.-b). Urban Area Quickfacts. Website accessed July 10, 2019: https://www.census.gov/programs-surveys/geography/guidance/geo-areas/urban-rural/ua-quickfacts.html

United Nations Office of Disaster Risk Reduction. (2015). Sendai Framework for Disaster Risk Reduction. 2015–2030. Accessed 15 July 2019: https://www.unisdr.org/files/43291_sendaiframeworkfordrren.pdf

US National Oceanic and Atmospheric Administration. (2013). State of the Coast Report Series, National Coastal Population Report: Population Trends from 1970 to 2020. Accessed 15 July 2019 at: https://coast.noaa.gov/digitalcoast/training/population-report.html

Vale, L. J., & Campanella, T. J. (2005). *The resilient city: How modern cities recover from disaster.* Oxford University Press.

Vancouver Climate Adaptation Strategy. (2012). https://vancouver.ca/files/cov/Vancouver-Climate-Change-Adaptation-Strategy-2012-11-07.pdf

Walker, K. E. (2016). Baby boomer migration and demographic change in US metropolitan areas. *Migration Studies, 4*(3), 347–372.

Weinkle, J., Landsea, C., Collins, D., et al. (2018). Normalized hurricane damage in the continental United States, 1900–2017. *Nature Sustainability, 1*(12), 808–813.

White, R. (2011). *Railroaded: The transcontinentals and the making of modern America.* Norton.

Wigtil, G., Hammer, R. B., Kline, J. D., Mockrin, M. H., Stewart, S. L., Roper, D., & Radeloff, V. C. (2016). Places where wildfire potential and social vulnerability coincide in the coterminous United States. *International Journal of Wildland Fire, 25*, 896–908.

Wisner, B., Blaikie, P., Cannon, T., & Davis, I. (2004). *At risk: Natural hazards, people's vulnerability and disasters* (2nd ed.). Routledge.

Environmental Migration in Latin America

11

Daniel H. Simon and Fernando Riosmena

Abstract

This chapter reviews prior empirical research on the climatic and environmental influences of human migration in Latin America. Through illustrative case studies, we focus on the prevailing sources of environmental and related stress -and, thus, scholarship- across four different sub-regions: Mexico, Central America, the Caribbean, and South America. In Mexico, we cover both internal and international migration following climatic variability mostly associated with drought. For Central America, our review focuses on research documenting migration following natural disasters, and land-use change as a result of migrant remittances. Next, we review literature from Caribbean nations, examining what is known about migration following rapid-onset climate events such as hurricanes

and earthquakes. We conclude our review in South America, drawing on past research to highlight the associations between migration and environmental conditions and climatic variability, as well as on migration related to broader land use changes. Upon reviewing environment-migration linkages in each sub-region, we reflect on gaps in the literature and goals for future research.

Introduction

In this chapter, we review prior empirical research on the ways in which different forms of climatic variability and environmental stressors are associated with international migration in Latin America. While our examination, like most of this kind, is not fully exhaustive of all environment-migration research in the region, we focus on the most important themes using several illustrative case studies.

Following the same spatial logic employed in this *Handbook* as a whole, we divide the chapter into sub-regions given that the types of weather shocks and migration regimes/responses vary across Latin America. In Mexico, the

D. H. Simon (✉)
Department of Sociology, CU Population Center, Institute of Behavioral Science, University of Colorado Boulder, Boulder, CO, USA
e-mail: Daniel.H.Simon@colorado.edu

F. Riosmena
Geography Department, CU Population Center, Institute of Behavioral Science, University of Colorado Boulder, Boulder, CO, USA

© Springer Nature Switzerland AG 2022
L. M. Hunter et al. (eds.), *International Handbook of Population and Environment*, International Handbooks of Population 10, https://doi.org/10.1007/978-3-030-76433-3_11

literature has primarily focused on the type of climate variability that produces drought, affecting crop yields in particular, which can elicit both internal and international migration responses (also see Simon, 2018). This body of scholarship documents the complex ways in which the climate-migration association varies according to the combined level of variability in temperature and rainfall, socioeconomic conditions, and migrant/social networks in sending areas (Nawrotzki et al., 2015a, b, 2016; Riosmena et al., 2018).

In Central America, we highlight two strengths from the literature that are not covered in the review of Mexico – natural disasters and land-use change as a result of migrant remittances. These two themes are particularly relevant given that tropical storms and severe weather events are increasing in intensity (Elsner et al., 2008; IPCC, 2013) and under some conditions are indeed associated with emigration to the United States (Mahajan & Yang, 2017) and to other parts of Latin America (Spencer & Urguhart, 2018). We further focus on the impacts of remittances on land use to examine how and when migration may worsen the vulnerability or bolster the resilience of communities in the region. For the Caribbean, we provide a review of work on migration following environmental shocks, particularly from disasters like hurricanes. Along these lines, in both the Central American and Caribbean sections, we also discuss the body of evidence on the migration implications of devastating earthquakes, like those affecting El Salvador in 2001, and Haiti and the Dominican Republic in 2010 (Alscher, 2011; Lu et al., 2012).

South America is a much larger and diverse region and, yet, much less has been written for English-language outlets on migration-environment linkages, perhaps due to the closer historical and demographic connections between the United States and Mesoamerica and the Caribbean. Still, there is a strong set of studies on the region, highlighting two major themes. First, as in the other cases, we discuss research on the association between land and environmental conditions and migration, mainly in Ecuador (Gray, 2009, 2010; Gray & Bilsborrow, 2013),

but also including some recent work examining climatic variability and migration across eight South American nations (Thiede et al., 2016). Finally, we highlight displacement as a result of dam construction in Brazil (Randell, 2016a, b, 2017, 2018) given that part of these movements have been driven by the (foreseen or unforeseen) environmental changes produced by these development projects.

After reviewing the environmental correlates of migration in each sub-region, we conclude by pulling together reflections on the current gaps in this literature, including some that may apply to environmental migration more generally across the globe.

Environmental Migration in Mexico

As most all nations in the Global South, Mexico has and will likely continue to exhibit a fair amount of vulnerability to climate and other environmental changes. Rapid-onset events such as flooding and earthquakes continue to occur (Saldaña-Zorrilla & Sandberg, 2009), and sea level rise will become an increasingly relevant issue along coastal areas. Yet, much of the country is especially vulnerable to heat-related shocks and low-rainfall spells, which can combine to produce drought, and challenge livelihoods in natural-resource and primary-sector dependent settings (Christensen et al., 2013; Collins et al., 2013; McSweeney et al., 2008). For this reason, research on environmental migration in Mexico has largely prioritized the study of rainfall and/or temperature variability and its influence on migration.

Indeed, research on these environmental dimensions in Mexico – especially those published in English-speaking outlets – has burgeoned over the last decade. Prior research clearly suggests that internal migration is a more likely and common response to environmental change relative to cross-border mobility (Black et al., 2011; Findlay, 2011). However, international climate migration may be a relatively viable option for families and communities with strong international migration networks and traditions

(Bardsley & Hugo, 2010), which includes many in rural Mexico (Fussell & Massey, 2004; Massey et al., 2002). We review these studies first, followed by those examining environmental internal migration. We also reflect on their similarities and differences, and the need for more comparative work in and beyond the Mexican context.

Climate Variability and International Migration

With the notable exception of Munshi's (2003) seminal study on rainfall and U.S. migration, earlier scholarship largely measured precipitation (and, in some cases, migration) at the very coarse state-level resolution (Chort & LaRupelle, 2016; Feng et al., 2010; Feng & Oppenheimer, 2012; Hunter et al., 2013; Nawrotzki et al., 2013). Recent scholarship has significantly advanced this early work by measuring migration and climate at the municipal- or locality-level. We will thus mostly refer to studies using these finer spatial resolutions.

This research suggests that some rural Mexicans may use U.S. migration as a livelihood strategy to cope with the impacts of variability in precipitation (Barrios-Puente et al., 2016; Munshi, 2003), temperature (Nawrotzki & De-Waard, 2016; Nawrotzki et al., 2015c), or both (Riosmena et al., 2018). Initially, studies mostly used precipitation estimates to measure climatic conditions and anomalies. While rainfall is an important factor in understanding drought, more recent studies have also examined anomalies in temperature, which helps further depict drought conditions given that temperatures affect the rate at which rain/moisture may be evapo-transpiring from the landscape. For example, temperature extremes have a strong negative impact on agricultural production and profit (Mueller et al., 2014), with corn – Mexico's primary crop – being extremely sensitive to high temperatures (Keleman et al., 2009). Indeed, taken together, studies examining both precipitation and temperature fluctuations indicate that the latter are more influential predictors of international moves than

the former (Jessoe et al., 2017; Nawrotzki et al., 2015c; Riosmena et al., 2018), likely as a result of the sensitivities described above.

In this way, temperature extremes are associated with increased U.S. migration from rural places, but not from urban locales (Nawrotzki et al., 2015b). Consistent with this finding, climate impacts on migration operate in tandem with reductions in local employment in rural areas (Jessoe et al., 2017), confirming the notion that the link occurs via the impact of climate on natural resource use and primary-sector types of activities. However, it is important to note that migration following slower-onset stressors like drought may be delayed as their impacts are not always immediate (Findlay, 2011; McLeman, 2011). Illustrating this, Nawrotzki and DeWaard (2016) find that migration following a year of drought-like conditions is initially quite low but increases and then peaks around three years after the shock (Nawrotzki & DeWaard, 2016).

It is important to note that almost every study on the topic finds relatively small or moderate impacts overall (e.g., in terms of estimated excess migrations occurring due to climatic variability) or relative to the impact of other factors such as migrant networks. Furthermore, one recent study employing data from the 2000 and 2010 Census finds that – in the vast majority of the Mexican territory – rainfall deficits and hot temperature spells are associated with *lower* U.S. migration (Riosmena et al., 2018), perhaps because the livelihood challenges produced by environmental shocks increase the opportunity costs of migrating, sometimes even trapping populations (Foresight, 2011). Yet, this study also finds a fair degree of heterogeneity in these associations, including positive instances that are likely consistent with many other studies' findings, but where the *average* effects differ due to differences in spatial and temporal coverage. On the spatial coverage, prior work had focused mainly on communities surveyed by the Mexican Migration Project (MMP) and on periods prior to the contemporary wave of Mexico-US migration (e.g., see Nawrotzki et al., 2015a,b,c, 2016; Nawrotzki & Dewaard, 2016). While MMP data represent a large diversity of settings within Mexico and have represented Mexico-US migrants fairly well for many years (Massey & Zenteno, 1999; Massey & Capoferro, 2004), the Census covers a somewhat broader set of migration and other conditions. Likewise, some prior work with both MMP and the Census used state level climate data (Feng & Oppenheimer, 2012; Hunter et al., 2013; Nawrotzki et al., 2013), a spatial resolution that only allows for a general and more indirect assessment of how climate may affect local economies and migration. Next, we review this and other recent work illustrating the many heterogeneous findings in the climate-migration nexus across the Mexico-US context.

An important question is whether climatic variability is associated in similar ways with undocumented migration relative to less irregular forms of international movement. Because of large permanent resident application backlogs and a low supply of temporary work visas, legal migration appears to be much less sensitive to economic shocks than undocumented movement (Papademetriou & Terrazas, 2009). In a similar fashion, undocumented migration from Mexico appears to be more sensitive to climatic variability than population movements through more regular channels (Nawrotzki et al., 2015a, c). As we detail at the end of the chapter, these findings are an important consideration for estimating how ongoing climate change might influence international (and, indirectly, internal) mobility in this and other contexts in the future.

More recent research on international migration from Mexico has also uncovered several important instances in which the climate-migration association varies across types of sending areas, particularly along three major lines. First, studies consistently find that climate-migration is more likely under particularly dry conditions (Leyk et al., 2017; Nawrotzki et al., 2013; Riosmena et al., 2018). Second, international climate-migration is also more likely in the least vulnerable communities in terms of both levels of marginalization (Riosmena et al., 2018) and the availability and use of irrigation for agriculture (Leyk et al., 2017). Third, and perhaps most importantly, the climate-migration nexus is more

clearly present in places with stronger migratory traditions (Hunter et al., 2013; Riosmena et al., 2018; but see also Nawrotzki et al., 2015c). In this way, climate-migration may be more likely in these places as a result of better-developed transnational networks from prior migrations, and these networks contain considerable social capital to facilitate subsequent migrations, particularly after economic shocks, some climate-related.

Climate Variability and Internal Migration

The bulk of the Mexican environmental-migration literature has focused on international movement, in part a reflection of the larger amount of research on international than internal migration in the Mexican setting (Simon, 2018). Given that most people prefer to remain in place or move short-distances to reduce the costs of migration (e.g., social, psychological, economic) (Bodvarsson & Van den Berg, 2009; Findlay, 2011), additional research on the environmental correlates of internal migration in Mexico is warranted.

Nonetheless, the smaller body of literature that does focus on internal climate migration shows that, as in the case of international migration overall, annual variability in weather patterns has similar negative impacts on rural labor markets and subsequent positive impacts on internal migration (Jessoe et al., 2017). Most notably, during years of hotter than average temperatures and – less clearly – rainfall deficits, movement from rural to urban areas becomes more likely (Jessoe et al., 2017).

The few studies that do compare both kinds of movement find that international migration is more strongly influenced by climate than domestic mobility in *relative* terms (i.e., in metrics like risk or odds ratios, Leyk et al., 2017; Nawrotzki et al., 2016). Yet, internal migration is much larger than international movement – increasingly so during the last 15 years with the slowdown of Mexican undocumented outmigration to the United States (Villarreal, 2014). While neither the estimates in Leyk et al. nor in Nawrotzki et al.

directly tackle whether climate variability produces a larger *number* of additional international than internal migrants, Jessoe et al. (2017) do confirm that climate shocks are associated with a larger absolute number of internal than international migrants. In our view, this research calls for more work on internal climate migration and more comparative research between environmental changes' impacts on both types of movement.

Finally, research on internal climate migration has also improved our understanding of non-linearities in the climate-migration association in Mexico. Using data from the 2000 and 2010 Mexican Census, Nawrotzki et al. (2017) find that higher-than-normal temperatures are associated with slightly lower rural-urban migration. However, this decrease in migration turns positive following 34 months of substantially higher-than-average temperatures during the 60-month migration and climate observation window preceding the Census, with subsequent heat shocks increasing rural-urban migration (Nawrotzki et al., 2017). As in the case of international migration discussed above, these findings suggest that internal climate migration may be more common after particularly intense shocks. Again, future research would benefit from comparing thresholds across domestic vs. international types of movement. Before further discussing the possible uses and implications of these additions, however, let us review other dimensions of the climate-migration nexus more common in other parts of Latin America.

Environmental Migration in Central America

Overview of Environment-Migration Connections in Central America

Compared to the body of scholarship on climate-migration in Mexico, there is far less research on Central America in large part due to much lower availability of data on migration. This is problematic given that all nations on the isthmus are vulnerable due to their higher exposure to natural disasters and severe weather events (Magrin

et al., 2014; Tucker et al., 2010) and, in most cases, due to their higher sensitivity and lower adaptive capacity. A reflection of the fact that Central America is among the least urbanized regions of Latin America, agriculture remains an important economic sector (Watkins, 2007). Lack of access or inadequate quality of land are also a major contributor to rural poverty (de Janvry & Sadoulet, 2000), with colonial and postcolonial practices concentrating the bulk of land among the wealthiest of landowners (Brockett, 1990, 1992). These practices, in turn, have been identified as drivers of internal (Lopez-Carr, 2012) and international migration (Jonas & Rodriguez, 2015; Menjívar, 2000) via their role in producing economic dispossession, but also conflict, violence, and displacement.

Against this backdrop, migration represents an important livelihood strategy for households seeking to mitigate the challenges associated with environmental as well as economic, social, and political shocks throughout the region. And while rural-urban migration is nontrivial, international mobility — particularly to the United States — remains a vital part of the social fabric for many Central American communities. Most notably, U.S.-bound migration from El Salvador and Guatemala accelerated in the 1980s as a result of civil war and violence (Jonas & Rodriguez, 2015; Menjívar, 2000). Also related to civil conflict, Nicaraguan migration to the United States rose considerably during that period (Lundquist & Massey, 2005).

Unfortunately, a new and different wave of violence, this time gang-related, continues to displace people today, especially from El Salvador, Honduras, and Guatemala. Between 2000 and 2010, migration out of Central America increased 50% (Patten, 2012), a situation that continues with a large number of Central Americans emigrating, many seeking asylum in the United States and other countries. While gang-related violence is the most important driver of these movements, recent journalistic reports suggest climate variability may also be displacing some people, particularly those hailing from the "dry corridor" spanning from Northern Guatemala to Northwestern Costa Rica and covering most of Guatemala, El Salvador, and Honduras (e.g., Blitzer, 2019). Yet, with very few exceptions, population research to assess the pervasiveness of and heterogeneity in these associations is missing due to the aforementioned data limitations.

Despite these shortcomings, research on the environmental correlates of migration in Central America complements the scholarship carried out in Mexico in two important ways. First, studies have largely focused on examining the impact of natural disasters on both internal and — to a lesser extent — international mobility. Second, this research investigates the ways in which demographic and economic changes brought by migration have impacted different forms of land-use change. We review each of these themes in turn below.

Migration Following Natural Disasters

Migration represents an important livelihood strategy in the face of both slow– and rapid-onset environmental shocks, though of course it may occur very differently after each of these (Black et al., 2011). Following this logic, Baez et al. (2017) examined internal migratory responses from both droughts and hurricanes using Census data from several rounds and countries in northern Latin America and the Caribbean, and which included Costa Rica, El Salvador, Nicaragua, and Panama. Baez and colleagues find that both droughts and hurricanes influence migration, particularly among rural youth who typically go to nearby towns to pursue off-farm employment and diversify income sources (Baez et al., 2017).

These general findings mask important heterogeneity however. In rural Nicaragua, Koubi et al. (2016) find that the perception of drought is associated with a reduction in internal out-migration while perceived sudden-onset events such as floods are associated with increased out-migration. Likewise, in Costa Rica, Robalino et al. (2015) find that "hydro-meteorological emergencies" (e.g., floods, excessive rainfall, strong winds, landslides) are associated with higher internal migration. However, the most severe sudden-onset events –defined as those resulting in the loss of life – reduce domestic migration, which might reflect both trapping in origin as well as disruptions in economic activity (and migrant labor demand) in the traditional migrant destinations.

The type or intensity of natural disasters also appears to have significantly different impacts on international out-migration from the region. Among the few scholarly works investigating the topic, Halliday (2006) examined the use of migration as a livelihood strategy in response to different environmental shocks (e.g., livestock and crop loss). Similar to the Costa Rican internal mobility case, adverse agricultural conditions increased U.S. migration. However, following devastating 7.6 and 6.6 magnitude earthquakes in early 2001 – which killed over 1000 people and left more than one million without shelter (Nicolas & Olson 2001) – Halliday found a sharp (40%) reduction in international migration. This suggests that disasters could disrupt local livelihoods too much for people to be able to move, despite potentially being willing to do so.

Indeed, disasters can lead to heightened international migration under some circumstances, as illustrated by the impacts of Hurricane Mitch – a catastrophic storm that hit the Caribbean side of Central America in 1998. Mitch resulted in nearly a year's worth of rainfall occurring in just one week in many communities, killing over 20,000 people and displacing roughly two million individuals. While the material damages of the storm were spread out across the region (Girot, 2002), its death toll and migration impacts were highly concentrated in Honduras. Almost 90% of deaths from the hurricane occurred there, where tens of thousands of people also sought refuge in makeshift camps in urban areas (McGirk, 2000). In addition to this internal displacement, international migration from Central America and Honduras, in particular, increased considerably in the following months (Oliver-Smith, 2009; UNFAO, 2001). Yet, as in the Mexican case, research on Nicaragua-Costa Rica flows suggest international movement was a response only among the least vulnerable, with the most vulnerable populations

Impacts of International Migration on Land-Use Change

Migration itself – environmental or otherwise – may also alter the vulnerability of households and communities to future environmental shocks. These impacts are particularly prone to occur in rural areas via changes in land use. On the one hand, outmigration lowers the supply of labor and can lead to lower deforestation or even to the reforestation of fallow lands. On the other hand, migrant remittances can also lead to extensification— when these monies are used to purchase, lease, or otherwise clear and use additional land for primary sector activities–and/or intensification— using different techniques to increase inputs on land already cleared. This can, in turn, render "migrant" households better able to cope with the impacts of environmental and other shocks, even if remittances could also lead to the concentration of land in fewer hands, rendering those selling/losing their lands more vulnerable.

Because these processes may be more accentuated by the longer absences and greater resources associated with international relative to internal migration, one notable strength of the environment-migration literature in Central America – particularly for Guatemala– is its focus on the former type of movement. In a context of deep historical inequality in the distribution of land and land scarcity for many in the country, ethnographic evidence in Guatemala illustrates how migrant households draw on remittances from abroad to purchase land and cattle, enabling and financing the conversion of land from rainforest to rangeland for pasture (Taylor et al., 2006). In the highlands region, migration reduces rural vulnerability not only by facilitating extensification but also by intensifying production: in addition to being better able to purchase additional land to devote to other vegetable crops (Moran-Taylor & Taylor, 2010), migrant households in the region also took advantage of infrastructure improvements and an expanding biofuel economy to convert their fields and cultivate African Palm (Taylor et al., 2016). Note, however, that work on other countries in the region finds that migration led to extensification but not to intensification (Davis & Lopez-Carr, 2014), suggesting again that these relationships are at least somewhat situated.

Environmental Migration in the Caribbean

Environmental Degradation, Natural Disasters, and Migration

Perhaps even more clearly than in the case of Central America, the Caribbean basin is known for its vulnerability to natural hazards – in particular hurricanes, often made worse by longstanding processes of deforestation and related land use change. Unfortunately, the extent of environmental migration stemming from these highly visible, impactful, and even fairly common natural disasters is not well documented in the Caribbean, despite calls for more research on the issue (Kaenzig & Piguet, 2014).

Nevertheless, existing research indicates that hurricanes are increasing in intensity (Elsner et al., 2008) and are associated with rising internal migration and displacement (e.g., in Haiti, Dominican Republic, and Jamaica, see Baez et al., 2017). As the impacts of these events cover a larger share of territory and the economies of small-island nations are often less diversified, international mobility may also rise in tandem with hurricanes (Spencer & Urguhart, 2018). As in the Mexican case, this response is particularly likely for those with more direct connections to migrants (for evidence on U.S. immigration from many nations, including some Caribbean ones, see Mahajan & Yang, 2017). This highlights, again, how prior migration can shape the adaptive capacity of those impacted by environmental shocks.

Also similar in some ways to the Mexican and Central American cases discussed above, differences in vulnerability across Caribbean nations may also shape migration responses to environmental and climate shocks. This differential vulnerability is perhaps clearest in the two countries sharing the island of Hispaniola, Haiti and the Dominican Republic. They are both frequently exposed to tropical storms, flooding, and heavy rainfall events (Alscher, 2011), but Haiti is much more vulnerable due to economic and political collapse, as well as land degradation (e.g., massive deforestation) putting Haitians at considerable additional risks (e.g., from mudslides) during storms. As a result, the impacts of environmental shocks have led to more internal and international displacement in Haiti than in the Dominican Republic following tropical storm Fay and hurricanes Gustav, Hanna, and Ike, all of which struck Hispaniola in the span of three weeks in 2008 (Alscher, 2011).

Puerto Rico provides another recent and striking example of how sudden-onset natural disasters can lead to outmigration. While generally less sensitive to these impacts than other Caribbean nations, Puerto Rico – a U.S. territory – was severely affected by Hurricane Irma in early September 2017, and – especially — Hurricane Maria a few weeks later. The hurricanes' impacts were exacerbated by social and economic vulnerabilities related to Puerto Rico's decade-long debt crisis (Cohn et al. 2014). Hurricane Maria alone caused 90 billion USD in damages in Puerto Rico, making it the third costliest ever in U.S. history (US NHC, 2018). In the subsequent days, failed emergency response systems and blackouts that undermined medical treatment of all kinds led to the deaths of more than 2000 people (Kishore et al., 2018).

While the full impacts on migration of the hurricanes are unclear as of this writing, there is descriptive evidence that movement to the U.S. mainland rose considerably as a response to Irma/Maria-related dislocations. Specifically, net migration during the year after the hurricanes (160,000) was twice as large as the two previous years combined (Hinojosa & Melendez, 2018). As more detailed analyses become available in the coming months and years, we will know more about the precise impact of Hurricane Maria on migration, including the ways in which disaster-related movement relates to social vulnerability.

As in the case of Central America, other types of environmental shocks have also altered mobility patterns in the Caribbean, particularly in Haiti. Most notably, in early 2010 a massive earthquake affecting the greater Port-au-Prince area killed between 65,000 and 300,000 people and left almost two million homeless overnight (Archibold, 2011; Schwartz, 2011). As a result, an estimated 23% of Port-au-Prince inhabitants – or 1.5 million people — moved in the days after the disaster (cf., IOM, 2012; Lu et al., 2012). And while these displacements were mostly directed to other parts of the country, a large number of Haitians sought refuge or work overseas in the months after the earthquake – particularly in Brazil (Miura, 2014) and the United States. Yet, it remains unclear whether the impact of the disaster was truly to increase international migration, or if an even larger number that would have emigrated under other circumstances did not after being rendered less able to do so by the impacts of the earthquake.

Environmental Migration in South America

Climatic Variability and Migration Across the Region, and in Comparative Perspective

Despite having more diversified, urban economies than Central America and the Caribbean, South America remains vulnerable to climatic changes, especially among rural and poor households (Cook and Vizy, 2008; Cox et al., 2004; Franchito et al., 2014). In many nations of the region, nontrivial portions of either the population, economic output, or both depend on primary-sector activities (World Bank, 2008, 2009, 2016).

As in other parts of Latin America, households in South American nations often employ migration to cope with livelihood challenges as a result of economic and environmental shocks (see Cerrutti & Parrado, 2015). For example, in Brazil, the semi-arid northeastern part of the country is highly prone to drought, which has severe consequences for rural agricultural livelihoods (Kahn & Campus, 1992). To cope with rainfall shortages, many of these households often send a migrant to the wealthier southern part of the country to diversify income sources (Capellini et al., 2011). These patterns are likely to persist, as future assessments project that climate change will continue to cripple the agricultural sector, serving as a push-factor for migration to urban areas in many other parts of the country too (Barbieri et al., 2010). Rapid rates of resettlement and urbanization in Brazil have also been tied to deforestation and resource extraction in rural places (Richard & VanWey, 2016), something we address in the next section.

The Andean region also faces many kinds of climate hazards (Urrutia & Vuille, 2009). As Andean glaciers retreat, many small low– and medium-altitude formations have been completely wiped out (Vuille et al., 2008). In Bolivia, for example, glacial retreat has led to decreased water supply for agricultural activities, drinking water, and energy production (Hoffman, 2008; Vergara, 2005). Drought and desertification have, in turn, increased migration among agricultural households to nearby urban areas or different rural regions in the country (Balderrama Mariscal et al., 2011) and placed additional stress on water provision in urban areas, further accentuated by privatization (e.g., in Ecuador, see Swyngedouw, 2004).

In the Ecuadorian Andes, Gray and colleagues (Gray, 2009, 2010; Gray & Bilsborrow, 2013) have also shown that changing environmental conditions and their impacts on agricultural production influence various types of migration. One of a handful of studies concurrently examining local, regional, and international mobility, Gray (2009) demonstrates how both types of domestic migration (local and regional) are more sensitive to climatic variability and its impacts, despite the fact that the region also has an important history of international movement to Spain and the United States. Consistent with prior work discussed above, the type of movement households engage in as a response to

environmental change is contingent on their level of vulnerability, with landless households most likely to migrate to local and regional destinations and land-rich ones more likely to engage in international movement (Gray, 2009). Later work builds on the idea that higher vulnerability does not always lead to displacement, and Gray and Bilsborrow (2013) show that environmental stressors are often associated with reductions in migration flows (an issue previously discussed in the Mexican context). These patterns are also highly gendered: Among men, access to land facilitated international moves and reduced internal migration, while women's international migration was not dependent on land tenure at all, suggesting different mechanisms of environmental migration for men and women (Gray, 2010).

Comparative work on the influences of climatic variability on internal migration has also highlighted heterogeneity in the impacts of environmental change on migration. Thiede et al. (2016) assessed the association between climate variability and inter-provincial migration across eight South American countries (Argentina, Bolivia, Brazil, Chile, Colombia, Ecuador, Paraguay, and Uruguay) between 1970 and 2011. Thiede and colleagues find that climate variability — especially in temperature — is associated with migration across the region in all periods. However, because these impacts vary in magnitude and, especially, direction, their findings also demonstrate the heterogeneity both within and across settings. To illustrate, temperature increases reduced inter-province moves in (more vulnerable) Bolivia while increasing them in Brazil and Uruguay. Additionally, exposure to high temperature months increased migration in Argentina and Uruguay, with negative temperature shocks associated with increased migration in Bolivia, Chile, Colombia, and Ecuador. While complex and heterogeneous, much of the climate-related internal moves were directed towards urban areas and extreme temperature shocks had the most consistent associations with migration in the region. Future research should investigate the factors driving these differences and examine whether these patterns generalize to international movement.

Dams and Displacement in Brazil

Renewable and cleaner energy sources represent a promising and increasingly efficient strategy to meet the demand for electric power while mitigating greenhouse gas emissions and other types of pollution. As a result, hydroelectric power generation — harnessing and controlling the flow of major riverine systems – has become a common part of sustainability and development activities across the world. Within Latin America, Brazil is at the forefront of these efforts as hydropower provides nearly 65% of the entire country's electricity (Alves et al., 2009), with 500 dams constructed since 1950 to meet the energy needs of the growing population and economy (International Rivers, 2012).

These projects are worth mentioning in this chapter as they directly displace people living in floodplains where water storage is directed (3.4 million hectares of land in the Brazilian case, with more than one million people displaced, see Zhouri & Oliveira, 2007). In addition, these projects change river flow and other environmental conditions in ways that can alter biodiversity, livelihoods, and future mobility patterns (see Winemiller et al., 2016).

These changes can also have far-reaching implications on land use. As research by Randell (2016a, b) on the Belo Monte dam in the Brazilian Amazon shows, those displaced from rural areas often resettle in other rural areas, purchasing and clearing new lands for agricultural use. Indeed, Randell finds that those engaging in this kind of mobility experienced much more favorable socioeconomic outcomes than families migrating to urban areas, which might be unique to this case but can also have important consequences for the acceleration of deforestation (Winemiller et al., 2016). Analogous to the case of research on migration and land use in Central America, this work highlights the ways in which migration changes land use which, when occurring at a massive scale, can increase the sensitivity of communities and – especially in the case of the Amazon – even the Earth System to climate and environmental change.

Conclusions

The work reviewed in this chapter illustrates the complex ways in which the environment is mediated by a host of factors in origin and destination communities to produce migration responses that differ in magnitude, spatial structure, and selectivity.

While some of the research reviewed here suggests that internal migration is the most likely response to environmental change (Black et al., 2011; Findlay, 2011), international migration remains a viable option for many Latin American families and communities given the strong tradition of international movement across and out of the region. International environmental migration may be particularly relevant for those with more social and financial capital, but who nonetheless experience direct or indirect damage from environmental shocks (e.g., Gray, 2009; Riosmena et al., 2018). Among those with more social and financial capital, families and places with long histories of prior migration stand out in facilitating international migration (Mahajan & Yang, 2017; Riosmena et al., 2018), likely due to the role of social capital contained within migrant networks in facilitating moves, particularly during emergencies. Many people and communities in Mesoamerica, the Caribbean, and – to a lesser extent –the Andean region possess such historical connections to other countries in the region, as well as the United States and other Global North nations (e.g., Sana & Massey, 2003; Cerruti & Parrado, 2015). As such, past, current, and even future U.S. immigration has been or will be driven at least in part by environmental change in Latin America. Future research should continue evaluating these links as well as directly comparing international and internal migration responses to climate.

While climate and environmental shocks will likely continue to influence migration patterns in and from Latin America, prior research suggests the effects are often quite complex and in ways that do not always suggest that increasing climatic variability will lead to more displacement, particularly internationally. For example, existing research reviewed in this chapter illustrates how these relationships are often non-linear, with other research also documenting that the *net* impact of climatic variability on international migration could be moderate (Jessoe et al., 2017) or even mildly negative, as environmental change may both increase the outmigration of some people while reducing the capacity of many others to engage in cross-border movement (Gray & Bilsborrow, 2013; Loebach, 2016; Riosmena et al., 2018). On this regard, scholarship on the environmental dimensions of migration in Latin America has played an important role in moving past earlier deterministic predictions that climate change will displace millions of "environmental refugees" internationally (Myers, 2002), and even in showing the heterogeneity of the climate-migration association beyond its average impacts, even if simpler, often-alarmist narratives still persist in national security and public discourses.

While social vulnerability is an important component in shaping who is better able to migrate, more intense exposure does tend to predict higher migration even if not invariably. For example, particularly intense episodes of high temperature and low precipitation are associated with slightly higher Mexico-U.S. migration (Riosmena et al., 2018). Likewise, internal and international migration are responsive to the impact of intense sudden-onset events in much of Central America and the Caribbean (Baez et al., 2017; Koubi et al., 2016; Mahajan & Yang, 2017). This notably includes increases in outmigration after particularly intense events in Honduras after Mitch in 1998 (Oliver-Smith, 2009), Puerto Rico after the passage of Irma and Maria in 2017 (see Hinojosa & Melendez, 2018), and Haiti after the devastating 2010 Earthquake. On the other hand, emigration decreased following other important disasters, including the 2001 Earthquake in El Salvador (Halliday, 2006), and among the most vulnerable in Honduras after Mitch (Loebach, 2016). Internal migration also decreased after

particularly intense hydro-meteorological events in Costa Rica (Robalino et al., 2015).

With these complexities in mind, future research is needed to better understand the heterogeneity in migration responses to climate shocks that differ in type and severity (which could take the form of threshold effects, e.g., Nawrotzki et al., 2017), and how these shocks may be affecting people differently according to their vulnerability and adaptive capacity. For example, comparative work examining across national and local contexts, time periods, and climate measures would provide more empirical leverage to understand what is driving the heterogeneous responses to environmental shocks, with a continued focus on concurrently understanding instances in which environmental change increases migration with those in which these associations are negative, and at what thresholds of stress. This should include more detailed assessments of whether shocks are trapping people in place (Foresight, 2011). Likewise, because comparisons of the magnitude of environmental migration across case studies are frequently complicated by differences in measurement, comparative examinations using consistent measures would also allow for a better understanding of how strong and likely migration responses to environmental shocks are in different settings. This also includes a more thorough examination of how social, economic, and environmental changes in potential destinations may also mediate environmental migration.

We would also encourage researchers to expand on the examination of temporality in both migration and environmental measures. Moreover, while both internal and international migration may be employed as an adaptation strategy in response to previous environmental stressors (i.e., *ex post*), or take the form of anticipatory risk mitigation strategies in order to protect against future livelihood challenges (i.e., *ex ante*) that will likely come with environmental change (Black et al., 2011; McLeman & Hunter, 2010), the studies discussed in this chapter overwhelmingly examined *ex post* responses. More research is needed in order to understand *ex ante* migration processes, and related behaviors aimed at lowering risk related to climate variability. Presumably, this is particularly important in settings with higher levels of circular mobility, where people may be migrating temporarily to meet a specific earnings target that allows for investments in physical or human capital back in the sending community. Likewise, a better understanding of the different pathways by which environmental shocks cause migration is necessary. In this way, future research might explore the timing of climate shocks (e.g., growing season measures), as well as differential migratory responses to direct losses (e.g., personal property) and indirect shocks (e.g., local economic opportunities). Future work would do well to also examine the ways in which climate and environmental shocks impact population movement to and from urban areas, particularly in places connected to and dependent on agricultural and primary-sector production.

Finally, regardless of whether people use migration as a concerted *ex ante* adaptation strategy or an *ex post* response, their remittances, savings and return migration behavior can change their own adaptive capacity – and, in some cases, the adaptive capacity of others around them. In Central America, we describe the impacts of migrants, and remittances, on land-use change, with scholarship in general agreement that migration does indeed alter the environment in sending communities as households invest in more land and agricultural infrastructure (Davis & Lopez-Carr, 2010, 2014; Moran-Taylor & Taylor, 2010; Taylor et al., 2006). Likewise, rural households displaced by development projects resettling in other parts of the rural (frontier) also engage in important land-use practices that are beneficial to them (Randell et al., 2016a, b) but can further contribute to land degradation in important ecosystems.

Taken together, the state of the literature on migration-environment linkages in Latin America is strong and has come a long way in the past decade. In the years and decades to come, we argue that when possible researchers should focus on understanding and explaining the variability in migration responses across climate events, regions, and communities, which may require additional data collection, or otherwise

more cross-national comparisons, such as the study by Thiede et al. (2016) in South America. Additionally, this work is well suited to quasi-experimental research designs that could help make sense of these complex associations and pathways. Finally, we urge scholars to consider non-linearities and threshold effects in both climate measures (e.g., rainfall deficits and excesses) and migration responses (i.e., trapped populations). In doing so, future research would bolster the scholarship reviewed in this chapter.

References

Alscher, S. (2011). Environmental degradation and migration on Hispaniola Island. *International Migration, 49*, e164–e188.

Alves, S. A., Nascimento, A. C., & Mesquita, H. A. (2009). Movimento dos Atingidos Por Barragens (MAB): Resistência popular ea construção de um novo modelo energético para o Brasil. *Revista Estudos Amazônidas: Fronteiras e Territórios, 1*(1), 635–650.

Archibold, R. C. (2011). Haiti: Quake's toll rises to 316,000. *NY Times*.

Baez, J., Caruso, G., Mueller, V., & Niu, C. (2017). Droughts augment youth migration in Northern Latin America and the Caribbean. *Climatic Change, 140*(3–4), 423–435.

Balderrama Mariscal, C., Tassi, N., Rubena Miranda, A., Aramayo Caned, L., & Cazorla, I. (2011). *Rural migration in Bolivia: The impact of climate change, economic crisis and state policy* (International Institute for Environment and Development (IIED) Human Settlements Working Paper, 45). : IIED.

Barbieri, A., Domingues, E., Queiroz, B., Ruiz, R., Rigotti, J., Carvalho, J., & Resende, M. (2010). Climate change and population migration in Brazil's Northeast: Scenarios for 2025–2050. *Population and Environment, 31*, 344–370.

Bardsley, D. K., & Hugo, G. J. (2010). Migration and climate change: Examining thresholds of change to guide effective adaptation decision-making. *Population and Environment, 32*(2–3), 238–262. https://doi.org/10.1007/s11111-010-0126-9

Barrios Puente, G., Perez, F., & Gitter, R. J. (2016). The effect of rainfall on migration from Mexico to the United States. *International Migration Review, 50*(4), 890–909.

Black, R., Adger, W. N., Arnell, N. W., Dercon, S., Geddes, A., & Thomas, D. S. (2011). The effect of environmental change on human migration. *Global Environmental Change-Human and Policy Dimensions, 21*, S3–S11. https://doi.org/10.1016/j.gloenvcha.2011.10.001

Blitzer, J. (2019). How climate change is fueling the US border crisis. The New Yorker. *New Yorker, 3*.

Bodvarsson, Ö. B., & Van den Berg, H. (2009). Hispanic immigration to the United States. In *The economics of immigration* (pp. 315–341). Berlin/Heidelberg: Springer.

Brockett, C. (1990). *Land, power, and poverty: Agrarian transformation and political conflict in Central America.*

Brockett, C. D. (1992). Measuring political violence and land inequality in Central America. *American Political Science Review, 86*(1), 169–176.

Capellini, N., Castro, M. C., & Gutjahr, E. (2011). Patterns of environmental migration in Brazil: Three case studies. In F. Gemenne, P. Brücker, & J. Glaser (Eds.), *The state of environmental migration 2010* (pp. 87–110). Paris: IDDRI Sciences Po and IOM (International Organization for Migration).

Cerrutti, M., & Parrado, E. (2015). Intraregional migration in South America: Trends and a research agenda. *Annual Review of Sociology, 41*, 399–421.

Chort, I., & De La Rupelle, M. (2016). Determinants of Mexico-US outward and return migration flows: A state-level panel data analysis. *Demography, 53*(5), 1453–1476.

Christensen, J. H., et al. (2013). Climate phenomena and their relevance for future regional climate change. In T. F. Stocker et al. (Eds.), *Climate change 2013: The physical science basis* (pp. 1217–1308). Contribution of Working Group I to the fifth assessment report of the Intergovernmental Panel on Climate Change. Cambridge University Press.

Cohn, D. V., Patten, E., & Lopez, M. H. (2014). Puerto Rican population declines on island, grows on US mainland. *Pew Research Center's Hispanic Trends Project.*

Cox, P. M., Betts, R. A., Collins, M., Harris, P. P., Huntingford, C., & Jones, C. D. (2004). Amazonian Forest dieback under climate-carbon cycle projections for the 21st century. *Theoretical and Applied Climatology, 78*, 137–156. https://doi.org/10.1007/s00704-004-0049-4

Collins, M., et al. (2013). Long-term climate change: Projections, commitments and irreversibility. In T. F. Stocker et al. (Eds.), *Climate change 2013: The physical science basis* (pp. 1029–1136). Contribution of working group I to the fifth assessment report of the Intergovernmental Panel on Climate Change. Cambridge University Press.

Cook, K. H., & Vizy, K. H. (2008). Effects of twenty-first-century climate change on the Amazon rain Forest. *Journal of Climate, 21*, 542–560. https://doi.org/10.1175/2007JCLI1838.1

Davis, J., & Lopez-Carr, D. (2010). The effects of migrant remittances on population–environment dynamics in migrant origin areas: International migration, fertility, and consumption in highland Guatemala. *Population and Environment, 32*(2–3), 216–237.

Davis, J., & Lopez-Carr, D. (2014). Migration, remittances and smallholder decision-making: Implications for land use and livelihood change in Central America. *Land Use Policy, 36*, 319–329.

de Janvry, A., & Sadoulet, E. (2000). Rural poverty in Latin America: Determinants and exit paths. *Food Policy, 25*(4), 389–409.

Elsner, J. B., Kossin, J. P., & Jagger, T. H. (2008). The increasing intensity of the strongest tropical cyclones. *Nature, 455*(7209), 92.

Feng, S., & Oppenheimer, M. (2012). Applying statistical models to the climate-migration relationship. *Proceedings of the National Academy of Sciences, 109*, E2915.

Feng, S., Krueger, A. B., & Oppenheimer, M. (2010). Linkages among climate change, crop yields and Mexico–US cross-border migration. *Proceedings of the National Academy of Sciences, 107*(32), 14257–14262.

Findlay, A. M. (2011). Migrant destinations in an era of environmental change. *Global Environmental Change, 21*, S50–S58.

Franchito, S. H., Fernandez, J. P. R., & Pareja, D. (2014). Surrogate climate change scenario and projections with a regional climate model: Impact on the aridity in South America. *American Journal of Climate Change, 3*(05), 474.

Foresight: Migration and Global Environmental Change Final Project Report (Government Office for Science). (2011). Available at http://go.nature.com/somswg

Fussell, E., & Massey, D. S. (2004). The limits to cumulative causation: International migration from Mexican urban areas. *Demography, 41*(1), 151–171.

Girot, P. O. (2002). *Environmental degradation and regional security: Lessons from hurricane Mitch.* International Institute for Sustainable Development.

Gray, C. L. (2009). Environment, land, and rural out-migration in the southern Ecuadorian Andes. *World Development, 37*(2), 457–468.

Gray, C. L. (2010). Gender, natural capital, and migration in the southern Ecuadorian Andes. *Environment and Planning A, 42*(3), 678–696.

Gray, C., & Bilsborrow, R. (2013). Environmental influences on human migration in rural Ecuador. *Demography, 50*(4), 1217–1241.

Halliday, T. (2006). Migration, risk, and liquidity constraints in El Salvador. *Economic Development and Cultural Change, 54*(4), 893–925.

Hinojosa, J., Román, N., & Meléndez, E. (2018). Puerto Rican post-Maria relocation by states. *Center for Puerto Rican Studies, 1*(1), 1–15.

Hoffmann, D. (2008). Consecuencias Cel Retroceso Glaciar En La Cordillera Boliviana. *Pirineos, 163*, 77–84.

Hunter, L. M., Murray, S., & Riosmena, F. (2013). Rainfall patterns and U.S. migration from rural Mexico. *International Migration Review, 47*, 874–909.

International Organization for Migration (IOM). (2012). *Haiti: From emergency to sustainable recovery* (IOM Haiti Two-Year Report (2010–2011), p. 56). Port-au-Prince: International Organization for Migration Haiti.

International Rivers. (2012). Questions and Answers about Large Dams. http://www.internationalrivers.org/node/570

IPCC. (2013). Technical summary. In T. F. Stocker, D. Qin, G.-K. Plattner, L. V. Alexander, S. K. Allen, N. L. Bindoff, F.-M. Bréon, et al. (Eds.), *Climate change 2013: the physical science basis. Contribution of Working Group I to the Fifth Assessment Report of the Intergovernmental Panel on Climate Change* (pp. 33–115). Cambridge University Press.

Jessoe, K., Manning, D. T., & Taylor, J. E. (2017). Climate change and labour allocation in rural Mexico: Evidence from annual fluctuations in weather. *The Economic Journal, 128*(608), 230–261.

Jonas, S., & Rodríguez, N. (2015). *Guatemala-US migration: Transforming regions.* University of Texas Press.

Kaenzig, R., & Piguet, E. (2014). Migration and climate change in Latin America and the Caribbean. In *People on the move in a changing climate* (pp. 155–176). Springer.

Keleman, A., Hellin, J., & Bellon, M. R. (2009). Maize diversity, rural development policy, and farmers' practices: Lessons from Chiapas, Mexico. *Geographical Journal, 175*, 52–70. https://doi.org/10.1111/j.1475-4959.2008.00314.x

Khan, A. S., & CAMPUS, R. T. (1992). Effects of drought on agricultural sector of Northeast Brasil. ICID.

Kishore, N., Marqués, D., Mahmud, A., Kiang, M. V., Rodriguez, I., Fuller, A., . . . Maas, L. (2018). Mortality in Puerto Rico after hurricane Maria. *New England Journal of Medicine, 379*(2), 162–170.

Koubi, V., Spilker, G., Schaffer, L., & Böhmelt, T. (2016). The role of environmental perceptions in migration decision-making: Evidence from both migrants and non-migrants in five developing countries. *Population and Environment, 38*(2), 134–163.

Leyk, S., Runfola, D., Nawrotzki, R. J., Hunter, L. M., Riosmena, F. (2017). Internal and international mobility as adaptation to climatic variability in contemporary Mexico: Evidence from the integration of census and satellite data. *Population, Space and Place.*

Loebach, P. (2016). Household migration as a livelihood adaptation in response to a natural disaster: Nicaragua and Hurricane Mitch. *Population and Environment, 38*(2), 185–206.

López-Carr, D. (2012). Agro-ecological drivers of rural out-migration to the Maya biosphere reserve, Guatemala. *Environmental Research Letters, 7*(4), 045603.

Lu, X., Bengtsson, L., & Holme, P. (2012). Predictability of population displacement after the 2010 Haiti earthquake. *Proceedings of the National Academy of Sciences, 109*(29), 11576–11581.

Lundquist, J. H., & Massey, D. S. (2005). Politics or economics? International migration during the Nicaraguan contra war. *Journal of Latin American Studies, 37*(1), 29–53.

Magrin, G. O., Marengo, J. A., Boulanger, J. P., Buckeridge, M. S., Castellanos, E., Poveda, G., Scarano, F. R., & Vicuña, S. (2014). Central and South America. *Climate Change,* 1499–1566.

Mahajan, P., & Yang, D. (2017). *Taken by storm: Hurricanes, migrant networks, and us immigration* (No. w23756). National Bureau of Economic Research.

Massey, D. S., & Capoferro, C. (2004). Measuring undocumented migration. *Internaitonal Migration Review, 38*(3), 1075–1102.

Massey, D. S., Durand, J., & Malone, N. J. (2002). *Beyond smoke and mirrors: Mexican immigration in an era of economic integration.* Russell Sage Foundation.

Massey, D. S., & Sana, M. (2003). Patterns of US migration from Mexico, the Caribbean, and Central America. *Migraciones Internacionales, 2*(2).

Massey, D. S., & Zenteno, R. M. (1999). The dynamics of mass migration. *Proceedings of the National Academy of Sciences, 96*(9), 5328–5335.

McGirk, J. (2000). Forgotten million still reeling from Hurricane Mitch, *The Independent.*

McLeman, R. A. (2011). Settlement abandonment in the context of global environmental change. *Global Environmental Change, 21,* S108–S120.

McLeman, R. A., & Hunter, L. M. (2010). Migration in the context of vulnerability and adaptation to climate change: Insights from analogues. *Wiley Interdisciplinary Reviews: Climate Change, 1*(3), 450–461.

McSweeney, C., New, M., & Lizcano, G. (2008). *UNDP climate change country profiles: Mexico.* United Nations Development Programme.

Menjívar, C. (2000). *Fragmented ties: Salvadoran immigrant networks in America.* Univ of California Press.

Miura, H. (2014). The Haitian migration flow to Brazil: Aftermath of the 2010 earthquake. The state of environmental migration 2014. A Review of 2013.

Moran-Taylor, M. J., & Taylor, M. J. (2010). Land and leña: Linking transnational migration, natural resources, and the environment in Guatemala. *Population and Environment, 32*(2–3), 198–215.

Mueller, V., Gray, C. L., & Kosec, K. (2014). Heat stress increases long-term human migration in rural Pakistan. *Nature Climate Change, 4*(3), 182–185. https://doi.org/10.1038/nclimate2103

Munshi, K. (2003). Networks in the modern economy: Mexican migrants in the US labor market. *The Quarterly Journal of Economics, 118*(2), 549–599.

Myers, N. (2002). Environmental refugees: A growing phenomenon of the 21st century. *Philosophical Transactions of the Royal Society of London Series B: Biological Sciences, 357*(1420), 609–613.

Nawrotzki, R. J., & DeWaard, J. (2016). Climate shocks and the timing of migration from Mexico. *Population and Environment, 38*(1), 72–100.

Nawrotzki, R. J., Riosmena, F., & Hunter, L. M. (2013). Do rainfall deficits predict U.S.-bound migration from rural Mexico? Evidence from the Mexican census. *Population Research and Policy Review, 32,* 129–158. https://doi.org/10.1007/s11113-012-9251-8

Nawrotzki, R. J., Riosmena, F., Hunter, L. M., & Runfola, D. M. (2015a). Undocumented migration in response to climate change. *International journal of population studies, 1*(1), 60.

Nawrotzki, R. J., Hunter, L. M., Runfola, D. M., & Riosmena, F. (2015b). Climate change as a migration driver from rural and urban Mexico. *Environmental Research Letters, 10*(11), 114023.

Nawrotzki, R. J., Riosmena, F., Hunter, L. M., & Runfola, D. M. (2015c). Amplification or suppression: Social networks and the climate change—Migration association in rural Mexico. *Global Environmental Change, 35*, 463–474.

Nawrotzki, R. J., Runfola, D. M., Hunter, L. M., & Riosmena, F. (2016). Domestic and international climate migration from rural Mexico. *Human Ecology, 6*(44), 687–699.

Nawrotzki, R. J., DeWaard, J., Bakhtsiyarava, M., & Ha, J. T. (2017). Climate shocks and rural-urban migration in Mexico: Exploring nonlinearities and thresholds. *Climatic Change, 140*(2), 243–258.

Nicolas, N., & Olson, R. (2001). *Final report "SUMA" and the 2001 El Salvador earthquakes: An independent external evaluation*. Pan-Health Organization.

Oliver-Smith, A. (2009). Understanding hurricane Mitch: Complexity, causality, and the political ecology of disaster. *The Legacy of Hurricane Mitch: Lessons from the Post-Disaster Reconstruction in Honduras, 1–21.*

Papademetriou, D. G., & Terrazas, A. (2009). *Immigrants and the current economic crisis: Research evidence, policy challenges, and implications*. Migration Policy Institute.

Patten. (2012). Statistical portrait of the foreign-born population in the United States, 2010 Pew Research Center, Washington.

Randell, H. (2016a). Structure and agency in development-induced forced migration: The case of Brazil's Belo Monte Dam. *Population and Environment, 37*(3), 265–287.

Randell, H. (2016b). The short-term impacts of development-induced displacement on wealth and subjective well-being in the Brazilian Amazon. *World Development, 87*, 385–400.

Randell, H. (2017). Forced migration and changing livelihoods in the Brazilian Amazon. *Rural Sociology, 82*(3), 548–573.

Randell, H. (2018). The strength of near and distant ties: Social capital, environmental change, and migration in the Brazilian Amazon. *Sociology of Development, 4*(4), 394–416.

Richards, P. D., & Van Wey, L. (2016). Farm-scale distribution of deforestation and remaining forest cover in Mato Grosso. *Nature Climate Change, 6*(4), 418.

Riosmena, F., Nawrotzki, R., & Hunter, L. (2018). Climate migration at the height and end of the Great Mexican Emigration Era. *Population and Development Review, 44*(3), 455.

Robalino, J., Jimenez, J., & Chacón, A. (2015). The effect of hydro-meteorological emergencies on internal migration. *World Development, 67*, 438–448.

Saldaña-Zorrilla, S. O., & Sandberg, K. (2009). Spatial econometric model of natural disaster impacts on human migration in vulnerable regions of Mexico. *Disasters, 33*(4), 591–607.

Schwartz, T. T., Pierre, Y. F., & Calpas, E. (2011). Building assessments and rubble removal in quake-affected Neighborhoods in Haiti: BARR survey: Final report: Draft. In *Building assessments and rubble removal in quake-affected neighborhoods in Haiti: BARR survey: Final report: Draft*. LTL Strategies; United States Agency for International Development (USAID).

Simon, D. H. (2018). Environmental migration in Mexico. *Routledge Handbook of Environmental Displacement and Migration, 257.*

Spencer, N., & Urquhart, M. A. (2018). Hurricane strikes and migration: Evidence from storms in Central America and the Caribbean. *Weather, Climate, and Society, 10*(3), 569–577.

Swyngedouw, E. (2004). *Social power and the urbanization of water: Flows of power.* Oxford University Press.

Taylor, M. J., Moran-Taylor, M. J., & Ruiz, D. R. (2006). Land, ethnic, and gender change: Transnational migration and its effects on Guatemalan lives and landscapes. *Geoforum, 37*(1), 41–61.

Taylor, M. J., Aguilar-Støen, M., Castellanos, E., Moran-Taylor, M. J., & Gerkin, K. (2016). International migration, land use change and the environment in Ixcán, Guatemala. *Land Use Policy, 54*, 290–301.

Thiede, B., Gray, C., & Mueller, V. (2016). Climate variability and inter-provincial migration in South America, 1970–2011. *Global Environmental Change, 41*, 228–240.

Tucker, C. M., Eakin, H., & Castellanos, E. J. (2010). Perceptions of risk and adaptation: Coffee producers, market shocks, and extreme weather in Central America and Mexico. *Global Environmental Change, 20*(1), 23–32.

UNFAO (2001). Analysis of the medium-term effects of hurricane Mitch on food security in Central America. .

United States National Hurricane Center (2018). Tropical Cyclones Tables Updated.

Urrutia, R., & Vuille, M. (2009). Climate change projections for the tropical Andes using a regional climate model: Temperature and precipitation simulations for the end of the 21st century. *Journal of Geophysical Research-Atmospheres, 114*(D2).

Vergara, W. (2005). Adapting to climate change: Lessons learned, work in progress, and proposed next steps for the World Bank in Latin America. In *Sustainable Development Working* (No. 25). The World Bank. Environmentally and Socially Sustainable Development Department.

Villarreal, A. (2014). Explaining the decline in Mexico-US migration: The effect of the Great Recession. *Demography, 51*(6), 2203–2228.

Vuille, M., Francou, B., Wagnon, P., Juen, I., Kaser, G., Mark, B. G., & Bradley, R. S. (2008). Climate change and tropical Andean glaciers: Past, present and future. *Earth-Science Reviews, 89*(3–4), 79–96.

Watkins, K. (2007). Human development report 2007/8. Fighting climate change: Human solidarity in a divided world. *Fighting Climate Change: Human Solidarity in a Divided World (November 27, 2007). UNDP-HDRO Human Development Report.*

Winemiller, K. O., McIntyre, P. B., Castello, L., Fluet-Chouinard, E., Giarrizzo, T., Nam, S., Baird, I. G., Darwall, W., Lujan, N. K., Harrison, I., & Stiassny, M. L. J. (2016). Balancing hydropower and biodiversity in the Amazon, Congo, and Mekong. *Science, 351*(6269), 128–129.

World Bank. (2008). *Rural population statistics.* Available via http://data.worldbank.org/indicator/SP.RUR.TOTL

World Bank. (2009). Mexico – Country note on climate change aspects in agriculture. *Country Note on Climate Change Aspects in Agriculture.* Washington, DC. © World Bank. https://openknowledge.worldbank.org/handle/10986/9478 License: CC BY 3.0 IGO.

World Bank. (2016). World development indicators. The World Bank, Washington, D.C. Available at: http://databank.worldbank.org/data/home.aspx

Zhouri, A., & Oliveira, R. (2007). Desenvolvimento, Conflitos Sociaise violencia no Brasil rural: O case dasusinas hidreletricas. *Ambiente e Sociedade, 10*(2), 119–135.

Part IV

Health and Mortality

Air Pollution, Health, and Mortality

12

Melissa LoPalo and Dean Spears

Abstract

Air pollution has emerged as a leading threat to population health. A recent *Lancet* Commission reported that 6.5 million people worldwide die each year from exposure to air pollution. Emerging evidence further suggests that health co-benefits from air pollution reductions could be an important justification for aggressive climate mitigation policy. Here we review the empirical literature that has established air pollution's quantitative importance to population health. Our review is organized around two themes. One theme is global inequality: although most of the literature has focused on richer countries, where environmental and demographic data are more likely to be available, exposure levels can be much greater in the developing world. Another theme is the relevance and irrelevance of gaps in the literature: although there are important gaps in data from developing countries, enough is known

M. LoPalo (✉)
Department of Agricultural Economics & Economics, Montana State University, Bozeman, MT, USA
e-mail: melissa.lopalo@montana.edu

D. Spears
Population Research Center and Economics Department, The University of Texas at Austin, Austin, TX, USA

to conclude that air pollution is a priority not only for environmental health policy, but also for public health and climate policy.

Introduction

Air pollution has emerged as a leading threat to population health. A recent Lancet Commission report suggests that 6.5 million people worldwide die each year from exposure to air pollution (Landrigan et al., 2018). Beyond being quantitatively important among causes of mortality and morbidity, air pollution is of increasing importance to population scientists for three key reasons. First, unlike other population-level determinants of mortality such as maternal health, sanitation, or infectious disease more broadly, exposure to air pollution is worsening for some large populations, such as India (Spears, 2019), although improving for others, such as China (Kahn & Zheng, 2016). Overall, evidence suggests that the number of air pollution deaths is increasing each year, and that air pollution poses the largest environmental threat to health (Landrigan et al., 2018). Second, exposure to air pollution is a driver of health disparities: research is increasingly moving beyond

© Springer Nature Switzerland AG 2022
L. M. Hunter et al. (eds.), *International Handbook of Population and Environment*, International
Handbooks of Population 10, https://doi.org/10.1007/978-3-030-76433-3_12

243

documentation of average effects to instead highlight inequalities, including substantial disparities between low and middle income (LMIC)[1] and high-income countries. Finally, as important as air pollution itself is, pollution further interacts with the policy challenges of climate change: evidence and models indicate that health co-benefits from reducing exposure to air pollution could be an important factor in designing the best policy response to climate change. If the health impacts of air pollution are as harmful as recent literature suggests, the costs of mitigating emissions enough to significantly manage warming could be justified for the health benefits of improved air quality alone (Markandya et al., 2018).

Here we review empirical literature that documents air pollution's quantitative importance to population health, with a core focus on particulate pollution. We begin with an overarching review of what we've learned about the impacts of air pollution on human health throughout the life cycle. Our subsequent review is organized around three themes. One theme is global inequality: although most of the literature has focused on more developed countries, where environmental and demographic data are more likely to be available, exposure levels can be much greater in the developing world. A second theme engages the active research question of the *shape* of the concentration-response curve: do additional damages increase or decrease at high levels of exposure? And finally, another theme is the relevance (or irrelevance) of gaps in our understanding. Since data can be scarce in LMIC, especially the sort of detailed data that would permit a full accounting of all sources of pollution, policymakers in those regions might conclude that it's not possible to know whether air pollution is a population health priority. These evidence gaps are important; however, the literature that does exist clearly indicates that air pollution should be a policy priority not only for environmental or energy policy-makers, but also for public health and climate policy as well.

Air Pollution and Health

Evidence from the multidisciplinary literature on air pollution is near univocal: air pollution is a major global cause of poor health and mortality. Recent estimates suggest that air pollution caused as much as 16 percent of global deaths in 2015 (Landrigan et al., 2018). A recent scoping review revealed upwards of 5000 relevant studies on air pollution and health outcomes from environmental science, medicine, and economics, among other fields. Of the 799 included in the review, over 95 percent presented a statistically significant association between air pollution and the chosen health outcome, ranging from asthma to cardiovascular events to cognitive disorders (Sun & Zhu, 2019).

Evidence on the health effects of air pollution is diverse in methodology, geographical context, and population of interest. The early literature on air quality in epidemiology studies the effects of pollutants from many sources, from second-hand smoke exposure to ambient air pollution. This research focuses on a range of human health outcomes, such as heart disease, lung cancer, acute lower respiratory infection, and other causes of human mortality that are directly affected by air pollution. Data from the Global Burden of Disease study in 2015 suggest that, despite declines in indoor air pollution, air pollution from all sources outstrips malnutrition and major communicable diseases as the fifth leading risk factor for death worldwide (Cohen et al., 2017).

The mechanisms through which air pollution impacts health are increasingly well understood, although the relative impacts of types of pollutants remain difficult to pin down. Excess deaths from respiratory and cardiovascular causes have long been associated with higher concentrations of particulate matter pollution (Seaton et al., 1995). Exposure to air pollution is also associated with increased susceptibility to infectious disease (Clay et al., 2018). While many types of pollutants have been shown to cause adverse health outcomes, fine particulate matter, or particles with a diameter of less than 2.5 micrometers, is understood to be

---

[1] Here we use World Bank terminology.

particularly harmful (Franck et al., 2011). Small particulates can infiltrate deeply into human lungs, instigating an inflammatory response that can cause or exacerbate respiratory conditions such as asthma. Fine particulate matter can also potentially enter the bloodstream, causing increased coagulability, which exposes individuals to adverse cardiovascular events (Seaton et al., 1995). Generalizations of impacts are difficult to make since particulate matter pollution varies in its chemical composition depending on the original source (Xing et al., 2016). However, particulate matter from sources such as vehicle emissions, biomass burning, and indoor sources is receiving increased attention due to both their increasing prevalence in densely populated areas such as India and to findings pointing to particularly large health risks (e.g. Nel, 2005).

In addition, certain sources of particulate matter pollution are predicted to increase as climate change progresses. Exposure to wildfire smoke, for example, is likely to increase and has been tied to adverse effects on respiratory infections and mortality in particular (Liu et al., 2015; Reid et al., 2016).

Physiological Mechanisms and the Life Cycle

The pathways through which air pollution harms human health are myriad and vary throughout the human life cycle. Overall, research suggests that the already-sick (especially those with lung and cardiovascular conditions), the elderly, and the very young are most vulnerable (Pope 3rd, 2000). Figure 12.1 summarizes the most well-established health effects by age group, from in utero through old age.

While early studies did not find consistent evidence pointing towards impacts on fetal health (Glinianaia et al., 2004), newer work suggests that pregnant women's exposure to air pollution puts stress on the fetus, leading to increased risk of low birthweight, premature birth, and neonatal death (e.g. Bell et al., 2007; Bobak, 2000; Wang et al., 1997). Stieb et al. (2012) review the literature on outdoor air pollution, birthweight, and preterm birth, finding that the evidence is especially decisive for exposure during the third trimester and exposure to carbon monoxide, nitrogen dioxide (NO_2), and particulate matter. The physiological mechanisms through which effects operate are relatively poorly understood for in utero exposure, but there is some evidence that exposure in utero is associated with a decrease in placental mitochondrial DNA (mtDNA) content, a measure linked to health at birth (Janssen et al., 2012). A study of a Boston birth cohort found that exposure to fine particulate matter during pregnancy is associated with reduced heart rate variability in response to stress, another risk factor for various health issues (Cowell et al., 2019). Shah and Balkhair (2011) review the literature on air pollution and birth outcomes, which points to impacts on low birth weight and preterm birth through mechanisms such as inflammatory reactions in the lungs and placenta. Some work also links in utero and early life exposure to the development of asthma later in childhood (Clark et al., 2010).

Exposure to air pollution is also strongly associated with infant mortality, according to a well-established and extensive literature. Chay and Greenstone (2003) find that most infant deaths in response to air pollution occur during the perinatal period, implying that in utero exposure plays a role in infant deaths. However, other evidence strongly ties post-neonatal deaths, including from Sudden Infant Death Syndrome (SIDS), to air pollution exposure after birth (e.g. Woodruff et al., 2008). Later in childhood, many studies tie exposure to air pollution with the development of asthma as well as acute exacerbations. Lleras-Muney (2010), for example, finds that ozone pollution has large impacts on child respiratory hospitalizations. Cohort analysis suggests that long-term exposure to traffic-related pollution in childhood leads to worsened lung-function and increased sensitivity to allergens (Nordling et al., 2008). Beatty and Shimshack (2014) find that exposure to carbon monoxide and ozone are associated with contemporaneous health treatments for children, and that carbon monoxide exposure in particular has lasting effects. Air pollution is also tied to worsened learning outcomes, impacts

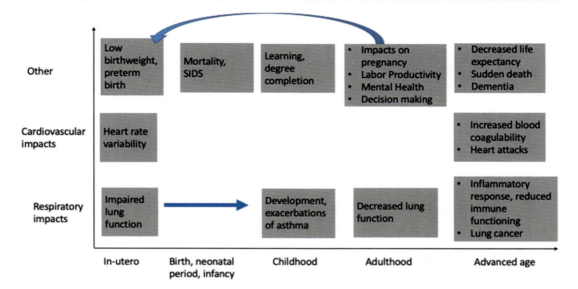

Fig. 12.1 This figure provides a visualization of selected adverse health and other outcomes that have been linked with exposure to air pollution in the environmental, economic, and other literatures

SIDS Sudden Infant Death Syndrome

which may persist into adulthood. For example, Bharadwaj et al. (2017), find that fetal exposure to air pollution has a detrimental effect on fourth-grade test scores in Chile.

Adulthood is generally associated with lower morbidity and mortality, including from air pollution, but there is some evidence that adverse effects of air pollution exist for prime-aged adults as well. While existing evidence does not find a consistent link between adult mortality and short- or long-term air pollution exposure (Anderson, 2020), evidence does tie air pollution to adverse health outcomes. The ESCAPE cohort study in Europe, for example, shows a relationship between chronic air pollution exposure and lung function (Adam et al., 2015). Schlenker and Walker (2016) find impacts of exposure to pollution from airports on adult hospitalizations. Furthermore, growing evidence links air pollution exposure to declining labor productivity in both physical tasks (e.g. Chang et al., 2019; Graff Zivin & Neidell, 2012) as well as among high-skilled workers such as judges (Kahn & Li, 2019) and investors (Meyer & Pagel, 2017). Voorheis (2017) finds impacts on schooling outcomes and incarceration. There is also evidence that air pollution affects decision-making in adults; Heyes et al. (2016) find evidence of impacts on investor behavior, potentially through the route of decreased risk tolerance. Finally, evidence from China suggests that short-term increases in particulate matter pollution in China are associated with increased incidence of several mental health issues (Chen et al., 2018b), as well as decreased self-reported happiness and increased depressive symptoms (Zhang et al., 2017).

In old age, cumulative exposure to air pollution is at its highest, as is vulnerability to health shocks. Work by Deryugina et al. (2019) suggests that vulnerability to air pollution among the elderly depends substantially on predicted life expectancy, with most of the health burden of air pollution concentrated in the population in the lowest quartile of life expectancy. Bishop et al. (2018) track 10 years of cumulative residential exposure to particulate matter pollution for Medicare patients and find a significant effect of long-term exposure on dementia diagnoses.

Methodological Challenges and Approaches

The epidemiological literature largely measures correlations between air pollution and health outcomes, generally by measuring population-wide associations between exposure to air pollution and adverse health outcomes. The early literature focuses on studies of single locations (e.g. Pope III et al., 1992; Schwartz & Dockery, 1992). More recently, epidemiological studies have moved towards repeated cross-sectional approaches using data from multiple locations due to concerns about generalizability of results from any single location.

Methodological challenges are posed by the fact that exposure to air pollution is not random. On the one hand, in many samples, more disadvantaged populations are exposed to higher levels of air pollution (Currie & Walker, 2011). The association between socioeconomic status and exposure to air pollution is one that has been repeatedly documented in various settings (we discuss the literature further in "Effects of Air Pollution in Developed Vs. Developing Countries"), and it implies that simple correlations between air pollution exposure and health outcomes may suffer from the confounding effect of lower socioeconomic status on health. This would tend to exaggerate the impacts of air pollution on health. On the other hand, individuals in worse health may be more likely to move away from heavily polluted areas (Currie, 2013). The resulting association between poor underlying health status and low air pollution exposure would mask the treatment effect for the most vulnerable individuals, biasing estimates of pollution on health downwards. In addition, measuring cumulative exposure to air pollution over the course of a person's life is challenging: individuals may experience varying exposure depending on where they work in addition to where they live. Migration poses an additional challenge: it is difficult to calculate cumulative exposure without having a complete register of all of the locations where an individual has lived and for how long, which is information that is rarely available in commonly-used datasets. Finally, it is difficult to determine the number of life years lost due to air-pollution related deaths, since the already-sick are likely most vulnerable (McMichael et al., 1998).

In the epidemiology literature, the more recent cohort and repeated cross-sectional studies have sidestepped some of these methodological issues: they often have richer data on potential confounding variables at the individual level, allowing for a more complete set of controls than studies that examine population-level outcomes (e.g. Dockery et al., 1993; Jerrett et al., 2009; Sheffield et al., 2011). These data allow researchers to compare health outcomes for individuals with varying exposure to air pollution, holding factors such as income, underlying health and health behaviors, and age constant.

Ultimately, however, the scope for omitted variable bias remains in these studies: in the absence of random assignment, the possibility that exposure to air pollution is correlated with other unobserved determinants of health is a threat to internal validity of results. In response to these challenges, recent work has increasingly incorporated tools of causal identification from observational data. Many of these studies are in the microeconometric literature and exploit sharp changes in air pollution that are plausibly unrelated to changes in other correlates of population health outcomes. Due to the focus on estimating cleanly-identified causal effects of air pollution, these studies have often focused on infant and child mortality, as these are less likely to indicate "harvesting," where an already-sick individual's death is hastened rather than being largely due to air pollution (Chay & Greenstone, 2003).

Mortality data can also be available in large, population-level datasets from censuses or vital registration systems; these permit the power necessary to estimate environmental effects where available, but of course such high-quality data are not available for all populations. Table 12.1 contains summaries of findings from a collection of these papers, putting them into the context of the local levels of the pollutant. This is not an exhaustive review, but rather aims to highlight overall trends in the literature and establish what the recent turn to causal identification has and has

Table 12.1 Review of effect sizes estimated in selected studies of the health impacts of particulate matter pollution

Study	Pollutant(s)	Setting	Average Pollution in Sample	Outcome Variable(s)	Effect Size
Chay & Greenstone, 2003	TSPs	U.S.	60–70 μg/m^3	IMR	1% decline in TSP → 0.35% decrease in IMR
Currie & Neidell, 2005	PM$_{10}$	CA	39.4 μg/m^3	IMR	No significant effect
Jayachandran, 2009	PM$_{2.5}$	Indonesia	-	Fetal, infant, and child mortality	Wildfire smoke → 15,600 estimated child deaths
Adhvaryu et al., 2016	PM$_{2.5}$	W. Africa	44 μg/m^3	IMR	100 μg/m^3 increase in in-utero exposure → 2.6 deaths/1000 births
Arceo-Gomez et al., 2016	PM$_{10}$	Mexico City	66.94 μg/m^3	IMR	100 μg/m^3 increase in PM$_{10}$ → 0.24 deaths/1000 births
Heft-Neal et al., 2018	PM$_{2.5}$	Africa	25 μg/m^3	IMR	10 μg/m^3 increase → 9% increase in IMR
Chen et al, 2013	TSPs	China	354.7 μg/m^3	Life expectancy	100 μg/m^3 → 3-year reduction in life expectancy

Notes: Table presents estimates from studies of the impacts of particulate matter pollution on mortality in a variety of settings
IMR Infant Mortality Rate, *TSPs* Total Suspended Particulates, *PM$_{2.5}$* Particulate matter less than 2.5 micrometers in diameter, *PM$_{10}$* Particulate matter less than 10 micrometers in diameter.

not clarified in this literature. These studies are also chosen to illustrate their methodology.

Studies in the microeconomic literature tend to take advantage of plausibly random variation in air pollution due to changes in policy, infrastructure, or economic activity. In developed countries, this literature has found robust evidence of adverse effects of air pollution at levels well below U.S. Environmental Protection Agency (EPA) standards. For example, Chay and Greenstone (2003) examined the impact of total suspended particulates (TSPs) on infant mortality rate by exploiting the association between local economic activity and ambient air pollution. They focus on the U.S. recession in the early 80's, which had substantially varying impacts in different geographical regions, and they find significant effects even at low levels of exposure.

Currie and Neidell (2005) similarly examine infant health, this time focusing on California in the 1990s, where average pollution was relatively low. The authors examine the impact of air pollution on the risk of death, controlling for a number of variables such as child age, characteristics of the mother and child, and place and time fixed effects. They find significant effects of carbon monoxide on infant mortality, even at low levels, as well as PM_{10} (particulate matter with a diameter of less than 10 $\mu g/m^3$) in some specifications.

Several papers set in the U.S. look at transportation-related pollutants from highways (Anderson, 2020; Currie & Walker, 2011) and airports (Schlenker & Walker, 2016). Each paper has a strategy for netting out non-random locational sorting near these large sources of pollution. Currie and Walker (2011) take advantage of the introduction of electronic toll collection in New Jersey, which sharply changed the level of pollution along highways at some points due to reduced idling traffic. They then compare the health of infants born to mothers close to a toll plaza to that of infants born to mothers also close to the highway, but relatively far from a toll plaza, finding large effects of E-ZPass adoption on probability of premature birth and low birthweight.

Schlenker and Walker (2016) examine the effect of pollution from Los Angeles International Airport, taking advantage of the large impact of flight delays on air pollution due to airplanes idling on the tarmac. To cope with the fact that delays may be related to local weather and other factors that may also influence health, they focus on delays that originated in large airports in other regions of the country, and they allow health effects of the pollution to vary by whether individuals live upwind or downwind of the airport. They examine the effect of this high-frequency variation in air pollution on hospitalizations throughout the age distribution, finding that hospitalizations increase for adults as well as infants and the elderly. The effects are driven by carbon monoxide.

Rather than using sharp variation in air pollution within a place to net out concerns over non-random exposure, Lleras-Muney (2010) uses variation in exposure coming from migration. She focuses on children who moved due to parents' military transfers, which are plausibly unrelated to other correlates of the child's underlying health. She examines the effects of several pollutants but only finds significant impacts of ozone, finding that ozone pollution has large impacts on child respiratory hospitalizations.

Another methodological challenge is posed by the dearth of available data on both health outcomes and air pollution in developing countries, which we explore in "Data Availability". The literature from high income countries has the advantage of relatively plentiful data on outcome variables such as deaths and hospitalizations but tends to estimate the effects of air quality at relatively low levels of air pollution. In many LMIC contexts, exposure to air pollution is many orders of magnitude more severe than in lower-income populations, making it challenging to extrapolate the results of this literature to these settings. Even a highly persuasive study of the effect of E-ZPass in New Jersey may tell us little about the effects of a new coal plant in Kanpur, Uttar Pradesh. Uttar Pradesh, however, does not have the sort of high-quality vital registration system that would allow a researcher to observe every infant death. Kanpur, moreover, only has one air pollution monitor – and the neighboring districts have none. In recent years, several papers have found ways to circumvent shortcomings in the availability of high-quality air pollution and health data in developing countries to add to the body of knowledge on the impacts of air pollution at higher levels. These papers often use variation in air pollution from weather or other natural phenomena that are observable using satellite data, which has grown in availability and quality in recent decades.

As an example, Jayachandran (2009) studies the impacts of smoke from severe wildfires in Indonesia. Lacking vital statistics data on infant mortality, Jayachandran looks for "missing" children in the 2000 Census, which was taken a few years after the wildfires. She controls for month and year of birth to net out average differences in survival by cohort and exploits geographic variation within Indonesia in the severity of smoke, as measured by satellite data. Using this strategy, she estimates that 1.2 percent of the affected birth cohorts are missing from the 2000 Census, which amounts to approximately 15,600 deaths.

Arceo-Gomez et al. (2016) examine the impacts of elevated pollution caused by thermal inversions in Mexico City. This weather phenomenon causes pollution to be trapped in the valley where the city is located due to a layer of hot air. They find that inversions significantly elevate local concentrations of PM_{10} and carbon monoxide, which each has large impacts on infant mortality. Adhvaryu et al. (2016) exploit wind patterns in West Africa, which carry large amounts of dust from the Sahara Desert to population centers near the coast. They use mortality data from Demographic and Health Surveys and satellite data on dust concentrations, finding large impacts of $PM_{2.5}$ on infant mortality. Rangel and Vogl (2019) examine the impacts of agricultural fires on infant health in Brazil, using changes in wind direction as well as fire location to separate the impacts of smoke from the association between agricultural fires and local economic activity. They find that birth weight, gestational length, and in utero survival are negatively impacted by in-utero exposure to smoke from fires.

Evidence on long-term effects of air pollution in developing countries is limited, largely because the natural phenomena used most often to identify effects in developing countries are transient. There are a few exceptions, however: Chen et al. (2013) make use of a long-term policy that created a difference in pollution levels in Northern versus Southern China to find large impacts on life expectancy. Cao et al. (2011) exploit unique cohort data from the National Hypertension Survey, finding strong associations between exposure to sulfur dioxide (SO2) and Total Suspended Particulates (TSP's) and mortality from causes such as cardiovascular mortality and lung cancer.

Global Distribution of Air Pollution Exposure and the Concentration-Response Curve

All populations are exposed to some degree of air pollution, and the studies reviewed in "Air Pollution and Health" find health impacts even at relatively low levels. However, a guiding question in the study of air pollution's population health impacts is the level at which harmful impacts can be sufficiently averted. Relatedly, it is important to understand the correlation between rising air pollution and population health. Incremental increases (or decreases) in pollution may have differential health impacts according to the baseline level of pollution. Such questions have to do with the shape of the concentration-response "curve" linking levels of air pollution to harmful health impacts.

This question, unresolved in the literature, has clear implications for policy decisions around air pollution. If the curve is "supra-linear," meaning that air pollution reductions would have the largest impacts in relatively clean places, then policymakers would achieve the largest reductions in health impacts by focusing on areas with low air pollution, making "blue skies bluer" (Marshall et al., 2015). If the curve is linear, on the other hand, then the benefits of pollution abatement for health would be the same everywhere. These possibilities are visualized in Fig. 12.2. Unfortunately, many regression-based empirical studies implicitly constrain models to a linear shape, ignoring this important question.

This issue is especially crucial to our understanding since exposure to air pollution varies tremendously by location. Figure 12.3 displays satellite-derived estimates (2016) of the

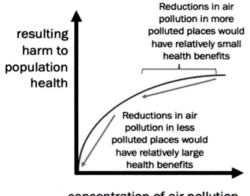 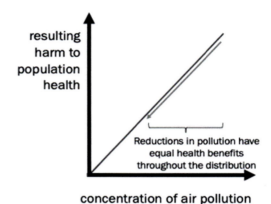

Fig. 12.2 This figure displays a few possible concentration-response curves. The graphs display the relationship between severity of air pollution exposure and adverse health impacts under the assumption that the concentration-response curve is supralinear (left) vs. linear (right) Source: Adapted from Spears et al. (2019)

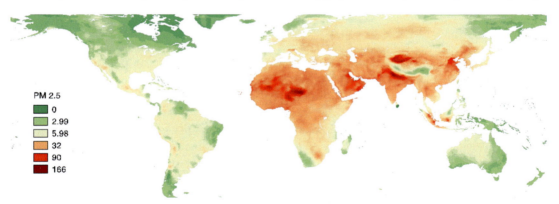

Fig. 12.3 This figure displays 2016 annual average $PM_{2.5}$ concentrations at a 0.1 × 0.1 degree spatial resolution globally. Estimates of particulate matter concentrations include dust and salt Source: Authors' computation from van Donkelaar et al. (2016)

average annual concentration of fine particulate matter ($PM_{2.5}$) air pollution for every 0.1 × 0.1-degree latitude/longitude grid point. The figure makes clear that exposure to air pollution varies in particular between the developed and developing world. The Americas, Europe, and Australia largely enjoy relatively clean skies, while Africa, the Middle East, and South and Southeast Asia are often exposed to levels of particulate matter pollution rarely experienced in places such as the U.S. Exposure to air pollution in developed countries has drastically fallen over the past half century, as the introduction of clean air legislation has prompted industrial sources to clean up and car manufacturers to produce more efficient vehicles, among other notable improvements (Chen & Kan, 2008; Sullivan et al., 2018).[2]

---

[2] However, there is some evidence that suggests that increases in wildfires and a trend towards environmental deregulation may be causing air pollution to tick back up in areas such as the United States (Clay & Muller, 2019).

Exposure varies significantly even within heavily polluted countries such as India; however, recent estimates by the WHO suggest that up to 92 percent of the global population are exposed to annual average air pollution that exceeds WHO healthy standards (WHO, 2016). The U.S. Environmental Protection Agency (EPA)'s standard for annual exposure to fine particulate matter air pollution is 12 $\mu g/m^3$. The highest reading in Fig. 12.3 is 166 $\mu g/m^3$, nearly 14 times the EPA's standard. Those who live in places such as the dusty Saharan desert and the Indo-Gangetic Plain of North India are exposed to annual averages well above 100 $\mu g/m^3$. Further, annual averages mask substantial variation on a month-to-month, day-to-day, and intraday basis, with much higher peak exposure. Levels of particulate matter pollution in Delhi, for example, have been known to reach 1000 $\mu g/m^3$ in the winter months, when smoke from agricultural fires combines with winter weather and pollution from winter heating.

Understanding the shape of the curve linking air pollution to health would help inform policymakers whether to focus on pollution abatement in North India or South India, China or California, as well as whether there would be value in further reducing air pollution in the U.S., where most places have annual average concentrations that meet EPA standards. Knowledge of this dose-response relationship is also a key step in quantifying the health co-benefits of climate mitigation policy. Scovronick et al. (2019), for example, compute that there would be large health co-benefits of climate mitigation in part because of the enormous opportunities to reduce exposure in India, where pollution levels are exceptionally high. In their baseline analysis, they assume the relationship between pollution exposure and health impacts to be linear; a supralinear concentration-response curve would likely significantly impact these results.

There are several reasons why quantifying the shape of the curve is difficult. Base mortality rates are very different in developed and developing countries. Age-specific mortality is difficult to clarify: mortality rates are more concentrated at young ages in LMIC, where pollution levels are more severe but age-specific mortality data are more difficult to come by. Furthermore, indoor exposures vary due to differences in cooking fuel and smoking behavior, as well as differences in the extent to which individuals are protected from outdoor pollution in their homes. These factors may amplify differences in exposure between developed and developing countries.

Additionally, as we discuss in "Data Availability", the data availability on health outcomes differs by place: developing countries tend to lack population-wide datasets on mortality. It is difficult to find a context where a researcher may compare the impacts of different levels of pollution from the same sources in order to credibly trace out the shape of the curve. "High" pollution levels in a study from the U.S. would be low levels of exposure in a study of South Asia.

The epidemiology literature has attempted to tackle this open question using cohort analysis: a few recent papers analyze lifetime smoking patterns and exposure to other sources of pollution and estimate the association with mortality from pollution-associated causes such as lung cancer and cardiovascular mortality. This evidence suggests that the shape of the dose response curve differs by cause of mortality, with lung cancer mortality risk rising linearly with exposure and cardiovascular mortality risk flattening out at very high levels of particulate matter pollution (Pope III et al., 2011; Pope III, Burnett, et al., 2009). Burnett et al. (2014) integrate studies on air pollution from ambient and indoor sources as well as smoking to construct a relative risk model that spans the distribution of global exposure to particulate matter air pollution. These methods necessarily make the assumption that health impacts of pollution exposure are the same regardless of the source (smoking versus ambient versus indoor).

This assumption is important in light of the fact that pollution in developing countries comes from different sources, and thus has a different chemical composition, than pollution in developed countries. Lelieveld et al. (2015), for example, estimate which source of pollution accounts for the largest mortality impacts in different areas of the world, finding that pollution from indoor sources, biomass burning, and natural sources are particularly prevalent in developing countries, whereas pollution from industrial sources, traffic, and agriculture are the biggest sources of harm in developed areas such as the U.S. and Europe. Given the evidence that morbidity and mortality effects vary by chemical composition (e.g. Peng et al., 2009) this poses a substantial threat to evidence extrapolated from developed to developing countries.

Lacking perfect settings to investigate the health impacts caused by the full range of possible air pollution concentrations, some researchers have focused on estimating the impacts of changes in air pollution within a place and testing for evidence of nonlinearity based on the variation in pollution exposure within their samples. Evidence from these exercises is mixed. As an example, Spears et al. (2019) use a panel of ambient particulate pollution data spanning 5 years and all of India's districts. The result is variation in exposure levels that spans from low, developed-country-like levels to north India's high winter levels. In several tests, they do not find statistically significant evidence that the shape of the curve is not linear. Arceo-Gomez et al. (2016), find similar effect sizes of pollution in infant mortality to Chay and Greenstone (2003) despite much higher average pollution levels, adding more evidence that the dose-response relationship may be linear. Chay and Greenstone themselves, however, do find evidence of nonlinearities within their sample.

Given the wide variety of pollution levels experienced by populations in different parts of the world, understanding the shape of the concentration response curve is crucial to determining where mitigation efforts would be most helpful. Issues with data availability and heterogeneity in pollution sources by place make quantification of the shape of this curve difficult, and existing evidence is quite mixed. However, in light of the importance of this question, this will be a crucial area for future research.

Effects of Air Pollution in Developed Vs. Developing Countries

There are several a priori reasons to believe that the health effects of air pollution might differ substantially in developing countries relative to developed countries. For one, average pollutant levels are often higher in developing countries, as was clear from Fig. 12.3. How much worse this is for human health depends on the shape of the curve linking quantities of air pollution to human health outcomes, as discussed in "Global Distribution of Air Pollution Exposure and the Concentration-Response Curve". Second, people in developing countries may have worse baseline health on average irrespective of exposure to air pollution, which may exacerbate the effects of air pollution, an effect likely reinforced by a lack of access to quality health care. Finally, people in developing countries may be less able to avoid the effects of air pollution if they work outdoors, do not have access to technologies such as air filters, or are exposed to more sources of air pollution.

The existing literature on health effects in developing countries is mixed on these differences. One type of comparison is simply to compare separate studies from separate populations, but here one may wonder whether other factors, such as differences in chemical composition of pollutants or exposure to indoor pollution, drive any differences in estimates. Arceo et al. (2016) find effect sizes for carbon monoxide that are larger than previous estimates from the U.S., but that the effect is similar for PM_{10}, despite large differences in average pollutant levels. Tanaka (2015) examines the impacts of Chinese regulations that imposed pollution standards in certain areas, finding effects on infant mortality that were much larger than effect sizes estimated in developed countries.

Alternatively, a few papers set in developing countries make comparisons *within* their sample, examining heterogeneity with respect to socioeconomic characteristics of individuals or populations. Adhvaryu et al. (2016) find that the health impacts of dust are smaller in more developed economies and in countries with greater health resources. Jayachandran (2009) finds larger health impacts of smoke in low-income areas of Indonesia and in areas where people cook more often with wood-burning stoves. On the other hand, Heft-Neal et al., (2018), find that the health impacts of $PM_{2.5}$ in Africa do not vary with household wealth.

Inequality in Exposure

The total health impacts of air pollution may vary based on income and wealth between individuals, communities, and economies for two reasons: inequality of exposure, or differences in health impacts conditional on exposure. Here we note briefly that the first reason is undeniable, both globally and locally. As discussed in "Global Distribution of Air Pollution Exposure and the Concentration-Response Curve", the burden of air pollution falls very disproportionately on developing countries and relatively disadvantaged populations within developed countries. Within countries, the less well-off tend to be located in more polluted areas (see O'Neill et al. (2003) for a review). This is true in developed countries as well. For example, Clark et al. (2014) find that residential exposure to nitrogen dioxide is much higher for nonwhite populations than for white populations in the U.S. and that this disparity in exposure accounts for 7000 excess heart disease deaths per year among nonwhite populations. Similarly, Mikati et al. (2018) find substantial disparities in $PM_{2.5}$ exposure by race in the U.S., with racial differences exceeding differences based on poverty status.

Inequality in Effect Conditional on Exposure

The second possible reason for inequality in total impact is inequality in the health effects of exposure: either because of difference in the protectiveness of environments or because air pollution may have larger effects on health at the higher levels found disproportionately in developing countries. The ability to avoid negative

health consequences of air pollution may also differ by place and by socioeconomic status. There are well-documented differences in the quality of health care between developing and developed countries, likely exacerbating the impacts on health in developed countries. Poorer people may also be less able to avoid exposure to air pollution, for example if they live in more poorly-constructed housing, which might do a poorer job of protecting them from outdoor pollutants. Vyas et al. (2016), for example, show that even in relatively high-socioeconomic-status apartments in an elite neighborhood of Delhi, commercially available air filters are unable to reduce indoor particulate pollution to healthy levels (or to prevent it from being correlated with outdoor ambient pollution levels) because houses are not built to construction standards that separate indoor air from outdoor air. For example, windows might simply be openings in walls, without panes of glass to interrupt airflow. There would be even less separation for poorer people's houses, although on the other hand less separation could be beneficial for dispersing pollution from indoor sources such as cooking.

Several studies investigate the scope for avoidance behavior by examining behavioral responses to elevated levels of air pollution, such as willingness to pay for avoidance technology or changes in time spent outdoors. However, the detailed data requirements of these studies mean they are focused on higher income populations' exposure levels. Neidell (2009) studies the response to smog alerts in Southern California using a regression discontinuity design, finding that attendance at the Los Angeles Zoo and Botanical Gardens and Griffith Park Observatory, two venues for outdoor recreation, fell on days where ozone was over the cutoff for a smog alert. However, Graff Zivin and Neidell (2009) find that this response becomes muted if the alert persists for more than one day. Chen et al., (2018a, 2018b) find that the health impacts of air pollution fall discontinuously on days when the level of pollutants qualifies for an air quality alert in Toronto. Moretti and Neidell (2011) exploit a source of variation in ozone that is likely unobserved by residents of Los Angeles:

the amount of boat traffic at the ports, to examine the impact of ozone on health. This, they argue, allows them to observe health responses to air pollution holding avoidance behavior fixed. They compare estimates using OLS to estimates using boat traffic as an instrument, interpreting the large difference in effect sizes as indicative of both avoidance behavior and measurement error.

Other studies estimate willingness to pay for clean air, which is an important parameter for understanding optimal environmental regulation (Greenstone & Jack, 2015). For example, Ito and Zhang (2020) use scanner data on air purifier purchases in China, taking advantage of product differentiation in the market. One of the product attributes has a strong impact on an air purifier's ability to effectively filter out air pollution; the authors study buyers' valuation of this feature. They find that the willingness to pay to reduce PM_{10} pollution by 1 $\mu g/m^3$ per year is \$1.34, and they also find strong evidence of that richer households have a higher willingness to pay for air quality. Zhang and Mu (2018) find that sales of particulate-filtering facemasks increase sharply during significant smog events, indicating nonlinearities in willingness to pay along the distribution of air pollution.

Freeman et al. (2019) follow Bayer et al. (2009) in using a residential sorting model combined with an instrumental variables approach to cope with the endogeneity of air pollution to measure willingness to pay, finding that the median household in China would pay \$21.70 to reduce $PM_{2.5}$ pollution by 1 $\mu g/m^3$ per year. This contrasts with Bayer et al.'s estimate of a willingness to pay of \$149–\$185 (1982–84 dollars) to reduce PM_{10} by one unit in the United States.

The ability to avoid the ill effects of air pollution through technologies such as air filters or medication may depress willingness to pay estimated through methods such as residential sorting models. Deschenes et al. (2017) estimate the impact of a reduction in air pollution due to a cap-and-trade market in California on medication expenditures, finding that defensive investments account for over a third of willingness to pay for air quality in their context. However, the ability

to avoid air pollution through expenditures on medicine or technology may vary by income and country context (Vyas et al., 2016). This has strong implications for both estimates of willingness to pay for air quality—people may not be willing to pay much for technologies that do not effectively mitigate health risks—and for health impacts in developing countries. Sun et al. (2017) find that richer people in China are more likely to purchase in air filters, which are more expensive and more effective than masks, widening socioeconomic gaps in exposure to air pollution.

We emphasize, however, that willingness to pay to avoid air pollution is only one factor to consider in policy-making. One reason is that even people exposed to very high levels of air pollution may not fully understand its harms or have the ability to pay to avoid them (Spears, 2019). Another reason is that air pollution causes *external* harms: when considering the costs and benefits of polluting activities, individuals and companies may not consider the resulting harms of air pollution to the health and productivity of others. These externalities fall both on individuals and on society as a whole, through the consequences of productivity and health for tax revenues and government budgets.

Co-Benefits & Climate Change: Population Health Consequences in the Long Run

It is clear that current levels of air pollution have large and meaningful impacts on human health and well-being globally. Air pollution is often a by-product, or externality, of economic production that affects people other than the polluters themselves. In other words, the social cost of air pollution exceeds the cost to the polluter. Under classic economic theory, this justifies government intervention: policymakers should intervene by discouraging pollution until the social cost of pollution reductions (from reduced economic activity) are just equal to the benefits from improved health outcomes. However, the benefits of pollution abatement go beyond the immediate impacts for today's populations. This type of intervention would make progress towards solving another classic externality problem: climate change.

The impacts of changes in climate and the associated increase in the frequency of extreme temperature days on human health and productivity are the subject of another well-established literature. Extreme temperatures have been found to have large impacts on mortality (e.g. Barreca et al., 2016; Deschenes & Moretti, 2009), economic growth (e.g. Burke et al., 2015; Hsiang, 2010), agricultural productivity (Burke & Emerick, 2016), conflict (Hsiang et al., 2013), and many other outcomes. Avoiding these impacts is one well-known rationale in favor of decreasing carbon emissions. But here, we emphasize another: health benefits associated with decreased air pollution that could occur as a by-product of reducing emissions. Shindell et al. (2018), for example, find that reducing emissions to cut warming to 1.5 degrees would avert 150 million untimely deaths due to air pollution, largely in Asia and Africa.

However, there is some debate about the net impact of air pollution abatement on climate change. This is because certain types of pollution cause *cooling* by blocking out the sun's rays. Therefore, thorough cost-benefit analyses of pollution mitigation must consider both the benefits of the economic activity that produces pollution and of the cooling effect of some pollution on climate against the impacts of pollution on health combined with the contribution of carbon emissions, which cause air pollution, to climate change. Scovronick et al. (2019) weigh both of these consequences. They find that the health benefits of air pollution abatement are larger than the cooling impacts, tilting the scale in favor of mitigation of emissions-producing technology. Further, the health benefits of emissions reductions are sufficiently large that mitigation would cause net benefits immediately, rather than only in the future when damages from climate change are averted. Therefore, the benefits of reducing exposure to climate change and air pollution reinforce each other, both in the short and long term – although these benefits, too, depend on the shape of the concentration-response curve.

Data Availability

In sections 2–4, we broadly surveyed the existing literature on air pollution, health and mortality, paying special attention to differences in exposure and impact conditional on exposure between high- and low-income people and places. These literatures often differ in their data sources, making tracing out the "shape of the curve" difficult. These studies diverge in both their sources of air pollution data—studies of high-income countries tend to use rich air quality monitor data where LMIC studies often use satellite data to infer pollution levels—and in the availability of outcome variables.

Independent Variables: Air Pollution Data

One of the key difficulties with measuring the impacts of air quality in developing countries is a lack of information on air quality itself. Air quality monitors provide the most accurate information on local levels of air pollution, which can vary sharply over fine geographic areas. For example, an air quality monitor attached to a rickshaw in Delhi revealed air pollution levels that changed significantly close to vehicular traffic (Apte et al., 2011). Air quality monitors do potentially suffer from issues such as politically-motivated placement: Grainger et al. (2018) find that monitors in the U.S. are sometimes placed in areas where pollution is lower, such as upwind of pollutant sources, in order to help local areas achieve compliance with EPA rules, for example. These types of manipulation could have serious consequences in terms of bias of estimated coefficients. However, air quality monitors are the only source of data that is measured at the surface, where humans are exposed to pollution.

Most developing countries do not have a comprehensive system of air quality monitors that span both urban and rural areas. India, for example, lacks monitoring coverage in rural areas (Pant et al., 2019), despite the fact that satellite data shows extreme levels of particulate matter pollution across North India, both rural and urban. Delhi, where richer people live, has many more monitors than Uttar Pradesh and Bihar, where many poorer people live.

This problem exists in developed countries as well, albeit to a lesser extent. Sullivan and Krupnick (2018) use satellite data to show that 24.4 million people in the U.S. live in places that should be classified as nonattainment areas, or areas with particulate matter pollution above the Environmental Protection Agency standards, but are not due to lack of monitor coverage. Similarly, Fowlie et al. (2019) show that satellite data imply errors in both directions: areas that should be nonattainment areas are classified as in attainment and vice versa. This lack of coverage makes measuring health impacts most difficult where it is likely most crucial.

Due to the lack of air quality monitor coverage, many studies in developing countries use satellite data to measure air quality. Satellites measure air pollution by measuring extent to which pollutant particles in a column of air absorb or scatter light. This provides a measure of the amount of air pollution in the atmospheric column between the satellite instrument and the surface of the Earth. These measurements, called Aerosol Optical Depth (AOD) and Aerosol Optical Thickness (AOT) have been shown to correlate strongly with ground measurements of fine particulate matter (Chu et al., 2003; Wang & Christopher, 2003). However, this correlation varies with other atmospheric conditions, such as temperature and humidity, potentially complicating analysis that uses satellite data. Some researchers have responded by controlling for weather conditions in empirical models, as weather conditions may also affect outcome variables of interest. But as Fowlie et al. (2019) note, there is a range of possible satellite readings for a given monitor reading, instilling conclusions using satellite data with a level of uncertainty that is not accounted for by most of the literature. In particular, they find that satellite data seem to be biased down at higher levels of air pollution.

However, the number and quality of satellite data products have increased in recent years, allowing researchers to take advantage of data at relatively fine geographic and temporal resolution. There are also data products available that allow researchers to track the location of pollutant sources such as fires and dust storms, which are increasingly used by researchers to estimate reduced-form effects of sources of pollution on health.

Dependent Variables: Mortality and Population Health

As discussed in "Air Pollution and Health", a large portion of the literature in developing countries examining the impacts of air pollution on health has focused on infant mortality as an outcome, in part because these deaths are relatively unlikely to solely reflect the deaths of the vulnerable being displaced forward in time. These studies rely on the high-quality vital registration systems available in countries like the U.S., which record all births and deaths in the population and thus provide a sufficiently large sample size to statistically detect determinants of mortality. Many developing countries do not have credible vital registration systems, so some deaths are simply not recorded (Setel et al., 2007). Some studies in developing countries have taken advantage of notable exceptions, such as Sao Paulo, Brazil (Rangel & Vogl, 2019), and Mexico City (Arceo et al., 2016) – although even these reflect the fact that Latin American populations are advantaged relative to sub-Saharan Africa or South Asia.

Other studies turn to household survey data, such as the Demographic and Health Survey (DHS) datasets. These types of survey data often have the advantage of containing rich health information beyond mortality. The DHS, for instance, contains complete reproductive histories for adult female respondents, including instances of infant mortality, but also other outcomes such as child height that may be affected by environmental factors such as air pollution. However, since household survey datasets represent small samples of the entire population, it can be difficult to obtain the statistical power necessary to detect effects. Furthermore, there is little in these types of datasets on adult mortality, and even less on the other health outcomes mentioned in Fig. 12.1, making it difficult to measure the impact of air pollution at all ages in developing countries.

Conclusion

Air pollution is a critical threat to population health. Unlike many determinants of mortality, for some large populations such as India, exposure to ambient air pollution is getting worse, not better (GBD Maps Working Group, 2018). Air pollution is also an important source of health inequalities (O'Neill et al., 2003), as exposure to air pollution, baseline health, and access to health care all vary with income both within and across countries. And it may be at the center of how policy-makers should think about climate mitigation: indeed, if health co-benefits are as important to climate policy as some recent estimates suggest, then the urgency of air pollution would overturn the conventional wisdom that climate mitigation would only help future generations.

Policy-makers know enough to act. Some steps, such as reducing exposure to pollution from burning coal in densely-populated north India, are clear (Gupta & Spears, 2017). And yet, important open questions remain. Many of these questions can only be understood at the population level: how do exposures and consequences differ between Bangladesh, on the one hand, and Belgium, on the other? For many policy questions, such as climate policy, these are the most important questions.

Simple, descriptive studies of population-level data can often help guide policy (sometimes the important question is simply where there is a problem) and can highlight inequalities. But for observational studies of cause and effect, these inequalities are exactly a core methodological challenge: the poor are more likely to be exposed to pollution, and are often worse off in other ways, too. Here, we have highlighted methodological innovations in the study of air pollution that focus on this challenge. Most of these studies are not at the population level, but instead learn what they

can from a context where data are available in a special situation that happens to be informative. This raises a question: how well do these studies address the most important policy-relevant questions? Notably, in writing this chapter, we struggled to make the quantitative estimates from different studies of different populations directly comparable to one another – in the end, we simply decided not to, rather than run the risk of producing a flawed comparison. The implication is that even motivated research teams may struggle to combine even high-quality studies to answer the most urgent policy questions. Ultimately, both types of studies are needed in the dialogue between population science and the policy response to air pollution.

References

Adam, M., Schikowski, T., Carsin, A. E., Cai, Y., Jacquemin, B., Sanchez, M., Vierkötter, A., Marcon, A., Keidel, D., Sugiri, D., & Al Kanani, Z. (2015, January 1). Adult lung function and long-term air pollution exposure. ESCAPE: A multicentre cohort study and meta-analysis. *European Respiratory Journal., 45*(1), 38–50.

Adhvaryu, A., Bharadwaj, P., Fenske, J., Nyshadham, A., & Stanley, R. (2016, February 20). *Dust and death: Evidence from the West African Harmattan*. Centre for the Study of African Economies, University of Oxford working paper.

Anderson, M. L. (2020 August 1). As the wind blows: The effects of long-term exposure to air pollution on mortality. *Journal of the European Economic Association., 18*(4), 1886–1927.

Apte, J. S., Kirchstetter, T. W., Reich, A. H., Deshpande, S. J., Kaushik, G., Chel, A., Marshall, J. D., & Nazaroff, W. W. (2011 August 1). Concentrations of fine, ultrafine, and black carbon particles in auto-rickshaws in New Delhi, India. *Atmospheric Environment., 45*(26), 4470–4480.

Arceo-Gomez, E., Hanna, R., & Oliva, P. (2016, January 12). Does the effect of pollution on infant mortality differ between developing and developed countries? Evidence from Mexico City. *The Economic Journal., 126*(591), 257–280.

Barreca, A., Clay, K., Deschenes, O., Greenstone, M., & Shapiro, J. S. (2016, February 1). Adapting to climate change: The remarkable decline in the US temperature-mortality relationship over the twentieth century. *Journal of Political Economy., 124*(1), 105–159.

Bayer, P., Keohane, N., & Timmins, C. (2009, July 1). Migration and hedonic valuation: The case of air quality. *Journal of Environmental Economics and Management., 58*(1), 1–4.

Beatty, T. K., & Shimshack, J. P. (2014, January 1). Air pollution and children's respiratory health: A cohort analysis. *Journal of Environmental Economics and Management., 67*(1), 39–57.

Bell, M. L., Ebisu, K., & Belanger, K. (2007, July). Ambient air pollution and low birth weight in Connecticut and Massachusetts. *Environmental Health Perspectives., 115*(7), 1118–1124.

Bharadwaj, P., Gibson, M., Zivin, J. G., & Neilson, C. (2017, June 1). Gray matters: Fetal pollution exposure and human capital formation. *Journal of the Association of Environmental and Resource Economists., 4*(2), 505–542.

Bishop, K.C., Ketcham, J.D., & Kuminoff, N.V. (2018, August 30). *Hazed and confused: The effect of air pollution on dementia.* National Bureau of Economic Research.

Bobak, M. (2000, February). Outdoor air pollution, low birth weight, and prematurity. *Environmental health perspectives., 108*(2), 173–176.

Burke, M., & Emerick, K. (2016, August). Adaptation to climate change: Evidence from US agriculture. *American Economic Journal: Economic Policy., 8*(3), 106–140.

Burke, M., Hsiang, S. M., & Miguel, E. (2015, November). Global non-linear effect of temperature on economic production. *Nature, 527*(7577), 235–239.

Burnett, R. T., Pope, C. A., III, Ezzati, M., Olives, C., Lim, S. S., Mehta, S., Shin, H. H., Singh, G., Hubbell, B., Brauer, M., & Anderson, H. R. (2014, April). An integrated risk function for estimating the global burden of disease attributable to ambient fine particulate matter exposure. *Environmental Health Perspectives., 122*(4), 397–403.

Cao, J., Yang, C., Li, J., Chen, R., Chen, B., Gu, D., & Kan, H. (2011, February 28). Association between long-term exposure to outdoor air pollution and mortality in China: A cohort study. *Journal of Hazardous Materials., 186*(2–3), 1594–1600.

Chang, T. Y., Graff Zivin, J., Gross, T., & Neidell, M. (2019, January). The effect of pollution on worker productivity: Evidence from call Center Workers in China. *American Economic Journal: Applied Economics., 11*(1), 151–172.

Chay, K. Y., & Greenstone, M. (2003, August 1). The impact of air pollution on infant mortality: Evidence from geographic variation in pollution shocks induced by a recession. *The Quarterly Journal of Economics., 118*(3), 1121–1167.

Chen, B., & Kan, H. (2008, March 1). Air pollution and population health: A global challenge. *Environmental Health and Preventive Medicine., 13*(2), 94–101.

Chen, H., Li, Q., Kaufman, J. S., Wang, J., Copes, R., Su, Y., & Benmarhnia, T. (2018a, January 1). Effect of air quality alerts on human health: A regression discontinuity analysis in Toronto, Canada. *The Lancet Planetary Health., 2*(1), e19–e26.

Chen, S., Oliva, P., Zhang, P. (2018b, June 7). *Air pollution and mental health: evidence from China.* National Bureau of Economic Research.

Chen, Y., Ebenstein, A., Greenstone, M., & Li, H. (2013, August 6). Evidence on the impact of sustained exposure to air pollution on life expectancy from China's Huai River policy. *Proceedings of the National Academy of Sciences., 110*(32), 12936–12941.

Chu, D. A., Kaufman, Y. J., Zibordi, G., Chern, J. D., Mao, J., Li, C., & Holben, B. N. (2003, November 16). Global monitoring of air pollution over land from the earth observing system-Terra moderate resolution imaging Spectroradiometer (MODIS). *Journal of Geophysical Research: Atmospheres, 108*(D21).

Clark, L. P., Millet, D. B., & Marshall, J. D. (2014, April 15). National patterns in environmental injustice and inequality: Outdoor NO2 air pollution in the United States. *PLoS One, 9*(4), e94431.

Clark, N. A., Demers, P. A., Karr, C. J., Koehoorn, M., Lencar, C., Tamburic, L., & Brauer, M. (2010, February). Effect of early life exposure to air pollution on development of childhood asthma. *Environmental health perspectives., 118*(2), 284–290.

Clay, K., Lewis, J., & Severnini, E. (2018). Pollution, infectious disease, and mortality: Evidence from the 1918 Spanish influenza pandemic. *The Journal of Economic History.* Cambridge University Press, *78*(4), 1179–1209.

Clay, K., & Muller, N. Z. (2019, October 17). *Recent Increases in Air Pollution: Evidence and Implications for Mortality.* National Bureau of Economic Research.

Cohen, A. J., Brauer, M., Burnett, R., Anderson, H. R., Frostad, J., Estep, K., Balakrishnan, K., Brunekreef, B., Dandona, L., Dandona, R., & Feigin, V. (2017, May 13). Estimates and 25-year trends of the global burden of disease attributable to ambient air pollution: An analysis of data from the global burden of diseases study 2015. *The Lancet., 389*(10082), 1907–1918.

Cowell, W. J., Brunst, K. J., Malin, A. J., Coull, B. A., Gennings, C., Kloog, I., Lipton, L., Wright, R. O., Enlow, M. B., & Wright, R. J. (2019, October 30). Prenatal exposure to PM 2. 5 and cardiac vagal tone during infancy: Findings from a multiethnic birth cohort. *Environmental Health Perspectives, 127*(10), 107007.

Currie, J. (2013, December). Pollution and infant health. *Child development perspectives., 7*(4), 237–242.

Currie, J., & Neidell, M. (2005, August 1). Air pollution and infant health: What can we learn from California's recent experience? *The Quarterly Journal of Economics., 120*(3), 1003–1030.

Currie, J., & Walker, R. (2011, January). Traffic congestion and infant health: Evidence from E-ZPass. *American Economic Journal: Applied Economics., 3*(1), 65–90.

Deryugina, T., Heutel, G., Miller, N. H., Molitor, D., & Reif, J. (2019, December). The mortality and medical costs of air pollution: Evidence from changes in wind direction. *American Economic Review., 109*(12), 4178–4219.

Deschenes, O., & Moretti, E. (2009, November 1). Extreme weather events, mortality, and migration. *The Review of Economics and Statistics., 91*(4), 659–681.

Deschenes, O., Greenstone, M., & Shapiro, J. S. (2017, October). Defensive investments and the demand for air quality: Evidence from the NOx budget program. *American Economic Review., 107*(10), 2958–2989.

Dockery, D. W., Pope, C. A., Xu, X., Spengler, J. D., Ware, J. H., Fay, M. E., Ferris, B. G., Jr., & Speizer, F. E. (1993, December 9). An association between air pollution and mortality in six US cities. *New England Journal of Medicine., 329*(24), 1753–1759.

Fowlie, M., Rubin, E., Walker, R. (2019, May). Bringing satellite-based air quality estimates down to earth. In *AEA Papers and Proceedings* (Vol. 109, pp. 283–88).

Franck, U., Odeh, S., Wiedensohler, A., Wehner, B., & Herbarth, O. (2011, September 15). The effect of particle size on cardiovascular disorders—The smaller the worse. *Science of the Total Environment., 409*(20), 4217–4221.

Freeman, R., Liang, W., Song, R., & Timmins, C. (2019, March 1). Willingness to pay for clean air in China. *Journal of Environmental Economics and Management., 94*, 188–216.

GBD MAPS Working Group. (2018, January). *Burden of disease attributable to major air pollution sources in India*. Special report. 21.

Glinianaia, S. V., Rankin, J., Bell, R., Pless-Mulloli, T., & Howel, D. (2004, January). Particulate air pollution and fetal health: A systematic review of the epidemiologic evidence. *Epidemiology, 1*, 36–45.

Graff Zivin, J., & Neidell, M. (2009, September 1). Days of haze: Environmental information disclosure and intertemporal avoidance behavior. *Journal of Environmental Economics and Management., 58*(2), 119–128.

Graff Zivin, J., & Neidell, M. (2012, December). The impact of pollution on worker productivity. *American Economic Review., 102*(7), 3652–3673.

Grainger, C., Schreiber, A., & Chang, W. (2018). *Do Regulators Strategically Avoid Pollution Hotspots when Siting Monitors? Evidence from Remote Sensing of Air Pollution*. Working Paper.

Greenstone, M., & Jack, B. K. (2015, March). Envirodevonomics: A research agenda for an emerging field. *Journal of Economic Literature., 53*(1), 5–42.

Gupta, A., & Spears, D. (2017, November 1). Health externalities of India's expansion of coal plants: Evidence from a national panel of 40,000 households. *Journal of Environmental Economics and Management., 86*, 262–276.

Heft-Neal, S., Burney, J., Bendavid, E., & Burke, M. (2018, July). Robust relationship between air quality and infant mortality in Africa. *Nature, 559*(7713), 254.

Heyes, A., Neidell, M., & Saberian, S. (2016, October 19). *The effect of air pollution on investor behavior: evidence from the S&P 500*. National Bureau of Economic Research.

Hsiang, S. M. (2010, August 31). Temperatures and cyclones strongly associated with economic production in the Caribbean and Central America. *Proceedings of the National Academy of sciences., 107*(35), 15367–15372.

Hsiang, S. M., Burke, M., & Miguel, E. (2013, September 13). Quantifying the influence of climate on human conflict. *Science, 341*(6151).

Ito, K., & Zhang, S. (2020, May 1). Willingness to pay for clean air: Evidence from air purifier markets in China. *Journal of Political Economy., 128*(5), 1627–1672.

Janssen, B. G., Munters, E., Pieters, N., Smeets, K., Cox, B., Cuypers, A., Fierens, F., Penders, J., Vangronsveld, J., Gyselaers, W., & Nawrot, T. S. (2012, September). Placental mitochondrial DNA content and particulate air pollution during in utero life. *Environmental health perspectives., 120*(9), 1346–1352.

Jayachandran, S. (2009, October 2). Air quality and early-life mortality evidence from Indonesia's wildfires. *Journal of Human Resources., 44*(4), 916–954.

Jerrett, M., Burnett, R. T., Pope, C. A., III, Ito, K., Thurston, G., Krewski, D., Shi, Y., Calle, E., & Thun, M. (2009, March 12). Long-term ozone exposure and mortality. *New England Journal of Medicine., 360*(11), 1085–1095.

Kahn, M.E., & Li, P. (2019, February 22). *The effect of pollution and heat on high skill public sector worker productivity in China*. National Bureau of Economic Research.

Kahn, M. E., & Zheng, S. (2016, May 17). *Blue skies over Beijing: Economic growth and the environment in China*. Princeton University Press.

Landrigan, P. J., Fuller, R., Acosta, N. J., Adeyi, O., Arnold, R., Baldé, A. B., Bertollini, R., Bose-O'Reilly, S., Boufford, J. I., Breysse, P. N., & Chiles, T. (2018, February 3). The lancet commission on pollution and health. *The Lancet., 391*(10119), 462–512.

Lelieveld, J., Evans, J. S., Fnais, M., Giannadaki, D., & Pozzer, A. (2015, September). The contribution of outdoor air pollution sources to premature mortality on a global scale. *Nature, 525*(7569), 367–371.

Liu, J. C., Pereira, G., Uhl, S. A., Bravo, M. A., & Bell, M. L. (2015, January 1). A systematic review of the physical health impacts from non-occupational exposure to wildfire smoke. *Environmental Research., 136*, 120–132.

Lleras-Muney, A. (2010, July 1). The needs of the army using compulsory relocation in the military to estimate the effect of air pollutants on children's health. *Journal of Human Resources., 45*(3), 549–590.

Markandya, A., Sampedro, J., Smith, S. J., Van Dingenen, R., Pizarro-Irizar, C., Arto, I., & Gonzáles-Eguino, M. (2018). Health co-benefits from air pollution and mitigation costs of the Paris agreement: A modeling study. *Lancet Planet Health, 2*, e126–e133.

Marshall, J. D., Apte, J. S., Coggins, J. S., & Goodkind, A. L. (2015, November 23). Blue skies bluer? *Environmental Science & Technology., 49*(24), 13929–13936.

McMichael, A. J., Anderson, H. R., Brunekree, B., & Cohen, A. J. (1998, June 1). Inappropriate use of daily mortality analyses to estimate longer-term mortality effects of air pollution. *International Journal of Epidemiology., 27*(3), 450–453.

Meyer S, Pagel M. Fresh Air Eases Work–The Effect of Air Quality on Individual Investor Activity. National Bureau of Economic Research; 2017, November 22.

Mikati, I., Benson, A. F., Luben, T. J., Sacks, J. D., & Richmond-Bryant, J. (2018, April). Disparities in distribution of particulate matter emission sources by race and poverty status. *American Journal of Public Health., 108*(4), 480–485.

Moretti, E., & Neidell, M. (2011, January 1). Pollution, health, and avoidance behavior evidence from the ports of Los Angeles. *Journal of Human Resources., 46*(1), 154–175.

Neidell, M. (2009, March 31). Information, avoidance behavior, and health the effect of ozone on asthma hospitalizations. *Journal of Human Resources., 44*(2), 450–478.

Nel, A. (2005, May 6). Air pollution-related illness: Effects of particles. *Science, 308*(5723), 804–806.

Nordling, E., Berglind, N., Melén, E., Emenius, G., Hallberg, J., Nyberg, F., Pershagen, G., Svartengren, M., Wickman, M., & Bellander, T. (2008, May). Traffic-related air pollution and childhood respiratory symptoms, function and allergies. *Epidemiology, 1*, 401–408.

O'Neill, M. S., Jerrett, M., Kawachi, I., Levy, J. I., Cohen, A. J., Gouveia, N., Wilkinson, P., Fletcher, T., Cifuentes, L., & Schwartz, J. (2003, December). Workshop on air pollution and socioeconomic conditions. Health, wealth, and air pollution: Advancing theory and methods. *Environmental Health Perspectives., 111*(16), 1861–1870.

Pant, P., Lal, R. M., Guttikunda, S. K., Russell, A. G., Nagpure, A. S., Ramaswami, A., & Peltier, R. E. (2019, January 8). Monitoring particulate matter in India: Recent trends and future outlook. *Air Quality, Atmosphere & Health., 12*(1), 45–58.

Peng, R. D., Bell, M. L., Geyh, A. S., McDermott, A., Zeger, S. L., Samet, J. M., & Dominici, F. (2009, June). Emergency admissions for cardiovascular and respiratory diseases and the chemical composition of fine particle air pollution. *Environmental Health Perspectives., 117*(6), 957–963.

Pope, C. A., 3rd. (2000, August). Epidemiology of fine particulate air pollution and human health: Biologic mechanisms and who's at risk? *Environmental Health Perspectives., 108*(suppl 4), 713–723.

Pope, C. A., III, Burnett, R. T., Krewski, D., Jerrett, M., Shi, Y., Calle E, et al. (2009). Cardiovascular mortality and exposure to airborne fine particulate matter and cigarette smoke: Shape of the exposure-response relationship. *Circulation, 120*, 941–94819720932.

Pope, C. A., III, Burnett, R. T., Turner, M. C., Cohen, A., Krewski, D., Jerrett, M., Gapstur, S. M., & Thun, M. J. (2011, November). Lung cancer and cardiovascular disease mortality associated with ambient air pollution and cigarette smoke: Shape of the exposure–response relationships. *Environmental Health Perspectives., 119*(11), 1616–1621.

Pope, C. A., III, Schwartz, J., & Ransom, M. R. (1992, June 1). Daily mortality and PM10 pollution in Utah Valley. *Archives of Environmental Health: An International Journal., 47*(3), 211–217.

Rangel, M. A., & Vogl, T. S. (2019, October). Agricultural fires and health at birth. *Review of Economics and Statistics., 101*(4), 616–630.

Reid, C. E., Brauer, M., Johnston, F. H., Jerrett, M., Balmes, J. R., & Elliott, C. T. (2016, September). Critical review of health impacts of wildfire smoke exposure. *Environmental Health Perspectives., 124*(9), 1334–1343.

Seaton, A., Godden, D., MacNee, W., & Donaldson, K. (1995, January 21). Particulate air pollution and acute health effects. *The Lancet., 345*(8943), 176–178.

Schlenker, W., & Walker, W. R. (2016, October 20). Airports, air pollution, and contemporaneous health. *The Review of Economic Studies., 83*(2), 768–809.

Schwartz, J., & Dockery, D. W. (1992, March). Increased mortality in Philadelphia associated with daily air pollution concentrations. *American Review of Respiratory Disease., 145*(3), 600–604.

Scovronick, N., Budolfson, M., Dennig, F., Errickson, F., Fleurbaey, M., Peng, W., Socolow, R. H., Spears, D., & Wagner, F. (2019, May 7). The impact of human health co-benefits on evaluations of global climate policy. *Nature Communications, 10*(1), 2095.

Setel, P. W., Macfarlane, S. B., Szreter, S., Mikkelsen, L., Jha, P., Stout, S., & AbouZahr, C. (2007, November 3). Monitoring of vital events (MoVE) writing group. A scandal of invisibility: Making everyone count by counting everyone. *The Lancet., 370*(9598), 1569–1577.

Shah, P. S., & Balkhair, T. (2011, February 1). Knowledge synthesis group on determinants of preterm/LBW births. Air pollution and birth outcomes: A systematic review. *Environment International., 37*(2), 498–516.

Sheffield, P., Roy, A., Wong, K., & Trasande, L. (2011, May 1). Fine particulate matter pollution linked to respiratory illness in infants and increased hospital costs. *Health Affairs., 30*(5), 871–878.

Shindell, D., Faluvegi, G., Seltzer, K., & Shindell, C. (2018, April). Quantified, localized health benefits of accelerated carbon dioxide emissions reductions. *Nature Climate Change., 8*(4), 291.

Spears, D. (2019). *Air: Pollution, climate change, and India's choice between policy and Pretence.* HarperCollins.

Spears, D., Dey, S., Chowdhury, S., Scovronick, N., Vyas, S., & Apte, J. (2019, December). The association of early-life exposure to ambient $PM_{2.5}$ and later-childhood height-for-age in India: An observational study. *Environmental Health, 18*(1), 62.

Stieb, D. M., Chen, L., Eshoul, M., & Judek, S. (2012, August 1). Ambient air pollution, birth weight and preterm birth: A systematic review and meta-analysis. *Environmental research., 117*, 100–111.

Sullivan, D.M., & Krupnick, A. (2018). *Using Satellite Data to Fill the Gaps in the US Air Pollution Monitoring Network.* Working Paper.

Sullivan, T. J., Driscoll, C. T., Beier, C. M., Burtraw, D., Fernandez, I. J., Galloway, J. N., Gay, D. A., Goodale,

C. L., Likens, G. E., Lovett, G. M., & Watmough, S. A. (2018, June 1). Air pollution success stories in the United States: The value of long-term observations. *Environmental Science & Policy., 84*, 69–73.

Sun, C., Kahn, M. E., & Zheng, S. (2017, January 1). Self-protection investment exacerbates air pollution exposure inequality in urban China. *Ecological economics., 131*, 468–474.

Sun, Z., & Zhu, D. (2019). Exposure to outdoor air pollution and its human health outcomes: A scoping review. *PLoS One, 14*(5), e0216550.

Tanaka, S. (2015, July 1). Environmental regulations on air pollution in China and their impact on infant mortality. *Journal of Health Economics., 42*, 90–103.

Van Donkelaar, A., Martin, R. V., Brauer, M., Hsu, N. C., Kahn, R. A., Levy, R. C., Lyapustin, A., Sayer, A. M., & Winker, D. M. (2016, March 24). Global estimates of fine particulate matter using a combined geophysical-statistical method with information from satellites, models, and monitors. *Environmental Science & Technology., 50*(7), 3762–3772.

Voorheis, J. (2017, May). *Air quality, human capital formation and the long-term effects of environmental inequality at birth. Center for Economic Studies*, US Census Bureau.

Vyas, S., Srivastav, N., & Spears, D. (2016, December 15). An experiment with air purifiers in Delhi during winter 2015–2016. *PLoS One, 11*(12), e0167999.

Wang, J., & Christopher, S. A. (2003, November). Intercomparison between satellite-derived aerosol optical thickness and PM2. 5 mass: Implications for air quality studies. *Geophysical Research Letters, 30*(21).

Wang, X., Ding, H., Ryan, L., & Xu, X. (1997, May). Association between air pollution and low birth weight: A community-based study. *Environmental health perspectives., 105*(5), 514–520.

WHO. Ambient air pollution: A global assessment of exposure and burden of disease. (2016). https://apps.who.int/iris/bitstream/handle/10665/250141/9789241511353-eng.pdf?sequence=1

Woodruff, T. J., Darrow, L. A., & Parker, J. D. (2008, January). Air pollution and postneonatal infant mortality in the United States, 1999–2002. *Environmental Health Perspectives., 116*(1), 110–115.

Xing, Y. F., Xu, Y. H., Shi, M. H., & Lian, Y. X. (2016, January). The impact of PM2. 5 on the human respiratory system. *Journal of Thoracic Disease, 8*(1), E69.

Zhang, J., & Mu, Q. (2018, November 1). Air pollution and defensive expenditures: Evidence from particulate-filtering facemasks. *Journal of Environmental Economics and Management., 92*, 517–536.

Zhang, X., Zhang, X., & Chen, X. (2017, September 1). Happiness in the air: How does a dirty sky affect mental health and subjective Well-being? *Journal of Environmental Economics and Management., 85*, 81–94.

Population and Water Issues: Going Beyond Scarcity

13

Stéphanie Dos Santos, Bénédicte Gastineau, and Valérie Golaz

Abstract

The growth in the number of human beings on Earth is undeniably a challenge for the planet, especially in terms of nonrenewable resources. However, scientific knowledge of the issues linking population and water should not be limited to only two variables – quantity of water and number of human beings. Other phenomena at play must be taken into account, and in particular at a micro-scale level. One of the central issues is certainly that of inequity in access to and the use of water, and its consequences. Population health, social and economic inequalities and their measures (including the questions of gender and poverty), governance of the hydro-socio-system, and in particular the participation of all stakeholders, are among the research priorities in the field of population, development, and the environment. This review's first section summarizes international-scale dialogues on the population water issue. It is followed by a review

of some empirical research demonstrating the associations between population health and water. Finally, we present some illustrations of the complex interrelations between water resources and socio-demographic characteristics of populations. The chapter concludes on the need for a mixed and interdisciplinary approach, a dialogue between the social sciences, biomedical sciences, and the life and earth sciences.

Difficult to purify, expensive to transport and with no possible substitute, water is essential for the survival of all species. On 28 July 2010, the United Nations recognized the universal right to water and to proper sanitation as essential for the achievement of full human rights (United Nations General Assembly, 2010). Water resources are also an important factor shaping the association between population dynamics and economic development. In this sense, water may be seen as one of the pillars of the population – environment relationship. Even so, population–water connections have received relatively little attention by scientists specializing in population issues, although the topic is increasingly of interest (Vörösmarty et al., 2000). In her 1998 Presi-

S. Dos Santos (✉) · B. Gastineau
IRD, AMU, LPED, Marseille, France
e-mail: stephanie.dossantos@ird.fr

V. Golaz
INED, LPED, Marseille, France

© Springer Nature Switzerland AG 2022
L. M. Hunter et al. (eds.), *International Handbook of Population and Environment*, International
Handbooks of Population 10, https://doi.org/10.1007/978-3-030-76433-3_13

dential Address to the Population Association of America, Anne Pebley mentioned a 1986 U.S. National Academy of Sciences report as having a major influence on demographers' interest in environmental issues generally, and on the link between demographic growth and water resources (in terms of quantity and quality) in particular.

Questions regarding water resources tend to be approached first in terms of quantity, especially through a Malthusian lens focused on the balance between population growth and water as a nonrenewable resource. The argument is simple, even simplistic: the more human beings there are on Earth, the greater the need for water. In the 1990s, the hydrologist Malin Falkenmark (Falkenmark & Widstrand, 1992) focused specifically on this link between demographic growth and water resources on the macro scale of the main world regions, and at the national scale, emphasizing the role of high fertility, rapid rural–urban migration and changing patterns of international population movement.

The work of Engelman and Leroy (Engelman & Le Roy, 1993, 1996), based on population projections and quantities of renewable water, is quite illustrative of the logic of Malthusian thought: the larger the population of a country, the less available water per capita. They also compare China and Canada. At that time, these two countries had roughly the same amount of renewable freshwater, but China already had 40 times the population of Canada. Therefore, each person in China had less than 3% of the water available to each person in Canada. Their work also makes use of the population projections provided by the United Nations and their reassessment: in their 1996 update, they state that depending on the speed of world population growth, there would be between 400 million and 1.5 billion fewer people than initially projected in water-scarce countries in 2050. Although they concede that living standards can influence water demand, they reaffirm that it is the number of people that is the most important human factor determining a country's water availability.

As with other population-environment connections, the demographic aspects of water supply and demand cannot be reduced to simple measures of population size. Of course population size alone does not determine water demand. Instead, issues of modes of production, of consumption, of distribution of goods and services, the impacts of these factors on the quality of available water, etc., illustrate the complexity of these interrelations. For example, more recently, the water footprint concept has given more consideration to the volume of water needed to produce the goods and services consumed by a country's inhabitants, including the volume of water used in other countries to produce imported goods and services (Hoekstra & Chapagain, 2007). Studying each nation of the world for the period 1997–2001, the authors show that the world average water footprint of national consumption is 1385 m^3 per capita per year, whereas the average citizens in China have 1.071 m^3 per year, and the water footprint reaches 2480 m^3 per capita per year in the USA. They conclude that consumption patterns (e.g., food diets and in particular high versus low meat consumption) and agricultural practices (in terms of water use efficiency) are among the most direct factors determining the water footprint of a country. Recently, the volume of polluted water (produced by the country's population/industries) was added to the concept of a water footprint. (Hoekstra & Mekonnen, 2012). Based on the water footprint of each country, it would be interesting to carry out in-depth analyses comparing various population indicators (in absolute and relative figures). In a simple intuitive way, the China/USA comparison shows that the most populated countries in the world, such as China, or countries where the average annual rates of population change are among the highest in the world during the same period, are not especially those with the highest water footprint.[1]

---

[1]The lowest water footprint of national consumption for countries with a population larger than 5 million, is that of the Democratic Republic of the Congo (a little more than 500 m^3/year/capita), a country that recorded an average annual rate of population change of 3.02%, among the highest over the period 2000–2005 (UNDESA, 2018) corresponding to the study period of Hoekstra and Mekonnen (2012). In the same vein, this water footprint in Angola (a little less than 1000 m^3/year/capita) or Chad (a little less than 1500 m^3/year/capita, the same as in the Netherlands),

The indispensable side of water can be illustrated through three important pathways. First, water is simply essential for life: for hydration, for food production and for the maintenance of health, dignity, and social equity. In this sense, water must be a central consideration within social and health policy. Second, water is essential for production, to generate energy, and more generally to support economic development and its concomitant improvements in social well-being. Third, water is essential for the survival and sustainability of the Earth's ecosystems, and is therefore of fundamental importance. These three pathways intersect to make water a central element of sustainable development and, therefore, an issue at the heart of international policy concerns for several decades. In this sense, this review's first section summarizes international-scale dialogues on the population water issue. It is followed by a review of some empirical research demonstrating the associations between population health and water. Finally, we present some illustrations of the complex interrelations between water resources and socio-demographic characteristics of populations.

International Recognition of the Population–Water Nexus

For more than four decades now, "the United Nations has long been addressing the global crisis caused by insufficient water supply to satisfy basic human needs and growing demands on the world's water resources to meet human, commercial and agricultural needs" (United Nations, 2019b). Indeed, this international recognition of the various dimensions of the water crisis dates back to the United Nations Mar del Plata conference in 1977 (United Nations, 1977), one of the largest international conferences ever held on water (Biswas, 2004). This conference was part of the cycle of major international conferences organized by the United Nations during the 1970s on various development topics: on the environment (Stockholm 1972), population (Bucharest 1974), food (Rome 1974), women (Mexico 1975), human settlements (Vancouver 1976), desertification (Nairobi 1977), and renewable energies (Nairobi 1979). In previous years, intellectuals around the world began to question the issue of economic development, and its antonym, the "underdevelopment." The concept of the Third World had been introduced by the demographer Alfred Sauvy, in 1952 (Véron, 1995). The debates then focused on economic growth and its compatibility with environmental questions, particularly from a neo-Malthusian perspective directly linked to population growth (Ehrlich, 1968) and more generally to the limits of growth (Meadows et al., 1972). Countries with high birth rates, often developing countries, were seen to be responsible for the ills of the planet, as for example, illustrated in one of René Dumont's works (Dumont, 1965). Parallel to this intellectual context, the beginning of the 1970s was also marked by very severe droughts in the Sahel (Nicholson, 2001), one of the factors of famine that received a lot of media attention in Western countries. The first objective of the Mar del Plata Conference was therefore to avoid a water crisis at the end of the century, in particular by considerably improving the efficient use of water in agriculture and industry and by enhancing international cooperation in all water sectors (assessment of water resources, institutional and financial arrangements, etc.). The Conference also led to the approval of a Plan of Action and, above all, to the launching of the "Decade for Drinking Water and Sanitation," with the objective that by 1990 all human beings should have access to drinking water in sufficient quantity and quality, as well as to basic sanitation facilities.

Two decades later, as the goals set during the Water Decade were still not being met, the Agenda 21 adopted at the Rio de Janeiro Earth Summit in 1992 incorporated emphasis on water issues (Chap. 18 of the Agenda 21) and highlighted the importance of maintaining adequate supplies of quality water for the global population while preserving the hydrological, biological, and chemical functions of ecosystems. By adopting

---

in countries that recorded average annual rates of population change of 3.02 and 3.8% respectively.

the first principle of the Rio declaration, that is, the recognition that human beings are at the core of concerns, and in particular that they have a right to a healthy life, the international community implicitly declared the social health aspect of water to be the most fundamental with regard to the water crisis. On a micro scale, the capacity of populations to satisfy their basic needs, those related to domestic uses, became the focus. In a 2003 report, UNESCO eloquently illustrated the impact of water-related diseases on human health, by comparing the number of people dying every day from diarrheal diseases with the daily crashing of 20 fully-loaded jumbo jets, with no survivors (UNESCO, 2003). During this period, population growth as the main cause of water scarcity was much less frequently highlighted in international reports. Levels of economic development are given greater prominence. Thus, in the same 2003 report, the UNESCO recognized the role of the mode of development in the consumption of water by stating that a child born in a developing country consumes almost 50 times fewer water resources than a child in a developed country.

Still, a decade later after the Rio Declaration, and noting the recurring issue of non-access to water for a large part of the world's population, the United Nations reformulated a new commitment in 2000 through the Millennium Development Goals (MDGs); Goal 7 being to "Halve, by 2015, the proportion of the population without sustainable access to safe drinking water and basic sanitation." At that time, this was seen as a real challenge: given the projections for demographic growth,[2] achieving this goal would require providing access to water to 2.2 billion people by 2015, which is on average 280,000 people more per day. In 2000, one sixth of the world population still did not have access to safe drinking water. Regions of particular challenge included Africa where 2 in 5 inhabitants do not have access to water, and Asia (1 in 4) with important variations from country to country on these two continents in relation to the level of urbanization (WHO/UNICEF, 2000). There has been some progress in the past few decades. Between 1990 and 2000, an estimated 820 million people had improved access to water for the first time.

During the 2012 World Water Forum held in Marseille, France, the United Nations announced that the goal had already been achieved, with only 11% of the world population lacking access to improved water supply.[3] Between 1990 and 2015, the number of nations in which more than 50% of the population did not have access to improved water supply fell from 23 to 3. Of course, these global trends mask important regional and spatial disparities. The African continent, the countries of the Caucasus, Central Asia, and Oceania fell short of the MDG target (Fig. 13.1).

Significant disparities were also apparent between rural and urban areas (Fig. 13.1). Globally, in 2015, 16% of the rural population, but only 4% of the urban population worldwide used an unimproved source of water. The rural/urban gap was particularly striking in several nations, including Angola, where 72% of the rural population did not have access to improved water supply in 2015 in contrast to 25% of urban dwellers. In other countries of the world, this observation can also be made, such as in Papua New Guinea, where the level was 67% for rural and 12% urban (WHO/UNICEF, 2015).[4] The cover rate of access to water had slightly increased in rural areas since 1990. However, it would appear to have declined in urban areas because of the pace of

---

[2]The world population was 2.5 billion inhabitants in 1950, 4.5 billion in 1980, 6 billion in 2000, 7.8 billion in 2020. Forecasts estimate 8.5 billion inhabitants in 2020 (UN/DESA 2018).

[3]By definition, a water supply referred to as 'improved' is a water supply that, because of its source and its utilization, provides adequate protection from outside contamination, in particular, from contact with fecal matter. This denomination includes household connection, public standpipes, boreholes, protected dug wells, protected springs, rainwater collection. Otherwise, the water supply is referred to as "unimproved".

[4]Despite significant progress in the worldwide extension of access to improved water supply sources, the accuracy of the figures published by the Joint Monitoring Program (JMP – the joint WHO and UNICEF monitoring system) may give rise to some scepticism owing to the difficulty of measuring water supply in the informal neighborhoods of major urban centers (Nganyanyuka et al., 2014) and the insufficient attention paid to health issues related to access to water and sanitation (Lim et al., 2012).

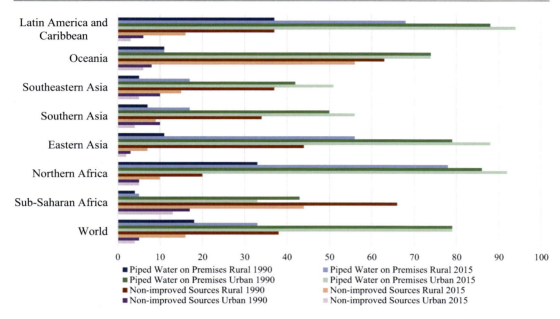

Fig. 13.1 Trends in non-improved access water and in piped water on premises as a percentage of total population, by region, 1990–2015. Source: WHO/UNICEF, 2015

urbanization. In Africa, for example, the urban population forecasts suggested that the public services involved in water provision would be facing a major challenge. To meet the target of the Millennium Development Goals (WHO/UNICEF, 2000), 210 million people should have had access to drinking water in urban areas between 1990 and 2015.

In 2015, the United Nations updated the MDGs to reflect new Sustainable Development Goals (SDGs) aiming to "transform our world" by 2030. Goal 6 works toward "the availability and the sustainable management of water and sanitation for all.". New, finer indicators have been generated to ensure that access to water is fair, safe, affordable, and appropriate, especially in terms of quality. The new goals also distinguish between the lack of access to an improved water supply (e.g., use of surface water) and limited access to an improved water source (e.g., collection time over 30 min) (WHO/UNICEF, 2016). Water quality is also more centrally integrated as "improved" sources are not necessarily healthy and free from contamination (Bain et al., 2014; Boateng et al., 2013; Martínez-Santos, 2017). Indicators also include handwashing facilities. In addition, official reports are expected to document socioeconomic inequalities, particularly inequalities of access between the richest and poorest households in the same cluster (country, region, etc.). Analyses are also carried out on the share of water purchases in the household budget, or on the burden of fetching water for women and girls.

At present, progress since the implementation of the SDGs is still difficult to measure owing to the lack of updated data: the last JMP report was published in 2017, with data from 2015. The most recent information dates from May 2019, published in a short report by the UN Economic and Social Council (United Nations, 2019a). Globally, the proportion of the population using safely managed drinking water services seems to have remained unchanged since 2015.

These levels of access are not without consequences for the health of human populations. This is the subject of the next section, based on a nonsystematic and even less exhaustive literature review. Then, in the final section, major demo-

Water-Related Challenges to Population Health

Numerous works have developed conceptual schemes that include issues linking water and population health. In the field of population sciences, Mosley and Chen's work on developing countries (Mosley & Chen, 1984) was probably the first to put environmental variables, and what they call "environmental contamination" (which includes a set of variables including water and hygiene) on an equal footing in the analysis alongside other groups of socio-economic determinants such as "maternal factors," "nutrient deficiency," and "injury."

Basically, there are four main pathways through which water plays a role in disease transmission to the human population (Cairncross & Feachem, 1993): 1. diseases caused by contaminated water; 2. diseases caused by poor personal hygiene; 3. diseases propagated by insects that breed in or near stagnant water and 4. diseases transmitted by aquatic organisms which penetrate skin (Table 13.1).

Today, infections transmitted by water, of which the most common symptom is diarrheal diseases, remain the ninth cause of mortality worldwide in general and the fourth cause of death in children less than 5 years of age. In absolute figures, this represents 1.3 million deaths annually, including 500,000 children of less than 5 years of age (GBD 2015 Diarrhoeal Diseases Collaborators 2017). Sub-Saharan Africa and South Asia are the two regions most affected by these diseases. For example, the two countries that record the highest rates of mortality from diarrheal diseases are Chad (594 deaths for 100,000 children under 5 years of age) and Niger (485 deaths for 100,000 children of less than 5 years of age).[5] Nigeria and India, because of their large populations, accounted for 42% of the global number of child deaths from diarrheal diseases in 2015 (GBD 2015 Diarrhoeal Diseases Collaborators, 2017).

Although water-related challenges to population health remain staggering, progress has been made. From 2005 to 2015, child deaths from diarrheal diseases declined nearly 35% (Fig. 13.2).

Yet, empirically, estimating the impact of improved water access conditions in the reduction of diarrheal diseases, for example, is still complex and controversial. In a meta-analysis covering 21 regions of the world, Lim et al. estimated the independent effects of 67 risk factors and clusters of risk factors between 1990 and 2010. They concluded that further epidemiological studies are needed to understand the real burden from water and sanitation (Lim et al., 2012).

Several water-related factors contribute to this diversity in the levels and patterns of water-related diseases around the world.

On the one hand, access to drinking water is the most effective means of combating this kind of water-related disease. Long considered the predominant factor with regard to health,[6] the quality of drinking water is still recognized for its role in the transmission of pathogenic elements. Infections may be of bacterial origin, such as typhoid or cholera, viral such as hepatitis A, or parasitic such as amoebas (Prost, 1996). The importance of good-quality drinking water is all the greater in that it plays a key role in epidemic episodes and in the persistence of endemic diseases (Payment & Hunter, 2001). Water quality is of paramount importance at the point of supply, hence the importance given to the type of access to water as "improved" or not. However, although the water leaving a production plant may meet the norms of quality for drinking water, the risk of pollution goes beyond that point, notably along the piping system when the network is old. Therefore, quality is also important at the point of

---

[5]These data for mortality are nevertheless estimations and only give an order of magnitude because of the bias inherent in data collection systems in developing countries.
[6]Relative to the quantities of water, in particular.

13 Population and Water Issues: Going Beyond Scarcity

Table 13.1 Classification of diseases related to water and their transmission pathways

	Transmission pathways	Diseases examples
Water-borne diseases	Infections transmitted through fecal contamination of drinking water	Cholera, *Cryptosporidium, E. -coli*, Hepatitis A, Typhoid
Water-washed diseases	Infections spread because of insufficient water for personal hygiene	Amoebic dysentery, Scabies, Trachoma
Water-related diseases	Infections spread by vector that need water for breeding	Malaria, Onchocerciasis, Trypanosomiasis
Water-based diseases	Infections spread by a pathogen that spends part of its lifecycle in an animal that is water based	Dracunculiasis, Schistosomiasis

use because water quality may deteriorate beyond the source of supply. It has been observed that the microbiological quality of the water in storage vessels within households is worse than at the common source of supply (Lindskog & Lindskog, 1988; Wright et al., 2004).

Two other types of diseases related to water quality result from the accumulation of physical and chemical pollution or from micro-pollutants and those linked to the excess or lack of certain elements in the water. These are forms of pollution and are the greatest cause for concern in the long term. Perhaps counter-intuitively, these risks are generally higher in developing countries because of the lack of investment in modern technologies and the inadequacy or even lack of legislation. Chemical pollution is relatively independent of domestic practices, in contrast to diseases caused by microbiological pollution, which may be the consequence of the management of water and its use within households. But the level of development expected in developing countries suggests that this type of pollution may have a very serious impact on morbidity and mortality in the future. Today, the association between the pollution of the environment and human health is still complex. Research on these issues remains very patchy, whereas pollution is already recognized as a major public health issue by the World Health Organization (WHO, 2007). Arsenic (Naujokas et al., 2013) and fluoride are among the chemical products that have been most widely studied, in particular in developing countries (Kjellstrom et al., 2006). The long latency, multiple exposure to different pollutants and their cumulative effects make the measurement of the impact all the more

complex (Briggs, 2003). The action of public authorities on risk prevention of diseases related to water is thus rather focused on the short term, i.e., on the diseases related to access to water, sanitation, and hygiene.

On the other hand, concerns were focused on water quality alone until the study by White et al. (1972). These authors showed that the quantities of water used, in particular for personal hygiene, had an impact on the health of individuals that was manifested by diseases such as severe conjunctivitis, trachoma, and dermatosis, as well as 10–20% of cases of diarrhea. Since this pioneering study, the links between the quantities of water and health have been extensively discussed. In a survey of the literature on trachoma, Prüss and Mariotti (2000) found only 2 studies out of 19 where quantities of water available to the household had an impact on this infection.

Going beyond the quality and quantity of water used within households, Hunter et al. (2010) described six determinants of the water supply that play a positive role in the maintenance of good health: quality, quantity, access (physical distance or socio-economic and cultural dimensions of access), reliability, cost, and ease of management (Table 13.2).

On the basis of a survey of the literature, Howard and Bartram (2003) drew up a survey table distinguishing four levels of access to water, defined by the distance from the source or the collection time for public supply sources or the type of connection if the dwelling disposes of a private tap or a shared tap in a communal yard. The quantities potentially collected and the associated health risk are dependent on this accessibility.

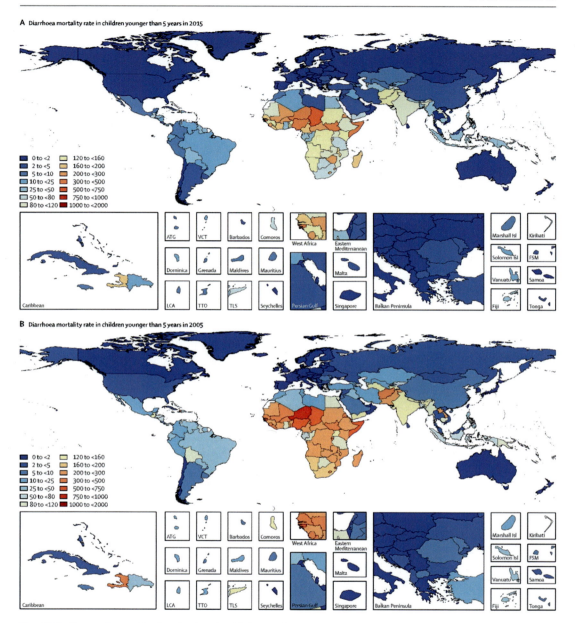

Fig. 13.2 Rates of mortality by diarrhea in children under 5 years of age (per 100,000) in 2015 (**a**) and 2005 (**b**). Source: GBD (2015), Diarrhoeal Diseases Collaborators (2017)

Their conclusion leads to the idea that access to piped water, which they refer to as intermediary access if the household has at its disposal a single tap in the yard or the dwelling, and optimal access if several taps are available, would appear to be the most suitable type of supply for the purpose of reducing health risks. Below this level of access, health risks are considerable. This is, however, based on the hypothesis of a continuous water supply service at the level of the infrastructure, without cuts. Yet, in African towns, interruptions in water supply are frequent (Thompson et al., 2001).

Table 13.2 Dimensions of water accessibility at the household level (adapted from Hunter et al., 2010)

Dimensions	Description	Examples
Quality	Pathogens Chemical constituents	Treatment and effectiveness
Quantity	Availability Uses	Quantities collected Quantities for each domestic purpose
Access	Geographical distance Time Social, economic and/or cultural differentials	Length and difficulty of the trip Waiting time Gender hierarchy
Reliability	Cuts Sustainability Maintenance	Drying up Breakdown
Cost	Cash tariff Time cost Drudgery	Part of the budget Time spent fetching water Means of carriage (physical challenges)
Ease of management	Payment Water reserves	Faculties and facilities Handling

To grasp the complexity of population health and water issues, it is relevant to go one step further, taking into account uses for hygiene purposes. In particular, the assumption that a given quantity of water necessarily implies certain types of hygiene-related practices, in the biomedical sense of the term, might be reconsidered. Research in East Africa revealed a 250% increase in water used for washing, dish washing, and laundry when household access improved through internal, piped water compared with reliance on outside sources (Thompson et al., 2001).

In addition, social perceptions of cleanliness and dirt and those related to the transmission of diseases are more likely to induce a change in behavior than simply the quantities of available water in the household (Dos Santos, 2011). In reality, if a minimum of 20–50 l water per person per day is necessary (Gleick, 1996) for the prophylactic practice of hygiene, this quantity is insufficient for it to be effective. For example, the management and treatment of water at the point of use is a determining factor with regard to the risk of water-related diseases, in particular from microbiological contamination (Fewtrell & Colford, 2004).

Other interesting opportunity costs may also have implications for populationhealth.

As an example, improved access to water may save time and energy that could be invested in child care, food production/preparation, or income-generating activities with potential positive impacts on household well-being (Esrey, 1994). In addition, the physical fatigue related to the collection of water may contribute to a quantitative and qualitative decline in milk production by a nursing mother, inducing a risk of malnutrition for the breast-fed child (Dufault, 1988).

Major Demographic Variations

Looking in more detail at the population–water relationship, particularly at the micro level, research has highlighted a number of population variables as being related in some way to water. These can be both explanatory, independent variables, which are factors in access to water or water-related diseases, or variables to be explained, which are dependent on water-related explanatory factors. It is these interactive loops between these two concepts, population and water, that make this field of research so complex and exciting to study.

Age

Among the demographic variables associated with access to water, age is undoubtedly among the most widely studied, in relation to its health dimension.

Literature has amply demonstrated that young children, especially those under 5 years of age, are disproportionately at risk for water-related diseases, mainly because of their less mature immune system (Simon et al., 2015). This phenomenon is nevertheless in inverse proportion to the intensity of breast-feeding, which, in contrast, decreases with age. Thus, the more the child grows, the less the diet is based solely on breast milk. The decline in the input of antibodies derived from breast milk and a greater risk of contamination by kitchen utensils and meals may thus result in an increase in exposure to the pathogens responsible for certain water-related diseases, in particular, diarrheal diseases (Mané et al., 2006).

Recent research has shown the relevance of studying children above the age of 5 as well. An analysis of diarrheal diseases carried out at Dakar in Senegal showed that the median age at one episode of diarrhea was 4 years, whereas the mean age in this sample of 1000 children aged 2–10 years was 4.79 years, owing to some episodes occurring up to the age of 7 years (Rautu et al., 2016).

Studying people at a high risk for diarrhea, living in an overcrowded housing with poor sanitation, unhealthy and unhygienic living condition, unsafe water use, and low income dwelling, Chowdhury et al. (2015) showed the inversely proportional relationship between age and risk of diarrhea. Their results show, however, that in a population with high-risk environmental conditions, the phenomenon of diarrheal diseases is not negligible after the age of 15 years either, with a prevalence rate of 12.8 per 1000 in both sexes combined (Fig. 13.3).

In the same vein, and specifically regarding endemic cholera in Kolkata (India) and Jakarta (Indonesia), Ali et al. (Ali et al., 2012) showed that the incidence is not negligible in children over 5 years of age as well as adults (although the risk is much higher in young children; children under 1 year old have an incidence seven times higher than people over 14 years old; Fig. 13.4).

Similarly, an analysis conducted in the Gaza Strip (Abuzerr et al., 2019) showed that the number of diarrhea cases among people aged 16 and over was not negligible (153 out of 3658 people were observed in this age group). Although this results in a prevalence rate four times lower than the rate among children under 5 years of age (4.2% and 15.8% respectively), it is also relevant to observe the phenomenon more closely.

However, studies on other older age groups are limited, especially because of the lack of accurate age-specific incidence rates. A gap in the research certainly needs to be filled in there as well.

Sex and Gender

Two points on the relationship between sex/gender and access to water can be discussed in particular: studies on the differential health consequences for males and females of limited access to water, and the gender impact of inequalities in access to water.

First, studies on the health consequences of access to water do not agree on a differential effect according to the sex of the child. Some studies show that little boys are more prone to diarrhea (Siziya et al., 2013), whereas others show no differences (Kandala et al., 2009). More specifically, Chowdhury et al. (2015) showed that the risk of diarrhea according to sex depends on age: before the age of 15, boys are at a higher risk than girls, a risk that then reverses to the detriment of girls (Fig. 13.3). The aforementioned analysis on the Gaza Strip (Abuzerr et al., 2019) also showed significant sex differentials in such prevalence: boys 5 years of age or younger had three times the prevalence of diarrhea compared with girls of that age; men 16 years of age and older had four times the prevalence of diarrhea compared with women in that age group.

At present, because the results do not converge, it is difficult to find a one-sided understanding of the relationship between gender and water-related diseases. Some people think that it

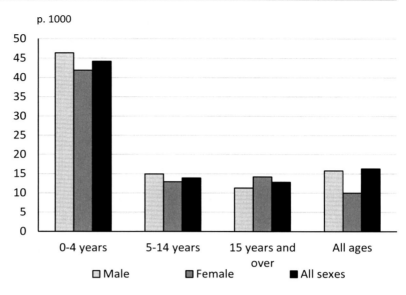

Fig. 13.3 Prevalence (per 1000) of diarrhea by age and sex groups among a high-risk population in Dhaka, Bangladesh. Source: Chowdhury et al. (2015))

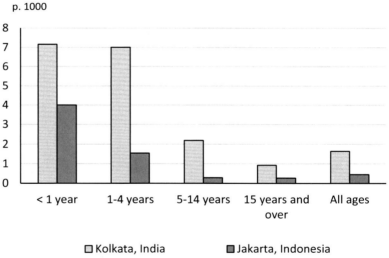

Fig. 13.4 Annual incidence rates (per 1000) of cholera by age group in Kolkata, India, and Jakarta, Indonesia. Source: Ali et al. (2012))

is a matter of differential gender-specific childcare behaviors, different lifestyle habits between men and women (boys/men more frequently being outside the home and therefore more at risk for multiple viruses and bacteria). Here again, we can only conclude that research needs to involve more detailed comprehensive analyses to better understand these phenomena, which seem to be highly dependent on local contexts.

Second, and until recently, the relation between issues related to access to water and gender was only cursorily considered in research programs on access to water on the scale of households and communities. Nevertheless, in the least developed countries, 67% of the population uses a water source (improved or not) outside the dwelling. Forty% use a source more than 30 min from their dwelling (WHO/UNICEF, 2015). In such cases, water collection is time-consuming. In many regions, water collection is gendered with socially constructed gender roles assigning it predominantly to women and girls (Dos Santos, 2012). In sub-Saharan Africa, it is estimated that women and girls spend around 40 billion hours a year in water transportation (PNUD, 2006). In parts of India, the participation of women in these activities is six times higher than that of men (Hirway & Jose, 2011).

Gender inequalities in water collection are particularly strong in the poorest areas (UN-Habitat, 2016). For example, a recent study carried out in some informal neighborhoods of Ouagadougou in Burkina Faso shows that women are responsible for the collection of water in 84 percent of the households sampled (Dos Santos & Wayack-Pambe, 2016). Men are more likely to collect water when standpipes are used, whereas women more often fetch water from wells and boreholes that require considerably more physical effort. These gender distinctions are principally shaped both by economic factors and by the distance. First, the price of a liter of water at a standpipe can be up to twice as high as at a borehole: women often favor wells and boreholes for this reason. Second, in informal settlements, boreholes are generally closer to the house. As women generally have less efficient means of transporting water (more often on foot or by bicycle) than men (more often by motorcycle), women tend to prefer the sources of supply that are the closest to their homes.

Literature studying the consequences of the difficulties of access to water for women is relatively extensive. In particular, from an economic perspective, the time spent collecting water is analyzed as a waste of time with an economic cost. There are two mainly studied types of opportunity costs for women: for younger women, attending school; for women of activity age, participating in income-generating activities (Sorenson et al., 2011).

Indeed, these activities constitute a strong constraint on women's use of time, which can be analyzed through the prism of the concept of time poverty (Garhammer, 1998). Fetching water then limits the ability to take advantage of economic opportunities, particularly the possibility of developing economic or training activities (Blackden and Wodon, 2006). Therefore, freeing up this time could allow women to devote more time to increasing their income (Sorenson et al., 2011).

Since the 1980s, international conferences on water have called for the integration of gender in the policies and programs. In this respect, Principle 3 of The Dublin Statement on Water and Sustainable Development states that "Women play a central part in the provision, management and safeguarding of water" (Dos Santos, 2012). Yet, despite strong recognition at an international level, additional research is needed to systematically document gender distinctions and the implications of resource-related gender roles.

Socio-economic Status

Socio-economic status has many relationships with the issue of access to water, and as is often the case with population–environment relations, in a feedback loop: the type of water source used is generally influenced by socioeconomic status, at a national level as well as at a household level; conversely, access to water can have a significant impact on socio-economic status; and socioeconomic status is one of the key determinants of the incidence of water-related diseases, through various direct or indirect pathways.

First, evidence from across the developing world has suggested that there is a strong correlation between socioeconomic status and water access on both a household and a national scale. In an analysis of data from 135 countries,[7] Gomez et al. (2019) demonstrated the positive association between certain variables measuring the economic level of the country (i.e., gross national income, female primary completion rate, or governance) and water access. For example, an increase of 1% in female primary completion rate increases access by 0.23%, for total improved water access and by 0.15% for piped water access, all countries combined. One of the explanations given by the authors is that educated women can exert greater pressure on their communities to have better access to water, themselves being the first to be affected by the burden of fetching water.

---

[7]31 countries classified as low-income, 51 as lower middle income and 53 as higher-middle-income countries, located in sub-Saharan Africa, South America and in Southeast Asia.

Second, it is also possible to think in the opposite direction, as the association found by Gomez et al. (2019) is not a causal relationship: countries with better access to water allow girls to devote more time to their school work, and therefore to perform better, as they are relieved of the burden of fetching water for their households. At the household level, Marcoux (1995) found in Mali that a water source within the concession increased children's school attendance probabilities by about 25%, after controlling for socioeconomic status.

Thus, poverty indicators and water access are also linked (Sullivan, 2002). In a regression analysis made on census data in South Africa, Dungumaro (2007) highlighted that the likelihood of living in a household with safe water is linked to the main source of income: households with income from wages or salaries are 5.2 times more likely to have access to safe water than households without income. Those with income from the sales of farm products are 1.4 times more likely to have access to safe water than those without income.

Because of this relationship between poverty and access to water, water-related indicators are often used to measure socioeconomic status or standard of living in the Global South (Filmer & Pritchett, 2001; Montgomery et al., 2000). For example, many studies include the type of access to water in a wealth index, households with access to improved water are considered to have a higher standard of living than those without (Garip, 2014; Yount, 2008). Comparing alternative approaches to welfare measurement, Filmer and Scott (2012), moreover, showed that asset indices including housing quality variables such as water access are sufficiently robust to measure the economic status of the household.

Finally, education, as a component of socioeconomic status, exhibits its own association with water access. To cite only one result, a study of more than 2000 Malagasy households found that an additional year of school increased the probability of using a public tap in rural areas by 2%. This effect was not significant in urban areas (Boone et al., 2010).

It is also widely recognized that education plays an important role in the acquisition of hygiene-related practices, by facilitating the acquisition of biomedical knowledge. Preston and Haines (1991) identified biomedical knowledge, i.e., acceptance of the germ theory in etiology and associated hygiene practices, as a determinant of the decline in mortality in the USA. This biomedical knowledge would be the product of mother's education (Preston, 1985). Education was found in a recent systematic review as a determinant of handwashing, with other studies most commonly reported factors of handwashing as knowledge, risk, gender, wealth, or infrastructure (White et al., 2020). In Kenya, Schmidt et al. found that lower education was associated with reduced rates of handwashing with soap among the poorest population groups (Schmidt et al., 2009). Nevertheless, studies have raised the question of whether it is mainly formal education that is determinant, or rather the knowledge acquired through experience and social interactions (Jalan & Ravaillon, 2003). Preston and Haines (1991) went even further in raising the hypothesis that it was public health programs rather than educational programs that had had a beneficial effect during the eighteenth century in North America, notably the development of water and sewerage infrastructure. Later, Cutler and Miller (Cutler & Miller, 2005) showed the impact of water chlorination and filtration on infant mortality between 1900 and 1936 in 13 US cities (47% of the decline in log mortality). More recently, a review of impact evaluations in 35 low- and middle-income countries during the past three decades highlighted that hygiene interventions, and provision of soap in particular, have led to a 31% relative reduction in child diarrheal morbidity (95% confidence interval) (Waddington & Snilsveit, 2009). To date, formal education is still most often used to address hygiene-related practices at an individual level (Snilstveit & Waddington, 2009).

Household Composition

Two types of relationships are discussed, depending on the water variable analyzed: household size and source of water; and household size and quantity of water used per capita.

First, several variables that approximate household composition were found to be associated with differential access to water. Here, we focus on the size of the household. A neo-Malthusian reading of this relationship should lead one to think of a negative correlation: the fewer members in the household, the better the access to water. The number of people in the household has been shown to be a determinant of the type of access to water in the household in South Africa. Dungumaro (2007) observed a negative association between the numbers of household members and the likelihood of having safe water access: households with one or two members are about four times more likely than households with more than ten members to access water from a safe source.

Second, other studies also demonstrated that the relationship between household composition and water is not so simple. In a study conducted in a rural area of Iran, a significant negative correlation between water consumption and household size was found (Keshavarzi et al., 2006): higher average water consumption per person was associated with smaller families, owing to the economies of scale of certain domestic water uses (e.g., house cleaning and washing dishes, outdoor water uses, etc.). This is consistent with findings in rural Benin where an inverted U-shaped curve was found between household size and per capita water consumption (Arouna & Dabbert, 2010). This refers to the conclusions of Martin (Martin, 1999) in Israel where the increase in the amount of domestic use was not explained by population growth as such, but rather by the increase in the number of household units.

Place of Residence

As we saw earlier, access to water varies greatly depending on the place of residence, with a very marked disadvantage for people living in rural areas all over the world, particularly in South Asia, sub-Saharan Africa, and Oceania.

Water provision in rural areas in developing countries is challenging. Lack of infrastructure, spare parts and services, as well as little monitoring, intersect to result in rural water issues. Throughout sub-Saharan rural Africa, thousands of water supply systems are developed each year, such as boreholes with motorized pumps, or manual or foot extraction methods. Nevertheless, it is estimated that 50,000 water supply systems throughout Africa do not work because of a lack of maintenance, mainly due to the lack of community participation (Graham, 2011). The complexity of rural water provision calls for a multidisciplinary analysis engaging the full range of stakeholders. In recent years, project initiatives based on the participation of communities, coupled with the public and private sectors, offer valuable lessons, although successes and failures have been observed. Notably, a project carried out in 400 rural communities in Peru, Bolivia, and Ghana highlighted the importance of the community management model and the demand-driven approach in the success of the project. Three to 12 years after the implementation of the project, all project hand pumps were still working in 90% of the villages in Ghana. In Peru and Bolivia, 95% of the households had operational taps among those included in the project (Whittington et al., 2009). Other examples may also show more mixed results undermined by weak capacity, such as in Kenya (Spaling et al., 2014).

Living in a city is a key factor in increasing the likelihood of improved water access, in comparison with rural residence (Pullan et al., 2014). In Zambia, households in rural areas were 70% less likely to have improved access to water than households in urban places (Mulenga et al., 2017).

However, despite optimistic figures for access to water in urban places, with many cities approaching 100% (Hopewell & Graham, 2014), access to water in urban areas presents major current and future challenges, owing to issues abundantly demonstrated in the literature, particularly on the measure of access (Martínez-Santos, 2017). Thus, although relative figures show an improvement in access to water in cities, absolute figures show an increasing number of urban dwellers without access to water in many cities. An analysis of DHS data from 2000 to 2012 on trends in water access in 31 sub-Saharan African cities highlighted that Abidjan, Antananarivo, Dakar, Douala, Mombasa, Kampala, Lagos, Maputo, Ouagadougou, or Yaoundé have made such little progress (between -0.5 and $+ 0.5$ annual percentage points) that the number of people without improved access has increased owing to urban growth (Hopewell & Graham, 2014).

In addition, the challenges of urban growth are putting a strain on access to water for city dwellers. Today's urbanization levels are unprecedented and constitute one of the largest demographic challenges facing efforts to improve water access and quality. It is projected that 2.7 billion people will be added to the world urban population by 2050 because of the combined effects of natural growth and migration. In absolute terms, the population will increase from 4.0 billion in 2015 to 6.7 billion in 2050 (UNDESA, 2018). Africa faces particular challenges. According to estimations, the urban population of sub-Saharan Africa will probably increase 3-fold, from 376 million to 1.3 billion between 2015 and 2050. This urban growth has generally led to a rapid expansion of informal settlements in poor urban and peri-urban informal settlements (UN Habitat, 2016), resulting from survival-driven land management practices and engendering even more challenges. In Malawi, where 70% of the urban residents live in formal areas, households deal with chronic water insecurity every day, in terms of quality, distribution affordability, and reliability (Adams, 2017).

Informal access to water is also a research topic that has emerged in recent years, but that needs to be treated in more depth, in particular in the light of the unprecedented growth of informal settlements in the cities of developing countries (Dos Santos et al., 2017). This field of research should focus not only on day-to-day practices regarding access to water in informal neighborhoods but also on the negotiation of measures of adaptation at the household and community levels. In this respect, research on the peri-urban zones of Ghana and Burkina Faso have shown how the households' strategies have developed to deal with water shortage at sources of supply classified as "improved." Some households among the least poor diversify their sources of water supply, in particular, by turning to informal suppliers, guaranteeing their security at the point of supply (often at "improved" sources). These parallel systems also allow them a certain flexibility (Dos Santos, 2015; Peloso & Morinville, 2014). And yet these modes of supply are still largely unknown, as they have been too rarely studied, whereas they constitute an alternative source that may be a major source for certain populations, notably in the informal urban neighborhoods. This is the case in Accra in Ghana where recent studies have also highlighted the negative perceptions that populations dwelling in informal neighborhoods have of the government's role in the management of crises in the city's water supply (Peloso & Morinville, 2014).

Conclusion

The population–environment–development debates within the academic community of human population scientists are relatively divided, between the most optimistic and the most pessimistic.[8] The debate on population and water is not exempt from strong positions/extreme views, given its importance: in one of the first reports dealing with the links between population and environment, Lori Hunter (2000) wrote of the interrelations between population and water as the second level of "common sense," linking population and the use of resources, after the question of food.

The growth in the number of humans on Earth is undeniably a challenge for the planet, especially in terms of nonrenewable resources. Mathematically speaking, the Malthusian vision is as correct as it is simplistic: the more humans are on the planet, the scarcer the water resources,

---

[8] Cf. the exchanges between Becker (2013) and Lam (2011, 2013) concerning the debate on population growth.

especially in the case of overexploitation of this largely nonrenewable resource. However, policy action, in the etymological sense of the term – management of the City – cannot be based on only two variables: quantity of water and number of humans. In order to understand the stakes, and provide solutions, the other phenomena at play must be taken into account.

Certainly, for populations worldwide, access to water resources raises questions, in particular, in terms of quantity and the availability of water on a macro scale, which is on the scale of the world, regions, or specific countries. The sustainable management of the resource mostly concerns the long term, the coming decades, as trends in water withdrawals largely depend on socio-economic scenarios (Shen et al., 2008).

However, at present, concerns arise on a more micro scale. Often, the populations living under poor water accessibility conditions are victims above all of poor governance, rather than undergoing a real shortage of the resource itself. The international community has been constantly committed to reducing or eradicating the number of people in the world who do not have access to this essential resource, without ever having been able to do so until now. This situation persists, despite the numerous solutions implemented in a more or less concerted manner with all stakeholders, and in particular, with the users themselves. The commodification of the provision of water services developed in some regions after the Dublin Conference is particularly exemplary (Dos Santos et al., 2017).[9] However, international agencies remain divided on this approach (Adams & Halvorsen, 2014). Research is also inconclusive, with little evidence to support the claim that the privatization of water services is more efficient than public sector water management (Prasad, 2006).

Finally, one of the central issues is certainly that of equity in access to and use of water, and its consequences. Population health, social and economic inequalities and their measures (including the questions of gender and poverty), governance of the hydro-socio system, and in particular, the participation of all stakeholders, are among the research priorities in the field of population, development, and the environment. These emerging themes highlight the necessity of exploring, in an empirical manner and in detail, on the micro scale of households and individuals, access to water in general. This is particularly crucial in view of contributing to the management of urban dynamics and human settlement. A social theory perspective could be highly relevant to study power between local, national, and international actors, to identify in-depth those who play a role in the decision-making and governance in the sector of water, and those who are excluded from such roles, etc. (Dos Santos et al., 2017). These research questions can, however, only be dealt with in an integrated way, with the aim of moving toward a holistic perspective on the population–water issue. This involves deploying mixed methodologies (combining discourse analysis and statistical models) and an interdisciplinary approach linking social sciences (demography, sociology, anthropology, economics, geography, political sciences, etc.) with earth sciences (hydrology, agronomy, climatology, ecology, etc.) as well as both life and biomedical sciences.

References

Abuzerr, S., Nasseri, S., Yunesian, M., Hadi, M., Mahvi, A. H., Nabizadeh, R., & Mustafa, A. A. (2019). Prevalence of diarrheal illness and healthcare-seeking behavior by age-group and sex among the population of Gaza strip: A community-based cross-sectional study. *BMC Public Health, 19*(1), 704. https://doi.org/10.1186/s12889-019-7070-0

Adams, E. A. (2017). Thirsty slums in African cities: Household water insecurity in urban informal settlements of Lilongwe, Malawi. *International Journal of Water Resources Development, 34*(6), 869–887. https://www.tandfonline.com/doi/ref/10.1080/07900627.2017.1322941?scroll=top

Adams, E. A., & Halvorsen, K. E. (2014). Perceptions of non-governmental organization (NGO) staff about water privatization in developing countries. *Human Geographies, 8*(2), 35.

---

[9]The most publicized example is undoubtedly that of the privatization of the water sector in the city of Cochabamba in Bolivia in 1999, the 35% increase in the price of water, the prohibition of the collection of rainwater, and massive social protests that followed (Castro, 2007).

Ali, M., Lopez, A. L., You, Y. A., Kim, Y. E., Sah, B., Maskery, B., & Clemens, J. (2012). The global burden of cholera. *The Bulletin of WHO, 90*(3), 157–244. https://doi.org/10.2471/BLT.11.093427

Arouna, A., & Dabbert, S. (2010). Determinants of domestic water use by rural households without access to private improved water sources in Benin: A seemingly unrelated Tobit approach. *Water Resources Management, 24*(7), 1381–1398. https://doi.org/10.1007/s11269-009-9504-4

Bain, R., Cronk, R., Wright, J., Yang, H., Slaymaker, T., & Bartram, J. (2014). Fecal contamination of drinking-water in low-and middle-income countries: A systematic review and meta-analysis. *PLoS Medicine, 11*(5), e1001644.

Becker, S. (2013). Has the world really survived the population bomb? (commentary on "how the world survived the population bomb: Lessons from 50 years of extraordinary demographic history"). *Demography, 50*(6), 2173–2181. https://doi.org/10.1007/s13524-013-0236-y

Biswas, A. K. (2004). From Mar del Plata to Kyoto: An analysis of global water policy dialogue. *Global Environmental Change, 14*, 81–88. https://doi.org/10.1016/j.gloenvcha.2003.11.003

Blackden C. M., Wodon Q. (2006) *Gender, time use, and poverty in sub-saharan Africa*, Washington, D.C., World Bank, Working Paper n. 73.

Boateng, D., Tia-Adjei, M., & Adams, E. A. (2013). Determinants of household water quality in the Tamale Metropolis, Ghana. *Journal of Environment and Earth Science, 3*, 70–77.

Boone, C., Glick, P., & Sahn, D. E. (2010). Household water supply choice and time allocated to water collection: Evidence from Madagascar. *Journal of Development Studies, 47*(12), 1826–1850.

Briggs, D. (2003). Environmental pollution and the global burden of disease. *British Medical Bulletin, 68*, 1–24. https://doi.org/10.1093/bmb/ldg019

Cairncross, S., & Feachem, R. (1993). *Environmental health engineering in the tropics: An introductory text.* Wiley.

Castro, J. E. (2007). Poverty and citizenship: Sociological perspectives on water services and public–private participation. *Geoforum, 38*, 756–771.

Chowdhury, F., Khan, I. A., Patel, S., Siddiq, A. U., Saha, N. C., Khan, A. I., Saha, A., Cravioto, A., Clemens, J., Qadri, F., & Ali, M. (2015). Diarrheal illness and healthcare seeking behavior among a population at high risk for diarrhea in Dhaka, Bangladesh. *PLoS One, 10*(6), e0130105. https://doi.org/10.1371/journal.pone.0130105

Cutler, D., & Miller, G. (2005). The role of public health improvements in health advances: The twentieth-century United States. *Demography, 42*(1), 1–22.

Dos Santos, S. (2011). Les risques sanitaires liés aux usages domestiques de l'eau. Représentations sociales mossi à Ouagadougou (Burkina Faso). *Natures, Sciences et Sociétés, 19*, 103–112.

Dos Santos, S. (2012). Le rôle des femmes selon la GIRE: regard sur le troisième principe de Dublin en Afrique au sud du Sahara. In: Julien, F. (ed) *La gestion intégrée des ressources en eau en Afrique subsaharienne. Paradigme occidental, pratiques africaines* (pp. 135–164), Presses de l'Université du Québec, Montréal.

Dos Santos, S. (2015). L'accès à l'eau des populations dans les quartiers informels de Ouagadougou: un objectif loin d'être atteint. In A. B. Soura, S. Dos Santos, & F. C. Ouédraogo (Eds.), *Climat et accès aux ressources en eau en zones informelles de Ouagadougou* (pp. 57–72). Presses universitaires de Ouagadougou.

Dos Santos, S., Adams, E. A., Neville, G., Wada, Y., de Sherbinin, A., Mullin Bernhardt, E., & Adamo, S. B. (2017). Urban growth and water access in sub-Saharan Africa: Progress, challenges, and emerging research directions. *Science of the Total Environment, 607–608*, 497–508. https://doi.org/10.1016/j.scitotenv.2017.06.157

Dos Santos, S., & Wayack-Pambe, M. (2016). Les Objectifs du Millénaire pour le Développement, l'accès à l'eau et les rapports de genre. *Mondes en développement, 44*(174), 63–78.

Dufault, A. (1988). Women carrying water: How it affects their health. *Waterlines, 3*(6), 23–25.

Dumont, R. (1965). *Chine surpeuplée: Tiers-monde affamé.* Éditions du Seuil Doullens, Sévin..

Dungumaro, E. W. (2007). Socioeconomic differentials and availability of domestic water in South Africa. *Physics and Chemistry of the Earth, 32*, 1141–1147. https://doi.org/10.1016/j.pce.2007.07.006

Ehrlich, P. (1968). *The population bomb.* Ballantine Books.

Engelman, R., & Le Roy, P. (1993). *Sustaining water: Population and the future of renewable water supplies.* Population Action International.

Engelman, R., & Le Roy, P. (1996). *Sustaining water, easing scarcity: A second update.* Population Action International.

Esrey, S. A. (1994). *Multi-country study to examine relationships between the health of children and the level of water and sanitation service, distance to water and type of water used.* Mc Gill University.

Falkenmark, M., & Widstrand, C. (1992). Population and water resources: A delicate balance. *Population Bulletin, 47*(3).

Fewtrell, L., & Colford, J. M. (2004). *Water, sanitation and hygiene: Interventions and diarrhoea a systematic review and meta-analysis.* World Bank.

Filmer, D., & Pritchett, L. (2001). Estimating wealth effects without expenditure data or tears: An application to educational enrolments in states of India. *Demography, 38*, 115–132.

Filmer, D., & Scott, K. (2012). Assessing asset indices. *Demography, 49*(1), 359–392.

Garhammer, M. (1998). Time poverty in modern Germany. *Society and Leisure, 21*, 327–352.

Garip, F. (2014). The impact of migration and remittances on wealth accumulation and distribution in rural Thailand. *Demography, 51*(2), 673–698.

GBD 2015 Diarrhoeal Diseases Collaborators. (2017). Estimates of global, regional, and national morbidity, mortality, and aetiologies of diarrhoeal diseases: A systematic analysis for the global burden of disease study 2015. *The Lancet Infectious Diseases, 388*, 1459–1544. https://doi.org/10.1016/S1473-3099(17)30276-1

Gleick, P. H. (1996). Basic water requirements for human activities: Meeting basic needs. *Water International, 21*(2), 83–92.

Gomez, M., Perdiguero, J., & Sanz, A. (2019). Socioeconomic factors affecting water access in rural areas of low and middle income countries. *Water, 11*(2), 202. https://doi.org/10.3390/w11020202

Graham, J. (2011). Tackling the water crisis: A continuing need to address spatial and social equity. In J. M. H. Selendy (Ed.), *Water and sanitation-related diseases and the environment* (pp. 1–15). Wiley.

Hirway, I., & Jose, S. (2011). Understanding women's work using time-use statistics: The case of India. *Feminist Economics, 17*(4), 67–92.

Hoekstra, A. Y., & Chapagain, A. K. (2007). Water footprints of nations: Water use by people as a function of their consumption pattern. *Water Resources Management, 21*(1), 35–48. https://doi.org/10.1007/s11269-006-9039-x

Hoekstra, A. Y., & Mekonnen, M. M. (2012). The water footprint of humanity. *Proceedings of the National Academy of Sciences, 109*(9), 3232. https://doi.org/10.1073/pnas.1109936109

Hopewell, M. R., & Graham, J. P. (2014). Trends in access to water supply and sanitation in 31 major sub-Saharan African cities: An analysis of DHS data from 2000 to 2012. *BMC Public Health, 14*(1), 208. https://doi.org/10.1186/1471-2458-14-208

Howard, G., & Bartram, J. (2003). *Domestic water quantity, service level and health*. WHO.

Hunter, L. M. (2000). *The environmental implications of population dynamics*. RAND Corporation.

Hunter, P. R., MacDonald, A. M., & Carter, R. C. (2010). Water supply and health. *PLoS Medicine, 7*(11), e1000361. https://doi.org/10.1371/journal.pmed.1000361

Jalan, J., & Ravaillon, M. (2003). Does piped water reduce diarrhea for children in rural India? *Journal of Econometrics, 112*, 153–173.

Kandala, N.-B., Emina, J. B., Nzita, P. D. K., & Cappuccio, F. P. (2009). Diarrhoea, acute respiratory infection, and fever among children in the Democratic Republic of Congo. *Social Science & Medicine, 68*(9), 1728–1736. https://doi.org/10.1016/j.socscimed.2009.02.004

Keshavarzi, A. R., Sharifzadeh, M., Kamgar Haghighi, A. A., Amin, S., Keshtkar, S., & Bamdad, A. (2006). Rural domestic water consumption behavior: A case study in Ramjerd area, Fars province, I.R. Iran. *Water Research, 40*(6), 1173–1178. https://doi.org/10.1016/j.watres.2006.01.021

Kjellstrom, T., Lodh, M., McMichael, T., Ranmuthugala, G., Shrestha, R., & Kingsland, S. (2006). Air and water pollution: Burden and strategies for control. In D. T. Jamison, J. G. Breman, & A. R. Measham (Eds.), *Disease control priorities in developing countries* (pp. 817–832). The International Bank for Reconstruction and Development/The World Bank/Oxford University Press.

Lam, D. (2011). How the world survived the population bomb: Lessons from 50 years of extraordinary demographic history. *Demography, 48*(4), 1231–1262.

Lam, D. (2013). Reply to Stan Becker, "has the world really survived the population bomb? (Commentary on "how the world survived the population bomb: Lessons from 50 years of extraordinary demographic history")". *Demography, 50*(6), 2183–2186.

Lim, S. S., Vos, T., Flaxman, A. D., Danaei, G., Shibuya, K., Adair-Rohani, H., AlMazroa, M. A., Amann, M., Anderson, H. R., Andrews, K. G., Aryee, M., Atkinson, C., Bacchus, L. J., Bahalim, A. N., Balakrishnan, K., Balmes, J., Barker-Collo, S., Baxter, A., Bell, M. L., Blore, J. D., Blyth, F., Bonner, C., Borges, G., Bourne, R., Boussinesq, M., Brauer, M., Brooks, P., Bruce, N. G., Brunekreef, B., Bryan-Hancock, C., Bucello, C., Buchbinder, R., Bull, F., Burnett, R. T., Byers, T. E., Calabria, B., Carapetis, J., Carnahan, E., Chafe, Z., Charlson, F., Chen, H., Chen, J. S., Cheng, A. T.-A., Child, J. C., Cohen, A., Colson, K. E., Cowie, B. C., Darby, S., Darling, S., Davis, A., Degenhardt, L., Dentener, F., Des Jarlais, D. C., Devries, K., Dherani, M., Ding, E. L., Dorsey, E. R., Driscoll, T., Edmond, K., Ali, S. E., Engell, R. E., Erwin, P. J., Fahimi, S., Falder, G., Farzadfar, F., Ferrari, A., Finucane, M. M., Flaxman, S., Fowkes, F. G. R., Freedman, G., Freeman, M. K., Gakidou, E., Ghosh, S., Giovannucci, E., Gmel, G., Graham, K., Grainger, R., Grant, B., Gunnell, D., Gutierrez, H. R., Hall, W., Hoek, H. W., Hogan, A., Hosgood, H. D., Hoy, D., Hu, H., Hubbell, B. J., Hutchings, S. J., Ibeanusi, S. E., Jacklyn, G. L., Jasrasaria, R., Jonas, J. B., Kan, H., Kanis, J. A., Kassebaum, N., Kawakami, N., Khang, Y.-H., Khatibzadeh, S., Khoo, J.-P., Kok, C., Laden, F., Lalloo, R., Lan, Q., Lathlean, T., Leasher, J. L., Leigh, J., Li, Y., Lin, J. K., Lipshultz, S. E., London, S., Lozano, R., Lu, Y., Mak, J., Malekzadeh, R., Mallinger, L., Marcenes, W., March, L., Marks, R., Martin, R., McGale, P., McGrath, J., Mehta, S., Memish, Z. A., Mensah, G. A., Merriman, T. R., Micha, R., Michaud, C., Mishra, V., Hanafiah, K. M., Mokdad, A. A., Morawska, L., Mozaffarian, D., Murphy, T., Naghavi, M., Neal, B., Nelson, P. K., Nolla, J. M., Norman, R., Olives, C., Omer, S. B., Orchard, J., Osborne, R., Ostro, B., Page, A., Pandey, K. D., Parry, C. D. H., Passmore, E., Patra, J., Pearce, N., Pelizzari, P. M., Petzold, M., Phillips, M. R., Pope, D., Pope, C. A., Powles, J., Rao, M., Razavi, H., Rehfuess, E. A., Rehm, J. T., Ritz, B., Rivara, F. P., Roberts, T., Robinson, C., Rodriguez-Portales, J. A., Romieu, I., Room, R., Rosenfeld, L. C., Roy, A., Rushton, L., Salomon, J. A., Sampson, U., Sanchez-Riera, L., Sanman, E., Sapkota, A., Seedat, S., Shi, P., Shield, K., Shivakoti, R., Singh, G. M., Sleet, D. A., Smith, E., Smith, K. R., Stapelberg, N. J.

C., Steenland, K., Stöckl, H., Stovner, L. J., Straif, K., Straney, L., Thurston, G. D., Tran, J. H., Van Dingenen, R., van Donkelaar, A., Veerman, J. L., Vijayakumar, L., Weintraub, R., Weissman, M. M., White, R. A., Whiteford, H., Wiersma, S. T., Wilkinson, J. D., Williams, H. C., Williams, W., Wilson, N., Woolf, A. D., Yip, P., Zielinski, J. M., Lopez, A. D., Murray, C. J. L., & Ezzati, M. (2012). A comparative risk assessment of burden of disease and injury attributable to 67 risk factors and risk factor clusters in 21 regions, 1990–2010: A systematic analysis for the global burden of disease study 2010. *The Lancet, 380*(9859), 2224–2260. https://doi.org/10.1016/S0140-6736(12)61766-8

Lindskog, R. U., & Lindskog, P. A. (1988). Bacteriological contamination of water in rural areas: An intervention study from Malawi. *Journal of Trocipal Medecine and Hygiene, 91*, 1–7.

Mané, N. B., Simondon, K. B., Diallo, A., Marra, A. M., & Simondon, F. (2006). Early breastfeeding cessation in rural Senegal: Causes, modes, and consequences. *American Journal of Public Health, 96*(1), 139–144. https://doi.org/10.2105/AJPH.2004.048553

Marcoux, R. (1995). Fréquentation scolaire et structure démographique des ménages en milieu urbain au Mali. *Cahiers des Sciences Humaines, 31*(3), 655–674.

Martin, N. (1999). *Population, households and domestic water use in countries of the Mediterranean Middle East (Jordan, Lebanon, Syria, the West Bank, Gaza and Israel)*. International Institute for Applied Systems Analysis Report, vol A-2361 Laxenburg, Austria.

Martínez-Santos, P. (2017). Does 91% of the world's population really have "sustainable access to safe drinking water"? *International Journal of Water Resources Development, 33*(4), 514–533. https://doi.org/10.1080/07900627.2017.1298517

Meadows, D. H., Meadows, D. L., Randers, J., & Behrens, W. W. (1972). The limits to growth. *New York, 102*(1972), 27.

Montgomery, M., Gragnolati, M., Burke, K. E., & Paredes, E. (2000). Measuring living standards with proxy variables. *Demography, 37*, 155–174. https://doi.org/10.2307/2648118

Mosley, W. H., & Chen, L. C. (1984). An analytical framework for the study of Child survival in developing countries. *Population and Development Review, 10*(Supplement), 25–45.

Mulenga, J. N., Bwalya, B. B., & Kaliba-Chishimba, K. (2017). Determinants and inequalities in access to improved water sources and sanitation among the Zambian households. *International Journal of Development and Sustainability, 6*(8), 746–762.

Naujokas, M. F., Anderson, B., Ahsan, H., Aposhian, H. V., Graziano, J. H., & Thompson, C. (2013). The broad scope of health effects from chronic arsenic exposure: Update on a worldwide public health problem. *Environmental Health Perspectives, 121*(3), 295–302. https://doi.org/10.1289/ehp.1205875

Nganyanyuka, K., Martinez, J., Wesselink, A., Lungo, J. H., & Georgiadou, Y. (2014). Accessing water services in Dar es Salaam: Are we counting what counts? *Habitat International, 44*, 358–366. https://doi.org/10.1016/j.habitatint.2014.07.003

Nicholson, S. (2001). Climatic and environmental change in Africa during the last two centuries. *Climate Research, 17*, 123–144.

Payment, P., & Hunter, P. R. (2001). Endemic and epidemic infections intestinal disease and its relationship to drinking water. In L. Fewtrell & J. Bartram (Eds.), *Water quality: Guidelines, standard and health* (pp. 61–88). IWA.

Peloso, M., & Morinville, C. (2014). 'Chasing for water': Everyday practices of water access in peri-urban Ashaiman, Ghana. *Water Alternatives, 7*(1), 121–139.

PNUD. (2006). *Rapport mondial sur le développement humain 2006. Au-delà de la pénurie: pouvoir, pauvreté et crise mondiale de l'eau*. ECONOMICA.

Prasad, N. (2006). Privatisation results: Private sector participation in water services after 15 years. *Development Policy Review, 24*(6), 669–692.

Preston, S. H. (1985). Resources, knowledge, and child mortality: A comparison of the U.S. in the late nineteenth century and developing countries today. In *International population conference*. International Union for the Scientific Study of Population (IUSSP).

Preston, S. H., & Haines, M. R. (1991). *Fatal years: Child mortality in late nineteenth-century America*. Princeton University Press.

Prost, A. (1996). L'eau et la santé. In F. Gendreau, P. Gubry, & J. Véron (Eds.), *Populations et environnement dans les pays du Sud* (pp. 231–251). Karthala-CEPED.

Prüss, A., & Mariotti, S. P. (2000). Preventing trachoma through environmental sanitation: A review of the evidence base. *Bulletin of WHO, 78*(2), 258–266.

Pullan, R. L., Freeman, M. C., Gething, P. W., & Brooker, S. J. (2014). Geographical inequalities in use of improved drinking water supply and sanitation across sub-Saharan Africa: Mapping and spatial analysis of cross-sectional survey data. *PLoS Medicine, 11*(4), e1001626. https://doi.org/10.1371/journal.pmed.1001626

Rautu, I., Dos Santos, S., & Schoumaker, B. (2016). Facteurs de risque pour les maladies diarrhéiques chez les enfants à Dakar: une analyse multi-niveaux avec variables latentes. *African Population Studies, 30*(1), 2201–2212.

Schmidt, W. P., Aunger, R., Coombes, Y., Maina, P. M., Matiko, C. N., Biran, A., & Curtis, V. (2009). Determinants of handwashing practices in Kenya: The role of media exposure, poverty and infrastructure. *Tropical Medicine & International Health, 14*(12), 1534–1541. https://doi.org/10.1016/j.ijheh.2020.113512

Shen, Y., Oki, T., Utsumi, N., Kanae, S., & Hanasaki, N. (2008). Projection of future world water resources under SRES scenarios: Water withdrawal/projection des ressources en eau mondiales futures selon les scénarios du RSSE: prélèvement d'eau. *Hydrological Sciences Journal, 53*(1), 11–33. https://doi.org/10.1623/hysj.53.1.11

Simon, A. K., Hollander, G. A., & McMichael, A. (2015). Evolution of the immune system in humans from infancy to old age. *Proceedings of the Royal Society. B, 282*(1821), e20143085.

Siziya, S., Muula, A. S., & Rudatsikira, E. (2013). Correlates of diarrhoea among children below the age of 5 years in Sudan. *African Health Sciences, 13*(2), 376–383. https://doi.org/10.4314/ahs.v13i2.26

Snilstveit, B., & Waddington, H. (2009). Effectiveness and sustainability of water, sanitation, and hygiene interventions in combating diarrhoea. *Journal of Development Effectiveness, 1*(3), 295–335.

Sorenson, S. B., Morssink, C., & Abril Campos, P. (2011). Safe access to safe water in low income countries: Water fetching in current times. *Social Science and Medicine, 72*, 1522–1526.

Spaling, H., Brouwer, G., & Njoka, J. (2014). Factors affecting the sustainability of a community water supply project in Kenya. *Development in Practice, 24*(7), 797–811. https://doi.org/10.1080/09614524.2014.944485

Sullivan, C. (2002). Calculating a Water Poverty Index. *World Development, 30*(7), 1195–1211.

Thompson, J., Porras, I. T., Tumwine, J. K., Mujwahuzi, M. R., Katui-Katua, M., Johnstone, N., & Wood, L. (2001). *Drawers of water II: 30 years of change in domestic water use and environmental health in East Africa*. IIED.

UNDESA. (2018). *World urbanization prospects: The 2018 revision*. UNDESA.

UNESCO. (2003). *Water for people, water for life: A joint report by the twenty-three UN agencies concerned with freshwater*. World Water Assessment Programme.

UN Habitat. (2016). World cities report 2016. Urbanization and development. Emerging futures. Retrieved from http://wcr.unhabitat.org/wp-content/uploads/sites/16/2016/05/WCR-percent20Full-Report-2016.pdf10.18356/d437cd7e-en

United Nations. (1977). Rapport de la Conférence des Nations Unies sur l'eau, Mar del Plata, 14–25 mars 1977. vol E/CONF.70/29.

United Nations. (2019a). Special edition: Progress towards the Sustainable Development Goals. vol E/2019/68.

United Nations. (2019b). *Water*. https://www.un.org/en/sections/issues-depth/water/

United Nations General Assembly. (2010). *The human right to water and sanitation*. https://www.un.org/en/ga/search/view_doc.asp?symbol=A/RES/64/292 Accessed 2 April 2019.

Véron, J. (1995). L'INED et le Tiers Monde. *Population, 6*, 1565–1578.

Vörösmarty, C. J., Green, P., Salisbury, J., & Lammers, R. B. (2000). Global water resources: Vulnerability from climate change and population growth. *Science, 289*, 284–288.

Waddington, H., & Snilstveit, B. (2009). Effectiveness and sustainability of water, sanitation, and hygiene interventions in combating diarrhoea. *Journal of Development Effectiveness, 1*(3), 295–335. https://doi.org/10.1080/19439340903141175

White, G. F., Bradley, D. J., & White, A. U. (1972). *Drawers of water*. Chicago University Press.

White, S., Thorseth, A. H., Dreibelbis, R., & Curtis, V. (2020). The determinants of handwashing behaviour in domestic settings: An integrative systematic review. *International Journal of Hygiene and Environmental Health, 227*, 113512. https://doi.org/10.1016/j.ijheh.2020.113512

Whittington, D., Davis, J., Prokopy, L., Komives, K., Thorsten, R., Lukacs, H., Bakalian, A., & Wakeman, W. (2009). How well is the demand-driven, community management model for rural water supply systems doing? Evidence from Bolivia, Peru and Ghana. *Water Policy, 11*(6), 696–718. https://doi.org/10.2166/wp.2009.310

WHO. (2007). *Chemical safetyof drinking-water: Assessing prioritiesfor risk management*. World Health Organization Geneva.

WHO/UNICEF. (2000). *Global water supply and sanitation assessment 2000 report*. WHO/UNICEF.

WHO/UNICEF. (2015). *Progress on sanitation and drinking water: 2015 update and MDG assessment*. WHO/Unicef.

WHO/UNICEF. (2016). *Safely managed drinking water*. http://www.prographic.com/wp-content/uploads/2016/11/UNICEF-SafelyMngDrinkWater-2016-11-18-web.pdf.

Wright, J., Gundry, S., & Conroy, R. (2004). Household drinking water in developing countries: A systematic review of microbiological contamination between source and point-of-use. *Tropical Medicine and International Health, 9*(1), 106–117.

Yount, K. M. (2008). Gender, resources across the life course, and cognitive functioning in Egypt. *Demography, 45*(4), 907–926.

Heat, Mortality, and Health

14

Heather Randell

Abstract

Climate change has led to an increased frequency, duration, and intensity of heat waves, and extreme heat is projected to worsen in severity over the coming decades. Exposure to extreme heat is a critical threat to population health and well-being, particularly in urban areas and among vulnerable groups including the elderly, pregnant women, individuals living in poverty, those who work outdoors, and people who are socially isolated. However, the relationship between heat exposure, health, and mortality remains understudied within demography. This chapter provides an overview of some of the direct and indirect health impacts of extreme heat including its effects on mortality, birth outcomes, nutrition and food security, and infectious disease transmission. The chapter concludes with a discussion of priority areas for future research as well as data needs to facilitate environmental demographers' investigation into the linkages between heat, mortality, and health. A demo-graphic perspective offers a unique lens into the population-level effects of extreme heat and can inform adaptation measures to mitigate the health effects of heat in a warming world.

Introduction

Extreme heat is a growing threat to population health. Heat waves have increased in frequency, duration, and intensity since the mid-twentieth century as a result of rising greenhouse gas emissions (Perkins et al., 2012). The number of people exposed to extreme heat grew by 125 million between 2000 and 2016 due to climate change, population growth, and changes in population distribution (World Health Organization, 2018). In 2016, the Paris Agreement was adopted with the goal of limiting global warming to 1.5 °C above pre-industrial levels—a threshold widely viewed as necessary to avoid catastrophic impacts on human and natural systems associated with rising global temperatures, including the effects of extreme heat (Hoegh-Guldberg et al., 2018; Rogelj et al., 2016). However, the world is currently on track to experience 3 to 5 °C of

H. Randell (✉)
Department of Agricultural Economics, Sociology, and Education, The Pennsylvania State University, University Park, PA, USA
e-mail: hrandell@psu.edu

© Springer Nature Switzerland AG 2022
L. M. Hunter et al. (eds.), *International Handbook of Population and Environment*, International Handbooks of Population 10, https://doi.org/10.1007/978-3-030-76433-3_14

warming by 2100 (World Meteorological Organization, 2019). This warming trend is predicted to lead to more numerous, severe, and widespread heat waves in the coming decades (Stocker et al., 2013). Further, tropical regions, where the majority of the world's poor live, are expected to experience unprecedented heat sooner than more temperate regions (Harrington et al., 2016).

Extreme heat negatively impacts health in a variety of direct and indirect ways. Exposure to high temperatures affects the body's ability to thermoregulate (maintain a constant core temperature), which can lead to heat cramps, heat stress, heat stroke, hyperthermia, and death (World Health Organization, 2018). People with existing health conditions including diabetes, hypertension, and cardiovascular disease are particularly vulnerable because these diseases are associated with a reduced ability to thermoregulate (Kenny et al., 2010). Further, heat can threaten maternal and fetal health by increasing the risk of preterm birth, low birthweight, and stillbirth (Kuehn & McCormick, 2017). Proposed mechanisms include decreased maternal and fetal nutrition, placental damage, and reduced uterine blood flow (Bakhtsiyarava et al., 2018; Basu et al., 2010, 2018). Heat also affects agricultural production, the transmission of infectious diseases, and labor productivity, which in turn impact the physical and mental health and well-being of exposed populations (Heal & Park, 2016; Lesk et al., 2016; Levy et al., 2016; Siebert & Ewert, 2014; World Health Organization, 2018). Altogether, these risks mean that a number of sub-populations are particularly vulnerable to the health impacts of extreme heat including the elderly, infants and young children, pregnant women, people who work outdoors, individuals with pre-existing health conditions, those who live in poverty, and people who are socially isolated (Benmarhnia et al., 2015; Mora, Counsell, et al., 2017; World Health Organization, 2018).

While the connections between heat, mortality, and health are well-documented in the public health and economics literatures, the topic remains understudied among demographers. For example, a search for the terms "heat" and "temperature" in *Demography*, the top journal in the field, returned one article related to health or mortality (Muhuri, 1996). Similarly, *Population and Environment,* the flagship environmental demography journal, returned only five papers on the topic (Cho, 2020; Davenport et al., 2020; Mueller & Gray, 2018; Xu et al., 2018; Yang & Jensen, 2017). Research on heat, health, and mortality could greatly benefit from an environmental demographic perspective. Demographers are interested in relationships at the population level, and understanding population-level (versus individual-level) effects of heat can help inform policies that reach large numbers of people. Demographers are also uniquely positioned to consider the interrelationships between population processes (fertility, mortality, and migration), how these processes are affected by extreme temperatures, and how this leads to a variety of health outcomes. As such, this chapter focuses on areas of heat-health research in which demographers can offer important contributions.

Section "Climate Change, Extreme Temperatures, and Heat Waves" provides a brief background on climate change, extreme temperatures, and heat waves. Section "Heat Waves and Mortality" focuses on the effects of heat waves on mortality, while Section "Prenatal Heat Exposure and Adverse Birth Outcomes" addresses linkages between heat exposure during pregnancy and adverse birth outcomes. Section "Heat, Food Security, and Undernutrition" discusses the relationship between heat, food security, and nutrition, and Section "Heat and Infectious Diseases" addresses links between heat and infectious diseases. Lastly, Section "Future Research Directions and Data Needs", concludes with a discussion of priority areas for future research as well as data needs to facilitate environmental demographers' investigation into the linkages between heat, mortality, and health.

Climate Change, Extreme Temperatures, and Heat Waves

Global temperatures have risen steadily over the past several decades, with 2016 and 2019 the two warmest years on record (World Meteorological Organization, 2020). In 2018, numerous locations around the world broke all time heat records including Riverside, California, which hit 118 °F (47.8 °C), and Ouargla, Algeria, which reached 124.3 °F (51.3 °C), the hottest temperature ever recorded in Africa (Samenow, 2018; Vives et al., 2018). In July 2019, Anchorage, Alaska reached 90 °F (32 °C) for the first time on record, breaking its previous heat record by 5 °F (Rathbun, 2019), and in August 2020, Death Valley, California reached 130 °F (54.4 °C), likely the hottest reliably recorded temperature on earth (Masters, 2020).

By the end of the twenty-first century, record-setting temperatures are expected to occur every year in nearly 60 % of the world if greenhouse gas emissions continue to rise (Power & Delage, 2019). The frequency of uncomfortably hot days is also predicted to rise with climate change. For example, the number of heat wave days in Houston is projected to increase from 5.2 per year in recent years to 58.2 per year in 2061–2080 under a medium-emissions scenario (RCP4.5) and 103.2 per year under a high-emissions scenario (RCP8.5) (Oleson et al., 2018).

Those living in cities are particularly vulnerable to the direct effects of extreme heat, as densely populated urban areas tend to be hotter than nearby locations that have fewer large buildings, less traffic, and more trees and green space. This phenomenon, called the urban heat island effect, occurs for a number of reasons: roads, concrete, and buildings absorb solar radiation; tall buildings and narrow streets trap hot air; cars, buildings, and industrial activities directly emit heat into the surrounding area; and a lack of vegetation reduces shading as well as cooling due to evapotranspiration (UCAR Center for Science Education, 2011). During the summer, dense urban neighborhoods can be as much as 18 °F (10 °C) warmer than nearby wooded areas (Kim, 1992).

Ambient temperature (the measured air temperature) is only one of several environmental factors that lead to heat stress. Additional factors include humidity and solar radiation (Epstein &

Moran, 2006). Mean radiant temperature, which factors in solar radiation, accounts for the surface temperatures of a person's surroundings (Thorsson et al., 2014). For example, on a hot and sunny day, an area surrounded by large buildings and dark pavement will have higher mean radiant temperature than a nearby area with fewer buildings and light-colored concrete due to the solar radiation absorbed by, and then emitted from, the buildings and pavement. Indeed, mean radiant temperature was found to be a better predictor of mortality than ambient temperature in urban areas (Thorsson et al., 2014). Humidity is also important to consider, as evaporative cooling through sweating is a primary means through which humans dissipate heat. As humidity increases the air becomes saturated with water vapor, which limits the ability for humans to release body heat through sweating. Hot weather accompanied by humid conditions therefore poses a higher risk to human health than hot and dry conditions (Sherwood, 2018).

Many studies on heat, mortality, and health rely on ambient air temperature to capture heat stress, likely due to the widespread availability of air temperature data. However, a number of heat indices, including the Universal Thermal Climate Index (UTCI) and the Discomfort Index (DI) have been created that take into account additional factors driving heat stress including humidity and mean radiant temperature (Blazejczyk et al., 2012; Epstein & Moran, 2006). Using these indices in studies of heat and health helps to more holistically capture the causes of heat stress in different environments.

There is no universal definition of heat waves, however a heat wave is generally defined as a period of consecutive days in which temperatures are considerably hotter than what is normal for a particular place at a given time of year (Perkins & Alexander, 2013). Thus, a span of 3 days in August in which the temperature reaches 95 °F (35 °C) may be typical in Las Vegas, which has a subtropical hot desert climate, but considered a heat wave in Boston, which has a humid continental climate characterized by warm summers and cold, snowy winters. Similarly, reaching 95 °F for 3 days in Las Vegas may be typical during June but considered a heat wave in January. Climate change is expected to increase heat wave severity, frequency, and duration over the short term (Stocker et al., 2013). Indeed, with each 1 °C of warming, between 4 and 34 additional heat wave days per summer are expected globally (Perkins-Kirkpatrick & Gibson, 2017). Further, the duration of individual heat waves is predicted to increase by 2 to 10 days per 1 °C of warming, with the greatest increases occurring in the tropics (Perkins-Kirkpatrick & Gibson, 2017).

Indeed, it is projected that even if warming is limited to 1.5 °C, there will be a significant increase in the severity and frequency of heat waves, particularly in Africa, Central and South America, and Southeast Asia (Dosio et al., 2018). With 2 °C of warming, the frequency of extreme heat waves is expected to double over much of the world, with the severest of heat waves occurring in large parts of Africa (Dosio et al., 2018). This suggests that people in many of the poorest parts of the world—particularly those living and working in urban areas—are likely to face the most severe impacts from the increased intensity and frequency of heat waves associated with climate change.

Heat Waves and Mortality

How Heat Kills

Extreme heat increases the risk of a number of adverse health outcomes, including mortality. Though cold-related deaths in the U.S. are about twice as common as heat-related deaths, heat deaths alone account for a larger total than deaths due to flood, storms, or lightning combined (Berko et al., 2014). Exposure to extreme heat for prolonged periods of time can cause mortality through a number of mechanisms, as hyperthermia (abnormally high body temperature) damages vital organs including the heart, brain, lungs, kidneys, and liver (Mora, Counsell, et al., 2017). For example, in hot conditions, blood vessels dilate to redirect blood from the core towards the skin in order to dissipate heat. This reduces blood flow to other organs, which can lead to cell damage, compromise organ function, and potentially cause organ failure and death (Mora, Counsell, et al., 2017). Another example is through heat-induced dehydration, which thickens the blood and constricts blood vessels, which can trigger a stroke.

Indeed, a study across 18 countries in North America, South America, Europe, and Asia found that heat waves[1] were linked with an increased risk of death in all places, and that the effects of heat waves were greater in moderately cold and moderately hot areas, versus cold and hot areas (Guo et al., 2017). Individuals with a reduced ability to thermoregulate (e.g., the elderly, young children, and sick people), socioeconomically disadvantaged individuals, and those involved in outdoor activities such as sports or strenuous labor are particularly vulnerable to heat-related mortality (Mora, Counsell, et al., 2017).

Some Notable Heat Waves

A number of heat waves in recent decades have led to considerable mortality. In 2003, Europe experienced its hottest summer since the 1500s, causing 70,000 heat-related deaths (Poumadère et al., 2005; Robine et al., 2008). France was hit especially hard, with temperatures in some locations rising to over 104 °F (40 °C) for over a week (Poumadère et al., 2005). Because summers in France tend to be mild, the country was not prepared for such a severe and unprecedented heat wave. Many homes lacked air conditioning, few residents recognized the severity of the situation, and the country's health system was ill-equipped to handle the increased rates of morbidity and mortality. Nearly 15,000 heat-related deaths occurred in

---

[1] The study used multiple community-specific heat wave definitions based on local average daily mean temperatures in the 400 study communities. Twelve categories were created based on temperatures exceeding the 90th, 92.5th, 95th, and 97.5th percentiles, as well as durations of 2, 3, and 4 days. Heat waves of all definitions were linked to mortality.

France alone, with elderly women, socially isolated individuals, urban dwellers, those of low socioeconomic status, and individuals with pre-existing conditions the most vulnerable (Poumadère et al., 2005).

Sociologist Eric Klinenberg (1999) examined the 1995 Chicago heat wave, which lasted 5 days and was characterized by unprecedented temperatures during both the day and nighttime, coupled with high levels of humidity (Semenza et al., 1996). The heat wave caused 739 deaths over the 5-day period. On the worst day of the heat wave, temperatures reached nearly 106 °F (41 °C), with indoor temperatures in some apartments that lacked air conditioning rising to 115 °F (46 °C), even with the windows open. Klinenberg found that deaths were concentrated among socially isolated elderly residents who lived in poorer, more violent, and more heavily African American neighborhoods of the city.

In addition to relatively cool places like France and Chicago, historically hot regions are also susceptible to heat wave mortality. For example, much of India has a hot, tropical climate with temperatures typically reaching 104 °F (40 °C) during the pre-monsoon season (Di Liberto, 2015). Yet in 2015, a heat wave occurred during which time temperatures exceeded 113 °F (45 °C) for multiple days, melting asphalt in Delhi and leading to over 2500 heat-related deaths (Wehner et al., 2016). The Indian heat wave was the fifth deadliest in world history, with mortality concentrated among the elderly, the young, and those who worked outdoors (Di Liberto, 2015).

Using Population-Level Data to Understand Links Between Heat and Mortality

In addition to the case studies noted above, researchers have linked mortality and climate data to better understand broader spatial and temporal patterns. For example, the timing of a heat wave has important implications for mortality. In the U.S. the first heat wave in a given summer is associated with a higher risk of mortality than subsequent heat waves, likely due to biophysical and/or behavioral adaptation as the summer progresses (Anderson & Bell, 2011). Further, cooler parts of the country (e.g., the Northeast and Midwest) tend to suffer more from heat wave mortality than warmer regions (e.g., the South) (Anderson & Bell, 2011; Barreca et al., 2015; Yang & Jensen, 2017). Similarly, a study of 16 Asian countries found that high temperatures significantly increased mortality and that this increase was larger in cooler nations, which tend to experience hot weather much less frequently (Deschênes, 2018). These studies highlight the importance of considering both spatial and temporal differences in vulnerability to extreme heat and provide empirical support for greater vulnerability in historically cooler areas that are not adapted to hot temperatures.

Much of the research on heat-related mortality focuses on urban areas given the urban heat island effect. However, the urban heat island is not the only risky context. In China, high mortality risk can be found in rural regions (Hu et al., 2019). Urban-rural socioeconomic and demographic disparities in age, poverty, education, occupational type, access to health care, and prevalence of air conditioning are the likely causes. In rural India, days above the 98th percentile for heat (>102 °F or 39 °C) were associated with a 33% increase in all-cause mortality and a 57% increase in non-infectious disease mortality (Ingole et al., 2015). Men and working-age individuals experienced a higher risk than women and the elderly, suggesting that those who are exposed to direct heat through outdoor labor (namely working-age men) are highly vulnerable to heat-related mortality.

Heat and Mortality among Infants and Children

Researchers have used demographic data to understand how heat affects mortality among adults as well as infants and young children. For example, Health and Demographic Surveillance System (HDSS) data can illuminate key relationships between heat waves and deaths.[2] Studies using HDSS data from Ghana, Burkina Faso, and Kenya found that high temperatures had strong impacts on mortality, particularly among children (Azongo et al., 2012; Diboulo et al., 2012; Egondi et al., 2012). For example, in Burkina Faso each additional 1 °C increase in daily maximum temperature was associated with a 2.6 times greater odds of short-term mortality among all age groups, and a 3.7 times greater odds among children under five (Diboulo et al., 2012). Given these findings, it is important to explore the relationship between heat and child mortality in greater depth to understand which populations and age groups are most at risk.

Infants are particularly vulnerable to extreme heat because their ability to thermoregulate is not fully developed (Ringer, 2013). However, research on heat and infant mortality remains extremely limited. Geruso and Spears (2018) used data from the Demographic and Health Surveys Program (DHS)[3] from 53 countries in Africa, Asia, and Latin America to examine the relationship between heat, humidity, and infant mortality. Hot and humid days led to large increases in infant mortality, with the strongest effects for high heat exposure occurring during the month of birth as opposed to in utero or during later infancy. This suggests that newborns are particularly vulnerable to the direct physiological impacts of extreme heat, likely due to their limited ability to thermoregulate. Further, the greatest impacts of heat on infant mortality occurred in Southeast Asia, a humid region, rather than in hotter parts of Sub-Saharan Africa, which tend to be dry. This indicates that extreme heat alone is less a threat to infant survival than hot and humid conditions.

Heat has been shown to affect infant mortality in high-income countries as well. A study in Montreal, Canada found that maximum daily temperatures above 84 °F (29 °C) were associated with a 2.8 times greater odds of mortality from Sudden Infant Death Syndrome (SIDS), possibly due to thermal stress in hot weather (Auger et al., 2015). Further, Deschênes and Greenstone (2011) found that if greenhouse gas emissions continue unabated, infant mortality in the United States is predicted to increase by 7.8% by the late twenty-first century.

Climate Change and Adaptation

A number of studies predict that climate change will intensify the risk of heat-related mortality globally over the coming decades. Approximately 30% of the global population currently lives in places where temperature and humidity conditions reach deadly levels at least 20 days per year (Mora, Dousset, et al., 2017). By 2100, this is predicted to increase to 48% under a low-emissions scenario (RCP2.6) and 74% under a high-emissions scenario (RCP8.5). In addition, under a medium-emissions scenario (RCP4.5), by 2100 densely populated areas of South Asia as well as parts of the Middle East are predicted to experience combined temperature-humidity levels that are extremely hazardous to human health (Im et al., 2017; Pal & Eltahir, 2016).

Mortality caused by extreme heat can be reduced—at least to a point—through adaptation measures. Heat-related deaths have declined in recent decades in high-income countries, likely due in part to improvements in infrastructure, air conditioning, and public health measures, (Gasparrini et al., 2015). In the U.S., mortality associated with daily temperatures above 80.6 °F (27 °C) declined by 75% over the course of the twentieth century, with nearly all of the decline attributable to the adoption of air conditioning in homes (Barreca et al., 2016). However, other studies debate the importance of air conditioning in heat-related mortality decline in the U.S., instead suggesting that general improvements in cardiovascular health among the elderly have made them more resilient to heat (Bobb et al., 2014).

---

[2]HDSS sites collect health and demographic data from entire communities in a number of low- and middle-income countries over extended periods of time. Data from some HDSS sites can be accessed through the INDEPTH Network (http://www.indepth-network.org/)

[3]The DHS program collects nationally representative data on population health and nutrition from over 90 countries. For more information, see www.dhsprogram.com

There are a number of other adaptation measures that can mitigate the adverse health impacts of extreme heat. Human bodies can adapt physiologically to hot weather over the course of days to weeks, for example through more efficient sweating (Hondula et al., 2015). City planners can reduce the urban heat island effect by installing green roofs, limiting automobile traffic, replacing surfaces with reflective materials, planting trees, and expanding parks (Akbari et al., 2015). And local governments can improve heat wave preparedness by developing early warning systems and improving access to cooling centers (Hondula et al., 2015). Adaptation measures such as these may lead to declines in heat-related mortality in places with the resources to implement them. However, predictions suggest that many populations globally will remain vulnerable to mortality associated with extreme heat (Mora, Dousset, et al., 2017).

Prenatal Heat Exposure and Adverse Birth Outcomes

Another topic important for investigation by environmental demographers is the link between extreme heat during pregnancy and adverse birth outcomes. Two common birth outcomes affected by conditions in utero are low birthweight (weighing under 2500 grams at birth) and preterm birth (being born before 37 weeks gestation). Globally, 20 million infants per year are characterized by low birthweight and 15 million are born preterm (World Health Organization, 2014). Over 60 % of preterm births occur in Africa and South Asia, however low birthweight and preterm birth are challenges in high-income countries as well. In the United States, approximately one in ten babies is born preterm and 8 % are low birthweight (CDC, 2017).

Birth outcomes such as these are critical determinants of long-term health and socioeconomic outcomes. Evidence from both poorer and wealthier countries suggests that low birthweight and preterm birth are linked to an increased risk of disability and are negatively associated with adult height, educational attainment, and wages (Behrman & Rosenzweig, 2004; Black et al., 2007; Lindstrom et al., 2007; Stein et al., 2013). This indicates that conditions experienced in utero can have lasting impacts on physical and cognitive development, which in turn affect well-being throughout the life course.

A number of studies in low- and middle-income countries have investigated the relationship between prenatal heat exposure and birthweight. Grace et al. (2015) used DHS data from 19 African countries and found that experiencing a greater number of days above 100.4 °F (38 °C), particularly during the first and second trimesters, was associated with a higher likelihood of low birthweight. Similarly, a study of 13 African countries found that a 10% increase in days above 104 °F (40 °C) during pregnancy decreased the likelihood of a healthy birthweight by 4.4% (Davenport et al., 2020). Among food cropping households in Kenya, the number of months in which the average temperature exceeded 95 °F (35 °C) was negatively associated with birthweight (Bakhtsiyarava et al., 2018), and in rural Colombia, exposure to heat waves during the third trimester of pregnancy was associated with lower birthweight (Andalón et al., 2016). Lastly, in Bolivia, Colombia, and Peru, experiencing temperatures above normal during pregnancy was associated with a reduction in birthweight, and this was mainly driven by temperatures during the first trimester (Molina & Saldarriaga, 2017).

In the U.S., Deschênes et al. (2009) used data from more than 37 million births and found that in utero exposure to extreme heat, particularly during the second and third trimesters, was associated with a higher probability of low birthweight. The link between heat and low birthweight was found to be strongest in the northeastern U.S. as well as an in other areas with cold climates (Sun et al., 2019). Similarly, in California, high temperatures during pregnancy were positively associated with low birthweight, and this was driven primarily by temperatures during the third trimester (Basu et al., 2018). Lastly, a study in New York City discovered that each additional day above 84 °F (29 °C) during pregnancy was associated with a 1.7 g reduction in birthweight, and the relationship was strongest for teen mothers (Ngo & Horton, 2016).

The above research provides strong evidence for the link between hot conditions during pregnancy and low birthweight. However, there is variation in critical periods of exposure although this variation is poorly understood. In some cases, the impact of heat is relatively stronger in the first trimester (Grace et al., 2015; Molina & Saldarriaga, 2017), although in other cases impacts appear greatest later in pregnancy (Andalón et al., 2016; Basu et al., 2018). There is clearly a need for clarification with regard to the contextual factors and mechanisms involved in the heat-birthweight relationship.

Less work has been conducted on the linkages between heat and preterm birth, but the findings are generally consistent with that of low birthweight. For example, in California during the warm season, a 10 °F (5.6 °C) increase in weekly average temperature was associated with an 8.6 increase in the likelihood of preterm delivery (Basu et al., 2010). Another U.S. study found that 25,000 infants per year are born early as a result of exposure to high temperatures (Barreca & Schaller, 2020). While a number of studies on heat and preterm birth exist in higher income regions including Europe, North America, and Australia, research in low- and middle-income countries is virtually absent (Carolan-Olah & Frankowska, 2014; Zhang et al., 2017). This is likely due in part to limited use of ultrasound technology to determine gestational age, particularly in rural areas (Wanyonyi et al., 2017).

Identifying the mechanisms underlying the effects of heat on adverse birth outcomes is difficult, but researchers have proposed two potential drivers. The first suggests that food production and income represent key mechanisms. Heat has been shown to reduce agricultural production (Zhao et al., 2017), which threatens food security, particularly in rural areas of low- and middle-income countries. This in turn may negatively impact maternal and fetal nutrition, leading to intrauterine growth restriction and low birthweight (Bakhtsiyarava et al., 2018; Hu & Li, 2019; Molina & Saldarriaga, 2017). The second is

a biological mechanism. Extreme heat may cause blood to be diverted from the vital organs to the skin, which damages the placenta and decreases fetal nutrition, leading to low birthweight (Basu et al., 2018). Heat exposure may also trigger early labor, potentially leading to preterm birth. A few mechanisms are hypothesized for the link between heat and early labor. Pregnant women exposed to extreme heat, particularly during the third trimester, may become dehydrated, decreasing blood flow to the uterus and inducing labor (Basu et al., 2010). Further, heat stress may trigger the release of hormones like cortisol, which also induce labor (Carolan-Olah & Frankowska, 2014).

If a food production and income mechanism is driving the relationship in a particular place, adaptation solutions would involve improving food security among pregnant women through, for example, the provision of heat-tolerant crop and livestock varieties, diversification into non-agricultural livelihood strategies to reduce dependence on agricultural income, and provision of food or monetary aid to populations whose food security is impacted by heat. If a biological mechanism drives the relationship in a particular context, solutions would center on reducing exposure to heat and encouraging sufficient hydration among pregnant women. Women who work outside or in places such as factories or warehouses, and those who do not have air conditioning in their homes, should be targeted for interventions.

Heat, Food Security, and Undernutrition

A third area of research centers on the linkages between heat, food security, and undernutrition. This issue is of particular relevance in low- and middle-income countries, where many people engage in agriculture and/or live in poverty. Globally, 10 % of the population experiences food insecurity, with rates as high as 34 % in Sub-Saharan Africa (FAO et al., 2018). While hunger has declined over the past several decades, there has been an uptick in recent years and climate change has been identified as a primary driver (FAO et al., 2018). A number of studies in Sub-Saharan Africa and Asia have found that exposure to droughts or severe flooding in early childhood can lead to undernutrition, as indicated by stunting (low height-for-age), wasting (low weight-for-height), and slower growth (Hoddinott & Kinsey, 2001; del Ninno & Lundberg, 2005; Rodriguez-Llanes et al., 2011; e.g., Chotard et al., 2011). However, there has been very little work on the relationship between heat and nutritional outcomes.

Extreme heat affects nutrition and food security primarily through its impacts on agriculture. As mentioned above, increasing temperatures are associated with declines in crop and livestock production (Lesk et al., 2016; Porter et al., 2014; Zhao et al., 2017). This in turn reduces food availability and income among farming and pastoralist households. Further, reduced agricultural productivity leads to increased food prices, which negatively impacts food security among those who purchase the majority of their food, especially the urban poor. Examining connections between heat and undernutrition among children is particularly important, as nutrition during early childhood has long-term effects on health and human capital formation. Early life undernutrition is linked to poorer cognitive development, lower educational attainment, shorter adult height, and lower wages and lifetime earnings (Alderman et al., 2006; Dewey & Begum, 2011; Glewwe & King, 2001; Hoddinott et al., 2008, 2013; Maluccio et al., 2009; Victora et al., 2008). Indeed, Randell and Gray (2019) discovered that in some parts of the global tropics, experiencing hotter-than-normal conditions during early childhood is associated with lower educational attainment in adolescence. This is likely driven in part by the negative effects of heat on agricultural production and child nutrition.

A few studies have used DHS data to examine the connections between climatic conditions and child undernutrition in Sub-Saharan Africa. A study in Kenya did not find a significant relationship between heat and stunting, arguing that rainfall was a more important determinant of nutrition in this context. (Grace et al., 2012). However, a recent study of 18 countries in Sub-Saharan Africa discovered that hot-and-dry conditions were negatively associated with weight-for-height among children under 5 years of age, which is an indicator of recent acute undernutrition (Thiede & Strube, 2019). This suggests that areas experiencing increased warming and drying trends may suffer from more frequent food shortages, which threatens short-term food security and nutrition among children.

Another study used data from the Mexican National Nutrition Survey to examine the effects of climatic conditions during the agricultural season on stunting (Skoufias & Vinha, 2012). While hotter temperatures were not associated with stunting among the entire sample, certain sub-populations were vulnerable to the effects of heat including boys, children aged 1 to 2 years at the time of the survey, and those with less educated mothers. Mueller and Gray (2018) examined the effects of heat on adult nutrition using the China Health and Nutrition Survey and discovered that hotter temperatures were associated with a greater likelihood of being underweight among adults over age 40. Heat was also associated with a higher probability of respiratory illnesses, gastrointestinal symptoms, headaches, and skin conditions among the elderly.

Taken together, this research indicates that rising global temperatures may lead to increased rates of undernutrition and poor health, particularly among two key demographic groups—young children and the elderly. Given that many parts of the world are experiencing rapid population aging (Bloom et al., 2015), and that nutrition during early childhood is a critical determinant of long-term health and well-being, it is essential to better understand the effects of heat on nutrition and food security, particularly among these populations. Potential solutions could involve increasing the provision of heat-tolerant crop and livestock varieties, promoting livelihood diversification to include non-agricultural sources of income that are not sensitive to heat (such as small businesses or handicraft production), and expanding social safety net programs to ensure the availability and affordability of food during and after periods of extreme heat.

Heat and Infectious Diseases

Heat is an important factor impacting the transmission of infectious diseases. Temperature influences the survival and reproductive rates of pathogens and disease vectors (Mordecai et al., 2019; Warrell & Gilles, 2002). Heat can also affect the geographic range of pathogens and vectors, as well as the seasonality of disease transmission (Morin & Comrie, 2013). In addition, temperature affects human behavior, which could modify the extent to which people are exposed to infectious diseases. For example, in hot weather people may swim in contaminated water to cool off, or may avoid sleeping under bed nets to keep cooler at night (von Seidlein et al., 2012).

Malaria, for example, infects nearly 220 million people and kills 435,000 per year worldwide, with over 90 of malaria deaths occurring in Africa (World Health Organization, 2019a). The disease is caused by parasites of the genus *Plasmodium* and is transmitted by *Anopheles* mosquitoes, both of which are sensitive to temperature. The life cycles of both parasite and mosquito speed up as temperatures increase from approximately 68 to 86 °F (20 to 30 °C), but survival is greatly hindered when temperatures exceed 95 °F (35 °C) (Warrell & Gilles, 2002). As global temperatures rise, malaria transmission is therefore expected to decrease in areas that are already hot— such as West Africa—and increase in cooler areas including the East African highlands (Caminade & Jones, 2016). Indeed, in recent decades, warmer years were associated with malaria cases extending to higher altitudes in highland areas of Ethiopia and Colombia (Siraj et al., 2014). Climate change is expected to continue this trend. In Africa, for example, the future risk of malaria is projected to increase in the East African highlands as well as in areas bordering the desert in West Africa (Endo et al., 2017).

Regions that are already near the optimal temperature range for malaria, such as parts of West Africa, may experience declines in malaria transmission with future climate change (Yamana et al., 2016). However, this decline is likely to be coupled with an increase in the transmission of other vector-borne diseases such as dengue and Zika viruses. These viruses, transmitted primarily by mosquitoes of the species *Aedes aegypti*, thrive in hotter weather than the parasites and mosquitoes that transmit malaria (Mordecai et al., 2019). Indeed, rising global temperatures are predicted to expand the geographic ranges of diseases such as dengue and Zika, with the greatest impacts in low- and middle-income countries (Ebi & Nealon, 2016).

Diarrheal diseases are also affected by temperature, as higher temperatures affect pathogen concentrations, rates of water consumption, and food spoilage (Levy et al., 2016). Diarrhea is the leading cause of undernutrition and second leading cause of death among children under five, killing approximately 525,000 children per year (World Health Organization, 2017). There is ample evidence that hotter temperatures in both low- and high-income countries are positively correlated with diarrheal diseases, particularly those caused by bacteria (Jiang et al., 2015; Levy et al., 2016). For example, a study using DHS data from 14 African countries discovered that a 1 °C increase in average monthly maximum temperature was associated with a one percentage point increase in the prevalence of diarrhea (Bandyopadhyay et al., 2012). Further, in Maryland, extreme heat exposure is linked to increases in *Salmonella* infections, the main cause of acute gastroenteritis worldwide (Jiang et al., 2015).

Future Research Directions and Data Needs

The evidence reviewed in this chapter indicates that extreme heat negatively impacts population health and mortality in several ways, and that vulnerability varies by geographic context as well as factors including age, gender, socioeconomic status, and occupation. However, in order to more comprehensively understand the heat-health relationship, environmental demographers should pursue new areas of research in addition to building upon the existing research discussed above. I highlight a few of these priority areas below.

One priority area for future research focuses on the relationship between heat and maternal morbidity and mortality. Existing research indicates that extreme heat is associated with higher rates of infant mortality, stillbirth, preterm birth, and low birthweight (Kuehn & McCormick, 2017). However, there are no studies that examine how heat affects health and mortality among mothers. Though much progress has been made in reducing maternal mortality, rates remain dangerously high in some countries. In 2017, global maternal mortality rates (MMR) averaged 211 per 100,000 births, ranging from 2 in high-income countries including Norway and Italy to 1150 in South Sudan (World Health Organization, 2019b). 295,000 women died from pregnancy- or childbirth-related complications in 2017 alone, with 86% of deaths occurring in Sub-Saharan Africa and Southern Asia (World Health Organization, 2019b). Further, there are stark disparities in maternal morbidity and mortality by race/ethnicity, geography, and education (Creanga et al., 2014; Curtin & Hoyert, 2017; Gao et al., 2017; Small et al., 2017). As heat waves increase in frequency, severity, and duration, it will be critical to understand the extent to which heat exposure creates added risks for mothers in addition to their newborns.

Moreover, while many studies rely on ambient temperature as an indicator of heat stress, factors such as humidity and solar radiation are important to consider as well (Epstein & Moran, 2006). To account for the direct health effects of heat stress more comprehensively, future research could incorporate heat indices such as the UTCI or DI. Collaborations between demographers and biometeorological researchers would encourage the appropriate use of heat indicators for particular health outcomes, resulting in better modeling of the direct linkages between heat, health, and mortality.

Additional research using demographic data would help further uncover links between heat and infectious disease transmission. For example, the DHS Program implements the Malaria Indicator Survey (MIS) in a number of African countries, and these data could be linked to climate data to explore linkages between heat and malaria transmission as well as prevention behaviors. Further, research utilizing DHS data as well as other population health data could help to uncover the mechanisms underlying how heat affects diarrheal disease transmission, determine which populations are most vulnerable, and identify the extent to which sanitation and access to clean water mitigate the links between heat and diarrheal diseases.

Moreover, there is very little qualitative research on the health impacts of extreme heat. One exception is Klinenberg's (1999) study of the 1995 Chicago heat wave, which combined field work and interviews with quantitative, geospatial, and historical data. This approach enabled Klinenberg to identify who was most vulnerable to heat wave mortality, and most importantly, to describe the individual- and neighborhood-level social factors that left certain Chicago residents at high risk of death. Additional qualitative data on how people respond to heat would help uncover some of the mechanisms underlying the heat-health relationship. Qualitative research would enable researchers to better understand which behaviors expose people to hot conditions, types of adaptation strategies people and communities use to avoid the negative health effects of heat, and the barriers faced by certain sub-populations which increase their vulnerability to heat.

Lastly, data availability issues present a challenge for furthering heat-health research, particularly in low- and middle-income countries. For example, data on birthweight and gestational length are often limited or inaccurate, particularly in rural areas of low-income countries where access to ultrasounds, prenatal care, and hospital births is less common. This constrains researchers' ability to understand how climatic conditions affect birth outcomes among some of the world's most vulnerable populations. In addition, inaccurate recall of birth dates affects the quality of child nutrition measures, particularly stunting, which relies on measuring the height of a child and indexing it to their exact age (Larsen et al., 2019). Misreported birth dates also lead to incorrect temporal linkage of children to shock events, such as heat waves.

One solution, albeit costly, would be to track a sample of women throughout the course of their pregnancy and follow their children through the first few years of life. This would provide accurate, high quality data that could be linked to climate data to better understand how heat affects birth outcomes as well as child nutrition. In addition, these data would enable researchers to examine fine-scale relationships between heat and health including identifying whether there are critical periods and durations of exposure as well as heat and humidity thresholds.

As discussed in this chapter, extreme heat is a critical threat to the health and well-being of populations across the world, from infants to the elderly, throughout rural and urban areas, and from low- to high-income countries. As climate change continues to increase the occurrence of record-breaking temperatures and expand the severity of heat waves, environmental demographers are poised to make key contributions to help better understand how extreme heat impacts population health and mortality. A demographic perspective—one that explicitly considers the dynamics of population processes and seeks to understand how these processes are impacted by external stressors such as heat—is critical to understand the population-level effects of extreme heat, to identify which populations are most vulnerable to heat waves, and to inform adaptation measures that mitigate the effects of heat in a warming world.

For example, policies aimed at reducing exposure to heat waves could be focused at individual-level solutions, such as offering subsidized air conditioners for households to cool their homes during a heat wave. This would help households that have the resources to purchase an air conditioner and pay for the increased energy usage, but would likely not reach the poorest households. In contrast, a population-level policy might focus on developing heat-resilient infrastructure in the most vulnerable neighborhoods such as opening cooling centers, planting trees, and designing buildings and roads that radiate less heat into nearby surroundings. Population-level thinking is critical in order to reach the greatest number of people and promote social justice in the face of a warming world.

References

Akbari, H., Cartalis, C., Kolokotsa, D., et al. (2015). Local climate change and urban heat island mitigation techniques – The state of the art. *Journal of Civil Engineering and Management, 22,* 1–16. https://doi.org/10.3846/13923730.2015.1111934

Alderman, H., Hoddinott, J., & Kinsey, B. (2006). Long term consequences of early childhood malnutrition. *Oxford Economic Papers, 58,* 450–474. https://doi.org/10.1093/oep/gpl008

Andalón, M., Azevedo, J. P., Rodríguez-Castelán, C., et al. (2016). Weather shocks and health at birth in Colombia. *World Development, 82,* 69–82. https://doi.org/10.1016/j.worlddev.2016.01.015

Anderson, G. B., & Bell, M. L. (2011). Heat waves in the United States: Mortality risk during heat waves and effect modification by heat wave characteristics in 43 U.S. communities. *Environmental Health Perspectives, 119,* 210–218. https://doi.org/10.1289/ehp.1002313

Auger, N., Fraser, W. D., Smargiassi, A., & Kosatsky, T. (2015). Ambient heat and sudden infant death: A case-crossover study spanning 30 years in Montreal, Canada. *Environmental Health Perspectives, 123,* 712–716. https://doi.org/10.1289/ehp.1307960

Azongo, D. K., Awine, T., Wak, G., et al. (2012). A time series analysis of weather variability and all-cause mortality in the Kasena-Nankana districts of northern Ghana, 1995-2010. *Global Health Action, 5,* 14–22. https://doi.org/10.3402/gha.v5i0.19073

Bakhtsiyarava, M., Grace, K., & Nawrotzki, R. J. (2018). Climate, birth weight, and agricultural livelihoods in Kenya and Mali. *American Journal of Public Health,* e1–e7. https://doi.org/10.2105/AJPH.2017.304128

Bandyopadhyay, S., Kanji, S., & Wang, L. (2012). The impact of rainfall and temperature variation on diarrheal prevalence in sub-Saharan Africa. *Applied Geography, 33,* 63–72. https://doi.org/10.1016/J.APGEOG.2011.07.017

Barreca, A., Clay, K., Deschênes, O., et al. (2015). Convergence in adaptation to climate change: Evidence from high temperatures and mortality, 1900–2004. *The American Economic Review, 105,* 247–251. https://doi.org/10.1257/aer.p20151028

Barreca, A., Clay, K., Deschenes, O., et al. (2016). Adapting to climate change: The remarkable decline in the US temperature-mortality relationship over the twentieth century. *Journal of Political Economy, 124,* 105–159. https://doi.org/10.1086/684582

Barreca, A., & Schaller, J. (2020). The impact of high ambient temperatures on delivery timing and gestational lengths. *Nature Climate Change, 10,* 77–82. https://doi.org/10.1038/s41558-019-0632-4

Basu, R., Malig, B., & Ostro, B. (2010). High ambient temperature and the risk of preterm delivery. *American Journal of Epidemiology, 172*, 1108–1117. https://doi.org/10.1093/aje/kwq170

Basu, R., Rau, R., Pearson, D., & Malig, B. (2018). Temperature and term low birth weight in California. *American Journal of Epidemiology, 187*, 2306–2314. https://doi.org/10.1093/aje/kwy116

Behrman, J. R., & Rosenzweig, M. R. (2004). Returns to birthweight. *The Review of Economics and Statistics, 86*, 586–601. https://doi.org/10.1162/003465304323031139

Benmarhnia, T., Deguen, S., Kaufman, J. S., & Smargiassi, A. (2015). Vulnerability to heat-related mortality: A systematic review, meta-analysis, and meta-regression analysis. *Epidemiology, 26*, 781–793. https://doi.org/10.1097/EDE.0000000000000375

Berko, J., Ingram, D.D., & Saha, S. (2014). Deaths attributed to heat, cold, and other weather events in the United States, 2006–2010.

Black, S. E., Devereux, P. J., & Salvanes, K. G. (2007). From the cradle to the labor market? The effect of birth weight on adult outcomes. *Quarterly Journal of Economics, 122*, 409–439. https://doi.org/10.1162/qjec.122.1.409

Blazejczyk, K., Epstein, Y., Jendritzky, G., et al. (2012). Comparison of UTCI to selected thermal indices. *International Journal of Biometeorology, 56*, 515–535. https://doi.org/10.1007/s00484-011-0453-2

Bloom, D. E., Canning, D., & Lubet, A. (2015). Global population aging: Facts, challenges, solutions & perspectives. *Daedalus, 144*, 80–92. https://doi.org/10.1162/DAED_a_00332

Bobb, J. F., Peng, R. D., Bell, M. L., & Dominici, F. (2014). Heat-related mortality and adaptation to heat in the United States. *Environmental Health Perspectives, 122*, 811–816. https://doi.org/10.1289/ehp.1307392

Caminade, C., & Jones, A. E. (2016). Malaria in a warmer West Africa. *Nature Climate Change, 6*, 984–985. https://doi.org/10.1038/nclimate3095

Carolan-Olah, M., & Frankowska, D. (2014). High environmental temperature and preterm birth: A review of the evidence. *Midwifery, 30*, 50–59. https://doi.org/10.1016/j.midw.2013.01.011

CDC. (2017). *Birthweight and gestation*. https://www.cdc.gov/nchs/fastats/birthweight.htm. Accessed 1 July 2019.

Cho, H. (2020). Ambient temperature, birth rate, and birth outcomes: Evidence from South Korea. *Population and Environment, 41*, 330–346. https://doi.org/10.1007/s11111-019-00333-6

Chotard, S., Mason, J. B., Oliphant, N. P., et al. (2011). Fluctuations in wasting in vulnerable child populations in the greater horn of Africa. *Food and Nutrition Bulletin, 32*, S219–S233. https://doi.org/10.1177/15648265100313S302

Creanga, A. A., Bateman, B. T., Kuklina, E. V., & Callaghan, W. M. (2014). Racial and ethnic disparities in severe maternal morbidity: A multistate analysis, 2008-2010. *American Journal of Obstetrics and Gynecology, 210*, 435.e1–435.e8. https://doi.org/10.1016/j.ajog.2013.11.039

Curtin, S. C., & Hoyert, D. L. (2017). *Maternal morbidity and mortality: Exploring racial/ethnic differences using new data from birth and death certificates* (pp. 95–113). Springer.

Davenport, F., Dorélien, A., & Grace, K. (2020). Investigating the linkages between pregnancy outcomes and climate in sub-Saharan Africa. *Population and Environment*. https://doi.org/10.1007/s11111-020-00342-w

del Ninno, C., & Lundberg, M. (2005). Treading water. The long-term impact of the 1998 flood on nutrition in Bangladesh. *Economics and Human Biology, 3*, 67–96. https://doi.org/10.1016/j.ehb.2004.12.002

Deschênes, O. (2018). Temperature variability and mortality: Evidence from 16 Asian countries. *Asian Development Review, 35*, 1–30. https://doi.org/10.1162/adev_a_00112

Deschênes, O., & Greenstone, M. (2011). Climate change, mortality, and adaptation: Evidence from annual fluctuations in weather in the US. *American Economic Journal: Applied Economics, 3*, 152–185. https://doi.org/10.1257/app.3.4.152

Deschênes, O., Greenstone, M., & Guryan, J. (2009). Climate change and birth weight. *American Economic Review: Papers and Proceedings, 99*, 211–217. https://doi.org/10.1257/aer.99.2.211

Dewey, K. G., & Begum, K. (2011). Long-term consequences of stunting in early life. *Maternal & Child Nutrition, 7*, 5–18. https://doi.org/10.1111/j.1740-8709.2011.00349.x

Di Liberto, T. (2015). *India heat wave kills thousands*. https://www.climate.gov/news-features/event-tracker/india-heat-wave-kills-thousands. Accessed 26 June 2019.

Diboulo, E., Sié, A., Rocklöv, J., et al. (2012). Weather and mortality: A 10 year retrospective analysis of the Nouna health and demographic surveillance system, Burkina Faso. *Global Health Action, 5*, 6–13. https://doi.org/10.3402/gha.v5i0.19078

Dosio, A., Mentaschi, L., Fischer, E. M., & Wyser, K. (2018). Extreme heat waves under 1.5 °c and 2 °c global warming. *Environmental Research Letters, 13*. https://doi.org/10.1088/1748-9326/aab827

Ebi, K. L., & Nealon, J. (2016). Dengue in a changing climate. *Environmental Research, 151*, 115–123. https://doi.org/10.1016/j.envres.2016.07.026

Egondi, T., Kyobutungi, C., Kovats, S., et al. (2012). Time-series analysis of weather and mortality patterns in Nairobi's informal settlements. *Global Health Action, 5*, 23–32. https://doi.org/10.3402/gha.v5i0.19065

Endo, N., Yamana, T., & Eltahir, E. A. B. (2017). Impact of climate change on malaria in Africa: A combined modelling and observational study. *Lancet, 389*, S7. https://doi.org/10.1016/s0140-6736(17)31119-4

Epstein, Y., & Moran, D. S. (2006). Thermal comfort and the heat stress indices. *Industrial Health, 44*, 388–398.

FAO, IFAD, UNICEF, et al., (2018), *The state of food security and nutrition in the world: Building climate resilience for food security and nutrition*. Rome.

Gao, Y., Zhou, H., Singh, N. S., et al. (2017). Progress and challenges in maternal health in western China: A countdown to 2015 national case study. *Lancet Global Health, 5*, e523–e536. https://doi.org/10.1016/S2214-109X(17)30100-6

Gasparrini, A., Guo, Y., Hashizume, M., et al. (2015). Temporal variation in heat–mortality associations: A multicountry study. *Environmental Health Perspectives, 123*, 1200–1207. https://doi.org/10.1289/ehp.1409070

Geruso M, Spears D (2018) Heat, humidity, and infant mortality in the developing world. .

Glewwe, P., & King, E. M. (2001). The impact of early childhood nutritional status on cognitive development: Does the timing of malnutrition matter? *World Bank Economic Review, 15*, 81–113. https://doi.org/10.1093/wber/15.1.81

Grace, K., Davenport, F., Funk, C., & Lerner, A. M. (2012). Child malnutrition and climate in sub-Saharan Africa: An analysis of recent trends in Kenya. *Applied Geography, 35*, 405–413. https://doi.org/10.1016/j.apgeog.2012.06.017

Grace, K., Davenport, F., Hanson, H., et al. (2015). Linking climate change and health outcomes: Examining the relationship between temperature, precipitation and birth weight in Africa. *Global Environmental Change, 35*, 125–137. https://doi.org/10.1016/j.gloenvcha.2015.06.010

Guo, Y., Gasparrini, A., Armstrong, B. G., et al. (2017). Heat wave and mortality: A multicountry, multicommunity study. *Environmental Health Perspectives, 125*. https://doi.org/10.1289/EHP1026

Harrington, L. J., Frame, D. J., Fischer, E. M., et al. (2016). Poorest countries experience earlier anthropogenic emergence of daily temperature extremes. *Environmental Research Letters, 11*, 055007. https://doi.org/10.1088/1748-9326/11/5/055007

Heal, G., & Park, J. (2016). Reflections-temperature stress and the direct impact of climate change: A review of an emerging literature. *Review of Environmental Economics and Policy, 10*, 347–362. https://doi.org/10.1093/reep/rew007

Hoddinott, J., Behrman, J. R., Maluccio, J. A., et al. (2013). Adult consequences of growth failure in early childhood. *The American Journal of Clinical Nutrition, 98*, 1170–1178. https://doi.org/10.3945/ajcn.113.064584

Hoddinott, J., & Kinsey, B. (2001). Child growth in the time of drought. *Oxford Bulletin of Economics and Statistics, 63*, 409–436. https://doi.org/10.1111/1468-0084.t01-1-00227

Hoddinott, J., Maluccio, J. A., Behrman, J. R., et al. (2008). Effect of a nutrition intervention during early childhood on economic productivity in Guatemalan adults. *Lancet, 371*, 411–416. https://doi.org/10.1016/S0140-6736(08)60205-6

Hoegh-Guldberg, O., Jacob, D., Taylor, M., et al., (2018). Impacts of 1.5° C global warming on natural and human systems. In: Delmotte, V.M.-, Zhai, P., Pörtner, H.O., et al., (eds) *Global Warming of 1.5°C. An IPCC Special Report on the Impacts of Global Warming of 1.5°C above Pre-Industrial Levels and Related Global Greenhouse Gas Emission Pathways*, in the Context of Strengthening the Global Response to the Threat of Climate Change,. In Press.

Hondula, D. M., Balling, R. C., Vanos, J. K., & Georgescu, M. (2015). Rising temperatures, human health, and the role of adaptation. *Current Climate Change Reports, 1*, 144–154.

Hu, K., Guo, Y., Hochrainer-Stigler, S., et al. (2019). Evidence for urban–rural disparity in temperature–mortality relationships in Zhejiang Province, China. *Environmental Health Perspectives, 127*, 037001. https://doi.org/10.1289/EHP3556

Hu, Z., & Li, T. (2019). Too hot to handle: The effects of high temperatures during pregnancy on adult welfare outcomes. *Journal of Environmental Economics and Management, 94*, 236–253. https://doi.org/10.1016/j.jeem.2019.01.006

Im, E.-S., Pal, J. S., & Eltahir, E. A. B. (2017). Deadly heat waves projected in the densely populated agricultural regions of South Asia. *Science Advances, 3*(e1603322), 2.

Ingole, V., Rocklöv, J., Juvekar, S., & Schumann, B. (2015). Impact of heat and cold on total and cause-specific mortality in Vadu HDSS—A rural setting in Western India. *International Journal of Environmental Research and Public Health, 12*, 15298–15308. https://doi.org/10.3390/ijerph121214980

Jiang, C., Shaw, K. S., Upperman, C. R., et al. (2015). Climate change, extreme events and increased risk of salmonellosis in Maryland, USA: Evidence for coastal vulnerability. *Environment International, 83*, 58–62. https://doi.org/10.1016/j.envint.2015.06.006

Kenny, G. P., Yardley, J., Brown, C., et al. (2010). Heat stress in older individuals and patients with common chronic diseases. *CMAJ, 182*, 1053–1060.

Kim, H. H. (1992). Urban heat island. *International Journal of Remote Sensing, 13*, 2319–2336. https://doi.org/10.1080/01431169208904271

Klinenberg, E. (1999). Denaturalizing disaster: A social autopsy of the 1995 Chicago heat wave. *Theory and Society, 28*, 239–295. https://doi.org/10.1023/A:1006995507723

Kuehn, L., & McCormick, S. (2017). Heat exposure and maternal health in the face of climate change. *International Journal of Environmental Research and Public Health, 14*, 853.

Larsen, A. F., Headey, D., & Masters, W. A. (2019). Misreporting month of birth: Diagnosis and implications for research on nutrition and early childhood in developing countries. *Demography, 56*, 707–728. https://doi.org/10.1007/s13524-018-0753-9

Lesk, C., Rowhani, P., & Ramankutty, N. (2016). Influence of extreme weather disasters on global crop production. *Nature, 529*, 84–87. https://doi.org/10.1038/nature16467

Levy, K., Woster, A. P., Goldstein, R. S., & Carlton, E. J. (2016). Untangling the impacts of climate change on waterborne diseases: A systematic review of relationships between diarrheal diseases and temperature, rainfall, flooding, and drought. *Environmental Science & Technology, 50*, 4905–4922.

Lindstrom, K., Winbladh, B., Haglund, B., & Hjern, A. (2007). Preterm infants as young adults: A Swedish National Cohort Study. *Pediatrics, 120*, 70–77. https://doi.org/10.1542/peds.2006-3260

Maluccio, J. A., Hoddinott, J., Behrman, J. R., et al. (2009). The impact of improving nutrition during early childhood on education among Guatemalan adults. *The Econometrics Journal, 119*, 734–763. https://doi.org/10.1111/j.1468-0297.2009.02220.x

Masters, J. (2020). Death Valley, California, may have recorded the hottest temperature in world history. In: *Yale Climate Connections*https://yaleclimateconnections.org/2020/08/death-valley-california-may-have-recorded-hottest-temp-in-world-history/. Accessed 10 Sept 2020.

Molina, O., & Saldarriaga, V. (2017). The perils of climate change: In utero exposure to temperature variability and birth outcomes in the Andean region. *Economics and Human Biology, 24*, 111–124. https://doi.org/10.1016/j.ehb.2016.11.009

Mora, C., Counsell, C. W. W., Bielecki, C. R., & Louis, L. V. (2017). Twenty-seven ways a heat wave can kill you: Deadly heat in the era of climate change. *Circulation. Cardiovascular Quality and Outcomes, 10*, e004233. https://doi.org/10.1161/CIRCOUTCOMES.117.004233

Mora, C., Dousset, B., Caldwell, I. R., et al. (2017). Global risk of deadly heat. *Nature Climate Change, 7*, 501–506. https://doi.org/10.1038/nclimate3322

Mordecai, E. A., Caldwell, J. M., Grossman, M. K., et al. (2019). Thermal biology of mosquito-borne disease. *Ecology Letters, 22*, 1690–1708.

Morin, C. W., & Comrie, A. C. (2013). Regional and seasonal response of a West Nile virus vector to climate change. *Proceedings of the National Academy of Sciences of the United States of America, 110*, 15620–15625. https://doi.org/10.1073/pnas.1307135110

Mueller, V., & Gray, C. (2018). Heat and adult health in China. *Population and Environment, 40*, 1–26. https://doi.org/10.1007/s11111-018-0294-6

Muhuri, P. K. (1996). Estimating seasonality effects on child mortality in Matlab, Bangladesh. *Demography, 33*, 98–110. https://doi.org/10.2307/2061716

Ngo, N. S., & Horton, R. M. (2016). Climate change and fetal health: The impacts of exposure to extreme temperatures in New York City. *Environmental Research, 144*, 158–164. https://doi.org/10.1016/J.ENVRES.2015.11.016

Oleson, K. W., Anderson, G. B., Jones, B., et al. (2018). Avoided climate impacts of urban and rural heat and cold waves over the U.S. using large climate model ensembles for RCP8.5 and RCP4.5. *Climatic Change, 146*, 377–392. https://doi.org/10.1007/s10584-015-1504-1

Pal, J. S., & Eltahir, E. A. B. (2016). Future temperature in Southwest Asia projected to exceed a threshold for human adaptability. *Nature Climate Change, 6*, 197–200. https://doi.org/10.1038/nclimate2833

Perkins, S. E., & Alexander, L. V. (2013). On the measurement of heat waves. *Journal of Climate, 26*, 4500–4517. https://doi.org/10.1175/JCLI-D-12-00383.1

Perkins, S. E., Alexander, L. V., & Nairn, J. R. (2012). Increasing frequency, intensity and duration of observed global heatwaves and warm spells. *Geophysical Research Letters, 39*. https://doi.org/10.1029/2012GL053361

Perkins-Kirkpatrick, S. E., & Gibson, P. B. (2017). Changes in regional heatwave characteristics as a function of increasing global temperature. *Scientific Reports, 7*, 12256. https://doi.org/10.1038/s41598-017-12520-2

Porter, J. R., Xie, L., Challinor, A. J., et al. (2014). Food security and food production systems. In C. B. Field, V. R. Barros, D. J. Dokken, K. J. Mach, M. D. Mastrandrea, T. E. Bilir, M. Chatterjee, K. L. Ebi, Y. O. Estrada, R. C. Genova, B. Girma, E. S. Kissel, A. N. Levy, S. MacCracken, P. R. Mastrandrea, & LLW (Eds.), *Climate change 2014: Impacts, adaptation, and vulnerability. Part A: Global and sectoral aspects. Contribution of working group II to the fifth assessment report of the intergovernmental panel on climate change* (pp. 485–533). Cambridge University Press.

Poumadère, M., Mays, C., Le Mer, S., & Blong, R. (2005). The 2003 heat wave in France: Dangerous climate change Here and now. *Risk Analysis, 25*, 1483–1494. https://doi.org/10.1111/j.1539-6924.2005.00694.x

Power, S. B., & Delage, F. P. D. (2019). Setting and smashing extreme temperature records over the coming century. *Nature Climate Change, 1*. https://doi.org/10.1038/s41558-019-0498-5

Randell, H., & Gray, C. (2019). Climate change and educational attainment in the global tropics. *Proceedings of the National Academy of Sciences, 116*, 8840–8845. https://doi.org/10.1073/pnas.1817480116

Rathbun, B. (2019). *90-degree heat stifles Anchorage for first time in its history as sweltering heat wave grips Alaska.* In: AccuWeather. https://www.accuweather.com/en/weather-news/90-degree-heat-stifles-anchorage-for-first-time-in-its-history-as-sweltering-heat-wave-grips-alaska/70008741. Accessed 7 July 2019.

Ringer, S. A. (2013). Core concepts: Thermoregulation in the newborn, part II: Prevention of aberrant body temperature. *NeoReviews, 14*, e221–e226.

Robine, J. M., Cheung, S. L. K., Le Roy, S., et al. (2008). Death toll exceeded 70,000 in Europe during the summer of 2003. *Comptes Rendus Biologies, 331*, 171–178. https://doi.org/10.1016/j.crvi.2007.12.001

Rodriguez-Llanes, J. M., Ranjan-Dash, S., Degomme, O., et al. (2011). Child malnutrition and recurrent flooding in rural eastern India: A community-based survey. *BMJ Open, 1*, e000109–e000109. https://doi.org/10.1136/bmjopen-2011-000109

Rogelj, J., den Elzen, M., Höhne, N., et al. (2016). Paris agreement climate proposals need a boost to keep warming well below 2 °C. *Nature, 534*, 631–639. https://doi.org/10.1038/nature18307

Samenow, J. (2018). *Red-hot planet: All-time heat records have been set all over the world during the past week*. In: Washington Post. https://www.washingtonpost.com/news/capital-weather-gang/wp/2018/07/03/hot-planet-all-time-heat-records-have-been-set-all-over-the-world-in-last-week/?utm_term=.ad4f7dbe3958. Accessed 24 June 2019.

Semenza, J. C., Rubin, C. H., Falter, K. H., et al. (1996). Heat-related deaths during the July 1995 heat wave in Chicago. *The New England Journal of Medicine, 335*, 84–90. https://doi.org/10.1056/NEJM199607113350203

Sherwood, S. C. (2018). How important is humidity in heat stress? *Journal of Geophysical Research – Atmospheres, 123*, 11,808–11,810.

Siebert, S., & Ewert, F. (2014). Future crop production threatened by extreme heat. *Environmental Research Letters, 9*, 041001. https://doi.org/10.1088/1748-9326/9/4/041001

Siraj, A. S., Santos-Vega, M., Bouma, M. J., et al. (2014). Altitudinal changes in malaria incidence in highlands of Ethiopia and Colombia. *Science, 343*(80), 1154–1158. https://doi.org/10.1126/science.1244325

Skoufias, E., & Vinha, K. (2012). Climate variability and child height in rural Mexico. *Economics and Human Biology, 10*, 54–73. https://doi.org/10.1016/j.ehb.2011.06.001

Small, M. J., Allen, T. K., & Brown, H. L. (2017). Global disparities in maternal morbidity and mortality. *Seminars in Perinatology, 41*, 318–322. https://doi.org/10.1053/J.SEMPERI.2017.04.009

Stein, A. D., Barros, F. C., Bhargava, S. K., et al. (2013). Birth status, chiled growth, and adult outcomes in low- and middle-income countries. *The Journal of Pediatrics, 163*, 1740–1746.e4. https://doi.org/10.1016/j.jpeds.2013.08.012

Stocker, T. F., Qin, D., Plattner, G.-K., et al. (2013). Technical summary. In T. F. Stocker, D. Qin, G.-K. Plattner, M. Tignor, S. K. Allen, J. Boschung, A. Nauels, Y. X. VB, & PMM (Eds.), *Climate change 2013: The physical science basis. Contribution of working group i to the Fifth assessment report of the intergovernmental panel on climate change*. Cambridge.

Sun, S., Spangler, K. R., Weinberger, K. R., et al. (2019). Ambient temperature and markers of fetal growth: A retrospective observational study of 29 million U.S. singleton births. *Environmental Health Perspectives, 127*, 067005. https://doi.org/10.1289/EHP4648

Thiede, B.C., & Strube, J. (2019). Climate Variability and Nutritional Security in Early Childhood: Findings from Sub-Saharan Africa. SocArXiv.

Thorsson, S., Rocklöv, J., Konarska, J., et al. (2014). Mean radiant temperature – a predictor of heat related mortality. *Urban Climate, 10*, 332–345. https://doi.org/10.1016/j.uclim.2014.01.004

UCAR Center for Science Education. (2011) .*Urban Heat Islands*. https://scied.ucar.edu/longcontent/urban-heat-islands. Accessed 25 June 2019.

Victora, C. G., Adair, L., Fall, C., et al. (2008). Maternal and child undernutrition: Consequences for adult health and human capital. *Lancet, 371*, 340–357. https://doi.org/10.1016/S0140-6736(07)61692-4

Vives, R., Kim, V., & Parvini, S. (2018). Southern California sets all-time heat records amid broiling conditions. In: *Los Angeles Times*. https://www.latimes.com/local/lanow/la-me-record-heat-20180706-story.html. Accessed 24 June 2019.

von Seidlein, L., Ikonomidis, K., Bruun, R., et al. (2012). Airflow attenuation and bed net utilization: Observations from Africa and Asia. *Malaria Journal, 11*, 200. https://doi.org/10.1186/1475-2875-11-200

Wanyonyi, S., Mariara, C., Vinayak, S., & Stones, W. (2017). Opportunities and challenges in realizing universal access to obstetric ultrasound in sub-Saharan Africa. *Ultrasound International Open, 3*, E52–E59. https://doi.org/10.1055/s-0043-103948

Warrell, D. A., & Gilles, H. M. (2002). *Essential malariology*. CRC Press.

Wehner, M., Stone, D., Krishnan, H., et al. (2016). The deadly combination of heat and humidity in India an d Pakistan in summer 2015 [in "explaining extremes of 2015 from a climate perspective"]. *Bulletin of the American Meteorological Society, 97*(supple), S81–S86. https://doi.org/10.1175/BAMS-D-16-0149

World Health Organization (2014) Global nutrition targets 2025 low birth weight policy brief. .

World Health Organization. (2017). Diarrhoeal disease. https://www.who.int/news-room/fact-sheets/detail/diarrhoeal-disease. Accessed 8 July 2019.

World Health Organization. (2018). Information and public health advice: heat and health. In: WHO. https://www.who.int/globalchange/publications/heat-and-health/en/. Accessed 24 June 2019.

World Health Organization. (2019a). Malaria. https://www.who.int/en/news-room/fact-sheets/detail/malaria. Accessed 8 July 2019.

World Health Organization. (2019b). *Trends in maternal mortality 2000 to 2017: Estimates by WHO, UNICEF, UNFPA, World Bank Group and the United Nations population division*. World Health Organization.

World Meteorological Organization. (2019). *WMO statement on the state of the global climate in 2018*. World Meteorological Organization (WMO).

World Meteorological Organization. (2020). WMO confirms 2019 as second hottest year on record | World Meteorological Organization. In: WMO Press Release Number 01/15/2020. https://public.wmo.int/en/media/press-release/wmo-confirms-2019-second-hottest-year-record. Accessed 10 Sept 2020.

Xu, Y., Wheeler, S. A., & Zuo, A. (2018). Will boys' mental health fare worse under a hotter climate in Australia? *Population and Environment, 40*, 158–181. https://doi.org/10.1007/s11111-018-0306-6

Yamana, T. K., Bomblies, A., & Eltahir, E. A. B. (2016). Climate change unlikely to increase malaria burden in West Africa. *Nature Climate Change, 6*, 1009–1013. https://doi.org/10.1038/nclimate3085

Yang, T. C., & Jensen, L. (2017). Climatic conditions and human mortality: Spatial and regional variation in the United States. *Population and Environment, 38*, 261–285. https://doi.org/10.1007/s11111-016-0262-y

Zhang, Y., Yu, C., & Wang, L. (2017). Temperature exposure during pregnancy and birth outcomes: An updated systematic review of epidemiological evidence. *Environmental Pollution, 225*, 700–712.

Zhao, C., Liu, B., Piao, S., et al. (2017). Temperature increase reduces global yields of major crops in four independent estimates. *Proceedings of the National Academy of Sciences, 201701762*. https://doi.org/10.1073/pnas.1701762114

Land Use Change and Health

15

William K. Pan and Gabrielle Bonnet

Abstract

In this chapter, we introduce a conceptual framework that identifies typologies of land use/land cover change and population dynamics connected with human health. We draw upon the preponderance of empirical and theoretical work related to vector-borne and zoonotic diseases as well as land change science to discuss how population-land typologies influence non-zoonotic communicable diseases, non-communicable diseases, and accidents and injuries. Given that population-land typologies have generally not been conceptualized in these health contexts, we present an abundance of research and case studies to demonstrate the linkages. We present these linkages as functions of the proximate determinants of land use change: agricultural expansion and intensification, infrastructure expansion, and

W.K. Pan (✉)
Nicholas School of Environment and Duke Global Health Institute, Duke University, Durham, NC, USA
e-mail: William.pan@duke.edu

G. Bonnet
London School of Hygiene & Tropical Medicine, London, UK
e-mail: gabrielle.bonnet@duke.edu

resource extraction. We conclude with a call for increased inter- and trans-disciplinary collaboration to improve understanding of complex environmental health challenges and develop stronger policies for sustainable development.

Introduction

The relationship between land use/land cover change (LULC) and human health is complex. Although this relationship has been described across many disciplines, the preponderance of research has focused on highlighting and unpacking coupled human-land use dynamics on *infectious diseases*, particularly vector-borne and zoonotic diseases (VBZD). What has not been thoroughly vetted is whether the coupled land-climate and population-land dynamics influencing animal and arthropod diseases similarly influence pathways connecting LULC to non-communicable diseases, accidents/unintentional injuries, or non-zoonotic infectious diseases. Current conceptual and empirical models of proximate and underlying determinants of human health outline a chain of causal events in which environmental drivers

© Springer Nature Switzerland AG 2022
L. M. Hunter et al. (eds.), *International Handbook of Population and Environment*, International
Handbooks of Population 10, https://doi.org/10.1007/978-3-030-76433-3_15

and risks play a key role in shaping disease patterns; however, population-environment factors related to health are often conceived as occurring within a black box that influences health throughout an individual's life (WHO, 2002; Lammle et al., 2013). In contrast, scientists specializing in infectious disease, land science, demography, public health, and other disciplines have actively debated the conceptualization of VBZD risk. For example, in 2002, a working group sponsored by the International Society for Ecosystem Health was convened to evaluate how land use change contributes to infectious disease emergence, focusing on changes to the biophysical environment, human and animal migration, urbanization and the land-water nexus, recommending a novel health policy framework that includes upstream linked land use and human behavioral determinants (Patz et al., 2004). Lambin and colleagues described principles of "landscape epidemiology" in which land use and habitat changes are conceptualized to occur at multiple scales (population, natural community, landscape, and macro-levels) that are associated with the spatial distribution of pathogens (Lambin et al., 2010). Along a similar vein, Parratt and colleagues (Parratt et al., 2016) describe disease transmission as a function of local landscape heterogeneity, such as land fragmentation and biodiversity, characterizing transmission as dependent on host and pathogen landscape dynamics, which is similar to the concept of pathogeography that describes the ecological constraints associated with the spatial limits, reproduction potential, and co-occurrence of infectious diseases (Murray et al., 2018). Other contributions include impacts of global environmental change and biodiversity on infectious diseases, which broadly refers to coupled land-climate interactions, LULC, urbanization and natural resource consumption as drivers of zoonotic disease transmission (Gage et al., 2008; Patz et al., 2008; Pan et al., 2014; Sandifer et al., 2015; Nava et al., 2017). These debates have helped stimulate research to begin untangling demographic, environmental, economic and other factors influencing infectious diseases, as well as address ontological differences in land change science and epidemiology that hamper synthesis of methods and data to properly quantify LULC contributions to human health (Messina & Pan, 2013). However, the question remains whether these relationships are generalizable to other health outcomes and how population characteristics, such as fertility, migration, and population structure, *ceteris paribus*, mediate the land-health dynamic.

The goal of this chapter is to propose a generalizable conceptual framework connecting LULC, population dynamics, and human health, drawing upon the wealth of research on VBZD. We discuss specific forms of LULC—agricultural expansion and intensification, urbanization and road development, and extractive industries—and how context-specific population and economic drivers of change can lead to multiple adverse health outcomes. Case studies are presented to highlight proposed causal pathways in both developed and developing country contexts. Due to the multiple routes through which LULC modifies human health, we do not intend to describe all the complexities involved; rather, we hope this chapter inspires policy innovations and critical thinking of the connectivity between LULC, population health, and human well-being beyond infectious disease.

Conceptual Model of Land Use Change and Health

VBZD models that include LULC as a factor in transmission recognize spatial and temporal scales, from macro to micro, as integral to explain variations in the evolution of disease. At macro scales, climate is a global forcing that predicts both human and vector population risks, such as the inter-annual variability of the El Nino-Southern Oscillation (ENSO); however, at a human population level, climate is coupled with landscape features that jointly mediate seasonal variability of both animal/vector distributions and human social and economic behavior. Although large-scale climate patterns may increase or decrease precipitation across a region over a long (or short) period of time, the coupling of land and climate results in areas with different vegetation cover experiencing different levels of problems such as excess floods or drought. These smaller-scale factors mediate how animal habitat and human decision-making evolve (i.e., where and when to farm, build roads, conduct business, etc.).

While the multiscale structure is critical to understanding how factors interact and evolve, the underlying causal relationships operating across scales and producing changes in human health are between land and human population. As described in several disease ecology approaches (see introduction), population and land interactions at the landscape and human scales are perhaps the most important factors to explain how and why specific disease outcomes occur. This insight stems from work in land use science that identifies the proximate drivers of land use change as infrastructure expansion, agriculture, and extractive industries (Geist & Lambin, 2002), which are mediated by underlying causes, namely, demographic, economic, technological, institutional, and cultural factors. Nuances of the relationship between proximate and underlying drivers of LULC have been studied and debated extensively by demographers and human geographers since the time of Malthus during the early 1800s (Boserup, 1981; Bilsborrow, 1987; Carr et al., 2006).

We begin the development of our conceptual land-health model with the understanding that proximal drivers of LULC are intimately and uniquely coupled with underlying population factors. For example, societal needs for expanded infrastructure, which includes transportation, markets, housing, and public services, are directly tied to population size (i.e., demand for basic services), age structure (i.e., young population demands educational infrastructure vs. working age demands inputs to private sector), stage of demographic transition (i.e., infrastructure needed to leverage the demographic dividend), and density. While population does not always lead to infrastructure investments—e.g., many African countries suffer from severe infrastructure deficits regardless of population growth as described by (Yepes et al., 2008)—infrastructure shortages often result from lack of technology or financial constraints, which are also important factors in mediating population-land relationships. For example, the choice between large scale agricultural production versus smallholder cultivation and cattle-ranching is likewise influenced by population size and structure, but access to technology plays a key role in a population's ability to clear land, cultivate, and enhance yields (i.e., through chemical inputs).

In addition, these emerging population-land relationships are not just unidirectional: infrastructure or agricultural expansion and resource extraction all influence seasonal and permanent migration flows, particularly as a function of local labor demand. The link between population-land relationships and health can become complex due to underlying causes of health risk. For example, agricultural expansion and oil/gas exploration cause deforestation in the tropics that fragments habitats and changes ecosystem services, both of which have been associated with vector-borne and zoonotic diseases (e.g. (Weigle et al., 1993; Coluzzi, 1994; Takken et al., 2003; Barbieri et al., 2005; Vittor et al., 2009; Castellanos et al., 2016). However, each population-land use interaction has a unique mix of underlying factors (i.e., technology, culture, etc.) that may map onto similar or mutually exclusive human health outcomes. Agricultural expansion, whether linked to smallholder or plantation farming, not only alters vector habitats and infectious disease risk, but widespread use of chemical fertilizer and pesticide inputs (an economic and social choice) has also been associated with endocrine disruption (McKinlay et al., 2008), kidney impairment (Khan et al., 2008; Patil et al. 2009), and occupationally-related cancers (Flower et al., 2004; Kim et al., 2017). In comparison, oil/gas exploration shares several non-infectious disease impacts similar to pesticide exposures (Colborn et al., 2011), but also includes impacts that are psychosocial (Adgate et al., 2014), developmental (McKenzie et al., 2014; Stacy et al., 2015), disruptive of overall indigenous health (Martinez et al., 2007; O'Callaghan-Gordo et al., 2018), and that affect animal health (Finer et al., 2008; Bamberger & Oswald, 2012). While these two land uses may result in similar physical landscape features (i.e., in terms of forest cover or fragmentation), the observed health outcomes may differ due to distinct population features that mediate the trajectory, scale, and structure of land use change.

Thus, we conceptualize the land, population and health relationship as a coupling of population characteristics that drive, directly or indirectly, environmental changes (Fig. 15.1). We identify three proximate causes of LULC change—agricultural expansion/intensification, expansion of extractive industries, and infrastructure development—and depict four population dynamics coupled to these changes—fertility and mortality, migration, and human behavior, which includes technology adoption. As described above, these population-land (or population-environment) relationships result in typologies that describe the habitat and interactions of vertebrate and invertebrate animals, food production, distribution of chemicals and pathogens in environmental media, and the natural and built environment. From a health perspective, human morbidity and mortality are a direct result of these population-environment typologies, which, *ceteris paribus*, directly influence human susceptibility to several health outcomes, including vector-borne and zoonotic diseases, non-zoonotic diseases, non-communicable diseases, and accidental injury or death.

There is an important nuance to our conceptual model that we alluded to above—similar typologies can result in multiple and unique health outcomes. Since different causal pathways connect land to human populations, factors such as structural land cover change (i.e., land mosaics), human-land use connectivity, and method of land clearing or management, differentially connect to health risks (Fig. 15.2). This has been shown repeatedly in vector-borne disease research and described as disease ecology (or landscape epidemiology) research. Drawing upon the agricultural expansion versus

oil/gas exploration example above, even if the two activities evolved with the exact same metrics of total land cleared, the process of clearing and the resulting landscape shape and structure will vary significantly, i.e., each having different compositions of land cover types (forest mixed with agriculture vs. mixed forest with desertification). These aspects of structure modify many disease processes, including pathogenic environments as described by Lambin and colleagues, how chemicals are transported, and how exogenous factors such as water and air quality are modified. Human-land use connectivity can either modify the physical or temporal distance between humans and specific land uses, or the way humans interact with the land (or both). Differential connectivity results in different disease risks for humans for different combinations of land cover. Finally, the mechanism to clear land—i.e., burning, clear-cutting, slashed and burned, slashed and mulched, and many others—predicts the types of pathogens or chemicals that might emerge to affect human health.

The remainder of this chapter provides examples and case studies of how population-land typologies are linked to differential and overlapping health effects. We present these typologies as functions of the three drivers of LULC change. As noted previously, vector-borne and zoonotic diseases have received considerable attention in parsing the nuances of land use change with specific disease outcomes. Thus, the majority of our case studies evaluate the connections between land use change and other major health outcomes.

Agricultural Extensification and Intensification

In 1798, Thomas Malthus highlighted agricultural production as the seminal relationship between humans and the environment. Since that time, scientists have debated the complex relationship between population, economic and social changes related to food demand, and land

15 Land Use Change and Health

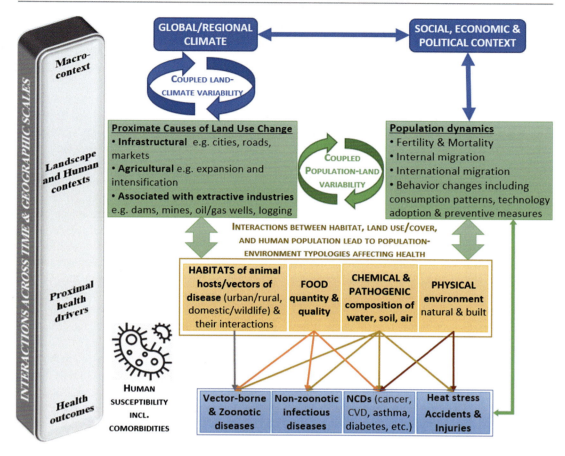

Fig. 15.1 Impact of coupled land use and land cover changes and population on health outcomes

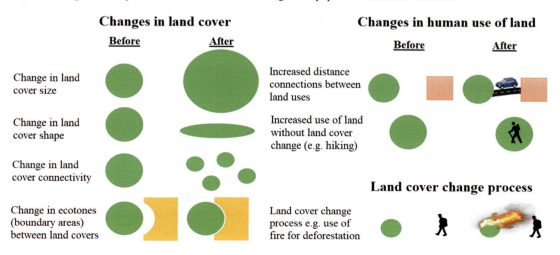

Fig. 15.2 Changes in land structure, use and change processes that link land use change to observed health impacts

conversion for agricultural production (Carr et al., 2006; Sherbinin et al., 2007; Fischer-Kowalski et al., 2014). What we have observed in the past half-century (between 1960 and 2015) is a 142% increase in global population and a 300% increase in food production (FAO, 2017; World Bank, 2019b). As a consequence, food security improved and hunger and malnutrition decreased overall, though many low-and-middle income countries, particularly in Sub-Saharan Africa, still have important leeway to make further progress. Food production increases have been obtained in two major ways: agricultural expansion (or extensification) and agricultural intensification. World Bank data from 1994 to 2014 suggest that there has been limited expansion of agricultural lands at the global level (just a 3.0% increase), while cereal yields have increased by 29% and the overall food production index by 67% (World Bank, 2019b). However, this overall result hides large disparities between countries and income groups. Low income countries, where population has grown the fastest (+73%), have expanded agricultural land by 30%, while yields have increased by 26%. In contrast, middle income countries have increased agricultural land by just over 2%, but yields by 48% and population by 29% (World Bank, 2019b).

Agricultural Expansion and Deforestation

The conversion of land to agricultural production is the largest driver of deforestation worldwide and is estimated to contribute 72% of deforestation-driven carbon emissions (Carter et al., 2018). The majority of this conversion occurs in vulnerable environments and marginal landscapes in developing countries that are not ideal for agricultural production, such as tropical forests with poor soil quality, marshes, and arid ecosystems. The risk of VBZD has been well-documented theoretically and empirically for this population-agriculture typology (Lambin et al., 2010). The mechanisms are primarily through ecosystem disturbances and introduction of susceptible human populations.

For example, deforestation creates culverts that expand the reproductive habitat of mosquitoes capable of transmitting malaria, increasing risk among people working near forest edges (Patz et al., 2004; de Castro et al., 2006; Vittor et al., 2009; Basurko et al., 2013). Forest fragmentation/patchiness has also been identified as a driver of emerging diseases such as Ebola or Lyme disease (Lantos, 2015; Olivero et al., 2017; Rulli et al., 2017).

Beyond infectious diseases, agricultural extensification has other important health impacts. Deforestation tends to increase the length of forest edges (interfaces between humans and the forest), which has been associated with higher risks of wildfire ignition in the Amazon forest (Luiz Eduardo et al., 2008) and in the western United States (Miller et al., 2008). Extensification along hillsides and mountain slopes has been associated with increased risk of landslides and avalanches (Glade, 2003; Mugagga et al., 2012; García-Hernández et al., 2017), and contributes to flood severity in some contexts through its impact on soil stability, permeability, and evapotranspiration[1] (Tollan, 2002; Ferreira & Ghimire, 2012; Adhakari, 2013; Lima et al., 2014; Ghimire et al., 2018). Finally, in some cases, land use change processes themselves are hazardous as seen with intentional fires for deforestation in Indonesia (Frankenberg et al., 2005).

Agricultural Intensification

In contrast to extensification, agricultural intensification has been associated with reductions in deforestation rates and increases in food production. Some countries have even managed concurrent increases in yields, food production and forest cover, and decreases in malnutrition (FAO, 2017; Pellegrini & Fernández, 2018). However, agricultural intensification comes with its own health consequences, most importantly, the coupled effects of soil erosion and hazardous agricultural inputs (pesticides and fertilizers). This coupling is

---

[1] At the global level, urbanization appears to be a more robust driver of flooding than deforestation.

perhaps the biggest of the threats to future human health conditions associated with land use/land cover changes. Indeed, mounting evidence suggests that agricultural intensification is threatening the long-term viability of soils to fulfill the needs of growing populations (Gomiero, 2016; Kopittke et al., 2019) and causing eutrophication of reservoirs and other bodies of water due to nitrogen runoff, leading to widespread fish kills and algal blooms (Anderson et al., 2002). These dual impacts on land and water have occurred globally following decades of intensification. For example, France experienced different pulses of agricultural intensification since the 1950s, with long-term effects associated with elevated sediment flux (erosion) and eutrophication of reservoirs and other bodies of water from fertilizers (Foucher et al., 2014, 2020). Similar findings were reported for the Chesapeake Bay in the United States (Boesch et al., 2001), Lake Erie (Baker et al., 2017), eastern Italy and the Adriatic Sea (Haddrill et al., 1983; Sangiorgi & Donders, 2004), and in China (Zhang & Shan, 2008), just to name a few. With a third of the world's agricultural land moderately or highly degraded, the use of agricultural intensification methods as a sustainable solution for maintaining food demands needs to be reevaluated (FAO, 2017).

In addition to the coupled impacts discussed above, the concentrated application of chemical fertilizers and pesticides as a land management strategy for agricultural intensification has serious human health consequences. Most pesticides operate by impairing some mechanism of nerve function, for example, organophosphates and carbamates impair neurotransmitters, while organochlorines and pyrethroids disrupt nerve-impulses (i.e., electrical impulses in the nervous system). Human exposures result from direct contact (usually by farmers and farm workers), atmospheric drift (residents of nearby farms), or through contaminated air, water and food. At the most local level, acute exposure to pesticides leads to permanent motor and cognitive impairment, particularly among children (Kofman et al., 2006). Chronic exposure, which results from atmospheric drift or environmental contamination, has been associated with an increase in mood

disorders, neurological diseases, autism and cancers, even at low-exposure levels, in both the individuals exposed and their offspring (Roberts et al., 2012).

Unfortunately, pesticide use is now increasing faster than food production (Rahman, 2010; Schreinemachers & Tipraqsa, 2012) and it is expected that the advantage of fertilizers over crop rotation will decline as weather conditions become increasingly variable due to climate change (FAO, 2017). With the largest growth in population and agricultural intensification expected to occur in developing countries, it is important to consider how pesticides and fertilizers may affect health in these vulnerable regions. Unfortunately, transnational chemical corporations have lobbied extensively to maintain product availability in many countries. In the US alone, 85 pesticides that are banned or phased out in the EU, China and Brazil are allowed in agricultural production (Donley, 2019). Acephate and atrazine, banned in the EU, are among the top pesticides used in Brazil and are still applied in the US (Daam et al., 2019). Chlorpyrifos, which is associated with developmental disabilities and low birthweight, is also banned in EU, but allowed in most states in the US (except Hawaii, California and New York) and throughout Latin America and Africa (Abbots & Morales, 2020). The effects of these chemicals in the environment are compounded due to dissipation in the atmosphere, which also leads to invertebrate death, including pollinators (Babut et al., 2013; Daam et al., 2019).

Case Study: Intensive Agriculture and the Gulf of Mexico "Dead Zone"
Prior to WWII, the use of chemical fertilizers was very low, less than 0.3 Tg of nitrogen per year; however, the Green Revolution, which began around 1950 in response to increasing population size, set off a massive agricultural technology boom, including improved irrigation, new seed varieties that increased nitrogen absorption, and an exponential increase in fertilizer use (van Grinsven et al., 2015; Cao et al., 2017).

(continued)

Between 1960 and 2015, fertilizer use in the US quadrupled, with the largest increases occurring between 1960 and 1980 due to both agricultural expansion and intensification, while the period 1980 to 2015 saw smaller increases in fertilizer use (Cao et al., 2017). Geographically, nearly all of the fertilizer used in the US occurs in the Mississippi River watershed (Cao et al., 2017; Lu & Tian, 2017; Nishina et al., 2017). This has resulted in a tripling of nitrates draining into the Gulf of Mexico since the 1950s, with 70% of total nutrient inputs into the Gulf coming from 31 states.

Nutrient runoff into water bodies can have a substantial impact on the local ecosystem: excess nutrient accumulation (eutrophication) favors the growth of certain algae or phytoplankton, increasing oxygen consumption. If the oxygen level decreases below a certain level (hypoxia), marine organisms start dying (Oviatt & Gold, 2005; Glibert, 2017). Further, some harmful algal blooms (HABs) can also affect human health directly and through consumption of contaminated seafood.

This scenario has been unfolding in the Gulf of Mexico, where the yearly hypoxic "dead zone" resulting from agricultural runoff and phytoplankton proliferation ranges between 5000 and 20,000 km^2 (Ménesguen, 2014). The increase in nutrients from agricultural, wastewater and industrial runoff, combined with climate change, is also believed to increase the recurrence and intensity of red algal blooms (Kudela et al., 2008; Lassus et al., 2016). Collectively, these environmental effects have led to large-scale fish deaths, which has major economic consequences as the Gulf of Mexico supplies close to 40% of domestic catches in the contiguous US (National Marine Fisheries Service, 2020). However, agricultural runoff is only part of the story.

Increases in coastal population, recreational land use (seaside tourism), and fish farming have increased human exposure to poor water quality and toxic algal blooms (Lassus et al., 2016). Surveillance of health effects is difficult given the short incubation time and clinical disease duration; however, exposure to HABs have been associated with fatigue, dizziness, numbness, abdominal pain and risk of emergency room visits for both respiratory and digestive distress (Backer et al., 2003; Hoagland et al., 2014). Away from coastlines, HABs can create respiratory issues in individuals, with more significant effects in individuals with underlying conditions such as asthma (Kirkpatrick et al., 2004; Fleming et al., 2011; Hoagland et al., 2014).

Fortunately, it is possible to reduce these effects. Studies have shown that a 12% decrease in fertilizer use in contributing states could cut nitrate runoff by 33% (Goolsby et al., 2001). This reduction could contribute to reduce HABs in the Gulf, and would also improve water quality throughout the Mississippi watershed.

CAFOs and Livestock Farming

Agricultural intensification does not solely apply to cropping, but also to livestock farming. Today, most of the global poultry and pig production takes place on industrial farms, or concentrated animal feeding operations (CAFOs). During the 1970s, farms in the United State raised an average of 60 pigs, but industry consolidation during the 1980s and 1990s has resulted in farms today producing over 4000 pigs annually. CAFOs have been implicated in major global epidemics, including the Nipah virus in Malaysia, which began in concentrated swine herds and workers (Liverani et al., 2013), and avian influenza which originated in poultry and poultry workers (WHO, 2004).

(continued)

In addition to infectious diseases, two key challenges of CAFOs are maintaining animal health and managing animal waste. Large quantities of antibiotics are often used to counter the risk of disease that is heightened by high densities of animals—these antibiotics end up in food and the environment and contribute to antimicrobial resistant bacterial infections (AMR) (Feingold et al., 2012; Liverani et al., 2013; Hoelzer et al., 2017), whose impact has been estimated at 700,000 deaths per year worldwide, and is considered one of the greatest threats to global health according to WHO and the Bill & Melinda Gates Foundation. Low-dose exposure to antibiotics in food has also been suspected to contribute to obesity (Riley et al., 2013).

Meanwhile, billions of tons of animal waste are produced annually from CAFOs, yet many countries either lack national policies to regulate or treat animal waste or have minimal treatment standards (Food and Water Watch). Management of animal waste from CAFOs varies, but includes placing waste in holding ponds, spraying onto agricultural fields as fertilizer, incineration, composting, or illegal dumping. The method of disposal determines health impacts, which result from air pollution, eutrophication, and heavy metal exposure, among others. For example, families living near concentrated livestock farms in the Netherlands suffered increased incidence of airway inflammation and obstruction following animal waste aerosolization for fertilizer (Borlée et al., 2017); in Nigeria, concentrated poultry production, which uses heavy metals as additives in feed, was associated with elevated levels of arsenic, iron and lead in drinking water (Oyewale et al., 2019); and living near swine CAFOs in North Carolina, USA, has been associated with increased multi-drug resistant staphylococcus (Schinasi et al., 2014; Feingold et al., 2019) and increased risk of H1N1 infection (Lantos et al., 2016).

Although agricultural extensification and intensification have successfully increased food quantities to keep pace with a growing global population; these solutions carry significant ecological and health risks. As climate change and major climatic events such as hurricanes, typhons and flooding become more common, agricultural production practices face significant sustainability challenges.

Infrastructure Expansion

Cities, markets and roads have expanded significantly over the past half-century, driven by global population growth, economic development, and broader societal changes. Infrastructure expansion changes the physical, chemical and biological properties of the local environment, which in turn can influence human health, including infectious diseases through modifications of vector and host habitats, respiratory diseases due to air pollution and increasing population density driven by migration and/or fertility, and obesity and chronic diseases by altering human behaviors such as exercise or nutrition. In addition, expansion of one form of infrastructure (e.g. roads) may promote the development of other forms (e.g. human settlements) or extractive industries (such as logging), which can have rippling effects that extend far beyond the original change. This section focuses on the multiple ways in which population-infrastructure expansion typologies affect disease risks, directly or indirectly.

City Expansion

Cities are expanding quickly. According to the World Urbanization Prospects (UN, 2018), 68% of the world population is expected to be living in urban areas by 2050, against 36% in 2000 (55% in 2018). Population growth and migration are expected to bring an additional 2.5 billion urban dwellers by 2050. This growth will be uneven: out of the 50 cities with over 300,000 inhabitants in 2018 projected to grow the fastest between now and 2035, all are located in low or middle-income countries with 46 in Sub-Saharan Africa and 49 in the tropics or subtropics. The majority of this urban expansion will involve increasing populations that inhabit slums with marginal access to public services, insecure property rights, and dependence on informal employment sectors (UN, 2018; World Bank, 2019b). New urban environments will experience increases in human population density, different behaviors affecting nutrition and exercise, different service availability, new vector habitats, and encroachment onto new land. Many of the risks depend on urban planning, which makes the trajectory of urban development in the fast-growing cities of the South particularly critical.

Communicable and Vector-Borne Diseases

High population density is an important predictor of infectious disease transmission. Although cities have been epicenters of past epidemics (e.g. cholera in London (Newsom, 2005), the 1918 influenza pandemic and COVID-19), cities in developed countries have strong health systems and good water, sanitation and hygiene services, hampering the spread of diseases. In developing country cities, such infrastructure (if available) is limited and they generally harbor large, growing informal settlements or slums (Utzinger & Keiser, 2006). Slum residents face a high communicable disease burden due to favorable conditions for direct human to human transmission of diseases such as tuberculosis and influenza, but also to the spread of vector-borne diseases. Indeed, stagnant water and poor sanitation infrastructure can provide breeding grounds for mosquitoes, while population density and limited protection promote human transmission. It is therefore not surprising that slums have been associated with high disease circulation, from cholera in Dar es Salaam (Penrose et al., 2010) and Ebola in West Africa (Snyder et al., 2014), to Zika in Brazil (Snyder et al., 2017) and dengue in tropical and subtropical city slums (Gubler, 2011; Brady & Hay, 2020). Finally, disease spread within an urban slum does not remain in urban slums: Sierra Leone's slums are believed to have played a significant role in the spread of Ebola (Snyder et al., 2014) and flu circulation in Indian slums influences cumulative and peak infection rates for influenza nationwide (Chen et al., 2016).

As cities and slums continue to expand, so does the availability of open markets. Markets are crowded areas, which can create additional risks of disease transmission. Wet markets, which remain common in certain parts of the world, including in large cities, have generated particular concern: they bring together live/freshly slaughtered wild and domestic animals, as well as humans, in crowded quarters, promoting the emergence of new zoonotic diseases. Public and rapid transport connecting wet markets to other parts of a city and to other cities can further rapidly disseminate potential pathogens. Wet markets have been held responsible for the emergence of SARS and influenza (Webster, 2004; Yuen, 2004) and COVID-19 (Aguirre et al., 2020). Rest days during which market are closed and cleaned have therefore been promoted to decrease disease risks (Woo et al., 2006).

> **Case Study: Dengue Fever in Puerto Maldonado, Peru**
>
> Dengue virus (DENV) is a *Flavivirus* that causes up to 400 million new infections per year and is spread by *Aedes aegypti* and *Aedes albopictus* mosquitos, primarily in urban areas (Brady & Hay, 2020). The distribution of DENV and its *Aedes* vectors have expanded globally, impacting regions throughout tropical and subtropical areas, aided by the anthropophilic behavior of *Aedes* and its adaptability to breed in a variety of water bodies or containers (for example flower pots, discarded tires or plastic bags). Puerto Maldonado (pop 85,025, INEI 2017) is the largest urban area in Madre de Dios, Peru, a region in the southern Peruvian Amazon. The city has experienced 5.5% annual population growth since 2007, the highest among any city or region in Peru and among the highest in Latin America. Prior to 2009, DENV infection was rare, cycling between 0 and 100 cases annually (Pan et al., 2014); however, between 2005 and 2010, construction of the Interoceanic Highway combined with expansion of gold mining led to a

(continued)

massive wave of labor migrants, resulting in expanded urban settlements surrounding Puerto Maldonado and a predominantly male working-age population (Perz et al., 2013; Pan et al., 2014; Caballero Espejo et al., 2018). With new urban settlements and a rapidly growing and mobile population, DENV cases increased from 45 in 2008 to 798 in 2009 and 2952 in 2010 (Direccion General de Epidemiologia, 2014). Since this emergence, DENV has averaged 2100 cases annually, with an annual incidence of 1681 per 100 K population (highest in Peru) and a peak of 7399 cases in 2019 (Centro Nacional de Epidemiologia Prevencion y Control de Enfermedades, 2020). Recent research has found that DENV infection in this region results in a significant loss of income and that emergence of DENV is associated with sex, population age structure, and newly formed settlements in the city (Salmon-Mulanovich et al., 2015, 2018). Unfortunately, due to the common vector and population-infrastructure typology, Puerto Maldonado is also experiencing emergence of other arboviruses, including Oropouche, Chikungunya, Mayaro and Zika viruses (Alva-Urcia et al., 2017).

Non-communicable Diseases (NCDs)

NCDs are the leading cause of death worldwide and the burden is growing (WHO, 2020). Cardiovascular disease, cancer, chronic respiratory disease, and diabetes caused the most deaths, with common epidemiological risk factors that include dietary choices, physical activity, air pollution, hypertension, obesity and smoking status (Peters et al., 2019). Population factors play a major role in the emergence of these risks in both developed and developing countries, primarily due to shifts in dietary behavior. Since the 1960s, the average calories consumed per person increased by 22%, with a 35% increase in developing countries where the percentage of calories from meat and sugar also increased by over 120% (Kearney,

2010). This was due primarily to urbanization, income, consumer attitudes, and the aggressive expansion of the food industry in establishing large supermarkets, fast-food options, and advertising (Thow, 2009; Thow & Hawkes, 2009; Kearney, 2010). Collectively, the shifts in diet due to rapid population growth and urbanization has been described as the *Nutritional Transition* or *Westernization of Diet* by nutritionists and demographers, and causally associated with large increases in obesity and diabetes rates worldwide (Popkin, 1999; Barría & Amigo, 2006; Kelly, 2016). This transition is spilling over into rural areas due to infrastructure expansion, particularly of roads and expanded food industry networks that facilitate rural dietary changes (Steyn et al., 2012; Harris et al., 2019; Pettigrew et al., 2019; Oestreicher et al., 2020).

As described by the Nutrition Transition, urban dwellers not only have higher caloric intake compared to their rural counterparts, they are experiencing lower average energy expenditure as well as hypertension (Popkin, 1999). Urbanization itself does not predispose someone to lower physical activity, but the structure of urban landscapes can alter access and participation to higher activity levels. Several studies have demonstrated that, within urban areas, mixed land use, street connectivity, access to recreational facilities and parks and the presence of public transport, can significantly increase the use of walking versus cars for transportation (Knuiman et al., 2014; Van Cauwenberg et al., 2018), which is associated with a healthy BMI (Brown et al., 2009). However, efforts to increase walkability in cities need to be contextualized with the local culture and the pace of population growth, as, in many contexts, cities cannot implement proper planning strategies to keep up with the demand for basic services.

In addition to dietary and activity changes, another major cause of the emergence of NCDs in urban environments has been worsening air quality. Air pollution is recognized as the leading cause of death worldwide, primarily due to its broad impact on NCD risk (Lelieveld et al., 2020). Household and ambient air pollution are disproportionately worse in cities located in low-

and middle-income countries due to continued reliance on biomass fuels as an energy source, increasing motorized vehicles, lack of air quality controls in manufacturing, and population growth (Ye, 2018). Air pollution will disproportionately impact developing nations as 97% of cities over 100,000 inhabitants in low- and middle-income countries have particulate matter levels above the WHO guidelines (WHO, 2018b). In addition to population size increasing demand for emission-related technologies, urban land use and structure have been recognized as key determinants of air pollution, with urban expansion associated, generally, with reductions in air pollution, while urban shape irregularities were associated with more polluted conditions (Hankey & Marshall, 2017; Zhou et al., 2018).

Not all population-infrastructure typologies are associated with worsening NCD burden. As noted earlier, proper city planning can encourage walking or cycling over driving, which both increased physical activity and reduced air pollution. In addition, dementia risk is often lower in urban areas (Nunes et al., 2010; Guerchet et al., 2013; Jia et al., 2014; Weden et al., 2018), possibly due to an educational gap (urban residents attaining significantly higher levels of education), greater access to work and leisure activities, and improved healthcare access and utilization.

Accidents and Injuries

After NCDs, mortality due to road injury is the most important cause of death worldwide and the leading cause of death among people aged 5–29. In addition, road accidents account for over 50 million injuries annually that often result in long-term chronic morbidity (WHO, 2018a). Road traffic deaths are not equally distributed—the majority occur in low- and middle-income countries and are a major impedance to economic development (World Bank, 2017). Road traffic deaths have been associated with rapid population growth, urbanization and road development, and the number of motor vehicles (WHO, 2018a; Sun et al., 2019); however, these trends do not hold in high income countries, where traffic fatalities are more likely in rural areas

(Cabrera-Arnau et al., 2020). Slums have a particularly high risk of deadly accidents due to inadequate street infrastructure as well as low access to health centers (WHO and UN-Habitat, 2016). In contrast, mixed neighborhoods tend to have decreased car use and increased biking and walking, changing the frequency of car-on-car, car-on-bicycle or car-on-pedestrian accidents (Wedagama et al., 2011). Street structure plays a critical role in the likelihood of accidents: cul-de-sacs and gardens are associated with lower child pedestrian accident rates, long and straight streets and on-street parking with more accidents (Petch & Henson, 2000).

Urban Heat Island (UHI) Effect

As the number of large metropolises is increasing worldwide and ground cover is rapidly changing, urban residents are increasingly experiencing heat stress due to the urban heat island (UHI) effect. The UHI effect is driven primarily by different rates of heating and cooling in cities and in the surrounding countryside related to increases in dark, impermeable surfaces, loss of vegetation, air circulation and pollution, thermal properties of urban structures, and anthropogenic "waste heat" (Oke, 1982; Arnfield, 2003; Kleerekoper et al., 2012; Cao et al., 2016). Population plays a key role in mediating each of these causes, particularly through its impact on city size as the magnitude of the UHI effect is proportional to city area (Roth, 2007).

Multiple studies have suggested that climate change will likely increase the nighttime UHI effect (Wilby, 2008; Sachindra et al., 2016). In fact, in fast growing cities, changes in the UHI effect associated with urban expansion can be similar to temperature increases driven by climate change (Ndiaye et al., 2017; USAID, 2017). In locations that experience high temperatures, the UHI effect magnifies the impact of extreme heat and increases the risk of heat stress, particularly for those with heightened vulnerability such as the elderly and deprived populations (Macintyre et al., 2018). It can also worsen local pollution and its health consequences (Zheng et al., 2018), and, by changing local climatic conditions, the UHI effect can affect the local ecology with potential impacts on the risk of vector-borne diseases such as malaria (Phelan et al., 2015). Finally, heightened temperatures also increase energy consumption (Ndiaye et al., 2017).

The interactions of UHI drivers with LULC, climate and human behavior suggest that mitigation efforts should be contextualized and that there is no one-size-fits-all solution. In this context, it is concerning that less than 20% of articles on the UHI effect appear to focus on tropical and subtropical regions (Roth, 2007) where the vast majority of fast-growing cities are located (UN, 2018). Developing increased understanding of the UHI phenomenon and ways to mitigate it in these regions is crucial.

Case Study: Urban Heat Island Effect in Ouagadougou, Burkina Faso

Since the 1960s, the city of Ouagadougou grew from 99,000 to 2.5 million inhabitants in 2018, a 25-fold increase, while the city area increased 14-fold (UN, 2018). Ouagadougou's population is forecasted to reach 5.5 million by 2035. Part of the city's rapid growth has been through poor informal settlements, which now occupy around 20% of the city's area (Holmer et al., 2013). While city growth may be hard to avoid, the way in which the city develops matters, particularly for urban land cover and land use.

Ouagadougou has a dry tropical climate (Ibrahim et al., 2014), and climate change is expected to further decrease rainfall and to increase risks of heat cramps, heat exhaustion and heat stroke (Sylla et al., 2018). Since 1984, average temperatures at Ouagadougou's airport have increased by around 1 °C (Bambara et al., 2018), but as shown in Fig. 15.3, studies examining different locations within Ouagadougou show large variability (Offerle et al., 2005; Lindén, 2011; Holmer et al., 2013). For example, while Lindén (2011) noted an in-

crease of 1.9 °C at night in the modern urban center, she also noted a 5.0 °C decrease at night in the forested area near an urban reservoir as compared to a reference dry, mostly bare area in the surrounding countryside.

The primary drivers influencing temperature changes in the city are the area of green vegetation and presence of open water. Green vegetation reduces daytime warming and enhances nighttime cooling in both urban and rural areas. Lindén (2011) found that the presence of open water contributes significantly to daytime cooling, but also that poverty-driven use of vegetation in the nearby countryside (for firewood) can create a bare and warm peri-urban landscape, broadening warming beyond city limits. Areas with a higher sky-view factor (unblocked view of the sky) also had an increase in nighttime cooling (Lindén, 2011; Holmer et al., 2013). Further, housing material also affects temperatures, e.g., local clay, brick, or thatch vs. modern concrete, with local material being better in some (Offerle et al., 2005) but not all studies (Lindén, 2011).

While lack of vegetation in both urban and rural areas has been identified as a crucial driver to the urban heat island effect in Ouagadougou, future urban population growth increases pressure for more urban sprawl, more poor informal housing, and continued reliance on rural vegetation for firewood. Mitigating the urban heat island effect in Ouagadougou will therefore require deliberate efforts in multiple areas including urban planning and support to alternative energy sources for poor city inhabitants.

(continued)

Urban expansion has therefore been associated with marked changes in human density, behaviors, pollution, infrastructural quality and urban ecosystems, and associated health impacts. These impacts often differ in developed and developing

Land Surface Temperature in Ouagadougou in January (left) and September (right)

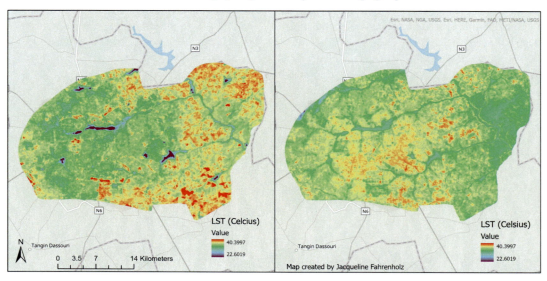

Fig. 15.3 Land Surface Temperatures in Ouagadougou, 2019 in January (dry season, left) and September (wet season, right). Variability in temperatures (heat islands) within the city are apparent in both seasons

countries, and within developing countries themselves, slum areas can face markedly different risks than high income areas: they face what has been termed a "triple threat" from violence and injury, NCDs and communicable diseases (WHO and UN-Habitat, 2016). Urban planning could help address some of these risks, but expansion in fast growing cities is often unplanned. Further, it is important to expand research in context-specific risks, particularly for subjects (such as the urban heat island effect) where most studies seem to concentrate in high-income areas while most urban expansion will take place in low-income regions of the world.

Road Development

Road development promotes the movement of people and goods, and is often considered as key to economic development by connecting people to cities and markets and reducing transport costs (Berg et al., 2016). According to (Meijer et al., 2018), an added 14–23% of road length will be built by 2050, driven by expected population and/or economic growth. The largest foreseen increases in road length, in absolute value, are in the USA and India. However, particular increases are expected in areas such as the Amazon, the Congo basin, and New Guinea. While roads per se cover a small percentage of the Earth's surface, they affect much larger areas through their indirect influence on humans, habitats, plants and the biochemical and physical environments. For example, in the United States, roads cover around 1% of land area, yet (Forman & Alexander, 1998) estimate that they have an ecological impact on 15–20% of the country's surface, including effects on biodiversity (e.g. through roadkills), transecting watersheds, and chemical runoff. In the Amazon, road development influences deforestation by making land accessible for settlement, agriculture, logging, or mining (Pan et al., 2007; Pfaff et al., 2007) hence creating large secondary impacts beyond the direct impact of roads. Thus, roads do not just influence the circulation of humans and goods, they are critical to shaping the distribution of city growth and population density, and, in vulnerable environments such as the

Amazon, they can affect ecosystems, promoting the circulation of animal or plant diseases and invasive species, hindering the movement of some of the local wildlife, or introducing pollutants in the environment. As some of the largest road developments are foreseen in areas currently with limited human influence and high biodiversity, it is critical to consider the direct and indirect impacts of roads in these contexts.

Roads, Increased Population Movement and Mixing, and Health

Increased population mixing and movement brought about by roads affects communicable disease as well as accident and injury risks. Highways for humans can become highways for diseases such as syphilis in North Carolina (Cook et al., 1999), avian influenza in Nigeria (Rivas et al., 2010), and HIV in Kenya, Uganda and the Brazilian Amazon (Morris & Ferguson, 2006; Donalisio et al., 2013). The effect of road development on disease circulation is particularly visible in areas undergoing massive demographic and infrastructural changes. For example, in Central Africa, the unprecedented increase in population and connectivity has worsened the severity of the Ebola epidemics (Munster et al., 2018). Using an index of road density as a proxy for human mobility, (Gomez-Barroso et al., 2017) have shown that a higher road density was associated with an increase in Ebola cases in Guinea, Sierra Leone and Liberia. With a growing, more connected population, a disease that may otherwise have remained relatively circumscribed can become a major epidemic.

As noted earlier, road development is also associated with road traffic accidents and dietary changes. The promotion of safer infrastructure (roads and roadsides) could significantly reduce accident risks, but the large majority of roads in low income countries still have low ratings with regarding to pedestrian, bicyclist and motorist safety, in contrast with the situation in high income countries (World Bank, 2019a). In the Peruvian Amazon, development of the Interoceanic Highway connecting urban and rural markets has been associated with an increased adoption of the Western diet and increased proportions of both stunted and overweight children (Ramalho et al., 2013; Pettigrew et al., 2019).

Also mentioned earlier in this chapter is the relationship between roads and vector-borne disease risk. By facilitating human movement, roads shape human settlements and urban centers, influencing the development of new farms, logging or mining sites, and tourism (Ferri, 2004; Christie et al., 2014; Ou et al., 2017; Levkovich et al., 2020). For example, in Tanzania, improved access to local cities has been associated with an increase in cropland and livestock in the Serengeti (Walelign et al., 2019), which may lead to the encroachment of pastures into protected areas. Similarly, road development has been associated with forest clearing and fragmentation in the Amazon, the Congo, and New Guinea, among others (Pfaff et al., 2007; Bryan et al., 2010; Li et al., 2015), which has been associated with outbreaks of several vector borne diseases, including malaria (Bauch et al., 2015), Ebola (Olivero et al., 2017; Rulli et al., 2017) and lymphatic filariasis (Melrose, 1999).

Roads can also bring about positive change. Human mobility promotes the movement of goods and ideas, fostering both economic and behavioral changes. In countries such as Nepal, China and Ethiopia, qualitative and quantitative improvements in the road network (density and year-round practicability) have been associated with lower poverty and improved nutritional status in children (Fan & Chan-Kang, 2005; Bucheli et al., 2018; Thapa & Shively, 2018). However, given the cost of motor vehicles and public transport, road development is not considered a panacea to increasing mobility in poor rural regions.

Impact of Road Development on the Local Physical and Chemical Environment and Ecosystem

Development of roads and the subsequent propagation of settlements and other land consuming activities also affect physical and chemical properties of an ecosystem, with ripple effects on human health. Proximity to roads has been associated with increased risk of several diseases, including, lung cancer among non-smokers (Puett et al., 2014), cardiovascular disease (Hoek et al., 2002), asthma prevalence among children and adults (Gowers et al., 2012; Gruzieva et al., 2013), obesity (Pyko et al., 2017) and dementia (Chen et al., 2017). These effects stem from contaminants during construction, maintenance, and use, such as road salt (in areas where it snows in winter) and heavy metals. Both impact ecosystems by altering water quality and aquatic life, magnifying exposures to heavy metals by increasing their bioavailability (Corsi et al., 2010; Novotny & Stefan, 2012; Schuler & Relyea, 2018).

Roads further contribute to the fragmentation of natural landscapes that result in corridors for certain plant and animal species, or, alternatively, create barriers that divide local animal habitat (Gippet et al., 2016; Habib et al., 2016; Bennett, 2017). These types of landscape modifications have important links to human health. For example, in Poland, structural changes in landscapes have resulted in more wildlife crossing roads, leading to an increase in car-wildlife collisions (Keken et al., 2016). Several infectious diseases have been shown to be exacerbated due to land fragmentation due to road development, for example, malaria in Peru and Brazil (Vittor et al., 2009; Bauch et al., 2015), bovine and human rabies in Brazil (de Andrade et al., 2016; Mastel et al., 2018), and Lyme disease throughout the US and Canada (Simon et al., 2014; Seukep et al., 2015).

The impact of road development is therefore complex and wide-reaching. Its association with economic growth and transport of goods and modern practices often has a welcome impact on health, but roads also bring new health challenges, as we have presented.

Extractive Industry

Natural resource extraction includes activities such as mining, logging, and oil/gas extraction, to excavate and recover natural materials for human use. It responds to the demands of an increased global population and economy, and plays a major economic role in many countries:

rents from the exploitation of oil, natural gas, coal, minerals and forests exceeded 10% of GDP in 47 countries in 2018 (World Bank, 2019b). Around 70 billion tons of non-metallic materials (mostly construction materials such as sand, gravel, crushed rocks or limestone), metallic ores (such as iron or gold), fossil fuels and wood are extracted every year (WU Vienna, 2019). Humans have significantly advanced technology to extract resources quickly and efficiently from the earth. However, when dealing with natural ecosystems, we remain relatively ignorant of how even small changes can be magnified. A recent review of hydraulic fracturing (fracking) presents the positive and negative benefits of this new technology, highlighting the importance of research to better understand the negative externalities and pathways through which fracking impacts human health and well-being (Jackson et al., 2014; Vengosh et al., 2014). Modern technologies of resource extraction are not exempt of risk, but in poor countries, primitive techniques of resource extraction are still used, with ensuing magnification of risks (see case study on artisanal small-scale gold mining).

For miners themselves, ecosystem disturbances brought about by mining can create fertile ground for interactions between humans, vectors and wildlife, elevating disease risks, as seen in the Amazon when mining increased incidence of malaria, hantavirus pulmonary syndrome, and leishmaniasis (Rotureau et al., 2006; Terças-Trettel et al., 2019). The relationship between vector-borne and zoonotic diseases with mining is multifactorial, namely, mining is a migration pull factor that draws occupational laborers often naïve to local circulating diseases, the land clearing process can create ecological niches for disease vectors to profligate, and mining activities result in chemical exposures that can reduce immune response to migrants and local residents. In some contexts, particularly mining regions in southern and central Africa, high density housing of migrant workers has also led to significantly elevated incidence of tuberculosis and HIV (Fielding et al., 2010).

However, natural resource extraction not only causes environmental changes that influence infectious disease risk, it is also related to acute and chronic exposure to toxic chemicals. For example, chemicals released into surface and groundwater from fracking (heavy metals, toluene, benzene, xylenes, polycyclic aromatic hydrocarbons, etc.) have been found in drinking wells of nearby residents and associated with neurotoxicity, neuroinflammation, and other neuro-developmental impacts (Weinberger et al., 2017; Webb et al., 2018). Mountaintop mining removal (a mining method where large portions of mountain are removed) can release contaminants naturally present in the ground into the air or waterways, increasing the risk of hypertension, pulmonary, heart and kidney disease and lung cancer (Palmer et al., 2010). Sand mining for glass production can produce contaminants in water (Kosior et al., 2015) and PM particulates in the air (Pierce et al., 2019). Contaminant pollution from mining is persistent and lasts long after the mines have closed: German regions with a history of mining have higher cadmium contamination in soy (Franzaring et al., 2019) and gold mining during the 1849 California Gold Rush has left a legacy of mercury pollution that continues to harm human health (Alpers et al., 2000). Soil disturbance associated with extractive industries (from mining to logging) can also release natural chemicals naturally present in soils: deforestation in the Amazon, for example, has been shown to contribute substantially to mercury pollution (Oestreicher et al., 2017).

Finally, accidental injury or death is, unfortunately, common in mining. Sinkholes (Karfakis, 1993) can appear in areas where former underground mines were located; flood risk can increase in areas where mountaintop removal mining is conducted due to large-scale removal of soils and topography (Palmer et al., 2010); and other unpredictable injuries can stem from machinery failure or land mismanagement. In fact, accidental failure of dams has become an important driver of unintentional injury in countries with minimal mining regulation. The failure of the Fundão tailings dam in Brazil killed 19 people (Guerra et al., 2017), while the failure of the Aznalcollar dam in Spain flooded around 46,000 hectares (Hudson-Edwards et al., 2003). Both accidents resulted in contaminated mining waste entering the environment with long-term risks due to increased exposure to dangerous metals.

Case Study: Multi-health Impacts of Artisanal Small-Scale Gold Mining (ASGM) in Madre de Dios, Peru

In ASGM, liquid elemental mercury (Hg^0) is added to excavated ground soil to bind with gold particles in the soil matrix to form a Hg-gold amalgam (Fig. 15.4). Burning the amalgam removes Hg, leaving purified gold, but much of the Hg is lost through evasion of gaseous Hg^0 to the atmosphere or disposal of contaminated waste soils near the excavation site. ASGM is the world's largest contributor of anthropogenic atmospheric Hg (UNEP, 2013) and is one of the largest threats to biodiversity and forest loss in the Amazon (Asner et al., 2013; El Bizri et al., 2016; Bebbington et al., 2018). A large portion of Hg entering the environment is converted into monomethylmercury (MeHg), a potent neurotoxicant, which biomagnifies in the aquatic food web and poses serious human health risks (Basu et al., 2011, 2015).

In Madre de Dios, Peru, ASGM has increased rapidly over the past 15 years (Swenson et al., 2011; Caballero Espejo et al., 2018), resulting in over 40% of the population experiencing levels of mercury exposure that exceed the WHO daily tolerable dose and almost 60% above the US EPA threshold (Wyatt et al., 2017; Feingold et al., 2020; Weinhouse et al., 2020). This exposure has resulted in increased levels of anemia, impaired immune function and deficits to cognitive function in children, just to name a few (Weinhouse et al., 2017; Wyatt et al., 2019; Reuben et al., 2020). Unfortunately, the health effects have extended well-beyond exposure-related outcomes. Vector-borne diseases, particularly malaria, dengue and leishmaniosis have all increased (Pan et al., 2014; Sanchez et al.,

(continued)

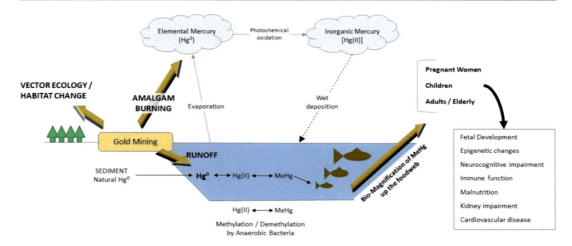

Fig. 15.4 Multi-health impacts of Artisanal and Small-Scale Gold Mining. Impacts, which are highlighted by gold arrows, include: (1) Disruption of habitat, which has been linked to vector-borne diseases; (2) Amalgam burning, which releases mercury to the atmosphere and contributes to mercury deposition in watersheds; (3) Runoff, which contributes to elemental mercury release into watersheds, which is converted into organic mercury; and (4) Bio-magnification of methylmercury (MeHg) up the foodweb. Exposure to humans via food consumption have multi-organ and multi-system impacts in humans

2017),[2] dietary changes in rural areas are occurring rapidly (Pettigrew et al., 2019), and chronic diseases such as chronic kidney disease, diabetes and hypertension are becoming increasingly prevalent in rural mining communities (Saxton et al., in review). A growing challenge is the recognition that ASGM-related deforestation is reinforcing mercury pollution, e.g., previously mined and deforested areas continue to release large amounts of mercury due to increases in soil erosion and lack of vegetation to absorb or prevent mercury from entering aquatic environments (Diringer et al., 2020). These conditions will result in continued damage to ecosystems and threaten human health.

ASGM is a rapidly growing economic sector across many low- and middle-income countries. Migrant families view mining as a pathway to relieve social, economic and health disparities; however, ASGM often fails to fulfil these expectations (Wilson et al., 2015), leading to organized

(continued)

crime, human trafficking, and social conflict (Heemskerk, 2002; Bury, 2004; Cortés-McPherson, 2019). Addressing policy options for ASGM or alternatives to mining is critical to improving population health and maintaining environmental viability.

---

[2] We note that malaria initially increased with the mining boom and concurrent road construction; however, malaria currently is not considered a major health risk in Madre de Dios. This evolution of malaria risk is an aberration compared to other mining areas in the Amazon, where ASGM is considered a leading indicator for the emergence of malaria. Vosti, S. A. (1990). "Malaria among gold miners in Southern Pará, Brazil: Estimates of determinants and individual costs." Social Science & Medicine 30(10): 1097–1105, Crompton, P., A. M. Ventura, J. M. de Souza, E. Santos, G. T. Strickland and E. Silbergeld (2002). "Assessment of mercury exposure and malaria in a Brazilian Amazon riverine community." Environmental research 90(2): 69–75, Barbieri, A. F. and D. O. Sawyer (2007). "Heterogeneity of malaria prevalence in alluvial gold mining areas in Northern Mato Grosso State, Brazil." Cadernos de saude publica/Ministerio da Saude, Fundacao Oswaldo Cruz, Escola Nacional de Saude Publica 23(12): 2878–2886.

Conclusions

The goal of this chapter was to introduce a conceptual framework that connects LULC and population with human health. Given the preponderance of collaborative work related to VBZD and land change science, we paid particular attention to descriptions of population-land typologies affecting other human health outcomes, such as non-zoonotic communicable diseases, NCDs, and accidents and injuries. In doing so, we presented several examples and case studies to demonstrate pathways connecting population, land and human health. While we have strived to present a comprehensive framework, we recognize many gaps and challenges in existing research. Most notably, successful and impactful research needs to evolve to become more inter- and trans-disciplinary. VBZD has made significant progress on this front, integrating knowledge and methods from different disciplines that demonstrate unity of intellectual frameworks beyond disciplinary perspectives. However, other areas of research have not achieved this sort of synthesis. How often do nutritional experts engage agricultural scientists, toxicologists or demographers to address issues related to food security? In 2015, the National Science Foundation began funding research teams to address *Innovations at the Nexus of Food, Energy, and Water Systems (INFEWS)*, which is an important step toward a more unified approach. In contrast, since 1980, the Environmental Protection Agency has funded the Superfund Program to protect human health and environment from toxic chemical exposure and pollution; however, among the 27 Superfund Projects funded today, none has active research programs or training cores led by faculty with expertise in agricultural or population sciences. Universities and funding agencies must move more quickly toward cultures fostering interdisciplinary collaboration.

Besides interdisciplinarity, there are some key areas for which better understanding is needed to link population-land typologies with health. First is migration—there are significant mischaracterizations in how migrants and migration are conceptualized in the epidemiological literature that contrast with how demographers and economists have described migration determinants and characteristics. Second, context relevant research. We have shown several examples where the relationships observed in high income country contexts do not necessarily apply to low- and middle-income countries. Population-land relationships matter everywhere, and research must focus on understanding geographical context to untangle the complexities. Third, food production is one of the most important activities for human survival and sustainable production and a priority and mission of FAO. However, in developing these sustainable solutions, we recommend greater integration and understanding of health consequences of different farming technologies and methods, both in the short- and long-term. Finally, as the world shifts to renewable energy, there will be greater pressure to mine the earth's precious and rare earth metals. Unfortunately, the total net costs of mining on human and environmental health are poorly quantified, particularly in low- and middle-income countries where mining regulation is limited and corruption is frequent. Research should focus on measuring the net impacts of several types of mining operations, both for local and distant populations.

We hope that this chapter contributes to shifting the interest towards the multi-scale population-land typologies that underlie the heterogeneities we observe in multiple human health outcomes. In many respects, we are encouraging the reader to build upon the traditional definitions of health ("a state of complete physical, mental and social well-being" according to WHO) and health care, bringing in land use, climate and demographic considerations. This would mean adopting a *pancentric* approach to health that encompasses aspects of individuals, populations, society, and the environment. Such an approach is becoming increasingly crucial if we are to achieve sustainable development.

References

Abbots, J., & Morales, M. (2020). *States lead to rid food crops of the neurotoxin chlorpyrifos*. Retrieved 11 Nov 2020, from https://www.sightline.org/2020/05/11/states-lead-to-rid-food-crops-of-the-neurotoxin-chlorpyrifos/

Adgate, J., Goldstein, B., & McKenzie, L. (2014). Potential public health hazards, exposures, and health effects from unconventional natural gas development. *Environmental Science & Technology, 48*, 8307–8320.

Adhakari, B. R. (2013). Flooding and inundation in Nepal Terai: Issues and concerns. *Hydro Nepal: Journal of Water, Energy and Environment, 12*, 59–65.

Aguirre, A. A., Catherina, R., Frye, H., & Shelley, L. (2020). Illicit wildlife trade, wet markets, and COVID-19: Preventing future pandemics. *World Medical and Health Policy, 12*(3), 256–265.

Alpers, C. N., Hunerlach, M. P., & Geological, S. (2000). *Mercury contamination from historic gold mining in California*. U.S. Department of the Interior, U.S. Geological Survey.

Alva-Urcia, C., Aguilar-Luis, M. A., Palomares-Reyes, C., Silva-Caso, W., Suarez-Ognio, L., Weilg, P., Manrique, C., Vasquez-Achaya, F., del Valle, L. J., & del Valle-Mendoza, J. (2017). Emerging and reemerging arboviruses: A new threat in Eastern Peru. *PLoS One, 12*(11), e0187897.

Anderson, D. M., Glibert, P. M., & Burkholder, J. M. (2002). Harmful algal blooms and eutrophication: Nutrient sources, composition, and consequences. *Estuaries, 25*(4), 704–726.

Arnfield, A. J. (2003). Two decades of urban climate research: A review of turbulence, exchanges of energy and water, and the urban heat island. *International Journal of Climatology, 23*(1), 1–26.

Asner, G. P., Llactayo, W., Tupayachi, R., & Luna, E. R. (2013). Elevated rates of gold mining in the Amazon revealed through high-resolution monitoring. *Proceedings of the National Academy of Sciences of the United States of America, 110*(46), 18454–18459.

Babut, M., Arts, G. H., Barra Caracciolo, A., Carluer, N., Domange, N., Friberg, N., Gouy, V., Grung, M., Lagadic, L., Martin-Laurent, F., Mazzella, N., Pesce, S., Real, B., Reichenberger, S., Roex, E. W. M., Romijn, K., Röttele, M., Stenrød, M., Tournebize, J., Vernier, F., & Vindimian, E. (2013). Pesticide risk assessment and management in a globally changing world—Report from a European interdisciplinary workshop. *Environmental Science and Pollution Research, 20*(11), 8298–8312.

Backer, L. C., Fleming, L. E., Rowan, A. D., & Baden, D. (2003). Epidemiology, public health and human diseases associated with harmful marine algae. In *Manual on harmful marine microalgae* (pp. 723–749). United Nations Educational, Scientific and Cultural Organization.

Baker, D. B., Johnson, L. T., Confesor, R. B., & Crumrine, J. P. (2017). Vertical stratification of soil phosphorus as a concern for dissolved phosphorus runoff in the Lake Erie Basin. *Journal of Environmental Quality, 46*(6), 1287–1295.

Bambara, D., Compaoré, H., & Bilgo, A. (2018). Évolution des températures au Burkina Faso entre 1956 et 2015: cas de Ouagadougou et de Ouahigouya. *Physio-Géo, 12*, 23–41.

Bamberger, B., & Oswald, R. E. (2012). Impacts of gas drilling on human and animal health. *New Solutions, 22*(1), 51–77.

Barbieri, A. F., & Sawyer, D. O. (2007). Heterogeneity of malaria prevalence in alluvial gold mining areas in Northern Mato Grosso State, Brazil. *Cadernos de saude publica /Ministerio da Saude, Fundacao Oswaldo Cruz, Escola Nacional de Saude Publica, 23*(12), 2878–2886.

Barbieri, A., Sawyer, D. O., & Soares, B. S. (2005). Population and land use effects on malaria prevalence in the Southern Brazilian Amazon. *Human Ecology, 33*(6), 847–874.

Barría, R. M., & Amigo, H. (2006). Nutrition transition: A review of Latin American profile. *Archivos Latinoamericanos de Nutrición, 56*(1), 3.

Basu, N., Nam, D. H., Kwansaa-Ansah, E., Renne, E. P., & Nriagu, J. O. (2011). Multiple metals exposure in a small-scale artisanal gold mining community. *Environmental Research, 111*(3), 463–467.

Basu, N., Clarke, E., Green, A., Calys-Tagoe, B., Chan, L., Dzodzomenyo, M., Fobil, J., Long, R. N., Neitzel, R. L., Obiri, S., Odei, E., Ovadje, L., Quansah, R., Rajaee, M., & Wilson, M. L. (2015). Integrated assessment of artisanal and small-scale gold mining in Ghana—Part 1: Human health review. *International Journal of Environmental Research and Public Health, 12*(5), 5143–5176.

Basurko, C., Demattei, C., Han-Sze, R., Grenier, C., Joubert, M., Nacher, M., & Carme, B. (2013). Deforestation, agriculture and farm jobs: A good recipe for Plasmodium vivax in French Guiana. *Malaria Journal, 12*(1), 90.

Bauch, S. C., Birkenbach, A. M., Pattanayak, S. K., & Sills, E. O. (2015). Public health impacts of ecosystem change in the Brazilian Amazon. *Proceedings of the National Academy of Sciences, 112*(24), 7414–7419.

Bebbington, A. J., Humphreys Bebbington, D., Sauls, L. A., Rogan, J., Agrawal, S., Gamboa, C., Imhof, A., Johnson, K., Rosa, H., Royo, A., Toumbourou, T., & Verdum, R. (2018). Resource extraction and infrastructure threaten forest cover and community rights. *Proceedings of the National Academy of Sciences of the United States of America, 115*(52), 13164–13173.

Bennett, V. J. (2017). Effects of road density and pattern on the conservation of species and biodiversity. *Current Landscape Ecology Reports, 2*(1), 1–11.

Berg, C. N., Deichmann, U., Liu, Y., & Selod, H. (2016). Transport policies and development. *The Journal of Development Studies, 53*(4), 465–480.

Bilsborrow, R. E. (1987). Population pressures and agricultural development in developing countries: A conceptual framework and recent evidence. *World Development, 15*(2), 183–203.

Boesch, D. F., Brinsfield, R. B., & Magnien, R. E. (2001). Chesapeake Bay eutrophication: Scientific understanding, ecosystem restoration, and challenges for agriculture. *Journal of Environmental Quality, 30*(2), 303–320.

Borlée, F., Yzermans, C. J., Aalders, B., Rooijackers, J., Krop, E., Maassen, C. B. M., Schellevis, F., Brunekreef, B., Heederik, D., & Smit, L. A. M. (2017). Air pollution from livestock farms is associated with airway obstruction in neighboring residents. *American Journal of Respiratory and Critical Care Medicine, 196*(9), 1152–1161.

Boserup, E. (1981). *Population and technological change: A study of long-term trends.* University of Chicago Press.

Brady, O. J., & Hay, S. I. (2020). The global expansion of dengue: How Aedes aegypti mosquitoes enabled the first pandemic arbovirus. *Annual Review of Entomology, 65*(1), 191–208.

Brown, B. B., Yamada, I., Smith, K. R., Zick, C. D., Kowaleski-Jones, L., & Fan, J. X. (2009). Mixed land use and walkability: Variations in land use measures and relationships with BMI, overweight, and obesity. *Health and Place, 15*(4), 1130–1141.

Bryan, J., Shearman, P., Ash, J., & Kirkpatrick, J. B. (2010). Impact of logging on aboveground biomass stocks in lowland rain forest, Papua New Guinea. *Ecological Applications, 20*(8), 2096–2103.

Bucheli, J. R., Bohara, A. K., & Villa, K. (2018). Paths to development? Rural roads and multidimensional poverty in the hills and plains of Nepal. *Journal of International Development, 30*(3), 430–456.

Bury, J. (2004). Livelihoods in transition: Transnational gold mining operations and local change in Cajamarca, Peru. *The Geographical Journal, 170*(1), 78–91.

Caballero Espejo, J., Messinger, M., Román-Dañobeytia, F., Ascorra, C., Fernandez, L., & Silman, M. (2018). Deforestation and forest degradation due to gold mining in the Peruvian Amazon: A 34-year perspective. *Remote Sensing (Basel, Switzerland), 10*(12), 1903.

Cabrera-Arnau, C., Prieto Curiel, R., & Bishop, S. R. (2020). Uncovering the behaviour of road accidents in urban areas. *Royal Society Open Science, 7*(4), 191739.

Cao, C., Lee, X., Liu, S., Schultz, N., Xiao, W., Zhang, M., & Zhao, L. (2016). Urban heat islands in China enhanced by haze pollution. *Nature Communications, 7*(1), 12509.

Cao, P., Lu, C., & Yu, Z. (2017). *Agricultural nitrogen fertilizer uses in the continental U.S. during 1850–2015: A set of gridded time-series data.* https://doi.org/10.1594/PANGAEA.883585

Carr, D. L., Suter, L., & Barbieri, A. (2006). Population dynamics and tropical deforestation: State of the debate and conceptual challenges. *Population and Environment, 27*(1), 89–113.

Carter, S., Herold, M., Avitabile, V., de Bruin, S., De Sy, V., Kooistra, L., & Rufino, M. C. (2018). Agriculture-driven deforestation in the tropics from 1990–2015:

Emissions, trends and uncertainties. *Environmental Research Letters, 13*(1), 014002.

Castellanos, A., Chaparro-Narváez, P., Morales-Plaza, C. D., Alzate, A., Padilla, J., Arévalo, M., & Herrera, S. (2016). Malaria in gold-mining areas in Colombia. *Memórias do Instituto Oswaldo Cruz, 111*, 59–66.

Centro Nacional de Epidemiologia Prevencion y Control de Enfermedades. (2020). *Situacion del Dengue en el Peru, 2020.* Ministerio de Salud.

Chen, J., Chu, S., Chungbaek, Y., Khan, M., Kuhlman, C., Marathe, A., Mortveit, H., Vullikanti, A., & Xie, D. (2016). Effect of modelling slum populations on influenza spread in Delhi. *BMJ Open, 6*(9), e011699.

Chen, H., Kwong, J. C., Copes, R., Tu, K., Villeneuve, P. J., van Donkelaar, A., Hystad, P., Martin, R. V., Murray, B. J., Jessiman, B., Wilton, A. S., Kopp, A., & Burnett, R. T. (2017). Living near major roads and the incidence of dementia, Parkinson's disease, and multiple sclerosis: A population-based cohort study. *The Lancet (British Edition), 389*(10070), 718–726.

Christie, I. T., Fernandes, E. A., Messerli, H. A., & Twining-Ward, L. A. (2014). *Tourism in Africa: Harnessing tourism for growth and improved livelihoods.* The World Bank.

Colborn, T., Kwiatkowski, C., Schultz, K., & Bachran, M. (2011). Natural gas operations from a public health perspective. *Human and Ecological Risk Assessment, 17*, 1039–1056.

Coluzzi, M. (1994). Malaria and the Afrotropical ecosystems: Impact of man-made environmental changes. *Parassitologia, 36*(1–2), 223–227.

Cook, R. L., Royce, R. A., Thomas, J. C., & Hanusa, B. H. (1999). What's driving an epidemic? The spread of syphilis along an interstate highway in rural North Carolina. *American Journal of Public Health, 89*(3), 369–373.

Corsi, S. R., Graczyk, D. J., Geis, S. W., Booth, N. L., & Richards, K. D. (2010). A fresh look at road salt: Aquatic toxicity and water-quality impacts on local, regional, and national scales. *Environmental Science & Technology, 44*(19), 7376–7382.

Cortés-McPherson, D. (2019). Labor trafficking of men in the artisanal and small-scale gold mining camps of Madre de Dios, a reflection from the "diaspora networks" perspective. In J. A. Winterdyk & J. Jones (Eds.), *The Palgrave international handbook of human trafficking* (pp. 1–18). Springer.

Crompton, P., Ventura, A. M., de Souza, J. M., Santos, E., Strickland, G. T., & Silbergeld, E. (2002). Assessment of mercury exposure and malaria in a Brazilian Amazon riverine community. *Environmental Research, 90*(2), 69–75.

Daam, M. A., Chelinho, S., Niemeyer, J. C., Owojori, O. J., De Silva, P. M. C. S., Sousa, J. P., van Gestel, C. A. M., & Römbke, J. (2019). Environmental risk assessment of pesticides in tropical terrestrial ecosystems: Test procedures, current status and future perspectives. *Ecotoxicology and Environmental Safety, 181*, 534–547.

de Andrade, F. A. G., Gomes, M. N., Uieda, W., Begot, A. L., Ramos, O. D. S., & Fernandes, M. E.

B. (2016). Geographical analysis for detecting high-risk areas for bovine/human rabies transmitted by the common hematophagous bat in the Amazon Region, Brazil. *PLoS One, 11*(7), e0157332.

de Castro, M. C., Monte-Mór, R. L., Sawyer, D. O., & Singer, B. H. (2006). Malaria risk on the Amazon frontier. *Proceedings of the National Academy of Sciences, 103*(7), 2452.

Direccion General de Epidemiologia. (2014). *Sala de Situacion de Salud 2014*. Ministerio de Salud.

Diringer, S. E., Berky, A. J., Marani, M., Ortiz, E. J., Karatum, O., Plata, D. L., Pan, W. K., & Hsu-Kim, H. (2020). Deforestation due to artisanal and small-scale gold mining exacerbates soil and mercury mobilization in Madre de Dios, Peru. *Environmental Science & Technology, 54*(1), 286–296.

Donalisio, M. R., Cordeiro, R., Lourenco, R. W., & Brown, J. C. (2013). The AIDS epidemic in the Amazon region: A spatial case-control study in Rondonia, Brazil. *Revista de Saúde Pública, 47*(5), 873–882.

Donley, N. (2019). The USA lags behind other agricultural nations in banning harmful pesticides. *Environmental Health, 18*(1), 44–44.

El Bizri, H. R., Macedo, J. C., Paglia, A. P., & Morcatty, T. Q. (2016). Mining undermining Brazil's environment. *Science, 353*(6296), 228.

Fan, S. and C. Chan-Kang (2005). Road development, economic growth, and poverty reduction in China.

FAO. (2017). *The future of food and agriculture—Trends and challenges*. Rome.

Feingold, B. J., Silbergeld, E. K., Curriero, F. C., van Cleef, B. A., Heck, M. E., & Kluijtmans, J. A. J. W. (2012). Livestock density as risk factor for livestock-associated methicillin-resistant Staphylococcus aureus, the Netherlands. *Emerging Infectious Diseases, 18*(11), 1841–1849.

Feingold, B. J., Augustino, K. L., Curriero, F. C., Udani, P. C., & Ramsey, K. M. (2019). Evaluation of methicillin-resistant Staphylococcus aureus carriage and high livestock production areas in North Carolina through active case finding at a tertiary care hospital. *International Journal of Environmental Research and Public Health, 16*(18), 3418.

Feingold, B. J., Berky, A., Hsu-Kim, H., Rojas Jurado, E., & Pan, W. K. (2020). Population-based dietary exposure to mercury through fish consumption in the Southern Peruvian Amazon. *Environmental Research, 183*, 108720.

Ferreira, S., & Ghimire, R. (2012). Forest cover, socioeconomics, and reported flood frequency in developing countries. *Water Resources Research, 48*(8), W08529.

Ferri, J. (2004). Evaluating the regional impact of a new road on tourism. *Regional Studies, 38*(4), 409–418.

Fielding, K. L., Grant, A. D., Hayes, R. J., Chaisson, R. E., Corbett, E. L., & Churchyard, G. J. (2010). Thibela TB: Design and methods of a cluster randomised trial of the effect of community-wide isoniazid preventive therapy on tuberculosis amongst gold miners in South Africa. *Contemporary Clinical Trials, 32*(3), 382–392.

Finer, M., Jenkins, C. N., Pimm, S. L., Keane, B., & Ross, C. (2008). Oil and gas projects in the Western Amazon: Threats to wilderness, biodiversity, and indigenous peoples. *PLoS One, 3*(8), e2932.

Fischer-Kowalski, M., Reenberg, A., Schaffartzik, A., Mayer, A., & SpringerLink. (2014). *Ester Boserup's legacy on sustainability: Orientations for contemporary research*. Springer Netherlands:Imprint: Springer.

Fleming, L. E., Kirkpatrick, B., Backer, L. C., Walsh, C. J., Nierenberg, K., Clark, J., Reich, A., Hollenbeck, J., Benson, J., Cheng, Y. S., Naar, J., Pierce, R., Bourdelais, A. J., Abraham, W. M., Kirkpatrick, G., Zaias, J., Wanner, A., Mendes, E., Shalat, S., Hoagland, P., Stephan, W., Bean, J., Watkins, S., Clarke, T., Byrne, M., & Baden, D. G. (2011). Review of Florida red tide and human health effects. *Harmful Algae, 10*(2), 224–233.

Flower, K. B., Hoppin, J. A., Lynch, C. F., Blair, A., Knott, C., Shore, D. L., & Sandler, D. P. (2004). Cancer risk and parental pesticide application in children of Agricultural Health Study participants. *Environmental Health Perspectives, 112*(5), 631–635.

Food and Water Watch Factory Farm Map: What's Wrong With Factory Farms? http://www.factoryfarmmap.org/problems/

Forman, R. T. T., & Alexander, L. E. (1998). Roads and their major ecological effects. *Annual Review of Ecology and Systematics, 29*(1), 207–231.

Foucher, A., Salvador-Blanes, S., Evrard, O., Simonneau, A., Chapron, E., Courp, T., Cerdan, O., Lefèvre, I., Adriaensen, H., Lecompte, F., & Desmet, M. (2014). Increase in soil erosion after agricultural intensification: Evidence from a lowland basin in France. *Anthropocene, 7*, 30–41.

Foucher, A., Evrard, O., Huon, S., Curie, F., Lefèvre, I., Vaury, V., Cerdan, O., Vandromme, R., & Salvador-Blanes, S. (2020). Regional trends in eutrophication across the Loire river basin during the 20th century based on multi-proxy paleolimnological reconstructions. *Agriculture, Ecosystems & Environment, 301*, 107065.

Frankenberg, E., McKee, D., & Thomas, D. (2005). Health consequences of forest fires in Indonesia. *Demography, 42*(1), 109–129.

Franzaring, J., Fangmeier, A., Schlosser, S., & Hahn, V. (2019). Cadmium concentrations in German soybeans are elevated in conurbations and in regions dominated by mining and the metal industry. *Journal of the Science of Food and Agriculture, 99*(7), 3711–3715.

Gage, K. L., Burkot, T. R., Eisen, R. J., & Hayes, E. B. (2008). Climate and vectorborne diseases. *American Journal of Preventive Medicine, 35*(5), 436–450.

García-Hernández, C., Ruiz-Fernández, J., Sánchez-Posada, C., Pereira, S., Oliva, M., & Vieira, G. (2017). Reforestation and land use change as drivers for a decrease of avalanche damage in mid-latitude mountains (NW Spain). *Global and Planetary Change, 153*, 35–50.

Geist, H. J., & Lambin, E. F. (2002). Proximate causes and underlying driving forces of tropical deforestation: Tropical forests are disappearing as the result of many pressures, both local and regional, acting in various combinations in different geographical locations. *Bioscience, 52*(2), 143–150.

Ghimire, G., Pokharel, L., Bhandari, D., Khadka, P., Kumal, B., & Uprety, M. (2018). *Nepal flood 2017: Wake up call for effective preparedness and response.* Practical Action.

Gippet, J. M. W., Mondy, N., Diallo-Dudek, J., Bellec, A., Dumet, A., Mistler, L., & Kaufmann, B. (2016). I'm not like everybody else: Urbanization factors shaping spatial distribution of native and invasive ants are species-specific. *Urban Ecosystem, 20*(1), 157–169.

Glade, T. (2003). Landslide occurrence as a response to land use change: A review of evidence from New Zealand. *Catena, 51*(3), 297–314.

Glibert, P. M. (2017). Eutrophication, harmful algae and biodiversity—Challenging paradigms in a world of complex nutrient changes. *Marine Pollution Bulletin, 124*(2), 591–606.

Gomez-Barroso, D., Velasco, E., Varela, C., Leon, I., & Cano, R. (2017). Spread of Ebola virus disease based on the density of roads in West Africa. *Geospatial Health, 12*(1), 552.

Gomiero, T. (2016). Soil degradation, land scarcity and food security: Reviewing a complex challenge. *Sustainability (Basel, Switzerland), 8*(3), 281.

Goolsby, D. A., David, M. B., Gertner, G. Z., & McIsaac, G. F. (2001). Eutrophication Nitrate flux in the Mississippi River. *Nature, 414*(6860), 166–167.

Gowers, A. M., Cullinan, P., Ayres, J. G., Anderson, H. R., Strachan, D. P., Holgate, S. T., Mills, I. C., & Maynard, R. L. (2012). Does outdoor air pollution induce new cases of asthma? Biological plausibility and evidence; A review. *Respirology (Carlton, Vic.), 17*(6), 887–898.

Gruzieva, O., Bergström, A., Hulchiy, O., Kull, I., Lind, T., Melén, E., Moskalenko, V., Pershagen, G., & Bellander, T. (2013). Exposure to air pollution from traffic and childhood asthma until 12 years of age. *Epidemiology (Cambridge, Mass.), 24*(1), 54–61.

Gubler, D. J. (2011). Dengue, urbanization and globalization: The unholy trinity of the 21(st) century. *Tropical Medicine and Health, 39*(4 Suppl), 3–11.

Guerchet, M., Mbelesso, P., Bandzouzi, B., Pilleron, S., Clément, J. P., Dartigues, J. F., & Preux, P. M. (2013). Comparison of rural and urban dementia prevalences in two countries of central Africa: The EPIDEMCA study. *Journal of the Neurological Sciences, 333*, e328–e328.

Guerra, M. B. B., Teaney, B. T., Mount, B. J., Asunskis, D. J., Jordan, B. T., Barker, R. J., Santos, E. E., & Schaefer, C. E. G. R. (2017). Post-catastrophe analysis of the Fundão tailings dam failure in the Doce River system, Southeast Brazil: Potentially toxic elements in affected soils. *Water, Air, & Soil Pollution, 228*(7), 1–12.

Habib, B., Rajvanshi, A., Mathur, V. B., & Saxena, A. (2016). Corridors at crossroads: Linear development-induced ecological triage as a conservation opportunity. *Frontiers in Ecology and Evolution, 4*, 132.

Haddrill, M. V., Keffer, R., Olivetti, G. C., Polleri, G. B., & Giovanardi, F. (1983). Eutrophication problems in Emilia Romagna, Italy: Monitoring the nutrient load discharged to the littoral zone of the Adriatic Sea. *Water Research (Oxford), 17*(5), 483–495.

Hankey, S., & Marshall, J. D. (2017). Urban form, air pollution, and health. *Current Environmental Health Reports, 4*(4), 491.

Harris, J., Chisanga, B., Drimie, S., & Kennedy, G. (2019). Nutrition transition in Zambia: Changing food supply, food prices, household consumption, diet and nutrition outcomes. *Food Security, 11*, 371–387.

Heemskerk, M. (2002). Livelihood decision making and environmental degradation: Small-scale gold mining in the Suriname Amazon. *Society & Natural Resources, 15*(4), 327–344.

Hoagland, P., Jin, D., Beet, A., Kirkpatrick, B., Reich, A., Ullmann, S., Fleming, L. E., & Kirkpatrick, G. (2014). The human health effects of Florida Red Tide (FRT) blooms: An expanded analysis. *Environment International, 68*, 144–153.

Hoek, G., Brunekreef, B., Goldbohm, S., Fischer, P., & van den Brandt, P. A. (2002). Association between mortality and indicators of traffic-related air pollution in the Netherlands: A cohort study. *The Lancet (British Edition), 360*(9341), 1203–1209.

Hoelzer, K., Wong, N., Thomas, J., Talkington, K., Jungman, E., & Coukell, A. (2017). Antimicrobial drug use in food-producing animals and associated human health risks: What, and how strong, is the evidence? *BMC Veterinary Research, 13*(1), 211–238.

Holmer, B., Thorsson, S., & Lindén, J. (2013). Evening evapotranspirative cooling in relation to vegetation and urban geometry in the city of Ouagadougou, Burkina Faso. *International Journal of Climatology, 33*(15), 3089–3105.

Hudson-Edwards, K. A., Macklin, M. G., Jamieson, H. E., Brewer, P. A., Coulthard, T. J., Howard, A. J., & Turner, J. N. (2003). The impact of tailings dam spills and clean-up operations on sediment and water quality in river systems: The Ros Agrio–Guadiamar, Aznalcóllar, Spain. *Applied Geochemistry, 18*(2), 221–239.

Ibrahim, B., Karambiri, H., Polcher, J., Yacouba, H., & Ribstein, P. (2014). Changes in rainfall regime over Burkina Faso under the climate change conditions simulated by 5 regional climate models. *Climate Dynamics, 42*(5), 1363–1381.

Jackson, R. B., Vengosh, A., Carey, J. W., Davies, R. J., Darrah, T. H., O'Sullivan, F., & Petron, G. (2014). The environmental costs and benefits of fracking. *Annual Review of Environment and Resources, 39*(39), 327–362.

Jia, J., Wang, F., Wei, C., Zhou, A., Jia, X., Li, F., Tang, M., Chu, L., Zhou, Y., Zhou, C., Cui, Y., Wang, Q., Wang, W., Yin, P., Hu, N., Zuo, X., Song, H., Qin, W., Wu, L., Li, D., Jia, L., Song, J., Han, Y., Xing, Y., Yang, P., Li, Y., Qiao, Y., Tang, Y., Lv, J., & Dong, X. (2014). The prevalence of dementia in urban and rural areas of China. *Alzheimer's & Dementia: The Journal of the Alzheimer's Association, 10*(1), 1–9.

Karfakis, M. G. (1993). Chapter 18—Residual subsidence over abandoned coal mines. In E. Hoek (Ed.), *Surface and underground project case histories* (pp. 451–476). Pergamon.

Kearney, J. (2010). Food consumption trends and drivers. *Philosophical Transactions: Biological Sciences, 365*(1554), 2793–2807.

Keken, Z., Kušta, T., Langer, P., & Skaloš, J. (2016). Landscape structural changes between 1950 and 2012 and their role in wildlife–vehicle collisions in the Czech Republic. *Land Use Policy, 59*, 543–556.

Kelly, M. (2016). The nutrition transition in developing Asia: Dietary change, drivers and health impacts. In P. Jackson, W. E. L. Spiess, & F. Sultana (Eds.), *Eating, drinking: Surviving: The international year of global understanding—IYGU* (pp. 83–90). Springer.

Khan, D. A., Bhatti, M. M., Khan, F. A., Naqvi, S. T., & Karam, A. (2008). Adverse effects of pesticides residues on biochemical markers in Pakistani tobacco farmers. *International Journal of Clinical and Experimental Medicine, 1*(3), 274–282.

Kim, K., Kabir, E., & Jahan, S. A. (2017). Exposure to pesticides and the associated human health effects. *Science of the Total Environment, 575*, 525–535.

Kirkpatrick, B., Fleming, L. E., Squicciarini, D., Backer, L. C., Clark, R., Abraham, W., Benson, J., Cheng, Y. S., Johnson, D., Pierce, R., Zaias, J., Bossart, G. D., & Baden, D. G. (2004). *Literature review of Florida red tide: Implications for human health effects* (Vol. 3, pp. 99–115). Elsevier B.V.

Kleerekoper, L., van Esch, M., & Salcedo, T. B. (2012). How to make a city climate-proof, addressing the urban heat island effect. *Resources, Conservation and Recycling, 64*, 30–38.

Knuiman, M. W., Christian, H. E., Divitini, M. L., Foster, S. A., Bull, F. C., Badland, H. M., & Giles-Corti, B. (2014). A longitudinal analysis of the influence of the neighborhood built environment on walking for transportation: The RESIDE study. *American Journal of Epidemiology, 180*(5), 453–461.

Kofman, O., Berger, A., Massarwa, A., Friedman, A., & Jaffar, A. A. (2006). Motor inhibition and learning impairments in school-aged children following exposure to organophosphate pesticides in infancy. *Pediatric Research, 60*(1), 88–92.

Kopittke, P. M., Menzies, N. W., Wang, P., McKenna, B. A., & Lombi, E. (2019). Soil and the intensification of agriculture for global food security. *Environment International, 132*, 105078.

Kosior, G., Samecka-Cymerman, A., Kolon, K., Brudzińska-Kosior, A., Bena, W., & Kempers, A. J. (2015). Trace elements in the Fontinalis antipyretica from rivers receiving sewage of lignite and glass sand mining industry. *Environmental Science and Pollution Research, 22*(13), 9829–9838.

Kudela, R. M., Lane, J. Q., & Cochlan, W. P. (2008). The potential role of anthropogenically derived nitrogen in the growth of harmful algae in California, USA. *Harmful Algae, 8*(1), 103–110.

Lambin, E. F., Tran, A., Vanwambeke, S. O., Linard, C., & Soti, V. (2010). Pathogenic landscapes: Interactions between land, people, disease vectors, and their animal hosts. *International Journal of Health Geographics, 9*(1), 54–54.

Lammle, L., Woll, A., Gert, B., Mensink, M., & Bos, K. (2013). Distal and proximate factors of health behaviors and their associations with health in children and adolescents. *International Journal of Environmental Research and Public Health, 10*(7), 2944–2978.

Lantos, P. M. (2015). Chronic Lyme disease. *Infectious Disease Clinics of North America, 29*(2), 325–340.

Lantos, P. M., Hoffman, K., Höhle, M., Anderson, B., & Gray, G. C. (2016). Are people living near modern swine production facilities at increased risk of influenza virus infection? *Clinical Infectious Diseases, 63*(12), 1558–1563.

Lassus, P., Chomérat, N., Hess, P., & Nézan, E. (2016). Toxic and harmful microalgae of the world ocean/Micro-algues toxiques et nuisibles de l'océan mondial. In *IOC manuals and guides, 68 (Bilingual English/French)*. International Society for the Study of Harmful Algae/Intergovernmental Oceanographic Commission of UNESCO.

Lelieveld, J., Pozzer, A., Pöschl, U., Fnais, M., Haines, A., & Münzel, T. (2020). Loss of life expectancy from air pollution compared to other risk factors: A worldwide perspective. *Cardiovascular Research, 116*(11), 1910–1917.

Levkovich, O., Rouwendal, J., & van Ommeren, J. (2020). The impact of highways on population redistribution: The role of land development restrictions. *Journal of Economic Geography, 20*(3), 783–808.

Li, M., Li, M., De Pinto, A., De Pinto, A., Ulimwengu, J. M., Ulimwengu, J. M., You, L., You, L., Robertson, R. D., & Robertson, R. D. (2015). Impacts of road expansion on deforestation and biological carbon loss in the Democratic Republic of Congo. *Environmental and Resource Economics, 60*(3), 433–469.

Lima, L. S., Coe, M. T., Soares Filho, B. S., Cuadra, S. V., Dias, L. C., Costa, M. H., Lima, L. S., & Rodrigues, H. O. (2014). Feedbacks between deforestation, climate, and hydrology in the Southwestern Amazon: Implications for the provision of ecosystem services. *Landscape Ecology, 29*(2), 261–274.

Lindén, J. (2011). Nocturnal Cool Island in the Sahelian city of Ouagadougou, Burkina Faso. *International Journal of Climatology, 31*(4), 605–620.

Liverani, M., Waage, J., Barnett, T., Pfeiffer, D. U., Rushton, J., Rudge, J. W., Loevinsohn, M. E., Scoones, I., Smith, R. D., Cooper, B. S., White, L. J., Goh, S., Horby, P., Wren, B., Gundogdu, O., Woods, A., & Coker, R. J. (2013). Understanding and managing zoonotic risk in the new livestock industries. *Environmental Health Perspectives, 121*(8), 873–877.

Lu, C., & Tian, H. (2017). Global nitrogen and phosphorus fertilizer use for agriculture production in the past half century: Shifted hot spots and nutrient imbalance. *Earth System Science Data, 9*(1), 181–192.

Luiz Eduardo, O. C. A., Malhi, Y., Barbier, N., Lima, A., Shimabukuro, Y., Anderson, L., & Saatchi, S. (2008). Interactions between rainfall, deforestation and fires during recent years in the Brazilian Amazonia. *Philo-*

sophical *Transactions of the Royal Society, B: Biological Sciences, 363*(1498), 1779–1785.

Macintyre, H. L., Heaviside, C., Taylor, J., Picetti, R., Symonds, P., Cai, X. M., & Vardoulakis, S. (2018). Assessing urban population vulnerability and environmental risks across an urban area during heatwaves—Implications for health protection. *Science of the Total Environment, 610–611*, 678–690.

Martinez, M., Napolitano, D., MacLennan, G., O'Callagham, C., Ciborawski, S., & Fabregas, X. (2007). Impacts of petroleum activities for the Achuar people of the Peruvian Amazon: Summary of existing evidence and research gaps. *Environmental Research Letters, 2*, 045006.

Mastel, M., Bussalleu, A., Paz-Soldán, V. A., Salmón-Mulanovich, G., Valdés-Velásquez, A., & Hartinger, S. M. (2018). Critical linkages between land use change and human health in the Amazon region: A scoping review. *PLoS One, 13*(6), e0196414.

McKenzie, L., Guo, R., Witter, R., Savitz, D., Newman, L., & Adgate, J. (2014). Birth outcomes and maternal residential proximity to natural gas development in rural Colorado. *Environmental Health Perspectives, 122*, 412–417.

McKinlay, R., Plant, J. A., Bell, J. N., & Voulvoulis, N. (2008). Endocrine disrupting pesticides: Implications for risk assessment. *Environment International, 34*(2), 168–183.

Meijer, J. R., Huijbregts, M. A. J., Schotten, K. C. G. J., & Schipper, A. M. (2018). Global patterns of current and future road infrastructure. *Environmental Research Letters, 13*(6), 064006–064011.

Melrose, W. (1999). *Deforestation in Papua New Guinea: Potential impact on health care.* James Cook University.

Ménesguen, A. (2014). Eutrophication of the marine environment. In P. Prouzet & A. Monaco (Eds.), *The land-sea interactions* (pp. 71–191). John.

Messina, J., & Pan, W. K. (2013). Different ontologies: Land change science and health research. *Current Opinion in Environmental Sustainability, 5*, 1–7.

Miller, J. D., Safford, H. D., Crimmins, M., & Thode, A. E. (2008). Quantitative evidence for increasing forest fire severity in the Sierra Nevada and Southern Cascade Mountains, California and Nevada, USA. *Ecosystems (New York), 12*(1), 16–32.

Morris, C. N., & Ferguson, A. G. (2006). Estimation of the sexual transmission of HIV in Kenya and Uganda on the trans-Africa highway: The continuing role for prevention in high risk groups. *Sexually Transmitted Infections, 82*(5), 368–371.

Mugagga, F., Kakembo, V., & Buyinza, M. (2012). Land use changes on the slopes of Mount Elgon and the implications for the occurrence of landslides. *Catena (Giessen), 90*, 39–46.

Munster, V. J., Bausch, D. G., de Wit, E., Fischer, R., Kobinger, G., Muñoz-Fontela, C., Olson, S. H., Seifert, S. N., Sprecher, A., Ntoumi, F., Massaquoi, M., & Mombouli, J.-V. (2018). Outbreaks in a rapidly changing Central Africa—Lessons from Ebola. *The New England Journal of Medicine, 379*(13), 1198–1201.

Murray, K. A., Olivero, J., Roche, B., Tiedt, S., & Guegan, J.-F. (2018). Pathogeography: Leveraging the biogeography of human infectious diseases for global health management. *Ecography, 41*(9), 1411–1427.

National Marine Fisheries Service. (2020). *Fisheries of the United States, 2018.* N. C. F. S. N. U.S. Department of Commerce.

Nava, A., Shimabukuro, J. S., Chmura, A. A., & Bessa Luz, S. L. (2017). The impact of global environmental changes on infectious disease emergence with a focus on risks for Brazil. *ILAR Journal, 58*(3), 393–400.

Ndiaye, A., Adamou, R., Gueye, M., & Diedhiou, A. (2017). Global warming and heat waves in West-Africa: Impacts on electricity consumption in Dakar (Senegal) and Niamey (Niger). *International Journal of Energy and Environmental Science, 2*(1), 16–26.

Newsom, S. W. B. (2005). The history of infection control: Cholera—John Snow and the beginnings of epidemiology. *British Journal of Infection Control, 6*(6), 12–15.

Nishina, K., Ito, A., Hanasaki, N., & Hayashi, S. (2017). Reconstruction of spatially detailed global map of NH4+ and NO3- application in synthetic nitrogen fertilizer. *Earth System Science Data, 9*(1), 149–162.

Novotny, E. V., & Stefan, H. G. (2012). Road salt impact on lake stratification and water quality. *Journal of Hydraulic Engineering, 138*(12), 1069–1080.

Nunes, B., Silva, R. D., Cruz, V. T., Roriz, J. M., Pais, J., & Silva, M. C. (2010). Prevalence and pattern of cognitive impairment in rural and urban populations from Northern Portugal. *BMC Neurology, 10*(1), 42–42.

O'Callaghan-Gordo, C., Flores, J. A., Lizárraga, P., Okamoto, T., Papoulias, D. M., Barclay, V., Orta-Martínez, M., Kogevinas, M., & Astete, J. (2018). Oil extraction in the Amazon basin and exposure to metals in indigenous populations. *Environmental Research, 162*, 226–230.

Oestreicher, J. S., Lucotte, M., Moingt, M., Bélanger, É., Rozon, C., Davidson, R., Mertens, F., & Romaña, C. A. (2017). Environmental and anthropogenic factors influencing mercury dynamics during the past century in Floodplain Lakes of the Tapajós River, Brazilian Amazon. *Archives of Environmental Contamination and Toxicology, 72*(1), 11–30.

Oestreicher, J. S., do Amaral, D. P., Passos, C. J. S., Fillion, M., Mergler, D., Davidson, R., Lucotte, M., Romaña, C. A., & Mertens, F. (2020). Rural development and shifts in household dietary practices from 1999 to 2010 in the Tapajós River region, Brazilian Amazon: Empirical evidence from dietary surveys. *Globalization and Health, 16*(1), 36.

Offerle, B., Jonsson, P., Eliasson, I., & Grimmond, C. S. B. (2005). Urban modification of the surface energy balance in the West African Sahel: Ouagadougou, Burkina Faso. *Journal of Climate, 18*(19), 3983–3995.

Oke, T. R. (1982). The energetic basis of the urban heat island. *Quarterly Journal of the Royal Meteorological Society, 108*(455), 1–24.

Olivero, J., Fa, J. E., Real, R., Márquez, A. L., Farfán, M. A., Vargas, J. M., Gaveau, D., Salim, M. A., Park, D., Suter, J., King, S., Leendertz, S. A., Sheil, D., & Nasi, R. (2017). Recent loss of closed forests is associated with Ebola virus disease outbreaks. *Scientific Reports, 7*(1), 14291–14299.

Ou, J., Liu, X., Li, X., Chen, Y., & Li, J. (2017). Quantifying spatiotemporal dynamics of urban growth modes in metropolitan cities of China: Beijing, Shanghai, Tianjin, and Guangzhou. *Journal of Urban Planning and Development, 143*(1), 4016023.

Oviatt, C. A., & Gold, A. J. (2005). Nitrate in coastal waters. In T. Addiscott (Ed.), *Nitrate, agriculture and the environment* (pp. 127–144). CABI Publishing.

Oyewale, A. T., Adesakin, T. A., & Aduwo, A. I. (2019). Environmental impact of heavy metals from poultry waste discharged into the Olosuru Stream, Ikire, Southwestern Nigeria. *Journal of Health & Pollution, 9*(22), 190607–190610.

Palmer, M. A., Bernhardt, E. S., Schlesinger, W. H., Eshleman, K. N., Foufoula-Georgiou, E., Hendryx, M. S., Lemly, A. D., Likens, G. E., Loucks, O. L., & Power, M. E. (2010). Mountaintop mining consequences. *Science, 327*(5962), 148–149.

Pan, W., Carr, D., Barbieri, A., Bilsborrow, R., & Suchindran, C. (2007). Forest clearing in the Ecuadorian Amazon: A study of patterns over space and time. *Population Research and Policy Review, 26*(5/6), 635–659.

Pan, W. K., Branch, O. H., & Zaitchik, B. (2014). Impact of climate change on vector-borne disease in the Amazon. In K. Pinkerton & W. Rom (Eds.), *Global climate change and public health*. Springer.

Parratt, S. R., Numminen, E., & Laine, A.-L. (2016). Infectious disease dynamics in heterogeneous landscapes. *Annual Review of Ecology, Evolution, and Systematics, 47*, 283–306.

Patil, J. A., Patil, A. J., Sontakke, A. V., & Govindwar, S. P. (2009). Occupational pesticides exposure of sprayers of grape gardens in western Maharashtra (India): Effects on liver and kidney function. *Journal of Basic and Clinical Physiology and Pharmacology, 20*(4), 335–355.

Patz, J., Daszak, P., Tabor, G. M., Aguirre, A. A., Pearl, M., Epstein, J., Wolfe, N. D., Kilpatrick, A. M., Foufopoulos, J., Molyneux, D., Bradley, D. J., & C. Members of the Working Group on Land Use, E. Disease, C. Working Group on Land Use, E. Disease and E. Members of the Working Group on Land Use Change Disease. (2004). Unhealthy landscapes: Policy recommendations on land use change and infectious disease emergence. *Environmental Health Perspectives, 112*(10), 1092–1098.

Patz, J., Campbell-Lendrum, D., Gibbs, H., & Woodruff, R. (2008). Health impact assessment of global climate change: Expanding on comparative risk assessment approaches for policy making. *Annual Review of Public Health, 29*, 27–39.

Pellegrini, P., & Fernández, R. J. (2018). Crop intensification, land use, and on-farm energy-use efficiency during the worldwide spread of the green revolution. *Proceedings of the National Academy of Sciences of the United States of America, 115*(10), 2335–2340.

Penrose, K., Castro, M. C. D., Werema, J., & Ryan, E. T. (2010). Informal urban settlements and cholera risk in Dar es Salaam, Tanzania. *PLoS Neglected Tropical Diseases, 4*(3), e631.

Perz, S. G., Qiu, Y., Xia, Y., Southworth, J., Sun, J., Marsik, M., Rocha, K., Passos, V., Rojas, D., Alarcón, G., Barnes, G., & Baraloto, C. (2013). Trans-boundary infrastructure and land cover change: Highway paving and community-level deforestation in a tri-national frontier in the Amazon. *Land Use Policy, 34*, 27–41.

Petch, R. O., & Henson, R. R. (2000). Child road safety in the urban environment. *Journal of Transport Geography, 8*(3), 197–211.

Peters, R., Ee, N., Peters, J., Beckett, N., Booth, A., Rockwood, K., & Anstey, K. J. (2019). Common risk factors for major noncommunicable disease, a systematic overview of reviews and commentary: The implied potential for targeted risk reduction. *Therapeutic Advances in Chronic Disease, 10*, 2040622319880392–2040622319880392.

Pettigrew, S. M., Pan, W. K., Berky, A., Harrington, J., Bobb, J. F., & Feingold, B. J. (2019). In urban, but not rural, areas of Madre de Dios, Peru, adoption of a Western diet is inversely associated with selenium intake. *The Science of the Total Environment, 687*, 1046–1054.

Pfaff, A., Robalino, J., Walker, R., Aldrich, S., Caldas, M., Reis, E., Perz, S., Bohrer, C., Arima, E., Laurance, W., & Kirby, K. (2007). Road investments, spatial spillovers, and deforestation in the Brazilian Amazon. *Journal of Regional Science, 47*(1), 109–123.

Phelan, P. E., Kaloush, K., Miner, M., Golden, J., Phelan, B., Silva, H., & Taylor, R. A. (2015). Urban heat island: Mechanisms, implications, and possible remedies. *Annual Review of Environment and Resources, 40*(1), 285–307.

Pierce, C., Fuhrman, E., Xiong-Yang, P., Kentnich, J., Husnik, P., Dahlen, J., Liang, R., & Awad, J. (2019). Monitoring of airborne particulates near industrial silica sand mining and processing facilities. *Archives of Environmental & Occupational Health, 74*(4), 185–196.

Popkin, B. M. (1999). Urbanization, lifestyle changes and the nutrition transition. *World Development, 27*, 1905–1916.

Puett, R. C., Hart, J. E., Yanosky, J. D., Spiegelman, D., Wang, M., Fisher, J. A., Hong, B., & Laden, F. (2014). Particulate matter air pollution exposure, distance to road, and incident lung cancer in the nurses' health study cohort. *Environmental Health Perspectives, 122*(9), 926–932.

Pyko, A., Eriksson, C., Lind, T., Mitkovskaya, N., Wallas, A., Ögren, M., Östenson, C.-G., Pershagen, G., Sahlgrenska, A., Göteborgs, U., A. F. S. O. F. E. F.

A.-O. M. Institutionen för medicin, Gothenburg, U., D. O. P. H. Institute of Medicine, S. O. O. Community Medicine, M. Environmental, & Sahlgrenska, A. (2017). Long-term exposure to transportation noise in relation to development of obesity—A cohort study. *Environmental Health Perspectives, 125*(11), 117005.

Rahman, S. (2010). Six decades of agricultural land use change in Bangladesh: Effects on crop diversity, productivity, food availability and the environment, 1948–2006. *Singapore Journal of Tropical Geography, 31*(2), 254–269.

Ramalho, A. A., Mantovani, S. A. S., Delfino, B. M., Pereira, T. M., Martins, A. C., Oliart-Guzmán, H., Brāna, A. M., Branco, F. L. C. C., Campos, R. G., Guimarães, A. S., Araújo, T. S., Oliveira, C. S. M., Codeço, C. T., Muniz, P. T., & da Silva-Nunes, M. (2013). Nutritional status of children under 5 years of age in the Brazilian Western Amazon before and after the interoceanic highway paving: A population-based study. *BMC Public Health, 13*(1), 1098.

Reuben, A., Frischtak, H., Berky, A., Ortiz, E. J., Morales, A. M., Hsu-Kim, H., Pendergast, L. L., & Pan, W. K. (2020). Elevated hair mercury levels are associated with neurodevelopmental deficits in children living near artisanal and small-scale gold mining in Peru. *GeoHealth, 4*(5), e2019GH000222.

Riley, L. W., Raphael, E., & Faerstein, E. (2013). Obesity in the United States—Dysbiosis from exposure to low-dose antibiotics? *Frontiers in Public Health, 1*, 69.

Rivas, A. L., Chowell, G., Schwager, S. J., Fasina, F. O., Hoogesteijn, A. L., Smith, S. D., Bisschop, S. P. R., Anderson, K. L., & Hyman, J. M. (2010). Lessons from Nigeria: The role of roads in the geo-temporal progression of avian influenza (H5N1) virus. *Epidemiology and Infection, 138*(2), 192–198.

Roberts, J. R., Karr, C. J., & H. Council On Environmental and H. Council On Environmental. (2012). Pesticide exposure in children. *Pediatrics (Evanston), 130*(6), e1765–e1788.

Roth, M. (2007). Review of urban climate research in (sub)tropical regions. *International Journal of Climatology, 27*(14), 1859–1873.

Rotureau, B., Joubert, M., Clyti, E., Djossou, F., & Carme, B. (2006). Leishmaniasis among gold miners, French Guiana. *Emerging Infectious Diseases, 12*(7), 1169–1170.

Rulli, M. C., Santini, M., Hayman, D. T. S., & D'Odorico, P. (2017). The nexus between forest fragmentation in Africa and Ebola virus disease outbreaks. *Scientific Reports, 7*(1), 41613.

Sachindra, D. A., Ng, A. W. M., Muthukumaran, S., & Perera, B. J. C. (2016). Impact of climate change on urban heat island effect and extreme temperatures: A case-study. *Quarterly Journal of the Royal Meteorological Society, 142*(694), 172–186.

Salmon-Mulanovich, G., Blazes, D. L., Lescano, A. G., Bausch, D. G., Montgomery, J. M., & Pan, W. K. (2015). Economic burden of dengue virus infection at the household level among residents of Puerto Maldon-

ado, Peru. *The American Journal of Tropical Medicine and Hygiene, 93*(4), 684–690.

Salmón-Mulanovich, G., Blazes, D. L., Guezala, M. C., Rios, V. Z., Espinoza, A., Guevara, C., Lescano, A. G., Montgomery, J. M., Bausch, D. G., & Pan, W. K. (2018). Individual and spatial risk of dengue virus infection in Puerto Maldonado, Peru. *The American Journal of Tropical Medicine and Hygiene, 99*(6), 1440–1450.

Sanchez, J. F., Carnero, A. M., Rivera, E., Rosales, L. A., Baldeviano, G. C., Asencios, J. L., Edgel, K. A., Vinetz, J. M., & Lescano, A. G. (2017). Unstable malaria transmission in the Southern Peruvian Amazon and its association with gold mining, Madre de Dios, 2001-2012. *The American Journal of Tropical Medicine and Hygiene, 96*(2), 304–311.

Sandifer, P. A., Sutton-Grier, A. E., & Ward, B. P. (2015). Exploring connections among nature, biodiversity, ecosystem services, and human health and well-being: Opportunities to enhance health and biodiversity conservation. *Ecosystem Services, 12*, 1–15.

Sangiorgi, F., & Donders, T. H. (2004). Reconstructing 150 years of eutrophication in the north-western Adriatic Sea (Italy) using dinoflagellate cysts, pollen and spores. *Estuarine, Coastal and Shelf Science, 60*(1), 69–79.

Saxton, A. T., Stanifer, J., Miranda, J., Ortiz, E., Taype-Rondan, A., & Pan, W. K. (in review). Prevalence and associated risk factors of diabetes, chronic kidney disease, and hypertension of adults in the Peruvian Amazon: The Amarakaeri Reserve Cohort study. *Preventing Chronic Disease*.

Schinasi, L., Wing, S., Augustino, K. L., Ramsey, K. M., Nobles, D. L., Richardson, D. B., Price, L. B., Aziz, M., MacDonald, P. D. M., & Stewart, J. R. (2014). A case control study of environmental and occupational exposures associated with methicillin resistant Staphylococcus aureus nasal carriage in patients admitted to a rural tertiary care hospital in a high density swine region. *Environmental Health, 13*(1), 54.

Schreinemachers, P., & Tipraqsa, P. (2012). Agricultural pesticides and land use intensification in high, middle and low income countries. *Food Policy, 37*(6), 616–626.

Schuler, M. S., & Relyea, R. A. (2018). A review of the combined threats of road salts and heavy metals to freshwater systems. *Bioscience, 68*(5), 327–335.

Seukep, S. E., Kolivras, K. N., Hong, Y., Li, J., Prisley, S. P., Campbell, J. B., Gaines, D. N., & Dymond, R. L. (2015). An examination of the demographic and environmental variables correlated with Lyme disease emergence in Virginia. *EcoHealth, 12*(4), 634–644.

Sherbinin, A. d., Carr, D., Cassels, S., & Jiang, L. (2007). Population and environment. *Annual Review of Environment and Resources, 32*(1), 345–373.

Simon, J. A., Marrotte, R. R., Desrosiers, N., Fiset, J., Gaitan, J., Gonzalez, A., Koffi, J. K., Lapointe, F.-J., Leighton, P. A., Lindsay, L. R., Logan, T., Milord, F., Ogden, N. H., Rogic, A., Roy-Dufresne, E., Suter, D., Tessier, N., & Millien, V. (2014). Climate change and habitat fragmentation drive the occurrence of Borrelia burgdorferi, the agent of Lyme disease, at the northeastern limit of its distribution. *Evolutionary Applications, 7*(7), 750–764.

Snyder, R. E., Marlow, M. A., & Riley, L. W. (2014). Ebola in urban slums: The elephant in the room. *The Lancet Global Health, 2*(12), e685–e685.

Snyder, R. E., Boone, C. E., Cardoso, C. A. A., Aguiar-Alves, F., Neves, F. P. G., & Riley, L. W. (2017). Zika: A scourge in urban slums. *PLoS Neglected Tropical Diseases, 11*(3), e0005287.

Stacy, S., Brink, L., Larkin, J., Saovsky, Y., Goldstein, B., Pitt, B., & Talbott, E. (2015). Perinatal outcomes and unconventional gas operations in Southwest Pennsylvania. *PLoS One, 10*(6), e0126425.

Steyn, N. P., Nel, J. H., Parker, W., Ayah, R., & Mbithe, D. (2012). Urbanisation and the nutrition transition: A comparison of diet and weight status of South African and Kenyan women. *Scandinavian Journal of Public Health, 40*(3), 229–238.

Sun, L.-L., Liu, D., Chen, T., & He, M.-T. (2019). Road traffic safety: An analysis of the cross-effects of economic, road and population factors. *Chinese Journal of Traumatology, 22*(5), 290–295.

Swenson, J. J., Carter, C. E., Domec, J. C., & Delgado, C. I. (2011). Gold mining in the Peruvian Amazon: Global prices, deforestation, and mercury imports. *PLoS One, 6*(4), e18875.

Sylla, M. B., Faye, A., Giorgi, F., Diedhiou, A., & Kunst-mann, H. (2018). Projected heat stress under 1.5°C and 2°C global warming scenarios creates unprecedented discomfort for humans in West Africa. *Earth's Future, 6*(7), 1029–1044.

Takken, W., de Tarso, P., Vilarinhos, R., Schneider, P., & dos Santos, F. (2003). Effects of environmental change on malaria in the Amazon region of Brazil. In W. Takken, P. Martens, & R. J. Bogers (Eds.), *Environmental change and malaria risk: Global and local implications* (pp. 113–123). Springer.

Terças-Trettel, A. C. P., Oliveira, E. C. D., Fontes, C. J. F., Melo, A. V. G. D., Oliveira, R. C. D., Guter-res, A., Fernandes, J., Silva, R. G. D., Atanaka, M., Espinosa, M. M., & Lemos, E. R. S. D. (2019). Malaria and hantavirus pulmonary syndrome in gold mining in the Amazon Region, Brazil. *International Journal of Environmental Research and Public Health, 16*(10), 1852.

Thapa, G., & Shively, G. (2018). A dose-response model of road development and child nutrition in Nepal. *Research in Transportation Economics, 70*, 112–124.

Thow, A. M. (2009). Trade liberalisation and the nu-trition transition: Mapping the pathways for public health nutritionists. *Public Health Nutrition, 12*(11), 2150–2158.

Thow, A. M., & Hawkes, C. (2009). The implications of trade liberalization for diet and health: A case study from Central America. *Globalization and Health, 5*(1), 5–5.

Tollan, A. (2002). Land-use change and floods: What do we need most, research or management? *Water Science and Technology: a Journal of the International Associ-ation on Water Pollution Research, 45*(8), 183–190.

UN. (2018). *World urbanization prospects*. United Na-tions, Department of Economic and Social Affairs, Pop-ulation Division. The 2018 Revision, Online Edition.

UNEP. (2013). *Global mercury assessment*. Nairobi, Kenya.

USAID. (2017). *Climate change risk in Senegal: Country risk profile*. Fact Sheet.

Utzinger, J., & Keiser, J. (2006). Urbanization and tropical health—Then and now. *Annals of Tropical Medicine and Parasitology, 100*(5–6), 517–533.

Van Cauwenberg, J., Nathan, A., Barnett, A., Barnett, D. W., Cerin, E., & E. Council on, G. Physical Activity -Older Adults Working, E. the Council on and G. Phys-ical Activity -Older Adults Working. (2018). Relation-ships between neighbourhood physical environmental attributes and older adults' leisure-time physical ac-tivity: A systematic review and meta-analysis. *Sports Medicine, 48*(7), 1635–1660.

van Grinsven, H. J. M., Bouwman, L., Cassman, K. G., van Es, H. M., McCrackin, M. L., & Beusen, A. H. W. (2015). Losses of ammonia and nitrate from agri-culture and their effect on nitrogen recovery in the European Union and the United States between 1900 and 2050. *Journal of Environmental Quality, 44*(2), 356–367.

Vengosh, A., Jackson, R. B., Warner, N., Darrah, T. H., & Kondash, A. (2014). A critical review of the risks to water resources from unconventional shale gas devel-opment and hydraulic fracturing in the United States. *Environmental Science & Technology, 48*(15), 8334–8348.

Vittor, A. Y., Pan, W., Gilman, R. H., Tielsch, J., Glass, G., Shields, T., Sanchez-Lozano, W., Pinedo, V. V., Salas-Cobos, E., Flores, S., & Patz, J. A. (2009). Linking de-forestation to malaria in the Amazon: Characterization of the breeding habitat of the principal malaria vec-tor, Anopheles darlingi. *American Journal of Tropical Medicine and Hygiene, 81*(1), 5–12.

Vosti, S. A. (1990). Malaria among gold miners in southern Pará, Brazil: Estimates of determinants and individual costs. *Social Science & Medicine, 30*(10), 1097–1105.

Walelign, S. Z., Nielsen, M. R., & Jacobsen, J. B. (2019). Roads and livelihood activity choices in the Greater Serengeti Ecosystem, Tanzania. *PLoS One, 14*(3), e0213089–e0213089.

Webb, E., Moon, J., Dyrszka, L., Rodriguez, B., Cox, C., Patisaul, H., Bushkin, S., & London, E. (2018). Neurodevelopmental and neurological effects of chem-icals associated with unconventional oil and natural gas operations and their potential effects on infants and children. *Reviews on Environmental Health, 33*(1), 3–29.

Webster, R. G. (2004). Wet markets—A continuing source of severe acute respiratory syndrome and influenza? *The Lancet, 363*(9404), 234–236.

Wedagama, D. M. P., Dissanayake, D., & Bird, R. (2011). Modelling the relationship between male and female pedestrian accidents and land use characteristics (Case

Study: Newcastle upon Tyne, UK). *Journal of the Eastern Asia Society for Transportation Studies, 9,* 1715–1730.

Weden, M. M., Shih, R. A., Kabeto, M. U., & Langa, K. M. (2018). Secular trends in dementia and cognitive impairment of U.S. rural and urban older adults. *American Journal of Preventive Medicine, 54*(2), 164–172.

Weigle, K. A., Santrich, C., Martinez, F., Valderrama, L., & Saravia, N. G. (1993). Epidemiology of cutaneous leishmaniasis in Colombia: A longitudinal study of the natural history, prevalence, and incidence of infection and clinical manifestations. *The Journal of Infectious Diseases, 168*(3), 699–708.

Weinberger, B., Greiner, L. H., Walleigh, L., & Brown, D. (2017). Health symptoms in residents living near shale gas activity: A retrospective record review from the Environmental Health Project. *Preventive Medical Reports, 8,* 112–115.

Weinhouse, C., Ortiz, E. J., Berky, A. J., Bullins, P., Hare-Grogg, J., Rogers, L., Morales, A. M., Hsu-Kim, H., & Pan, W. K. (2017). Hair mercury level is associated with Anemia and micronutrient status in children living near artisanal and small-scale gold mining in the Peruvian Amazon. *The American Journal of Tropical Medicine and Hygiene, 97*(6), 1886–1897.

Weinhouse, C., Gallis, J. A., Ortiz, E., Berky, A. J., Morales, A. M., Diringer, S. E., Harrington, J., Bullins, P., Rogers, L., Hare-Grogg, J., Hsu-Kim, H., & Pan, W. K. (2020). A population-based mercury exposure assessment near an artisanal and small-scale gold mining site in the Peruvian Amazon. *Journal of Exposure Science & Environmental Epidemiology.* https://doi.org/10.1038/s41370-020-0234-2

WHO. (2002). *The world health report 2002: Reducing risks, promoting health life.* World Health Organization.

WHO. (2004). *Avian Influenza A (H5N1)—Update 31.* World Health Organization.

WHO. (2018a). *Global status report on road safety 2018.* World Health Organization.

WHO. (2018b). *WHO global ambient air quality database (update 2018).* World Health Organization.

WHO. (2020). *World health statistics 2020: Monitoring health for the SDGs.* World Health Organization. License: CC BY-NC-SA 3.0 IGO.

WHO and UN-Habitat. (2016). *Global report on urban health: Equitable healthier cities for sustainable development.* World Health Organization.

Wilby, R. L. (2008). Constructing climate change scenarios of urban Heat Island intensity and air quality. *Environment and Planning. B, Planning & Design, 35*(5), 902–919.

Wilson, M. L., Renne, E., Roncoli, C., Agyei-Baffour, P., & Tenkorang, E. Y. (2015). Integrated assessment of artisanal and small-scale Gold Mining in Ghana—Part 3: Social sciences and economics. *International Journal of Environmental Research and Public Health, 12*(7), 8133–8156.

Woo, P. C. Y., Lau, S. K. P., & Yuen, K.-y. (2006). Infectious diseases emerging from Chinese wet-markets: Zoonotic origins of severe respiratory viral infections. *Current Opinion in Infectious Diseases, 19*(5), 401–407.

World Bank. (2017). *The high toll of traffic injuries: Unacceptable and preventable.* World Bank.

World Bank. (2019a). *Guide for road safety opportunities and challenges: Low- and middle-income countries country profiles.* World Bank.

World Bank (2019b). *World Bank open data.* https://data.worldbank.org/

WU Vienna. (2019). *Global domestic extraction in 2017, by material group.* Visualisations based upon the UN IRP Global Material Flows Database. V. U. o. E. a. Business: materialflows.net/visualisation-centre/raw-material-profiles

Wyatt, L., Ortiz, E., Feingold, B., Berky, A., Diringer, S., Morales, A., Jurado, E., Hsu-Kim, H., & Pan, W. (2017). Spatial, temporal, and dietary variables associated with elevated mercury exposure in Peruvian riverine communities upstream and downstream of artisanal and small-scale gold mining. *International Journal of Environmental Research and Public Health, 14*(12), 1582.

Wyatt, L., Permar, S. R., Ortiz, E., Berky, A., Woods, C. W., Amouou, G. F., Itell, H., Hsu-Kim, H., & Pan, W. (2019). Mercury exposure and poor nutritional status reduce response to six expanded program on immunization vaccines in children: An observational cohort study of communities affected by gold mining in the Peruvian Amazon. *International Journal of Environmental Research and Public Health, 16*(4), 1–22.

Ye, M. F. (2018). 4.2 Causes and consequences of air pollution in Beijing, China. In K. A. Clark, T. R. Shaul, & B. H. Lower (Eds.), *Environmental science bites.* Ohio State University Press.

Yepes, T., Pierce, J., & Foster, V. (2008). *Making sense of Sub-Saharan Africa's infrastructure endowment: A benchmarking approach* (Working Paper 1, Africa Infrastructure Country Diagnostic). World Bank.

Yuen, K. Y. (2004). SARS: Market, toilet, hospital, and laboratory. *Hong Kong Medical Journal = Xianggang yi xue za zhi, 10*(3), 148.

Zhang, H., & Shan, B. (2008). Historical records of heavy metal accumulation in sediments and the relationship with agricultural intensification in the Yangtze–Huaihe region, China. *The Science of the Total Environment, 399*(1–3), 113–120.

Zheng, Z., Ren, G., Wang, H., Dou, J., Gao, Z., Duan, C., Li, Y., Ngarukiyimana, J. P., Zhao, C., Cao, C., Jiang, M., & Yang, Y. (2018). Relationship between fine-particle pollution and the urban heat island in Beijing, China: Observational evidence. *Boundary-Layer Meteorology, 169*(1), 93–113.

Zhou, C., Li, S., & Wang, S. (2018). Examining the impacts of urban form on air pollution in developing countries: A case study of China's megacities. *International Journal of Environmental Research and Public Health, 15*(8), 1565.

Health and Mortality Consequences of Natural Disasters

16

Mark VanLandingham, Bonnie Bui, David Abramson, Sarah Friedman, and Rhae Cisneros

Abstract

We review, summarize, and critically evaluate recent theoretical and empirical research focusing on the health and mortality consequences of disasters. We begin with a general conceptual framework to outline key causal mechanisms linking disasters with health-related outcomes. We highlight recent substantive and methodological advances in the field and identify gaps and shortcomings. We conclude that the impacts of disasters on health and mortality are large, widely distributed, and unevenly distributed; are often spread out over a long period of time; and are conceptually complex. Our

principal substantive recommendation for moving the field forward is a shift from the current primary emphasis on the health and mortality consequences of disasters that occur during the immediate aftermath of the event to a broader time horizon encompassing the medium and long term effects. Such a longer-term perspective would allow for more research on trajectories of recovery rather than reliance on point-in-time estimates. Our principal methodological recommendation is a shift to longitudinal approaches in order to avoid the many pitfalls associated with cross-sectional designs. Leveraging research already underway when a disaster strikes oftenmakes available pre-disaster data that

---

M. VanLandingham (✉)
Department of Social, Behavioral, and Population Sciences, Tulane University School of Public Health and Tropical Medicine, New Orleans, LA, USA
e-mail: mvanlan@tulane.edu

B. Bui
Department of Global Community Health and Behavioral Sciences, Tulane University School of Public Health and Tropical Medicine, New Orleans, LA, USA

D. Abramson
Department of Social and Behavioral Sciences, Program on Population Impact, Recovery, and Resilience, New York University School of Global Public Health, New York, NY, USA

S. Friedman
New York University School of Global Public Health, New York, NY, USA

R. Cisneros
Public Health Region 6/5S Preparedness and Response Program, Texas Department of State Health Services, Austin, TX, USA

© Springer Nature Switzerland AG 2022
L. M. Hunter et al. (eds.), *International Handbook of Population and Environment*, International Handbooks of Population 10, https://doi.org/10.1007/978-3-030-76433-3_16

Introduction

Natural disasters[1] disrupt communities in profound and myriad ways. The loss of life, event-related injuries, declines in health, and spread of disease are among the most noteworthy and devastating of these impacts. Such health and mortality impacts of disasters are also among the most difficult to assess. One source of complexity with regard to conceptualization results from the fact that the health and mortality impacts from disasters are inextricably linked to a wide array of circumstances that exist prior to, during, and after the disaster. A source of difficulty in assessment results from the scarcity of high-quality data on these pre- and post-disaster circumstances as well as post-disaster health and mortality outcomes, especially when these occur subsequent to the immediate aftermath of the event (Wood & Bourque, 2018).

These difficulties and complexities in assessing the effects of natural disasters on health and mortality in turn complicate important policy and program decisions regarding how to prepare for and mitigate the consequences of such disasters. The importance of addressing these scientific and policy challenges is becoming increasingly urgent since the intensity and frequency of weather-related disasters are expected to increase over the next few decades (Van Aalst, 2006; Field et al., 2012; Goldstein et al., 2016). In addition, population density in areas especially vulnerable to natural disasters is expected to increase (Neu-

---

[1] By "natural disaster" we mean an event that has a principal basis in nature, *i.e.*, an event with a significant environmental basis. We acknowledge significant human contributions to these events, such as the influence of fossil fuel consumption on global climate change, engineering failures leading to the collapse of the federal levee system after Hurricane Katrina struck New Orleans, *etc.* We employ the term to exclude purposeful events such as war, terrorist attacks, and mass shootings, although we occasionally draw on studies of such events when the results seem especially relevant for an understanding of the health consequences of natural disasters. In this review paper, we use the term disaster to mean natural disaster.

mann et al., 2015). In this chapter, we summarize and synthesize the existing state-of-knowledge on the health and mortality consequences of natural disasters. We begin with a conceptual framework positioning the health consequences of disasters within the context of disaster *recovery* more generally. We then review the recent empirical literature that assesses how disasters affect health and follow this with an exploration of a set of more specific issues related to the mortality consequences of disasters. We conclude with recommendations for moving this area of research forward with an aim to inform the development of responsive policy and programs.

A Conceptual Framework

The International Federation of Red Cross and Red Crescent Societies (IFRC) defines a disaster as "a sudden, calamitous event that seriously disrupts the functioning of a community or society and causes human, material, and economic or environmental losses that exceed the community's or society's ability to cope using its own resources" (IFRC website, accessed July 9, 2019). A natural disaster (hereafter, disaster) is defined as " a sudden and terrible event in nature (such as a hurricane, tornado, or flood) that usually results in serious damage and many deaths (Merriam-Webster Dictionary website, accessed July 16, 2019).

Often, the precipitating hazardous event sets in motion a cascading sequence of subsequent events with manifold health consequences. For example, the Great East Japan Earthquake of 2011 triggered a massive tsunami and led directly to the failure and partial meltdown of the Fukushima Daiichi nuclear plant which resulted in the radiological exposure of human and animal populations. The 2010 Deepwater Horizon oil spill in the Gulf of Mexico led to over 170 million gallons of oil dispersed into the Gulf ecosystem, threatening human health through direct contact, inhalation, and ingestion of contaminated seafood. The defining features of such hazardous events that turn into disasters are the scale of destruction or disruption and the systemic effects. During the Gulf oil spill, for example, the initial environmental event was followed almost immediately by massive economic consequences, as both commercial fishing and oil exploration, two principal sources of regional revenue, were suspended. In these and other disasters, the initial events produce systemic effects, both in terms of the secondary consequences (*e.g.*, population evacuation and displacement) and in terms of system disruptions (*e.g.*, disrupted supply chains). Large-scale disasters can degrade or destroy any number of overlapping and inter-connected systems associated with population health and well-being, including local markets; civic and institutional sectors including health and human services, educational and faith-based services, and governmental services; and critical lifelines such as energy, communications, transportation, and water and wastewater systems, among others. Population health is highly dependent upon the routine functioning of these overlapping and inter-connected systems.

Our starting point for this review of health, mortality and natural disasters is the Socioecological Model of Disaster Recovery, developed by Abramson and colleagues – see the Figure (Abramson et al., 2010b). This model allows for consideration of complex exposures to disasters and their consequences, while also integrating pre-existing vulnerabilities (*e.g.*, social determinants of health related to socio-economic class or neighborhood levels of poverty) and multiple moderating effects (*e.g.*, the role of formal and informal social supports, networks, and capital). Health is a principal outcome of interest. As the diagram illustrates, several important interactions shape population health outcomes of disasters. First, as illustrated in the far right section of the Figure, health outcomes co-vary with other key factors related to housing, economics, and social functioning. Well-conceived studies of the health and mortality consequences of disasters will include a wide range of indicators of recovery in order to place the health findings in proper context. Second, the extent and severity of health impacts will depend on several factors in place at around the time of the disaster and during the subsequent recovery period (center of the Figure).

Disaster exposure may contain both a primary exposure, such as the initial hurricane, as well as a secondary exposure, such as a subsequent population displacement. Formal Help indicates assistance which can be timely and ample, or late and inadequate. Post-Event Modifiers account for the fact that some communities will be well-positioned to bounce back from a major setback, while others will never fully recover (Fig. 16.1).

Pre-Event Factors

As illustrated in the far left section in the Figure, the extent and severity of these health-related impacts depends on several sets of characteristics that were already in place prior to the disaster event and recovery period (Chronic Stressors and Pre-Event Moderators). This set of pre-event factors influences not only the course of recovery for individuals who are affected by a disaster, but also the likelihood of disaster in the first place. Low income families often seek housing in flood-prone areas, where prices are low. Pre-event chronic stressors and moderators can worsen the consequences of disaster impacts. Such stressors can operate at collective levels (Bronfenbrenner, 1986), such as poverty and economic disparities. Stressors can also be located within households, such as family dysfunction; or within individuals, such as chronic co-morbidities or pre-existing mental illness.

These critically important pre-disaster characteristics are among the most vexing to anticipate, measure, and model in efforts to understand the health and mortality impacts of disasters. That pre-disaster health and general well-being would positively correlate with post-disaster health accords with intuition (good begets good; bad begets bad) and also with perspectives focusing on fundamental causes (Link & Phelan, 1995) and upstream variables (McKinlay, 1993). What makes these predictors difficult to anticipate and measure is that disasters by their very nature are unpredictable and thus data collection often takes place after the event. Victims of disasters—or anyone else—are often not able to report with accuracy and precision

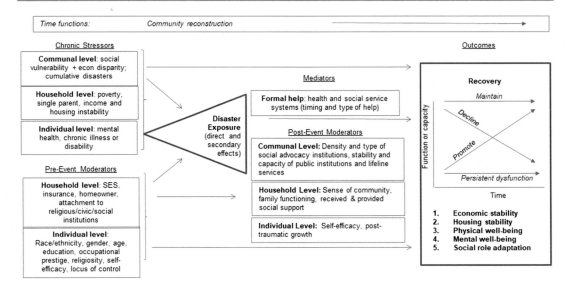

Fig. 16.1 Socioecological model of disaster recovery

how they were faring at some point in the past, *i.e.*, before an event occurred. A related empirical challenge relates to the measurement and modeling of health, which is generally considered an outcome variable. Yet as we point out above, health can also be a moderating variable: pre-disaster health may moderate (influence) the relationship between the experience of a disaster and subsequent health and well-being. Health can also be a mediating variable: long-term displacement may result in epigenetic changes that in turn may lead to increases in chronic morbidity.

Factors Related to the Disaster Event

To understand the level and distribution of disasters' health and mortality consequences, studies need to include comparable measures of impact, such as housing damage and length of displacement. Several of these factors are illustrated in the middle part of the Figure. All other things being equal, a family that loses their home and are displaced for an extended period of time can expect to experience worse health outcomes than a household that experiences only minor damage and short term displacement. Unfortunately, families that experience similar levels of damage do not necessarily receive similar levels of formal and informal help, and this variability in assistance will need to be considered in any model seeking to predict health outcomes. Studies of resilience are in an early stage, but this early work makes clear that some families – or communities – will face a disaster with a greater store of resilience than others (VanLandingham, 2017), a source of variability that also needs to be considered.

Long-Term Consequences and Invisible Factors

Another challenge to proper conceptualization and measurement of the health and mortality consequences of disasters—and one that is illustrated to some extent in the top of the Figure — involves time. While most assessments of the health and mortality consequences of disasters focus on the immediate aftermath, in fact many consequences will not become manifest until much later, when most assessments have ended. Individuals who appear at first to have weathered the event well may subsequently suffer from an earlier death than would have been the case had they not suffered this setback, or from disaster-related negative health consequences that were simply delayed in their manifestation. Following a cohort of affected families for even a few years can be challenging and expensive, and a longitudinal study lasting for a decade or more can be especially so.

A fifth and final challenge is that health itself is multi-faceted, and its measurement usually involves multiple layers of subjectivity and error, not only from self-reported results but also from clinician judgment and biological marker measurement and interpretation. Furthermore, while good health certainly has intrinsic value, it is often of interest as a critical steppingstone to economic and social engagement or productivity for adults, and to education, growth and development for children. Such life-course outcomes, which build upon good physical and mental health, are central to the interests of many health researchers and social scientists. Linking health to such outcomes also presents additional measurement challenges.

The health consequences from a disaster are both complex and manifold. Disasters impact health both directly and through other disruptions to society. To begin to unravel the complexity of disasters' impact on health, a wide set of characteristics must be considered, from pre-event factors which influence the level of impact that individuals and communities will experience from the disaster; to variations in impact, such as housing damage and length of displacement; and to time, since some effects will be delayed. Next we turn to examining the current research on disasters, health and mortality, in order to better understand the current state of the science in this area and to assess where the field needs to turn.

Current State of the Field – Disasters and Health

The following review of the empirical literature on the health consequences of disasters first examines the most common impacts at each level of social organization at which they occur. Next, we turn to a discussion of common impacts at various time points over the course of a disaster.

Finally, this section concludes with a discussion of disparities in outcomes.

Levels of Impact

Community-Level Impacts

The impacts of a disaster upon the health of the affected population occur at multiple levels, including community, household, and individual. At the community level, damage to health facilities and more general community infrastructure can make access to care much more difficult than it was before the disaster (Berggren & Curiel, 2006). Such impacts may be especially severe in areas where the health infrastructure and access to care were already strained before the disaster occurred (Chandra et al., 2011; Runkle et al., 2012). These impacts may also be especially acute for sub-groups who were already facing serious health challenges. For instance, chronic disease requiring regular care is very common among older adults (Aldrich & Benson, 2007) and interruptions in care are, therefore, especially consequential (Ngo, 2001).

If a disaster is severe enough, medical providers may choose to permanently leave the area, resulting in a shortage of healthcare professionals (Berggren & Curiel, 2006; Rudowitz et al., 2006). Damage to the electrical grid can make activities of daily living more difficult and dangerous, *e.g.*, for chronically ill individuals who rely on electricity-dependent medical equipment such as ventilators, or for elderly populations who are vulnerable to extremes in temperature without air conditioning or heat. Water may become polluted and dangerous to drink.

Household-Level Impacts

At the household level, disasters can have multi-faceted impacts including financial strains due to interruption of income flows, repair expenses from damages to the home, displacement costs, *etc.* These financial strains may lessen a household's ability to afford health care or purchase nutritious food (Stehling-Ariza et al., 2012). Strain can also impact exercise and sleep, as well as yielding poor mental health outcomes, especially among those in middle age (Sastry & Vanlandingham, 2009; Vu & Vanlandingham, 2011) who shoulder much of the burden for household well-being (Norris et al., 2002).

In addition to the direct effects that disasters can have on the health of individuals through these community and family-level impacts, there are other potential effects that are more indirect. For instance, displacement complicates access to health care, goods, and other services that contribute to well-being. Access to social networks is also hindered which inhibits knowledge about employment, health care, schools, and other dimensions of daily life. Changes to one's built environment can lead to unexpected obstacles such as decreased walkability and, in turn, worse physical health (Arcaya et al., 2014a).

Individual-Level Impacts

The effects of disasters on health operate directly on individuals, too. For instance, direct experience of fear and trauma related to rising water, crumbling buildings, injuries to oneself or witnessing injuries or death to others can have short-term or long-lasting effects on the physical and mental health of persons who experience a disaster first-hand, whether they are living in affected areas or are first responders to the event (Benedek et al., 2007). Research on the mental health consequences of disasters is well-developed (Norris et al., 2002; Goldmann & Galea, 2014). There are several measures of mental distress that are commonly utilized in the disaster research literature. Post-traumatic stress disorder, or PTSD, is a mental health condition recognized in the Diagnostic and Statistical Manual of Mental Disorders. PTSD can result from exposure to a traumatic event. A person experiencing PTSD continues to experience terror or horror at what they've seen or experienced previously and remains in an especially vigilant state, sometimes called "hyper-arousal." Post-traumatic stress symptoms (PTS or PTSS) and post-traumatic stress reactivity (PTSR) are characterized by symptoms that are severe and intense but do not reach the threshold of a clinical disorder. General psychological distress (PD) is characterized by symptoms of mood disorders and anxiety. PD is a major outcome in the National Health Interview Survey and is measured using the K6 scale of non-specific mental illness (Kessler et al., 2002). Major depressive disorder (MDD) is frequently measured in studies of the mental health consequences of disasters, and elevated rates post-disaster are common (Norris et al., 2002). These constructs of mental health challenges that are common after a disaster are not mutually exclusive; more than one may occur simultaneously (Goldmann & Galea, 2014). The widely-used Short Form 36 (SF-36), developed by RAND Corporation and J. E. Ware and colleagues (Ware & Sherbourne, 1992), has three measures of mental health; one of the most commonly used scales is the mental health component summary (MCS) score. One major advantage of this measure is that it can assess variations in mental well-being within and across healthy populations and across time; it is not limited to discerning cases with extreme outcomes.

Galea and colleagues (Galea et al., 2007) documented PTSD rates of about 30% in the area affected by Katrina. Kessler and colleagues, using the K6, found elevated rates of PD post-Katrina: both serious mental illness and mild-moderate mental illness were nearly twice as high in the Katrina-affected area compared to levels measured before Katrina (Kessler et al., 2006). Using the SF-12, a shortened version of the SF-36, Karaye and colleagues (Karaye et al., 2019) found that 3 months after Hurricane Harvey made landfall along the Texas Gulf Coast, affected residents reported mental health scores well below (worse than) the national average.

Timing of Impacts

Short Term Impacts

Impacts are often temporally clustered and occur at various time points after a disaster. In the immediate aftermath of a disaster, first responders and local clinics and hospitals deal with injury, trauma, exposure to the elements, and lack of food and water. While concerns about increased spread of infectious disease (*e.g.*, cholera) and coming into contact with extremely toxic flood waters in the aftermath of a disaster are generally inflated compared to the actual impacts—at least for recent hurricanes in the U.S.—acute gastroenteritis was very common among Katrina evacuees in Louisiana, Texas, Louisiana, and Mississippi during the weeks immediately following the storm. Norovirus gastroenteritis is often found among sheltered populations in congregate post-disaster settings. The Centers for Disease Control and Prevention (CDC) also reported increases in acute respiratory illnesses in the Katrina-affected area immediately after the storm (Bourque et al., 2006). In massive catastrophes such as the 2010 earthquake in Haiti, in addition to widespread mortality and morbidity in the immediate aftermath of such a disaster, a breakdown of law and order can lead to increases in sexual assault and other violent crime (Kolbe et al., 2010). Worry and concern about the whereabouts and well-being of loved ones—especially children—are a source of major anxiety and influence subsequent post-disaster psychological adjustment (Lowe et al., 2011).

Medium Term Impacts

In the medium term (Years 1–5), issues such as previously-undiagnosed trauma often becomes apparent, along with more chronic stressors related to making one's way in a new environment and putting the family's life back together (Uscher-Pines, 2009). Paxson and colleagues found that among members of their Resilience in the Survivors of Katrina (RISK) cohort, while both PD and PTSS declined between Waves 1 (at around Year 1) and Wave 2 (at around Year 4), at Wave 2 rates were still elevated compared to what they were before Katrina (Paxson et al., 2012). Similarly, while Mississippi residents who experienced Katrina showed small improvements in mental health between two early post-Katrina waves (during Year 1 and near Year 2), more than half of the overall Gulf Coast Child and Family Health (GCAFH) cohort of residents of Mississippi and Louisiana who were significantly impacted by Katrina

continued to report significant mental health distress 2 years after the event (Abramson et al., 2008). In a related study focusing on the well-being of children in the GCAFH cohort described above, depression, anxiety, or behavior disorders were found to be widespread during the first 4 years after Katrina. Children who were exposed to Hurricane Katrina were five times more likely to show serious emotional disturbance than were their counterparts from a pre-Katrina cohort (Abramson et al., 2010a). For the young women respondents of RISK cohort study of Hurricane Katrina, experiencing an unstable housing situation post-disaster or relocation to a new community both led to poor mental health outcomes relative to returning to one's pre-disaster community (Fussell & Lowe, 2014).

Longer Term Impacts

While many studies have examined the health implications of disasters in the near term and a fair number of studies have followed victims through the medium term, few studies have taken a longer-term view. Opportunities for long-term follow up of a cohort with pre-disaster measures are especially rare, and even where they do exist, keeping such a cohort intact for an extended study is difficult and expensive. Where pre-disaster data do not exist, data collected around the time of the event ("peri-event data") can be predictive of subsequent post-disaster outcomes. In a longitudinal study of Hurricane Ike survivors, Lowe and colleagues (Lowe et al., 2015) leverage such peri-event data to predict mental health and general wellness at later points in time. An exploratory study of World Trade Center survivors suggests that disaster-related PTSD may lead to new-onset diabetes years later (Miller-Archie et al., 2014). Thomas, Frankenberg, and colleagues using more compelling biomarker and longitudinal data (including data collected before the disaster) find that middle-age men living in communities heavily damaged by the 2004 Indian Ocean tsunami had higher levels of diabetes 12 years later than did their counterparts who had been living in undamaged areas at the time of the tsunami (Thomas et al., 2018). More positively, Bui and colleagues find that even after taking into account pre-disaster health and other characteristics and disaster-related losses, having high quality social support soon after a disaster is positively associated with good mental and physical health outcomes later on (Bui et al., 2020).

Disparities in Post-Disaster Health Impacts

The Influence of Pre-Disaster Circumstances

The issue of timing discussed in the previous section has vast implications for post-disaster health disparities. Specifically, post-disaster health impacts are shaped in large part by pre-disaster circumstances. In pre-Katrina New Orleans, for example, many older adults, individuals with disabilities, people experiencing poverty, and ethnic minorities were pre-disposed to poor post-Katrina health outcomes because of poor access to transportation for evacuation and limited resources to finance a displacement. These pre-disaster circumstances made their evacuation from the ensuing flooding much more difficult than it was for their counterparts who were more advantaged on these characteristics (Elliott & Pais, 2006; Zoraster, 2010). Also, those with lower income were more likely to live in low-lying areas of the city that were destined to flood extensively after the federal levee system collapsed the day after Katrina made landfall (Logan, 2006). High levels of housing damage, in turn, explains much of the higher levels of psychological distress among Katrina-impacted black residents 1 year post-Katrina (Sastry & Vanlandingham, 2009) and also likely contributed to displacement and delays in return for black families (Fussell et al., 2010) which, in turn negatively impacted post-disaster mental health and well-being (Uscher-Pines, 2009). Pre-disaster social connectivity contributes to post-disaster health disparities, too: poor-quality pre-disaster social support is predictive of post-disaster psychological distress (Green et al., 2012).

Wealth, socioeconomic status, prior health, and privilege loom large in the literature on the distribution of health and mortality impacts of disasters. For instance, neighborhood affluence was negatively associated with mortality during the 1995 Chicago heat wave; areas that experienced commercial decline also experienced structural disadvantages in collective efficacy and suffered from an unequal distribution of community-based resources that proved lethal for many residents of those neighborhoods (Browning et al., 2006). Females, minorities, the poor, and persons who had mental health challenges before a disaster have been shown to experience poor mental health outcomes after a disaster (Norris et al., 2002; Neria et al., 2008). General health status as a pre-disaster characteristic can influence a wide range of post-disaster outcomes. For example, poor pre-Katrina health status among individuals displaced from New Orleans is associated with resettlement into neighborhoods with higher levels of poverty[2] (Arcaya et al., 2014b).

Low-income areas and rural communities experience delayed resumption of medical care after a disaster (Davis et al., 2010). Communities that were medically underserved before a disaster might be especially vulnerable to disaster-related disruptions in the limited health care services that do exist. For example, enclaves of recent immigrants lacking language skills in English often have a limited supply of health care providers who can communicate effectively with them. This was certainly the case for the Vietnamese community in New Orleans before Katrina. Do and colleagues (Do et al., 2009) find that after Hurricane Katrina, this community experienced a steep decrease in routine health care. A major review concludes that much more research is needed on how communities that were medically underserved before a disaster fare afterwards (Davis et al., 2010).

---

[2] Health and neighborhood poverty were not associated for these respondents at baseline (before Katrina).

The Influence of Circumstances During the Disaster

What happens to affected individuals around the time of the disaster is also consequential for disparities in later health outcomes. In the Lowe et al. study cited above (Lowe et al., 2015), the team finds that lower emotional reactions around the time of the event and higher levels of community collective efficacy bode well for subsequent mental health, while loss of pets and financial setbacks bode poorly for subsequent general wellness. Other studies focusing on Katrina also find that disaster-related financial losses are strongly related to the onset and duration of negative mental health outcomes (Galea et al., 2008). Merdjanoff (2013) concludes from her analysis that having a home rendered uninhabitable by Katrina is even worse for psychological well-being than having a completely destroyed home; renters fared especially poorly on the emotional distress outcomes.

Elsewhere, Frankenberg and colleagues (Frankenberg et al., 2008) find that traumatic events, loss of kin, and property damage were significantly associated with higher PTSR scores among residents of coastal Ache and North Sumatra experiencing the 2004 Indian Ocean tsunami. Indeed, disruptions and dispersion of social networks and social support have a wide range of negative health consequences for disaster victims (Norris et al., 2002). For instance, Morris and Deterding (2016) find that dispersion of respondents' social networks after Katrina is associated with post-Katrina post-traumatic stress. Key social-psychological mechanisms identified are a lack of deep belonging and an inability to fulfill important social obligations. In an analysis of several studies, Chan and Rhodes (2014) find that the following threats and situations had strong associations to both GPD and PTS: threat to physical integrity of self and others, a lack of basic necessities, and loss of pet.

Exposure to the effects of a disaster seem especially consequential for subsequent health outcomes for children. In the 2010 analysis of GCAFH data (cited earlier), exposure to the effects of Katrina had negative impacts on subsequent mental health outcomes for children (Abramson et al., 2010a). Children seem particularly vulnerable to PTSD (Neria et al., 2008). Disasters may also have implications for the subsequent physical health of affected children. A study of the effects of heat waves experienced early in life for children in Mexico suggests a negative impact on subsequent adult height (Agüero, 2014). And disasters often take a disproportionate mortality toll on children: when Sri Lanka was struck by the 2004 Indian Ocean tsunami, the death rate of young children living in a camp for displaced persons was 32%, more than 4 times the death rate for young adults (Nishikiori et al., 2006).

Older persons are also especially vulnerable to negative health consequences, including mortality (Peek, 2013), after experiencing a disaster (Ngo, 2001). In Katrina's aftermath, older persons over age 60—and especially over age 74—were substantially overrepresented among those who lost their lives (Bourque et al., 2006). The same is true for the Sri Lankan camp struck by the 2004 Indian Ocean tsunami (Nishikiori et al., 2006).

Resilience

Recently, some research teams have addressed the issue of disparities in post-disaster outcomes by focusing on individuals or groups with positive outcomes: why are some more resilient to disasters than others? Bonanno et al. (2006) find that even among those highly exposed to the September 2001 terrorist attacks in New York City, those exhibiting resilience to trauma exceeded a third of the sample. In a subsequent article related to the same study, the authors find associations between post-disaster resilience and a wide range of socioeconomic categories and previous life stressors (Bonanno et al., 2007). Chan and Rhodes (2013) report that positive religious coping, such as seeking spiritual guidance, predicts post-traumatic growth, *i.e.*, subjective positive psychological changes resulting from a traumatic event. Focusing on a Vietnamese American community that was heavily flooded by Hurricane Katrina, VanLandingham concludes that resilience can be a group-level attribute as well as an individual attribute, resulting from a shared history and a resulting set of shared narratives that emerge from that shared history (VanLandingham, 2017). Norris et al. find much lower rates of PTSD among Vietnamese Americans who experienced the effects of Hurricane Katrina compared to levels reported by similarly-affected populations (Norris et al., 2009). Other recent work has focused on the interplay between one's genetic endowment and experience in determining resilience. Dunn et al. find preliminary evidence in recent work for gene-environment interactions driving disparities in PTG (Dunn et al., 2014). Clearly, resilience is a fertile field for further exploration.

We've organized this review of recent empirical work on the health consequences of disasters to parallel our initial conceptual framework. We first consider level of impact (community, household, or individual) and then consider the timing of impact (short-term, medium-term, and longer-term). We concluded this section with some attention to disparities in impact, and highlighted some new exciting fields of inquiry.

Current State of the Field – Disasters and Mortality

That research on the mortality consequences of disasters is vast is not surprising. Mortality is an event of profound consequence not just to the individual who dies but also to loved ones and associates. Death is also a clear outcome. Unlike health or socioeconomic outcomes, death is an event that is unambiguous to categorize, and given its importance, governments devote significant resources to documenting occurrences. In fact, the estimated death toll in the immediate aftermath of a disaster is often regarded as the principal metric of disaster magnitude.

Some recent major disasters have had staggering estimated death tolls, *e.g.*, over 300,000 from the 2010 earthquake in Haiti and over 200,000 from the 2004 "Indian Ocean" tsunami caused by the Sumatra-Andaman Islands earthquake (Wood & Bourque, 2018). These very large initial estimates are often revised substantially upward or downward after the fact, but given the immediate requirements of massive burials, the needs of survivors, and the general lack of resources in many disaster-prone countries, more precise figures often remain elusive. Even in affluent societies like the United States and Japan, estimating the death toll from a major disaster is fraught with difficulties. During the chaos of post-Katrina New Orleans, it was difficult to ascertain whether missing persons were displaced or deceased. Similarly for Puerto Rico after Hurricane Maria, the initial modest death toll of 64 seemed implausibly low, and was perhaps motivated in part by political interests motivated by a desire to demonstrate an effective government response. More systematic approaches have resulted in much higher estimates of Maria-related deaths such as an independent assessment using a representative survey and baseline mortality rates which estimates 4645 deaths (Kishore et al., 2018). Santos-Burgoa et al. (2018) adopted a different strategy, conducting a time-series analysis based on actual death records from September 2017 through February 2018, but accounting for differential base populations as residents migrated from and returned to the island. The team estimated an excess mortality of 1191 deaths associated with Hurricane Maria. In a third approach to ascertain a more accurate death toll from Hurricane Maria, Santos-Lozada and Howard (2018) used monthly death counts from vital records from January 2010 through December 2016 to establish expected monthly deaths and a 95% confidence interval (CI). For September through December 2017, they calculated the difference between the number of deaths and the upper 95% CI bound, estimating 1139 excess deaths, much higher than the official death toll of 64 which does not account for indirect deaths. Notwithstanding the differences in methods and estimates, all three studies concluded that Maria-related mortality was much higher than the original estimate of 64 deaths. While these much higher figures include many omitted deaths that occurred during the immediate aftermath of Maria, the larger numbers are also due to the inclusion of deaths that resulted from Maria but occurred later. For example, injuries, disease, and hardship resulting from a disaster may lead to a pre-mature death at a later point, as will the interruption of access to health care for pre-existing conditions. Assessing whether mortality is disaster-related or not is complicated by the fact that the precise proximate cause of death can often be difficult to pinpoint. Five years after the Indonesian tsunami, older adults who had experienced property loss from that event continued to experience elevated mortality risk (Ho et al., 2017). Ten years out, mortality was elevated for older men with poor post-tsunami psychosocial health and for older women whose spouse died in the tsunami (Frankenberg et al., 2020).

To date, there is no standard method to properly attribute deaths to a disaster although some efforts are underway. Combs and colleagues (Combs et al., 1999) have developed a flow chart to help coroners and others make data-based judgments as to whether a decedent's death was caused by a disaster. Such data would include the location and timing of the disaster in question, the circumstances of death, and the cause and manner of death. The instrument results in a tally of deaths "directly" attributable to the disaster by "environmental forces"; deaths "indirectly" attributable to the disaster by "loss or disruption of service"; and deaths "indirectly" attributable to the disaster by "personal loss or lifestyle disruption." Uscher-Pines proposes further development of this framework to include more precise timeframes during which deaths resulting indirectly from a disaster may occur and by also adding a third category for deaths that are eventually "triggered" by the disaster (Uscher-Pines, 2007). More recently, the U.S. Centers for Disease Control (CDC) has developed and issued a reference guide to help officials and researchers further refine estimates of disaster-related mortality. This reference guide employs a flow chart similar to the two earlier efforts discussed above, and provides examples for proper death certificate notation (NCHS, 2017).

Compared to the numerous challenges of measuring health outcomes related to a disaster, in many ways enumerating deaths is more straightforward. Nevertheless, important methodological challenges face researchers as they attempt to calculate the full mortality consequences of disasters.

Conclusions and Recommendations for Moving Forward the Field on Health, Mortality, and Natural Disasters

This critical review of recent conceptual and empirical work focusing on the impacts of disasters on health and mortality makes clear that such impacts are large, widely distributed, and unevenly distributed; are often spread out over a long period of time; and are conceptually complex. One part of this complexity is related to timing and sequence. The distribution and severity of impact is driven in part by factors that are in play before the disaster occurs and by other factors that unfold during or soon after the disaster. These features of timing and sequence highlight the importance of longitudinal approaches that incorporate pre-event characteristics and long-term follow-up. A second part of this complexity is related to causal pathways. Some mortality and morbidity from disasters will result from the direct consequences of exposure, *e.g.*, death by drowning or injury from flying debris. More often, the health and mortality consequences are mediated by intervening factors – for example, a decline in long-term mental health due to the destruction of one's home – or moderated by interacting factors that confer resilience or vulnerability – *e.g.*, a bolstering of mental health recovery resulting from strong social support soon after the disaster. These features of causal pathways highlight the importance of sophisticated analytical approaches and a theoretical grounding in the social sciences.

Our recommendations for moving this body of research forward include topics of substance and methods. Regarding substance, we recommend a shift from the current primary emphasis on the health and mortality consequences of disasters that occur during the immediate aftermath of the event to a broader time horizon encompassing the medium (1–5 year) and long term (5 years and beyond) effects. Such a longer-term perspective would allow for more research on trajectories of recovery rather than reliance on point-in-time estimates. This shift in perspective will provide a more accurate and complete understanding of how the recovery of health status unfolds over time (Lowe & Rhodes, 2013; Goldmann & Galea, 2014). We also recommend a shift from the current primary emphasis on negative health and mortality outcomes to a broader perspective that would include more emphasis on the positive end of the distribution of outcomes. We're excited by the expansion of research on resilience and post-traumatic growth and encourage funding agencies to support these new initiatives.

Regarding methods, we conclude that a longitudinal approach, while expensive, is essential. Following a cohort of affected individuals and families over time avoids many of the pitfalls of cross-sectional designs. Given the importance of pre-disaster data for understanding post-disaster outcomes, supporting studies that attempt to leverage data collection enterprises already underway in the field when a disaster occurs (Frankenberg et al., 2008; Rhodes et al., 2010; Vu & Vanlandingham, 2011; Heid et al., 2016) will likely pay substantial benefits for moving this field forward. We also recommend more work on attributable cause with regard to the mortality impacts of disasters. Finally, standard methods for estimating disaster-related mortality are sorely needed so that researchers can provide officials and policy makers with complete, accurate, comparable, and objective results.

References

Abramson, D., Stehling-Ariza, T., Garfield, R., & Redlener, I. (2008). Prevalence and predictors of mental health distress post-Katrina: Findings from the Gulf coast child and family health study. *Disaster Medicine and Public Health Preparedness, 2*(2), 77–86.

Abramson, D. M., Park, Y. S., Stehling-Ariza, T., & Redlener, I. (2010a). Children as bellwethers of recovery: Dysfunctional systems and the effects of parents, households, and neighborhoods on serious emotional disturbance in children after hurricane Katrina. *Disaster Medicine and Public Health Preparedness, 4*, S17–S27.

Abramson, D. M., Stehling-Ariza, T., Park, Y. S., Walsh, L., & Culp, D. (2010b). Measuring individual disaster recovery: A socioecological framework. *Disaster Medicine and Public Health Preparedness, 4*, S46–S54.

Agüero, J. M. (2014). *Long-term effect of climate change on health: Evidence from heat waves in Mexico* (IDB working paper series). Inter-American Development Bank.

Aldrich, N., & Benson, W. F. (2007). Disaster preparedness and the chronic disease needs of vulnerable older adults. *Preventing Chronic Disease, 5*(1), A27–A27.

Arcaya, M., James, P., Rhodes, J. E., Waters, M. C., & Subramanian, S. (2014a). Urban sprawl and body mass index among displaced hurricane Katrina survivors. *Preventive Medicine, 65*, 40–46.

Arcaya, M. C., Subramanian, S. V., Rhodes, J. E., & Waters, M. C. (2014b). Role of health in predicting moves to poor neighborhoods among hurricane Katrina survivors. *Proceedings of the National Academy of Sciences of the United States of America, 111*(46), 16246–16253.

Benedek, D. M., Fullerton, C., & Ursano, R. J. (2007). First responders: Mental health consequences of natural and human-made disasters for public health and public safety workers. *Annual Review of Public Health, 28*, 55–68.

Berggren, R. E., & Curiel, T. J. (2006). After the storm—Health care infrastructure in post-Katrina New Orleans. *The New England Journal of Medicine, 354*(15), 1549–1552.

Bonanno, G. A., Galea, S., Bucciarelli, A., & Vlahov, D. (2006). Psychological resilience after disaster: New York City in the aftermath of the September 11th terrorist attack. *Psychological Science, 17*(3), 181–186.

Bonanno, G. A., Galea, S., Bucciarelli, A., & Vlahov, D. (2007). What predicts psychological resilience after disaster? The role of demographics, resources, and life stress. *Journal of Consulting and Clinical Psychology, 75*(5), 671–682.

Bourque, L. B., Siegel, J. M., Kano, M., & Wood, M. M. (2006). Weathering the storm: The impact of hurricanes on physical and mental health. *The Annals of the American Academy of Political and Social Science, 604*, 129–151.

Bronfenbrenner, U. (1986). Ecology of the family as a context for human development: Research perspectives. *Developmental Psychology, 22*(6), 723–742.

Browning, C. R., Wallace, D., Feinberg, S. L., & Cagney, K. A. (2006). Neighborhood social processes, physical conditions, and disaster-related mortality: The case of the 1995 Chicago heat wave. *American Sociological Review, 71*(4), 661–678.

Bui, B., Anglewicz, P., & VanLandingham, M. (2020). *The impact of hurricane Katrina on physical and mental health within New Orleans' Vietnamese-American community: The role of social support* (CSDP working paper series). T. University.

Chan, C. S., & Rhodes, J. E. (2013). Religious coping, posttraumatic stress, psychological distress, and post-traumatic growth among female survivors four years after hurricane Katrina. *Journal of Traumatic Stress, 26*(2), 257–265.

Chan, C. S., & Rhodes, J. E. (2014). Measuring exposure in hurricane Katrina: A meta-analysis and an integrative data analysis. *PLoS One, 9*(4), e92899.

Chandra, A., Acosta, J., Stern, S., Uscher-Pines, L., Williams, M. V., Yeung, D., Garnett, J., & Meredith, L. S. (2011). Building community resilience to disasters: A way forward to enhance national health security. *Rand Health Quarterly, 1*, 6.

Combs, D. L., Quenemoen, L. E., Parrish, R. G., & Davis, J. H. (1999). Assessing disaster-attributed mortality: Development and application of a definition and classification matrix. *International Journal of Epidemiology, 28*, 1124–1129.

Davis, J. R., Wilson, S., Brock-Martin, A., Glover, S., & Svendsen, E. R. (2010). The impact of disasters on populations with health and health care disparities. *Disaster Medicine and Public Health Preparedness, 4*(1), 30–38.

Do, M. P., Hutchinson, P. L., Mai, K. V., & Vanlandingham, M. J. (2009). Disparities in health care among Vietnamese New Orleanians and the impacts of hurricane Katrina. *Research in the Sociology of Health Care, 27*, 301–319.

Dunn, E. C., Solovieff, N., Lowe, S. R., Gallagher, P. J., Chaponis, J., Rosand, J., Koenen, K. C., Waters, M. C., Rhodes, J. E., & Smoller, J. W. (2014). Interaction between genetic variants and exposure to hurricane Katrina on post-traumatic stress and post-traumatic growth: A prospective analysis of low income adults. *Journal of Affective Disorders, 152-154*, 243–249.

Elliott, J. R., & Pais, J. (2006). Race, class, and hurricane Katrina: Social differences in human responses to disaster. *Social Science Research, 35*(2), 295–321.

Field, C. B., Barros, V., Stocker, T. F., & Dahe, Q. (2012). *Managing the risks of extreme events and disasters to advance climate change adaptation: Special report of the intergovernmental panel on climate change.* Cambridge University Press.

Frankenberg, E., Friedman, J., Gillespie, T., Ingwersen, N., Pynoos, R., Rifai, U., Sikoki, B., Steinberg, A., Sumantri, C., Suriastini, W., & Thomas, D. (2008). Mental health in Sumatra after the tsunami. *American Journal of Public Health, 98*(9), 1671–1677.

Frankenberg, E., Sumantri, C., & Thomas, D. (2020). Effects of a natural disaster on mortality risks over the longer term. *Nature Sustainability, 3*, 614.

Fussell, E., & Lowe, S. R. (2014). The impact of housing displacement on the mental health of low-income parents after hurricane Katrina. *Social Science & Medicine, 113*, 137–144.

Fussell, E., Sastry, N., & VanLandingham, M. (2010). Race, socioeconomic status, and return migration to New Orleans after hurricane Katrina. *Population and Environment, 31*(1–3), 20–42.

Galea, S., Brewin, C. R., Gruber, M., Jones, R. T., King, D. W., King, L. A., McNally, R. J., Ursano, R. J., Petukhova, M., & Kessler, R. C. (2007). Exposure to hurricane-related stressors and mental illness after hurricane Katrina. *Archives of General Psychiatry, 64*, 1–8.

Galea, S., Tracy, M., Norris, F., & Coffey, S. F. (2008). Financial and social circumstances and the incidence and course of PTSD in Mississippi during the first two years after hurricane Katrina. *Journal of Traumatic Stress, 21*(4), 357–368.

Goldmann, E., & Galea, S. (2014). Mental health consequences of disasters. *Annual Review of Public Health, 35*(1), 169–183.

Goldstein, R. B., Smith, S. M., Chou, S. P., Saha, T. D., Jung, J., Zhang, H., Pickering, R. P., Ruan, W. J., Huang, B., & Grant, B. F. (2016). The epidemiology of DSM-5 posttraumatic stress disorder in the United States: Results from the national epidemiologic survey on alcohol and related conditions-III. *Social Psychiatry and Psychiatric Epidemiology, 51*(8), 1137–1148.

Green, G., Lowe, S. R., & Rhodes, J. E. (2012). What can multiwave studies teach us about disaster research: An analysis of low-income hurricane Katrina survivors. *Journal of Traumatic Stress, 25*(3), 299–306.

Heid, A. R., Christman, Z., Pruchno, R., Cartwright, F. P., & Wilson-Genderson, M. (2016). Vulnerable, but why? Post-traumatic stress symptoms in older adults exposed to hurricane Sandy. *Disaster Medicine and Public Health Preparedness, 10*(3), 362–370.

Ho, J. Y., Frankenberg, E., Sumantri, C., & Thomas, D. (2017). Adult mortality five years after a natural disaster. *Population and Development Review, 43*(3), 467–490.

Karaye, I. M., Ross, A. D., Perez-Patron, M., Thompson, C., Taylor, N., & Horney, J. A. (2019). Factors associated with self-reported mental health of residents exposed to hurricane Harvey. *Progress in Disaster Science, 2*, 100016.

Kessler, R. C., Andrews, G., Colpe, L. J., Hiripi, E., Mroczek, D. K., Normand, S.-L., Walters, E. E., & Zaslavsky, A. M. (2002). Short screening scales to monitor population prevalences and trends in non-specific psychological distress. *Psychological Medicine, 32*(6), 959–976.

Kessler, R. C., Galea, S., Jones, R. T., & Parker, H. A. (2006). Mental illness and suicidality after hurricane Katrina. *Bulletin of the World Health Organization, 84*, 930–939.

Kishore, N., Marqués, D., Mahmud, A., Kiang, M. V., Rodriguez, I., Fuller, A., Ebner, P., Sorensen, C., Racy, F., Lemery, J., Maas, L., Leaning, J., Irizarry, R. A., Balsari, S., & Buckee, C. O. (2018). Mortality in Puerto Rico after hurricane Maria. *New England Journal of Medicine, 379*(2), 162–170.

Kolbe, A. R., Hutson, R. A., Shannon, H., Trzcinski, E., Miles, B., Levitz, N., Puccio, M., James, L., Noel, J. R., & Muggah, R. (2010). Mortality, crime and access to basic needs before and after the Haiti earthquke: A random survey of Port-au-Prince households. *Medicine, Conflict, and Survival, 26*(4), 281–297.

Link, B. G., & Phelan, J. (1995). Social conditions as fundamental causes of disease. *Journal of Health and Social Behavior, 1995*, 80–94.

Logan, J. R. (2006). *The impact of Katrina: Race and class in storm-damaged neighborhoods.* Spatial Structures in the Social Sciences, Brown University.

Lowe, S. R., & Rhodes, J. E. (2013). Trajectories of psychological distress among low-income, female survivors of hurricane Katrina. *The American Journal of Orthopsychiatry, 83*(2 Pt 3), 398–412.

Lowe, S. R., Chan, C. S., & Rhodes, J. E. (2011). The impact of child-related stressors on the psychological functioning of lower-income mothers after hurricane Katrina. *Journal of Family Issues, 32*(10), 1303–1324.

Lowe, S. R., Joshi, S., Pietrzak, R. H., Galea, S., & Cerdá, M. (2015). Mental health and general wellness in the aftermath of hurricane Ike. *Social Science & Medicine (1982), 124*, 162–170.

McKinlay, J. B. (1993). The promotion of health through planned sociopolitical change: Challenges for research and policy. *Social Science & Medicine, 36*(2), 109–117.

Merdjanoff, A. A. (2013). There's no place like home: Examining the emotional consequences of hurricane Katrina on the displaced residents of New Orleans. *Social Science Research, 42*(5), 1222–1235.

Miller-Archie, S. A., Jordan, H. T., Ruff, R. R., Chamany, S., Cone, J. E., Brackbill, R. M., Kong, J., Ortega, F., & Stellman, S. D. (2014). Posttraumatic stress disorder and new-onset diabetes among adult survivors of the world trade center disaster. *Preventive Medicine, 66*, 34–38.

Morris, K. A., & Deterding, N. M. (2016). The emotional cost of distance: Geographic social network dispersion and post-traumatic stress among survivors of hurricane Katrina. *Social Science & Medicine, 165*, 56–65.

NCHS. (2017). *A reference guide for certification of deaths in the event of a natural, human-induced, or chemical/radiological disaster. U. S. D. o. H. a. H. Services.* National Center for Health Statistics, Centers for Disease Control and Prevention.

Neria, Y., Nandi, A., & Galea, S. (2008). Post-traumatic stress disorder following disasters: A systematic review. *Psychological Medicine, 38*(4), 467–480.

Neumann, B., Vafeidis, A. T., Zimmermann, J., & Nicholls, R. J. (2015). Future coastal population growth and exposure to sea-level rise and coastal flooding-a global assessment. *PLoS One, 10*(3), e0118571.

Ngo, E. B. (2001). When disasters and age collide: Reviewing vulnerability of the elderly. *Natural Hazards Review, 2*(2), 80–89.

Nishikiori, N., Abe, T., Costa, D. G. M., Dharmaratne, S. D., Kunii, O., & Moji, K. (2006). Who died as a result of the tsunami? — Risk factors of mortality among internally displaced persons in Sri Lanka: A retrospective cohort analysis. *BioMed Central Public Health, 6*, 1.

Norris, F. H., Friedman, M. J., Watson, P. J., Byrne, C. M., Diaz, E., & Kaniasty, K. (2002). 60,000 disaster victims speak: Part I. An empirical review of the empirical literature, 1981–2001. *Psychiatry: Interpersonal and Biological Processes, 65*(3), 207–239.

Norris, F. H., VanLandingham, M. J., & Vu, L. (2009). PTSD in Vietnamese Americans following hurricane Katrina: Prevalence, patterns, and predictors. *Journal of Traumatic Stress, 22*(2), 91–101.

Paxson, C., Fussell, E., Rhodes, J., & Waters, M. (2012). Five years later: Recovery from post traumatic stress and psychological distress among low-income mothers affected by hurricane Katrina. *Social Science & Medicine, 74*(2), 150–157.

Peek, L. (2013). Age. In D. S. K. Thomas, B. D. Phillips, W. E. Lovekamp, & A. Fothergill (Eds.), *Social vulnerability to disasters* (pp. 167–198). CRC Press.

Rhodes, J., Chan, C., Paxson, C., Rouse, C. E., Waters, M., & Fussell, E. (2010). The impact of hurricane Katrina on the mental and physical health of low-income parents in New Orleans. *The American Journal of Orthopsychiatry, 80*(2), 237–247.

Rudowitz, R., Rowland, D., & Shartzer, A. (2006). Health care in New Orleans before and after hurricane Katrina. *Health Affairs, 25*(5), w393–w406.

Runkle, J. D., Brock-Martin, A., Karmaus, W., & Svendsen, E. R. (2012). Secondary surge capacity: A framework for understanding long-term access to primary care for medically vulnerable populations in disaster recovery. *American Journal of Public Health, 102*(12), e24–e32.

Santos-Burgoa, C., Sandberg, J., Suárez, E., Goldman-Hawes, A., Zeger, S., Garcia-Meza, A., Pérez, C. M., Estrada-Merly, N., Colón-Ramos, U., Nazario, C. M., Andrade, E., Roess, A., & Goldman, L. (2018). Differential and persistent risk of excess mortality from hurricane Maria in Puerto Rico: A time-series analysis. *The Lancet Planetary Health, 2*(11), e478–e488.

Santos-Lozada, A. R., & Howard, J. T. (2018). Use of death counts from vital statistics to calculate excess deaths in Puerto Rico following hurricane Maria. *JAMA, 320*(14), 1491–1493.

Sastry, N., & Vanlandingham, M. J. (2009). One year later: Mental illness prevalence and disparities among New Orleans residents displaced by hurricane Katrina. *American Journal of Public Health, 99*, 725–731.

Stehling-Ariza, T., Park, Y. S., Sury, J. J., & Abramson, D. (2012). Measuring the impact of hurricane Katrina on access to a personal healthcare provider: The use of the National Survey of Children's Health for an External Comparison Group. *Maternal and Child Health Journal, 16*, S170–S177.

Thomas, D., Frankenberg, E., Seeman, T., & Sumantri, C. (2018). Effect of stress on cardiometabolic health 12 years after the Indian Ocean tsunami: A quasi-experimental longitudinal study. *The Lancet Planetary Health, 2*, S8.

Uscher-Pines, L. (2007). "But for the hurricane": Measuring natural disaster mortality over the long term. *Prehospital and Disaster Medicine, 22*(2), 149–151.

Uscher-Pines, L. (2009). Health effects of relocation following disaster: A systematic review of the literature. *Disasters, 33*(1), 1–22.

Van Aalst, M. K. (2006). The impacts of climate change on the risk of natural disasters. *Disasters, 30*(1), 5–18.

VanLandingham, M. (2017). *Weathering Katrina: Culture and recovery among Vietnamese-Americans.* Russell Sage Foundation.

Vu, L., & Vanlandingham, M. J. (2011). Physical and mental health consequences of Katrina on Vietnamese immigrants in New Orleans: A pre- and post-disaster assessment. *Journal of Immigrant and Minority Health, 14*(3), 386–394.

Ware, J. E., & Sherbourne, C. D. (1992). The MOS 36-item short-form health survey (SF-36). Conceptual framework and item selection. *Medical Care, 30,* 473.

Wood, M. M., & Bourque, L. B. (2018). Morbidity and mortality associated with disasters. In H. Rodríguez, W. Donner, & J. E. Trainor (Eds.), *Handbook of disaster research* (pp. 357–383). Springer International Publishing.

Zoraster, R. M. (2010). Vulnerable populations: Hurricane Katrina as a case study. *Prehospital and Disaster Medicine, 25*(1), 74–78.

Part V

The Influence of Demographic Dynamics on the Environment

Cities and Their Environments

17

Mark R. Montgomery, Jessie Pinchoff, and Erica K. Chuang

Abstract

If it is properly managed with attention not only to human well-being but also to the well-being of the other species that share our common spaces, urbanization can be a powerful force for good. This analytic survey of cities and their environments begins with a review of commonly misunderstood features of the demography of urban population growth and the geography of urban spatial expansion. The principal theme that runs through the remainder of the chapter is the need for unified governance approaches to highly heterogenous spaces, which span core urban areas, the urban fringe, and an outer envelope of rural land, water basins, and a variety of ecological habitats. Some of these critical spaces are large geographic units that extend far beyond the boundaries of municipal jurisdictions as such. The need for some integrating governance mechanism has been apparent for decades in low- and

M. R. Montgomery (✉)
Economics department, Stony Brook University and Population Council, Stony Brook, NY, USA

J. Pinchoff
Population Council, New York, NY, USA

E. K. Chuang
University of California San Diego, San Diego, CA, USA

high-income countries alike. The all-too-obvious absence of such joined-up authorities, especially in poor countries, testifies to the legal, administrative, budgetary, technical, and scientific challenges entailed in creating such sustainable governance systems.

Introduction

For many decades, the force of urbanization has been reshaping the spaces in which humans and other species live, profoundly altering the terms on which we all co-exist. City- and town-dwellers inhabit spaces that are relatively compact, dense, and diverse by comparison with those of rural areas. If it is effectively anticipated and managed, this increasing concentration of the human population has the potential to improve access to essential services and raise living standards, by tapping the scale economies and positive externalities that stem from agglomeration (Panel on Urban Population Dynamics 2003; World Bank 2009). Well-managed urban concentration would also benefit rural dwellers, for whom city economies provide markets, services, and (via migration) both employment and significant flows of remittances. In these ways, urbanization—if accompanied

© Springer Nature Switzerland AG 2022
L. M. Hunter et al. (eds.), *International Handbook of Population and Environment*, International Handbooks of Population 10, https://doi.org/10.1007/978-3-030-76433-3_17

by a dense network of rural–urban economic and social connections—can be a powerful force for inclusive growth (UNFPA 2007). The sustainability of such growth, however, depends crucially on the environmental resources on which cities depend and influence over time. Some of these ecosystem resources are akin to public goods; others are affected by negative externalities from human activities; and in general, it cannot be assumed that the full value of such resources is factored into private economic decisions.

By virtue of their on-the-ground focus, local urban and rural governments are in a position to take an integrated and inclusive view of all their constituents' needs. But each such government has a limited jurisdictional space within which it can act. Only higher-level governments can be expected to perceive cross-jurisdiction connections and inequities and take effective action to address them. Effective governance in an urbanizing world thus requires complex mutually-supportive arrangements among the tiers and units of government that oversee different spaces. Exactly how to fashion such arrangements is a significant challenge, and no one optimal model has yet emerged (Panel on Urban Population Dynamics 2003, Chapter 9). Furthermore, although an appreciation of their roles is no doubt growing, it is probably still rare that the non-human inhabitants of these common spaces are given adequate consideration by any level of government. Much is being learned—sometimes at great cost—about the protections that natural barriers to flooding can provide, for example, and about the threats to habitats and species survival that arise from ill-considered human actions.

In the lead-up to the Sustainable Development Goals, the High-Level Panel (United Nations 2013a) argued forcefully for an holistic view of urban–rural–environmental systems, and warned of the risks of narrowly urban-specific or rural-specific perspectives:

> The post-2015 agenda must be relevant for urban dwellers. Cities are where the battle for sustainable development will be won or lost. Yet the Panel also believes that it is critical to pay attention to rural areas, where three billion near-poor will still be living in 2030. The most pressing issue is not urban versus rural, but how to foster a local, geographic approach to the post-2015 agenda.

In seeking insights into such integrated approaches, this chapter provides an introduction and brief guide to the literature on cities and their immediate environments. The relevant literature spans many scientific disciplines, and we cannot hope to do it justice here. To bring a measure of order to our discussion, we will adopt a framing device that was put to good effect by Ros-Tonen et al. (2015), as depicted in Fig. 17.1. They distinguish three broad spaces of governance: a core urban region marked by dense connections and within-urban externalities both positive and negative; a peri-urban transition or *desakota* zone having a mixed urban–rural character; and an outer rural envelope, which both influences and is influenced by the other two zones.[1] As with the early and still under-appreciated analysis of McGranahan et al. (2004), these authors place externalities, governance, and both local and far-flung spatial effects at the center of the story.

The urban core can be likened to an "urban metabolism" that is fed via supplies of fresh water, nutrients, raw materials and energy from the transitional and rural zones, and is protected to an extent by their natural barriers and absorptive capacities. If not properly regulated, this urban organism will emit untreated wastewater into the transitional zone, along with quantities of air, light, and even noise pollution. Like the outer rural envelope, the transitional zone can be subject to deforestation, over-exploitation of resources, and fragmentation of natural habitats, damages that stem from economic activities underway in the core (as well as the international economic system) that take little heed of un-priced environmental externalities. In some treatments, a portion of the boundary between the transitional zone and the outer envelope is termed the wildland–urban interface, a phrase that draws attention to the possible transmission of risks in both directions.

To draw out the implications of this wide-lens perspective, our chapter is organized as follows. In section "Demography, Geography, and Sus-

---

[1] The term *desakota* originated with McGee (1987, 1991).

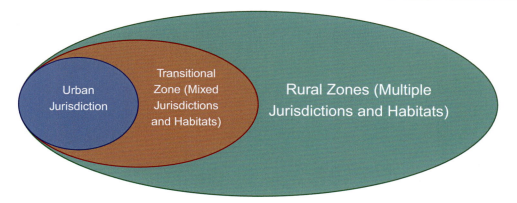

Fig. 17.1 Three broad spaces of urban and rural governance. Adapted from Ros-Tonen et al. (2015)

tainability", we sketch important demographic and geographic features of urbanization that must be borne in mind in any discussion of environmental contexts. Section "Environmental Risks Facing Urban Dwellers" examines the risks that face inhabitants of the urban core; section "Environmental Effects of Urbanization: The Urban–Rural Interface" explores how urban-core activities affect the fringe and outer rural zones. Section "Conclusion" concludes.

Demography, Geography, and Sustainability

To situate cities and towns in their wider contexts, Fig. 17.2 provides a view of the spaces surrounding the core urban zone of Mumbai, one of the world's largest cities located on the coast of Maharashtra state. The top image of the figure gives a land-cover perspective: it shows areas of built-up land in what is known as the Mumbai Metropolitan Region (MMR), with the more built-up land shown in brighter colors. In an ideal world, the MMR would function as a set of linked areas constituting an integrated economic–environmental planning zone. The middle image displays one of the challenges that stands between that ideal and current realities: it depicts the boundaries of the many local government jurisdictions that oversee the spaces of this zone. The boundaries of statutory urban governments are indicated in blue (where available, within-city wards are also shown). Census towns, which in India have urban characteristics but whose governments are legally rural, are shown in gray. In light red are the multitude of "revenue villages" in the MMR, which are sets of villages proper that are combined for administrative purposes. As can be seen, the jurisdictional space of the MMR is exceedingly complex. To add to the challenges, some of these jurisdictions have a hand in safeguarding nature reserves and protected areas, shown in green in the bottom image, which include Sanjay Gandhi National Park.

Not evident in these figures, but a significant presence nevertheless, is the state government of Maharashtra (which has considerable authority in the legal designation of land and settlements as urban or rural), the national government, and both Indian and international non-governmental actors, which are important especially in the technical definition and support given to protection of natural reserves. If the holistic vision of the MMR is to materialize, then somehow these diverse actors and presences must find a way of coordinating across their human and non-human constituencies in an appropriate and sustainable plan of action.

Fig. 17.2 Mumbai metropolitan region (MMR): (**a**) Top-most image shows percentage built-up; (**b**) middle image depicts boundaries of statutory urban settlements (in blue, within-urban wards are shown where available); census towns (gray); and rural villages (light red); and (**c**) bottom image shows three protected areas (in green) that intersect the MMR. Sources: Authors' calculations drawing on data from Corbane et al. (2019); ML Infomap Pvt. Ltd. (2014); UNEP-WCMC and IUCN (2019)

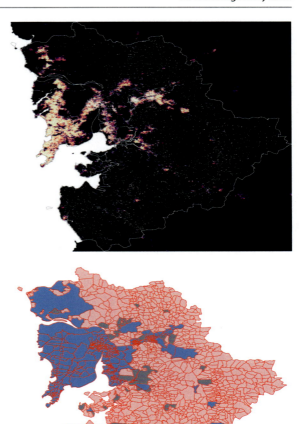

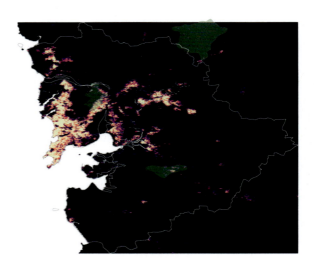

Empirical Regularities in Urbanization

To appreciate how urbanization presents challenges to low- and high-income countries more generally, five of its important features need to be appreciated. First, the amount of land taken up by core urban areas is very small when set against the area of the transitional and outer rural zones. In an analytic review of the literature, Liu et al. (2014) note that estimates of the percentage of urban land typically range from 1 to 3%.[2] They distinguish three types of such land: that covered by impervious surfaces (roads, sidewalks, driveways, parking lots, and rooftops, which are usually estimated to account for less than 1% of all land); built-up land which includes impervious surfaces but also includes buildings and the industrial facilities that may not take the form of buildings as such; and all remaining land contained in urban jurisdictions (administrative areas). It is difficult to be very precise about these calculations, since much rests on an agreed-upon definition of urban (a contested concept), but there seems to be general consensus on the 1–3% range.

Second, the population that will be packed into these relatively compact spaces is expected to grow enormously in the coming decades. As indicated in Fig. 17.3 (top panel), the urban percentages of national populations are rising across the board, in the least developed countries, other developing countries, and high-income developed countries alike. By 2050, according to the United Nations estimates and forecasts shown here, urban-dwellers will be in the majority in all but the least-developed countries; and even in those countries, nearly half of the population will be urban. The expected numbers of additional urban and rural dwellers are shown in the bottom panel, set against a year-2000 baseline. In the other-developing countries alone, nearly 2.5 billion urban residents are expected to be added by 2050 to the year-2000 totals, while total rural populations are projected to be on the decline except in the least-developed countries. These trends have been in the making for many decades,

and although urban percentages will eventually approach an asymptote short of 100%, no reversal of the upward trends is now in sight.

Third, perhaps because the largest urban places are so vivid in the mind's eye and present such distinctive governance challenges, many observers have imagined that the urban residents of low- and middle-income countries are concentrated in huge agglomerations. This is very far from being the case. As Fig. 17.4 shows, among all developing-country urban dwellers, only 10–12% live in mega-cities above 10 million in population. By contrast, 40–50% live in smaller cities and towns under 300,000 in population. Despite their numerical predominance and the distinctive positions they occupy in linking rural areas and to the networks of larger cities, these smaller places continue to be overlooked both in the scientific literature and in urban policies.

The fourth and often-misunderstood feature of urbanization has to do with the relative roles of migration and urban natural increase (the excess of urban births over deaths) in driving urban and city population growth. Chen et al. (1998) found that as a rule, *natural increase accounts for a greater share of urban population growth than the combination of migration and spatial expansion*, with (typically) the share of natural increase being about 60% of urban growth. To be sure, there are many exceptions to this rule (China being the most prominent) and the relative shares of city population growth attributable to the three proximate sources—natural increase, migration, and spatial expansion—are likely to vary over time even for a single city. We emphasize natural increase in part because urban fertility can be reduced through voluntary urban family planning and reproductive health programs (Panel on Urban Population Dynamics 2003), whereas few such effective tools are available to manage migration and spatial expansion.

The fifth noteworthy feature of the urban transformation is the most recent to have been documented: the geographic expansion of urban places relative to their population growth. Angel et al. (2005, 2011) have shown that cities in poor countries tend to become less dense as they grow

---

[2]Exclusive of the Arctic, Antarctica, and Greenland.

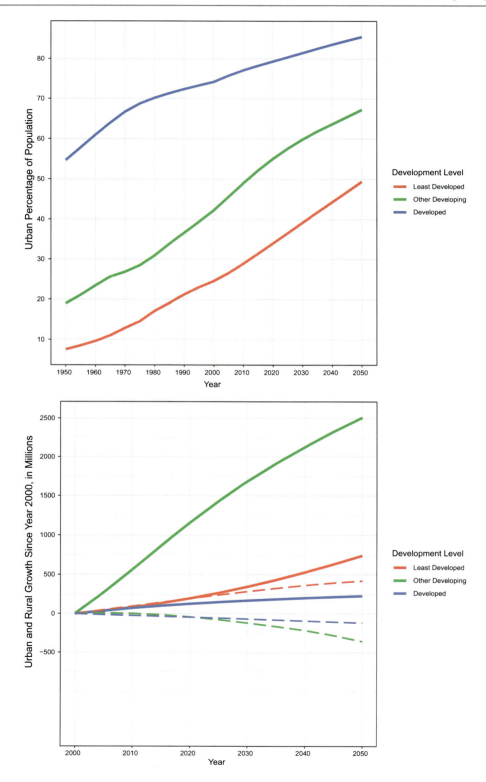

Fig. 17.3 Urban percentage of national populations, by development level (top panel **a**); and total estimated and projected growth in urban and rural populations since 2000, also by development level (bottom panel **b**; urban in solid lines, rural in dashed lines). Source: Authors's calculations from data provided in United Nations (2014)

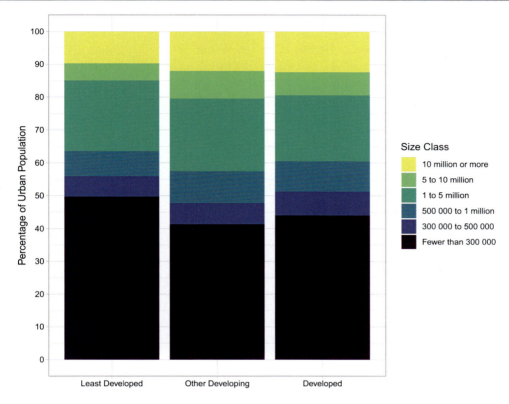

Fig. 17.4 Distribution of urban population by city size and development level in 2015. Authors' calculations based on estimates in United Nations (2014)

in population—to "lose their density" as United Nations (2013b) puts it. This empirical regularity is important for at least three reasons: (1) the compactness of cities is a fundamental determinant of transport and therefore of greenhouse gas emissions; (2) where some spatial expansion is desirable, it should be planned rather than haphazard, appropriately guided so that future residents enjoy equitable access to city services and are provided with adequate infrastructure to meet their basic needs; and (3) spatial expansion brings urban demands for land into conflict with rural needs and can compromise the protection of natural habitats and ecosystems that benefit both rural and urban dwellers. Even if the percentage taken by urban land remains on the order of 1–3%, the apparent trend toward lower density may well present a cause for concern.

Local Governments

Urban and rural local governments, as well as regional authorities and a variety of other entities, operate within systems that link governments at multiple levels from the national to the local. As Martínez-Vázquez and Smoke (2010, 73) write:

> Local governments play increasingly more critical roles in delivering basic infrastructure services, such as roads, transportation and water, and social services, such as education and health. …Citizens look more to local governments for those public services that improve daily living conditions. Central governments depend on local governments to support priority development and poverty reduction goals. Private firms increasingly rely on local governments to deliver infrastructure and other services that support production and stimulate job creation.

Although when taken together smaller cities form outsize presence in urban populations as a whole, their local governments are often severely constrained in terms of authority, budgets, and both technical and managerial expertise (Tanner et al. 2009; World Bank 2011; Jha et al. 2012). Local authorities are often overwhelmed by the demands of day-to-day management and generally lack the capacity to foresee and properly prepare for the longer term or to anticipate emergent threats posed by environmental degradation and climate change.

Environmental Risks Facing Urban Dwellers

This section explores the implications of local environmental conditions and resources for the well-being of urban-dwellers. Later, in section "Environmental Effects of Urbanization: The Urban–Rural Interface", we address the consequences of urban growth for their surrounding environments. We must confess here at the outset that to divide up the subject according to these two causal directions is more than a little artificial. If ill-planned development and depletion of groundwater in the coastal environs of a city causes land subsidence and thus exposes city residents to sea-level rise and storm surges, both directions of effect are at work, operating through complex feed-backs. These sorts of effects are examined in some detail from the urban ecological perspective in Douglas et al. (2011), a remarkable compilation which we would recommend to the interested reader. Also, a full exploration of the issues at hand would examine the international environmental implications of the levels and composition of consumption in a given rapidly growing large cities. Within the confines of this chapter, we cannot hope to document and trace through the implications of such distant (but undoubtedly significant) spatial and economic linkages.

The literature on urban risks has been ably summarized in three recent publications by the Intergovernmental Panel on Climate Change: IPCC (2012), Revi et al. (2014), and Bazaz et al. (2018). In a marked departure from the early IPCC reports that tended to overlook cities and towns, urban concerns have been highlighted in the fourth and fifth IPCC assessments. In particular, IPCC (2012, 234, 247) underscores a critical feature of urban risks—exposure and vulnerability are unevenly distributed *within* individual cities and towns: "The most vulnerable populations include urban poor in informal settlements, refugees, internally displaced people, and those living in marginal areas. …Occupants of informal settlements are typically more exposed to climate events with no or limited hazard-reducing infrastructure. The vulnerability is high due to very low-quality housing and limited capacity to cope due to a lack of assets, insurance, and marginal livelihoods, with less state support and limited legal protection." As we now proceed to discuss, many of the chronic and extreme-event risks facing today's urban residents are likely to become amplified in severity or frequency as global climate change takes hold (Bazaz et al. 2018, 11).

Inadequate Access to Fresh Water

Drawing upon an extensive literature on water scarcity, McDonald et al. (2011) show that as cities grow in population, the quantities of water they need grow as well. The increase in total municipal water demand can compel cities to seek out new water sources, leading to the creation of sometimes quite complex and far-flung systems of urban water infrastructure (McDonald et al. 2014). In addition to population, water scarcity is shaped by fundamental geographic limitations—some cities are situated in arid climates or located far from water sources adequate to serve their populations. Municipal finance also matters, in that richer cities will often have access to sources of funds and financing mechanisms that enable the construction of more robust water infrastructure.

Figure 17.5 illustrates these constraints with reference to Pakistan, studied by McDonald et al. (2011). The spatial extents of Karachi and Hyderabad (estimated via night-time lights) are depicted in red in the figure, and United Nations sources provide the approximate population of

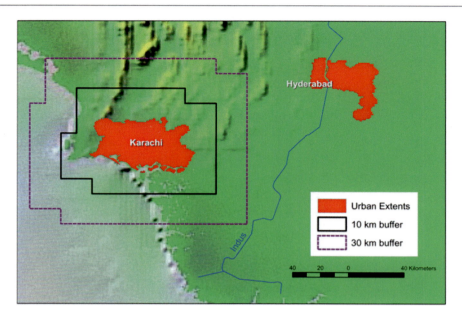

Fig. 17.5 Source: McDonald et al. (2011), supplementary material

these urban agglomerations. Hyderabad (shown in the upper right) can meet its basic water needs (100 liters per person per day) within its urban extents by drawing from the Indus River. By contrast, Karachi cannot find adequate supplies within its own extents, and would still fall short even at 10 km beyond those extents. Investments extended to 30 km distance, however, would very nearly reach the Indus and in theory would enable Karachi's basic needs to be met. McDonald et al. (2011) apply this theoretical logic to compute the implications of future city population growth for water scarcity, assuming that water infrastructure is confined within a fixed set of distance buffers. Looking out to the year 2050, they find that over this period expected urban population growth is likely to drive up water scarcities (especially on a seasonal basis) to a much greater extent than any anticipated effects from climate change. In follow-up research that exploits data from water utilities, reflecting the infrastructural investments that large cities have actually been able to undertake, McDonald et al. (2014) identify a number of water-stressed cities in high- and low-income countries alike; see Fig. 17.6.

Detailed models of water basins by Hofste et al. (2019) supply estimates of baseline water stress around the world.[3] The city of Chennai in Tamil Nadu—seen in the upper right of Fig. 17.7—illustrates the results. This city is situated in a basin classified as having "extremely high" levels of water stress (Hofste et al. 2019). Indeed, in mid-June of 2019, Chennai experienced a catastrophic failure of its water supply system. The proximate cause was that the monsoon rains failed to replenish the reservoirs on which Chennai's system relies. But a significant contributing factor was the inadequacy of regulatory and management efforts: the rivers from which Chennai might draw adequate water are simply too polluted to use. Figure 17.7 is suggestive of the scale of the governance challenge in securing water: the basins on which Chennai and other cities of Tamil Nadu depend not only span the jurisdictions of multiple local governments, but extend well into neighboring states.

Over the longer term, global warming is likely to exacerbate water scarcity, albeit with consider-

---

[3]Baseline water stress is measured as the ratio of total water withdrawals (from domestic, industrial, irrigation, and livestock uses) to available renewable surface and groundwater supplies, taking into account the impact of upstream consumptive water users and dams on downstream water availability.

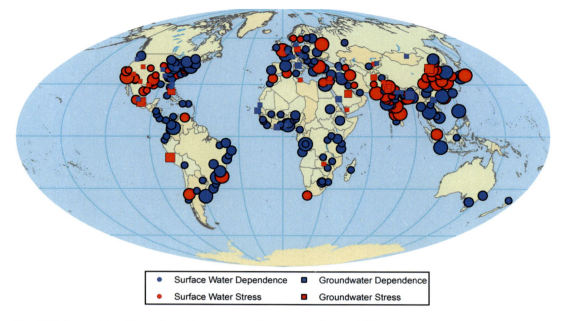

Fig. 17.6 Large cities affected by surface and groundwater stress. Source: McDonald et al. (2014)

able variation by region. According to Bazaz et al. (2018, 11) as global average temperatures move from the now-inevitable 1.5°C increase to reach an increase of 2.0°C, the total population living under conditions of water scarcity is expected to double. Insights into the urban implications can be gleaned from the Hofste et al. (2019) basin-specific projections.

Drought

Approximately 70% of the world's population lives in areas that are drought-prone, with Sub-Saharan Africa being perhaps the most afflicted region (Mishra and Singh 2010). Drought can reduce accessible quantities of both surface and groundwater, and can also impede electricity generation, affecting the economies of cities dependent on hydropower. In Africa, for example, the cities of Lusaka, Windhoek, Kampala, Addis Ababa, and Cairo all rely on surface and groundwater sources for drinking water and power (Calow et al. 2010).

Other economic effects of drought on urban-dwellers operate on an indirect basis, through increases in food prices and food insecurity, and potentially through increased rural–to–urban migration. On an individual and household level, the role of water as an input into domestic production means that water scarcity, either in terms of quantity or quality, can hinder household income-generating activities (Baker 2012; Calow et al. 2010).[4] Tracing through the full effects of drought on urban well-being is exceedingly difficult, requiring detailed data on food prices and quantities consumed (which may be affected by conditions in domestic and even international economic markets) as well as close attention to the role of urban agriculture in providing a buffer against food price shocks for low-income city residents.

Long-term global trends in drought frequency and severity are uncertain at present, owing to natural variability, scientific disputes about appropriate measurement, and the gaps and deficiencies that afflict precipitation data. If the global prospects remain unclear, convincing evidence of greater drought frequency and

---

[4]Drought can compromise water quality through changes in lake chemistry or the lack of nutrients that are typically transported by water runoff (Mishra and Singh 2010).

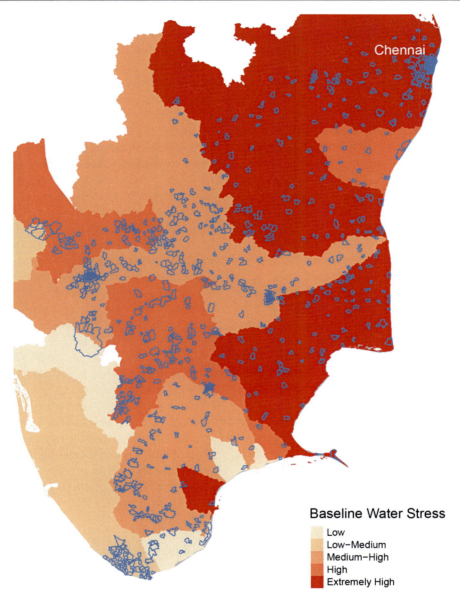

Fig. 17.7 Levels of baseline water stress in the hydro-basins of Tamil Nadu, India, estimated by Hofste et al. (2019) using the AQUEDUCT 3.0 model. Blue areas represent the boundaries of statutory towns in the state; the city of Chennai is in the upper north-east, located in an "extremely high" water stress basin. Source: Authors' calculations based on data supplied by Hofste et al. (2019)

severity has emerged for selected regions—the Mediterranean, North Africa and Middle East, many parts of sub-Saharan Africa, Central China, India, other regions of east and south Asia, the southern Amazon, portions of North America, and eastern Australia (IPCC 2019, Chapter 2 of full report, page 2-21).

Urban Heat and Air Pollution

Urban Heat Islands

Buildings in urban areas store heat and release it at slower rates than would rural structures or natural ecosystems, leading what is known as UHI, the "urban heat island" effect (Arnfield 2003;

Depietri et al. 2012). Elevated urban temperatures also arise from the lack of porous surfaces and vegetation cover (Broadbent et al. 2018). In addition, the flow of winds is hindered by urban buildings, which act as obstacles to wind–paths and induce drag that reduces the transfer of heat (Britter and Hanna 2003; Ongoma et al. 2013). Recent IPCC assessments confirm that mean annual surface air temperature in cities and in their surroundings are higher by as much as 0.19°C–2.60°C, affecting night-time temperatures more than daytime. Heatwaves, which are sustained periods of abnormal heat, are also typically more intense in cities by 1.22°C–4°C, with their effects likewise more pronounced at night (IPCC 2019, Chapter 2 of full report, page 2-74).

The health effects of heat and heat stress are well documented in both low- and middle-income countries (Revi et al. 2014, 556). In high-income countries the elderly appear to be more vulnerable to heat-related mortality. Among the urban poor of low-income countries, selected studies have shown that temperatures in informal neighborhoods can significant exceed those of higher-income neighborhoods (e.g., Scott et al. 2017, for Nairobi) and although more research is needed for confirmation, some studies suggest that extreme temperatures increase mortality in these settings (Egondi et al. 2015, also for Nairobi). Poor quality, badly ventilated housing is associated with the risk of heat-related illness and death, especially among women, the elderly, and the poor (Baker 2012). Workers in outdoor occupations—a common mode of employment for poor men—can suffer from prolonged exposure to heat-related risks (Kovats and Akhtar 2008).

Ambient Air Pollution

The Urban levels of exposure to air pollution can be seen in Fig. 17.8 for the Indian cities of Mumbai and Agra. The estimates shown in the figure are annual averages of fine particulate (PM2.5) concentrations, derived from a combination of remotely-sensed data and (limited) on-the-ground monitoring. The levels of exposure in these two Indian cities are far in excess of international safe standards (see WHO 2018; UNICEF 2016, for extensive compilations). Exposure to risk is not only an urban problem: In India and elsewhere, urban emissions of pollutants can threaten rural residents, and likewise for rural emissions and urban residents (Cusworth et al. 2018).

A recent review summarizes the health impacts of ambient air pollution, which has been convincingly linked to cardiovascular and respiratory disease, lung cancer, diabetes, low birth weight, neurodegenerative disease, and premature mortality (USAID et al. 2019). Most of the health damage from air pollution is thought to result from long-term exposure to fine particulates (PM2.5) but gases such as ozone, sulfur dioxide and nitrogen dioxide are also implicated. Worldwide, most air pollution–related premature deaths (approximately three-quarters) occur in poor countries, as do the vast majority of such deaths among children under 5 years of age. (In part this is because young children breathe more quickly than adults and take in more air in relation to their body weight.) Pregnant women are also at risk as the fetus—for reasons that are not yet firmly established—appears to be highly susceptible to air pollution toxins. Increased air temperatures and excessive heat can further compromise air quality by stimulating the formation of photochemical oxidants and increasing the concentration of pollutants such as ozone, with corresponding threats to human health (IPCC 2019, chapter 2 of full report, page 2–75).

Flooding

Flooding presents significant risks to lives, health, and livelihoods in cities and towns around the world. Urban areas share a number of characteristics linked to flooding: impermeable surfaces that cause water run-off; a general scarcity of parks and other green spaces to absorb such flows; drainage systems that are often clogged by solid waste and which in any case can be quickly overloaded with water; and the ill-advised development of natural drainage channels, marshlands and other natural buffers (Douglas et al. 2008; Depietri et al. 2012; Romero Lankao 2010; Benson-Lira et al. 2016). Coastal cities face additional dis-

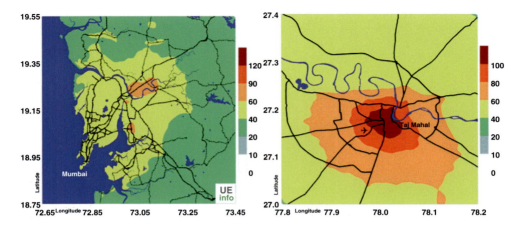

Fig. 17.8 Modeled annual average PM2.5 concentrations in Mumbai (2018, left) and Agra (2015, right), expressed in milligrams per cubic meter. For urban Mumbai, the average PM2.5 concentration over the year was $51.8 \pm 24.8\,\mu g/m^3$, over five times the WHO guideline level. For Agra, the average PM2.5 concentration was $89.0 \pm 12.2\,\mu g/m^3$, more than eight times the WHO guideline. See http://www.urbanemissions.info/india-apna/apna-leaflets for details of data and methods

tinctive risks from storm surges and the winds that accompany tropical cyclones, and in parts of the world as different as Houston (United States) and Jakarta (Indonesia), daunting long-term threats are emerging in the combination of sea-level rise and land subsidence (Ingebritsen and Galloway 2014; Abidin et al. 2011).

When urban flooding takes place, chemicals and fecal and other hazardous materials contaminate flood waters and spill into open wells, elevating the risks of water-borne disease (Kim and Kannan 2007). Comprehensive accounts of the health consequences include Ahern et al. (2005); Shultz et al. (2005); ActionAid International (2006); Kovats and Akhtar (2008); Portier et al. (2010); Jha et al. (2012), with an authoritative summing-up in the 2012 IPCC report on extreme events (IPCC 2012). Urban floods are thus aptly described as "complex emergencies". The urban poor are often more exposed than others to flooding hazards, because the housing they can afford is less durable and tends to be located in more flood-prone areas (IPCC 2012, 248).

Figure 17.9—based on work with colleagues Deborah Balk and Zhen Liu—may help in communicating the enormity of the flood risk challenge in Asia alone. It depicts the estimated percentage of urban populations at risk of inland flooding (top panel) and coastal flooding (bottom panel). Many of Asia's largest cities are located in the flood-plains of major rivers (the Ganges–Brahmaputra, Mekong, and Yangtze rivers) and in coastal areas that have long been cyclone-prone. Mumbai saw massive floods in 2005, as did Karachi in 2007 and again in 2010. Flooding and storm surges also present a threat in coastal African cities (e.g., Port Harcourt, Nigeria, and Mombasa, Kenya) and in Latin America (e.g., Caracas, Venezuela). Elsewhere, large cities at risk include Guangzhou, Shanghai, Miami, Ho Chi Minh City, Kolkata, New York, Osaka-Kobe, Alexandria, Tokyo, Tianjin, Bangkok, Dhaka, and Hai Phong. In general, the economic consequences of flooding can be severe, as Revi et al. (2014, 556) note for the 2007 floods in the city of Villahermosa, Mexico, and the 2011 flooding of the Chao Phraya River which damaged many companies and several industrial estates in Bangkok and disrupted global supply chains. Revi et al. (2014, 555) emphasize the risks to cities with extensive port facilities and large-scale petro-chemical and energy-related industries, which were vividly demonstrated in the massive 2017 floods of Houston.

The cross-sectoral challenges of adaptation may be appreciated by reference to the situation

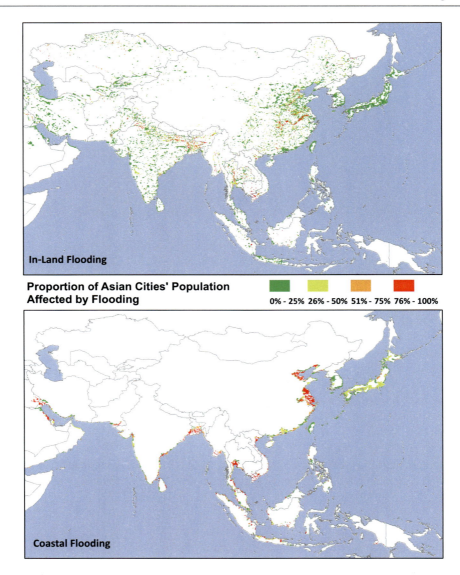

Fig. 17.9 For Asian cities, percentage of population at risk of coastal or inland flooding. Source: Calculations by Balk, Liu, and Montgomery

of Accra, Ghana. Here, annual flooding is produced by the heavy rains that occur nearly every June and July, causing loss of life, damaging property and, especially in poor neighborhoods, threatening health by contaminating water with a toxic brew of solid waste, sewage, and industrial waste (Dickson and Benneh 1996; Afeku 2005; Rain et al. 2011; Nyarko 2002; International Federation of Red Cross and Red Crescent Societies 2009). Both natural and constructed drains can be found across Accra, but the city's drainage network remains woefully undersized relative to the population it should serve, with most of the channels being so choked with silt and debris that storm water cannot be conveyed to points of outlet.

Slum residents here and elsewhere often have a good understanding of why their neighborhoods flood, and can point to specific blockages and features of local development that are important proximate causes. Durable, long-term solutions to flood risks require investments and collective action not only at the neighborhood level, but also on the part of local and other units of government, which in Accra have long neglected drainage problems and which have acquiesced in ill-advised development schemes that have eliminated natural protective buffers. To be sure, the municipal governments of such cities typically function without access to detailed local population data; their land-use maps are rarely adequate even at the city level, to say nothing of within-city neighborhoods; and hydrological data and models of flood risk, when these are available at all, are likely to be studied only in universities and specialized government departments, hence rarely analyzed from a health equity point of view.

A useful perspective on meeting flood risks is set out in Table 17.1, in which protective efforts are categorized in terms of "keeping water away from people," and "keeping people away from water". This way of viewing adaptation makes it clear that effective risk reduction requires cross-sectoral linkages on a grand scale, not only within urban governments and their neighborhoods, but also across multiple levels of government. Yet even in a richly-resourced, high-income country such as the United States, the necessary coordination often fails to materialize. As Galloway et al. (2018, 5) write on the U.S. case, "While primary responsibility for mitigation of urban flooding rests with local governments, the division of responsibilities among federal, state, regional, local, and tribal governments for urban flood and stormwater management [is] not clearly defined. Responsibilities are diffused and lack the collaboration and coordination necessary to address the technical and political challenges that must be faced. …At the federal level, there is no agency charged with oversight of federal support of urban flood mitigation related activities". Much the same could be said for cities and towns around the world.

One important type of flood occupies its own category—flash floods that are common occurrences in mountainous regions, where they are often mixed with mud, gravel, and rock and are thus similar to precipitation-induced landslides. The distinction between flash floods and standing-water floods is important: Effective remote-sensing methods for detection of standing-water floods have been developed and new methods are being tested for urban areas—where buildings, streets, and alleyways present daunting remote-sensing challenges (see NASA 2018)—but much methodological work remains before similarly effective detection of flash floods becomes possible.

Landslide Risks

Another way in which floods and landslides warrant joint attention is that for both hazards, significant international scientific collaborations are being forged to improve early warning and detection, and to create databases of verified events. For example, the U.S. space agency NASA has developed a world-wide landslide-risks raster (Stanley and Kirschbaum 2017) and is using it with nearly real-time rainfall data, to produce spatially-specific forecasts of heightened risks. The aim is to provide an early-warning system to alert national civil defense and disaster authorities. Figure 17.10 shows an excerpt from the NASA risk estimates model for Mexico.

Significant risks from precipitation-triggered landslides are evident across a great range of countries and regions—throughout much of Central America and the Andean regions of South America, and in Cameroon, India, and Indonesia (Kirschbaum et al. 2015; Che et al. 2011; Rumbach and Follingstad 2019; Cepeda et al. 2010). As Rumbach and Follingstad (2019, 99) write with reference to the Darjeeling district in West Bengal (India), situated in hilly terrain between Nepal and Bhutan, "the steep topography and protected land uses in the region, like tea estates and national forests, means that developable land is at a premium. As the towns and cities in the region have grown, urban development has moved from the relatively flat and stable areas like ridgelines onto steeper and more hazardous land." With reference to Cameroon, Che et al. (2011) make a similar point, observing that "most slope instabilities

Table 17.1 Summary of flood-protection strategies. Adapted from Svetlosakova and Jha (2012)

Keeping water away from people	Keeping people away from water
Hard-engineered:	*Increased preparedness*:
Flood conveyance	Awareness campaigns
Flood storage	Urban management
Urban drainage systems	*Flood avoidance*:
Ground water management	Land-use planning
Flood-resilient building design	Resettlement
Flood defenses	*Emergency planning and management*:
Eco-system management:	Early warning systems and evacuation
Utilizing wetlands	Critical infrastructure
Creating environmental buffers	*Speeding up recovery*:
	Building back safer

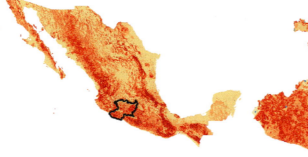
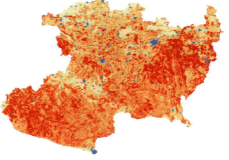

Fig. 17.10 NASA estimates of landslide risks in Mexico (left panel) and the state of Michoacán (right panel). Areas of very high estimated landslide risk are shown in red; yellow areas have low risk. The urban localities of Michoacán are depicted in blue. Source: Authors' calculation based on data from Kirschbaum and Stanley (2018)

within the study area are associated with and appear to be exacerbated by man-made factors such as excavation, anarchical construction, and deforestation of steep slopes." In other words, urban growth not only places populations in the pathway of landslide risk, but in some cases, by destabilizing rock and hillsides, has created new risks.

Although rainfall-induced landslides are known to be important to urban- and rural-dwellers in mountainous and hilly terrain, a proper accounting for specifically urban risk exposure has proven difficult. One promising approach is being tested in Brazil, one of the few middle-income countries with both spatially detailed population census data and similarly detailed geological surveys (de Assis Dias et al. 2018). Here it is possible to overlay precise spatial information on urban settlements with the geological and hydrological variables that are associated with landslides, and through this combination of data, to develop profiles of urban exposure to risk. In a test case of the methodology focused on three municipalities in the mountainous regions of Rio de Janeiro state, de Assis Dias et al. (2018) find that 19–30% of the population of these municipalities reside in landslide risk zones; those exposed to risk can be further characterized by age, sex, and many other socioeconomic factors on which data are collected in the Brazilian census.

Environmental Effects of Urbanization: The Urban–Rural Interface

Following the theory set out by Ros-Tonen et al. (2015) that was illustrated in Fig. 17.1, we now shift the discussion away from core urban areas to the urban fringe and surrounding rural land, which can be affected by economic activities and various spill-overs from the core. The existence of such fringe spaces, where urban and rural features are thoroughly mixed, has long been recognized by geographers. Terry McGee, who worked in Indonesia in the 1980s, coined the term *desakota zone* to describe what he saw then as a newly emerging form of heterogeneous land cover and varieties of economic activity (McGee 1987, 1991).

More recently, Gupta et al. (2015) and especially Ros-Tonen et al. (2015), have written penetrating accounts of the governance challenges. The municipal governments that oversee the urban core may not wield any formal authority in these outside spaces, and because interactions here often involve externalities and public goods that are not priced into the decisions of private economic actors, there exists significant potential for over-exploitation and degradation of natural habitats (Güneralp et al. 2017). In such imperfect economic markets, it is certainly not a given that the conversion of rural land to urban uses is properly guided by prices in the manner seen in textbook economic models. Indeed, as Güneralp et al. (2017, 4) note, it is probably more common than not for land speculation by the wealthy to occur in the urban fringe, a practice no doubt abetted by weak land-use planning and regulatory controls. Nor are public investments necessarily placed here with any consideration of ecosystem services: Urban-oriented transport routes typically thread through fringe and rural spaces, risking fragmentation of habitats.

Vector-Borne Disease Risks

Especially at the fringes of core urban areas, there can be a great deal of on-going construction and encroachment on previously unused land, a chaotic process that inserts human activities, structures, and waste into terrain that formerly might have supplied habitats for various kinds of wildlife. When ecological habitats are significantly disturbed in these ways, a range of effects can follow: changes in the number and

distribution of disease vectors and their breeding sites; niche invasions or interspecies host transfers; changes in biodiversity; and over time, disease transmission between domestic and wild animals and human-induced genetic changes affecting resistance (Patz et al. 2005; McDonald et al. 2009). Tropical countries are apt to be more affected because they have greater exposure to vectors and lack the resources to anticipate and regulate environmental modifications (Patz et al. 2005).

Recent surges in the prevalence of arboviruses such as dengue, chikungunya, and the Zika virus, illustrate some of the implications. The vector for all three of these viruses is the Aedes mosquito, which prefers to find its home in dirty stagnant water in close proximity to dense urban populations. In the urban fringe, poor sanitation systems and dense disorganized housing can create multiple breeding sites: an old tire, an open water drum, or even a flower pot can all serve the purpose. The potential scale of the problem is enormous: It is believed that more than half of the world's population now lives in areas infested with Aedes mosquitoes; see Baud et al. (2017) and https://www.who.int/neglected_diseases/vector_ecology/mosquito-borne-diseases/en.

Cropland Loss

Although the total amount of core urban land is relatively small, as we have discussed, the urban fringe can be the most active zone of land conversion, accounting for a disproportionate share of all land transitions. Bren d'Amour et al. (2017) project that on the order of 2% of global cropland (excluding pastures) will be converted to urban uses by 2030, with some 80% of the cropland losses expected to take place in Africa and Asia. There is some evidence that on average, the converted land is substantially more productive than other agricultural land—the land lost to urban expansion is estimated to be 1.77 times as productive as the global average (Bren d'Amour et al. 2017, 8940).

It is difficult to know just how to interpret this disconcerting finding. Perhaps there are transport advantages in such converted land that had formerly benefited agriculture and which encouraged farm investment, or perhaps such sites enjoyed easier access to agricultural inputs due to their proximity to urban services and markets. To understand the issues, focused research in selected regions would seem to be of high priority: in sub-Saharan Africa, in particular, Nigeria and the region around Lake Victoria; Egypt in North Africa; and in Asia, areas of concern include Java and the Yangtze River Delta in China.

Threats to Biodiversity

As urban outward expansion continues, in some cases it is likely to put local biodiversity at risk (Mcdonald et al. 2019). An innovative study by Seto et al. (2012) explores the intersection between areas of probable urban expansion and biodiversity "hot-spots". In the year 2000, less than 1% of biodiversity hot-spot area was urbanized; these authors project that by 2030, a further 1.8% of such areas will be caught up in urban expansion. The biodiversity hot-spots likely to be most affected (in terms of area) are in the Guinean forests of West Africa, Japan, the Caribbean Islands, the Philippines, the Western Ghats (India), and Sri Lanka.

Huang et al. (2018) extend the earlier analysis by using country-level GDP and indicators of governance, factors that help to identify resource-constrained countries with inadequate institutions and regulatory systems. Newly developed, spatially-specific maps of mammal, bird and amphibian biodiversity are employed to sharpen the focus on diversity (Jenkins et al. 2013). The authors adopt the phrase "land governance" to underscore the importance of land-use planning and effective government regulation to biodiversity conservation. For example, Namibia and Malaysia are cited as countries in which urban expansion might well pose a threat to biodiversity, but in which governance capacities are potentially strong enough to hold off such threats. In Indonesia, Brazil, and Nigeria, however, urban spatial growth similarly threatens diversity and there are doubts about whether effective land governance is possible amid sharply contending interests and political volatility. Drawing on similar materials, Güneralp et al. (2017) reflect on the prospects for successful conservation efforts in Africa. Research on urban land use and species richness has also focused on North America (Olden et al. 2006). Here, as Bradley and Altizer (2007) write, "urbanization dramatically alters the composition of wildlife communities, leading to biodiversity loss and increases abundance of species that thrive in urban areas". Franks et al. (2018) and Booker and Franks (2019) have synthesized lessons learned from a number of detailed recent studies of the social and economic challenges facing protected nature reserves, with a focus on the implications for local governments in giving their poor residents a stake in the success of the reserve, and in the regulation, vigilant monitoring, and enforcement of agreements among all interested parties.

Given the importance of low-elevation coastal zones to urbanization, a small literature has examined whether economic and recreational activities in these zones may bring about reductions in coastal biodiversity, increase the populations of invasive species, induce hypoxia in local water ecosystems which can result in toxic phytoplankton blooms, and introduce trace metals and toxin concentrations in biota that could harm food sources (McGranahan et al. 2007; Perkol-Finkel et al. 2012). Von Glasow et al. (2013) organize their review of potential coastal effects around mega-cities (conventionally defined as having 10 million inhabitants or more). As they explain, the science of coastal effects is dauntingly complex and conclusions are heavily dependent on local characteristics. Where emissions and run-offs of phytoplankton nutrients, fixed nitrogen and phosphorus are concerned, an important general point is that in mega-cities "the localized scale of discharge relative to local dilution capacity is often particularly large with the potential for nonlinearities in the ecosystem response". Industrial activities and shipping within megacities and their vicinity can also produce discharges of trace metals and potentially toxic organic compounds. The local hydrogeography is key in determining whether such outputs of coastal cities are flushed away or recirculated in a manner that compounds the damage. For example, many estuaries are tidal and pollutant emissions can be effectively trapped by complex estuarine hydrodynamics, producing significant damages to intertidal systems such as salt marshes and mangrove forests.

Fire Risks Along the Wildland–Urban Interface

According to IPCC (2019, chapter 2, page 2-27), the frequency of fires is likely to increase by some 27% globally, in part owing to global warming. The fire weather season has significantly lengthened across fully one-quarter of the Earth's vegetation-covered land. An important contributing factor is the expansion of urban and peri-urban fringe settlement into what had formerly been wildland. As Radeloff et al. (2018) put it, "When houses are built close to forests or other types of natural vegetation, they pose two problems related to wildfires. First, there will be more wildfires due to human ignitions. Second, wildfires that occur will pose a greater risk to lives and homes, they will be hard to fight, and letting natural fires burn becomes impossible". As more people move to areas situated in the wildland–urban interface, as has been the case in the western United States, exposure to fire risk increases unless protective measures (e.g., clearance of flammable material from a 30-foot zone around each home) are adopted to counteract the effects of exposure (Hammer et al. 2009; Shafran 2008). When such fires occur, ash and fine particulates can be dispersed over wide areas, affecting air quality tens or even hundreds of miles away (Bein et al. 2008).

In the United States, where the wildland–urban interface is home to an estimated 99 million people, detailed assessments of risk exposure in the wildland–urban interface have been made by the U.S. Forest Service, involving precise definitions of *intermix* and *interface* areas (Martinuzzi et al. 2015). An interface is created when a block of set-tlement lies adjacent to wild land; an intermix, by contrast, involves tendrils or isolated settlements (not necessarily urban in any conventional sense) situated amid wild land. Even for communities that are not themselves adjacent to wildland, there can be significant risks from fires that originate in such spaces. Figure 17.11 shows the path of a September 8–9 2020 fire in southern Oregon (United States) during which ferociously strong winds drove embers from nearby burning forests along a corridor of low-income housing, utterly destroying the poor communities of Talent and Phoenix. The fire was advancing northward upon Medford, a city of some 80,000 residents, when the winds finally shifted.[5] At that time, many other communities in Oregon and northern California were also being threatened and seriously damaged by outbreaks on a scale that has rarely been seen.

Subsidence in Deltaic and Coastal Areas

Cities and towns situated in river deltas often face the problem of land subsidence. The directly affected populations are large: Edmonds et al. (2017) estimate that while deltas cover less than 1% of total habitable area, they contain over 4% of the world's population, with a projected 322 million inhabitants in 2020. Subsidence is a complex phenomenon that can stem not only from local development activities (including over-extraction of groundwater and removal of the natural barriers that would have reduced erosion and land loss), but also from upstream disruptions and blockages (whether from dams, hydroelectric power-generation infrastructure, artificial reservoirs, or disconnections of key links in upstream wetlands) which inhibit the natural processes of silting and replenishment of downstream deltaic land. This combination of local and regional effects—the latter can extend for hundreds of kilometers—can stand as an instructive example of the mutual dependence of both urban

---

[5]See https://www.washingtonpost.com/nation/2020/10/20/oregon-almeda-fire.

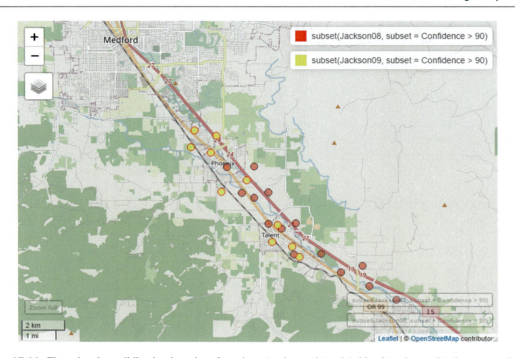

Fig. 17.11 Fires in the wildland–urban interface in southern Oregon, United States, from 8–9 September 2020. Fires detected by VIIRS satellite sensors (resolution of 350 m) on September 8 are shown in red; those detected on September 9 are shown in yellow. Between Medford (to the north) and Ashland (to the south), the communities of Talent and Phoenix were burned to the ground. Source: Authors' calculations using VIIRS active fires data made available globally at http://viirsfire.geog.umd.edu, University of Maryland, Department of Geographical Sciences

and rural populations on regional environmental resources, whereby activities in one part of a large interconnected system affect the well-being of humans and other species living nearby and downstream.

When subsidence occurs in a deltaic environment, it can substantially enlarge the land area and population that is exposed to coastal flooding and storm surge, given the typically gentle elevation gradient that characterizes many deltas. Taken in combination with off-shore increases in sea levels, subsidence can increase the rate of relative sea-level rise, opening new pathways for the intrusion of seawater and risking salinization of drinking water and agricultural soils. It also places city infrastructure at risk, as uneven subsidence causes roads, canals, building foundations, and sub-surface structures (sewerage, gas pipes, and the like) to tilt, crack, and buckle, thereby raising the costs of maintenance and replacement (Brown and Nicholls 2015; Tessler et al. 2015; Szabo et al. 2016; Tessler et al. 2016).

Although many factors are at work, to judge from the Erkens et al. (2015) study of cities as various as Jakarta, Ho Chi Minh City, Bangkok, Dhaka, Shanghai, and Tokyo, the principal cause of subsidence is the extraction of groundwater for both domestic consumption and industrial uses. (In some deltas, notably that of the Mississippi River, oil and gas extraction also plays a significant role.) Over-extraction to secure drinking water is especially prevalent when the surface water alternatives are seriously polluted (as has been the case in Jakarta and Dhaka). The failure of governments to regulate surface water pollution has arguably driven residents and firms to overexploit groundwater resources, thereby setting in motion a series of unanticipated consequences. In this way, one unaddressed externality spawns a host of others.

Unusually severe subsidence has been taking place in Jakarta, where rates of subsidence have recently been the order of an astonishing 3–10 cm per year, with a total subsidence of up to 4 m (from 1974 to 2010) having been observed in several locations within the urban area (Abidin et al. 2015). In addition to groundwater extraction, other contributing factors in Jakarta include the natural consolidation of the area's soft alluvial soils, the pressures exerted by construction on such highly compressible material, and residual effects attributable to tectonic activity. Things have reached the point where there is now serious consideration being given to the relocation of the capital from Jakarta to Kalimantan, in part because the subsidence underway in the northern portions of the metropolitan region appears to present a long-term threat to livability.

Subsidence is not only a coastal phenomenon: In parts of Mexico City, which is land-locked but sits atop the vast Basin of Mexico aquifer, mean subsidence rates in excess of 350 mm/year have been seen (Carrera-Hernández and Gaskin 2007; Cabral-Cano et al. 2008). In Mexico City, much as in coastal urban areas, the unevenness with which soils are compacted presents the most significant risks for urban infrastructure.

Conclusion

The principal theme that runs through this survey of cities and their environments is the need for unified governance approaches to highly heterogenous spaces, which span core urban areas, the urban fringe, and an outer envelope of rural land, water basins, and a variety of ecological habitats. Some of the critical spaces—watersheds, for example—are large geographic units that extend far beyond the boundaries of municipal jurisdictions. The need for some integrating governance mechanism has been apparent for decades (Panel on Urban Population Dynamics 2003, Chapter 9), but the all-too-obvious absence of such joined-up authorities, especially in poor countries, testifies to the legal, administrative, budgetary, technical,

and scientific difficulties that confront their construction.

This chapter began, in Fig. 17.2, with a depiction of the many local governments with jurisdiction over the spaces of the Mumbai Metropolitan Region, which includes Sanjay Gandhi National Park. Just east of this park the Mithi River begins its journey to the Arabian Sea; while it flows through the park its waters are said to be clear and clean in keeping with the river's name ("sweet" in Hindi). But upon leaving the park the Mithi proceeds through a number of poorly regulated urban communities where it becomes contaminated by pollutants and choked with plastics, debris, and other solid wastes.[6] These serial failures of governance are then transported downstream to threaten the health and well-being of the Mumbai poor living in the slums of Morarji Nagar, Bhim Nagar, and Dharavi, especially where the Mithi stagnates and its waters become a breeding-ground for malaria-transmitting mosquitos. At this point in its journey, the Mithi is typically so overloaded with waste that it cannot always absorb the monsoon rains and usher them safely toward the sea; instead, the rainwaters overtop the river's banks and inundate the poor communities living nearby. Although local, regional, and state governments have begun to address these complex and varied problems, their efforts to date appear to have been halting, underfunded, and so far inadequate to the task.

Writing with reference to the climate adaptation efforts that will be urgently needed in the coming decades, Revi et al. (2014, 575–577) express very well what is at stake in such governance challenges:

> Many aspects of adaptation can be implemented only through what urban governments do, encourage, allow, support, and control. This necessarily involves overlapping responsibilities and authority across other levels of government as well. …The capacity of local authorities to work effectively, alone or with other levels, is constrained by limited funding and technical expertise, institutional mechanisms, and lack of information and leadership. …While urban governments have authority for many relevant adaptation decisions, they can

---

[6]See the account in https://www.indiawaterportal.org/articles/mithi-recounting-rivers-apathetic-journey.

be enabled, bounded, or constrained by national, subnational, or supranational laws, policies, and funding and land use and infrastructure planning decisions. …Action has to be coordinated and harmonized across multiple urban jurisdictions …Water authorities, for instance, may operate at water-basin level, representing both national and local interests while operating independently of urban authorities. Failing to ensure consistent alignment and integration in risk management can lock in outcomes that raise the vulnerability of urban populations, infrastructure, and natural systems.

What is needed, in short, is a program to build new institutions that span and link multiple jurisdictions. The need for such institutions has long been recognized, and, looking across countries, one sees a number of successful examples that could prove to be internationally instructive. Each country must of course find its own way toward solutions that fit its requirements and constraints, but surely there is much to be learned from cross-country exchanges of experience on how to create and support systems of integrated human–environmental governance.

References

Abidin, H. Z., Andreas, H., Gumilar, I., Fukuda, Y., Pohan, Y. E., & Deguchi, T. (2011). Land subsidence of Jakarta (Indonesia) and its relation with urban development. *Natural Hazards, 59*(3), 1753. https://doi.org/10.1007/s11069-011-9866-9.

Abidin, H. Z., Andreas, H., Gumilar, I., & Wibowo, I. R. R. (2015). On correlation between urban development, land subsidence and flooding phenomena in Jakarta. *Proceedings of the International Association of Hydrological Sciences, 370*, 15–20. https://doi.org/10.5194/piahs-370-15-2015

ActionAid International (2006). Unjust waters: Climate change, flooding and the protection of poor urban communities: Experiences from six African cities. Action-AID International report

Afeku, K. (2005). Urbanization and flooding in Accra, Ghana. Master's thesis, Miami, FL: Department of Geography, Miami University.

Ahern, M. R., Kovats, S., Wilkinson, P., Few, R., & Matthies, F. (2005). Global health impacts of floods: Epidemiological evidence. *Epidemiologic Reviews, 27*(1), 36–45.

Angel, S., Sheppard, S. C., & Civco, D. L. (2005). The Dynamics of Global Urban Expansion. World Bank, Washington, D.C.: Transport and Urban Development Department.

Angel, S., Parent, J., Civco, D. L., & Blei, A. M. (2011). Making room for a planet of cities. Policy Focus Report PF027, Massachusetts: Lincoln Institute of Land Policy.

Arnfield, A. J. (2003). Two decades of urban climate research: A review of turbulence, exchanges of energy and water, and the urban heat island. *International Journal of Climatology, 23*(1), 1–26.

Baker, J. L. (2012). Climate change, disaster risk, and the urban poor: Cities building resilience for a changing world. Washington, D.C.: The World Bank.

Baud, D., Musso, D., Vouga, M., Alves, M. P., & Vulliemoz, N. (2017). Zika virus: A new threat to human reproduction. *American Journal of Reproductive Immunology, 77*(2), e12614. https://onlinelibrary.wiley.com/doi/abs/10.1111/aji.12614.

Bazaz, A., Bertoldi, P., Buckeridge, M., Cartwright, A., de Coninck, H., Engelbrecht, F., Jacob, D., Hourcade, J. C., Klaus, I., de Kleijne, K., Lwasa, S., Markgraf, C., Newman, P., Revi, A., Rogelj, J., Schultz, S., Shindell, D., Singh, C., Solecki, W., Steg, L., & Waisman, H. (2018). Summary for urban policymakers—what the IPCC Special Report on 1.5c means for cities. Indian Institute for Human Settlements (IIHS), Bengaluru 560 080, India.

Bein, K. J., Zhao, Y., Johnston, M. V., & Wexler, A. S. (2008). Interactions between boreal wildfire and urban emissions. *Journal of Geophysical Research: Atmospheres, 113*(D7). https://doi.org/10.1029/2007JD008910.

Benson-Lira, V., Georgescu, M., Kaplan, S., & Vivoni, E. R. (2016). Loss of a lake system in a megacity: The impact of urban expansion on seasonal meteorology in Mexico City. *Journal of Geophysical Research: Atmospheres, 121*(7), 3079–3099.

Booker, F., & Franks, P. (2019). Governance assessment for protected and conserved areas (GAPA): Methodology manual for GAPA facilitators. London: International Institute for Environment and Development. http://pubs.iied.org/17655IIED

Bradley, C. A., & Altizer, S. (2007). Urbanization and the ecology of wildlife diseases. *Trends in Ecology & Evolution, 22*(2), 95–102. https://doi.org/10.1016/j.tree.2006.11.001. http://www.sciencedirect.com/science/article/pii/S0169534706003648

Bren d'Amour, C., Reitsma, F., Baiocchi, G., Barthel, S., Güneralp, B., Erb, K. H., Haberl, H., Creutzig, F., & Seto, K. C. (2017). Future urban land expansion and implications for global croplands. *Proceedings of the National Academy of Sciences, 114*(34), 8939–8944. https://doi.org/10.1073/pnas.1606016114. https://www.pnas.org/content/114/34/8939. https://www.pnas.org/content/114/34/8939.full.pdf

Britter, R. E., & Hanna, S. R. (2003). Flow and dispersion in urban areas. *Annual Review of Fluid Mechanics, 35*(1), 469–496.

Broadbent, A. M., Coutts, A. M., Tapper, N. J., & Demuzere, M. (2018). The cooling effect of irrigation on urban microclimate during heatwave conditions. *Urban Climate, 23*, 309–329.

Brown, S., & Nicholls, R. (2015). Subsidence and human influences in mega deltas: The case of the Ganges–Brahmaputra–Meghna. *Science of the Total Environment, 527–528*, 362–374. https://doi.org/10.1016/j.scitotenv.2015.04.124. http://www.sciencedirect.com/science/article/pii/S0048969715300589

Cabral-Cano, E., Dixon, T. H., Miralles-Wilhelm, F., Díaz-Molina, O., Sánchez-Zamora, O., & Carande, R. E. (2008). Space geodetic imaging of rapid ground subsidence in Mexico City. *GSA Bulletin, 120*(11–12), 1556–1566. https://doi.org/10.1130/B26001.1. https://pubs.geoscienceworld.org/gsabulletin/article-pdf/120/11-12/1556/3398058/i0016-7606-120-11-1556.pdf

Calow, R. C., MacDonald, A. M., Nicol, A. L., & Robins, N. S. (2010). Ground water security and drought in Africa: Linking availability, access, and demand. *Groundwater, 48*(2), 246–256.

Carrera-Hernández, J., & Gaskin, S. (2007). The Basin of Mexico aquifer system: Regional groundwater level dynamics and database development. *Hydrogeology Journal, 15*, 1577–1590. https://doi.org/10.1007/s10040-007-0194-9

Cepeda, J., Smebye, H. C., Vangelsten, B., Nadim, F., & Muslim, D. (2010). Landslide risk in Indonesia. Background paper prepared for the 2011 Global Assessment Report on Disaster Risk Reduction. Prepared by the International Centre for Geohazards, Norwegian Geotechnical Institute. Geneva, Switzerland.

Che, V. B., Kervyn, M., Ernst, G. G. J., Trefois, P., Ayonghe, S., Jacobs, P., Van Ranst, E., Suh, C. E. (2011). Systematic documentation of landslide events in Limbe area (Mt Cameroon volcano, SW Cameroon): Geometry, controlling, and triggering factors. *Natural Hazards, 59*(1), 47–74. https://doi.org/10.1007/s11069-011-9738-3

Chen, N., Valente, P., & Zlotnik, H. (1998). What do we know about recent trends in urbanization? In R. E. Bilsborrow (Ed.) Migration, Urbanization, and Development: New Directions and Issues, United Nations Population Fund (UNFPA), New York (pp. 59–88).

Corbane, C., Pesaresi, M., Kemper, T., Politis, P., Florczyk, A. J., Syrris, V., Melchiorri, M., Sabo, F., & Soille, P. (2019). Automated global delineation of human settlements from 40 years of Landsat satellite data archives. *Big Earth Data, 3*(2), 140–169. https://doi.org/10.1080/20964471.2019.1625528

Cusworth, D. H., Mickley, L. J., Sulprizio, M. P., Liu, T., Marlier, M. E., DeFries, R. S., Guttikunda, S. K., & Gupta, P. (2018). Quantifying the influence of agricultural fires in northwest India on urban air pollution in Delhi, India. *Environmental Research Letters, 13*(4), 044018. https://doi.org/10.1088%2F1748-9326%2Faab303

de Assis Dias, M. C., Saito, S. M., dos Santos Alvalá RC, Stenner, C., Pinho, G., Nobre C. A., de Souza Fonseca, M. R., Santos, C., Amadeu, P., Silva, D., Lima, C. O., Ribeiro, J., Nascimento, F., Corréa, C. O. (2018). Estimation of exposed population to landslides and floods risk areas in Brazil, on an intra-urban scale. *International Journal of Disaster Risk Reduc-*

tion, 31, 449–459. https://doi.org/10.1016/j.ijdrr.2018.06.002. http://www.sciencedirect.com/science/article/pii/S2212420918300359

Depietri, Y., Renaud, F. G., & Kallis, G. (2012). Heat waves and floods in urban areas: A policy-oriented review of ecosystem services. *Sustainability Science, 7*(1), 95–107.

Dickson, K. B., & Benneh, G. (1996). A new geography of Ghana. London: Longmans.

Douglas, I., Alam, K., Maghenda, M., McDonnell, Y., McLean, L., & Campbell, J. (2008). Unjust waters: Climate change, flooding, and the urban poor in Africa. *Environment and Urbanization, 20*(1), 187–205.

Douglas, I., Goode, D., Houck, M. C., Wang, R. (Eds.) (2011). The Routledge handbook of urban ecology. Routledge: London.

Edmonds, D., Caldwell, R., Baumgardner, S., Paola, C., Roy, S., Nelson, A., & Nienhuis, J. (2017). A global analysis of human habitation on river deltas. In: EGU general assembly conference abstracts, EGU general assembly conference abstracts (p. 10832).

Egondi, T., Kyobutungi, C., Rocklöv, J. (2015). Temperature variation and heat wave and cold spell impacts on years of life lost among the urban poor population of Nairobi, Kenya. *International Journal of Environmental Research and Public Health, 12*(25739007), 2735–2748. https://www.ncbi.nlm.nih.gov/pmc/articles/PMC4377929/

Erkens, G., Bucx, T., Dam, R., de Lange, G., & Lambert, J. (2015). Sinking coastal cities. *Proceedings of the International Association of Hydrological Sciences, 372*, 189–198. https://doi.org/10.5194/piahs-372-189-2015. https://proc-iahs.net/372/189/2015/

Franks, P., Small, R., & Booker, F. (2018). Social assessment for protected and conserved areas (SAPA): Methodology manual for SAPA facilitators. London: International Institute for Environment and Development (IIED). http://pubs.iied.org/14659IIED

Galloway, G. E., Reilly, A., Ryoo, S., Riley, A., Haslam, M., Brody, S., Highfield, W., Gunn, J., Rainey, J., & Parker, S. (2018). The growing threat of urban flooding: A national challenge. University of Maryland, Center for Disaster Resilience, and Texas A&M University, Galveston Campus, Center for Texas Beaches and Shores. College Park: A. James Clark School of Engineering.

Güneralp, B., Lwasa, S., Masundire, H., Parnell, S., & Seto, K. C. (2017). Urbanization in Africa: Challenges and opportunities for conservation. *Environmental Research Letters, 13*(1), 015002. https://doi.org/10.1088/1748-9326/aa94fe

Gupta, J., Pfeffer, K., Verrest, H., Ros-Tonen, M. (Eds.) (2015). Geographies of Urban governance: Advanced theories, methods and practices. Berlin: Springer. https://doi.org/10.1007/978-3-319-21272-2

Hammer, R. B., Stewart, S. I., & Radeloff, V. C. (2009). Demographic trends, the wildland–urban interface, and wildfire management. *Society and Natural Resources, 22*(8), 777–782.

Hofste, R. W., Kuzma, S., Walker, S., Sutanudjaja, E. H., Bierkins, M. F., Kuijper, M. J., Sanchez, M. F., van Beek, R., Wada, Y., Rodríguez, S. G., & Reig, P. (2019). AQUEDUCT 3.0: Updated decision-relevant global water risk indicators. Technical Note, Washington DC: World Resources Institute. https://www.wri.org/publication/aqueduct-30

Huang, C. W., McDonald, R. I., & Seto, K. C. (2018). The importance of land governance for biodiversity conservation in an era of global urban expansion. *Landscape and Urban Planning, 173*, 44–50. https://doi.org/10.1016/j.landurbplan.2018.01.011. http://www.sciencedirect.com/science/article/pii/S0169204618300331

Ingebritsen, S. E., & Galloway, D. L. (2014). Coastal subsidence and relative sea level rise. *Environmental Research Letters, 9*(9), 091002. https://doi.org/10.1088/1748-9326/9/9/091002

International Federation of Red Cross and Red Crescent Societies (2009). Ghana: Floods. DREF Operation. http://www.reliefweb.int/rw/RWFiles2009.nsf/FilesByRWDocUnidFilename/JBRN-7TYJ32-full_report.pdf/$File/full_report.pdf

IPCC (2012) Managing the risks of extreme events and disasters to advance climate change adaptation. In A special report of working groups I and II of the intergovernmental panel on climate change. Cambridge: Cambridge University.

IPCC (2019) IPCC Special Report on Climate Change, Desertification, Land Degradation, Sustainable Land Management, Food Security, and Greenhouse Gas Fluxes in Terrestrial Ecosystems: Summary for Policymakers. http://www.ipcc.ch/, approved draft.

Jenkins, C. N., Pimm, S. L., & Joppa, L. N. (2013). Global patterns of terrestrial vertebrate diversity and conservation. *Proceedings of the National Academy of Sciences, 110*(28), E2602–E2610. https://doi.org/10.1073/pnas.1302251110. https://www.pnas.org/content/110/28/E2602. https://www.pnas.org/content/110/28/E2602.full.pdf

Jha, A. K., Bloch, R., & Lamond, J. (2012). Cities and flooding: A guide to integrated urban flood risk management for the 21st Century. Washington, DC: World Bank.

Kim, S. K., & Kannan, K. (2007). Perfluorinated acids in air, rain, snow, surface runoff, and lakes: Relative importance of pathways to contamination of urban lakes. *Environmental Science & Technology, 41*(24), 8328–8334.

Kirschbaum, D. B., & Stanley, T. (2018). Satellite-based assessment of rainfall-triggered landslide hazard for situational awareness. *Earth's Future, 6*(3), 505–523. https://doi.org/10.1002/2017EF000715. https://agupubs.onlinelibrary.wiley.com/doi/abs/10.1002/2017EF000715. https://agupubs.onlinelibrary.wiley.com/doi/pdf/10.1002/2017EF000715

Kirschbaum, D. B., Stanley, T., & Simmons, J. (2015). A dynamic landslide hazard assessment system for Central America and Hispaniola. *Natural Hazards and Earth System Sciences, 15*, 2257–2272. https://doi.org/10.5194/nhess-15-2257-2015

Kovats, R. S., & Akhtar, R. (2008). Climate, climate change and human health in Asian cities. *Environment and Urbanization, 20*(1), 165–175.

Liu, Z., He, C., Zhou, Y., & Wu, J. (2014). How much of the world's land has been urbanized, really? A hierarchical framework for avoiding confusion. *Landscape Ecology, 29*(5), 763–771. https://doi.org/10.1007/s10980-014-0034-y

Martínez-Vázquez, J., & Smoke, P. (2010). Local Government Finance: The Challenges of the 21st Century, United Cities and Local Governments, Barcelona, Spain, chap Conclusion (pp 73–93). www.cities-localgovernments.org

Martinuzzi, S., Stewart, S. I., Helmers, D. P., Mockrin, M. H., Hammer, R. B., & Radeloff, V. C. (2015). The 2010 wildland-urban interface of the conterminous United States. Research Map NRS-8., U.S. Department of Agriculture, Forest Service, Northern Research Station., Newtown Square, PA (US). https://www.nrs.fs.fed.us/data/WUI/

Mcdonald, R., Mansur, A., Ascensão, F., Colbert, M., Crossman, K., Elmqvist, T., Gonzalez, A., Güneralp, B., Haase, D., Hamann, M., Hillel, O., Huang, K., Kahnt, B., Maddox, D., Pacheco, A., Seto, K., Simkin, R., Walsh, B., & Ziter, C. (2019). Research gaps in knowledge of the impact of urban growth on biodiversity. *Nature Sustainability, 3*(1), 1–9. https://doi.org/10.1038/s41893-019-0436-6

McDonald, R. I., Forman, R. T. T., Kareiva, P., Neugarten, R., Salzer, D., & Fisher, J. (2009). Urban effects, distance, and protected areas in an urbanizing world. *Landscape and Urban Planning, 93*(1), 63–75. https://doi.org/10.1016/j.landurbplan.2009.06.002. http://www.sciencedirect.com/science/article/pii/S0169204609001133

McDonald, R. I., Green, P., Balk, D., Fekete, B., Revenga, C., Todd, M., & Montgomery, M. R. (2011). Urban growth, climate change, and freshwater availability. In Proceedings of the National Academy of Sciences (PNAS) Published 28 March 2011 in Online Early Edition, manuscript 2010-11615R.

McDonald, R. I., Weber, K., Padowski, J., Flörke, M., Schneider, C., Green, P. A., Gleeson, T., Eckman, S., Lehner, B., Balk, D., Timothy Boucher, G. G., & Montgomery, M. (2014). Water on an urban planet: Urbanization and the reach of urban water infrastructure. *Global Environmental Change, 27*, 96–105.

McGee, T. G. (1987). Urbanization or Kotadesasi—the emergence of new regions of economic interaction in Asia. Honolulu: East-West Center.

McGee, T. G. (1991). The emergence of *desakota* regions in Asia. Expanding a hypothesis. In: N. Ginsburg, B. Koppel, T. G. McGee (Eds.) The Extended Metropolis. Settlement Transition in Asia. Honolulu: University of Hawaii (pp. 3–25).

McGranahan, G., Satterthwaite, D., & Tacoli, C. (2004). Rural–urban change, boundary problems and environmental burdens. International Institute for Environment and Urbanization (IIED), Working Paper 10, Working Paper Series on Rural–Urban Interactions and Livelihood Strategies, London.

McGranahan, G., Balk, D., & Anderson, B. (2007). The rising tide: Assessing the risks of climate change to human settlements in low-elevation coastal zones. *Environment and Urbanization, 19*(1), 17–37.

Mishra, A. K., & Singh, V. P. (2010). A review of drought concepts. *Journal of Hydrology, 391*(1–2), 202–216.

ML Infomap Pvt Ltd (2014) Village points in Arunachal Pradesh, 2011. http://www.mlinfomap.com

NASA (2018) Monitoring urban floods using remote sensing. In Applied Remote Sensing Training Program, 25 July and 1 August 2018. Training materials downloaded September 2018. https://arset.gsfc.nasa.gov

Nyarko, B. K. (2002). Application of a rational model in GIS for flood risk assessment in Accra. *Journal of Spatial Hydrology, 2*(1), 1–14.

Olden, J. D., Poff, N. L., & McKinney, M. L. (2006). Forecasting faunal and floral homogenization associated with human population geography in North America. *Biological Conservation, 127*(3), 261–271. https://doi.org/10.1016/j.biocon.2005.04.027. http://www.sciencedirect.com/science/article/pii/S0006320705003575

Ongoma, V., Muthama, N. J., & Gitau, W. (2013). Evaluation of urbanization influences on urban winds of Kenyan cities. *Ethiopian Journal of Environmental Studies and Management, 6*(3), 223–231.

Panel on Urban Population Dynamics (2003) Cities transformed: Demographic change and its implications in the developing world. Washington, DC: National Academies Press. M. R. Montgomery, R. Stren, B. Cohen, & H. E. Reed (Eds.)

Patz, J. A., Confalonieri, U. E., Amerasinghe, F. P., Chua, K. B., Daszak, P., Hyatt, A. D., Molyneux, D., Thomson, M., Yameogo, D. L., Mwelecele-Malecela-Lazaro, Vasconcelos, P., Rubio-Palis, Y., Campbell-Lendrum, D., Jaenisch, T., Mahamat, H., Mutero, C., Waltner-Toews, D., & Whiteman, C. (2005). Human health: Ecosystem regulation of infectious diseases. In R. Hassan, R. Scholes, N. Ash (Eds.) Ecosystems and human well-being: Current state and trends: Findings of the Condition and Trends Working Group, Island Press, chap 14 (pp. 391–415).

Perkol-Finkel, S., Ferrario, F., Nicotera, V., & Airoldi, L. (2012). Conservation challenges in urban seascapes: Promoting the growth of threatened species on coastal infrastructures. *Journal of Applied Ecology, 49*(6), 1457–1466. http://doi.wiley.com/10.1111/j.1365-2664.2012.02204.x

Portier, C. J., Thigpen, T. K., Carter, S. R., Dilworth, C. H., Grambsch, A. E., Gohlke, J., Hess, J., Howard, S. N., Luber, G., Lutz, J. T., Maslak, T., Prudent, N., Radtke, M., Rosenthal, J. P., Rowles, T., Sandifer, P. A., Scheraga, J., Schramm, P. J., Strickman, D., Trtanj, J. M., & Whung, P. Y. (2010). A human health perspective on

climate change: A report outlining the research needs on the human health effects of climate change. Research Triangle Park, NC: Environmental Health Perspectives/National Institute of Environmental Health Sciences. https://doi.org/10.1289/ehp.1002272. www.niehs.nih.gov/climatereport

Radeloff, V. C., Helmers, D. P., Kramer, H. A., Mockrin, M. H., Alexandre, P. M., Bar-Massada, A., Butsic, V., Hawbaker, T. J., Martinuzzi, S., Syphard, A. D., & Stewart, S. I. (2018). Rapid growth of the US wildland-urban interface raises wildfire risk. *Proceedings of the National Academy of Sciences, 115*(13), 3314–3319. https://doi.org/10.1073/pnas.1718850115. https://www.pnas.org/content/115/13/3314. https://www.pnas.org/content/115/13/3314.full.pdf

Rain, D., Engstrom, R., Ludlow, C., & Antos, S. (2011). Accra Ghana: A city vulnerable to flooding and drought-induced migration. Case study prepared for *Cities and Climate Change: Global Report on Human Settlements 2011.* United Nations Human Settlements Programme. http://www.unhabitat.org/grhs/2011

Revi, A., Satterthwaite, D. E., Aragón-Durand, F., Corfee-Morlot, J., Kiunsi, R. B., Pelling, M., Roberts, D. C., & Solecki, W. (2014). Urban areas. In C. Field, V. Barros, D. Dokken, K. Mach, M. Mastrandrea, T. Bilir, M. Chatterjee, K. Ebi, Y. Estrada, R. Genova, B. Girma, E. Kissel, A. Levy, S. MacCracken, P. Mastrandrea, L. White (Eds.) Climate change 2014: Impacts, adaptation, and vulnerability. Part A: Global and sectoral aspects. Contribution of working group II to the fifth assessment report of the intergovernmental panel on climate change: Cambridge: Cambridge University Press (pp. 535–612)

Romero Lankao, P. (2010). Water in Mexico city: What will climate change bring to its history of water-related hazards and vulnerabilities? *Environment and Urbanization, 22*(1), 157–178.

Ros-Tonen, M., Pouw, N., & Bavinck, M. (2015). Governing beyond cities: The urban-rural interface. In J. Gupta, K. Pfeffer, H. Verrest, M. Ros-Tonen (Eds.) Geographies of Urban Governance. Berlin: Springer, chap 5 (pp. 85–105). https://doi.org/10.1007/978-3-319-21272-2_5

Rumbach, A., & Follingstad, G. (2019). Urban disasters beyond the city: Environmental risk in India's fast-growing towns and villages. *International Journal of Disaster Risk Reduction, 34*, 94–107. https://doi.org/10.1016/j.ijdrr.2018.11.008. http://www.sciencedirect.com/science/article/pii/S2212420918306617

Scott, A. A., Misiani, H., Okoth, J., Jordan, A., Gohlke, J., Ouma, G., Arrighi, J., Zaitchik, B. F., Jjemba, E., Verjee, S., & Waugh, D. W. (2017). Temperature and heat in informal settlements in Nairobi. *PLoS One, 12*(11), 1–17. https://doi.org/10.1371/journal.pone.0187300

Seto, K. C., Güneralp, B., & Hutyra, L. R. (2012). Global forecasts of urban expansion to 2030 and direct impacts on biodiversity and carbon pools. *Proceedings of the National Academy of Sciences, 109*(40), 16083–16088. https://doi.org/10.1073/pnas.1211658109. https://

www.pnas.org/content/109/40/16083. https://www.pnas.org/content/109/40/16083.full.pdf

Shafran, A. P. (2008). Risk externalities and the problem of wildfire risk. *Journal of Urban Economics, 64*(2), 488–495.

Shultz, J. M., Russell, J., & Espinel, Z. (2005). Epidemiology of tropical cyclones: The dynamics of disaster, disease, and development. *Epidemiologic Reviews, 27*, 21–35.

Stanley, T., & Kirschbaum, D. B. (2017). A heuristic approach to global landslide susceptibility mapping. *Natural Hazards, 87*, 145–164. https://doi.org/10.1007/s11069-017-2757-y

Svetlosakova, Z., & Jha, A. K. (2012). Cities and flooding: Urban flood risk management. In Presentation to "Urban Resilience and Adaptation to Climate Change" Learning Event, World Bank.

Szabo, S., Brondizio, E., Renaud, F. G., Hetrick, S., Nicholls, R. J., Matthews, Z., Tessler, Z., Tejedor, A., Sebesvari, Z., Foufoula-Georgiou, E., da Costa, S., & Dearing, J. A. (2016). Population dynamics, delta vulnerability and environmental change: Comparison of the Mekong, Ganges-Brahmaputra and Amazon delta regions. *Sustainability Science, 11*(4), 539–554. https://doi.org/10.1007/s11625-016-0372-6

Tanner, T. M., Mitchell, T., Polack, E., & Guenther, B. (2009). Urban governance for adaptation: Assessing climate change resilience in ten Asian cities. IDS Working Paper 315, Brighton: Insitute for Development Studies at the University of Sussex.

Tessler, Z. D., Vörösmarty, C. J., Grossberg, M., Gladkova, I., Aizenman, H., Syvitski JPM, Foufoula-Georgiou E (2015) Profiling risk and sustainability in coastal deltas of the world. *Science, 349*(6248), 638–643. https://doi.org/10.1126/science.aab3574. https://science.sciencemag.org/content/349/6248/638. https://science.sciencemag.org/content/349/6248/638.full.pdf

Tessler, Z. D., Vörösmarty, C. J., Grossberg, M., Gladkova, I., & Aizenman, H. (2016). A global empirical typology of anthropogenic drivers of environmental change in deltas. *Sustainability Science, 11*(4), 525–537. https://doi.org/10.1007/s11625-016-0357-5

UNEP-WCMC, IUCN (2019) Protected planet: The world database on protected areas (WDPA)/the global database on protected areas management effectiveness (gd-pame)] [on-line]. downloaded july 2019. Cambridge, UK: UNEP-WCMC and IUCN. https://www.protectedplanet.net

UNFPA (2007). State of world population 2007: Unleashing the potential of urban growth. New York: United Nations Population Fund. George Martine, lead author.

UNICEF (2016). Clear the air for children: The impact of air pollution on children. https://www.unicef.org/publications/files/UNICEF_Clear_the_Air_for_Children_30_Oct_2016.pdf

United Nations (2013a). A new global partnership: Eradicate poverty and transform economics through sustainable development. In The report of the High-Level Panel of Eminent Persons on the post-2015 development agenda. http://www.post2015hlp.org/wp-content/uploads/2013/05/UN-Report.pdf

United Nations (2013b). TST issues brief: Sustainable cities and human settlements. The Technical Support Team (TST) is co-chaired by the Department of Economic and Social Affairs and the United Nations Development Programme. This issues brief was co-led by UN-Habitat and UNEP with the participation of ECLAC, ESCAP, IFAD, ILO, UNDP, UNFPA, UNICEF, UNISDR, UN-Women, WHO, WMO, and the World Bank.

United Nations (2014). World Urbanization Prospects. United Nations: Department of Economic and Social Affairs. Population Division.

USAID, Mailman School of Public Health Columbia University, Vital Strategies (2019). LMIC urban air pollution solutions: Technical document.

von Glasow R, Jickells TD, Baklanov A, Carmichael GR, Church TM, Gallardo L, Hughes C, Kanakidou M, Liss PS, Mee L, Raine R, Ramachandran P, Ramesh R, Sundseth K, Tsunogai U, Uematsu M, Zhu T. Megacities and large urban agglomerations in the coastal zone: interactions between atmosphere, land, and marine ecosystems. Ambio. 2013 Feb;42(1):13–28. https://doi.org/10.1007/s13280-012-0343-9. Epub 2012 Oct 18. PMID: 23076973; PMCID: PMC3547459.

WHO (2018). Air pollution and child health: Prescribing clean air. https://www.who.int/ceh/publications/air-pollution-child-health/en/ (Advance copy—Under review).

World Bank (2009). World Development Report 2009: Reshaping Economic Geography. Washington, DC: The World Bank.

World Bank (2011). Urban Risks Assessments: An Approach for Understanding Disaster & Climate Risk in Cities. Washington, DC: World Bank. Urban Development & Local Government Unit, Finance, Economics and Urban Department.

Population and Agricultural Change

18

Richard E. Bilsborrow

Abstract

This chapter assesses the state-of-the art pertaining to studies on the relationships between population and land use over the last half century, focusing on developing countries. The theoretical perspectives drawn upon are from Malthus (population increase leads to more land brought into agricultural use, or land extensification) and Boserup (population increase tends to induce more intensive use of existing land). But both are widely misinterpreted, and a more nuanced view shows each had much more to say. The extensive literature is reviewed in two sections, pre-2000 and since 2000, showing less focus since 2000 on whether population factors are affecting (or not) changes in land use or use of inputs but more on whether land intensification is occurring and how to stimulate it. Empirical studies of the available cross-country data find major differences in overall trends over time in land use and input use between the least developed countries and other developing countries, associated with both the level of development and rates of population growth (and hence fertility): developing countries at higher incomes and already low or declining fertility and rural population growth are experiencing little increase in land extensification (except some in South America) and instead advances in land intensification, while the least developed countries, which continue to have high fertility and rural population growth, continue to have substantial increases in the agricultural land area and little increase in land intensification. Malthus thus continues to be relevant for the latter. The chapter concludes with recommendations for further research and policy improvements.

Introduction

Over 30 years ago, several colleagues and I conducted research for the UN Food and Agricultural Organization (FAO) on whether population change affects agricultural change and land use

R. E. Bilsborrow (✉)
Carolina Population Center, University of North Carolina, Chapel Hill, NC, USA
e-mail: richard_bilsborrow@unc.edu

© Springer Nature Switzerland AG 2022
L. M. Hunter et al. (eds.), *International Handbook of Population and Environment*, International Handbooks of Population 10, https://doi.org/10.1007/978-3-030-76433-3_18

in developing countries. The aim of this chapter is to bring this research assessment up to date, focusing on the views of Malthus and Boserup, clarifying their views, and misperceptions, and providing new evidence.

In section "The theories, the grand debate, and evidence up to the 1990's" following since it is immediately below, I look at the state of the art about the population-agriculture debate at that time 30–40 years ago, assessing both the quantitative empirical evidence and a number of the most important case studies up to the 1990s. This is followed in section "Reassessment of the theory and definitions" by a pause to reflect on studies of Brookfield and others which raise fundamental questions about definitions, Malthusian and Boserupian theory and the direction of causation. In the largest section (3part III), I summarize results from a meta-study of the more recent scientific literature based mainly on journal articles of country-case studies but also including multi-country quantitative analyses. The final section "More recent evidence" provides a summary of where we are now in terms of anticipated future changes in land use, approaches for future research that take advantage of data becoming increasingly available, and policy considerations.

A brief caveat is desirable before embarking on this journey. First, there are a host of ways of envisaging population in relationship to agriculture. The focus here is on *population change* in relationship to the land area in agricultural use: (a) if the land area is viewed as expanding (at least partly) in response to population increase, we refer to this as *land extensification;* and (b) if the existing land is being used more intensely, this is *land intensification*. Effects on land use of changes in the *composition* of the population (according to age, sex, education, ethnicity, etc.) or changes in population *distribution* due to migration will generally not be considered, nor will studies focusing on demand-side factors associated consumer demands for food or other agricultural products, nor on changes in market prices, climate change, etc., although such factors are often in the background and affecting land use.

The Theories, the Grand Debate, and Evidence Up to the 1990's

Malthus and Boserup

Relationships between population and agricultural land use have been the stuff of philosophical conjecture for several thousand years, perhaps since the first domestication of seeds into growing grains around 10,000 years ago—wheat in the Tigris-Euphrates valleys of Mesopotamia or present-day Iraq, and maize or corn in the central valley of Mexico. By several centuries BC,[1] Greek and other scholars were speculating about links between population and food supply, including about the optimum population of the planet in the known world; this included the Zoroastrians around 325 BC, the Indian sage Kautilya around 300 BC, Aristotle in his *Politics,* and the Bible (Parsons, 1992: 3–6). But it is the economist and parson Thomas Malthus who is generally considered the first to develop a comprehensive theoretical approach, albeit with distinct threads that have made it both fascinating and subject to criticism from the beginning. But there has been much misunderstanding of Malthus as well, particularly among contemporary social scientists, who continue to debate both its science and its policy implications while most ecologists (not just Ehrlich, 1968, the best known) are convinced of the negative impacts of large and growing populations on the natural environment and biodiversity. Mortality started to decline in most of the world after 1492 linked to the vast expansion in global trade in foodstuffs and transfers of plants and seeds across the oceans (the so called "Columbian exchange": Crosby, 1972), facilitating population

---

[1]Recent accumulating archeological research is providing increasing evidence of substantial human impacts on the planet as early as 1000 BC (*NY Times,* August 12, 2018) on research led by Ellis of Cornell University, and by Stephens et al. (2019). They cite results of the ArchaeoGlobe Project to synthesize knowledge from studies across the planet, coinciding with a new journal since 2012, *The Anthropocene Review*, with Anthropocene meaning the time since human activities began to alter the face of the planet, seen now as not 1000 but at least 3000 years ago.

to grow in the world Malthus knew, in Europe. This led to the thesis in Malthus most commonly cited, that human populations tend to grow geometrically over time as 1,2,4,8 ... while the "means of subsistence" (food to feed them) tend to grow only arithmetically as 1,2,3,4 ..., the former therefore tending to surpass the latter over time (Malthus 1798, *First Essay,* p. 9ff: 1960 version). Malthus saw agriculture as tending to grow through the expansion of the agricultural area, or via "agricultural extensification". This assumption that agriculture tends to expand through an expansion of its area is still associated with Malthus by the "Neo-Malthusians", who see the expansion occurring at the expense of the natural environment (clearing of forests, grasslands, depletion and damage to water supplies, etc.), in turn causing a decline in biodiversity.

It is important to examine these implied linkages and what Malthus *actually* wrote in more detail. First, he thought that people would recognize (learn) differences in the quality of soils and would hence tend to use the lands with better soils first for agriculture, so that over time agricultural expansion would tend to involve less productive soils brought into use at the "extensive margin" (*op. cit.*, 35ff, 151ff). To the extent this were true, the tendency would indeed be for the population to grow faster than the food supply to feed it, with diminishing returns to land expansion. The concept of diminishing returns was elaborated by a contemporary economist of Malthus, David Ricardo, who noted that if one factor of production is fixed (e.g., land), then increasing applications of labor eventually result in smaller increases in output per additional worker—a fundamental principle in classical and neo-classical economics. But was no basis for assuming that lands at the extensive margin would *necessarily* be less fertile.

Malthus's theory implies that population growth tends to exceed the growth in food to sustain it, provoking a "Malthusian crisis", of increased mortality from starvation, or so-called "positive checks" (158 ff; 165, Ch. III). To alleviate this, Parson Malthus urged people to reduce fertility and population growth by postponing marriage and decrease fertility within marriage by controlling the "passions within the sexes" (p. 8)—his so-called "preventative checks". Malthus' predictions of cycles of famine and immiseration have been disproved by history–since his time the world population has risen from one billion to nearly eight billion—but he could not have foreseen the vast changes in agricultural technology that were to occur. However prospects for population crises in England or more broadly in Europe in the nineteenth century were much alleviated by emigration to the New World, Australia and New Zealand, which Malthus anticipated:

> There are many parts of the globe .... hitherto uncultivated (*op. cit.*, 155) .... [so] in the case of a redundant population in the more cultivated parts of the world, the natural and obvious remedy.... is emigration to those parts.... uncultivated (Chap. 4 of *Second Essay,* 1983, p. 346 and *passim*)... but such a solution is only 'temporary' until the whole earth is settled. (p. 352)

Another observation of Malthus relevant to contemporary linkages between population growth and rising demands on the world's land is his noting that private land tenure is not a solution, that land inequality is a problem, and that the growing demand for meat may be a significant future problem:

> ... the inclosure of common fields [through private property] has frequently had a contrary effect... large tracts of land, which formerly produced great quantities of corn, by being converted to pasture both employ fewer hands and feed fewer mouths than before their inclosure. It is an acknowledged truth, that pastureland produces a smaller quantity of human subsistence than corn land of the same natural fertility ... [contributing to] the diminution of human subsistence. (*op. cit.* 1983, p. 113)[2]

---

[2]Malthus' theory of cycles of immiseration led to economics being dubbed the "dismal science": whenever living (food) conditions improve, the population would grow faster than the food supply, resulting in insufficient food and hence misery and deaths until the population declines to equilibrate again with the food supply. But this should be tempered by his later conclusions about the future–that education, an increasing role of government, and "moral restraints" (his preventative checks) could lead to an "improvement in human society in despair", facilitated by "the brilliant career of physical discovery ... in which mankind will be influenced by its progress and partake

A century and a half later, the Danish economist Ester Boserup (1965) postulated that population growth and rising population density could itself stimulate changes in the intensity of land use that would make it possible for the land to support the growing population—interpreted by "cornucopians" such as Simon (1977, 1981) as solving its own problem. But before returning to this, it is desirable to indicate her long view of history on how land use has been evolving from hunter-gathering through five stages of increasing intensity: (1) *forest-fallow*, in which plots of land are cleared and planted for a year or two, then left fallow for the forest (soil) to regenerate for 20–30+ years; (2) *bush-fallow*, with the fallow period reduced to 6–10 years, by which time the regenerating land is covered in bushes and small trees; (3) *short-fallow*, a system with a fallow of only 1–2 years during which wild grasses germinate and provide limited nutrients; (4) *annual cropping*, in which the land is not cultivated for only a few months between the harvest and the next planting; and finally (5) *multi-cropping*, in which the plot is planted in two or more sequential crops within a year, with no real fallow (Boserup, 1965: 12–13). She sees these stages as characterizing the phases of evolution of agriculture from prehistoric times to the present: "Even if we cannot be sure that systems of extensive land use have preceded the intensive ones in every part of the world, there seems to be little reason to doubt that the typical sequence of development of agriculture has been a gradual change—more rapid in some regions than in others—from extensive to intensive types of land use" (Boserup, 1965, pp. 17–18; also Boserup, 1981).

Thus she saw population growth and increasing density as tending to stimulate the evolution from one stage to the next over time, with most of the world's agriculture reaching stages 4 and 5 by the late twentieth century, by the time of her writings. She saw this as a logical response to there being less land per person so that fallow times had to shrink, and simultaneously more labor was available to work the land to increase land productivity (e.g., by weeding, small scale irrigation, etc.). Besides reducing fallow time, and thereby raising the labor/land ratio, Boserup also foresaw, though did not focus on, other ways to increase output per unit of land, including use of natural and chemical fertilizers and irrigation, which also happen to involve increased applications of labor. These forms of agricultural intensification are discussed in Bilsborrow (1987) and Bilsborrow and Geores (1994), along with another dimension of land intensification, the replacement of crops (e.g., grains) which require more land than labor by higher-value more labor-intensive crops, such as vegetables, which require more labor than land for the same value of output. Further discussion of intensification is found in section "Reassessment of the theory and definitions" below.

Although Boserup saw population growth and increasing density as *potentially stimulating* or contributing to an increase in the intensity of land use, she never wrote that such a response would automatically occur, or would be sufficient to keep production *per person* (and therefore income per capita and living conditions) from declining as population growth continued–only that in *some circumstances* it could stimulate sufficient changes.[3] For example, a study of Lele and Stone (1989) on different ethnic groups in six Sub-Saharan African countries examined how they were coping with high population growth. The authors concluded that in three countries the responses were positive while in three others they were not, so there, population growth was contributing to serious development problems. The positive responses were traced to national policies facilitating effective responses and/or to aspects of the local culture favorable to undertaking risk and innovation.

---

of its success" (*Second Essay:* 593–4). That is, he was aware of the ongoing industrial revolution and improving technology in manufacturing, but not of future possible improvements in agriculture.

[3] Personal discussion with Boserup at a conference in Switzerland in 1975, and consistent with her 1981 book on the urgency of expanding women's economic activities to achieve development in low-income countries.

Extensions and Alternative Theoretical Approaches to Analyzing the Determinants of Extensification vs. Intensification

Boserup was likely affected by the thinking of the famous British historian, Toynbee, who wrote up the history of the world's great empires in 12 volumes (in 1934–61), viewing the successful ones as responding to great challenges—the challenge-response theory of history. This influenced Kingsley Davis to develop his "theory of the multiphasic response", postulating that population growth and large families put pressures on living standards that stimulate one or more responses, and that *the more one response occurs, the less pressure there is for the other responses* to occur. In considering the historical processes in Japan and Western Europe, the declines in mortality led to high population growth and pressures on household living standards which Davis (1963) saw as occurring through three means: postponement of marriage, reduction of fertility within marriage, and international migration—all three curiously also mentioned by Malthus, albeit here without the religious overtones. Davis also postulated that any combination of the responses could occur and that there could be trade-offs. Bilsborrow (1987) expanded upon this theory by adding four additional, primarily economic, responses, all relevant to the discussion here: land extensification (Malthusian), land intensification (Boserupian), local off-farm employment, and rural-urban migration. Again, the more one "response" occurs, the less pressure there is for the other responses, but there may be multiple, simultaneous responses. Which of the seven responses occur depends on the conditions in the particular household, its local community, region and country—socio-economic, demographic, cultural, environmental, and political.

This theory of the multiphasic response provides a broad conceptual framework for examining the factors affecting the responses of rural households in developing countries to population increase (or other sources of economic pressures), *ceteris paribus*. Drawing on Bilsbor-row (1987) and Bilsborrow and Geores (1994), these responses are illustrated in Fig. 18.1, which also shows some of the factors that condition the likelihood of each type of response. The main responses of interest here are land extensification and land intensification. In traditional societies, the former is the most likely, depending on "the availability and accessibility of untapped, potentially cultivable land" (Bilsborrow and Geores, *op. cit.* p. 177), and both it and increased applications of labor to a fixed land endowment avoid disruptions to family life, community ties and customs which occur with the other responses (migration, reducing fertility, etc.). Among the factors that affect the likelihood of responses are access to roads and markets (von Thünen, 1826), which even in modern times continue to play crucial roles in deforestation (e.g., Pichon, 1997; Nepstad et al., 2001). Roads are the most common mode of access, so government road-building projects, especially in frontier regions or near forests and sparsely settled savanna or grasslands, can make this a likely response. Where holdings are communal, population growth can lead to a "tragedy of the commons" (Hardin, 1968) in the absence of effective "community property" controls (Ostrom, 1990). And a host of other factors that affect the demand for agricultural products may stimulate land extensification, intensification or both–not only population increase but increases in incomes, and associated changes in tastes toward more consumption of meat, ongoing in low-income countries. The latter increases the demand for pastureland, which has been stimulating the expansion of the agricultural land area (see "The cattle are eating the forest" of DeWalt 1985, on Honduras; Hecht, 1985; Bilsborrow & Carr, 2001; Keller et al., 2009). Other factors also affect land intensification at the household level, the community level (leadership, culture, location) and higher contextual or policy-making levels, such as government policies. At the household level, more education may facilitate use of chemicals, and also may be associated with higher consumption aspirations (Pichon & Bilsborrow, 1999). Younger household heads are more likely to try chemical fertilizers than their parents accustomed to natural fertiliz-

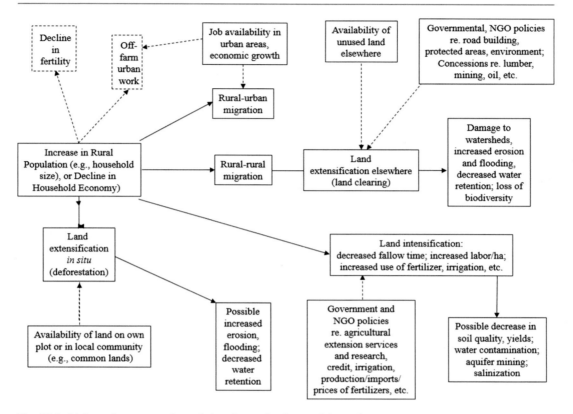

Fig. 18.1 Linkages between rural population change, land use and the environment

ers. Households with higher incomes can afford the expense of fertilizers (Sellers & Bilsborrow, 2020), which if successful in increasing yields widens inequality across farm households. Government policies can create a positive environment for improving technology, via funding for agricultural research and extension services to farmers and training in the use of new higher productivity or pest-resistant crops, and subsidizing input costs, lowering the cost of land intensification. Government construction or promotion of small- or large-scale irrigation programs can improve the availability and reliability of water for crops. Credit programs have often been found crucial in facilitating farmers' access to chemical inputs (*ibid.*) Finally, aspects of culture, access to education and health facilities, existence of agricultural cooperatives, and the extent to which households collaborate in community activities such as house construction and forest clearing can influence the extent to which farm households adopt modern technology (Blaikie & Brookfield, 1987; Brookfield, 2001).

Finally, there is a substantial body of relevant research from agricultural economics, including models of the agricultural household (Barnum & Squire, 1979; Singh et al., 1986), in which agricultural production on the supply side and household consumption on the demand side are simultaneously modeled in a "utility" maximization model, taking into account household land, human capital, and constraints. While these models are complex and can require strong assumptions, they have been used to study household size and other impacts on land extensification and intensification when the necessary data are available—a significant challenge. Another approach, induced innovation theory, focuses on the supply side and sees technology-induced increases in productivity per unit of land being independent of population pressures (Binswanger & Ruttan, 1978; Pingali & Binswanger, 1984). It sees technology change as mostly exogenous– developed outside the case study population in question. Thus much of the innovations in the past half century or so is linked to international organizations, following the success of the Green Revolution promoted by the Rockefeller Foundation, which facilitated Borlaug's development of dwarf rust-resistant varieties of wheat in Mexico and India in the 1950's and 1960's, setting off the Green Revolution. In addition, some developing countries have developed and funded their own national agricultural research centers, to improve seed varieties and promote more crop rotation and nitrogen-fixing legumes, etc. But the main sequelae of Borlaug is the support of the World Bank and others since 1994 for a family of CGIAR (Consultative Group on International Agricultural Research) international research centers, most on a single key crop (such as rice, wheat, corn, potatoes, sorghum) or agricultural system (viz., agroforestry). These have played a major role in developing and promoting the distribution of highly productive seeds/plants to farmers, free or subsidized.

Some Evidence Up to the 1990s, from Case Studies

Much of the "earlier" work, i.e., latter part of last century, on land extensification and intensification was on a particular ethnic group or community in a single country, often involving rich discussions of land use in cultural and geographic context. Most of this was *case studies based on small numbers of households* and descriptive rather than aimed at gauging the relative impacts of factors on land use. Through the 1980s, the vast majority of this was by anthropologists and geographers, as reported in Bilsborrow (1987), Blaikie and Brookfield (1987), Little et al. (1987), Turner et al. (1993), and others. The little by economists was by agricultural economists, with very little by demographers or other social scientists. This stimulating literature may be heuristically (if arbitrarily) distinguished into that focusing on land extensification and that on land intensification. Here and throughout this chapter, items were selected based on overall quality and inclusion of population in some meaningful fashion.

We begin with extensification, though occasionally it is examined together with intensification when both occurred in the same reference period. Often other important factors were also involved—consistent with multi-phasic response theory–but are not elaborated here due to space limitations.

Land Extensification

1. In a famous, long-term historical study on population and change in cultivated area, Perkins (1969) compiled data for China from 1400 to 1957, showing the population growing from 75 million to 647 million while the cultivated area rose as well in each century, from 25 million hectares (ha hereafter) to 112 million ha in 1957. Population per ha almost exactly doubled, from 2.90 to 5.78, almost all rural population. He traced agricultural practices back to the fifth century AD, when most land use was slash- and-burn, shifting cultivation, with some bush and short fallow, following Boserup. By 1400 most people lived in stable settlements and practiced annual cropping. Increases in acreage were said by Perkins to account for 55% of the increase in output during the period 1400–1957 (*op. cit.*: 33, cited in Bilsborrow, 1987, p. 187), with increases in yields accounting for about 40% due to rising labor inputs per ha, small gradual farmer-managed improvements in seeds, replacement of millet by rice, growing multi-cropping of rice, reduced fallow times, and planting wheat and barley over the winter in rice growing areas. In addition, there was some introduction of corn after 1914, more irrigation, improved plows, and more use of animal fertilizer and (human) night soil–all stimulated by population growth, according to Perkins. There was thus both extensification on a large scale but also many small forms of intensification.

2. In another classic study, of Geertz (1963) on Java, increasing density and subdivision of plots led to increased inputs of labor per ha but eventually the plots were so small that more labor was spread over smaller plots,

reducing production per person and household income and leading to a decline in living standards, captured in the term "involution" for the failure to intensify land use and instead acceptance of a decline in living conditions. Since then, however, most of Java intensified its agriculture, while millions of people were moved to other islands through the Transmigration Program or migrated voluntarily (Whitten, 1987; World Bank, 1988). Crowded conditions in Java also provided fertile ground for one of the first successful family planning programs in rural areas of the developing world, beginning in the 1960's and leading to an early fertility decline, as part of a truly multiphasic response.

3. In Latin America, especially, there have been many studies on the effects of population growth as well as migration to frontier forest regions on deforestation of sub-tropical and tropical rainforests. Studies on Central American countries based mostly on qualitative data and small household surveys include Stonich (1989) and DeWalt (1985) on different parts of Honduras, and Joly (1989) and Heckandon and McKay (1984) on Panama. Stonich observed population growth leading to ever smaller landholdings of small farms due to partiable inheritance from one generation to the next in southern Honduras, given that most good land was already tied up in large farms in the fertile valleys, forcing growing households to move up into the highlands to clear forests to establish small farms for subsistence. Meanwhile, DeWalt (1985) found the expansion of cattle pastures the main proximate factor causing deforestation elsewhere in Honduras. In southern Panama, new roads opened up tropical forests to migrant colonists, who established cattle ranches by clearing forests, i.e., land extensification.

4. Similar processes have occurred in southeast Asia. For example, in the Philippines, Cruz and Cruz (1990) and Cruz (1999) describe a situation like that of southern Honduras in which the combination of a rapidly growing population linked to high rural fertility and a very inegalitarian land distribution in the valleys forced small farmers experiencing smaller and smaller landholdings to migrate to colonize nearby mountain slopes, clearing forests to establish family farms, which led to erosion and soil depletion. 30% of the total Philippine population was living on uplands by the 1980s. During 1960–80, land under crops rose by 94% while natural forests decreased by 42%, indicating substantial land extensification.

5. A host of studies have been undertaken on Brazil on the causes of deforestation since the 1970s, with Blaikie and Brookfield (1987, Ch. 9) providing historic context. The Amazon basin first began to be opened up for colonization via roads in the 1960s and 1970s, with an early demographic study based on household survey data for Ouro Preto and Gy Parana (in the southern Amazon) by Henriques (1985), and later by Walker et al. (2000) in the eastern Amazon. Henriques found small farmer colonization of plots for agriculture occurred via clearing forests, with migrants coming from the densely populated, poverty-stricken Northeast; this was then followed by high fertility and further forest clearing for crops. On the other hand, Walker observed vast cattle ranch expansion, as cattle farmers often cheaply bought up small farms cleared by migrant farmers who depleted the poor quality oxisoils of nutrients after only a few years. The resulting consolidation of landholdings into cattle ranches thus constituted the second wave of land extensification, which unlike the first one, had little to do with population increase.

6. In Ecuador, following the first detailed household survey of 418 migrant settler farms in the northeast Amazon in 1990, analyses of the determinants of land use and land clearing, starting with my Colombian doctoral student at UNC (Pichon, 1997; Bilsborrow & Pichon, 1999) were among the first multivariate studies on the determinants of land use. These studies modelled simultaneously the factors affecting not only

deforestation but also land areas in pasture, in perennial crops, and in annual/subsistence crops (see also Pan & Bilsborrow, 2005). These studies found the number of adult males in the household a significant factor in land clearing, suggesting some Malthusian effects, though other factors such as location/road access had stronger effects.

7. One of the more complex studies linking population growth to deforestation is on Guatemala by Bilsborrow and Stupp (1997), which took advantage of publicly available data from population and agricultural censuses (which could easily be done in many other developing countries which have similar data sets but appears to not have been done). Thus using data from a series of population and agricultural censuses from the 1950s to the 1980s (three of each), they showed how high rural population growth in the Altiplano led over time to smaller and smaller plots there, most too small to support a typical family according to the National Planning Agency, SEGEPLAN. Meanwhile, rural employment conditions were also deteriorating, and there was high migration occurring into the Peten rainforest region, resulting in rapid deforestation by migrant colonists. We linked the two to attribute the deforestation in one region (Peten) to high population growth in areas of origin which forced people to out-migrate. Later work of Carr (2008) and Carr et al. (2017) found that this was an oversimplification, as migrants to the Peten had rural origins all over Guatemala.

8. A number of relevant case studies were carried out by the 1990s in Sub-Saharan Africa, finding evidence of population-induced agricultural extensification as countries became independent, and research increased (compiled in a report to the FAO by Bilsborrow & Stupp, 1987). The Kisii highlands of Kenya (Bilsborrow & Okoth-Ogendo, 1992) were characterized by fertile soils and good rainfall with modest levels of population density up to the 1940s. But rapid population growth since then led

to erosion of the traditional Gusil clan-based land tenure system, fragmentation of plots and encroachment into forests and woodlands, and out-migration–processes widespread by the 1980s in much of rural Kenya. In the Kisii, fertile soils led also to some agricultural intensification via inter-cropping. Mosley (1963) had shown earlier that significant land extensification was occurring in both Kenya and Zimbabwe, along with some innovations such as use of better plows.

9. The Maasai are such a well-known and evolving population in East Africa that they warrant a separate point, which also ties in the impacts of prior colonial and post-colonial policies, both of which seriously circumscribed the area allowed to the Maasai compared to their traditional lands (Sindiga, 1984). First, the Maasai were gathered into native reserves in 1904 and 1911, leading to herds too large for the available land, resulting in overgrazing and soil erosion. Then in the 1940s game reserves were established, and in the 1960s post-independence governments preserved these reserves to promote tourism, along with providing private land titles to some but not most Maasai. With continuing high fertility and population growth, we concluded even then that "the Maasai population is now too large relative to its remaining lands and herd sizes and is no longer economically self-sufficient" (Bilsborrow & Stupp, p. 39). Indeed, in 1970–72, 90% of the livestock was lost due to drought, followed by another loss of 83% of the cattle in the 1983–84 drought (Talbot, 1986).

10. In the Sahel countries, as populations have grown, cultivators were pushed to move north into increasingly marginal low-rainfall areas, increasing conflicts with transhumant pastoralists (e.g., Niger, cited in Noronha, 1985). Similarly, in the Sudan, migration in the other direction (south) by pastoralists was increasing conflicts over land with sedentary farmers (*ibid.*). Other cases of increasing conflicts between pastoralists and sedentary

farmers are found in Little et al. (1987). In Nigeria, Adepoju (1983) describes how the growth of populations was leading to fragmentation of landholdings and increasing landlessness, while Diarra (cited in Monod, 1975) describes increasing conflicts between Fulani pastoralists and Hausa farmers in northern Nigeria.

11. An important and unique place of land *extensification* is Puerto Rico. For centuries, agricultural land expanded as the rural population grew, as elsewhere in the Caribbean. Starting about 1930, both cities, with manufacturing and services, and tourism attracted rural-urban migrants, leading to much farmland being abandoned and then regenerated by forests: so long-term land extensification linked to growth of population was followed by major *reforestation* linked to rural population decline and abandonment of land. Key studies include Del Mar Lopez et al. (2001), Rudel et al. (2000), and Helmer (2004).

Intensification of Land Use

At the same time, there were many case studies finding evidence of *intensification* (some below draw on reviews in Bilsborrow, 1987; Bilsborrow & Stupp, 1987; Geores & Bilsborrow, 1991; and Brookfield, 2001).

1. A historical study of native populations in Papua New Guinea detected many innovations over centuries as the population grew (Blaikie & Brookfield, 1987, Ch. 8). By the early 1800s, the population had been growing without problems due to the introduction of the sweet potato from the Philippines. During the century prior to the 1980s, new techniques of soil tillage especially cultivating mounds with ditches for drainage were developed by small farmers on their own, with just a supportive local social structure. However, over time, the growing population expanded up the highland slopes to where the mound technology was not sufficient to control erosion, so erosion started to occur, and some areas experienced declining land productivity.

2. In the densely populated area around Kano, Nigeria, population growth led to changes in both land tenure and land use, from shifting cultivation to permanent crops with stable land tenure, but smaller plots per person, leading to increases in labor inputs and manure use and shorter fallow periods (Mortimore, 1967). Since the manure needed was more than was available from local animal raising, it had to be purchased from Kano, necessitating men to engage in wage work to pay for it and reducing growing of subsistence crops in favor of cash crops (*viz.,* groundnuts, in Nigeria, peanuts). Over time, it was a multi-phasic response, involving land and labor intensification, increased use of natural fertilizer, diversifying livelihoods to include urban work, and switching land to higher value crops. In addition, in another area in the Southeast with high rural density, the Ibo adapted from short fallow to annual crops on smaller and smaller plots as the population grew and as plots were subdivided among multiple heirs due to their high fertility, while there was also much out-migration, both internal and international (World Bank, 2010, *African Migration Project*).

3. Brookfield (2001, p. 205) describes the Bambara village of Kala in Northern Mali, in which large households (typically 10+ members) adapted to their high fertility and population growth in multiple ways, including agricultural intensification, by investing labor in digging deep water wells by hand for irrigating crops and maintaining soil fertility. Other "phases" of adaptation (observed over time from the early 1980s to 1990s by Scoones & Toulmin, 1999) included off-farm work and out-migration followed by remittances received. Important institutional aspects of the society were the stability of land and household water rights, and the increasing participation of women in farm work as men migrated. Nomadic Fulani would visit seasonally to use their fields, corralling cattle on plots to provide manure in exchange for their cattle to have access to grass and water. After most households had wells by 1983,

this expanded. Besides increasing manuring, a second innovation of the Bambara was using plows with ox teams to replace manual plowing. Moreover, the village blacksmith developed a lighter plow that could be used for weeding between furrows. The Bambara increased their own cattle herds over time, especially on medium and larger farms, with larger households also benefitting by being more able to diversify livelihoods (wage employment, hunting) and exercise more political power.

4. The Koyfar of the northern central plains of Nigeria have been studied since the 1960s by prominent anthropologists (*viz.*, Netting, 1968; Netting & Stone, 1996). Over time they had migrated down from the hills (where they maintained terraces for growing rice, etc., but with population growth were depleting the trees), to the lowland plains where, however, they had to increase labor effort in farming and reduce leisure. As population density increased and more and more land was brought into use, farmers developed a new crop rotation scheme that helped conserve soil fertility, rotating commercial yams with sorghum, millet, and groundnuts. Household sizes increased to provide the increased labor needed to work the fertile ground, so hired labor. While Netting and colleagues viewed the experience as one of successful land intensification, Brookfield (2001, p. 206ff) saw it as also due to growing nearby urban market demands for yams and other products. Such an effect was also noted by Goldman and Smith (1995) for India as well as Nigeria, though their focus was on the adoption of the Green Revolution technology of the new rust-resistant, dwarf wheat seeds developed with great effort by Norman Borlaug, in Mexico and India.

5. Machakos, Kenya, is often cited as a counterexample to Malthusian changes in Sub-Saharan Africa–proof that land extensification does not have to occur as populations grow (Tiffen et al., 1994). Located close to the main road between Mombasa and Nairobi, the population of its Akamba people rose seven times between 1900 and 2000, but the population adapted via successful land intensification. The first change was from cattle and goat grazing and shifting cultivation–extensive and low-productivity forms of land use–to sedentary agriculture with terraced fields, ox-ploughs for plowing, adding tree crops, and integrating crops with livestock (feeding crops to livestock and applying the manure on the cropped fields). But Brookfield states (pp. 211ff) that even in colonial times, the community was well located and had contracts to sell to colonists, then later benefitted from early government and international research on improved seeds for maize. It also switched to stall feeding for cattle, to maximize collection of manure, which was starting to be supplemented by chemical fertilizers at the end of the millennium. While erosion was controlled, threats to soil fertility were evident two decades ago (*ibid.*, 214). But not enough can be said about the fortuitous location of Machakos in terms of market access.

6. In the southern highlands of Peru, indigenous farmers over time have developed a well-thought-out technology/survival strategy as a response to rising populations. As microclimates (temperatures and rainfall) as well as soils can vary widely across montane environments at different altitudes, risk factors for different crops (risks of crop failure) vary with location. An approach therefore evolved called "field scattering" (Winterhalder & Thomas, 1978) to take advantage of these differences as well as in the annual vicissitudes (and risks) of weather; thus each household has plots in different micro-climate areas, so whenever some are damaged by the vicissitudes of weather, others are not, ensuring their subsistence. Curiously in the lowland Peruvian Amazon, studies by Denevan and Padoch (1987) and Padoch and de Jong (1991) in the community of Santa Rosa near Iquitos found similar forms of adaptation plus intensification: traditional *ribereño* farmers living near the flood plain have both agricultural crops and tree crops, with plots or *chacras* both along the river, to

benefit from its rich alluvial, volcanic soils, and on higher ground for security in case of more than usual flooding (e.g., in these headwaters the floodplain is occasionally as high as 10 meters above normal: (Brookfield, 2001, p. 173). There is also no fine line between use of a plot and its abandonment, as a plot is traditionally planted in multiple species and used for crops for only a few years before the soil becomes depleted. By then there are plantain/fruit trees starting to produce, which attracts small animals that are hunted on visits to the plot to collect fruit, so the plots are not fully abandoned until after 10–12 years. Padoch and deJong's studies were based on only 46 households but were found engaging in 39 distinct combinations of land use in 1985. In my own visits to indigenous communities in the Ecuadorian Amazon, I observed similar "poly-culture" practices with long use of plot, but since 2001 and even before, they are coming to be replaced by monocropping of coffee, cacao, or African palm, which yield higher incomes per ha.

7. In India, traditional management of rural lands is mostly at the community level, with substantial lands as Common Property Resources (CPRs). It is crucial to not ignore these since up to ¾ of a billion persons in rural areas of much of Asia as well as in tribal communities in Africa and Latin America live in communities with significant CPRs, which are especially drawn upon for survival by the poor. Jodha (1985) collected data over 4 years from 80 villages in 20 districts in 7 states of India in 1982–84—data at the village level supplemented by household visits, data from informants, participant observation and monitoring. He also found archival data for the villages in 1950–52, allowing comparisons over time. With some privatization and increasing inequality in land ownership, the percentage of total land in CPRs in the villages fell from 55 to 31. Meanwhile, management of CPRs deteriorated, as traditional social and religious sanctions imposed when people failed to pay grazing taxes and provide labor to maintain the CPRs declined, leading to overuse of the "commons" and degradation of CPRs. This counter-example shows that CPRs, *well managed* (Ostrom et al., 1994) can help as a form of adaptation to population growth, in lieu of the "tragedy of the commons" (Hardin, 1968)– an institutionalized form of innovation in land use important to include as a possible form of intensification. But in this case, the opposite is occurring, Moreover, the case study also describes how in South Asia the dominance of CPR management by the elite (non-scheduled castes) is often unravelling it.

As a rough generalization, there is a tendency for extensification to still dominate in Latin America, intensification in Asia, and more of a mix to occur in Africa—in each case, consistent with contemporary person-land ratios. The examples above provide a rich set of case studies, with different results, some showing land extensification associated with (not necessarily caused or stimulated by) population growth and increasing population density, while others provide examples of successful intensification of land use, though very often related to other facilitating factors, notably access to roads and markets and sometimes leading to land degradation, raising questions about long-run sustainability. Some land extensification is successful in using unused land and avoiding deforestation or expanding onto marginal (e.g., semi-arid) lands or wetlands, while other expansion destroys those ecosystems. Land intensification is similar, if successful, it can increase the value of output per unit of land, avoid soil erosion or deterioration in soil fertility, and improve livelihoods; but in other cases, it has already had such effects, and/or chemical poisoning of the soil from overuse of modern chemical inputs, or of the waterways and water supply, impacting human, animal and plant health. But the effects on land degradation, and what the case studies above may have since led to, are beyond the scope of this chapter, though illustrated in Fig. 18.1.

Evidence Up to the 1990s: Cross-Country Quantitative Evidence

Several colleagues and I experimented with using country-level data to explore relating changes in population to changes in land-use and other factors, including in a larger context of multi-phasic responses, as in Fig. 18.1. This was first done at a continental level for low-income countries (Bilsborrow, 1987), for the period from 1961–65 (when the UN Food and Agricultural Organization first started to collect systematic data) to 1971 and 1976. Data were compiled from FAO *Production Yearbooks,* with countries grouped by FAO as (1) Africa (excluding North Africa and South Africa, still under apartheid), (2) Latin America (best regional data coverage), (3) Near East (Libya eastward to Afghanistan), (4) China, and (5) Other Asia. Data availability and quality for countries was uneven. Regarding the expansion of "arable land plus land under permanent crops" (not perfect, but the best available), it was found that Latin America's agricultural land grew by 23%, Africa's by 10%, and each of the other regions by 7%–9%, showing that land extensification was still occurring broadly, with the differences consistent with *a priori* expectations based on land endowments relative to population. In terms of land intensification at the continent level, data were compiled on fertilizer use, irrigation, and use of tractors, but found missing for many countries or implausible, so are not further discussed here.

That first effort indicated it would be more useful to look at data at the country level. A decade later, far more data were available for developing countries, making it possible to study whether changes in population tended to be linked more with changes in (a) land extensification, or (b) land intensification (Bilsborrow & Geores, 1994). Data were compiled from the FAO and the UN Population Division (on rural population) for 18 countries in Asia, 10 in the Middle East (including North Africa), 21 in Latin America, and 34 in Sub-Saharan Africa. The goal was to look at changes from the 1960s to the 1980s, but there were still data gaps, such as no rural population for a quarter of the countries. We also looked at indices of change in fertilizer use 1976–86, in the fraction of agricultural land irrigated, and in a land productivity index (1985/1965). What did we find?

First, on land extensification, we used median country values for each region, to take out the impact of outliers compared to means. Land per agricultural worker is the indicator of rural population density (its inverse), falling due to population growing faster than the agricultural land area in most countries: in Asia it fell from 0.38 ha/person in 1965 to 0.30 in 1985 and in Africa from 1.44 to 1.35; it rose in Latin America from 2.3 to 2.7 and in the Middle East from 1.17 to 1.39, likely due mostly to high rural-urban migration. Country values were plotted on x-y axes (% change in land area on y-axis and % change in population density on the x-axis, but simple correlations were not computed as it was clear they would be determined by an outlier in each region (Cuba, in Latin America, Thailand in Asia, Rwanda in Africa), each being the only country with a change over 10% in agricultural land, and Cuba with negative population growth. The bottom line is there appeared only a weak positive correlation.

For intensification, country data was again plotted in scatterplots, including identifying regions. For fertilizer use, changes in kilograms of fertilizer use were plotted against percent change in rural population, finding a definite positive relationship, as most countries experienced positive changes in both. But this was found to be due to a few outliers, with a clustering at 0 for the many countries reporting no change (partly due to, at least in Latin America, few countries implementing censuses of agriculture in the 1980s and hence having no new data to report). Outliers again dominated, with weak positive relationships for Latin America linked to a big increase in Venezuela, for Asia to China, and for the Middle East to Egypt. For Sub-Saharan Africa, a lower but more consistently positive line is driven by two countries with negative population growth. The overall Pearson correlation coefficient is positive but only marginally significant ($p = 0.09$), although for Africa the coefficient was 0.24 and significant at the 0.03 level. There was also a weak positive relationship between percent change in irrigated land area and rural population growth.

Finally, multi-phasic response theory implies a likelihood of *tradeoffs* between land extensification and land intensification, so we plotted absolute changes in fertilizer use against absolute changes in agricultural land: while there was a curvilinear, convex from below curve, it was again mostly due to three large countries, two with low or no change in agricultural land on the horizontal axis and the biggest increases in fertilizer use on the y-axis (Egypt and China) and one country with a huge change in arable and permanently cropped land (or A & P land) and little increase in fertilizer use along the horizontal axis (Brazil), with no evident pattern among the other countries. So the jury was still out on what might appear with more recent data. Our conclusion (*op. cit.*, p. 200) at that time (1994) was that "cross-country data are too aggregated for a proper investigation of the relationships", that research should be carried out based on *household level* data (p. 202), which can take into account differences in household characteristics as well as in the local context, and (I add now) get away from idiosyncratic country-level effects (discussed further in section "Reflections and moving forward").

Two other studies that analyzed the determinants of land use change based on country-level data before 2000 should be mentioned. One was by Allen and Barnes (1985), using data from FAO initially for 60 low-income countries. But by the time they drop those with obvious errors, no change in forest area (real or not, unable to determine), or missing data on explanatory variables (change in cultivated land, population growth, growth in GNP, change in wood production per capita, and wood exports), the final sample was 39 countries. The only statistically significant variable found then was that of population growth, but its t-statistic of 1.48 was barely significant at the weak 0.1 level.

Given the problems of data quantity and quality with country-level data, it is useful to further disaggregate and look at the innovative approach of Turner et al. (1977) based on pooling data

from studies based on small samples of households from communities in various countries. A multiple regression was then estimated based on community mean values, the dependent variable being intensity of agricultural land use (A), measured as the mean percent of time land units are in use vs. fallow. Thus, a unit used 1 year which is then left fallow for 9 years has an intensity of 0.1, one used for 5 years then not for five would be 0.5, and one planted in two crops within a year would be 2.0. The range across the 27 study populations (when converted to percent of time in use) was from 5 to 150. Data used were from studies of households in 27 communities/areas of 10 countries, on four continents plus two Pacific Islands. Population density (P), defined as persons per km^2 of area planted in crops, varied from 1.0 in Amazonian areas of Peru and Brazil to 200 on the Tonga Islands and the island of Uhara, in Lake Victoria—coincidentally, the very island Boserup initially studied! A simple regression of the log of intensity A on P resulted in a strongly statistically significant positive coefficient at the 1% level. To test its robustness, additional explanatory variables were added, including perceived soil quality (S), length of dry season (D), and whether the population had livestock (L), expecting the signs to all be positive. Rainfall was also tested but dropped when it was not significant. The final model was:

$$\text{Log A} = 0.79 + \underset{(0.024)}{0.109}\,\text{D} + \underset{(0.068)}{0.215}\,\text{S} + \underset{(0.0008)}{0.0042}\,\text{P}$$

$$+ \underset{(0.001)}{0.003}\,\text{LP} - \underset{(0.0003)}{0.0007}\,\text{DP}\ (R^2 = 0.87)$$

The results for population density remained strong even when the other significant variables were added, all with expected signs, with the effect of P reduced a bit with a longer dry season. While interesting, only 9 of the 27 study populations had (mean) values computed from actual household data, the others being just rough estimates of authors of the case studies. Evidently, the sample size is also very small and contains extreme outliers (discussed further in Bilsborrow & Geores, 1992, p. 80). One would also expect a *shorter* dry (agricultural) season to be associated

18 Population and Agricultural Change

with higher intensity of land use, suggesting the sample of sites is biased toward tropical rainforest areas. Nevertheless, pooling data from many communities is an intriguing approach and worth looking into further (see section "Reflections and moving forward" at end of chapter text below).

Reassessment of the Theory and Definitions

Often there is confusion about the basic definitions, as well as the relationships between the processes postulated by Malthus and Boserup. Are they really in direct opposition, as appears in many discussions?

Clearing Up Definitional Ambiguities

First, what should be meant by "agricultural land extensification"? By "agricultural land intensification"?

Agricultural land extensification refers to the expansion of the land area in agriculture into areas not being used at the time, whether tropical forests (most recent studies look at this, as it has implications for loss of biodiversity) secondary forests, temperate forests, grasslands, wetlands, arid or semi-arid lands, frozen or peat lands (which becomes more common with global warming). If any of these lands had been used before, but were not in use at the time, that doesn't matter—when they come into use again, that is extensification of agriculture. Nor does it matter *how* the new lands are used, whether for annual crops, perennial crops, forage crops, pasture for current or future use for livestock (cattle, horses, camels, etc.) or smaller animals (sheep or goats); nor if it is for annual use, multi-cropping or use only for few years, with a long period of fallow to follow. According to Malthus, population increase tends to push farmers to expand the agricultural land area, if there is unused land, whether their own land, nearby public/common lands, or land far away which may require migration.

A curious anomaly is what happens if population per unit of agricultural land is *declining*? That does not necessarily lead to a reverse relationship, that is, less land in use, since it could be due to out-migration of people. Land in use could then decline or even be abandoned due to factors other than population decline, such as declining soil fertility, depredation of crops due to wild animals (crop raiding), plant or insect damage, loss of water or climate change, or appropriation of land by expanding urban centers or infrastructure (roads, oil wells, etc.). Land in use may also fall if small farms are consolidated into large ones, as large landholders do not feel the same pressures to work the land as small landholders, just to survive. On the other hand, the agricultural land area could increase due to many factors besides a growing rural population, notably demand side factors, such as increasing local, national or international demands for locally produced products. Our concern in this chapter is on supply side factors, whether population growth is contributing or not to land extensification as postulated by Malthus.

A practical issue is how to *measure* the agricultural land area: it can be difficult on the ground, such as in dense tropical rainforests, as our teams of UNC field workers found in the Ecuadorian Amazon, trekking to isolated agricultural plots in dense forests on evanescent paths to measure their size with GPS under tree cover or during heavy rains. Logistic difficulties also exist on steep mountain slopes and in arid areas. While great advances are occurring in the quality and availability of *satellite imagery* and in the skill of physical geographers to better identify types of land use on the ground from satellites, including forest cover vs. agriculture vs. other land uses, challenges remain to distinguish primary forest from secondary forests from agroforestry from small perennial trees such as cacao and coffee; and agricultural areas with crops from fallow fields from barren or built-up areas. In addition, in multivariate statistical studies of the determinants of land clearing or areas of land in agricultural use, care must be taken in the way variables are formulated so they match, e.g., the absolute amounts of agricultural land with the area of forest lost in the same time period, or the % changes in both, which would require also other variables measured in a parallel way, such as % change in population, per capita income, distance to nearest road or market, etc.

An intriguing question is how to measure *land extensification* in indigenous communities which practice shifting cultivation and have no fixed agricultural land plots: is clearing a new plot extensification if an equivalent plot is for future use after 10 or 30 years for the soil to regenerate? If the new plot is larger than the old one, the answer is clear. But if the two are the same, and the time frame is sufficient for the fallow land to regenerate, there would be no net land extensification. However, if the *number* of indigenous households practicing that shifting agriculture in a fixed community area is rising due to population growth, then at the *community* level the forest area as a whole is being used more *extensively* since there are more total plots and total area in use, though the *intensity* of use of all the plots may not have changed. This is land extensification at the *community level.*

Moving on to *agricultural land intensification*, it is more multi-faceted. The measurement problems begin with the difficulties of collecting accurate data on it from satellite imagery, in contrast to land extensification. Gaps in data exist for countries in FAO files as many countries do not collect or publish reliable data on land use by crop, nor can it be determined from satellites whether fertilizer is being used, though irrigation can often be detected, so agricultural surveys are more important sources of data. Intensification is also more multi-dimensional than land extensification, with some measures based on inputs per unit of land and others not. Boserup's initial view of population-induced intensification was based on her five stages of increasing intensity of land use (from forest fallow to multi-cropping—see above), which measures intensity by the mean times a plot of land (of a household, country . . .) is planted in a crop or in use (including by cattle) in a specific time period, usually a calendar year or 12-month time interval. So a decline in fallow times or an increase in multiple cropping are equivalent measures of Boserupian intensifica-

tion of agriculture. Boserup recognized that with more people, there would be more labor available to be applied per plot, meaning a higher labor/-land ratio. This could be applied for weeding, creating ditches for irrigation or fences/other barriers to control animal depredation, etc., each of which would increase productivity, or net output per unit of land. However, this would not necessarily be sufficient to keep production *per person-hour* from declining.

Besides increases in labor inputs, there are also other forms of land intensification, some ancient (use of irrigation, manure), others mostly since the twentieth century and involving modern technology manifest especially in use of chemicals. These are described in Bilsborrow (1987), Bilsborrow and Geores (1992), and elsewhere, and include: increased use per unit of land of natural fertilizers (animal manure, also night soil, leaves, coffee/corn husks or other crop residues, food waste, etc.); of chemical fertilizers, pesticides, or herbicides; of water, from small- or large-scale irrigation, ground water, or even from carrying buckets of household wastewater to plots. A question exists regarding the application of physical capital and machinery–is it extensification or intensification, if tractor, thresher, etc., hours increase per unit of land? While it is intensification of one form of input, it usually involves a parallel *decline* in labor inputs per unit of land, so would not be intensification according to Boserup, nor to me.

Yet there is more to consider: improved varieties of (hybrid) seeds increase production per ha, and are hence considered a form of intensification, but may or may not involve more labor hours per ha (but even if time for planting and weeding does not change, time for harvesting a larger crop per ha should rise). Also, farmers may change from crops producing a low value of output per ha to others with higher value per ha, e.g., from grains to vegetables, or pasture to crops. In these cases, the land is used more intensively in terms of labor time per ha, so it is land intensification for that reason, not necessarily because it yields higher output value/ha (e.g., hybrid seeds). Thus these changes *usually* require additional complementary inputs that require additional labor per ha, such as to apply more fertilizer/ha for vegetables compared to grains. For example, Borlaug developed hybrid seeds for dwarf wheat from working years in the fields of Sonora, Mexico, and India, which were successful in increasing land productivity (wheat/ha), then had to confront the dilemma of the Green Revolution, as the new GR seeds required large amounts of both fertilizer and (irrigation) water to be successful. In northern India, once irrigation and tractor use were established, many fewer agricultural workers were needed, displacing much of its labor force and worsening inequality although it reduced hunger and famine.

Brookfield (2001) stated that intensification refers to increases in *any inputs of labor, capital or skills* intended to increase land productivity (p. 199–201): "intensification . . . [is] the addition of inputs up to the economic margin . . . in regard to land . . . must be measured by inputs only of capital, labor and skills against constant land" (quoting himself from 1972). The new element here is *skills*. In fact, the 2001 book is a masterful account of the importance of and neglect of smallholder farmers and the management skills they have developed over centuries and pass on, adapting to their local environment, resources, and capabilities. He presents concrete examples for Peru, Nigeria and Kenya, cited above in case studies. This includes technical skills and improving basic implements for local adaptation (such as a hoe, plow or digging stick); recognition of the complementarity of growing crops (to feed animals) and raising animals (providing manure for crops); adaptations in the social and community organization of land (CPRs) and labor (exchange labor); knowledge of local soils, vegetation, animals, and seasons to know when to plant and how to intercrop and rotate crops appropriate for the context. Despite this, due to a history of abuse of smallholders by both colonial and post-colonial governments, large landholders and politicians, and due to lack of access to credit or crop insurance, smallholders tend to be risk averters (*ibid., xiii and passim up to his last page 286*), chary of adopting new crops or methods. As a result, Brookfield bemoans, they are incorrectly viewed at not only too conservative but dumb. Identifying

such forms of intensification can rarely be done without time-intensive field-based case studies, which is rarely done.

Linkages Between Malthus and Boserup?

Brookfield (2001) pointed out that both Malthus and Boserup postulate deterministic, unidirectional, mostly linear models. By expanding his discussion, Malthus' theory can be illustrated as follows: if A = Agricultural land, P = Population (size or density), L = rural labor force, Q = agricultural output, and T = technology, then with T constant, $\uparrow P \rightarrow \uparrow A \rightarrow \downarrow Q/P$, as long as additional usable land A is available (assumed by Malthus to be less productive than existing land in use). Meanwhile, Boserup postulated that $\uparrow P \rightarrow \uparrow L/A \rightarrow \uparrow T \rightarrow \uparrow Q/A$ which may in the right circumstances raise Q/P at least to its original level, and if not $\downarrow Q/P$, as in Malthus. But *both* Malthusian and Boserupian effects could occur simultaneously, or neither in response to population change, depending on circumstances in the household, community and country and mediating factors (Fig. 18.1 above), as other multi-phasic responses may occur, and the more one occurs, the less the pressure for the others.

Equations and curves illustrating and relating the two theories, which economists favor, are found in Robinson and Schutjer (1984), with an introduction to the production possibility curves they use for land, labor and agricultural output in Bilsborrow and Geores (1992, p. 51–54). An intriguing mathematical exploration is in Cuffaro (2001), but further discussion is beyond the scope of this review.

Let us now move on to section "More recent evidence" for an assessment of the more recent literature, from about 2000 to 2019. This will mainly be via a meta-analysis of the articles in journals, rather than an analytical one based on attempting to identify the "best" statistical studies and assessing their models and coefficients.[4] Sec-

---

[4]Hoffmann et al. (2020) just published an excellent article doing such an analysis of 30 published statistical studies

tion "More recent evidence" will also include a summary of more recent cross-country analyses.

More Recent Evidence

A Meta-appraisal of the Literature in Journal Articles Since About 2000

This section looks at the academic literature since the turn of the millennium[5] to assess the newer research –what have we learned about extensification and intensification of agriculture since 2000? The studies cited can only be sketched in this meta-assessment, so it is not possible to get into closely related topics such as land degradation, migration or climate change, their linkages with changes in population, nor the other main exogenous drivers of extensification or intensification in each study. Most case studies are on a single study site, region, or country. In this review of recent literature, case studies are grouped not only by the extensification-intensification dichotomy but also by geographic region—Africa, Asia and Latin America, following standard UN classifications.

Focusing on Land Extensification

Africa
Not many studies on extensification were found in this time period on countries in Africa. In a study on land use change in Burkina Faso, where the population was growing at 3% per year, Knauer et al. (2017) analyzed (500) Landsat and (3000) Modis satellite scenes for the period 2001–2014. They observed that the rainfed agricultural area grew by 91%, mostly at the expense of forests, including protected areas. In the Zam-

bezi valley of Zimbabwe, Baudron et al. (2012) found smallholder farmers were expanding the land area in agriculture rather than intensifying its use via Conservation Agriculture practices being promoted (see Brown et al., 2018). Finally, Kreidenweis et al. (2018) examine the expected loss of biodiversity in Africa and Latin America based on projections of the likely land area needed to meet anticipated increases in demand for agricultural goods from 2015 to 2050. They expect the majority of the increase in land area to be in pasture for livestock, with some improvement in the use of more land-intensive methods of pasture in Africa but not in Latin America, due to the latter's greater stock of unused potentially arable land. This implies there will be a greater loss of biodiversity in Latin America. The main policy implication is that governments should develop stronger policies to protect forests.

Asia
1. While most of the selected relevant studies on Asia are on Southeast Asia, several of note on other prominent countries are worth citing first. For the Tarim River basin in Western China, Liu and Chen (2006) studied the link between growth of population and increase in agricultural land area, finding that massive draining of water from the river for irrigation led to water shortages and pollution, salinization and desiccation, and dust storms.

2. In Pakistan, Mahmood et al. (2010) use a "convenience sample" of 120 farmers in one district to study the effects of out-migration on stimulating agricultural development via the investment of remittances in expanding farmland and livestock.

3. In Papua New Guinea, Ningal et al. (2008) investigated land use change from 1975 to 2000, during which time population grew at 2.3% per year to 5.2 million, with 85% dependent on subsistence agriculture. In Morobe province, based on census and Landsat TM data, population grew by 99% and the land area by 58%– found to be correlated across districts—most of the expansion from primary forest, which they expect to continue "in the absence of improved farming systems."

---

on the relationships between environmental change and migration.

[5]An important caveat is needed for this meta review as well as the whole chapter. Journals searched are those that publish in English, with few exceptions. I am indebted to the Director of Information Services at the Carolina Population Center, Lori Delaney, for excellent assistance carrying out the computerized search and obtaining abstracts as well as PDFs of many dozens of articles.

4. Dutch and Belgian economists have conducted much interesting research modeling the linkages between agriculture, development, and changes in land use in low-income countries, especially in Indonesia. Maertens et al. (2006) studied agricultural expansion and deforestation in Central Sulawesi province, along with addressing the key question in the environment and planning literatures about the effects of intensification when it eventually kicked in: was it "forest-saving" (increasing land productivity which reduced pressures for more clearing) or "forest-clearing" (higher productivity leading to further clearing to take advantage of it). They note that in most of South-East Asia, the successful adoption of high yield varieties has been usually in rice, grown in lowlands, with most parallel land extensification occurring in the uplands. They elaborated a model (p. 198ff) of the agricultural household, unusual in being a two-sector model but not of the usual rural-urban type (Lewis, 1954; Fei & Ranis, 1964) but for lowlands-uplands sectors, in which most households have land plots in both sectors. The model has production functions with initially fixed labor and limited capital (per ha) inputs, both assumed to be mobile between the uplands and lowlands. Households maximize utility as a function of income and leisure, with farm output consumed or sold, and no credit available. Land (and thus labor) is allocated optimally for agricultural use in each sector: when output prices rise or technology improves, households tend to reduce leisure to increase labor (substitution effect from price changes) and increase the cultivated area. But at the same time, the rise in income may lead to more leisure and cultivating less land (income effect). Population growth enters explicitly as it tends to increase the agricultural area unless the land constraint is binding. The key question is what kind of technical change occurs—whether labor-intensive or labor-saving. The authors collected data from a random sample of 77 villages in the rural Lore Lindu region in 2001, with 1980 (baseline) land use obtained from interviews with farmers with memories stimulated by old maps, census data and road maps, and satellite images. The main crop is paddy lowland rice, supplemented by cacao and coffee cash crops and other crops. In the study period 1980–2001 the population grew by 60% (varying from 32 to 177% across the five subdistricts) while the agricultural area grew overall by 56%, also varying widely. Meanwhile, the area in rice grew by 30% and the area in perennials almost tripled, mostly in uplands at the forest margins. A von Thünen model (1826) was incorporated to incorporate the cost of access (transport) to the nearest (urban) market rising with distance. The two simultaneously determined dependent variables were areas in agricultural land in each sector; data averaged at the village level (n = 77) were then used to estimate the coefficients of endogenous explanatory variables (log of population, years of small-scale irrigation since 1980, number of hand tractors); exogenous explanatory variables (1980 area in agriculture, travel time to provincial capital Palu, whether land flat or sloped); and exogenous instruments (initial population size and density in 1980; years village had electricity, health center, and a junior high school in 1980–2001; mean elevation; and density of rivers). Instruments were needed to estimate a 3SLS (Three Stage Least Squares) simultaneous equations model to estimate the determinants of land in agricultural use in the two sectors. They found a 1% increase in population linked to a 0.69% increase in lowland agricultural land and a 0.56% increase in upland area in use. The larger effect in lowlands was surprising given less unused land was available in 1980 and attributed to the dual effect of increased food demand (for rice) combined with an increased labor supply, both from population growth. In addition, the more hand tractors became available as a new technology, the more the land clearing, which was also, as expected, the result of roads providing better market access, differentially across villages. Policy implications include the importance of better planning the location of roads and road improvements (in already densely settled areas

rather than where populations are dispersed), promoting yield-increasing technologies (e.g., better seeds) and taxing tractors to reduce agricultural expansion at forest margins.

Latin America

In Latin America, much of the research has been on the extensification of agriculture and deforestation in the Amazonian lowlands, especially in Brazil and Ecuador. But first several studies in Mexico and Central America will be described.

1. In an intriguing application of the Population Pressure Hypothesis inspired by Boserup, Pascual and Barbier (2006) studied the implications of growing population pressure on a shifting cultivation population in Yucatan, Mexico -does it lead to the "fallow crisis" of fallow time declining so much that soil fertility of plots does not have time to regenerate? A utility maximization model was developed for shifting cultivation households based on production of the food staple (maize, from milpa plots), subject to the usual income/budget and technology constraints, tracking soil fertility and taking into account increasing labor required for clearing land more frequently, endogenously tied to the agricultural cycle (p. 157). A key factor was the number of households relative to the area of common property lands available. As crop yields declined due to shortening fallow times, farmers extensified production more into common *ejido* lands. The study area was two communities in one of the 106 municipalities of the Yucatan, with a dry climate and deciduous forest, and shifting milpa cultivation on ejido lands. 60% of the (1035) households in the two communities were engaged in shifting cultivation, from which data were collected for 74 for estimating the model. Households were divided into the 21% poorest (only 15!) and the rest ("least poor"), with the optimum behavior of the poorest found to be to continue expanding into the common lands, while for the others it is to become more engaged in off-farm work. Unfortunately, there is no discussion of the effects of other household

characteristics that would favor off-farm work (e.g., more educated, living closer to roads?), and the policy implication seem obvious, to improve opportunities for local off-farm work (see also Reardon & Vosti, 1995) -the *deus ex machina*?

2. A growing theme in recent research is the linkage between international migration and land use in areas of origin, whether or not the initial migration was stimulated by population pressures on the land. Taylor et al. (2016) examine long-run impacts of international migration and subsequent receipt of remittances from the US in the *municipio* of Ixcan, Guatemala, in the Alta Verapaz lowlands bordering Mexico. Based on panel quantitative and qualitative data for 1986–2012, they first found out-migration led to remittances, used mostly to invest in pastureland and cattle. Some households failed at cattle-raising, some instead went into off-farm work, and others got government subsidies to preserve the forest, allowing some forest recovery. Later, road expansion led to converting maize to African palm, which involves clearing any remaining trees and heavy use of chemicals, seriously degrading the soil.

3. Among the studies in Brazil, Alves et al. (2003) for Rondonia in the southwestern Brazilian Amazon, used Landsat imagery to follow 2.5 km^2 level pixels in 1985–95 to measure deforestation, finding it often over the 50% legal limit, especially close to roads, and linked more to pasture expansion than population increase. Barretto et al. (2013) used aggregate census and other data to study the intensification and extensification of agriculture in the Amazon. While improvements in technology led to some contraction in agricultural areas in the mature southern and southeastern regions, in the frontier agricultural regions of central and northern Brazil they instead led to even more land extensification over the period 1975–2006, indicating stronger income than substitution effects. Additionally, Metzger (2002) studied fallow times of shifting cultivators at older areas of the agricultural

frontier, using Landsat-TM satellite images from 1985 to 1996. He selected three study areas of 250 ha each with short-fallow times (2–4 years) and three with long-fallow times (10–12 years). In the short fallow areas, secondary vegetation kept growing less between cycles while new areas of land continued to be cleared, the land area expanding at 3% per year. In contrast, long fallow plots were found to be sustainable, which was consistent with indigenous land use long fallow practices. This is found also by Lintemani et al. (2020) who studied soil samples at depths of 0–5, 5–10 and 10–20 cm in areas of slash-and-burn agriculture, measuring soil fertility after varying fallow durations between plantings. They found 15 or more years was needed to restore soil fertility.

4. Ecuador also continues to receive attention. Siren (2007) studied an indigenous community in the Amazon, where the population growth was 1.6%/year, with agricultural land growing at only 0.4%/year, or very low extensification. But over time the need for more agricultural plots to feed the growing population implied increased distance and time to walk to the plots, plus closer plots began to be fallowed for shorter times leading to more weeds and more labor for frequent clearing. In a larger study on some 500 households in 32 indigenous villages in the Ecuadorian Amazon, Gray and Bilsborrow (2020) use panel data to study factors related to the area cleared and in agricultural use in 2001–2012. The mean total area cleared per household was about 3.9 ha, and did not change, although the total mean area in forest of the communities declined, but by only 1.5% to 91.7%. Besides ethnicity, household wealth, travel time to nearest town, and household size were statistically significant in explaining differences across households in area in agricultural use, as one would expect from Malthus (household size effect). The same factors were also found to affect changes over time, albeit modest. In other studies on the Ecuadorian Amazon based on 418 to 687 migrant settler households, multivariate

models of the determinants of land in crops vs. remaining in forest found number of male adults positively linked to forest clearing, after controlling for a host of other factors (Pichon, 1997; Pan & Bilsborrow, 2005). Overall, there appears to be Malthusian land expansion, albeit modest.

Focusing on Intensification

Africa

It is curious why there should be so many recent interesting studies on intensification in Africa and so few on extensification, when there continues to be much extensification continuing. One reason could be the predominance of small farmers, usually poor, and the concern of scholars to find positive success stories that might be possible to replicate. The number of cases on intensification left after filtering here and in the other regions is sufficient for numbering them for the reader.

1. In Ethiopia, Abebe et al. (2013) examined the adoption of high yielding, disease-resistant varieties of potatoes, finding it low. He noted, following Brookfield, this is commonly attributed to small farmers' conservatism and risk aversion, but found they prefer local varieties due to perceived easier crop management and better quality when cooked in local stews. He used survey data on 346 potato farmers to study the factors affecting the adoption of higher yield varieties, finding it positively linked to household head's education, presence of a radio, land ownership, access to credit, and having a mobile phone. Also in Ethiopia, Mekonnen et al. (2018) found social networks had statistically significant effects on the adoption of row vs. scatter planting, especially helpful among women.

2. Among the studies in West Africa, in inland Benin, Djagba et al. (2019) note that crop diversification is very limited, with 83% of farmers monocropping rice, using large amounts of chemical fertilizers, with little use of crop residues and organic manure. They see it as a challenge to get farmers to understand the implications for human health and the environ-

ment. In Burkina Faso, Gray (2005) noted that intensification has been uneven, resulting in social costs in the form of increased inequality: Using data from villages in the southwest, he shows that, contrary to common views, it is not the poor who resort to degrading the environment for survival but rather the better off farmers who farm larger areas, maintain fewer trees in their fields, and use more animal traction for plowing, lowering soil fertility. Moreover, poorer farmers tend to be excluded from key local institutions that control access to resources. In the northern Nigeria savanna, Okike et al. (2005) observe population pressure stimulating better integration of crop-livestock activities, including increases in livestock per ha, in applications of manure on cropped areas, and higher labor inputs per ha. Finally, in Ghana, Nin-Pratt and McBride (2014) found no evidence of higher population density linked to higher input (fertilizer) intensity and were disappointed at the lack of adoption of Green Revolution intensification.

3. Using data from a nationally representative household-level panel survey for Malawi, Ricker-Gilbert et al. (2014) explore how increases in population density affect land intensification and household well-being. They use a household utility maximizing model to study the determinants of household decisions regarding its agricultural area and demand for fertilizer based on household and community traits, including local agricultural wage rates and maize prices. Data were used from a panel of 1375 households visited in the 2003/4, 2006/7 and 2008/9 growing seasons. Areas of higher density had smaller farm sizes, lower agricultural wage rates, and higher maize prices (as found in other models). Fertilizer demand per ha was not significantly related to population density but is negatively linked to fertilizer price and the local wage, and positively to household assets (ability to pay), government subsidies for fertilizer, and existence of a farm credit organization in the village. However, intensification success was minimal as maize yield per ha was lower at higher densities. And though total household income per capita was positively related to density, the relationship became weaker at higher densities, where people increasingly sought off-farm work despite lower wage rates. Thus there seemed to be a rural population density threshold beyond which further intensification did not yield further increases in yields, forcing people to seek outside work, as also found in other studies (see Kenya studies below).

4. In Tanzania, Bullock et al. (2014) studied the adoption of soil replenishment practices in the East Usambaras—a tropical biodiversity hotspot threatened by agricultural expansion—seeking how to promote intensification. Data from household surveys were used with logistic regression to identify the factors linked to the adoption of adequate fallow and organic fertilizer: being married, larger household size, having access to credit or remittances, and tenure security. Their main policy recommendation was to improve access to markets and agricultural extension services.

5. Part of the Kigezi district in southwest Uganda was studied over a 50-year period (Carswell, 2002). Despite being considered on the verge of an environmental disaster around 1920, a survey of land use carried out in 1945 found this was not observed despite the population doubling since 1920. Then it grew 2.5 times again by the time of later data collection in 1996 to reach one million. So what happened? The small original 1945 survey covered (only!) 34 farm households living along 14 transects at different altitudes and was repeated for 50 households, in the mountainous area above 1500 m, with regular rainfall and good soils. Instead of the expected Boserupian decline in fallowing over time, there was an *increase* in the percent of farms fallowing (from 19% to 32%) along with a slight increase in fallow time, with small decreases in the areas cultivated and in pasture. Four things made this possible: some small land extensification, mainly clearing the 7% of land that was in swamp in 1945; intensification via increased inter-cropping, crop rotation, and mulching, resulting in higher output per

ha cultivated, with soil quality not seriously suffering; a meaningful increase in the area covered by woodlots from 4% to 9%; and households that did not benefit from draining the swamp became more dependent on off-farm income as casual laborers. This reflected the increasing inequality in land and livestock ownership and effective landlessness. The future of the area was said to be unclear given that all extensification and most intensification possibilities had been exhausted (p. 139).

6. There seem to be more studies on Kenya than any other country in the region. Eckert et al. (2017) use images from Google Earth and Landsat to study land use change on the Mt. Kenya foothills, where population growth is associated with increases in both rainfed and irrigated agriculture at the expense of clearing forests. The period 1987–2000 was primarily one of land extensification, which flipped to intensification in 2000–2016, with more land irrigated, increased conflicts with pastoralists, and depletion of water from aquifers due to high withdrawals. Successful examples of land intensification with population growth are observed by Kiprono and Matsumoto (2018) in southwest Kenya in 2004–2012 from road expansion facilitating access to hybrid seeds, fertilizers, and marketing of milk and grains. And as roads were improved in remote areas, the benefits were pro-poor. Similarly, in the Embu District, Oduol and Tsuji (2005) report that increasing land scarcity from population increase led to increased use of modern inputs (fertilizers and pesticides), while the desire to increase incomes led also to the replacement of annual food crops by higher value cash crops, as in Nigeria (see above).

7. In Ethiopia, Josephson et al. (2014) examine the effects of growing rural population density (RPD) on agricultural intensification, based on panel data from six rounds of the Ethiopia Rural Household Survey between 1989 and 2009 (1293 households, but only 15

villages, and with no pastoralist households, both limiting generalizability, as the authors note). The authors describe the unusual political history of government-controlled and forced land-redistribution under Marxist and subsequent governments, but in 1995 anyone who wanted to farm could receive only usufruct rights to land, maintaining tenure insecurity and leading to rural-urban migration. They developed a household utility-maximization model influenced by Boserup to estimate the impact of RPD on landholding size, fertilizer demand, etc. A key equation is land in use = f (RPD, household characteristics including assets, education of head and household size, oxen owned, and community variables including rainfall, distance to nearest paved road and nearest agricultural cooperative, perceived soil quality, cooperatives in region, etc.). The impact of RPD on input prices is said to come from the theory of induced innovation (see above): population growth leads to land becoming scarce relative to labor, so the relative price of labor falls relative to land, stimulating land intensification. Seemingly unrelated regression is used to estimate coefficients for equations on the determinants of input and output prices, with RPD affecting input demand through direct and indirect pathways, the former including via information flow lowering transaction costs, and the latter to the greater existence of institutions that facilitate intensification in areas with high RPD. They also estimate an equation for fertilizer demand (kg/ha) which finds a strong direct effect of RPD. The only two significant determinants of farm income per ha were land size and fertilizer price, both negative, with RPD not significant. The authors conclude that farmers in high density areas tend to use more fertilizer per ha but do not get enough of an increase in output value to justify the cost, so income per ha is falling, with dim prospects for the future.

8. In the dry Rift Valley of Kenya, Greiner and Mwaka (2016) describe its evolution under high population growth from a mostly pastoralist to a sedentary farming lifestyle, sugarcoated with honey! The study region is East Pokot, whose population grew from 40,000 to 63,000 from the 1980s to the late 1990s, and then 133,189 according to the 2009 census. Mixed methods of data collection were used in 2010–2011 (survey on 271 randomly selected households, plus qualitative interviews with another 102 household informants, district leaders, merchants, etc.) and satellite imagery for 1986, 2000 and 2010. While small changes in livelihoods occurred (trading, making beer and charcoal, etc.), opportunities for non-agricultural work or labor out-migration remained minimal. In addition, cattle remained at 100,000 despite rapid population growth, while the number of sheep and goats rose five-fold to 700,000. But two factors led to successful intensification: a big expansion in rainfed cultivation (mostly maize) and the quick spread of a new crop, honey production, based on the traditional log hive (log cut in half, inside cleaned out and cow dung smeared there to attract bees, then the halves are joined with wire to make a cylinder). While new modern hives became available in the market, they were too expensive for most. Meanwhile, many Pokot households moved, partly due to rapidly growing population causing increased pressures on the land to the less-dry highlands, converting commons grazing lands to private farms. People developed more diversified livelihoods, including still raising animals but fewer cattle, growing corn, and producing honey for local markets – a sustainable intensification of land use adaptation. However, both access to markets and superior uplands agroecological areas were facilitating factors.

9. Another high-quality statistical study on Kenya was carried out on the effects of rising population density, focusing on smallholder farms (Muyanga & Jayne, 2014). With 40% of the rural population on 5% of the farmland and population growing at 2.7%/year due to a total fertility rate (mean births per woman in her lifetime) of 4.1 still in 2017 (UN), the question of what happens as rural density rises is a key one. They formulate an agricultural household utility maximization model (based on production of agricultural goods for own consumption, of purchased goods, and leisure), with first-order conditions yielding factor supply and demand functions. They had data from five-panels starting with 1300 households over the period 1997–2010, selected randomly from villages selected from 24 districts to represent diverse agroecological zones and production modes (1146 households followed across all panels). Population density is measured at the village level, as population/km^2 of potentially arable land, with other factors measured at the household level: price of maize, means of transport available, socioeconomic characteristics, demand for inputs and production. Households were organized by population density quintile, with mean landholdings ranging from a low of 3.8 persons/ha to a high of 1.0, falling by a third in the 13 years, with fallow land also declining by 19% and wage rates by 20%, and prices of maize and rent rising by 2%. Models were estimated of the determinants of demand for fertilizer and other purchased inputs based on population density, price of maize, and household factors (size, headship, education, land size, elevation, distance to road, etc.). Purchased inputs contributing to land intensification increased per cultivated hectare with density up to 600–700 persons/km and then declined above that. Thus the net value of farm production also rose up to that 75% density level (although not necessarily value per person). The authors concluded that *continuing population growth is not sustainable*, due to its effects on reducing fallow time so that purchased inputs could no longer sustain soil fertility beyond that. In a second paper by mostly the same authors (Willy et al., 2019), scientific soil samples were analyzed from two different study sites

studied by previous investigators (Machakos and Kisii counties, cited above) to examine whether Boserupian improvements in soil quality continued under increasing population density. Soil samples were collected from the largest plot in maize for 290 households, along with other household survey data in 2014. Measures included organic carbon, total nitrogen, phosphorus, micro-nutrients, soil Ph, soil texture, etc. Maize yield in kg/ha was estimated as a function of plot size, labor/ha, seed quality, kg fertilizer/ha, and soil quality attributes, along with household size, age and education of head, management practices, village population density, and other community-level factors such as distance to extension service and nearest town, price of fertilizer, altitude, and security of land tenure. Results showed that measures of soil quality and maize yields per ha improved with population density up to about 600 persons/km^2, beyond which soil quality declined. The authors concluded (p. 108): "endogenous technological change seems sufficient to drive agricultural growth at low population density but may be limited at higher densities." This was seen as due to a virtual disappearance of fallow, insufficient replenishment of organic matter, declining phosphorus, and rising soil acidity as density rose, causing yields and incomes/ha to decline. They recommend policies to make low-cost soil testing available to farmers to ensure the right nutrients are added rather than relying on whatever fertilizers happen to be sold locally in the private sector.

Some of the most important policy implications from the review of African studies include improving access to roads and markets for both products and off-farm employment, strengthening social networks for technology dissemination (see Horton et al., 2010 on a Participatory Chain Approach to marketing), and government interventions via promoting research, subsidizing fertilizer imports, and technical assistance/extension services. Without these, farm households sustained livelihoods by migration and receiving remittances (Greiner & Mwaka on Kenya Bullock; Oyekale & Adepoju, 2012 on South East Nigeria).

Asia

1. Indonesia is a large, fascinating country which has experienced over time complex long-term linkages between population and land use. In three areas of Sulawesi in the century from 1850 to 1950, Henley (2005) noted that periods of land intensification (increased labor for weeding, transplanting rice, and shortening fallow times) occurred when the rural population grew, but this seemed to be induced more often by price increases as "farmers play close attention to market signals" (Brookfield, 2001, p. 215). This also led to a question about the direction of causation, as once output and incomes rise, the land can support a larger population, facilitating population increase (the original Malthusian view!). Big increases in the export of tree products for export (copra, rubber, fruits . . . .) also occurred here and throughout Southeast Asia, facilitating increases in population. Henley recently observed in the three areas of Sulawesi continuing land extensification associated with areas of *either* high or low population density, so both extensification and intensification have been occurring over time and continue. Based on more recent village and household level data from Central Sulawesi, Grimm and Klasen (2015) found increasing population pressure on the land from in-migration led to endogenous changes to formalize land titles, which in turn led to investments in improving the land -planting trees, terraces, ditches and irrigation systems- and overall successful agricultural land intensification.

2. In a northern hill region of Vietnam, Tachibana et al. (2001) observed that significant land clearing had occurred over time historically due to shifting cultivation and lack of land titles. Commune-level data for 1978, 1987 and 1994 showed that establishing individual land rights in the late 1980s led to land intensification and forest restoration. In the same region, how farmers could use satellite imagery com-

bined with ground data to better plan land use was studied in two villages (Folving & Christensen, 2007). They used Sloping Agricultural Land Technologies (SALT) to detect changes in vegetation succession and fallow at the time of clearance based on high resolution satellite imagery, confirmed by ground observation and overlaid with a digital elevation model (DEM). Farmers were interviewed to determine their willingness and ability to use SALT to identify changes in attitudes, in the context of ongoing changes in livelihood diversification. The study shows the value of taking into account local views and traditional farming systems in ethnic communities and using remote sensing imagery to analyze changes in fallow length and environmental sustainability of shifting cultivators.

3. Hunt (2000) revisited Boserup to study whether the intensification of agriculture led to a decline in labor productivity, based on farm data on rice production in Sarawak and Thailand. Data were collected on five swidden farms (term in Asia for shifting cultivation) in Sarawak and 40 swidden farms in Thailand in 1960, on the total amounts of time involved in clearing forest, planting, walking to the plots to weed, repair irrigation canals, etc. This was then compared with the time requirements for sedentary or "sawah" rice farming of 57 households in Thailand in 1975. Mean labor requirements in days/ha/year fell from 175 to 100, while rice output rose from 1.3 to 2.7 million tons/ha, showing this change from a long fallow system to annual cropping with irrigation led to a more than tripling in labor productivity.

4. Li et al. (2013) use the New Economics of Labor Migration model to study the impacts of rural out-migration on origin area households in three townships of northwest China, using a 3-stage least squares model. They find the decline in multi-cropping from the loss of labor offset by the investment of migrant remittances in capital-intensive cash crops. Similarly, Liu et al. (2016) use panel data from households in five mountain cities in Guangdong Province from 1966 to 2012 to observe a initial decline in land use intensity from outmigration compensated for later by increased use of fertilizers and pesticides, which unfortunately had negative impacts later on the environment and food security.

5. The importance of small farmer access to agricultural (crop) insurance has been discussed for decades in the international aid community; Hosseini et al. (2017) describe how it led to greater willingness to invest in innovation in Iran, based on primary data from 2012–13, from which they developed a model of the willingness to invest, based on household and community variables.

6. For Nepal, Dahal et al. (2009) studied the switch from cereal crops to vegetables and other cash crops in the Himalaya hills region in the Ansikhola watershed, accompanying rising farmers' aspirations from food sufficiency to improving living conditions. This contributed to agricultural land intensification and market sales, improving incomes and wealth, and increasing rural-urban migration.

7. For six villages in northern India, Deb et al. (2013) describes how population increase over time led to a variety of small, local innovations and adaptations that both increased productivity and controlled land degradation: increased use of cover crops, retention of trees, better management of weeds, use of poles and logs for soil conservation, fallow management, and integration of cash and subsistence crops.

8. For West Bengal, India, Reddy et al. (2020) review previous studies and analyzes primary data from two villages to assess whether ongoing agricultural intensification is sustainable. They find little evidence for sustainability, as farmers continually have to respond to pressing needs for income. And while farmers are encouraged by public programs to use more organic fertilizer, this is overwhelmed by business promotions to use chemical fertilizers, which they cannot afford, showing the lack of coordination of institutions working at cross purposes. There is thus a need for them to coordinate with the local self-governing village institutions (traditional *Panchayati Raj*).

9. A quite different type of analysis of the effects of population pressure on agricultural intensification is provided by Ali (2007) for Bangladesh over the period 1975–2000. Bangladesh was already considered densely populated by 1975, with over 80% of its rural households being smallholders (<1 ha) engaged in subsistence farming with minimal market sales and major environmental constraints and risks, notably from flooding. So what happened when the population grew from 72 million to 134 million? Ali proposes a model of rural system change induced by population growth and growing market demands, mediated through government policies, environmental constraints, institutional supports, and changes in technology. A path model is described (but not presented), yielding coefficients to determine the direct and indirect effects (and total effects) of a host of factors on rural system change. The model is estimated from data at the *district* level (not household or village level, which is more appropriate but challenging) for 64 rural districts, drawing also on data from population and agricultural censuses, administrative data, market studies of prices, etc. Many indices of change of interest for present purposes were computed, including of cropping intensity (including multiple cropping) and technology (use of chemical fertilizers, irrigation). A complex dependent variable "rural system change" was created to include not only measures of agricultural change (including in the total value added by the 21 main crops grown in the country including fish) but also improvements in rural markets, banks, roads, etc. (economic infrastructure), social improvements (schools, hospitals, rural literacy), and rural living conditions. Factor analysis sorted out all these into 5 factor loadings. Then this index was regressed on various independent variables, via three paths: path 1 representing demand factors mediated by political economy (increase in population density, market prices), path 2 adding an environmental constraint index, and path 3

adding technological changes (in more canal excavations, provision of credit, increased area under irrigation, and increase in use of fertilizers). Of all the independent variables, the percent increase in population density had the biggest overall effect on rural system change, followed by changes in market prices and credit. Structural path models can be quite useful for identifying key direct and indirect causal pathways, with appropriately lagged data (e.g., see Wang et al., 2020), but cannot be well understood with such composite artificial variable constructs.

Latin America

1. In Brazil, Spera (2017) uses household data for 2000–2016 from Mato Grosso and Goias states on the border of the Amazon and the *Cerrado* savannah to show how incentives for intensive agriculture and improving degraded pastures can increase production while conserving the rest of the Cerrado. Dias et al. (2016) provide an overall view of land use change in 1940–2012 in the Amazon, based on changes in vegetation at the level of pixels at the 1 km^2 level from a series of satellite images. They show a marked shift from land extensification (deforestation) in the twentieth century to agricultural land intensification in the twenty-first century up to 2012.

2. Rocha (2011) studied the causes of land intensification in a peasant community in the north-central Andes of Peru based on Systems Ecology theory, drawing on hypotheses from anthropology and development economics. Household size (persons living in the household) relative to landholding size was found associated with the intensity of use of labor and technology for the 37 households studied. But this community is quite isolated – 4 miles walk to the nearest road plus another 30 min by truck to the nearest market town. Therefore, exogenous factors that might stimulate intensification are absent. The household density effect was interpreted as showing that what agricultural intensification there exists is forced upon households by diminishing returns to their labor and accomplished by bor-

rowing and increased debt to acquire the new technology. Some households instead abandon agriculture to migrate to urban centers.

3. In Bolivia, Zimmerer (2013) reports on an international research project to promote the integration of biodiversity conservation into smallholder farms' agricultural intensification. He notes there had been major, long-term (2000–2010) advances in maize production, combined with growing peach trees. The maize fields in 2010 still had high biodiversity, contributing to biodiversity on over 31% of the maize fields of Bolivia. This success was also linked to off-farm migration; access to resources of both traditional and non-traditional growers; use of local adapted varieties *(landrace)* of corn for food; and farmers' knowledge of local conditions and skills in adapting to them.

4. On Central America, Davis and Carr (2014) use the New Economics of Labor Migration theory as a framework to study the effects of international out-migration followed by remittances on land use in origin households. Using data for Costa Rica, El Salvador, Guatemala and Nicaragua (from the Latin America Migration Project, at Princeton University), multivariate regression analyses find that neither time of a household member spent abroad nor remittances received was associated with land intensification or increased sale of farm products, instead increasing the row cropped area and pasture.

5. Jakovac et al. (2016) in a study in the central Amazon of Brazil observed that the increased demand for the locally grown staple cassava (or flour, *farinha*) due to rising populations resulted in reduced fallow times under shifting cultivation. Ethnographic and biophysical surveys in a main cassava growing area collected data on management practices, including fallow cycles, weed infestation and weeding effort. With shorter cycles, weeding effort increased, and secondary growth became bushes rather than small trees, with soil fertility and plot productivity in cassava declining. With very limited access to fertilizers, herbicides and technical assistance, and market demand

limited to the single product, adaptation was minimal. This indicates a need for technical assistance to improve and broaden consumer's diets along with producers' market opportunities and diversification of crops.

Some Conclusions Regarding Extensification vs. Intensification Studies Since 2000 and the Role of Population

1. There is far *more research on intensification than extensification*, consistent with the declining role of extensification as a means for increasing agricultural output from unused but potentially arable lands as such lands continue to decline, while intensification possibilities expand with improving technology and global/national efforts to improve seeds, advances in biology, genetics, etc. (see below).

2. Most of the relevant research continues to be carried out by agricultural economists, geographers and anthropologists, not demographers, and usually *deals with population in a minimalist or simplistic way (viz.,* total population per ha) rather than taking into account its composition, qualities, or dynamics. While many studies cite the problems posed by population growth in leading to deforestation from land extensification and difficulties with unsustainable intensification (e.g., land degradation), almost none mentions reducing fertility and population growth as a relevant policy consideration, consistent with the point made about ecological economics by O'Sullivan (2020). An exception is Muyanga and Jayne (2014).

3. Sophisticated models of utility maximization of the agricultural household have often been used in research but with either *tiny samples of households* (fewer than 100), parameters estimated not for households at all but for aggregations of households in sample clusters or villages, often also with small sample sizes, and covering very small populations, raising questions about generalizability. These models require heroic assumptions and then estimate the equations providing results and conclusions about farm household behavior and policy implications that seem obvious *a*

18 Population and Agricultural Change

priori, and which could be reached via smaller more qualitative, intensive methods if well-constructed and applied.

4. Few studies look *simultaneously* at extensification and intensification as possible consequences of population growth leading to increased population density, and hence do not examine possible trade-offs or complementarities in responses over time. Indeed, none examines the full range of possible responses over time as postulated in multi-phasic response theory, logically due to the discipline orientation of the investigators as well as data demands, so more *collaborative interdisciplinary research* is needed (see also Jayne and Heady, 2014, in section "Reflections and moving forward" at the end of the chapter below).

5. Few studies are based on *panel data* of households, which is highly desirable for studying causality, i.e., of the effects of demographic changes on changes in land use over time, in contrast to studies based on a single cross-section.

A Broader Sweep of the Data for 130 Developing Countries, 1961–2008

This expands upon the earlier analysis of section "Evidence up to the 1990s: Cross-country quantitative evidence" in five ways: (a) uses a longer series of data, notably on land use for 1961–2008, and better measures of rural population growth (UN Population Division); (b) this makes it possible to divide the reference period into three time intervals to ascertain if changes are accelerating or decelerating; (c) additional measures of intensification are available (d) correlation coefficients are used to test for links between changes in population and changes in land use, and (e) countries were divided into the Least Developed Countries and Other Developing Countries, following UN usage (full report available from the authors: Bilsborrow & Salinas, 2012). Adequate data were available for the whole period for 49 Least Developed Countries (LDCs) and 67 Other Developing Countries (ODCs), for

three time periods: 1961/65–1980, 1980–1995, 1995–2008. The LDCs comprise 32 countries in Africa, 16 in Asia and one (Haiti) in Latin America. The ODCs are the rest of Latin America (28) plus 20 countries in Africa and 21 in Asia. 2/3 of the LDCs are in Africa, accounting for most gaps in data, particularly for the first year and therefore the first time period.

Data are presented in Appendix Table 18.1a and 18.1b on agricultural land area (arable and permanently cropped land, or A & P land) at the country level for all three sub-periods and for the whole period of 40+ years, separately for least developed countries (1a) and other developing countries (1b). Space is not available here to present total or rural population levels and growth rates, to explain the choice of using rural population growth rates in the analysis, nor to show the country-level data on measures of land intensification (available on request from the author). Further details on the data and their shortcomings are in Bilsborrow and Salinas (2012) and FAOSTAT. The two issues to be explored here are whether there appear to be linkages between rural population change, on the one hand, and changes in land use or in the intensification of land use, based on country level data, based on the longer time series.

We begin with (Malthusian) *land extensification*, first looking briefly at the data themselves, beginning with rural population growth. Median annual rates of *total population growth* for the 49 available LDCs were 2.4, 2.7, and 2.5 for the three time periods, 1961/65–1980, 1980–1995, and 1995–2008, respectively (the rate over the whole 40+ year period being 2.5), associated with median annual *rural population growth rates* of 1.8, 2.1, and 2.0 (overall 1.8). For the 67 ODCs, median rates of total population growth were 2.1, 2.2, and 1.5, while for rural populations they were 1.6, 0.9 and 0.4 (overall 0.9)—not only significantly lower (indeed half the rates overall) than those of the LDCs but, in contrast to the LDCs, consistently declining over time. Table 18.1a and 18.1b provides the data on land extensification, including a few gaps and implausible figures (also data for Eritrea and Ethiopia were separate starting in 1995 but combined before under Ethiopia PDR). For the LDCs, median annual % rates of expansion in agricultural land were 0.56, 0.85, and 1.03, in the three time periods, with the overall % rate of expansion over the period 1961/65–2008 being 0.75. These data show that land extensification was occurring throughout the 45-year period and was in fact accelerating early in the twenty-first century, suggesting that significant A & P land remains in parts of Africa and that population pressures could be factors in further expansion. Nine of the 49 countries had agricultural land expanding at 2% or more over the whole period, meaning a doubling of A & P land in 35 years or less. In the ODCs, in contrast, median rates of expansion in A & P land were 0.7, 0.6 and 0 in the three periods, with an overall median of 0.49 -a third lower than among the LDCs despite the fact that many of the ODCs are in South America, which is known to have the most unused, potentially arable land available.[6]

Are there correlations in either set of countries between population growth and land extensfication? For the LDCs, simple Pearson correlation coefficients were only 0.15, 0.29 and 0.13, in the three time periods, with an overall period value of 0.14. Thus the *relationships are indeed positive*, as Malthusian theory would expect— weak but providing some support, given the data vicissitudes. Among ODCs, the same correlations were 0.08, 0.42 and −0.05, with the overall being 0.28–almost statistically significant ($r = 0.31$ being criterion 0.05 value). Note for the middle period statistically significant or nearly significant correlations exist for both sets of countries. Ideally it would be desirable to take into account country endowments of *potentially usable land* (including forest land, grasslands, etc.) to control for the extent to which actual land use is pushing up against the country limits, which should push them more towards intensification. The final col-

umn in Appendix Table 18.1a and 18.1b provides the data available from FAO, which shows the difficulty in using these data.

From the discussion in section "Reassessment of the theory and definitions" above, it is evident that there are many ways to measure *land intensification*. In the time period since the 1960s, populations not engaged in at least annual cropping are few and small, and mostly isolated from markets, so looking at examining linkages between population increases and reductions in fallow times based on the Boserup stages is a thing of the past. In the absence of reliable data on multiple cropping in most countries, we present evidence only on the relationships between changes in rural population, on the one hand, and changes in fertilizer use (chemical) and area of land irrigated, on the other (other changes are explored in the original document).

Fertilizer (chemical) use is the clearest, available indicator of land intensification in modern times, but It is disappointing that the data are not more complete. The FAO collects data, sending out annual questionnaires to Ministries of Agriculture and statistical offices of countries all over the world; their source of data is usually a recent agricultural census or survey, but if none had been implemented recently–often the case, as large data collection efforts are very often the first to be postponed in times of fiscal problems– figures are reported only from an earlier source, or sometimes by mere extrapolations. Thus for the LDCs, 20 of the 49 countries did not have data for all of the 4 years used (1961–2002), as of this study in 2011. Working with countries which had data for *any three* years left 38 countries, for which we classified 23 as having none or modest increases in fertilizer use (less than doubling), with 15 others increasing by doubling or much more. The (Boserupian) hypothesis is that countries with higher population growth should be more induced to increase fertilizer use. Due to wildly different values for countries, we chose not to use correlations but rather set up 3 × 3 cross-tabulations to classify countries according to low (1–1.9), medium (1.9–2.8) and high (2.8+) rates of annual population growth against the increase in fertilizer use (none, small, large). We observed

---

[6]The FAO data for Ecuador show no change in the agricultural land area, despite heavy human migration and colonization of the Amazon accompanied by deforestation beginning about 1970 (Bilsborrow et al., 2004; Mena et al., 2006; "Deforestation in Ecuador" on web, accessed 9/14/20). Data on Brazil vary greatly with government policies (with land extensification from fires for land clearing currently high again due to Presidential policies).

18 Population and Agricultural Change

a weak positive relationship for the LDCs. For the ODCs, data were available for all 4 years for 66 of the 69 countries, with 32 experiencing low increases and 34 high ones, using the same criterion. Even with enough data to cross-classify countries by 5-categories of population growth against 3 categories of increase in fertilizer use, there was no expected with positive association (higher population growth, more rise in fertilizer use). As already noted, overall, ODCs had significantly higher median increases in fertilizer use than LDCs: a smaller proportion of LDCs experienced a doubling or more in fertilizer use between 1980 and 2002 than of the ODCs even though their initial levels of use were very low. We concluded that fertilizer use levels and growth are determined primarily by economic factors, including ability to pay, making it all the more difficult to identify the smaller effects of demographic factors.

It is therefore curious that there did seem to be a positive relationship between rural population increase and increase in the area of irrigated farmland: after dropping countries with poor data, 36 of the 49 LDCs were left, for which little or no change occurred among 20, a significant increase of 50–100% among 11, and a more than doubling in five. Among the 63 ODCs, there was little change in 34, a significant increase among 19, and a large one in 10 more, or more overall than among LDCs, as with fertilizer use. Frequency cross-tabulations explored whether there was a relationship with rural population growth. For LDCs, a 4 (population growth) x 3 (change in irrigation) table found no relationship, but for ODCs, a 3×3 table showed a tendency for irrigation increases to be associated with higher population growth. The rejection of the null hypothesis was confirmed by a Chi-square statistic computed for the 63 cases, found to be 19.93, highly significant at the 0.001 level.

A summary of the 2011 results shows that the division into LDCs and ODCs was useful, as growth rates of rural population started to diverge significantly by the latter period, 1995–2008. Increases in the land area used in agri-

culture occurred in most countries over the full observation period but in 1995–2008 were much *smaller* than before for ODCs while they actually *rose* for LDCs, providing a *prima facie* case that *higher population growth stimulated more land to be brought into use in the low-income countries*, suggesting that Malthus is still relevant in Sub-Saharan Africa. Regarding *intensification of agriculture*, data gaps and inconsistencies contribute to difficulties in analyzing patterns between change in rural population and change in fertilizer use at the country level, but there was a clear tendency for ODC countries with higher population growth to have greater increases in irrigated area. This work could be improved by designing and implementing multivariate statistical models of change in land use and input use to incorporate other country-level factors that may affect land use change, but faces additional data challenges (see 5.4 below).

One other recent multi-country study should be mentioned here. Using FAOSTAT and other data, including developed countries but dropping the many developing countries which did not have all the data needed (on energy use, mechanization, and distance for transporting agricultural products from the main agricultural area to main consumers in the country), Pellegrini and Fernandez (2018) end up working with 58 countries which they say account for 95% of marketed agricultural products in the world, but excluding most small farms which produce mostly for subsistence (p. 2335). They provide aggregated data on agricultural land and input changes from 1961 to 2014 for the 58 countries, showing cultivated land rising by a factor of only 1.1 (10%) globally, while use of nitrogen, phosphorus and potassium fertilizer rose by 9, 4 and 5 times, respectfully, and machinery and fuel inputs each doubled. Developing countries with 30% or more of their agricultural land irrigated in 2014 were excluded from the continent-level estimates for Africa (Egypt excluded), Latin America (Chile, Peru) and Asia (Bangladesh, China, India, Nepal, Pakistan)—with those excluded constituting "high irrigation countries". This left minimal countries representing Africa (9 countries), Asia (11), and Latin America (9). For these three regions, the three types of fertilizer inputs rose over the 50 years from 15 times (for nitrogen) to 3 times for Africa, with little increase in machinery or fuel inputs; 30 to 13 times for Asia, with both machinery and fuel rising hugely by about 40 times; and 25 to 40 times for Latin America, with machinery and fuel inputs both rising about 5 times. This shows huge differences by continent, though exaggerated by use of the very low initial 1961 values when data were first being collected by FAO, and perhaps underestimated, leading to likely inflated percentage increases up to 1980, so it would have been more useful to see changes computed from 1980 instead. In any case, increases in inputs would still have been far greater than in cultivated land, which were themselves large for all three "continents"—58% for Africa, 40% for Asia, and 135% for Latin America. It is important to bear in mind that the Asia data exclude the largest countries (China, India, etc.) as "high irrigation countries", which in fact increased their total agricultural areas by only 6% while the irrigated area doubled.

The main purpose of the Pellegrini and Fernandez paper was to study whether the total energy value of the outputs produced has been rising faster than the cost of inputs, to determine energy use efficiency (EUE). They indeed find the EUE ratio rising over time, due to improved genetic varieties and agronomic practices, "fertilization" of the air by rising CO_2, and improved water use efficiency, so are optimistic about future intensification being able to meet future world food needs for 2050 (as was Tilman et al., 2011). But it was the developed countries that had vast increases in use of inputs with little change in agricultural land, while most developing countries achieved smaller increases in inputs with considerable continuing land clearing, suggesting the stage of development has had much to do with trends in land extensification vs. land intensification. How much of this relates to the main issue in this chapter—growing divergences in population growth—cannot be answered definitively,

but it is consistent with the data above for ODCs vs. LDCs. The UN projects population growth to fall to zero in this twenty-first century even in the LDCs, and rural populations to decline virtually everywhere, so interest in the impact of population growth on land extensification vs. intensification is likely to wan albeit not yet for Sub-Saharan Africa. In addition, the larger issues of both land extensification (Malthusian or not) and land intensification (Boserupian or not) will continue to be of paramount concern for the world given their potential implications for soil degradation, other environmental damage, and loss of biodiversity.

Reflections and Moving Forward

Quick Summary of Main Points from the Literature Review and Update

1. Malthus continues to be widely misunderstood, with too much focus on the simplistic idea of arithmetic increases in food production vs. exponential growth in population leading to a Malthusian food crisis. More important is "Malthusian land extensification", which continues and is linked endogenously to rising populations, often at considerable environmental cost.

2. Boserup is also often misunderstood. She did not say that population increase *would* necessarily (endogenously) stimulate increased intensity of land use and adoption of innovations, only that it could under favorable conditions. So how can such conditions be created or at least facilitated? Some ways are discussed below.

3. Research up to the 1990s focused on whether population growth and rising density in rural areas would stimulating more land extensification or more land intensification of agriculture, under what conditions and with what mediating factors. There has been little of this

kind of research since, the focus being on what kinds of land intensification are occurring and the environmental consequences, with little research on land extensification, even on Africa.

4. In contrast, data show that land extensification is continuing, in much of Africa and Latin America, slower than before *except in the Least Developed Countries* where both fertility and population growth remain high. Elsewhere, in the Other Developing Countries with lower population growth and higher economic growth, land intensification dominates.

5. While increasing population pressures are often involved in land intensification and innovation, other factors are more important, including household livelihood decisions, access to markets, agricultural research/extension, and policies of governments and international agencies.

6. With *surprisingly few exceptions* (e.g., Muyanga & Jayne, 2014, on Kenya), the recent literature in section "More recent evidence" here often struggles to find plausible policy changes to confront the continuing problems of food security, poverty, and land/environmental degradation linked to continuing land extensification and/or insufficient intensification includes *no mention of the value of reducing population growth,* viz., fertility. This also reflects the fact that demographers have rarely led or even been involved in this research. By now there is widespread evidence that development policies to reduce fertility can be more successful at modest cost compared to policies to achieve development and environmental benefits via land intensification innovations.

Continuing albeit slowing population growth in developing countries, especially in Sub-Saharan Africa, will contribute to both land extensification and land intensification and thereby further land degradation and other impacts on the environment and biodiversity loss unless major policy changes occur.

Policy Implications

Those studies[7] pertaining to land extensification from this meta-analysis are summarized first, followed in more detail by those relating to land intensification. As is evident from Bilsborrow and Salinas (2012) and Pellegrini and Fernandez (2018), there continues to be significant agricultural land extensification ongoing in the developing countries, especially Africa and Latin America. While this process was slowing in most of Latin America by the new millennium, this has not been evident yet in most of Africa, which seems implausible to separate from its high fertility and population growth. The major concern about continuing land extensification is its implications for biodiversity, as noted in studies reviewed here, including Brookfield (2001), Pellegrini and Fernandez (2018), Knauer et al. (2017), and ecologists in *Nature, Science*, etc.[8] Governments especially in tropical low-income countries need to create more protected areas and effectively protect them, with assistance from high income countries, and better plan roads, to limit extensification and loss of biodiversity (Nepstad et al., 2001; Thais et al., 2020).

Meanwhile, the vast majority of the increase in agricultural output needed to feed the world's projected population of 9.8 billion people 30 years from now (about 2 billion more than the present population according to UN projections) must come from increases in output from existing agricultural land, or land intensification. Hunter et al. (2017) project that it is possible for productivity per ha to rise by 50% between now

---

[7]Many other policy discussions exist, including in documents of international agencies such as the World Bank, the UN Environment Program, FAO, WRI (2019), the UN Development Program, etc., as well as at conferences and panels of scientists and government officials (e.g., the UK Foresight Report).

[8]An oft-cited estimate of ongoing species extinctions is Kolbert (2014), at 0.4 to 0.6% per year, with the expansion of agricultural land cited at the biggest cause (WRI, 2019, p.19).

and 2050, which would be sufficient to meet the projected world population's needs, but to avoid increasing greenhouse emissions from expanded agriculture and maintain ecosystem functions will require new methods, calling for new directions in research and policy. Much has been learned in about the possibilities as well as limitations of Borlaug's Green Revolution hybrid seeds and the vast increases in fertilizer and irrigation they tend to require (Brookfield, 2001), contributing to land degradation and contamination of water supplies. Moreover, Beckmann et al. (2019), in a meta-analysis of 449 sites covered in 115 studies, found land intensification successful in increasing yields by a mean of 20% but resulting in a mean biodiversity loss of 9% (as well as higher inequality: Clay, 2018).

New technologies thus need to be developed, but existing ones can be drawn upon more widely as well, including farmers using satellite imagery to improve planning land use; but in the case of the Indian Digital Land Initiative, this was limited by the low digital literacy of farmers so Lele and Goswami (2017) point out the need to promote digital literacy. Another technology is mobile phones which provide access to information on better agricultural practices, new crops, weather forecasts for planning planting/harvesting, etc.: Mekonnen et al. (2015) in a study of 85 developing countries in 2004–11 found mobile phone coverage correlated with more efficient agricultural production. Genetic research on plants and using genomics to implant desirable traits in seeds and plants continues, along with use of greenhouses and hydroponics. Parallel to the concept in medicine, "precision agriculture" is evolving to identify optimum inputs for plants, along with "regenerative agriculture" to restore soil from overuse and nutrient depletion (reported on US National Public Radio, Jan. 20, 2021 also Yost et al. 2019). Better methods of management are evolving, including minimum tillage, leaving crop residues in fields, microdosing fertilizer, gene editing via CRISPR, pest management, and improving livestock and pasture productivity

with changes in feeding, grass quality and genetics (WRI, 2019, Chs. 10–13). Rockstrom et al. (2017) state we need a "paradigm shift in how we view world agriculture" – as not only the source of food... but also as a "key contributor to a global transition to a sustainable world."

Research Needs, Macro Level Going Forward

Can cross-country studies be improved? Several possibilities exist, one to revisit the analyses as more data become available in the 2020s, with a longer time series, to see if the diverging trends in land extensification of LDCs in contrast to the ODCs continue or not. This could be extended to multivariate statistical analysis to try to separate out the effects on land use of rural population growth/rising density from those of other factors. For land extensification, continuing improvements in spatial imagery from satellites will lead to more accurate measurement of changes in vegetation and other forest cover (e.g., Hunt et al., 2019). Also, IPUMS TERRA "integrates population and environmental data across disciplinary scientific domains, enabling research into dramatic transformations of human populations, the environment, and their interactions." (September 15, 2020): it combines data from population censuses; land cover classified from satellite imagery (the most disaggregated so far limited to 1 km grids using Modis imagery, 2001–13); and climate data from Worldclim, etc.

Research Needs, Micro Level Going Forward

Following Turner et al. (1977), it could be useful to collect and pool data from household surveys in multiple countries/study sites that use similarly measured variables to implement a cross-site statistical analysis of differences in household land extensification and land intensifica-

18 Population and Agricultural Change

tion across households in many countries. This would simultaneously control for differences in explanatory variables at the household, community (to the extent data are available for communities) and region or country level to better sort out and measure the determinants of differences in land use associated with a host of demographic and other factors at the household, community, and higher levels (using fixed or random effects models, depending on the availability of comparable data at the higher contextual levels).

But even most desirable – though more difficult to generalize from until many studies are implemented- for ascertaining whether and in what circumstances Malthus, Boserup or both have contemporary relevance in the twenty-first century are *studies that look at representative panel or longitudinal samples of households*, if possible, from different regions of a country, to allow analysis of the effects of factors that change *over time* on changes in land use at the level of households within communities (and even within regions). Explanatory variables could include household demographics (including household size, fertility, migration) and the evolution of other factors that may affect livelihood-seeking behaviors. Panel data facilitate sorting out endogeneity and teasing out causality. Thus, the case studies reviewed in this chapter based on household data implicitly control for (i.e., ignore) most non-household factors since they are similar for all households in the limited study area, though they vary across populations within a country as well as more across countries. In addition, few of the many studies reviewed collected enough breadth and depth of data on household demographics, land use, and livelihoods to allow adequate identification of the differences in the effects of factors affecting land use even in the single cross-sectional data available. As more panel data sets become publicly available in the future, not only will path-breaking analyses of individual

country/site household panels become feasible but so will the possibility of pooling data sets across countries/study sites to test for effects of contextual factors at the higher country level, providing more policy relevant insights about factors affecting land use change. Finally, this could be done in the context of a multi-phasic theoretical model (Bilsborrow, 1987), which has yet to be applied to adequate *household panel data* in one country much less for pooled country study sites.

Returning to this topic was inspired by a review article in a special issue of *Food Policy* on the evolution of farming systems in Africa, by Jayne et al. (2014). They describe the diverse responses to rural population growth leading to rising pressures on land across the continent, including land extensification and intensification, off-farm work, migration and fertility decline, "in the context of unsustainable agricultural intensification" (p. 1). They note the ongoing shrinkage in the size of smallholder farms and more continuous cropping (reduction of fallow) leading to land degradation. While a half dozen countries (out of 40 studied) still have plentiful land for extensification, only five combined account for 76% of all the non-forested unutilized land: Democratic Republic of Congo, Angola, Congo, Zambia, and Cameroon. They note the need to increase both off-farm employment opportunities and rural-urban migration to absorb the rapid increase in rural labor and contrast the limited success of intensification in Africa with that of Asia with the Green Revolution. They also note that women in areas of high population density in Africa say they want fewer births (p. 9), so propose expanded family planning as part of a full set of policies to stimulate a multi-phasic response to improve livelihoods.

Appendix

Appendix Table 18.1a Land in arable and permanent crops[a], Least Developed Countries (000 ha)

	Growth rate 1961–80 (annual %)	Growth rate (%)	Growth rate 1980–95 (annual %)	Growth rate (%)	Growth rate 1995–2008 (annual %)	Growth rate (%)	Growth rate 1961–2008 (annual %)	Growth rate (%) 1961–2008	% of potent. arable land in use[b]
Afghanistan	7700	0.23	8049	−0.25	7753	0.15	7910	0.06	265.0
Angola	3170	0.37	3400	0.19	3500	0.41	3690	0.32	4.0
Bangladesh	8880	0.16	9158	−0.78	8148	0.50	8700	−0.04	103.1
Benin	1000	2.42	1585	1.45	1970	2.83	2845	2.23	19.3
Bhutan	113	1.42	148	0.48	159	−0.15	156	0.69	744.4
Burkina Faso	2139	1.39	2785	1.43	3450	4.71	6360	2.32	17.5
Burundi	905	1.70	1250	0.31	1310	−0.12	1290	0.75	83.5
Cambodia	2938	−1.84	2070	4.09	3820	0.46	4055	0.69	31.4
Central African Rep.	1738	0.59	1945	0.25	2020	−0.04	2010	0.31	4.2
Chad	2900	0.44	3150	0.61	3450	1.75	4330	0.85	9.9
Comoros	80	0.90	95	1.56	120	0.91	135	1.11	–
Congo DR	6950	0.47	7600	0.09	7700	−0.25	7450	0.15	0.7
Equatorial Guinea	210	0.48	230	0.00	230	−0.85	206	−0.04	14.0
Eritrea	–	–	–	–	440	3.26	672	–	88.0
Ethiopia	–	–	–	–	10,500	2.49	14,513	–	25.6
Ethiopia PDR	11,486	0.94	13,715	–	–	–	–	0.59	–
Gambia	121	1.44	159	1.01	185	5.83	395	2.52	21.9
Guinea	1060	0.39	1142	1.85	1506	5.53	3090	2.28	5.5
Guinea-Bissau	278	0.45	303	1.95	406	2.34	550	1.45	14.7
Haiti	1160	−0.28	1100	0.12	1120	1.15	1300	0.24	107.6
Kiribati	39	−0.14	38	−0.18	37	−0.65	34	−0.29	–
Lao PDR	650	1.13	806	0.74	900	3.09	1345	1.55	15.3
Lesotho	361	−1.05	296	0.60	324	0.79	359	−0.01	88.4
Liberia	583	−0.06	576	−0.94	500	1.63	618	0.12	6.0
Madagascar	2145	1.84	3040	0.90	3480	0.15	3550	1.07	8.7
Malawi	1360	2.00	1990	1.33	2430	3.07	3622	2.08	25.1
Maldives	4	2.95	7	0.89	8	0.00	8	1.48	–
Mali	1698	1.07	2080	3.31	3419	2.89	4980	2.29	9.4

(continued)

Appendix Table 18.1a (continued)

	Growth rate 1961–80 (annual %)	Growth rate (%)	Growth rate 1980–95 (annual %)	Growth rate (%)	Growth rate 1995–2008 (annual %)	Growth rate (%)	Growth rate 1961–2008 (annual %)	Growth rate (%) 1961–2008	% of potent. arable land in use[b]
Mauritania	272	−1.26	214	5.79	510	−1.66	411	0.88	15.1
Nepal[c]	1831	1.20	2299	0.28	2399	0.24	2475	0.64	103.7
Niger	11,500	−0.62	10,220	2.10	14,000	0.29	14,536	0.50	35.1
Rwanda	615	2.64	1015	−0.51	940	3.95	1570	1.99	156.8
Samoa	70	1.02	85	−2.10	62	0.12	63	−0.22	–
Sao Tome and Prin.	34	0.30	36	1.34	44	1.58	54	0.98	–
Senegal	2947	0.34	3141	−0.35	2980	1.36	3554	0.49	17.7
Sierra Leone	410	1.25	520	1.17	620	8.74	1930	3.30	13.7
Solomon Islands	50	0.40	54	1.44	67	0.97	76	0.89	12.8
Somalia	905	0.53	1000	0.36	1056	−0.21	1027	0.27	42.8
Sudan	10,840	0.73	12,450	1.79	16,274	1.93	20,906	1.40	15.0
Tanzania	6000	2.13	9000	0.70	10,000	0.70	10,950	1.28	5.2
Timor-Leste	80	2.51	129	2.58	190	1.30	225	2.20	–
Togo	1870	0.44	2035	0.82	2300	1.03	2630	0.73	56.6
Tuvalu	2	0.00	2	0.00	2	−0.81	1.8	−0.22	–
Uganda	4018	1.82	5680	1.40	7010	0.92	7900	1.44	48.0
Vanuatu	80	1.33	103	1.29	125	1.14	145	1.27	–
Yemen	1337	0.47	1463	1.14	1736	−0.60	1606	0.39	30900.0
Zambia	2065	−0.58	1851	0.73	2064	1.11	2384	0.31	9.0
MEDIAN		**0.56**		**0.85**		**1.03**		**0.75**	

[a] Source: FAOSTAT | © FAO Statistics Division 2010 | URL: http://faostat.fao.org/DesktopDefault.aspx?PageID=377&lang=en#ancor. Accessed 15 December 2010.

[b] Source: Terrastat| © FAO Statistics Division 2011 | URL: http://www.fao.org/ag/agl/agll/terrastat/wsrout.asp?wsreport=5&region=1&region=2&region=3&region=4&region=5&region=6&region=7&search=Display+statistics+%21. Accessed 15 December 2010

[c] Data on agricultural area irrigated and data on arable land and permanent crops refer to agricultural holdings operated by households. Agricultural activities undertaken by governmental organizations, business, etc. are excluded

Appendix Table 18.1b Land in arable and permanent crops, Other Developing Countries (000 ha)

	Growth rate 1961–80 (annual %)	Growth rate (%)	Growth rate 1980–95 (annual %)	Growth rate (%)	Growth rate 1995–2008 (annual %)	Growth rate (%)	Growth rate 1961–2008 (annual %)	Growth rate (%) 1961–2008	% of potent. arable land in use[b]
Algeria	7066	0.32	7509	0.45	8029	0.37	8424	0.37	62.7
Argentina	19,472	1.72	26,981	0.25	28,020	1.26	33,000	1.12	30.0
Belize	42	1.12	52	4.09	96	0.47	102	1.89	5.8
Bolivia	1442	1.88	2062	1.71	2665	2.77	3819	2.07	3.8
Botswana	400	0.05	404	−1.03	346	−2.44	252	−0.98	4.6
Brazil	28,396	3.27	52,864	1.43	65,500	0.34	68,500	1.87	9.2
Cameroon	5510	1.21	6930	0.22	7160	0.00	7163	0.56	4.2
Cape Verde	40	0.00	40	0.78	45	3.18	68	1.13	–
Chile	3836	0.29	4050	−3.49	2400	−2.55	1722	−1.70	127.7
China	105,248	−0.26	100,219	1.87	132,715	−0.61	122,543	0.32	47.5
Colombia	4970	0.23	5192	−1.06	4430	−1.90	3461	−0.77	8.3
Congo	540	−0.15	525	−0.06	520	0.32	542	0.01	4.7
Costa Rica	480	0.28	506	0.05	510	−0.15	500	0.09	44.0
Côte d'Ivoire	2680	2.43	4255	3.13	6800	0.28	7050	2.06	14.1
Cuba	1650	3.70	3330	1.52	4184	−0.40	3970	1.87	45.0
Dominica	15	0.66	17	−0.83	15	2.59	21	0.72	–
Dominican Rep.	990	1.90	1420	−0.10	1400	−0.57	1300	0.58	68.2
Ecuador	2510	−0.10	2462	1.32	3001	−1.41	2500	−0.01	23.6
Egypt	2568	−0.26	2445	1.97	3283	0.58	3542	0.68	2892.6
El Salvador	648	1.11	800	0.44	855	0.52	915	0.73	84.5
Fiji	154	0.52	170	2.96	265	−0.36	253	1.06	77.2
French Guiana	3	1.51	4	7.86	13	2.11	17.1	3.70	0.2
Gabon	145	5.98	452	0.61	495	−0.32	475	2.53	2.6
Ghana	3300	0.46	3600	1.78	4700	3.33	7250	1.68	23.6
Guatemala	1536	0.69	1750	0.58	1910	1.32	2268	0.83	51.5
Guyana	360	1.68	495	0.12	504	−0.96	445	0.45	3.7
Honduras	1480	0.90	1757	0.70	1950	−2.40	1428	−0.08	59.3

Appendix Table 18.1b (continued)

	Growth rate 1961–80 (annual %)	Growth rate (%)	Growth rate 1980–95 (annual %)	Growth rate (%)	Growth rate 1995–2008 (annual %)	Growth rate (%)	Growth rate 1961–2008 (annual %)	Growth rate (%) 1961–2008	% of potent. arable land in use[b]
India	160,986	0.23	168,255	0.06	169,750	−0.02	169,320	0.11	82.2
Indonesia	26,000	0.00	26,000	1.04	30,387	1.54	37,100	0.76	42.0
Iran	15,271	−0.57	13,713	2.07	18,708	0.04	18,770	0.44	384.8
Iraq	4700	0.77	5439	0.30	5690	−0.33	5450	0.32	130.5
Jamaica	276	−0.74	240	0.74	268	−1.01	235	−0.34	140.4
Jordan	294	0.72	337	−0.28	323	−2.60	230.5	−0.52	71.9
Kenya	3900	0.49	4280	2.16	5918	−0.16	5800	0.84	28.5
Korea DPR	2330	0.30	2465	0.36	2600	0.84	2900	0.47	55.1
Korea Rep.	2095	0.25	2196	−0.67	1985	−0.98	1747	−0.39	52.3
Lebanon	262	0.73	301	0.18	309	−0.60	286	0.19	113.8
Libya	1970	0.29	2080	0.42	2215	−0.60	2050	0.09	88.1
Malaysia	3980	0.99	4800	3.07	7604	−0.02	7585	1.37	4.7
Mauritius	92	0.80	107	−0.13	105	−1.10	91	−0.02	–
Mexico	23,745	0.21	24,700	0.67	27,300	0.06	27,500	0.31	47.4
Mongolia	624	3.36	1182	0.75	1322	−3.38	852	0.66	745.8
Morocco	6970	0.75	8030	1.29	9749	−0.63	8981	0.54	75.7
Namibia	642	0.12	657	1.48	820	−0.11	808	0.49	5.6
Nicaragua	1180	0.28	1245	2.71	1870	1.00	2130	1.26	22.9
Nigeria	28,800	0.28	30,385	0.53	32,909	1.60	40,500	0.73	49.4
Pakistan	16,881	0.97	20,300	0.40	21,550	−0.13	21,200	0.49	392.3
Panama	564	−0.09	555	1.10	655	0.46	695	0.44	28.1
Papua New Guinea	425	2.33	662	1.04	774	1.33	920	1.64	2.9
Paraguay	811	4.00	1735	2.91	2685	3.62	4300	3.55	10.5
Peru	1956	3.14	3550	0.90	4064	0.68	4440	1.74	9.5
Philippines	6901	1.75	9628	0.04	9685	0.47	10,300	0.85	98.4
Saint Kitts and Nevis	16	−0.70	14	−3.73	8	−5.14	4.1	−2.90	–

Appendix Table 18.1b (continued)

	Growth rate 1961–80 (annual %)	Growth rate (%)	Growth rate 1980–95 (annual %)	Growth rate (%)	Growth rate 1995–2008 (annual %)	Growth rate (%)	Growth rate 1961–2008 (annual %)	Growth rate (%) 1961–2008	% of potent. arable land in use[b]
Saint Lucia	14	1.02	17	0.74	19	−4.94	10	−0.72	–
Saint Vincent & Gren.	9	0.56	10	0.00	10	−1.72	8	−0.25	–
Seychelles	5	0.00	5	0.00	5	−1.72	4	−0.48	–
South Africa	12,878	0.15	13,254	1.18	15,825	−0.18	15,450	0.39	46.9
Sri Lanka	1538	1.14	1910	−0.08	1886	1.19	2200	0.76	50.7
Suriname	35	1.77	49	2.19	68	−1.49	56	1.00	0.7
Swaziland	126	2.13	189	0.04	190	0.08	192	0.90	23.7
Syrian Rep.	6381	−0.61	5684	−0.22	5502	0.23	5666	−0.25	98.1
Thailand	11,393	2.49	18,298	0.73	20,410	−0.61	18,850	1.07	64.6
Tonga	26	0.75	30	−0.23	29	−0.55	27	0.08	–
Tunisia	4250	0.53	4701	0.25	4878	0.25	5041	0.36	149.6
Turkey	25,167	0.65	28,479	−0.33	27,115	−0.78	24,505	−0.06	109.7
Uruguay	1383	0.25	1449	−0.55	1335	1.74	1673	0.41	9.2
Venezuela	3482	0.28	3670	−0.55	3380	−0.07	3350	−0.08	7.1
Viet Nam	6020	0.46	6570	0.18	6751	2.56	9415	0.95	60.2
Zimbabwe	1985	1.43	2605	1.39	3210	1.40	3850	1.41	11.7
MEDIAN		**0.67**		**0.58**		**−0.02**		**0.49**	

References

Abebe, G. K., Bijman, J., Pascucci, S., & Omta, O. (2013). Adoption of improved potato varieties in Ethiopia: The role of agricultural knowledge and innovation system and smallholder farmers' quality assessment. *Agricultural Systems, 122*, 22–32.

Adepoju, A. (1983). *Population and space in Saharan Africa*. Economic Commission for Africa.

Ali, A. M. S. (2007). Population pressure, agricultural intensification and changes in rural systems in Bangladesh. *Geoforum, 38*(4), 720–738.

Allen, J., & Barnes, D. (1985). The causes of deforestation in developing countries. *Annals of the Association of American Geographers, 75*(2), 163–184.

Alves, D. S., Escada, M., Pereira, J., & Linhares, C. (2003). Land use intensification and abandonment in Rondonia, Brazilian Amazonia. *International Journal of Remote Sensing, 24*(4), 899–903.

Barnum, H., & Squire, L. (1979). An econometric application of the theory of the farm-household. *Journal of Development Economics, 6*(1), 79–102.

Barretto, A., Berndes, G., Sparovek, G., & Wirsenius, S. (2013). Agricultural intensification in Brazil and its effects on land-use patterns: An analysis of the 1975-2006 period. *Global Change Biology, 19*(6), 1804–1815.

Baudron, F., Andersson, J. A., Corbeels, M., & Giller, K. E. (2012). Failing to yield? Ploughs, conservation agriculture and the problem of agricultural intensification: An example from the Zambezi Valley, Zimbabwe. *Journal of Development Studies, 48*(3), 393–412.

Beckmann, M., Gerstner, K., Akin-Fajiye, M., Ceausu, S., Kambach, S., Kinlock, N. L., Phillips, H. R. P., Verhagen, W., Gurevitch, J., Klotz, S., Newbold, T., Verburg, P. H., Winter, M., & Seppelt, R. (2019). Conventional land-use intensification reduces species richness and increases production: A global meta-analysis. *Global Change Biology, 25*(6), 1941–1956.

Bilsborrow, R. E. (1987). Population pressures and agricultural development in developing countries: A conceptual framework and recent evidence. *World Development, 15*(2), 1–18.

Bilsborrow, R. E., & Stupp, P. (1987, June 12). *Demographic effects on rural development in Sub-Saharan Africa: An assessment of the literature and recommendations* (Report to the UN Food and Agricultural Organization (Rome) and International Labour Office (Geneva), p. 188).

Bilsborrow, R. E., & Okoth-Ogendo, H. (1992). Population-driven changes in land use. *Ambio, 21*(1), 37–45.

Bilsborrow, R. E., & Geores, M. (1992). *Rural population dynamics and agricultural development: Issues and consequences observed in Latin America*. Cornell University Population and Development Program and Cornell International Institute for Food, Agriculture and Development.

Bilsborrow, R. E., & Geores, M. (1994). Population change and agricultural intensification in developing countries. In L. Arizpe et al. (Eds.), *Population and the environment: Rethinking the debate* (pp. 171–207). Westview Press, for Social Science Research Council.

Bilsborrow, R. E., & Stupp, P. (1997). Demographic processes, land, and the environment in Guatemala. In A. Pebley & L. Rosero-Bixby (Eds.), *Demographic diversity and change in the central American Isthmus* (pp. 581–623). The Rand Corporation.

Bilsborrow, R. E., & Carr, D. (2001). Population, agricultural land use, and the environment in developing countries. In D. R. Lee & C. B. Barrett (Eds.), *Tradeoffs or synergies? Agricultural intensification, economic development and the environment* (pp. 35–56). CABI.

Bilsborrow, R. E., Barbieri, A., & Pan, W. (2004). Changes in population and land use over time in the Ecuadorian Amazon. *Acta Amazonica, 34*(4), 635–647.

Bilsborrow, R. E., & Salinas, R. V. (2012). *Population growth and agricultural change: Diverging paths for least developed countries and other developing countries?* (p. 177). United Nations Population Fund, project report.

Binswanger, H., & Ruttan, V. (1978). *Induced innovation: Technology, institutions and development*. Johns Hopkins University Press.

Blaikie, P., & Brookfield, H. (1987). *Land degradation and society*. Methuen.

Boserup, E. (1965). *The conditions of agricultural growth: The economics of agrarian change under population pressure*. Aldine Publishing Company.

Boserup, E. (1981). *Population and technological change: A study of long-term trends*. The University of Chicago Press.

Brookfield, H. (2001). *Exploring agrodiversity*. Columbia University Press.

Brown, B., Nuberg, I., & Llewellyn, R. (2018). Research capacity for local innovation: The case of conservation agriculture in Ethiopia, Malawi and Mozambique. *Journal of Agricultural Education & Extension, 24*(3), 249–262.

Bullock, R., Mithofer, D., & Vihemaki, H. (2014). Sustainable agricultural intensification: The role of cardamom agroforestry in the East Usambaras, Tanzania. *International Journal of Agricultural Sustainability, 12*(2), 109–129.

Carr, D. (2008). Migration to the Maya Biosphere Reserve, Guatemala: Why place matters. *Human Organization, 67*(1), 37–48.

Lopez-Carr, D., Martinez, A., Bilsborrow, R., & Whitmore, T. (2017). Geographical and individual determinants of rural out-migration to a tropical forest protected area: The Maya Biosphere Reserve, Guatemala. *European Journal of Geography, 8*(2), 78–106.

Carswell, G. (2002). Farmers and fallowing: Agricultural change in Kigezi District, Uganda. *Geographical Journal, 168*, 130–140.

Clay, N. (2018). Seeking justice in Green Revolutions: Synergies and trade-offs between large scale and smallholder agricultural intensification in Rwanda. *Geoforum, 97*, 352–362.

Crosby, A. (1972). *The columbian exchange: Biological and cultural consequences of 1492*. Praeger.

Cruz, M. C. (1999). Population pressure, economic stagnation, and deforestation in Costa Rica and the Philippines. In R. Bilsborrow & D. Hogan (Eds.), *Population and deforestation in the humid tropics* (pp. 99–121). International Union for the Scientific Study of Population.

Cruz, W. D., & Cruz, M. C. (1990). Population pressure and deforestation in the Philippines. *ASEAN Economic Bulletin, 7*(2), 200–212.

Cuffaro, N. (2001). *Population, economic growth and agriculture in less developed countries*. Routledge.

Dahal, B. M., Nyborg, I., Sitaula, B. K., & Bajracharya, R. M. (2009). Agricultural intensification: Food insecurity to income security in a mid-hill watershed of Nepal. *International Journal of Agricultural Sustainability, 7*(4), 249–260.

Davis, J., & Lopez-Carr, D. (2014). Migration, remittances and smallholder decision-making: Implications for land use and livelihood change in Central America. *Land Use Policy, 36*, 319–329.

Davis, K. (1963). The theory of change and response in modern demographic history. *Population Index, 294*, 345–366.

Deb, S., Lynrah, M., & Tiwari, B. K. (2013). Technological innovations in shifting agricultural practices by three tribal farming communities of Meghalaya, northeast India. *Tropical Ecology, 54*(2), 133–148.

Del Mar Lopez, T., Aide, T., & Thomlinson, J. (2001). Urban expansion and the loss of prime agricultural lands in Puerto Rico. *Ambio, 30*(1), 49–54.

Denevan, W., & Padoch, C. (1987). *Swidden-fallow agroforestry in the Peruvian Amazon* (Advances in Economic Botany) (Vol. 5). New York Botanical Garden.

De Walt, B. (1985). Microcosmic and macrocosmic processes of agrarian change in Southern Honduras: The cattle are eating the forest. In B. De Walt & P. Pelto (Eds.), *Micro and macro levels of analysis in anthropology: Issues in theory and research*. Westview Press.

Dias, L. C. P., Pimenta, F. M., Santos, A. B., Costa, M. H., & Ladle, R. J. (2016). Patterns of land use, extensification, and intensification of Brazilian agriculture. *Global Change Biology, 22*(8), 2887–2903.

Djagba, J. F., Zwart, S. J., Houssou, C. S., Tente, B. H. A., & Kiepe, P. (2019). Ecological sustainability and environmental risks of agricultural intensification in inland valleys in Benin. *Environment Development and Sustainability, 21*(4), 1869–1890.

Eckert, S., Kiteme, B., Njuguna, E., & Zaehringer, J. G. (2017). Agricultural expansion and intensification in the foothills of Mount Kenya: A landscape perspective. *Remote Sensing, 9*(8), 20.

Ehrlich, P. (1968). *The population bomb*. Sierra Club/Ballantine Books.

Fei, J., & Ranis, G. (1964). *Development of the labor surplus economy: Theory and policy*. Irwin.

Folving, R., & Christensen, H. (2007). Farming system changes in the Vietnamese uplands – Using fallow length and farmers' adoption of Sloping Agricultural Land Technologies as indicators of environmental sustainability. *Geografisk Tidsskrift-Danish Journal of Geography, 107*(1), 43–58.

Geertz, C. (1963). *Agricultural involution: Processes of ecological change in Indonesia*. University of California Press.

Geores, M., & Bilsborrow, R. E. (1991, August 28). *Population, environment and sustainable agricultural development in Asia: A review of the recent literature* (Report to the UN Food and Agricultural Organization, Rome, p. 40).

Goldman, A., & Smith, J. (1995). Agricultural transformations in India and northern Nigeria: Exploring the nature of Green Revolutions. *World Development, 23*(2), 243–263.

Gray, C., & Bilsborrow, R. E. (2020). Stability and change within indigenous land use in the Ecuadorian Amazon. *Global Environmental Change, 63*.

Gray, L. C. (2005). What kind of intensification? Agricultural practice, soil fertility and socioeconomic differentiation in rural Burkina Faso. *Geographical Journal, 171*, 70–82.

Greiner, C., & Mwaka, I. (2016). Agricultural change at the margins: Adaptation and intensification in a Kenyan dryland. *Journal of Eastern African Studies, 10*(1), 130–149.

Grimm, M., & Klasen, S. (2015). Migration pressure, tenure security, and agricultural intensification: Evidence from Indonesia. *Land Economics, 91*(3), 411–434.

Hardin, G. (1968, December 13). The tragedy of the commons. *Science, 162*(3859).

Hecht, S. B. (1985). Environment, development and politics: Capital accumulation and the livestock sector in eastern Amazonia. *World Development, 13*(6), 663–684.

Heckandon, M. S., & McKay, A. (Eds.). (1984). *Colonización y Destrucción de Bosques en Panamá: Ensayos sobre un Grave Problema Ecológico*. Asociación Panameña de Antropología.

Helmer, E. H. (2004). Forest conservation and land development in Puerto Rico. *Landscape Ecology, 19*(1), 29–40.

Henley, D. (2005). Agrarian change and diversity in the light of Brookfield, Boserup and Malthus: Historical illustrations from Sulawesi, Indonesia. *Asia Pacific Viewpoint, 46*(2), 153–172.

Henriques, M. H. (1985). The demographic dynamics of a frontier area: The state of Rondonia, Brazil. In R. E. Bilsborrow & P. DeLargy (Eds.), *Impact of rural development projects on demographic behavior* (pp. 133–168). United Nations Fund for Population Activities.

Hoffmann, R., Dimitrova, A., Muttarak, R., Crespo Cuaresma, J., & Peisker, J. (2020, September 14). A meta-analysis of country-level studies on environmental migration. *Nature Climate Change*. https://doi.org/10.1038/s41558-020-0898-6

Horton, D., Akello, B., Aliguma, L., et al. (2010). Developing capacity for agricultural market chain innovation experience with the 'PMCA' in Uganda. *Journal of International Development, 22*(3), 367–389.

Hosseini, S. M., Dourandish, A., Ghorbani, M., & Kakhki, M. D. (2017). Agricultural insurance and intensification investment: Case study of Khorasan Razavi Province. *Journal of Agricultural Science and Technology, 19*(1), 1–10.

Hunt, M. L., Blackburn, G. A., & Rowland, C. S. (2019). Monitoring the sustainable intensification of arable agriculture: The potential role of Earth observation. *International Journal of Applied Earth Observation and Geoinformation, 81*, 125–136.

Hunt, R. C. (2000). Labor productivity and agricultural development: Boserup revisited. *Human Ecology, 28*(2), 251–277.

Hunter, M. C., Smith, R. G., Schipanski, M., Atwood, L., & Mortensen, D. (2017). Agriculture in 2050: Recalibrating targets for sustainable intensification. *Bioscience, 67*(4), 385–390.

Jakovac, C. C., Pena-Claros, M., Mesquita, R. C. G., Bongers, F., & Kuyper, T. W. (2016). Swiddens under transition: Consequences of agricultural intensification in the Amazon. *Agriculture Ecosystems & Environment, 218*, 116–125.

Jayne, T., Chamberlin, J., & Headey, D. (2014). Land pressure, the evolution of farming systems, and development strategies in Africa: A synthesis. *Food Policy, 48*, 1–17.

Jodha, N. (1985). Population growth and the decline of common property resources in Rajasthan, India. *Population and Development Review, 11*(2), 247–264.

Joly, L. G. (1989). The conversion of rain forests to pastures in Panama. In D. Schumann & W. Partridge (Eds.), *The human ecology of tropical land settlement in Latin America* (pp. 89–110). Westview Press.

Josephson, A. L., Ricker-Gilbert, J., & Florax, R. (2014). How does population density influence agricultural intensification and productivity? Evidence from Ethiopia. *Food Policy, 48*, 142–152.

Keller, M., Bustamante, M., Gash, I., & Silva Dias, P. (Eds.). (2009). *Amazonia and global change* (Geophysical Monograph) (Vol. 186). American Geophysical Union.

Kiprono, P., & Matsumoto, T. (2018). Roads and farming: The effect of infrastructure improvement on agricultural intensification in South-Western Kenya. *Agrekon, 57*(3–4), 198–220.

Knauer, K., Gessner, U., Fensholt, R., Forkuor, G., & Kuenzer, C. (2017). Monitoring agricultural expansion in Burkina Faso over 14 years with 30 m resolution time series: The role of population growth and implications for the environment. *Remote Sensing, 9*(2), 25.

Kolbert, E. (2014). *The sixth extinction: An unnatural history*. Henry Holt.

Kreidenweis, U., Humpenoder, F., Kehoe, L., Kuemmerle, T., Bodirsky, B. L., Lotze-Campen, H., & Popp, A. (2018). Pasture intensification is insufficient to relieve pressure on conservation priority areas in open agricultural markets. *Global Change Biology, 24*(7), 3199–3213.

Lele, U., & Stone, S. (1989). Population pressure, environment and agricultural intensification: variations on the Boserup Hypothesis. In *Managing agricultural development in Africa/MADIA* (Paper 4). World Bank.

Lele, U., & Goswami, S. (2017). The fourth industrial revolution, agricultural and rural innovation, and implications for public policy and investments: A case of India. *Agricultural Economics, 48*, 87–100.

Lewis, W. A. (1954). Economic development with unlimited supplies of labour. *The Manchester School of Economic and Social Studies, 22*(2), 139–191.

Li, L., Wang, C. G., Segarra, E., & Nan, Z. B. (2013). Migration, remittances, and agricultural productivity in small farming systems in Northwest China. *China Agricultural Economic Review, 5*(1), 5–23.

Lintemani, M., Loss, A., Mendes, C., et al. (2020). Long fallows allow soil regeneration in slash-and-burn agriculture. *Journal of the Science of Food and Agriculture, 100*(3), 1142–1154.

Little, P., Horowitz, M., & Nyerges, A. (Eds.). (1987). *Lands at risk in the third world: Local-level perspectives* (Monographs in Development Anthropology). Westview Press.

Liu, G. S., Wang, H. M., Cheng, Y. X., Zheng, B., & Lu, Z. L. (2016). The impact of rural out-migration on arable land use intensity: Evidence from mountain areas in Guangdong, China. *Land Use Policy, 59*, 569–579.

Liu, Y. B., & Chen, Y. N. (2006). Impact of population growth and land-use change on water resources and ecosystems of the and Tarim River Basin in Western China. *International Journal of Sustainable Development and World Ecology, 13*(4), 295–305.

Maertens, M., Zeller, M., & Birner, R. (2006). Sustainable agricultural intensification in forest frontier areas. *Agricultural Economics, 34*(2), 197–206.

Mahmood, S., Khan, I. A., Maann, A., Shahbaz, B., & Akhtar, S. (2010). Role of international migration in agricultural development and farmers' livelihoods: A case study of an agrarian community. *Pakistan Journal of Agricultural Sciences, 47*(3), 297–301.

Malthus, T. R. (1960). *On population (First Essay on Population, 1798, and Second Essay on Population, 1803)*. Modern Library, for Random House.

Mekonnen, D., Gerber, N., & Matz, J. A. (2018). Gendered social networks, agricultural innovations, and farm productivity in Ethiopia. *World Development, 105*, 321–335.

Mekonnen, D., Spielman, D. J., Fonsah, E., & Dorfman, J. (2015). Innovation systems and technical efficiency in developing-country agriculture. *Agricultural Economics, 46*(5), 689–702.

Mena, C. F., Bilsborrow, R. E., & McClain, M. E. (2006). Socioeconomic drivers of deforestation in the Northern Ecuadorian Amazon. *Environmental Management, 34*(10), 1831–1849.

Metzger, J. P. (2002). Landscape dynamics and equilibrium in areas of slash-and-burn agriculture with short and long fallow period (Bragantina region, NE Brazilian Amazon). *Landscape Ecology, 17*(5), 419–431.

Monod, T. (1975). *Pastoralism in tropical Africa*. Oxford University Press, for the International African Institute.

Mortimore, J. (1967). Land and population pressure in the Kano close-settled zone, Northern Nigeria. *The Advancement of Service, 23*(118), 677–686.

Mosley, P. (1963). *The settler economies: Studies in the economic history of Kenya and Southern Rhodesia, 1900–1963*. Cambridge University Press.

Muyanga, M., & Jayne, T. S. (2014). Effects of rising rural population density on smallholder agriculture in Kenya. *Food Policy, 48*, 98–113.

Nepstad, D., et al. (2001). Road paving, fire regime feedbacks, and the future of Amazon forests. *Forest Ecology and Management, 154*(1), 395–407.

Netting, R. M. (1968). *Hill farmers of Nigeria: Cultural ecology of the Kofyar of the Jos Plateau*. University of Washington Press.

Netting, R. M., & Stone, M. P. (1996). Afro-diversity on a farming frontier: Kofyar smallholders on the Benue plains of central Nigeria. *Africa, 66*(1), 52–70.

Ningal, T., Hartemink, A. E., & Bregt, A. K. (2008). Land use change and population growth in the Morobe province of Papua New Guinea between 1975 and 2000. *Journal of Environmental Management, 87*(1), 117–124.

Nin-Pratt, A., & McBride, L. (2014). Agricultural intensification in Ghana: Evaluating the optimist's case for a Green Revolution. *Food Policy, 48*, 153–167.

Noronha, R., 1985. A review of the literature of land tenure systems in Sub-Saharan Africa. World Bank Discussion Paper No. ARU 43. .

Oduol, J. B. A., & Tsuji, M. (2005). The effect of farm size on agricultural intensification and resource allocation decisions: Evidence from smallholder farms in Embu District, Kenya. *Journal of the Faculty of Agriculture Kyushu University, 50*(2), 727–742.

Okike, I., Jabbar, M. A., Manyong, V. M., & Smith, J. W. (2005). Ecological and socio-economic factors affecting agricultural intensification in the West African savannas: Evidence from Northern Nigeria. *Journal of Sustainable Agriculture, 27*(2), 5–37.

Ostrom, E. (1990). Commons: The Evolution of Institutions for Collective Action. Cambridge University Press.

Ostrom, E., Gardner, R., & Walker, J. (1994). *Rules, games and common-pool resources*. University of Michigan Press.

O'Sullivan, J. (2020). The social and environmental influences of population growth rate and demographic pressure deserve greater attention in ecological economics. *Ecological Economics, 172*, 106648.

Oyekale, A. S., & Adepoju, A. O. (2012). Determinants of agricultural intensification in Southwest Nigeria. *Life Science Journal-Acta Zhengzhou University Overseas Edition, 9*(3), 370–376.

Padoch, C., & de Jong, W. (1991). The house gardens of Santa Rosa: Diversity and variability in an Amazonian agricultural system. *Economic Botany, 45*(2), 166–175.

Pan, W. K.-Y., & Bilsborrow, R. E. (2005). A multilevel study of fragmentation of plots and land use in the Ecuadorian Amazon. *Global and Planetary Change, 47*, 232–252.

Parsons, J. (1992, September 9–11). Population growth as a factor leading to conflict over land and other natural resources. In *Presented at British Society for Population Studies, annual conference on population & environment*.

Pascual, U., & Barbier, E. B. (2006). Deprived land-use intensification in shifting cultivation: the population pressure hypothesis revisited. *Agricultural Economics, 34*(2), 155–165.

Pellegrini, P., & Fernandez, R. J. (2018). Crop intensification, land use, and on-farm energy-use efficiency during the worldwide spread of the green revolution. *Proceedings of the National Academy of Sciences of the United States of America, 115*(10), 2335–2340.

Perkins, D. (1969). *Agricultural Development in China, 1368–1968*. Chicago: Aldine Publishing Co.

Pichón, F. (1997). Settler households and land-use patterns in the Amazon frontier: Farm-level evidence from Ecuador. *World Development, 25*(1), 67–91.

Pichón, F., & Bilsborrow, R. E. (1999). Land tenure and land use systems, deforestation, and associated demographic factors: Farm-level evidence from Ecuador. In R. E. Bilsborrow & D. Hogan (Eds.), *Population and deforestation in the humid tropics* (pp. 175–207). International Union for the Scientific Study of Population (IUSSP).

Pingali, P., & Binswanger, H. (1984). *Population density and agricultural intensification: A study of the evolution of technologies in tropical agriculture* (World Bank Discussion Paper ARU 22). Washington, DC.

Reardon, T., & Vosti, S. (1995). Links between rural poverty and the environment in developing countries: Asset categories and investment poverty. *World Development, 23*(9), 1495–1506.

Reddy, V. R., Chiranjeevi, T., & Syme, G. (2020). Inclusive sustainable intensification of agriculture in West Bengal, India; Policy and institutional approaches. *International Journal of Agricultural Sustainability, 18*(1), 70–83.

Ricker-Gilbert, J., Jumbe, C., & Chamberlin, J. (2014). How does population density influence agricultural intensification and productivity? Evidence from Malawi. *Food Policy, 48*, 114–128.

Robinson, W., & Schutjer, W. (1984). Agricultural development and demographic change: A generalization of the Boserup model. *Economic Development and Cultural Change, 32*, 355–366.

Rocha, J. M. (2011). Agricultural intensification, market participation, and household demography in the Peruvian Andes. *Human Ecology, 39*(5), 555–568.

Rockstrom, J., Williams, J., Daily, G., Noble, A., et al. (2017). Sustainable intensification of agriculture for human prosperity and global sustainability. *Ambio, 46*(1), 4–17.

Rudel, T., Perez-Lugo, M., & Zichal, H. (2000). When fields revert to forest: Development and spontaneous reforestation in post-war Puerto Rico. *The Professional Geographer, 52*(3), 386–397.

Scoones, I., & Toulmin, C. (1999). *Policies for fertility management in Africa*. Department for International Development.

Sellers, S., & Bilsborrow, R. E. (2020). Agricultural technology adoption among migrant settlers and indigenous populations of the Northern Ecuadorian Amazon: Are differences narrowing? *Journal of Land Use Science.* https://doi.org/10.1080/1747423X.2020.1719225

Simon, J. (1977). *The economics of population growth.* Princeton University Press.

Simon, J. (1981). *The ultimate resource.* Princeton University Press.

Sindiga, L. (1984). Land and population problems in Kajiado and Narok, Kenya. *African Studies Review, 27*(1), 23–39.

Singh, I., Squire, L., & Strauss, J. (1986). A survey of agricultural household models: recent findings and policy implications. *The World Bank Economic Review, 1*(1).

Siren, A. H. (2007). Population growth and land use intensification in a subsistence-based indigenous community in the Amazon. *Human Ecology, 35*(6), 669–680.

Spera, S. (2017). Agricultural intensification can preserve the Brazilian Cerrado: Applying lessons from Mato Grosso and Goias to Brazil's last agricultural frontier. *Tropical Conservation Science, 10*, 7.

Stephens, L., et al. (2019). Archaeological assessment reveals Earth's early transformation through land use. *Science, 365*(6456), 897–902.

Stonich, S. (1989). The dynamics of social processes and environmental destruction: A Central American case study. *Population and Development Review, 15*, 269–295.

Tachibana, T., Nguyen, T. M., & Otsuka, K. (2001). Agricultural intensification versus extensification: A case study of deforestation in the northern-hill region of Vietnam. *Journal of Environmental Economics and Management, 41*(1), 44–69.

Talbot, L. M. (1986). Demographic factors in resource depletion and environmental degradation in East African rangeland. *Population and Development Review, 12*(3), 441–451.

Taylor, M. J., Aguilar-Stoen, M., Castellanos, E., Moran-Taylor, M. J., & Gerkin, K. (2016). International migration, land use change and the environment in Ixcan, Guatemala. *Land Use Policy, 54*, 290–301.

Thais, V., Harb, A., Bruner, A., da Silva Arruda, V., Ribeiro, V., Costa Alencar, A., Escobedo Grandez, A., Laina, A., & Botero, R. (2020). A better Amazon road network for people and the environment. *Proceedings of the National Academy of Sciences, 117*(13), 7095–7102.

Tiffen, M., Mortimore, M. J., & Gichuki, F. (1994). *More people, less erosion: Environmental recovery in Kenya.* Wiley.

Tilman, J., Balzer, C., Hill, J., & Befort, B. (2011). Global food demand and the sustainable intensification of agriculture. *Proceedings of National Academy of Sciences, 108*, 20260–20264.

Turner, B. L., Hanham, R., & Portararo, A. (1977). Population pressure and agricultural intensity. *Annals of the Association of American Geographers, 67*(3), 384–396.

Turner, B. L., Hyden, G., & Kates, R. (Eds.). (1993). *Population growth and agricultural change in Africa.* University of Florida Press.

von Thunen, J. (1826). *The isolated state.* Translated to English, 1966.

Walker, R., Moran, E., & Anselen, L. (2000). Deforestation and cattle ranching in the Brazilian Amazon: External capital and household processes. *World Development, 28*(4), 683–699.

Wang, Y., Bilsborrow, R., Tao, S., Chen, X., Sullivan-Wiley, K., Huang, Q., Li, J., & Song, C. (2020). Effects of payments for ecosystem services programs in China on rural household labor allocation and land use: Identifying complex pathways. *Land Use Policy.* In Press.

Whitten, A. J. (1987). Indonesia's transmigration program and its role in the loss of tropical rain forests. *Conservation Biology, 1*(3), 239–246.

Willy, D. K., Muyanga, M., & Jayne, T. (2019). Can economic and environmental benefits associated with agricultural intensification be sustained at high population densities? A farm level empirical analysis. *Land Use Policy, 81*, 100–110.

Winterhalder, B. P., & Thomas, R. B. (1978). *Geoecology of southern highland Peru: A human adaptation perspective.* University of Colorado, Institute of Arctic and Alpine Research.

World Bank. (1988). *Indonesia: The transmigration program in perspective* (A World Bank Country Study). The World Bank.

World Bank (2010). *Africa Migration Project.* microdata.worldbank.org.mrs

World Resources Institute. (2019, July). *Creating a sustainable food future: A menu of solutions to feed nearly 10 billion people by* 2050 (Final Report). .

Yost, M. A., Sudduth, K., Walthall, C., & Kitchen, N. (2019). Public-private collaboration toward research, education and innovation opportunities in precision agriculture. *Precision Agriculture, 20*(1), 4–18.

Zimmerer, K. (2013). The compatibility of agricultural intensification in a global hotspot of smallholder agrobiodiversity (Bolivia). *Proceedings of the National Academy of Sciences, 110*(8), 2769–2774.

Population and Energy Consumption/Carbon Emissions: What We Know, What We Should Focus on Next

19

Brantley Liddle and Gregory Casey

Abstract

This paper surveys the literature on the demographic determinants of energy use and carbon emissions. The literature demonstrates that demographic patterns are important drivers of environmental outcomes. In particular, age structure, household composition, household size and population density all correlate strongly with energy use and, thus, carbon emissions. In addition, life-cycle and cohort effects appear to be important (e.g., millennials tend to consume less fuel from driving; younger homeowners are more likely to invest in technologies like rooftop solar than older homeowners are). In addition to reviewing existing work, we also make recommendations for future research. Our recommendations include empirical best practices, suggestions for new variables like

education, and a greater focus on the supply of energy. The existing literature is already quite extensive. Thus, we also stress the need for future work to integrate existing findings with broader models to help inform policy decisions and improve prediction analyses. Finally, we stress the possibility of a feedback from energy use and carbon emissions to demographic outcomes.

Introduction

Not surprisingly, the main way that people contribute to anthropogenic carbon emissions is by consuming energy services that use carbon-based fuels (e.g., gasoline/diesel and electricity that is generated from coal or natural gas). Demographic factors like age structure, household composition and size, and population density can impact energy consumption demand—particularly in the residential and transport sectors. Probably the most widely understood/studied path for demography to influence such demand is through life cycle effects, i.e., people have different consumption behaviors at different times in their lives (e.g., young adults, parents, retirees).

B. Liddle (✉)
Energy Studies Institute, National University Singapore, Singapore, Singapore

Bethesda, MD, USA
e-mail: btliddle@alum.mit.edu

G. Casey
Department of Economics, Williams College, Williamstown, MA, USA
e-mail: gpc2@williams.edu

© Springer Nature Switzerland AG 2022
L. M. Hunter et al. (eds.), *International Handbook of Population and Environment*, International
Handbooks of Population 10, https://doi.org/10.1007/978-3-030-76433-3_19

Perhaps less understood, but potentially important (particularly so for projections) are cohort effects. The idea of cohort effects is that adults of similar age, but who were born at different times, might exhibit different behavior because they were exposed to different external factors. Age and cohort effects could be important for two reasons: (1) age groups/cohorts could have different attitudes towards the consumption of energy services; and (2) age groups/cohorts could have different attitudes towards the adoption of new, more energy efficient or renewable technologies. So, in addition to influencing carbon emissions through energy demand/consumption (i.e., reason 1), people have the ability to impact the energy mix at a more grass-roots level (i.e., reason 2) by, for example, investing in rooftop solar panels for their homes or by purchasing an electric vehicle (if the national electricity grid already employs substantial non-fossil fuels).

The following section (section "Demographic determinants of energy use/carbon emissions") discusses the current state of knowledge on how demographic variables/processes impact energy consumption and carbon emissions. We focus on several lines of inquiry of demographic determinants. We start by considering work on how several demographic factors impact energy demand, namely: (i) age structure and household composition; (ii) population density; (iii) household size; and, (iv) cohort and life-cycle effects. We conclude section "Demographic determinants of energy use/carbon emissions" by summarizing research on how those demographic factors may influence the adoption of energy-savings/carbon-reducing technologies.

After surveying the existing literature, we make numerous suggestions for future research in section "Moving forward". We begin this discussion by reviewing empirical best practices. Most importantly, we argue that the STIRPAT (Stochastic Impacts by Regression on Population, Affluence, and Technology) regression framework has largely outlived its usefulness. In particular, it is well understood that under certain conditions the elasticity of carbon emissions (or energy use) with respect to population size should be one. Also, ample empirical evidence suggests that the elasticity of carbon emissions with respect to income is less than one. On their own, further documentation of these facts contributes little to the state of existing knowledge.[1]

Most of our suggestions on *what to do* in future work focus on moving beyond the standard regressions used in the literature.[2] Given the ample existing literature, researchers already understand well the partial correlations embedded in standard regressions. However, relatively little work has been done to integrate the energy-demography literature with other literatures in order to inform broader questions. Given the convincing evidence that demographic variables affect energy use and carbon emissions, understanding demographic forces is important when designing energy policy or predicting future energy use. Such analyses are likely to require combining existing work with integrated assessment models (IAMs) or undertaking multiple equation modeling. It is also likely to require greater attention to the supply of energy, instead of just demand for energy, which is the focus of the existing literature.

The existing literature focuses almost exclusively on how demographic variables affect energy use/carbon emissions. However, there is good reason to believe that energy use—and resulting carbon emissions—will affect demographic variables. These feedback loops are important for fully understanding the relationship between demography and energy use.

We draw some final conclusions in section "Conclusion". We hope that this paper will be of value to future researchers in extending our knowledge of the complex relationships between demography and energy use.

---

[1] Based on existing studies, we also argue that population density is a better predictor of energy use than urbanization.

[2] We do suggest that education has been an overlooked demographic variable in the existing literature.

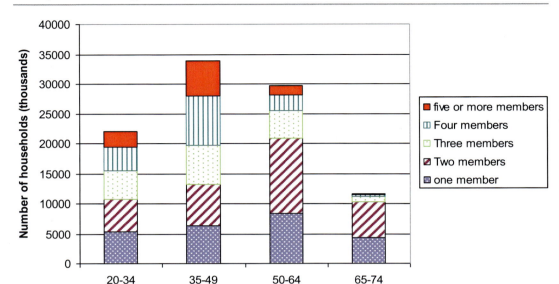

Fig. 19.1 Number of households in five different household-size groups by age of household head for the US in 2007. Data are from the US Census Bureau. (Reprinted by permission from Springer Population and Environment Age-structure, urbanization, and climate change in developed countries: Revisiting STIRPAT for disaggregated population and consumption-related environmental impacts, B. Liddle and S. Lung 2010.)

Demographic Determinants of Energy Use/Carbon Emissions

Age Structure and Household Composition[3]

In general, age structure matters because: (i) people in different age groups or at different stages of life have different levels of economic activity and resulting energy consumption; and (ii) the age of household head is associated with size of household (younger and older/retired-age adults typically have smaller households), and larger households consume more energy in aggregate, but less per person than smaller households. Early work demonstrating such relationships employed micro- (i.e., household-) level data to show that activities like transport and residential energy consumption vary according to age structure and household size.

For example, O'Neill and Chen (2002) showed how both residential and transportation energy consumption per capita differ nonlinearly when the age of householder is decomposed at 5-year intervals for US data. Transportation follows an inverted-U type shape, whereas residential energy consumption tends to increase with the age of householder—but at a non-constant rate. To some degree these consumption patterns reflect (i) the association of age of household head with size of household, and (ii) the fact that large households consume more energy in aggregate, but less per person, than smaller households. Figure 19.1 illustrates the first point, showing the breakdown of the number of households of various sizes by age of household head for the US in 2007. (The second point will be discussed further below and illustrated in Fig. 19.2.) Figure 19.1 indicates that

---

[3] Text from this section has been reprinted by permission from Springer Population and Environment Age-structure, urbanization, and climate change in developed countries: Revisiting STIRPAT for disaggregated population and consumption-related environmental impacts, B. Liddle and S. Lung 2010 as well as from Springer Population and Environment Impact of population, age structure, and urbanization on greenhouse gas emissions/energy consumption: Evidence from macro-level, cross-country analyses, B. Liddle, 2014.

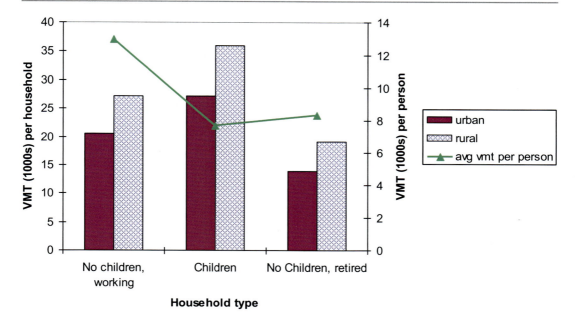

Fig. 19.2 Vehicle Miles Travelled (VMT) for three household types: (1) those working but without children, (2) those with children, and (3) those retired and without children. Left axis shows *VMT (in thousands) per household* and also differentiates between urban and rural households (VMT per urban/rural household is indicated by the bars). Right axis shows *VMT (again in thousands) per person* for the three household types (VMT per person is indicated by the line). Data are from US Department of Energy, Energy Information Agency, 2001. (Reprinted by permission from Springer Population and Environment Age-structure, urbanization, and climate change in developed countries: Revisiting STIRPAT for disaggregated population and consumption-related environmental impacts, B. Liddle and S. Lung 2010.)

large households (4 people or more) are predominately headed by people in the 35–49 age cohort, and that the vast majority of households headed by those aged 50 and older are either single or two-person households (the estimated[4] average household size for the four different household head age groupings shown in the figure are 2.7, 3.1, 2.2, and 1.8, respectively).

Liddle (2004), like O'Neill and Chen considering US data, showed that average miles driven per person decline as the number of household members increases, and, in small households (one to two people) at least, when controlling for the size of household, 20–30 year-olds drive more per person than other age-groups. Prskawetz et al. (2004) demonstrated that similar relationships exist for Austria. Figure 19.2 shows average vehicle miles traveled (US data from 2001), both per household (left axis) and per person within a household (right axis) for three household types: (i) working adults without children, (ii) households with children, (iii) retired adults without children.[5] The figure also differentiates between urban and rural households. Figure 19.2 illustrates a number of important generalizations: (i) households with children drive more—because they are larger—but drive less per person than smaller households; (ii) among households without children (typically one or two adults), younger, working-age households drive more; and (iii) keeping household types constant, rural households drive more than urban ones.

---

[4]This number is estimated because the last household size category supplied in the data is "seven or more" members, i.e., the number of households with exactly eight, nine, etc., members is not explicitly known from the data.

[5]The working or retired designation is merely to distinguish between two household types that do not include children. The data set used does not otherwise allow for disaggregations by employment status.

More recently, studies using cross-country, macro-level data have shown a similar age-structure relationships. For example, Liddle (2011) determined that for transport energy consumption, young adults (20–34) were intensive consumers, whereas the other age groups had negative coefficients; yet, for residential electricity consumption, age structure had a U-shaped impact: the youngest and oldest age groups had positive coefficients, while the age middle groups had negative coefficients. Using different methods, Liddle and Lung (2010) similarly found that, compared to younger ones, older age groups had a lower elasticity for CO_2 emissions from transport (i.e., negative for ages 35–64 but positive for ages 20–34), yet a higher elasticity for residential electricity use (i.e., negative for ages 35–49, but positive for ages 50–64). Okada (2012), using different methods and models than Liddle (2011), confirmed the result that a larger share of population over 65 is associated with lower CO_2 emissions from road transport.

Yet, age structure is less likely to directly impact national, aggregate carbon dioxide emissions; instead, those emissions should be heavily influenced by the size, structure, and energy intensity of the macro-economy (e.g., the presence and size of sectors like iron and steel and aluminum smelting), and by the technologies used to generate electricity (i.e., coal vs. nuclear). Indeed, macro-level studies that have considered age structure typically have used the World Bank definitions/data, i.e., the share of people aged less than 15, aged 15–64, and aged over 64, and typically have examined aggregate, economy-wide carbon emissions. Such studies mostly have found those age structure variables to be insignificant (see Liddle, 2014, Table 1). The finding of insignificance perhaps is not so surprising given the high level of aggregation for emissions and that it is not clear that such a broad definition of age structure (15–64) should have much explanatory power.

A few papers that have considered levels of disaggregation that approximate life-cycle behavior, such as family or household size, uncovered significant, complex relationships even for aggregate emissions. For example, Liddle and Lung (2010) uncovered a positive elasticity for young adults (aged 20–34) and a negative elasticity for older adults (aged 35–64). Menz and Welsch (2012), also analyzing aggregate carbon emissions, estimated differential age-structure elasticities too, with the middle ages (30–59) having negative elasticities. Menz and Welsch considered cohort effects as well, and determined that people born after 1960 are associated with increased carbon emissions. It is worth noting that the vast majority of research on age structure and household competition has focused on OECD countries, and it is not yet clear whether these findings will generalize countries with different levels of development.

Population Density[6]

Higher urban density[7] leads to smaller housing units, inter-unit insulation/space-conditioning savings, shorter commuting distances, and increased attractiveness of public transport and non-motorized personal transport (walking, biking). According to Rickwood et al. (2008) the impact of urban form on residential energy use is much less understood than the impact of urban form on transport energy use. Much of the earlier work on urban density focused on its relationship with lower levels of transport energy consumption (e.g., Newman & Kenworthy, 1989; Kenworthy & Laube, 1999). Some other studies, also employing city-level data, determined a relationship between urban density and lower levels of electricity consumption in buildings (e.g., Lariviere & Lafrance, 1999), and

---

[6]Some text from this section has been reprinted by permission from Springer Population and Environment Impact of population, age structure, and urbanization on greenhouse gas emissions/energy consumption: Evidence from macro-level, cross-country analyses, B. Liddle, 2014.

[7]Ewing and Cervero (2001) argued that urban form—i.e., the co-location of housing, employment, and services—is more important in reducing transport demand than density. Yet measures of urban form are hard to collect; hence, density is the default measure of many studies (Rickwood et al. 2008).

lower levels of greenhouse gas emissions (e.g., Marcotullio et al., 2012).[8]

While most work on density has employed data at a localized level, a few national-level studies have considered population density. Among those few studies, Hilton and Levinson (1998) found a significant, negative relationship between national population density and gasoline use in a study of 48 (developed and developing) countries. Similarly, Liddle (2004) found a significant, negative relationship between national population density and per capita road energy use in OECD countries. By contrast, early studies on electricity or energy consumption per capita found a small positive to insignificant effect for national population density (Jones, 1989; Burney, 1995; Parikh & Shukla, 1995). It is important to note that national-level population density is only weakly correlated ($\rho = 0.35$) with urban density (Liddle, 2013).

While the process of urbanization has led to a higher percentage of people living in cities, much of this growth has generated sprawl (and associated issues like congestion and fragmentation), rather than create benefits from agglomeration (Shaker, 2015; Kyatta et al., 2013).[9] Some recent work on environmental impact has recognized this diversity of density (i.e., highly dense urban, suburban, rural). For example, Jones and Kammen (2014) calculated household carbon footprints for zip codes, cities, counties, and metropolitan areas from US national survey data. In general, population density lowers carbon footprints by reducing the size of homes and lowering vehicle ownership. However, Jones and Kammen (2014) uncovered an inverted-U relationship between population density and household carbon footprints where emissions increase with density until a population density of 3000 persons per square mile is reached. At lower population density, a positive correlation between density and emissions is found, in part, because suburban areas tend to be wealthier than rural areas and so those suburban areas produce more emissions.

Otsuka (2017, 2018) considered Japanese data and used stochastic frontier analysis to estimate residential energy and electricity demand functions and to analyze the determinants of energy efficiency in those sectors. Stochastic frontier analysis is a cross-sectional based, econometric method that has been widely applied to measure relative efficiency in a variety of settings. Otsuka found a substantial role for population density for improving efficiency in both analyses. He found as well a small, positive relationship between household size and efficiency, which confirmed the idea that sharing common spaces (e.g., living rooms, kitchens) leads to per capita energy savings. For the case of residential energy, Otsuka (2018) found a nonlinear relationship between population density and energy efficiency, whereby a certain threshold of density must be reached for population agglomeration to improve efficiency. This last finding is in line with the work of Shaker (2015), Kyatta et al. (2013), and Jones and Kammen (2014).

Most of the studies on cities and transport energy/emissions have focused (effectively) on intra-city travel (i.e., denser cities reduced the need for individual motorized transport). Lim et al. (2019) considered the impact of national-level spatial distribution of cities on the resulting intercity transport-related CO_2 emissions by employing NASA's Gridded Population of the World data. They found that among urbanized countries, countries with several medium-sized cities have lower national transportation CO_2 emissions than spatially polarized urbanized countries (e.g., ones that have only a few large cities)—a potentially important finding given the proclivity in many developing countries to form mega-cities.

---

[8] Studies employing city-level data, e.g., Kenworthy and Laube (1999); Marcotullio et al. (2012), are among the few examples where information from non-OECD countries has been considered.

[9] Indeed, Liddle (2013) determined that the national-level share of people living in urban areas was *negatively* correlated with the density of large cities located in those respective countries.

Household Size

MacKellar et al. (1995) advocated for the consideration of the household as the unit of analysis. Outlining the importance of such a consideration, Bradbury et al. (2014) noted that household size is decreasing in all countries, and that in nearly every nation for which they had data, the number of households grew faster than the population. Yet, household size can be a difficult variable to collect for an empirical panel analysis; thus, few macro-level/cross-country studies have heeded MacKellar et al.'s advice— two exceptions are Cole and Neumayer (2004) and Liddle (2004). Liddle (2004) found that larger households were associated with lower levels of per capita road energy use in OECD countries, while Cole and Neumayer (2004) found that larger households were associated with lower levels of aggregate carbon emissions in both developed and developing countries.

Some recent work has used the language of sharing to compare the impacts of household size and population density. For example, Underwood and Fremstad (2018) used US household expenditure data and calculated the carbon emissions savings of sharing economies for both household size and population density. They found that household size economies come primarily from savings on residential energy. The savings for suburban areas come from transport since people living in such areas have similar sized houses as people living in rural areas, but suburbanites drive less than their rural counterparts. People living in dense urban areas save about equally on residential energy and transport, suggesting that in addition to having smaller living spaces, people in dense urban areas benefit from inter-household energy savings, too.

Fremstad et al. (2018) combined that understanding (of shared economies/savings) with the US trends of declining household size and increased urbanization, to estimate a net increase in per capita carbon emissions of 6% from those two competing forces. Hence, they argued that if household size continues to decline in rural and urban settings, dense urbanization will need to occur at an accelerated pace to offset the carbon impacts of declining household economies.

Cohort vs Life-Cycle Effects

One needs to decompose age effects into life cycle effects—people at similar stages of life behave similarly—and cohort effects—sharing generation-specific experiences leads to similar behavior. While both macro and micro data can be used to demonstrate an age effect on energy use, micro data is better suited to separate life cycle from cohort effects.

Chancel (2014) analyzed French and US household data to consider both residential and transport energy use. He found a cohort effect for France (the 1930–1955 cohort consumed more than others) but not for the US. Bardazzi and Pazienza (2018) focused on private transport in Italy. While they found that energy consumption falls with age, they determined an inverted-U cohort effect: those born (relatively soon) after WWII had the highest fuel consumption; and Millennials had lower car-related energy consumption, perhaps because Millennials have greater concern for climate change and greater interest in ride-sharing alternatives to private ownership. The Bardazzi and Pazienza (2018) and Chancel (2014) findings (at least for Europe) are in line with the ideas that different generations/cohorts develop different energy cultures, and that the younger generations may be developing less energy intensive cultures (Stephenson et al., 2015; Tilley, 2017).

The importance of cohort or generational effects are often invoked when discussing Millennials (or adults born from 1980 on). Indeed, much has been written, often in popular press and often based on anecdotal evidence, about Millennials being different from previous generations.[10] For example, Millennials prefer public transport and

---

[10]Such examples include: "Why Car Companies Can't Win Young Adults," Fortune 2013; "NOwnership, No Problem: Why Millennials Value Experience Over Owning Things," Forbes 2015; "The Many Reasons Millennials are Shunning Cars," Washington Post 2014; "Social

ride sharing to car ownership, perhaps because of a desire to engage with social media while traveling. While Millennials may indeed be different (much like the Baby Boomers—those born between 1946 and 1964—are/have been different from the generations they succeeded), it is possible, too, that the timing of events like the Great Recession have caused Millennials merely to delay decisions like marriage and car ownership.

Several recent papers have employed US household survey data in an attempt to shed light on this question. Klein and Smart (2017) analyzed expenditure data and determined that today's young adults (Millennials) have lower rates of car ownership than previous generations did at the same age. However, they found that today's young adults that are economically independent of their parents have slightly higher rates of car ownership, after controlling for income and wealth. Similarly, Kurz et al. (2018) found that there is little difference in the consumption and spending patterns of Millennials compared to earlier generations once age, income, and family size and marital status are considered. Lastly, Knittel and Murphy (2019) determined that after controlling for confounding variables (e.g., age and wealth), Millennials had more vehicle miles traveled than Baby Boomers. While Knittel and Murphy found evidence that Millennials are more likely to live in urban areas and less likely to marry before 35, after controlling for age, Millennials tended to have larger families. The net effect on vehicle ownership of these life-choice differences was about 1%.

The takeaway in much of this recent work is to temper the belief that Millennials will continue to have different consumption patterns from previous generations. Hence, this provides an example of the importance in taking a deeper dive into cohort/generational differences when using demographic trends to project energy consumption. Yet, it seems likely that this life-cycle vs. cohort effect issue will continue to be worth investigat-

ing. It is certainly worthwhile to consider, for example, that people currently over 65 or those who will be in the future may behave differently from those who were previously in that age group. This consideration is suggested as an area of continued research since such potential behavioral differences will be only just emerging.

Demographic Factors and the Adoption of Energy-Savings/Carbon-Reducing Technologies

There is a developing, related literature that seeks to determine whether demographic factors (e.g., life-cycle or cohort based) and various other factors influence the adoption of certain energy technologies. These technologies can be energy efficiency/energy savings or could impact the energy mix through factors like rooftop solar installations. For example, Barnicoat and Danson (2015) argued that the elderly, as people more likely to stay at home than younger people, are important candidates for demand side management approaches like smart metering and in-home displays. Yet, they found that many older people in the UK had no interest in switching to nor little knowledge of smart technology.

In addition to home-owners and wealthier people being more likely to invest in home energy technologies, there is evidence that for such investments age and family size matter as well. Older household heads may be less likely to adopt energy efficient technologies because the expected rate of return is lower than for households with younger heads. This line of reasoning is supported by the findings of Curtis et al. (1984), Walsh (1989), Poortinga et al. (2003) and Mahapatra and Gustavsson (2008). On the other hand, younger households may be more likely to move and hence be less inclined to invest in energy efficiency improvements, in particular, if these measures become an integral part of the built environment. Combining these perspectives, middle aged households should

---

Media Trumps Driving Among Today's Teens," Forbes 2012.

be most likely to adopt capital-intensive energy efficiency measures (e.g. Mills and Schleich, 2010), particularly if the technologies are structurally linked to the building.

There is evidence that the impact of age on technology adoption/investment depends on the technology. For example, Willis et al. (2011) used UK household survey data and found that households with members over 65 years of age are much less likely to adopt discretionary micro-generation technologies—e.g., solar thermal, solar photovoltaic, and wind power as supplements to the primary energy system(s) of the house—compared to the rest of the population. Ameli and Brandt (2015) analyzed OECD survey data to determine that investments in light bulbs, heat thermostats, thermal insulation, and energy-efficient windows correlate positively with age, while the likelihood of investing in heat pumps falls with age.

Ameli and Brandt (2015) also found that family size was positively related to the likelihood of investing in solar panels and light bulbs, not necessarily in other technologies. Relatedly, Mills and Schleich (2012) determined that families with young children were more likely to adopt energy-efficient technologies and practice conserving energy in the home. Lastly, Mills and Schleich (2012) found that more elderly households were motivated to save electricity merely for financial reasons rather than to reduce carbon emissions.

Araujo et al. (2019) studied electric vehicle (EV) and solar technology diffusion in New York State using geospatial data and techniques. They found a positive correlation between the adoption of solar energy technology and two age groups, 30–44 and over 59, and a negative correlation between EV ownership and the over 59 age group. Araujo et al. could not determine whether those findings reflected life-cycle-type effects or a generational/cultural cohort tendency. They noted a negative correlation between EV's and larger households sizes, which they attributed to range anxiety, i.e., larger households because of their greater demand for mobility were more concerned about the limited range that electric vehicles may have.

Moving Forward

What Not to Do[11]

Before making positive recommendations for future work, we discuss two areas of improvement that involve not repeating weaknesses of the past. The first suggestion focuses on the future/continued value of the well-used STIRPAT modeling framework, the second focuses on the suitability of the over-used variable share of people living in urban areas (i.e., urbanization).

What Not to Do—STIRPAT

Analyses examining population's impacts on the environment often employ Dietz and Rosa's (1997) STIRPAT (Stochastic Impacts by Regression on Population, Affluence, and Technology) framework. STIRPAT builds on the IPAT/impact equation of Ehrlich and Holdren (1971):

$$I = P \times A \times T \tag{19.1}$$

where I is aggregate environmental impact, P is total population, A is affluence or consumption per capita, and T is technology or impact per unit of consumption. Dietz and Rosa (1997) proposed a flexible, log-linear, regression framework that allows for hypothesis testing:

$$\ln I_{it} = \alpha_i + \gamma_t + \beta_1 \ln P_{it} + \beta_2 \ln A_{it} + \beta_3 \ln T_{it} + \varepsilon_{it} \tag{19.2}$$

where subscripts it denote the ith cross-section and tth time period. The constants α and γ are the country or cross-sectional and time fixed effects, respectively, and ε is the error term. Eq. 19.2 allows the coefficient on the population term to be different from unity—a feature that distinguishes the STIRPAT approach from other reduced form

[11] Some text from this section has been reprinted by permission from Springer Population and Environment Impact of population, age structure, and urbanization on greenhouse gas emissions/energy consumption: Evidence from macro-level, cross-country analyses, B. Liddle, 2014.

socio-economic environmental models that have the dependent variable (e.g., emissions, energy consumption) in per capita terms and do not include aggregate population among the explanatory variables.

We might not expect, however, population's elasticity to be different from one anyway, since as O'Neill et al. (2012) argued, "... if all other influences on emissions are controlled for, and indirect effects of population on emissions through other variables are excluded, then population can only act as a scale factor[,] and its elasticity should therefore be 1." After examining a large number of STIRPAT analyses, Liddle (2014) made two observations regarding the estimation of population's coefficient (β_1 in Eq. 19.2). First, analyses employing time-variant data produced a great variance in population coefficient estimations—sometimes significantly greater than one, sometimes significantly less than one, and other times insignificant (see Table 1, Liddle 2014). Second, by contrast, analyses of pure cross-sections typically estimated population elasticities very near one—a finding that was true even for studies considering different dependent variables (e.g., fuelwood consumption by Knight and Rosa 2012 as opposed to carbon emissions or energy consumption) or different units/scales of analysis (e.g., US county-level data in Roberts, 2011; international city-level data in Liddle, 2013 as opposed to national-level data).

Motivated by the question of the "true" population coefficient, Liddle (2015) performed a substantial robustness exercise on the STIRPAT framework and determined that the estimated (mean) population coefficient was greater than one and was unstable/inconsistent—i.e., it varied considerably depending on the panel (OECD vs. non-OECD), method (long-run vs. short-run/first difference estimation), and time-span considered; however, its accompanying standard errors were large, and so, the elasticity was typically not statistically different from one, nor statistically different between developed and developing countries. Furthermore, the lack of stability of the population coefficient over time was not evidence that the coefficient had changed—the

sensitivity analysis revealed no evidence that the size, significance, or sign of the population coefficient may have changed over-time (e.g., from 1970–1990 to 1990–2006). Rather, the more extreme estimated values (i.e., particularly large or insignificant estimations) typically occurred whenever the time span was shortest (e.g., 1971–1990, 1975–1995, 1980–2000, and 1985–2006).[12] Jorgenson and Clark (2010), using different methods, similarly concluded that their population coefficient estimations did not change over time.

Some researchers employing the STIRPAT framework reported a significant positive coefficient for population as a substantial finding. Yet, that total population would be highly, positively correlated with aggregate emissions or energy consumption seems expected and does not need to be repeatedly confirmed. Another question STIRPAT is used to answer is whether population or income has the larger effect. If we accept the idea that unity is a reasonable/expected coefficient for population, the answer to this question is well understood. Liddle (2015) estimated a carbon emissions-income relationship of statistically significantly less than one for OECD countries, and the energy-GDP elasticity has been estimated to be less than one for both OECD and non-OECD countries in several recent studies (Burke & Csereklyei, 2016; Csereklyei et al., 2016; Liddle & Huntington, 2020). In other words, increases in population led to more or less proportional increases in emissions/energy consumption, whereas, increases in income led to less than proportional increases.

Liddle (2015) hypothesized that one explanation for the population coefficient instability finding in that paper and the general wide dispersion of results found elsewhere is the challenging time series properties of population. Variables like GDP per capita, carbon emissions per capita, and energy consumption per capita may only need to be detrended to be transformed into single/constant mean variables; however, population likely

---

[12]See Liddle (2015) for details of the models, estimators, and regression diagnostics.

needs to be first difference, i.e., its growth rate may have a constant mean. Hence, Liddle (2015) recommended that modelers divide Eq. 19.2 by population, and thus, convert the dependent variable into a per capita term.

Another related issue with the STIRPAT approach is how to represent the T term. This term is often proxied by economy-wide energy intensity or energy consumption/GDP (when the dependent variable is carbon emissions). In addition to being in the spirit of the IPAT identity, the inclusion of energy intensity sometimes is supported by the idea that it represents energy efficiency. Yet, energy intensity has been demonstrated to be a poor proxy of energy efficiency (e.g., Filippini & Hunt, 2011). (Indeed, economy-wide energy intensity is a function of underlying energy efficiency and the structure of the economy among relatively energy intensive activities and relatively less intensive activities.) However, more problematic is the fact that carbon emissions are not observed, but rather are estimated from energy consumption data (specifically, fossil fuel consumption data). The relationship between fossil fuels and carbon emissions is known, well-understood, and explained by chemistry—it is of no interest for socio-economic modelling. In addition to including as an explanatory variable, one that is related to the dependent variable by data design, having energy/GDP on the right-hand-side has been demonstrated to introduce further statistical/modeling problems (Jaforullah & King, 2017).

Dietz and Rosa's STIRPAT framework has unquestionably helped to spur a substantial amount of scholarship. However, the STIRPAT approach has likely outlived its usefulness in further elucidating population-environment interactions. As discussed in the 'What to do' section, moving to multiple equation modeling or using integrated assessment models can achieve similar goals with a stronger theoretical underpinning.

What Not to Do—Urbanization

The *process* of urbanization—the reorganization of consumption and production in densely populated cities rather than in sparsely populated rural areas—is undoubtedly important. However, the *variable* urbanization—the share of a population living in areas classified as urban—likely adds little to our understanding of carbon emissions or energy consumption once income/GDP per capita and/or population density is controlled for. Unfortunately, urbanization continues to be among the most examined demographic-based variables despite the suggestion in Liddle (2014) that the share of people living in urban areas had little potential to further our understanding of energy demand/carbon emissions. Indeed, a simple keyword search on ScienceDirect (conducted on June 7, 2019) for "urbanization, carbon emissions" found 8601 papers from 2015. (Of course not all of these papers were regressing carbon emissions on urbanization.)

Urbanization and GDP per capita not only are highly correlated, they probably are interrelated; however, urbanization is likely an indicator/result of GDP growth/development/progress rather than a catalyst (Henderson, 2003). According to Liddle (2014), urbanization, like GDP per capita, typically is found to have a positive relationship with carbon emissions/energy consumption. Yet, that correlation between urbanization and emissions/energy is not necessarily causal, but rather, both are influenced by the same factors (e.g., the transformation from an agricultural economy to one dominated by industry and services). In other words, people move from rural areas to urban areas as agriculture becomes mechanized, and manufacturing and services—which tend to be located in urban areas—become the major employer. So, industrialization causes urbanization/rural-to-urban migration—at the same time that industrialization is fuelled by consumption of modern energy. Relatedly, Liddle and Lung (2014) found more evidence that electricity consumption Granger-caused urbanization rather than the other way around.

What to Do

The energy-demography literature has made considerable progress on correctly signing and estimating the partial correlations embedded in regressions models. Future work can build on this solid foundation to help inform policy and prediction analyses. While regression coefficients are helpful tools, they are often insufficient on their own for answering important questions in environmental social science. Combining insights from the energy-demography literature with integrated assessment models (IAMs) or moving to multiple equation modeling is likely to be especially fruitful. It may also be beneficial to consider education as part of the demographic structure and to incorporate feedbacks from energy use to demographic change.

What to Do – Prediction

Existing results from the economic-demographic literature contain many insights for predicting future carbon emissions and understanding the relationship between demography and energy use. In particular, a significant amount of effort has gone into predicting future demographic trends, with considerable success (e.g., Raftery et al., 2012; Lutz & Samir, 2011). When combining predicted trends with regression coefficients from the energy-demography literature, it is possible to (a) predict future emissions and (b) better understand how different demographic trajectories affect carbon emissions.

O'Neill et al. (2010, 2012) provided the first example of this type of analysis. They examined how different demographic trends affect carbon emissions using the Population-Environment-Technology (PET) model, a dynamic integrated assessment model (IAM).[13] The demographic scenarios are developed by the United Nations. The authors found that "slowing population growth could account for 16%-29% of the emissions reductions suggested to be necessary by 2050 to avoid dangerous levels of climate change."

Casey and Galor (2017) provided a complementary analysis that focused on the co-evolution of environmental and economic outcomes. They also examined the impact of moving between different UN scenarios for future demographic

trends. As noted above, STIRPAT regressions suggest that the partial impact of population size on carbon emissions is much larger than the partial effect of income. The authors combined this insight with an economic-demographic model for Nigeria developed by Ashraf et al. (2014). The model is specifically designed to incorporate all of the channels through which changes in demography affect economic outcomes. Combining this model with the STIRPAT regressions, Casey and Galor (2017) examined the impact of moving between the medium and low population projections from the United Nations. They found that slower population growth can simultaneously increase income per capita and decrease carbon emissions. Based on these results, the authors argued that demographic policies should be a part of climate change mitigation strategies.

There is much work to be done to expand on these results. First, Casey and Galor (2017) used a detailed model of economic outcomes, but only studied results in Nigeria. O'Neill et al. (2010, 2012) examined a much wider range of geographic outcomes, but in a less detailed economic model. Combing these two results would be an important next step in understanding how different demographic trends contribute to climate change. Of course, it is also necessary to examine alternate modelling assumptions and their impact on these results. As explained in the next section, it is especially important to extend these analyses to explicitly consider the impact of policy, rather than differences between exogenous demographic scenarios.

What to Do – Policy

The papers discussed above are explicitly interested in the effect of demographic *policies* on energy use and carbon emissions. Policy can be thought of as interventions – by government, international organizations or other entities – that move an economy from one demographic path to another. The papers discussed above compare exogenous demographic paths, creating opportunities for future work that explicitly focus on policy. Indeed, influencing energy and climate

---

[13] Integrated assessment models (IAMs) study the relationship between the economy and the climate.

change policy is an important 'end goal' for the energy-demography literature.[14]

There are several important steps that are necessary to move from comparing different demographic paths to analyzing the effects of specific policies. In all cases, these steps are likely to require combining empirical evidence from the energy-demography literature with IAMs or employing multiple equation modeling. The most obvious future step is matching specific policies to future demographic trends. For example, if contraceptives are subsidized, how much will this reduce fertility?[15] Will fertility reductions come from households with high or low energy use per person? How will this affect the future age distribution and population density of a society? In addition to understanding the consequences of policy, it is also important to think about costs. How might a government pay for subsidized contraceptives? How will any new taxes affect energy use and demographic structure?

When weighing the costs and benefits of policy, it will also be helpful to consider non-energy-related benefits. When considering climate change mitigation policy, economic outcomes may be especially important, since objections to climate change mitigation policy often focus on negative economic costs. Based on economic theory, there is good reason to believe that decreases in population growth – and concurrent changes in population structure – may boost income per capita, even as they decrease carbon emissions and energy use (Ashraf et al., 2014; Casey & Galor, 2017). Thus, demographic approaches to limiting energy use and carbon emissions may be beneficial.

What to Do – The Supply of Energy

As discussed above, the energy-demography literature generally focuses on the demand for energy. When thinking about prediction or policy analyses, however, consideration of energy supply is likely to be especially important. Most empirical papers in the energy-demography literature estimate cross-sectional regressions or panel regressions with time fixed effects. In other words, the identifying variation comes from comparing different countries at a particular point in time. Since energy prices are partially determined at the global level, these empirical designs essentially hold energy prices fixed. When predicting future energy use and carbon emissions, however, this assumption is no longer valid. Since energy prices respond to total energy use, changes in demography are likely to affect both the supply and the demand for future energy use. Thus, any prediction exercises or policy analyses will need to incorporate these effects.

Understanding the impact of demographic changes on the cost of energy is one reason why using IAMs with detailed representations of the energy sector is likely to be especially important (e.g., O'Neill et al. (2010, 2012). There is also scope to better understand this interaction using multiple equation modeling. For example, existing analyses of the impact of population structure on energy demand could be augmented with estimates of the impact of demographic structure and energy use on energy prices.

Relatedly, there is much we do not understand about the interaction between energy prices and demographic structure on the demand side. How strong are demographic predictors of energy use when controlling for energy prices?[16] Are some demographic groups more sensitive to changes in energy prices? There is also an existing literature on the impacts of demographic structure on overall economic productivity and technology adoption (e.g., Feyrer, 2007, 2008; Acemoglu & Restrepo, 2017). It may be particularly interesting to investigate the effect of demographic structure on the productivity of the energy extraction sector and the overall energy efficiency of the economy.

---

[14]It is important to note that the papers mentioned here are explicitly interested in *voluntary* changes in fertility and their impact on demographic structure and population size. See Das Gupta (2013) for a related discussion.

[15]The UN projections employed in existing work are not meant to correspond to different policy scenarios, although existing work uses those projections as proxies for differing levels of policy intervention.

[16]This issue is discussed in Liddle (2014).

What to Do – Multiple Equation Modeling

As explained above, the standard STIRPAT identity has been studied extensively, and it is no longer clear that future refinements of the standard single-equation regression will yield new insights. However, there is still a significant amount that could be learned from multiple equation extensions of the standard regressions. For example, the coefficient on population should be equal to one when "fully" controlling for technology (T). At the same time, relatively little is known about how population levels – or other demographic variables – affect technology. With multiple equation modeling, it would be possible to simultaneously estimate the standard STIRPAT regression, but also estimate how different demographic factors, including total population, affect T. This insight is crucial for determining how different demographic trajectories or policies will affect total energy use and carbon emissions. For example, if a policy affects fertility, age structure and total population will be altered. This will affect carbon emissions directly through the demographic change, but also indirectly through technology. The standard STIRPAT equation only gives the direct effect and is therefore incomplete.

What to Do – Expanding the Geographic Scope

As noted in section "Demographic determinants of energy use/carbon emissions", the study of demographic influences on energy demand and technology adoption tends to focus on developed countries. This is especially true outside of the STIRPAT literature. It is not clear whether the patterns related to age effects, cohort effects, population density, household size, and household composition are universal. They may depend on the level of development or differ across geographic regions. Given that much future demographic change will occur in developing regions, understanding the impact of demographic characteristics on energy use in developing countries may have significant consequences for global outcomes.

As a result, there would be considerable benefit to future studies that test whether the patterns discussed in this literature are similar in developing countries. This broader perspective would help inform both prediction and policy analyses. At the moment, such analyses require the assumption that demographic effects are similar for all countries. Analyses for developing countries are often hampered by a lack of available data. While the costs of undertaking this type of research are significant, the benefits are even greater.

What to Do – Education

In addition to the demographic variables already studied in this literature, future research can also examine the role of education. Education plays a large role in demographic research. For example, educated mothers tend to have fewer children (Lutz & Samir, 2011). Given the role that education plays in the wider demographic literature, it is somewhat surprising that the energy-demography literature has overlooked the role that education may play in energy consumption.[17]

Unlike the demographic variables discussed throughout this paper, the effect of education on energy use is likely to come through the production side of the economy. For example, educated individuals are more likely to work in service-based industries (Herrendorf et al., 2014), which tend to use less energy per unit of output (Schäfer, 2005). On the other hand, educated workers are more productive, leading to an increase in overall production.

On the demand side, it may be the case that educated individuals are more aware of the costs of consuming energy, especially fossil fuels (Zarnikau, 2003). Conversely, since educated individuals are likely to be wealthier, they may consume goods, like air travel, that are energy- or carbon-intensive. Given that education levels are predicted to rise arise the world (Roser and Nagdy, 2013), the potential for conflicting effects on energy use and carbon emissions makes future research in this area particularly valuable.

---

[17]For example, a recent study suggested that education is the true driver of 'demographic dividends,' an important concept in the study of economic growth and sustainability (Lutz et al., 2019).

What to Do – A Reverse Effect?

The energy-demography literature tends to focus on the impact of demography on energy use. However, there is also reason to believe that energy use – and especially fossil fuel energy use – will have a meaningful impact on demographic patterns. Climate change is likely to have widespread impact on both health and the economy (Auffhammer, 2018), Both health and the economy, in turn, have important impacts on fertility, mortality, and migration, key aspects of population growth and demographic change (e.g., Raftery et al., 2012). For example, significant empirical evidence shows that changes in temperature affect fertility, likely through channels related to reproductive health (Lam & Miron, 1996; Barreca et al., 2018). Similarly, a growing literature suggests that changes in temperature also affect migration (e.g., Cattaneo & Peri, 2016). These results suggest that climate change is likely to impact demographic structure. No existing work integrates these findings with the energy-demography literature discussed here. This feedback is likely to be particularly important in understanding the impacts of policy. As a result, this integration is a particularly important area for future research.

In addition to the direct impacts, adaptation to climate change might also involve changes in demographic structure. Casey et al. (2019) combined an existing economic-demographic model with estimates of the impact of climate change to show that climate change is likely to increase fertility and decrease education in countries located near the equator (Galor & Mountford, 2008; Desmet & Rossi-Hansberg, 2015). According to their theory, climate damages will lead to food scarcity, raising prices and wages. As the return to working in agriculture increases, workers reallocate towards this sector. Since agriculture tends to reward educational investments less, the return to education decreases. This incentivizes parents to invest less in the education of their children and to use the saved resources to increase fertility. Shayegh (2017) extended this model to account for migration. These papers are explicitly preliminary. There is much future work to be done to fully understand how climate change and energy use affect demographic structure. Moreover, as in the case of direct impacts, there is much work to be done to integrate these insights with the existing energy-demography literature.

The small existing literature mostly focuses on how climate change affects demographic structure. There is ample evidence, however, that household income also affects demographic structure. Changes in energy prices and the implementation of climate change policies may also affect demographic structure, since they impact effective household income. Further understanding this feedback is a fruitful area for future research as it will inform both prediction exercises and policy analyses.

Conclusions

An extensive literature has documented that demographic factors are important determinants of energy use and carbon emissions. We summarize this literature and provide suggestions for future work. At the most basic level, carbon emissions and energy use increase one-for-one with the size of the population, holding all other variables constant. Age structure, household size, and population density all appear to be strong correlates of energy use at the micro level, although it can sometimes be difficult to detect effects at more aggregate levels. Despite all of the work that has already been done, there is considerable opportunity for future research to further explore the relationship between energy use and demography. Relatively straightforward examples include (i) investigating the relationship between education and energy demand, (ii) understanding how energy prices and demographic factors interact in determining energy demand, and (iii) disentangling age and cohort effects.

On a larger scale, there is much work to be done pushing beyond the estimation of regression coefficients, which has been the primary focus of the literature. For example, there is good reason to believe that the insights developed in this literature can help design effective climate change mitigation and energy security policies. For example, how would subsidizing contraception affect carbon emissions? Answering this question (and related questions) requires fully understanding the relationship between demography and energy use. Such analyses are likely to require multiple equation modeling or the use of integrated assessment models. Among other things, these tools will be helpful in uncovering the relationship between demography, income, and technology, as well as the relationship between demography and energy supply.

The existing literature focuses almost exclusively on how demography drives energy use. However, there is reason to believe that energy use and carbon emissions will also exert influence over demographic structure. At the moment, very little is known about this 'feedback.'

For all of these reasons, we think that there is much to learn from the existing literature on the energy-demography nexus, and there is much more exciting work to be done in this field.

References

Acemoglu, D., & Restrepo, P. (2017). Secular stagnation? The effect of aging on economic growth in the age of automation. *American Economic Review, 107*(5), 174–179.

Ameli, N., & Brandt, N. (2015). Determinants of household's investment in energy efficiency and renewables: Evidence from the OECD survey on household environmental behaviour and attitudes. *Environmental Research Letters, 10.*

Araujo, K., Boucher, J., & Aphale, O. (2019). A clean energy assessment of early adopters in electric vehicle and solar photovoltaic technology: Geospatial, political and socio-demographic trends in New York. *Journal of Cleaner Production, 216*, 99–116.

Ashraf, Q. H., Weil, D. N., & Wilde, J. (2014). The effect of fertility reduction on economic growth. *Population and Development Review, 39*(1), 97–130.

Auffhammer, M. (2018). Quantifying economic damages from climate change. *Journal of Economic Perspectives, 32*(4), 33–52.

Bardazzi, R., & Pazienza, M. (2018). Ageing and private transport fuel expenditure: Do generations matter? *Energy Policy, 117*, 396–405.

Barnicoat, G., & Danson, M. (2015). The ageing population and smart metering: A field study of householders' attitudes and behaviours towards energy use in Scotland. *Energy Research and Social Science, 9*, 107–115.

Barreca, A., Deschenes, O., & Guldi, M. (2018). Maybe next month? Temperature shocks and dynamic adjustments in birth rates. *Demography, 55*(4), 1269–1293.

Bradbury, M., Peterson, M., & Liu, J. (2014). Long-term dynamics of household size and their environmental implications. *Population and Environment, 36*(1), 73–84.

Burke, P., & Csereklyei, Z. (2016). Understanding the energy-GDP elasticity: A sectoral approach. *Energy Economics, 58*, 199–210.

Burney, N. (1995). Socioeconomic development and electricity consumption. *Energy Economics, 17*, 185–195.

Casey, G., & Galor, O. (2017). Is faster economic growth compatible with reductions in carbon emissions? The role of diminished population growth. *Environmental Research Letters, 12*, 014003.

Casey, G., Shayegh, S., Moreno-Cruz, J., Bunzl, M., Galor, O., & Caldeira, K. (2019). The impact of climate change on fertility. *Environmental Research Letters, 14*, 054007.

Cattaneo, C., & Peri, G. (2016). The migration response to increasing temperatures. *Journal of Development Economics, 122*, 127–146.

Chancel, L. (2014). Are younger generations higher carbon emitters than their elders? Inequalities, generations and CO2 emissions in France and in the USA. *Ecological Economics, 100*, 195–207.

Cole, M., & Neumayer, E. (2004). Examining the impact of demographic factors on air pollution. *Population and Environment., 26*(1), 5–21.

Csereklyei, Z., Rubio Varas, M., & Stern, D. (2016). Energy and economic growth: The stylized facts. *Energy Journal, 37*(2), 223–255.

Curtis, F., Simpson-Housley, P., & Drever, S. (1984). Household energy conservation. *Energy Policy, 12*, 452–456.

Das Gupta, M. (2013). *Population, Poverty, and Climate Change* (World Bank Policy Research Working Paper) (Vol. 6631). The World Bank.

Desmet, K., & Rossi-Hansberg, E. (2015). On the spatial economic impact of global warming. *Journal of Urban Economics, 88*, 16–37.

Dietz, T., & Rosa, E. A. (1997). Effects of population and affluence on CO_2 emissions. *Proceedings of the National Academy of Sciences – USA, 94*, 175–179.

Ehrlich, P., & Holdren, J. (1971). The impact of population growth. *Science, 171*, 1212–1217.

Ewing, R., & Cervero, R. (2001). Travel and the build environment: A synthesis. *Transport Research Record Journal of Transport Research Board, 1780*, 87–114.

Feyrer, J. (2007). Demographics and productivity. *Review of Economics and Statistics, 89*(1), 100–109.

Feyrer, J. (2008). Aggregate evidence on the link between age structure and productivity. *Population and Development Review, 34*, 78–99.

Filippini, M., & Hunt, L. (2011). Energy demand and energy efficiency in the OECD countries: A stochastic demand frontier approach. *The Energy Journal, 32*(2), 59–80.

Fremstad, A., Underwood, A., & Zahran, S. (2018). The environmental impact of sharing: Household and urban economies in CO_2 emissions. *Ecological Economics, 145*, 137–147.

Galor, O., & Mountford, A. (2008). Trading population for productivity: Theory and evidence. *Review of Economic Studies, 75*(4), 1143–1179.

Henderson, V. (2003). The urbanization process and economic growth: The so-what question. *Journal of Economic Growth, 8*, 47–71.

Herrendorf, B., Rogerson, R., & Valentinyi, Á. (2014). Growth and structural transformation. In *Handbook of economic growth* (Vol. 2, pp. 855–941). Elsevier.

Hilton, F., & Levinson, A. (1998). Factoring the environmental Kuznets curve: Evidence from automotive lead emissions. *Journal of Environmental Economics and Management, 35*, 126–141.

Jaforullah, M., & King, A. (2017). The econometric consequences of an energy consumption variable in a model of CO_2 emissions. *Energy Economics, 63*, 84–91.

Jones, D. (1989). Urbanization and energy use in economic development. *The Energy Journal, 10*(4), 29–44.

Jones, C., & Kammen, D. (2014). Spatial distribution of US household carbon footprints reveals suburbanization undermines greenhouse gas benefits of urban population density. *Environmental Science and Technology, 48*, 895–902.

Jorgenson, A., & Clark, B. (2010). Assessing the temporal stability of the population/environment relationship in comparative perspective: A cross-national panel study of carbon dioxide emissions, 1960–2005. *Population and Environment, 32*, 27–41.

Kenworthy, J., & Laube, F. (1999). Patterns of automobile dependence in cities: An international overview of key physical and economic dimensions with some implications for urban policy. *Transportation Research Part A, 33*, 691–723.

Klein, N., & Smart, M. (2017). Millennials and car ownership: Less money, fewer cars. *Transport Policy, 53*, 20–29.

Knight, K., & Rosa, E. (2012). Household dynamics and fuelwood consumption in developing countries: A cross-national analysis. *Population and Environment, 33*, 365–378.

Knittel, C., & Murphy, E. (2019). *Generational trends in vehicle ownership and use: Are millennials any different?* (National Bureau of Economic Research Working Paper) (Vol. 25674).

Kurz, C., Li, G., & Vine, D. (2018). *Are millennials different? Finance and economics discussion series 2018– 080*. Board of Governors of the Federal Reserve System.

Kyatta, M., Broberg, A., Tzoulas, T., & Snabb, K. (2013). Towards contextually sensitive urban densification: Location-based softGIS knowledge revealing perceived residential environmental quality. *Landscape and Urban Planning, 113*, 30–46.

Lam, D. A., & Miron, J. A. (1996). The effects of temperature on human fertility. *Demography, 33*(3), 291–305.

Lariviere, I., & Lafrance, G. (1999). Modelling the electricity consumption of cities: Effect of urban density. *Energy Economics, 21*(1), 53–66.

Liddle, B. (2004). Demographic dynamics and per capita environmental impact: Using panel regressions and household decompositions to examine population and transport. *Population and Environment, 26*(1), 23–39.

Liddle, B. (2011). Consumption-driven environmental impact and age-structure change in OECD countries: A cointegration-STIRPAT analysis. *Demographic Research, 24*, 749–770.

Liddle, B. (2013). Urban density and climate change: A STIRPAT analysis using city-level data. *Journal of Transport Geography, 28*, 22–29.

Liddle, B. (2014). Impact of population, age structure, and urbanization on greenhouse gas emissions/energy consumption: Evidence from macro-level, cross-country analyses. *Population and Environment, 35*, 286–304.

Liddle, B. (2015). What are the carbon emissions elasticities for income and population? Bridging STIRPAT and EKC via Robust Heterogeneous Panel Estimates. *Global Environmental Change, 31*, 62–73.

Liddle, B., & Huntington, H. (2020). Revisiting the income elasticity of energy consumption: A heterogeneous, common factor, dynamic OECD & non-OECD country panel analysis. *The Energy Journal, 41*(3).

Liddle, B., & Lung, S. (2010). Age structure, urbanization, and climate change in developed countries: Revisiting STIRPAT for disaggregated population and consumption-related environmental impacts. *Population and Environment, 31*, 317–343.

Liddle, B., & Lung, S. (2014). Might electricity consumption cause urbanization instead? Evidence from heterogeneous panel long-run causality tests. *Global Environmental Change, 24*, 42–51.

Lim, J., Kang, M., & Jung, C. (2019). Effect of national-level spatial distribution of cities on national transport CO_2 emissions. *Environmental Impact Assessment Review, 77*, 162–173.

Lutz, W., & Samir, K. C. (2011). Global human capital: Integrating education and population. *Science, 333*(6042), 587–592.

Lutz, W., Cuaresma, J. C., Kebede, E., Prskawetz, A., Sanderson, W. C., & Striessnig, E. (2019). Education rather than age structure brings demographic dividend. *Proceedings of the National Academy of Sciences*.

Lutz, W., Cuaresma, J. C., Kebede, E., Prskawetz, A., Sanderson, W. C., & Striessnig, E. (2019). Education rather than age structure brings demographic dividend. *Proceedings of the National Academy of Sciences, 116*(26), 12798–12803.

MacKellar, F. L., Lutz, W., Prinz, C., & Goujon, A. (1995). Population, households, and CO_2 emissions. *Population and Development Review, 21*(4), 849–865.

Mahapatra, K., & Gustavsson, L. (2008). An adopter-centric approach to analyze the diffusion patterns of innovative residential heating systems in Sweden. *Energy Policy, 36*, 577–590.

Marcotullio, P., Sarzynski, A., Albrecht, J., & Schulz, N. (2012). The geography of urban greenhouse gas emissions in Asia: A regional analysis. *Global Environmental Change, 22*, 944–958.

Menz, T., & Welsch, H. (2012). Population aging and carbon emissions in OECD countries: Accounting for life-cycle and cohort effects. *Energy Economics, 34*, 842–849.

Mills, B., & Schleich, J. (2010). What's driving energy efficient appliance label awareness and purchase propensity? *Energy Policy, 38*, 814–825.

Mills, B., & Schleich, J. (2012). Residential energy-efficient technology adoption, energy conservation, knowledge, and attitudes: An analysis of European countries. *Energy Policy, 49*, 616–628.

Newman, P., & Kenworthy, J. (1989). *Cities and automobile dependence: An international sourcebook*. Gower Technical.

O'Neill, B. C., & Chen, B. S. (2002). Demographic determinants of household energy use in the United States. *Population and Development Review, 28*, 53–88.

O'Neill, B. C., Dalton, M., Fuchs, R., Jiang, L., Pachauri, S., & Zigova, K. (2010). Global demographic trends and future carbon emissions. *Proceedings of the National Academy of Sciences, 107*(41), 17521–17526.

O'Neill, B., Liddle, B., Jiang, L., Smith, K., Pachauri, S., Dalton, M., & Fuchs, R. (2012). Demographic change and carbon dioxide emissions. *The Lancet, 380*(9837), 157–164.

Okada, A. (2012). Is an increased elderly population related to decreased CO_2 emissions from road transportation? *Energy Policy, 45*, 286–292.

Otsuka, A. (2017). Determinants of efficiency in residential electricity demand: Stochastic frontier analysis on Japan. *Energy, Sustainability and Society, 7*, 31.

Otsuka, A. (2018). Regional determinants of energy efficiency: Residential energy demand in Japan. *Energies, 11*, 1557.

Parikh, J., & Shukla, V. (1995). Urbanization, energy use and greenhouse effects in economic development. *Global Environmental Change, 5*(2), 87–103.

Poortinga, W., Steg, L., Vlek, C., & Wiersma, G. (2003). Household preferences for energy-savings measures: A conjoint analysis. *Journal Economic Psychology, 24*, 49–64.

Prskawetz, A., Leiwen, J., & O'Neill, B. (2004). Demographic composition and projections of car use in Austria. *Vienna Yearbook of Population Research, 2004*, 247–326.

Raftery, A. E., Li, N., Ševčíková, H., Gerland, P., & Heilig, G. K. (2012). Bayesian probabilistic population projections for all countries. *Proceedings of the National Academy of Sciences, 109*(35), 13915–13921.

Rickwood, P., Glazebrook, G., & Searle, G. (2008). Urban structure and energy – A review. *Urban Policy and Research, 26*(1), 57–81.

Roberts, T. (2011). Applying the STIRPAT model in a post-Fordist landscape: Can a traditional econometric model work at the local level? *Applied Geography, 31*, 731–739.

Roser, M., & Nagdy, M. (2013). *Projections of future education*. Published online at OurWorldInData.org. Retrieved from: https://ourworldindata.org/projections-of-future-education

Schäfer, A. (2005). Structural change in energy use. *Energy Policy, 33*(4), 429–437.

Shaker, R. (2015). The well-being of nations: An empirical assessment of sustainable urbanization for Europe. *International Journal of Sustainable Development and World Ecology, 22*(3), 375–387.

Shayegh, S. (2017). Outward migration may alter population dynamics and income inequality. *Nature Climate Change, 7*(11), 828.

Stephenson, J., Barton, B., Carrington, G., Doering, A., Ford, R., Hopkins, D., Lawson, R., McCarthy, A., Rees, D., Scott, M., Thorsnes, P., Walton, S., Williams, J., & Wooliscroft, B. (2015). The energy cultures framework: Exploring the role of norms, practices and material culture in shaping energy behaviour in New Zealand. *Energy Research & Social Science, 7*, 117–123.

Tilley, S. (2017). Multi-level forces and differential effects affecting birth cohorts that stimulate mobility change. *Transport Reviews, 37*(3), 344–364.

Underwood, A., & Fremstad, A. (2018). Does sharing backfire? A decomposition of household and urban economies in CO_2 emissions. *Energy Policy, 123*, 404–413.

Walsh, M. (1989). Energy tax credits and housing improvement. *Energy Economics, 11*, 275–284.

Willis, K., Scarpa, R., Gilroy, R., & Hamza, N. (2011). Renewable energy adoption in an ageing population: Heterogeneity in preferences for microgeneration technology adoption. *Energy Policy, 39*, 6021–6029.

Zarnikau, J. (2003). Consumer demand for 'green power' and energy efficiency. *Energy Policy, 31*(15), 1661–1672.

Part VI

Other Arenas

Environment and Fertility

20

Sam Sellers

Abstract

Environmental shocks and stressors, such as natural disasters or interannual temperature and precipitation variability can affect fertility outcomes through a variety of pathways, including those characterized by socioeconomic, nutrition, or health indicators. This chapter broadly examines the research linking environmental correlates with outcomes related to human fertility. These outcomes include: union formation, fertility desires, family planning use, early pregnancy loss, live births, preterm birth, low birth weight, and stillbirth. In general, climatic conditions favorable for agriculture or natural resource based-livelihoods increase the likelihood of giving birth to healthy babies. However, there is a great deal of spatial and temporal variability underlying environment-fertility interactions, indicating the strong importance of understanding biological, social, economic, and cultural mechanisms that can influence such relationships.

S. Sellers (✉)
Independent Scholar, Bainbridge Island, WA, USA
e-mail: samsellers@gmail.com

Introduction

Within the past several decades, a growing body of literature has emerged centering on the linkages between environmental shocks and stressors and human fertility outcomes. This trend has been catalyzed in recent years by increasing concern among academics and the general public regarding the effects of global climate change, which disproportionately burden individuals in low- and middle-income settings, and are likely to have significant impacts on human fertility outcomes (Grace, 2017). Developing a stronger understanding of how environmental factors impact fertility can provide researchers and policymakers with the tools and guidance needed to develop mechanisms to improve the resilience of individuals to the effects of environmental change, particularly in contexts where vulnerability is high.

Fertility has long been a central motivation of population-environment research, largely due to concerns associated with human population growth. Much of the early academic interest in population-environment linkages more broadly was spurred by neo-Malthusian concerns related to the effects of human population growth on environmental conditions, as typified by Ehrlich's *The Population Bomb* (Ehrlich, 1968). More re-

© Springer Nature Switzerland AG 2022
L. M. Hunter et al. (eds.), *International Handbook of Population and Environment*, International
Handbooks of Population 10, https://doi.org/10.1007/978-3-030-76433-3_20

cent work in this area has emphasized the importance of feedbacks between high rates of population growth and adverse environmental changes (O'Neill et al., 2001). This line of inquiry is illustrated by Potts et al. (2013) who highlight how population growth rates in the Sahel are unsustainable because of deteriorating resource conditions as a result of growing consumption patterns and the effects of climate change.

However, while we acknowledge that the relationship between fertility and the environment is bidirectional, this chapter emphasizes linkages from environmental conditions to fertility outcomes. This is in part due to challenges associated with study design in population-environment research, particularly when it comes to understanding causality. It is difficult to attribute changes in environmental conditions to population growth, as these conditions are influenced by myriad, often unobserved, factors. This in turn hampers the ability of researchers to generate credible claims about specific aspects of these relationships, even if it is plausible that population growth contributes to environmental degradation. By contrast, because fertility outcomes are typically measured at the individual, household, or community levels, environmental factors can be more easily isolated and tested in statistical models allowing for stronger causal inferences. As this linkage has often been underexplored in earlier summaries of the population-environment literature, we center this chapter's focus on research seeking to explain fertility behaviors resulting from environmental causes.

This analysis complements recent reviews of the environment-fertility literature (de Sherbinin et al., 2008; Grace, 2017) that have examined the bidirectional relationships linking environmental conditions and fertility, noting the close linkages between environmental change (including climate change), household conditions, and changes in fertility. These reviews draw upon literature concerning livelihoods, defined as the collection of strategies that individuals and households apply in order to earn a living, which often depend on local environmental conditions (Ellis, 2000). As will be discussed later in this chapter, these livelihood strategies are closely related to fertility outcomes. The discussion below is organized by the different mechanisms through which environmental shocks and stressors can shape fertility outcomes, highlighting the realities that environmental exposures can influence a variety of fertility determinants, and that temporal variability of exposures during pregnancy can have differing effects on outcomes.

Literature linking environmental shocks and stressors to fertility outcomes is only a small set of the wide array of research exploring human fertility outcomes. A departure point for this review is the seminal work from Bongaarts (1978) on the determinants of fertility. Bongaarts outlines a simple model for understanding live birth outcomes as a function of "indirect" determinants of fertility, which include an array of socioeconomic, cultural, and environmental variables that impact "proximate" or "intermediate" determinants of fertility. Such proximate determinants include factors such as marriage rate, contraception use, and postpartum infecundity.[1] In turn, the levels of and relationships between these proximate determinants can be used to effectively estimate live birth outcomes.

This review centers on a subset of the pathways discussed by Bongaarts and other early scholars of fertility, namely those associated with the natural environment. We focus on research that explores the relationships between environmental factors and proximate determinants of fertility, live births, as well as adverse pregnancy outcomes. Section "Environment and Fertility: Methods, Understandings, and Theories" provides a brief summary of the lenses, theories, and methods found in environment-fertility literature. Sections "Union Formation and Fertility Desires", "Factors Affecting Conception and Early Pregnancy Loss Probability", "Live Births", and "Adverse Pregnancy Outcomes"

---

[1] In the referenced article, Bongaarts discusses eight proximate determinants of human fertility. However, some of these determinants are not discussed in this review because of limited evidence linking them with the environment. Determinants not discussed include induced abortion, coital frequency, sterility, and duration of the fertile period.

explore the effects of environmental stressors on specific fertility outcomes. Section "Population, Health, and Environment Programs" provides a brief summary of findings from studies of population, health, environment projects designed to address environmental drivers of fertility.

Environment and Fertility: Methods, Understandings, and Theories

The literature linking the environment to fertility outcomes uses a variety of methods and lenses in conceptualizing the environment and fertility, as well as the pathways through which these two concepts are related. How these ideas are operationalized varies considerably between studies, which has important ramifications for the conclusions that can be drawn from the universe of environment-fertility research. This section provides a very brief overview of some of the main framings and theories undergirding environment-fertility literature. While this summary is intended to orient readers towards many of the main characteristics and themes in the literature, we acknowledge that this brief overview only scratches the surface of the nuance and complexity underlying many of these ideas.

Conceptualizing the Environment: Shocks Vs. Stressors

Like much of the population-environment literature, the literature exploring fertility engages a variety of different understandings of the natural environment in order to assess fertility effects. Some studies explore environmental shocks—discrete events such as a natural disaster or annual crop failure. Others examine environmental stressors, longer-term changes such as interannual temperature or precipitation variability. Stressors can be measured discretely (i.e. number of days above a temperature threshold during pregnancy) or continuously (i.e. total precipitation volume or z-

scores compared to long-term averages). Both environmental shocks and stressors inform the discussion of fertility outcomes below. We adopt the phrase "environmental stimuli" in this chapter when referring to shocks and stressors jointly.

Conceptualizing the Environment: Human Vs. Nature

Understandings of the environment in population-environment research are often quite expansive, and also vary among scholars. This chapter focuses on stimuli from the natural environment, and the ways in which these can affect fertility outcomes. Moreover, this chapter is constrained through its emphasis on hydrometeorological shocks and stressors, as opposed to other environmental drivers of fertility outcomes (e.g., pollution).[2] Additionally, while they can have substantial effects on fertility outcomes, pathways originating in the human environment, such as a terrorist attack or an economic crisis, are not explored here (Rodgers et al., 2005; Sobotka et al., 2011). Even so, stimuli from the natural environment often impact fertility through socioeconomic processes such as changes in household income or maternal nutrition conditions due to crop losses. Such pathways are explored below.

Conceptualizing Pathways: Behavioral Vs. Physiological

Environmental shocks and stressors can affect behavioral choices related to fertility, such as how many children an individual desires or whether contraception is used. However, environmental factors have also been shown

---

[2]Brief mention is made in Section "Adverse Pregnancy Outcomes" of air pollution and adverse pregnancy outcomes because of the potential for temperature to increase pollution concentrations. However, there is a large body of literature exploring pollution and adverse pregnancy outcomes that is not centered on environmental change. In order to narrow the scope of this review to a manageable size, such literature is not discussed here.

to have physiological impacts on humans, affecting fertility outcomes. For instance, ovarian function can be adversely affected by seasonal undernutrition and/or increases in energy expended, which can lead to a lower likelihood of conceptions (Vitzthum, 2009). Behavioral and physiological mechanisms are often complementary in explaining fertility outcomes and both mechanism types are discussed below.

Conceptualizing Pathways: Direct Vs. Indirect

Many of the relationships linking environmental stimuli with fertility are mediated through at least one other variable, typically an indicator associated with livelihood activities and/or socioeconomic status. For instance, a study from rural Nepal found that fertility decline was mediated through the adoption of new agricultural technologies, which increased incomes and reduced the need for farm labor. This in turn discouraged births (Bhandari & Ghimire, 2013). Some relatively direct relationships between environmental stimuli and fertility exist, such as heat-related physiological impacts that appear to reduce the likelihood of a live birth (Barreca et al., 2018). However, in general, the relationships discussed below are contingent on specific livelihood contexts that affect fertility in particular ways, which limit the applicability of the findings in many environment-fertility studies to specific spatiotemporal domains.

Conceptualizing Fertility: Pre-Conception Vs. Post-Conception Outcomes

As shown in Fig. 20.1, there are a variety of steps linking environmental stimuli to fertility outcomes. However, the environment and fertility literature is diverse in terms of how "fertility outcomes" are defined. Predictors closely related to birth outcomes, such as the number of children a woman desires and current contraceptive use, are often included in this literature, along with post-conception outcomes such as live births or stillbirths. Adverse pregnancy outcomes, such as low birth weight and preterm birth, can increase the risk of infant mortality, affecting the probability of future births. For purposes of this chapter, fertility outcomes are those which are directly related to the behaviors and impacts associated with or the result of giving birth, including whether or not a woman wants a child in the future, union formation, current use of contraception, live births, stillbirths, low birth weight, and preterm birth.

Conceptualizing Research: Observational Vs. Intervention-Centered Studies

As in the population and environment field more broadly, most of the environment-fertility literature centers on observational studies, whereby inferences about relationships between environmental stimuli and fertility outcomes are made by assessing patterns within demographic, economic, agricultural, or land use datasets. Such studies are designed to capture extant conditions, and not to assess the effects of particular interventions. A key exception to this is the literature on population, health, environment (PHE) programs, discussed at the end of this chapter, which explores the effects of interventions designed to improve environmental and fertility outcomes in settings with high fertility rates and substantial dependence on natural resources for livelihoods.

Much of the environment-fertility literature uses one or more of several key frameworks in order to explain decision-making processes around fertility. These frameworks often provide bidirectional understandings, explaining both environmental behaviors resulting from fertility decisions, as well as fertility decisions resulting from environmental behaviors. Moreover, they generally seek to explain fertility decision-making through assumptions about the costs and benefits that individuals and couples perceive when making childbearing decisions. Importantly, these frameworks are not necessarily independent of each other. Each highlights specific factors affecting environment-fertility

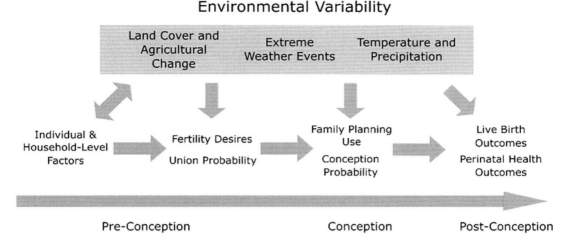

Fig. 20.1 Conceptual model of environmental variability on fertility outcomes

relationships, but as such relationships are complex and context-dependent, more than one framework may provide a useful lens for understanding environment-fertility relationships in a particular setting.

Multiphasic Response

Davis (1963) highlights how individuals seek to maintain relative socioeconomic prosperity despite population growth or environmental stimuli adversely affecting resource endowments. This in turn spurs multiple types of demographic change, albeit at different time lags from experiencing a stressor. He notes in particular the historic role of declining land-labor ratios, namely how a relatively fixed land base coupled with growing human populations require alternative economic arrangements in order to improve well-being.[3] In addition to adopting different livelihood strategies, often through rural outmigration, Davis notes the importance of slower population growth in order to maintain relative household prosperity given the growing costs of childrearing. This was achieved in Western Europe and Japan in the early-to-mid twentieth Century through a combination of delayed marriage and increased contraceptive uptake, ultimately reducing total fertility rates, a response which many countries in Asia and Latin America have since emulated (Davis, 1963; Lesthaeghe, 2014).

Life Cycle Hypotheses

Related to multiphasic response is the concept of life cycles, which are often used to help understand fertility and environment relationships in rural settings. Household demographic characteristics (e.g. number of adults and children in a household and age of parents) are sometimes associated with both land use choices and future fertility (McCracken et al., 2002; Perz, 2001). The study of household life cycles is largely credited to Chayanov (1986), who observed a close relationship between the number of working-age individuals on a farm and land use practices, with more labor-intensive practices taking place among households with a greater share of working-age individuals. By contrast, households with older children who have left the home may adopt less labor-intensive practices, such as cattle ranching.

---

[3] These ideas have been expanded upon by Bilsborrow (1987) who explores demographic cases where land availability is not a limiting factor, leading to an expansion of the area covered by human farms and settlement, with often delayed demographic responses as a result.

These household life cycle patterns parallel farm life cycle patterns in frontier regions. Typically, when a family settles in a forested plot, some land is initially cleared to support small-scale agriculture, followed by investments in more financially lucrative agricultural practices and further deforestation that are often supported through additional (family) labor and the accumulation of financial capital over time (Walker, 2003). A complicating factor is the relationship between households, farms, and broader markets. As noted by Vanwey et al. (2007) classical household life cycle hypotheses have key limitations including the absence of labor and capital markets that can affect household decision-making. Thus, as farm households become more closely integrated into off-farm labor markets, parents may choose to invest in educational opportunities and/or have children seek work in nearby cities, which can diminish the linkage between household labor supply and land use choices (Barbieri et al., 2005; Pan et al., 2007). A current area of research is determining when household life cycle effects predominate as opposed to farm life cycle effects, with some evidence suggesting a weaker relationship between household demographic cycles and land use choices as livelihood patterns become more diversified (Barbieri et al., 2020).

Fertility and Household Endowments

Becker and Lewis (1973) argue that resource constraints for households affect preferences for children, much as they do for any other type of good. Specifically, couples face a tradeoff between the number of children they want to have (quantity) versus the investment they make in each child through schooling, healthcare, etc. to improve that child's well-being (quality). Factors affecting future household endowments, including land availability, agricultural labor needs, and perceptions of environmental risks can all affect the tradeoffs households make to maximize their utility. Thus, an endowments approach illustrates the importance of understanding fertility decisions as emerging from a variety of household preferences and constraints (including environmental constraints) and the ties between these decisions and household consumption choices. However, there are limits to the utility of household endowments to understand fertility decision-making, as a result of variation in preferences between household members, limited insights into the social, religious, and cultural mechanisms that influence fertility decisions, and the importance of public policies that promote family planning and women's empowerment (de Sherbinin et al., 2008).

Risk Diversification

In areas with poor environmental conditions, a lack of diversified livelihood mechanisms and limited or no income support from government or other institutions, children may be a means for households to avoid slipping further into poverty after experiencing a shock or stressor that affects their livelihoods. Cain (1981, 1983) uses the example of Bangladesh to demonstrate that the benefits of a child to parental income security outweigh the short-term costs of childrearing, especially for rural households living in areas with substantial environmental hazards. While the risk diversification benefits associated with childrearing may be most likely to accrue to parents during old age, Cain notes that this mechanism can also provide economic benefits to parents well before then depending on local social and environmental conditions.

Child Survival and Replacement

Couples exposed to significant levels of environmental risk may be inclined to have more children than they otherwise would if exposed to lower levels of risk (Preston, 1978). As noted by Scrimshaw (1978), two closely related hypotheses exist concerning fertility responses in settings with substantial environmental risk. The first concerns child *survival*, whereby couples assess the risk of children surviving to adulthood based on their household and community circumstances and opt not to reduce their fertility when they otherwise would because risks to child survival are high. This mechanism understands high fertility as providing a form of insurance intended to ensure a desired number of children survive to adulthood. The second concerns child *replacement*, whereby couples opt to have additional births to replace children who die. While the motivations behind both theories are complex, child survival prospects and child replacement decisions can be impacted by environmental conditions and variability, among other causes (Mosley & Chen, 1984).

Vicious Circle Model

In settings where natural resource collection is a key livelihood strategy and barriers exist to adopting alternative livelihood strategies, environmental degradation and resource scarcity can incentivize fertility, in turn worsening environmental conditions in future time periods. Dasgupta (1995) details this vicious circle model (VCM), noting that many key commodities necessary for household sustenance, such as fuelwood or water, can be collected by children from a young age. As human populations grow and consumption levels increase, these resources often become increasingly scarce, requiring household members (particularly women and children) to expend more time and effort in future periods (O'Neill et al., 2001). This in turn can spur households to try for additional children, increasing the amount of labor available to individual households, but also increasing the total supply of labor competing for a shrinking pool of resources, creating a vicious cycle of worsening resource scarcity and higher fertility (Dasgupta, 1995).

The definitions and theories above inform the empirical literature on environment and fertility, examined below. We begin at the earliest stages of the fertility process, exploring research concerning union formation and fertility desires, followed by discussions of conception and early pregnancy loss, live births, and adverse pregnancy outcomes.

Union Formation and Fertility Desires

Despite its importance in affecting fertility outcomes, the evidence exploring the linkages between variation in environmental conditions and union formation is less robust. An analysis using historical data from the Netherlands finds that marriage probability increases during periods of lower-than-normal temperatures and flood events, both of which disrupted agricultural production. One explanation is that marriage may have been a mechanism for individuals to cope with poor environmental conditions. Marriage can increase household economies of scale, allowing formerly single adults to pool resources together (Jennings & Gray, 2017). However, in other settings, the effect of climate on marriage differs. In Nepal, better agricultural conditions, as measured by the amount of land devoted to agriculture, is positively associated with marriage probability, even after controlling for access to nonfarm employment opportunities (Yabiku, 2006). By contrast, there are more robust patterns suggesting linkages between intraannual climate variability and marriage timing. For instance, various European studies use robust historical marriage data to illustrate how marriage patterns correlated to specific times in the growing cycle and associated fluctuations in farm labor needs (Coppa et al., 2001; Dribe & Van de Putte, 2012; González-Martín, 2008). In industrialized societies, seasonal marriage patterns persist into the modern day. Current patterns of marriage seasonality, with weddings most often occurring in the summer months, are attributable in part to couples seeking to avoid parts of the year when inclement weather is more probable and to accommodate the school calendar (Rault & Régnier-Loilier, 2016; Ruiu & Gonano, 2015)

The effects of environmental stimuli on fertility desires vary substantially across settings, but this variability can largely be understood through differences in the socioeconomic and ecological contexts that underlie these relationships. Evidence from a longitudinal study in the Chitwan Valley, Nepal broadly supports VCM hypotheses, whereby individuals who are dependent on natural resources for their livelihoods increase the number of children desired if the labor needs to harvest those resources increase, despite the additional scarcity this could induce over time. In Nepal, the number of children a woman desires has been observed to increase as the amount of time devoted to fodder and fuelwood collection increases (Biddlecom et al., 2005; Brauner-Otto & Axinn, 2017). Similarly in Malawi, a setting where food scarcity is common, experiencing a food shortage increased the likelihood of young women wanting another child soon (Sennott & Yeatman, 2012).

Evidence from other contexts where agricultural livelihoods are common suggests that prolonged exposure to drought or higher-than-normal temperatures is often associated with reduced fertility desires. For example, after experiencing periods of above average temperatures that adversely affect agricultural conditions, women on farms in Indonesia were significantly less likely to want a child (Sellers & Gray, 2019). In sub-Saharan Africa, ideal family size and desire to have another child were reduced during periods of high temperatures, with long periods of lower-than-normal precipitation also reducing ideal family size (Eissler et al., 2019). Despite the disruption they may bring to individual livelihoods, there is less evidence, however, that environmental shocks exhibit similar effects, provided child mortality associated with the event is low. For instance, individuals with greater exposure to the 2010 earthquake in Haiti were no more likely to want a child following the event than individuals with less exposure (Behrman & Weitzman, 2016).

Factors Affecting Conception and Early Pregnancy Loss Probability

The probability of conception leading to a live birth can be affected by a variety of factors. Fecundity can vary seasonally, particularly in settings where there are substantial changes in resource availability during the year, which can in turn be one cause of seasonal patterns of fertility (see Section "Live Births"). Additionally, conception probability can be negatively affected by the use of modern contraceptives, with access to and use of contraceptives impacted by a variety of factors including environmental ones. Rates of early pregnancy loss may also be sensitive to environmental factors (Vitzthum, 2009).

Various studies have explored differences in ovarian function across human female populations and how this variability is associated with environmental factors. Ellison et al. (2005) note that a key underlying factor affecting ovarian function is energy balance, that is, energy intake less energy expended. Energy balance is thus a function of both nutritional status as well as workload, both of which can vary seasonally, particularly in rural settings. For instance, in rural Nepal, heavy workloads among women during the monsoon season can lead to temporary weight loss, reducing ovarian function during certain months of the year (Panter-Brick et al., 1993). Similar effects were observed among Lese populations in Northeastern Democratic Republic of the Congo (formerly Zaire) that experience seasonal food shortages (Bailey et al., 1992).

Additionally, nutritional factors can influence the duration of postpartum amenorrhea, which plays a significant role in birth spacing in many natural fertility settings. Numerous observational studies have found that better nourished women (Huffman et al., 1987; Peng et al., 1998; Tracer, 1996), as well as those that are able to supplement infant feeding with other foods (Dewey et al., 1997; Kurz et al., 1993) resume menstruation more quickly following pregnancy. More recent evidence suggests that both the intensity of suckling stimulus as well as energy balance can affect amenorrhea duration, with shorter durations generally associated with improving environmental conditions that can allow women to invest energy in future reproduction (Valeggia & Ellison, 2009).

Environmental shocks have also been linked with changes in family planning use, generally as clinics and commodity supply chains are damaged in the wake of the event, which can increase unwanted pregnancies. In both Indonesia and Haiti, overall contraceptive prevalence was largely stable among individuals exposed to earthquakes, but some minor changes were reported in terms of method mix, with slight decreases reported in injectable use (Behrman & Weitzman, 2016; Djafri et al., 2015; Hapsari et al., 2009). After Hurricane Katrina struck the southern United States in 2005, contraceptive access was significantly disrupted, affecting the rate of use after the event (Kissinger et al., 2007). Moreover, disasters can exacerbate social inequalities in contraceptive access. Following Hurricane Ike, which struck the southern United States in 2008, black women indicated a much greater likelihood of having difficulty accessing contraception, as did individuals who were forced to evacuate due to the storm (Leyser-Whalen et al., 2011).

Environmental stressors can affect agricultural and other livelihood activities, which in turn can change family planning decisions. There is growing evidence that many couples choose to adopt contraception following environmental stressors that adversely affect household economic conditions in order to reduce unwanted fertility. This behavior is in line with the evidence outlined in Section "Union Formation and Fertility Desires" suggesting that fertility desires may be lowered during periods of sustained adverse environmental conditions. For instance, in Tanzania, a longitudinal analysis found that family planning use significantly increases following crop failure, although much of this increase is associated with use of less-effective options such as the rhythm method (Alam & Pörtner, 2018). In Uganda, couples are significantly more likely to use family planning after periods of lower-than-average rainfall (Abiona, 2017). In Indonesia, farm households exposed to prolonged periods of above-average temperatures are significantly more likely to use family planning (Sellers & Gray, 2019). By contrast, in Nepal, areas of poorer environmental quality (as measured by plant diversity), are associated with lower use of contraceptives, lending further support for a VCM hypothesis in that setting (Brauner-Otto, 2014).

Environmental stimuli may also impact the risk of early pregnancy loss, although evidence is limited due to the difficulty of accurately measuring early pregnancy losses since most are unreported. Various studies have found seasonal patterns in early pregnancy loss, albeit with variation between studies in terms of the period of greatest risk (Czeizel et al., 1984; Kallan & Enneking, 1992; Weinberg et al., 1994). In rural Bolivia, conceptions occurring at times of high energy demand and low nutrient intake are more vulnerable to early pregnancy loss than conceptions during other parts of the year, in line with findings regarding diminished ovarian function during these periods (Vitzthum et al., 2009). Additionally, higher temperatures throughout pregnancy may marginally increase the risk of early pregnancy loss, although the evidence base for this relationship is currently limited (Asamoah et al., 2018).

Live Births

Compared to the topics covered earlier, far more research has been published exploring the effects of environmental shocks and stressors on live births. Several environmental mechanisms have been identified that can impact live births. First, environmental stimuli may affect household economic outcomes, particularly when households significantly depend on natural resources for their livelihoods. These stimuli can worsen nutritional outcomes (due to crop failures or lower incomes to buy food) or affect household labor patterns and in turn diminish the probability of a live birth (e.g. Sasson & Weinreb, 2017; Sellers & Gray, 2019). Second, environmental shocks often disrupt day-to-day household activities and livelihoods, which can have implications for access to family planning, coital frequency, and/or fertility desires, in turn affecting the likelihood of a live birth (e.g. Evans et al., 2010; Nobles et al., 2015). Third, a variety of physiological and socioeconomic mechanisms associated with intraannual environmental variability contribute to patterns of live birth seasonality (e.g. Cowgill, 1966; Martinez-Bakker et al., 2014). Finally, there is a growing set of studies centered around the effects of heat on human physiology and the probability of live births (e.g. Barreca et al., 2018; Lam & Miron, 1996). The literature described in this section is grouped by the primary mechanism driving the results found, with the caveat that because of the complexity associated with processes leading to live births, more than one of these processes may be reflected in a given study.

Household Economic Impacts

Household economic shocks can have significant effects on births, particularly in rural settings where households are highly dependent on natural resource-based livelihood activities. This literature often explores the relationship between various proxies for agricultural or natural resource conditions, including land cover change, or variability in temperature and precipitation, and live births. As noted above, increased need for household labor may stimulate births. Moreover, healthier women are generally less likely to experience fetal loss. It should be noted that most study designs do not permit researchers to clearly disentangle these competing hypotheses (nutrition/health vs. labor supply), and both mechanisms may be occurring simultaneously. Literature focusing on household economic impacts tends to use longitudinal analyses, focusing on interannual variability in meteorological or land use variables in a particular setting and seeking to understand the relationship between this variability and the magnitude of live birth outcomes. This is distinguished from research on seasonal patterns of fertility below, which explores the relationship between intraannual variability in birth rates and socioeconomic and/or physiological factors.

In general, unfavorable climate conditions for agriculture or natural resource harvesting are linked to reduced fertility, particularly in communities with little wealth or alternative livelihood activities. By contrast, demand for farm labor may increase fertility. For instance, in Indonesia, a delay in the monsoon rainfall necessary for planting rice (a key staple crop) is associated with reduced fertility the following year (Sellers & Gray, 2019). Crop failure in Tanzania is also associated with lower fertility in the years immediately following the event (Alam & Pörtner, 2018). Sasson and Weinreb (2017) found evidence for a VCM mechanism taking place in parts of sub-Saharan Africa where declines in local natural resources resulted in higher fertility, especially in communities with low baseline resource levels. However, in communities with higher baseline resource levels, declines in resource availability resulted in lower fertility. This may be because the latter set of communities had additional forms of wealth or economic opportunities to help buffer the impact of poorer resource conditions, and/or had less of a need for additional labor (Sasson & Weinreb, 2017).

Studies adopting land use and agricultural metrics to measure the impact of resource conditions on live births have yielded mixed results across settings. In Nepal, more land area devoted to agriculture (a sign of more favorable agricultural conditions) is associated with higher odds of a first birth (Ghimire & Axinn, 2010). In the same setting, researchers examined the role of technology adoption on births. Farm households that accumulate capital can substitute technology, such as tractors, for farm-related tasks, in turn reducing household labor needs. This phenomenon is associated with lower fertility in Nepali households that adopted labor-saving agricultural technologies (Bhandari & Ghimire, 2013). By contrast, a different set of patterns was found in the Ecuadorian Amazon, where fertility rose more quickly when households increased land in pasture (which generally requires less labor but is more profitable), as well as among households that lacked land titles (and thus faced greater risk when investing in their farms). This suggests that fertility was motivated by both higher incomes as well as a desire to diversify household risks (Pan & López-Carr, 2016).

Disruption of Livelihoods in Disasters

Various studies exploring the aftermath of natural disasters examine changes in the number of live births. Generally, only small effects emerge with the direction of association varying between settings. For example, evidence from the southeastern United States suggests that more severe storm events, where individuals are more likely to evacuate, reduce fertility. On the other hand, when individuals exposed to storms are able to stay in place, there is a slightly greater likelihood of having a child 9 months following the event (Evans et al., 2010). Cohan and Cole (2002) use data from hurricanes in South Carolina to argue that stress associated with extreme weather events may strengthen the emotional bonds among couples. The authors argue these bonds contributed to a short-term increase in fertility following the storms. Demonstrating another pathway to increased fertility, births rose following the 2010 Haiti earthquake, which researchers attribute to an increased unmet need for family planning services (Behrman & Weitzman, 2016). However, in line with the work in southeastern U.S., flooding in North Dakota, which forced widespread evacuations, was associated with decreases in fertility in the most affected counties in the years immediately following the event (Tong et al., 2011).

The perceived risks of child mortality have long been established as a predictor of fertility (Cain, 1981) and recent research has examined the environmental aspects of this association. For example, women that lost children as a result of the Indian Ocean tsunami in 2004 were more likely to have children in the years after the tsunami than those which did not lose a child. Moreover, childless women in communities that experienced greater mortality as a result of the event were more likely to give birth in the years following the event (Nobles et al., 2015).

If demographic subgroups are differentially affected by natural hazards, due to preexisting socioeconomic conditions, differences in hazard risk, or other causes, post-disaster fertility patterns will likely be impacted as well. Indeed, black residents experienced greater adverse impacts of Hurricane Katrina compared to white residents, which affected birth outcomes following the event. While white residents saw a slight increase in fertility, black residents experienced a significant decline (Seltzer & Nobles, 2017). By contrast, following the 2001 earthquake in Gujarat, India, birth rates rose significantly, particularly among less educated women and those from minority ethnic groups (Nandi et al., 2018).

However, while there is some evidence suggesting a short-term impact on births associated with natural hazards, there is less support for changes in longer-term fertility trends. After Hurricane Mitch struck Nicaragua, the most affected areas experienced an increase in births in the years immediately following the storm. However, a follow-up study roughly 5 years after the event found these differences had largely attenuated (Davis, 2017). Nevertheless, if the risk of an event persists over time, this may have lasting fertility effects. In Guatemala, women living in communities with a greater risk of hurricanes (as measured over a 118-year period) and who were in households that owned land had more children on average than women in communities with lower hurricane risks (Pörtner, 2008).

Environmental Contributors to Birth Seasonality

Closely integrated throughout this discussion is literature addressing the phenomenon of birth seasonality. Seasonal patterns in births have been documented in the demographic literature for over 50 years in a variety of settings (Cowgill, 1966; Lam & Miron, 1991; Udry & Morris, 1967). However, birth seasonality is a complex phenomenon, involving biological, social, and environmental factors (Ellison et al., 2005). In modern-day industrialized countries, seasonality in births tends to be more strongly associated with sociocultural factors as opposed to environmental ones, including factors such as school enrollment cut-offs (Dahlberg & Andersson, 2019), expectations about future economic security (Caleiro, 2010), or the timing of religious holidays (Friger et al., 2009; Herteliu et al., 2015; Polašek et al., 2005).

However, there are numerous historical and modern examples of environmental factors being associated with seasonal patterns in births. Some of these pathways may be indirect. For instance, certain proximate biological determinants of conception have been shown to exhibit seasonal patterns, although the extent to which these patterns result in seasonal birth outcomes remains unclear. Such determinants include sperm quality (Levine, 1999; Santi et al., 2016) and conception rates (Rizzi & Dalla-Zuanna, 2007; Rojansky et al., 2000).

Additionally, more direct pathways exist, such as the association between temperature or precipitation timing and growing/harvest periods or vector-borne disease cycles. This in turn is associated with both maternal nutritional/health status and household labor activity, affecting coital frequency and ovarian function, as discussed in Section "Factors Affecting Conception and Early Pregnancy Loss Probability". In agrarian European societies, marriage seasonality has been closely tied with agricultural cycles (Dribe & Van de Putte, 2012; Sanna & Danubio, 2008), and this seasonality likely played a strong role in influencing the timing of first births in particular (Grech et al., 2003). However, as European societies began to industrialize, these seasonal birth fluctuations have largely attenuated (Cancho-Candela et al., 2007; Régnier-Loilier, 2010; Ruiu & Breschi, 2019). The United States has experienced similar patterns as Europe, with most of the country exhibiting a seasonal birth pattern throughout much of the twentieth Century, albeit decreasing over time (Martinez-Bakker et al., 2014).

Consistent with physiological explanations discussed earlier, seasonality of births is particularly pronounced in rural populations that experience significant variability in environmental conditions throughout the year and have limited access to outside markets. Among the Turkana peoples of northwest Kenya, a region that experiences a lengthy dry season, conceptions strongly correlate with food quantity and quality. Conceptions are highest at the beginning of the dry season when food is most abundant, which translates into more than half of all births in this population occurring in a 4 month window (Leslie & Fry, 1989). Similar significant seasonal patterns in births have been documented among agrarian or pastoral populations in other regions where there is strong intraannual climatic variability, including Central Africa (Bailey et al., 1992) and South Asia (Becker et al., 1986; Panter-Brick, 1996).

Despite increases in modern contraceptive use and greater access to health services, seasonal birth patterns continue to be found in more recent literature exploring populations that rely on natural resources for their livelihoods, but are more closely integrated into growing regional or global economies. For instance, in Mali, conceptions among farmers and agropastoralists tend to coincide with annual harvest periods, when food is most plentiful (Grace & Nagle, 2015). By contrast, conceptions in Mali tend to decrease when malaria cases peak and seasonal outmigration is at its height (Philibert et al., 2013). A multicounty analysis using data from 1987–2008 notes seasonal trends in conceptions aligned with harvest periods throughout much of Central and East Africa (Dorélien, 2016).

Heat-Related Physiological Mechanisms

There is also a small but growing literature exploring how temperature may affect birth outcomes, although this is still an area of ongoing research in terms of understanding both the relationships between temperature and birth outcomes as well as the physiological mechanisms underlying these trends. This literature is distinct from research discussed above on agricultural livelihoods, as most of the studies examining temperature and births have been conducted in high-income settings and emphasize the role of physiology. Using a large sample of births from the United States, Barreca et al. (2018) found that births decline 8–10 months following extremely hot days, although this is somewhat compensated for by a slight increase in births 11–13 months after the event. Earlier findings indicate that extremely high temperatures may reduce concep-

20 Environment and Fertility

tions which contributes to this pattern (Lam & Miron, 1996). Temperature may also have effects on delivery dates for women due to give birth. Among women in Montreal, significant increases in deliveries occurred 1 day following extremely high temperatures, with opposite effects found on colder days (Benmarhnia et al., 2015). However, these effects reversed for lags of 4 or 5 days, suggesting that the effect of extreme heat on deliveries is short-term in nature. Similar findings were found in Montreal for early term (37–38 weeks gestation) pregnancies, where risk of delivery went up following extreme heat events (Auger et al., 2014).

Adverse Pregnancy Outcomes

Adverse pregnancy outcomes can result in an elevated risk of infant and child mortality, which in turn has ramifications on future fertility decision-making and the degree to which women can achieve their ideal family size. This section discusses environmental linkages with a variety of maternal and perinatal health outcomes, including low birth weight (LBW), preterm birth (PTB), and stillbirths.[4] As these outcomes are often related to similar drivers, including maternal stress and nutritional deficiencies, they are addressed here collectively, with a discussion of the effects of environmental shocks on adverse pregnancy outcomes, followed by a discussion of the effects of environmental stressors.

Environmental Shocks and Adverse Pregnancy Outcomes

Scholars have linked a variety of extreme weather events with adverse pregnancy outcomes, including PTB and LBW. These effects are often attributed to the stress associated with being ex-

posed to an event during pregnancy and the effect of this maternal stress response on a developing fetus (Bussieres et al., 2015), although event timing and the magnitude of the stress mediate this relationship. Additionally, the relationships for LBW appear to be somewhat more consistent in the literature than relationships for PTB (Harville et al., 2010).

Examples of environmental shocks affecting adverse pregnancy outcomes include: a multi-decade study from Quebec, which found that women exposed to a severe ice storm during the first or second trimester were more likely to give birth to LBW babies than women exposed in the third trimester or who were unexposed (Dancause et al., 2011); a study of pregnancy outcomes following North Dakota flooding, noting an increased risk of LBW or PTB outcomes among women in the most flood-exposed counties (Tong et al., 2011); studies of hurricane exposure in the United States and substantial increases in LBW and PTB outcomes among the most exposed women particularly when that exposure is during the second or third trimester (Currie & Rossin-Slater, 2013; Xiong et al., 2008); a study of women who experienced a 2005 Chilean earthquake, finding increased PTB and LBW outcomes, particularly among women exposed during the first trimester (Torche, 2011); and a study of women exposed to wildfires in California experiencing elevated risks of PTB and LBW outcomes (Holstius et al., 2012). Differential disaster impacts between demographic subgroups may also affect adverse pregnancy outcomes. For instance, after Hurricane Katrina struck the southern United States in 2005, PTB and LBW rates unexpectedly declined in most jurisdictions impacted by the storm, which scholars speculate may have resulted from reductions in live births among black women, who are more likely than white women to experience adverse pregnancy outcomes (Harville et al., 2020). Collectively, this literature illustrates that adverse pregnancy outcomes can result from many types of environmental shocks, and that shocks may result in adverse outcomes regardless of when they occur during pregnancy.

---

[4]While definitions vary across studies, typical definitions of these terms are as follows: LBW (births <2500 g); PTB (births <37 weeks gestation), stillbirth (fetal death at or after 20 weeks gestation).

Environmental Stressors and Adverse Pregnancy Outcomes

In addition to acute events, environmental health scholars have explored the relationship between environmental stressors, such as temperature or precipitation variability, and pregnancy outcomes. The most commonly explored environmental stressor linked to PTB, LBW, and stillbirth is extreme heat, with many studies detailing significant positive associations between high temperatures and these adverse pregnancy outcomes, largely in developed country contexts. For example, in Belgium, a maximum temperature at the 95th percentile 1 day before delivery is associated with a 9.6% increase in the risk of PTB versus a temperature at the 50th percentile (Cox et al., 2016). In California, higher temperatures are linked to significantly higher rates of stillbirth, particularly when temperatures are elevated in the week prior to fetal death being recorded (Basu et al., 2016) and LBW, particularly when temperatures are elevated during the third trimester (Basu et al., 2018).

Similar patterns linking elevated temperatures at various points of pregnancy to adverse pregnancy outcomes include studies from Europe (Arroyo et al., 2016; Schifano et al., 2013; Vicedo-Cabrera et al., 2014), East Asia (Guo et al., 2018; Son et al., 2019; Weng et al., 2018), North America, (Auger et al., 2017; Basu et al., 2010; Ha et al., 2017), and Australia (Mathew et al., 2017; Wang et al., 2013, 2019), with the bulk of this literature examining PTB as opposed to LBW or stillbirth (Chersich et al., 2020) An important caveat is that methods vary widely across studies, which can complicate developing more generalizable conclusions. Studies differ in lag structures (time from when exposure to an environmental stressor occurs to when a pregnancy outcome is measured) as well as in how environmental stressors are measured. For instance, research measuring heat exposure uses a variety of indicators, including maximum temperature, mean temperature, or binary indicators for heat waves, just to name a few. This complicates efforts to develop more generalizable conclusions about specific environmental changes and adverse pregnancy outcomes (Beltran et al., 2014; Chersich et al., 2020).

Additionally, elevated temperatures are generally associated with higher concentrations of some ambient air pollutants, such as ozone or PM10 (Kalisa et al., 2018; Stathopoulou et al., 2008), although meteorological phenomena such as inversions can complicate this relationship (Gehrig & Buchmann, 2003). While the association between air pollutant concentrations and adverse pregnancy outcomes is robust but not uniform (Shah & Balkhair, 2011), this literature suggests air pollution may be a key additional mechanism linking high temperatures with adverse pregnancy outcomes (Díaz et al., 2016; Li et al., 2018; Wang et al., 2018).[5]

Because of its impact on nutrition and livelihoods associated with natural resource use, temperature variability can also affect obstetric outcomes for individuals in lower-income settings engaged in rural livelihoods activities. As is true in industrialized settings, exposure to extreme heat appears to increase the likelihood of adverse birth outcomes. For instance, across sub-Saharan Africa, the number of days over 37.8 °C is significantly associated with significantly lower birth weights, particularly when those extremely hot days occur during the first and second trimesters (Grace et al., 2015). Interestingly, while higher levels of rainfall (generally favorable for agriculture) are associated with slightly higher birth weights, temperature effects predominated when placed in a regression model together with precipitation. Similarly, in Kenya and Mali, high levels of precipitation are significantly associated with higher birth weights, but only among households engaged in food cropping or pastoralism (as opposed to cash-cropping households), while the results for temperature are less robust (Bakhtsiyarava et al., 2018). These patterns also appear to

---

[5] As noted earlier, this review only addresses pollution impacts associated with environmental change on pregnancy outcomes. However, there is a broad and fast-growing air pollution and adverse pregnancy outcomes literature. Interested readers may wish to consult recent reviews (Klepac et al., 2018; Stieb et al., 2012; Vrijheid et al., 2016) which contain more thorough discussions of the developing science in this field.

extend beyond the African context. In Colombia, exposure to heat waves during the third trimester reduced average weight at birth by roughly three grams (Andalón et al., 2016), while in Brazil, increased rainfall before birth is associated with higher birth weights (Rocha & Soares, 2015).

While this literature is diverse in terms of the environmental stimuli and pregnancy outcomes measured, environmental changes associated with climate change, including higher temperatures and an increased frequency of extreme weather events, are often associated with a higher risk of adverse pregnancy outcomes. While some of these effects may be partly offset in some geographies due to higher rainfall, current trends point to greater environmental risks for adverse pregnancy outcomes in the coming decades.

Population, Health, and Environment Programs

A small, but growing component of the population-environment literature centers on interventions designed to address environmental drivers of fertility. These initiatives, often referred to as population, health, environment (PHE) programs, provide activities addressing natural resource management challenges that can affect fertility behavior, as well as gaps in access to health care, particularly the provision of high-quality family planning services (Oglethorpe et al., 2008). PHE initiatives are typically implemented in rural communities where there is a high dependence on natural resources for local livelihoods. While the motivation and structure of each project differs depending on local contexts, in general, PHE projects address factors related to unsustainable natural resource management, which typically include demographic factors (particularly rapid population growth, often due to high levels of fertility), social factors (unequal access to natural resource-based livelihood activities in many settings among women and men), and institutional factors (poor management

of community forests and fisheries resulting in overharvesting) (De Souza, 2014).

The PHE approach is favored by some development practitioners because of the potential for synergistic outcomes to result from conducting both environment and health activities under the auspices of a single intervention, including greater recognition among community members about the linkages between human health and environmental conditions (D'Agnes et al., 2010; De Souza, 2014). To date, there is evidence in both the peer-reviewed and grey literatures that PHE projects result in positive health and environmental outcomes, including outcomes likely resulting from these synergies (Lopez-Carr & Ervin, 2017; Yavinsky et al., 2015). Examples of success from PHE projects include an increase in fish catches in the Philippines (D'Agnes et al., 2010), greater family planning use in Nepal and Ethiopia (Hahn et al., 2011; Sinaga et al., 2015), reduced cases of childhood diarrhea in Cameroon (Lopez-Carr & Ervin, 2017), and improved resilience to environmental change in western Tanzania (Hardee et al., 2018). However, PHE projects are often subject to short funding cycles and many environmental indicators in particular require longer durations than projects are funded for in order to show improvement. Thus, the long-term sustainability and impact of PHE projects remains an important area of ongoing research (Sellers, 2019).

Conclusion

Environmental stimuli have significant effects on all stages of the fertility process, including union formation and fertility preferences, conception and early pregnancy loss, live births, and various adverse pregnancy outcomes. In cases where environmental stimuli present socioeconomic challenges to households, there is growing evidence that individuals with the ability to control their fertility opt to exercise this control by reducing fertility desires and increasing family planning use. In general, live birth and adverse pregnancy outcome responses are similar to those of other health outcomes; environmental stimuli that adversely affect socioeconomic and/or health status tend to be associated with reduced rates of live births and higher rates of adverse pregnancy outcomes. However, researchers have identified significant spatial and temporal heterogeneity in responses to environmental stimuli, necessitating additional investigation to further refine our understandings of the mechanisms linking environmental factors with fertility outcomes.

A key theme emanating from the above discussion is the robust interdisciplinary nature of this line of research. While much of the literature discussed in this chapter, particularly research exploring household-level responses, has been generated by environmental demographers, sociologists, economists, or geographers, essential contributions have also been made by scholars identifying as biodemographers or biological anthropologists, as well as those with backgrounds in epidemiology, environmental health, or medicine. Harnessing this diversity of perspectives through new and creative scholarly collaborations is essential to fostering more holistic understandings of how environmental stimuli shape fertility outcomes at different spatial and temporal scales on a rapidly changing planet, as well as how environmental effects are modified by biological, social, and cultural variables.

As the discussion above illustrates, the literature linking environmental factors with fertility outcomes is growing rapidly. Because of the increasingly important role of global climate change as a driver of new public policies, as well as of adverse health outcomes (including those associated with reproductive health), environment-fertility research is likely to become ever more important for shaping policy debates. Crafting research questions with an eye towards informing individual- and system-level efforts to respond to environmental change will be essential for ensuring that couples can achieve the fertility outcomes they desire. Such an approach will be essential to demonstrate the continued relevance of the field in the coming years.

References

Abiona, O. (2017). The impact of unanticipated economic shocks on the demand for contraceptives: Evidence from Uganda. *Health Economics, 26,* 1696–1709.

Alam, S. A., & Pörtner, C. C. (2018). Income shocks, contraceptive use, and timing of fertility. *Journal of Development Economics, 131,* 96–103. https://doi.org/10.1016/j.jdeveco.2017.10.007

Andalón, M., Azevedo, J. P., Rodríguez-Castelán, C., et al. (2016). Weather shocks and health at birth in Colombia. *World Development, 82,* 69–82. https://doi.org/10.1016/j.worlddev.2016.01.015

Arroyo, V., Díaz, J., Ortiz, C., et al. (2016). Short term effect of air pollution, noise and heat waves on preterm births in Madrid (Spain). *Environmental Research, 145,* 162–168.

Asamoah, B., Kjellstrom, T., & Östergren, P.-O. (2018). Is ambient heat exposure levels associated with miscarriage or stillbirths in hot regions? A cross-sectional study using survey data from the Ghana Maternal Health Survey 2007. *International Journal of Biometeorology, 62,* 319–330. https://doi.org/10.1007/s00484-017-1402-5

Auger, N., Fraser, W. D., Smargiassi, A., et al. (2017). Elevated outdoor temperatures and risk of stillbirth. *International Journal of Epidemiology, 46,* 200–208. https://doi.org/10.1093/ije/dyw077

Auger, N., Naimi, A. I., Smargiassi, A., et al. (2014). Extreme heat and risk of early delivery among preterm and term pregnancies. *Epidemiology, 25,* 344–350.

Bailey, R. C., Jenike, M. R., Ellison, P. T., et al. (1992). The ecology of birth seasonality among agriculturalists in Central Africa. *Journal of Biosocial Science, 24,* 393–412.

Bakhtsiyarava, M., Grace, K., & Nawrotzki, R. J. (2018). Climate, birth weight, and agricultural livelihoods in Kenya and Mali. *American Journal of Public Health, 108,* S144–S150.

Barbieri, A. F., Bilsborrow, R. E., & Pan, W. K. (2005). Farm household lifecycles and land use in the Ecuadorian Amazon. *Population and Environment, 27,* 1–27. https://doi.org/10.1007/s11111-005-0013-y

Barbieri, A. F., Guedes, G. R., & Onofre dos Santos, R. (2020). Land use systems and livelihoods in demographically heterogeneous frontier stages in the amazon. *Environmental Development, 100587.* https://doi.org/10.1016/j.envdev.2020.100587

Barreca, A., Deschenes, O., & Guldi, M. (2018). Maybe next month? Temperature shocks and dynamic adjustments in birth rates. *Demography, 55,* 1269–1293. https://doi.org/10.1007/s13524-018-0690-7

Basu, R., Malig, B., & Ostro, B. (2010). High ambient temperature and the risk of preterm delivery. *American Journal of Epidemiology, 172,* 1108–1117. https://doi.org/10.1093/aje/kwq170

Basu, R., Rau, R., Pearson, D., & Malig, B. (2018). Temperature and term low birth weight in California. *American Journal of Epidemiology, 187,* 2306–2314.

Basu, R., Sarovar, V., & Malig, B. J. (2016). Association between high ambient temperature and risk of stillbirth in California. *American Journal of Epidemiology, 183*, 894–901. https://doi.org/10.1093/aje/kwv295

Becker, G. S., & Lewis, H. G. (1973). On the interaction between the quantity and quality of children. *Journal of Political Economy, 81*, S279–S288.

Becker, S., Chowdhury, A., & Leridon, H. (1986). Seasonal patterns of reproduction in Matlab, Bangladesh. *Population Studies, 40*, 457–472. https://doi.org/10.1080/0032472031000142356

Behrman, J., & Weitzman, A. (2016). Effects of the 2010 Haiti earthquake on women's reproductive health. *Studies in Family Planning, 47*, 3–17. https://doi.org/10.1111/j.1728-4465.2016.00045.x

Beltran, A. J., Wu, J., & Laurent, O. (2014). Associations of meteorology with adverse pregnancy outcomes: A systematic review of preeclampsia, preterm birth and birth weight. *International Journal of Environmental Research and Public Health, 11*, 91–172.

Benmarhnia, T., Auger, N., Stanislas, V., et al. (2015). The relationship between apparent temperature and daily number of live births in Montreal. *Maternal and Child Health Journal, 19*, 2548–2551.

Bhandari, P., & Ghimire, D. (2013). Rural agricultural change and fertility transition in Nepal. *Rural Sociology, 78*, 229–252.

Biddlecom, A. E., Axinn, W. G., & Barber, J. S. (2005). Environmental effects on family size preferences and subsequent reproductive behavior in Nepal. *Population and Environment, 26*, 583–621. https://doi.org/10.1007/s11111-005-1874-9

Bilsborrow, R. E. (1987). Population pressures and agricultural development in developing countries: A conceptual framework and recent evidence. *World Development, 15*, 183–203.

Bongaarts, J. (1978). A framework for analyzing the proximate determinants of fertility. *Population and Development Review, 4*, 105–132.

Brauner-Otto, S. R. (2014). Environmental quality and fertility: The effects of plant density, species richness, and plant diversity on fertility limitation. *Population and Environment, 36*, 1–31.

Brauner-Otto, S. R., & Axinn, W. G. (2017). Natural resource collection and desired family size: A longitudinal test of environment-population theories. *Population and Environment, 38*, 381–406. https://doi.org/10.1007/s11111-016-0267-6

Bussieres, E.-L., Tarabulsy, G. M., Pearson, J., et al. (2015). Maternal prenatal stress and infant birth weight and gestational age: A meta-analysis of prospective studies. *Developmental Review, 36*, 179–199.

Cain, M. (1981). Risk and insurance: Perspectives on fertility and agrarian change in India and Bangladesh. *Population and Development Review, 7*, 435–474. https://doi.org/10.2307/1972559

Cain, M. (1983). Fertility as an adjustment to risk. *Population and Development Review*, 688–702.

Caleiro, A. (2010). Exploring the peaks and valleys in the number of births in Portugal. *Human Ecology, 38*, 137–145. https://doi.org/10.1007/s10745-009-9294-6

Cancho-Candela, R., Andrés-de Llano, J. M., & Ardura-Fernandez, J. (2007). Decline and loss of birth seasonality in Spain: Analysis of 33 421 731 births over 60 years. *Journal of Epidemiology & Community Health, 61*, 713–718.

Chayanov, A. V. (1986). *The theory of peasant economy*. Manchester University Press.

Chersich, M. F., Pham, M. D., Areal, A., et al. (2020). Associations between high temperatures in pregnancy and risk of preterm birth, low birth weight, and stillbirths: Systematic review and meta-analysis. *BMJ, 371*, m3811.

Cohan, C. L., & Cole, S. W. (2002). Life course transitions and natural disaster: Marriage, birth, and divorce following Hurricane Hugo. *Journal of Family Psychology, 16*, 14–25. https://doi.org/10.1037/0893-3200.16.1.14

Coppa, A., Di Donato, L., Vecchi, F., & Danubio, M. E. (2001). Seasonality of marriages and ecological contexts in rural communities of Central-Southern Italy (Abruzzo), 1500–1871. *Collegium Antropologicum, 25*, 403–412.

Cowgill, U. M. (1966). The season of birth in man. *Man, 1*, 232–240.

Cox, B., Vicedo-Cabrera, A. M., Gasparrini, A., et al. (2016). Ambient temperature as a trigger of preterm delivery in a temperate climate. *Journal of Epidemiology and Community Health, 70*, 1191–1199.

Currie, J., & Rossin-Slater, M. (2013). Weathering the storm: Hurricanes and birth outcomes. *Journal of Health Economics, 32*, 487–503. https://doi.org/10.1016/j.jhealeco.2013.01.004

Czeizel, A., Bognár, Z., & Rockenbauer, M. (1984). Some epidemiological data on spontaneous abortion in Hungary, 1971–80. *Journal of Epidemiology & Community Health, 38*, 143–148.

D'Agnes, L., D'Agnes, H., Schwartz, J. B., et al. (2010). Integrated management of coastal resources and human health yields added value: A comparative study in Palawan (Philippines). *Environmental Conservation, 37*, 398–409. https://doi.org/10.1017/s0376892910000779

Dahlberg, J., & Andersson, G. (2019). Fecundity and human birth seasonality in Sweden: A register-based study. *Reproductive Health, 16*, 87.

Dancause, K. N., Laplante, D. P., Oremus, C., et al. (2011). Disaster-related prenatal maternal stress influences birth outcomes: Project Ice Storm. *Early Human Development, 87*, 813–820. https://doi.org/10.1016/j.earlhumdev.2011.06.007

Dasgupta, P. (1995). *An inquiry into Well-being and destitution*. Clarendon Press.

Davis, J. (2017). Fertility after natural disaster: Hurricane Mitch in Nicaragua. *Population and Environment, 38*, 448–464. https://doi.org/10.1007/s11111-017-0271-5

Davis, K. (1963). The theory of change and response in modern demographic history. *Population Index, 29*, 345–366.

de Sherbinin, A., VanWey, L. K., McSweeney, K., et al. (2008). Rural household demographics, livelihoods and the environment. *Global Environmental Change, 18*, 38–53.

De Souza, R. M. (2014). Resilience, integrated development and family planning: Building long-term solutions. *Reproductive Health Matters, 22*, 75–83. https://doi.org/10.1016/s0968-8080(14)43773-x

Dewey, K. G., Cohen, R. J., Rivera, L. L., et al. (1997). Effects of age at introduction of complementary foods to breast-fed infants on duration of lactational amenorrhea in Honduran women. *The American Journal of Clinical Nutrition, 65*, 1403–1409.

Díaz, J., Arroyo, V., Ortiz, C., et al. (2016). Effect of environmental factors on low weight in non-premature births: A time series analysis. *PLoS One, 11*, e0164741.

Djafri, D., Chongsuvivatwong, V., & Geater, A. (2015). Effect of the September 2009 Sumatra earthquake on reproductive health services and MDG 5 in the city of Padang, Indonesia. *Asia Pacific Journal of Public Health, 27*, NP1444–NP1456.

Dorélien, A. M. (2016). Birth seasonality in sub-Saharan Africa. *Demographic Research, 34*, 761–796.

Dribe, M., & Van de Putte, B. (2012). Marriage seasonality and the industrious revolution: Southern Sweden, 1690–1895 1. *The Economic History Review, 65*, 1123–1146.

Ehrlich, P. (1968). *The population bomb*. Ballantine Books.

Eissler, S., Thiede, B. C., & Strube, J. (2019). Climatic variability and changing reproductive goals in sub-Saharan Africa. *Global Environmental Change, 57*, 101912.

Ellis, F. (2000). *Rural livelihoods and diversity in developing countries*. Oxford University Press.

Ellison, P. T., Valeggia, C. R., & Sherry, D. S. (2005). Human birth seasonality. In D. K. Brockman & C. P. van Schaik (Eds.), *Seasonality in Primates: Studies of living and extinct human and non-human Primates* (pp. 379–399). Cambridge University Press.

Evans, R. W., Hu, Y., & Zhao, Z. (2010). The fertility effect of catastrophe: U.S. hurricane births. *Journal of Population Economics, 23*, 1–36. https://doi.org/10.1007/s00148-008-0219-2

Friger, M., Shoham-Vardi, I., & Abu-Saad, K. (2009). Trends and seasonality in birth frequency: A comparison of Muslim and Jewish populations in southern Israel: Daily time series analysis of 200 009 births, 1988–2005. *Human Reproduction, 24*, 1492–1500.

Gehrig, R., & Buchmann, B. (2003). Characterising seasonal variations and spatial distribution of ambient PM10 and PM2.5 concentrations based on long-term Swiss monitoring data. *Atmospheric Environment, 37*, 2571–2580. https://doi.org/10.1016/S1352-2310(03)00221-8

Ghimire, D. J., & Axinn, W. G. (2010). Community context, land use, and first birth. *Rural Sociology, 75*, 478–513.

González-Martín, A. (2008). Ecological and cultural pressure on marriage seasonality in the Principality of Andorra. *Journal of Biosocial Science, 40*, 1–18.

Grace, K. (2017). Considering climate in studies of fertility and reproductive health in poor countries. *Nature Climate Change, 7*, 479–485.

Grace, K., Davenport, F., Hanson, H., et al. (2015). Linking climate change and health outcomes: Examining the relationship between temperature, precipitation and birth weight in Africa. *Global Environmental Change, 35*, 125–137. https://doi.org/10.1016/j.gloenvcha.2015.06.010

Grace, K., & Nagle, N. N. (2015). Using high-resolution remotely sensed data to examine the relationship between agriculture and fertility in Mali. *The Professional Geographer, 67*, 641–654. https://doi.org/10.1080/00330124.2015.1032899

Grech, V., Savona-Ventura, C., Agius-Muscat, H., & Janulova, L. (2003). Seasonality of births is associated with seasonality of marriages in Malta. *Journal of Biosocial Science, 35*, 95–105. https://doi.org/10.1017/S0021932003000956

Guo, T., Wang, Y., Zhang, H., et al. (2018). The association between ambient temperature and the risk of preterm birth in China. *Science of the Total Environment, 613*, 439–446.

Ha, S., Liu, D., Zhu, Y., et al. (2017). Ambient temperature and early delivery of singleton pregnancies. *Environmental Health Perspectives, 125*, 453–459.

Hahn, S., Anandaraja, N., & D'Agnes, L. (2011). Linking population, health, and the environment: An overview of integrated programs and a case study in Nepal. *Mount Sinai Journal of Medicine, 78*, 394–405. https://doi.org/10.1002/msj.20258

Hapsari, E. D., Widyawati, N. W. A., et al. (2009). Change in contraceptive methods following the Yogyakarta earthquake and its association with the prevalence of unplanned pregnancy. *Contraception, 79*, 316–322. https://doi.org/10.1016/j.contraception.2008.10.015

Hardee, K., Patterson, K. P., Schenck-Fontaine, A., et al. (2018). Family planning and resilience: Associations found in a population, health, and environment (PHE) project in Western Tanzania. *Population and Environment, 40*, 204–238.

Harville, E., Xiong, X., & Buekens, P. (2010). Disasters and perinatal health: A systematic review. *Obstetrical & Gynecological Survey, 65*, 713–728. https://doi.org/10.1097/OGX.0b013e31820eddbe

Harville, E. W., Xiong, X., David, M., & Buekens, P. (2020). The paradoxical effects of Hurricane Katrina on births and adverse birth outcomes. *American Journal of Public Health, 110*, 1466–1471. https://doi.org/10.2105/AJPH.2020.305769

Herteliu, C., Ileanu, B. V., Ausloos, M., & Rotundo, G. (2015). Effect of religious rules on time of conception in Romania from 1905 to 2001. *Human Reproduction, 30*, 2202–2214. https://doi.org/10.1093/humrep/dev129

Holstius, D. M., Reid, C. E., Jesdale, B. M., & Morello-Frosch, R. (2012). Birth weight following pregnancy during the 2003 Southern California wildfires. *Environmental Health Perspectives, 120*, 1340–1345.

Huffman, S. L., Ford, K., Allen, J., Hubert, A., & Streble, P. (1987). Nutrition and fertility in Bangladesh: Breastfeeding and post partum amenorrhoea. *Population Studies, 41*, 447–462.

Jennings, J. A., & Gray, C. L. (2017). Climate and marriage in the Netherlands, 1871–1937. *Population and Environment, 38*, 242–260.

Kalisa, E., Fadlallah, S., Amani, M., et al. (2018). Temperature and air pollution relationship during heatwaves in Birmingham, UK. *Sustainable Cities and Society, 43*, 111–120. https://doi.org/10.1016/j.scs.2018.08.033

Kallan, J. E., & Enneking, E. A. (1992). Seasonal patterns of spontaneous abortion. *Journal of Biosocial Science, 24*, 71–76.

Kissinger, P., Schmidt, N., Sanders, C., & Liddon, N. (2007). The effect of the Hurricane Katrina disaster on sexual behavior and access to reproductive care for young women in New Orleans. *Sexually Transmitted Diseases*, 34. 883–886. https://doi.org/10.1097/OLQ.0b013e318074c5f8

Klepac, P., Locatelli, I., Korošec, S., et al. (2018). Ambient air pollution and pregnancy outcomes: A comprehensive review and identification of environmental public health challenges. *Environmental Research, 167*, 144–159. https://doi.org/10.1016/j.envres.2018.07.008

Kurz, K. M., Habicht, J.-P., Rasmussen, K. M., & Schwager, S. J. (1993). Effects of maternal nutritional status and maternal energy supplementation on length of postpartum amenorrhea among Guatemalan women. *The American Journal of Clinical Nutrition, 58*, 636–642.

Lam, D. A., & Miron, J. A. (1991). Seasonality of births in human populations. *Social Biology, 38*, 51–78. https://doi.org/10.1080/19485565.1991.9988772

Lam, D. A., & Miron, J. A. (1996). The effects of temperature on human fertility. *Demography, 33*, 291–305. https://doi.org/10.2307/2061762

Leslie, P. W., & Fry, P. H. (1989). Extreme seasonality of births among nomadic Turkana pastoralists. *American Journal of Physical Anthropology, 79*, 103–115.

Lesthaeghe, R. (2014). The second demographic transition: A concise overview of its development. *Proceedings of the National Academy of Sciences of the United States of America, 111*, 18112. https://doi.org/10.1073/pnas.1420441111

Levine, R. J. (1999). Seasonal variation of semen quality and fertility. *Scandinavian Journal of Work, Environment & Health*, 34–37.

Leyser-Whalen, O., Rahman, M., & Berenson, A. B. (2011). Natural and social disasters: Racial inequality in access to contraceptives after Hurricane Ike. *Journal of Women's Health, 20*, 1861–1866. https://doi.org/10.1089/jwh.2010.2613

Li, Q., Wang, Y., Guo, Y., et al. (2018). Effect of airborne particulate matter of 2.5 μm or less on preterm birth: A national birth cohort study in China. *Environment International, 121*, 1128–1136.

Lopez-Carr, D., & Ervin, D. (2017). Population-health-environment (PHE) synergies? Evidence from USAID-sponsored programs in African and Asian core conservation areas. *European Journal of Geography, 8*, 92–108.

Martinez-Bakker, M., Bakker, K. M., King, A. A., & Rohani, P. (2014). Human birth seasonality: Latitudinal gradient and interplay with childhood disease dynamics. *Proceedings of the Royal Society B: Biological Sciences, 281*, 20132438. https://doi.org/10.1098/rspb.2013.2438

Mathew, S., Mathur, D., Chang, A., et al. (2017). Examining the effects of ambient temperature on preterm birth in Central Australia. *International Journal of Environmental Research and Public Health, 14*, 147.

McCracken, S. D., Siqueira, A. D., Moran, E. F., et al. (2002). Land use patterns on an agricultural frontier in Brazil. In *Deforestation and land use in the Amazon* (pp. 162–192). University Press of Florida.

Mosley, W. H., & Chen, L. C. (1984). An analytical framework for the study of child survival in developing countries. *Population and Development Review, 10*, 25–45. https://doi.org/10.2307/2807954

Nandi, A., Mazumdar, S., & Behrman, J. R. (2018). The effect of natural disaster on fertility, birth spacing, and child sex ratio: Evidence from a major earthquake in India. *Journal of Population Economics, 31*, 267–293.

Nobles, J., Frankenberg, E., & Thomas, D. (2015). The effects of mortality on fertility: Population dynamics after a natural disaster. *Demography, 52*, 15–38. https://doi.org/10.1007/s13524-014-0362-1

O'Neill, B. C., MacKellar, F. L., & Lutz, W. (2001). *Population and climate change*. Cambridge University Press.

Oglethorpe, J., Honzak, C., & Margoluis, C. (2008). *Healthy people, healthy ecosystems: A manual on integrating health and family planning into conservation projects*. World Wildlife Fund.

Pan, W. K., Carr, D., Barbieri, A., et al. (2007). Forest clearing in the Ecuadorian Amazon: A study of patterns over space and time. *Population Research and Policy Review, 26*, 635–659.

Pan, W. K., & López-Carr, D. L. (2016). Land use as a mediating factor of fertility in the Amazon. *Population and Environment, 38*, 21–46. https://doi.org/10.1007/s11111-016-0253-z

Panter-Brick, C. (1996). Proximate determinants of birth seasonality and conception failure in Nepal. *Population Studies, 50*, 203–220.

Panter-Brick, C., Lotstein, D. S., & Ellison, P. T. (1993). Seasonality of reproductive function and weight loss in rural Nepali women. *Human Reproduction, 8*, 684–690.

Peng, Y.-K., Hight-Laukaran, V., Peterson, A. E., & Perez-Escamilla, R. (1998). Maternal nutritional status is inversely associated with lactational amenorrhea in sub-Saharan Africa: Results from demographic and health surveys II and III. *The Journal of Nutrition, 128*, 1672–1680.

Perz, S. G. (2001). Household demographic factors as life cycle determinants of land use in the Amazon. *Population Research and Policy Review, 20*, 159–186. https://doi.org/10.1023/A:1010658719768

Philibert, A., Tourigny, C., Coulibaly, A., & Fournier, P. (2013). Birth seasonality as a response to a changing rural environment (Kayes region, Mali). *Journal of Biosocial Science, 45*, 547–565.

Polašek, O., Kolčić, I., Vorko-Jović, A., et al. (2005). Seasonality of births in Croatia. *Collegium Antropologicum, 29*, 249–255.

Pörtner, C.C. (2008). Gone with the wind? Hurricane risk, fertility and education. Working Paper UWEC-2006-19-R.

Potts, M., Henderson, C., & Campbell, M. (2013). The Sahel: A Malthusian challenge? *Environmental & Resource Economics, 55*, 501–512. https://doi.org/10.1007/s10640-013-9679-2

Preston, S. H. (1978). *The effects of infant and child mortality on fertility*. Academic.

Rault, W., & Régnier-Loilier, A. (2016). Seasonality of marriages, past and present. *Population, 71*, 675–680.

Régnier-Loilier, A. (2010). Changes in the seasonality of births in France from 1975 to the present. *Population, 65*, 145–185.

Rizzi, E. L., & Dalla-Zuanna, G. (2007). The seasonality of conception. *Demography, 44*, 705–728. https://doi.org/10.1353/dem.2007.0040

Rocha, R., & Soares, R. R. (2015). Water scarcity and birth outcomes in the Brazilian semiarid. *Journal of Development Economics, 112*, 72–91. https://doi.org/10.1016/j.jdeveco.2014.10.003

Rodgers, J. L., John, C. A. S., & Coleman, R. (2005). Did fertility go up after the Oklahoma City bombing? An analysis of births in metropolitan counties in Oklahoma, 1990–1999. *Demography, 42*, 675–692.

Rojansky, N., Benshushan, A., Meirsdorf, S., et al. (2000). Seasonal variability in fertilization and embryo quality rates in women undergoing IVF. *Fertility and Sterility, 74*, 476–481. https://doi.org/10.1016/S0015-0282(00)00669-5

Ruiu, G., & Breschi, M. (2019). Intensity of agricultural workload and the seasonality of births in Italy. *European Journal of Population*, 1–29.

Ruiu, G., & Gonano, G. (2015). Seasonality of marriages in Italian regions: An analysis from the formation of the Italian kingdom to the present. *Rivista italiana di economia, demografia e statistica, 69*, 135–143.

Sanna, E., & Danubio, M. E. (2008). Seasonality of marriages in Sardinian pastoral and agricultural communities in the nineteenth century. *Journal of Biosocial Science, 40*, 577–586.

Santi, D., Vezzani, S., Granata, A. R., et al. (2016). Sperm quality and environment: A retrospective, cohort study in a Northern province of Italy. *Environmental Research, 150*, 144–153.

Sasson, I., & Weinreb, A. (2017). Land cover change and fertility in West-Central Africa: Rural livelihoods and the vicious circle model. *Population and Environment, 38*, 345–368.

Schifano, P., Lallo, A., Asta, F., et al. (2013). Effect of ambient temperature and air pollutants on the risk of preterm birth, Rome 2001–2010. *Environment International, 61*, 77–87. https://doi.org/10.1016/j.envint.2013.09.005

Scrimshaw, S. C. (1978). Infant mortality and behavior in the regulation of family size. *Population and Development Review, 4*, 383–403.

Sellers, S. (2019). Does doing more result in doing better? Exploring synergies in an integrated population, health and environment project in East Africa. *Environmental Conservation, 46*, 43–51. https://doi.org/10.1016/j.worlddev.2019.02.003

Sellers, S., & Gray, C. (2019). Climate shocks constrain human fertility in Indonesia. *World Development, 117*, 357–369. https://doi.org/10.1016/j.worlddev.2019.02.003

Seltzer, N., & Nobles, J. (2017). Post-disaster fertility: Hurricane Katrina and the changing racial composition of New Orleans. *Population and Environment, 38*, 465–490. https://doi.org/10.1007/s11111-017-0273-3

Sennott, C., & Yeatman, S. (2012). Stability and change in fertility preferences among young women in Malawi. *International Perspectives on Sexual and Reproductive Health, 38*, 34.

Shah, P. S., & Balkhair, T. (2011). Air pollution and birth outcomes: A systematic review. *Environment International, 37*, 498–516. https://doi.org/10.1016/j.envint.2010.10.009

Sinaga, M., Mohammed, A., Teklu, N., et al. (2015). Effectiveness of the population health and environment approach in improving family planning outcomes in the Gurage, Zone South Ethiopia. *BMC Public Health, 15*, 1123. https://doi.org/10.1186/s12889-015-2484-9

Sobotka, T., Skirbekk, V., & Philipov, D. (2011). Economic recession and fertility in the developed world. *Population and Development Review, 37*, 267–306. https://doi.org/10.1111/j.1728-4457.2011.00411.x

Son, J.-Y., Lee, J.-T., Lane, K. J., & Bell, M. L. (2019). Impacts of high temperature on adverse birth outcomes in Seoul, Korea: Disparities by individual-and community-level characteristics. *Environmental Research, 168*, 460–466.

Stathopoulou, E., Mihalakakou, G., Santamouris, M., & Bagiorgas, H. S. (2008). On the impact of temperature on tropospheric ozone concentration levels in urban environments. *Journal of Earth System Science, 117*, 227–236. https://doi.org/10.1007/s12040-008-0027-9

Stieb, D. M., Chen, L., Eshoul, M., & Judek, S. (2012). Ambient air pollution, birth weight and preterm birth: A systematic review and meta-analysis. *Environmental Research, 117*, 100–111. https://doi.org/10.1016/j.envres.2012.05.007

Tong, V. T., Zotti, M. E., & Hsia, J. (2011). Impact of the Red River catastrophic flood on women giving birth in North Dakota, 1994–2000. *Maternal and Child Health Journal, 15*, 281–288. https://doi.org/10.1007/s10995-010-0576-9

Torche, F. (2011). The effect of maternal stress on birth outcomes: Exploiting a natural experiment. *Demography, 48*, 1473–1491.

Tracer, D. P. (1996). Lactation, nutrition, and postpartum amenorrhea in lowland Papua New Guinea. *Human Biology*, 277–292.

Udry, J. R., & Morris, N. M. (1967). Seasonality of coitus and seasonality of birth. *Demography, 4*, 673–679. https://doi.org/10.2307/2060307

Valeggia, C., & Ellison, P. T. (2009). Interactions between metabolic and reproductive functions in the resumption of postpartum fecundity. *American Journal of Human Biology: The Official Journal of the Human Biology Association, 21*, 559–566.

VanWey, L. K., D'Antona, Á., & Brondízio, E. (2007). Household demographic change and land use/land cover change in the Brazilian Amazon. *Population and Environment, 28*, 163–185. https://doi.org/10.1007/s11111-007-0040-y

Vicedo-Cabrera, A. M., Iñíguez, C., Barona, C., & Ballester, F. (2014). Exposure to elevated temperatures and risk of preterm birth in Valencia, Spain. *Environmental Research, 134*, 210–217. https://doi.org/10.1016/j.envres.2014.07.021

Vitzthum, V. J. (2009). The ecology and evolutionary endocrinology of reproduction in the human female. *American Journal of Physical Anthropology, 140*, 95–136.

Vitzthum, V. J., Thornburg, J., & Spielvogel, H. (2009). Seasonal modulation of reproductive effort during early pregnancy in humans. *American Journal of Human Biology: The Official Journal of the Human Biology Association, 21*, 548–558.

Vrijheid, M., Casas, M., Gascon, M., et al. (2016). Environmental pollutants and child health—A review of recent concerns. *International Journal of Hygiene and Environmental Health, 219*, 331–342. https://doi.org/10.1016/j.ijheh.2016.05.001

Walker, R. (2003). Mapping process to pattern in the landscape change of the Amazonian frontier. *Annals of the Association of American Geographers, 93*, 376–398.

Wang, J., Tong, S., Williams, G., & Pan, X. (2019). Exposure to heat wave during pregnancy and adverse birth outcomes: An exploration of susceptible windows. *Epidemiology, 30*. https://doi.org/10.1097/EDE.0000000000000995

Wang, J., Williams, G., Guo, Y., et al. (2013). Maternal exposure to heatwave and preterm birth in Brisbane, Australia. *BJOG: An International Journal of Obstetrics & Gynaecology, 120*, 1631–1641. https://doi.org/10.1111/1471-0528.12397

Wang, Q., Benmarhnia, T., Zhang, H., et al. (2018). Identifying windows of susceptibility for maternal exposure to ambient air pollution and preterm birth. *Environment International, 121*, 317–324.

Weinberg, C. R., Moledor, E., Baird, D. D., & Wilcox, A. J. (1994). Is there a seasonal pattern in risk of early pregnancy loss? *Epidemiology*, 484–489.

Weng, Y.-H., Yang, C.-Y., & Chiu, Y.-W. (2018). Adverse neonatal outcomes in relation to ambient temperatures at birth: A nationwide survey in Taiwan. *Archives of Environmental & Occupational Health, 73*, 48–55.

Xiong, X., Harville, E. W., Buekens, P., et al. (2008). Exposure to Hurricane Katrina, post-traumatic stress disorder and birth outcomes. *The American Journal of the Medical Sciences, 336*, 111–115.

Yabiku, S. T. (2006). Land use and marriage timing in Nepal. *Population and Environment, 27*, 445–461.

Yavinsky, R. W., Lamere, C., Patterson, K. P., & Bremner, J. (2015). *The impact of population, health, and environment projects: A synthesis of evidence*. The Evidence Project.

Gender, Population and the Environment

21

Jessica Marter-Kenyon, Sam Sellers, and Maia Call

Abstract

Gender mediates a variety of population-environment interactions, affecting natural resource use behaviors as well as the individual and household impacts of environmental shocks and stressors. This chapter provides an introduction to research on gender, population, and the environment. Following an overview of intellectual developments in gender and environment theory, we survey existing population-environment scholarship on gender as an intervening factor in four key areas of inquiry: perception of environmental change and risk; human impacts on the environment; adaptation to environmental change; and well-being outcomes following environmental stress. We also highlight key methodological considerations and critiques, and conclude with a discussion of research gaps and future directions for scholarship to better understand how gender mediates

J. Marter-Kenyon (✉)
University of Georgia, Athens, Georgia
e-mail: jsmk@uga.edu

S. Sellers · M. Call
United States Agency for International Development
(USAID), Washington, DC, USA
e-mail: ssellers@usaid.gov; mcall@usaid.gov

the impacts of increasingly disruptive environmental changes.

Introduction

Gender is one of the primary identities around which human societies are constructed, and through which human behavior is constrained. Consequently, gender plays a significant role in mediating the nature and outcomes of population-environment processes, cutting across each of the themes articulated in this anthology: from research methodologies, to migration decisions and health outcomes, to environmental impacts such as deforestation and agricultural change.

In the context of population-environment research, gender typically refers to sociocultural differences in roles, rights, norms, access, privileges, values, and experiences between men/boys and women/girls (Parpart et al., 2000; van Dijk & Bose, 2016). Gender is a social construct, meaning that differences between groups of men and women are not (only) biological and therefore 'natural' in origin, but develop through sociological processes (Bezner-Kerr, 2014; McStay & Dunlap,

© Springer Nature Switzerland AG 2022
L. M. Hunter et al. (eds.), *International Handbook of Population and Environment*, International
Handbooks of Population 10, https://doi.org/10.1007/978-3-030-76433-3_21

1983). Gender is one identity that categorizes individuals and their activities; others include class, age, and talent (Lorber, 1994). Roles and responsibilities are assigned to men and women based on sociocultural assumptions about their inherent differences, resulting in gendered experiences, behaviors, and perspectives (West & Zimmerman, 1987). These constructs have material consequences (Jarviluoma et al., 2003); social proscriptions about the relative abilities and appropriate behaviors of men and women have led to widespread, entrenched inequities in many spheres, including in the rural livelihood activities that are the focus of much population-environment research. Although social norms have largely led to women occupying a subordinate status, gender identities are fluid and dynamic (Lorber, 1994); their production, meaning, and influence change over time, place, person, and situation (Jackson, 1993).

Gender-focused research on population and the environment begins from the understanding that socially-constructed differences between genders influence how people affect environmental change and natural resource management and use, as well as how people experience and respond to environmental change (Meinzen-Dick et al., 2014). The relationships between gender and the natural environment are complex and contextual (McDowell, 1992). Scholars have predominantly focused on women and girls, in part because the sociopolitical marginalization of females often renders them particularly reliant on natural resources and vulnerable to environmental change (Carr & Thompson, 2014). While it is important to resist tendencies to cast women's environmental relationships as either naturally vulnerable or naturally virtuous (Arora-Jonsson, 2011), and while the mechanisms that precipitate differential responses vary between contexts (Goh, 2012), there are consistent patterns across many settings where women are disadvantaged relative to men in agriculture and natural resource management. Yet gender also includes men, boys, and masculinity; along with

a continued focus on women and interactions between genders, population-environment, scholars are increasingly investigating male experiences (Brandth & Haugen, 2016; Enarson & Pease, 2016; Pearce et al., 2011).

In this chapter, we primarily survey the interaction of gender and the environment within populations reliant on natural resource-based livelihoods in low- and middle-income countries. While gender-environment relationships in cities and wealthier countries are growing areas of research (Howard, 2009; Leichenko, 2011), we opt to focus on the rural populations that have historically been at the heart of population-environment scholarship, and which are emphasized throughout this anthology. Similarly, while a large body of research investigates the ways that policies and institutions influence, and are influenced by, gendered population-environment interactions (e.g. Forsyth, 2003; Peet & Watts, 2004; Sen & Grown, 1987; Carney, 1996), due to space limitations we do not emphasize this literature. Nonetheless, it is important to highlight that inequities are further entrenched when policy fails to adequately consider gender (Rocheleau et al., 1996). It is also useful to note that other chapters in this volume touch on gender as a mediating factor in population-environment relationships (see, for example Chap. 17 on fertility).

We begin with a brief overview of intellectual developments in gender and environment theory and describe how gender has specifically been theorized within population-environment research. Next, we survey existing population-environment scholarship on gender as a mediating factor in four key areas of inquiry: risk perception; human impact on the environment; adaptation to environmental change; and well-being outcomes following environmental stress. We conclude the chapter by highlighting some methodological considerations and critiques followed by a discussion of research gaps and future directions for scholarship on gender, population, and the environment.

How Does Gender Shape Population-Environment Linkages?

Gender and Environment Theory

Scholarly interest in the relationship between gender and the environment accelerated in the 1970s and 1980s, inspired by global social and environmental justice movements (Meinzen-Dick et al., 2014; Rico, 1998). Over time, several theories were developed to explain how gender shapes the ways in which individuals affect, and are affected by, their environment. Although these narratives have waned and waxed in influence, the ideas discussed below all remain present in contemporary population-environment research and policy (Resurreccion, 2017).

Early work, falling under the umbrella of *'ecofeminism'*, centered on women and their 'close relationships' with nature (i.e. the 'Earth Mother' vision) (Rico, 1998). As originally conceived by Vandana Shiva, ecofeminism posited that: (1) women have a special emotional and spiritual connection to nature, in part due to their biological functions as mothers and (2) the patriarchal domination of nature is interlinked with the subordination of women (Merchant, 1992; Mies & Shiva, 1993; Shiva, 1989). A modified version of ecofeminism known as *"Women, environment and development" (WED)* followed, gaining traction with NGOs and policymakers (Resurreccion, 2017). WED scholars argued women should be the target of environment and development work because they make better environmental stewards and are more reliant on natural resources, and will therefore most effectively contribute to, and benefit from, interventions (Rodda, 1993; Sontheimer, 1991). Ecofeminist approaches were ultimately criticized for essentializing women and 'nature' – overlooking heterogeneity within both concepts – and for framing women as victims, rather than complex actors who facilitate and negotiate environmental change (Jackson, 1993; Leach, 2007; Leach et al., 1995). Critics noted that relationships of power determining and regulating gendered

positions in socio-environmental systems were generally not discussed and that WED ideas had the potential to burden already-overworked women with additional unpaid responsibilities for environmental management (Rico, 1998). Moreover, men, other than as agents of patriarchy, were largely ignored (Rao, 1991).

Theoretical developments over the past thirty years have led to a broader, more inclusive understanding of gender and the environment (Meinzen-Dick et al., 2014). Conceptually, most scholars have moved away from a 'single-axis' focus on gender as binary groups of men and women with static and homogeneous characteristics (May, 2014). *Feminist environmentalism*, advanced in the 1990s by Bina Agarwal (a prominent critic of ecofeminism) was particularly important to this shift (Meinzen-Dick et al., 2014). Agarwal observed that women's environmental encounters are conditioned by class relationships and struggles over the division of labor and power, both outside and, crucially, within the household (Agarwal, 1992). Feminist environmentalism drew attention to the contextual, dynamic, and contingent nature of gender-environment relationships (Resurreccion, 2017). More recent insights from ethnic, queer, and feminist studies have led to a widespread recognition of the *'intersectionality'* of gender, finding that gender is "constituted through other kinds of social differences and axes of power such as race, sexuality, class and place" (Resurreccion & Elmhirst, 2008: 5) and therefore one of several interacting identities mediating human relationships with the environment (Resurreccion, 2017). For instance, marriage produces changes in women's resource use (Gururani, 2002b), yet those relationships are mediated by class (Nightingale, 2006); age affects both men and women's ability to adopt adaptive strategies (Ravera et al., 2016), while high status men may benefit from outsized control over environmental resources (Nightingale, 2001).

The other major advancement in gender-environment theory is our understanding of power (Jackson, 1993). *Feminist political ecology (FPE)*, in particular, has elucidated the structural underpinnings of gender-environment relationships. Rather than focusing on women's roles, FPE emphasizes gender relations (Rico, 1998), prioritizing: (1) how power dynamics between men and women influence and are influenced by the environment (Rocheleau et al., 1996); (2) how gendered environmental impacts, and gendered experience of these changes, are co-constituted with political, economic, and social structures (Castree & Braun, 2001); and (3) how gender operates in environmental politics and justice movements (Harris, 2015; Schroeder, 1999). Likewise, *intrahousehold economics*, while not always explicitly focused on gender, sheds light on the unequal distribution of resources and bargaining power between men and women (McElroy, 1990). Population-environment researchers build on this work (Sect. 21.2.2), further exploring the influence of these inequities on resource use and environmental vulnerability (Agarwal, 1997).

Major Areas of Research

These intellectual developments have furnished researchers with diverse pathways for empirically investigating gender as a mediating factor in population-environment relationships. Here, we briefly discuss how gender influences four major areas of population-environment scholarship.

Access to Resources Human relationships with the environment are mediated through the availability and use of productive resources including land and water, time, education, information, social networks, and finance (Meinzen-Dick et al., 2014). The distribution of, access to, and control of these resources are governed by complex systems of formal and informal rules, which are often pervasively gendered, resulting in profound differences in the way men and women interact with the environment (Nightingale, 2006). In most low- and middle-income settings, women are substantially less likely to own or have control over land, productive assets, and capital (Deere et

al., 2012; Doss et al., 2015; Fletschner & Kenney, 2014). Gendered differences in access to environmental knowledge and information (regarding, for instance, knowledge regarding productive assets, natural resource management, and environmental changes) stem from a number of factors, including barriers to women's education (Campos et al., 2014), biases against women in agricultural extension services (Mudege et al., 2017; Ragasa et al., 2013), and gendered social networks (Magnan et al., 2013). Women's household responsibilities (childcare, cleaning, and cooking) on top of their livelihood activities results in time constraints that affect the activities that women engage in and the natural resources they manage (Ahmed & Laarman, 2000; Gueye, 2000).

Gendered Power Dynamics Gendered differences in decision-making and bargaining power at varied scales influence the motivations, means, and opportunities to exploit environmental resources, cope with environmental change, and benefit from technological innovations in resource productivity or sustainability (Meinzen-Dick et al., 2014). Women are traditionally less likely to be involved in groups that govern communal resources, in turn affecting the benefits they receive (Coleman & Mwangi, 2013; Kleiber et al., 2018; Wutich, 2012) as well as the environmental impacts of management decisions (Sultana & Thompson, 2008). In households where land is owned jointly by women and men, the latter are often deferred to in key land-use, labor, and income allocation decisions (Doss, 2014). Gendered power differentials also frequently result in the social devaluation of women's environmental work and perspectives (Jackson, 1998; Orlove & Caton, 2010; Sachs, 1996). In the Andes, for example, researchers have found that women's involvement in local water management was hindered by the fact that their water use (and associated knowledge) is largely centered on supplying household needs as opposed to agricultural needs (the latter being a traditionally male domain), a sphere undervalued relative to agriculture (a traditionally male domain) (Trawick, 2003); there is evidence from East Africa that both men and women exhibit lower concern for forest degradation when the primary risk it poses is to women's time use rather than men's income generation (Quinn et al., 2003).

Sociospatial Environmental Relationships Social norms and the sexual division of labor in many societies mean that men and women move in different spaces and therefore engage with different components of the local environment (Meinzen-Dick et al., 2014). For instance, topography, perceptions of physical strength, and gender roles may interact to proscribe women's and men's spaces and natural resource use (Harman, 2013; Zwarteveen & Neupane, 1996). These sociospatial differences in environmental engagement consequently affect the kinds of knowledge that men and women have access to and prioritize. For example, women's responsibilities for household and care work can result in the development of expert knowledge of local varieties and medicinal plants; men, by contrast, may have more information about species found further from the home, and plants used for construction (Camou-Guerrero et al., 2008; Chambers & Momsen, 2007).

Environmental Preferences and Priorities These gendered differences in roles, responsibilities, power, and access subsequently inform differences between men and women regarding their short and long-term environmental perceptions, priorities, and risk preferences (Sect. 21.3). Women in the Chitwan Valley of Nepal, for example, are especially burdened by environmental change because they are responsible for the provision of firewood, fodder, water, and other household needs that come directly from surrounding natural resources (Bohra-Mishra & Massey, 2010). Though femininity has been linked to a proclivity to care about or 'nurture' the environment (Mies & Shiva, 1993), there is also evidence that women prioritize short-term household survival needs in times of scarcity, which can lead to unsustainable practices (Agarwal, 2000; Yang & Xi, 2001).

Gender and Perceptions of Environmental Change and Risk

Human perceptions of environmental change, and assessments of associated risk, are important factors in determining natural resource use and management (Sect. 21.4), as well as responses to environmental change (Sect. 21.5). Population-environment research has focused on gendered perceptions of impact and risk stemming from environmental changes involving the climate (e.g. Haq & Ahmed, 2016; Lazrus, 2015; Lujala et al., 2014; McKinley et al., 2016), deforestation and biodiversity loss (Campos et al., 2014; Flaherty & Flipchuck, 1993), soil erosion (Momsen, 1993), and water access (Quinn et al., 2003). This literature indicates that, while perceptions of environmental change and risk are strongly influenced by gender (Momsen, 2000), dynamics and outcomes vary widely across study contexts and generalizable conclusions are difficult to make (Bee, 2016). Individual characteristics, roles, social norms, institutional arrangements and socioeconomic circumstances interact with hazards in complex ways to produce variations in environmental perceptions, assessments and preferences (Patt & Schroter, 2008). Population-environment researchers have explored a number of mechanisms to explain observed differences between men and women regarding their perceptions of, and attitudes toward, environmental change and risk; these are briefly surveyed below:

Hazards-Based Approaches connect risk perception with the type, frequency, and severity of environmental hazards and the degree to which people and their livelihoods are *exposed and sensitive* to them (Boissiere et al., 2013; Kasperson & Kasperson, 1996). Since livelihoods and natural resource use are often gendered, men and women may differentially observe environmental change or prioritize policies to mitigate adverse environmental changes (Carr, 2008) as a result of differences in their relative exposure and sensitivity. For example, in rural Rajasthan, Singh et al. (2018) found that changes in frost and soil moisture were most frequently observed by men, attributed to the fact that men's responsibility for agricultural production made them more aware of these environmental changes (some women in the study reported never leaving the home).

Most evidence, however, indicates that hazards-based approaches are insufficiently explanatory for understanding the relationship between gender and environmental perception (Bee, 2016). For instance, while women's foraging activities in rural Thailand are highly exposed and sensitive to environmental degradation, the wild plants they gather from are a supplementary, rather than staple, part of their household diet (Flaherty & Flipchuck, 1993). As a result, women failed to connect local deforestation with the disappearance of the plants they use for foraging. In Indonesia, men's greater support for conservation programs is connected not to gender-specific hazard experiences, but rather to their historical interactions with environmental NGOs and researchers (Villamor et al., 2014). Thus, the **everyday material realities and lived experiences** of men and women matter for the perception and prioritization of environmental change and risk (Bee, 2016).

The literature also demonstrates that **socio-cultural norms** influence gendered differences in perceptions and tolerance of environmental change (Bee, 2016; Lazrus, 2015; McStay & Dunlap, 1983). A sizeable group of studies (typically employing national survey data) has found that women often exhibit higher degrees of environmental concern, and are more likely to engage in pro-environmental action than men (Goldsmith et al., 2012). These researchers link women's sensitivity to environmental threats with their socialization in caregiving roles (Davidson & Freudenburg, 1996; Goldsmith et al., 2012; Zelezny et al., 2000). Gender norms sometimes have the opposite effect, though, particularly in low-income agrarian and pastoralist populations. Case studies in rural Colombia (Eitzinger et al., 2018), Vietnam (Cullen & Anderson, 2017) and

several African countries (e.g. Bunce et al., 2010; Cullen et al., 2018; Daoud, 2016; Kristjanson et al., 2015; Quinn et al., 2003), find men are more likely to report perceived risks stemming from extreme weather, pasture degradation, livestock disease, and declines in soil fertility. In these communities, gendered patterns of production and income-generation (Quinn et al., 2003), and responsibility for relocation and repairs following extreme weather events (Daoud, 2016) appear to increase men's perception of certain types of environmental change.

Gendered perceptions of environmental change and risk can also be driven by **differences in knowledge and access to information** (Sect 21.2.2) (Bee, 2016; Kasperson & Kasperson, 1996). For instance, in Nigeria, men (relative to women) had greater access to information about environmental change from a greater diversity of formal and informal sources. In turn, receipt of this information was positively concerned with concern about climate change (Badmos et al., 2018). Meanwhile, in Egyptian Bedouin communities, sedentarization processes have shifted responsibility for livestock management to women, leading them to acquire new knowledge about local plants suitable for grazing sheep and goats. The acquisition of this knowledge, and associated information about the increased scarcity of these pastoral resources, is part of the reason many women in these communities exhibit substantial interest in adaptation through small-scale cultivation (Briggs et al., 2003).

Finally, research has touched on the influence of **socio-cognitive** factors on gendered perceptions of environmental change and risk (Cvetkovic et al., 2018). As Susan Cutter notes, "if women perceive and experience the world differently from men, then it stands to reason that they would also perceive risks differently" (Cutter et al., 2006: 181). Some studies find women underestimate their knowledge about environmental change, perhaps due to gendered differences in perceived self-efficacy (Gururani, 2002a; McCright, 2010). Social identities — how people view themselves in relation to others — also influence perceptions about

21 Gender, Population and the Environment

the credibility and legitimacy of information about environmental risk (Frank et al., 2011). Scholars working in rural sub-Saharan Africa and Latin America note that local people often mistrust environmental information relayed by government officials and foreign scientists (Frank et al., 2011; Patt & Schroter, 2008); while these particular studies do not explore gendered effects, differences in institutional (Davidson & Freudenburg, 1996) and interpersonal trust (Oakley & Momsen, 2005) between men and women have been observed in other contexts.

Gender and Environmental Impact

There is a varied and growing literature on gendered environmental impacts. For example, research investigates gender and pro-environmental attitudes (Blocker & Eckberg, 1997; McCright, 2010; Zelezny et al., 2000), as well as the relationship between gender, environmental impact and consumer behavior (Carlsson-Kanyama et al., 2003; McGuckin & Murakami, 1999; Räty & Carlsson-Kanyama, 2010; Shi, 2019). However, much of that work focuses on populations living in urban areas and relatively wealthy countries. Here, in line with this chapter's focus, we survey research on gender as a mediator of environmental impact in two areas of particular relevance in low-income, rural places: natural resource-based livelihoods, and household fuel use.

Natural Resource-Based Livelihoods Across the globe, over 2.5 billion rural people depend on natural resources to sustain their livelihoods (FAO, 2008). This section briefly summarizes the results of a recent review which elaborates on gendered natural resource management and its relationship to climate change adaptation and vulnerability (Call & Sellers, 2019). Gender has been observed to shape the environmental impacts of three major categories of natural resource-based livelihoods: agriculture, fisheries, and forestry.

Agriculture As noted above, men and women typically play distinct roles in the smallholder agricultural production process (Carr, 2008). Fur-

ther, women often lack access to land and other resources that may facilitate environmentally sustainable land management practices, such as the building and maintenance of stone bunds to prevent runoff or the adoption of zero tillage agricultural approaches (Jost et al., 2016). Though these and other environmentally sustainable land management approaches may have long term environmental and social benefits, they require an initial and ongoing investment of resources, labor, and time (Call & Sellers, 2019). As such, it is unsurprising that research consistently finds women are less likely than men to use these management strategies (e.g. Karamba & Winters, 2015; Theriault et al., 2017). However, some studies have found that when women have the time and other resources necessary to engage in sustainable agricultural management, they are more likely than their male counterparts to use approaches that improve their soil fertility as well as strategies that reduce the greenhouse gas emissions attributable to their crops (Sapkota et al., 2018; Snapp et al., 2018).

Fisheries Though fishing is often viewed as a male-dominated livelihood strategy, women across the globe are heavily involved in the management and utilization of fisheries (Call & Sellers, 2019). Women are often engaged in the processing and selling of fish, but may also harvest invertebrates and near-shore species for home consumption or sale (Harper et al., 2013; Kleiber et al., 2015). However, due to their lack of resources, women are more likely than men to engage in environmentally destructive fishing practices, including the use of mosquito nets to harvest fish or the trampling of seagrass beds during near shore fishing (Hauzer et al., 2013). Further, women have been observed to be less likely than men to follow environmental rules set up for marine protected areas, possibly due to the fact that women rarely get to engage in the rule-setting process. However, by breaking rules governing the management of marine protected areas, women may be creating long-term challenges for their local marine environment (Rohe et al., 2018).

Forests As with the agriculture and fisheries sectors, forest activities are gendered, with women generally collecting fuelwood and edible forest products (e.g. plants, fruit) while men engage in timber harvesting and hunting (Sunderland et al., 2014). Studies have found that when forest user communities have a greater share of women, this has the potential to improve forest growth, likely because of the different ways in which women and men utilize the forest growth (Agarwal, 2009). However, due to cultural gender power dynamics and women's extensive responsibilities, researchers have observed that women-dominated forest user groups tend to be less effective at forest monitoring and the sanctioning of rule-breakers (Sun et al., 2011).

Household Fuel Use Gender also shapes environmental impact through household energy consumption. Some 2.8 billion people around the world, mainly living in low-income, rural areas, rely on solid biomass energy and other 'dirty fuels' such as coal and kerosene for cooking, lighting and heating (OECD, 2017). The environmental impacts of household fuel use are significant for both local environmental degradation and global climate change (Edwards et al., 2004).

Research demonstrates that the gender composition of households influences the type of fuel (e.g. modern vs. traditional) used, although the evidence is mixed. Some studies find female-headed households prefer modern (e.g. clean burning) fuels to traditional fuels, likely because women's responsibilities for cooking means they are the ones most affected by the air pollution caused by dirtier traditional fuels (Farsi et al., 2007; Narasimha Rao & Reddy, 2007). Conversely, other research has found that the gender of the household head does not significantly affect fuel choice (Abebaw, 2007; Ouedraogo, 2006). In Nepal, research suggests that households with more women may use a greater proportion of traditional fuels, which are inexpensive and typically collected by women (Link et al., 2012); however, this relationship is not significant in Guatemala (Heltberg, 2005). Finally, there is evidence from Bolivia suggesting that households where more women work outside the home may elect to use modern fuels, which

tend to be costlier than traditional fuels but are less time-intensive to obtain (Israel, 2002). On the other hand, Indian households with large numbers of working women do not exhibit a shift in fuel type (Gupta & Köhlin, 2006). The heterogeneity of these findings suggests that access to resources, cultural context, and labor demands on women are the major factors determining gender variability in household fuel choice.

Gender and Adaptation

Research also investigates how responses to environmental change are mediated by gender. While humans take advantage of positive environmental changes, there is an imbalance in the literature towards adaptation to shocks and stressors, commensurate with the deteriorating conditions observed all around the world. Adaptation refers to the actions people take to reduce negative outcomes in the face of environmental change (Smit & Wandel, 2006). As Bee (2016: 71) notes, "adaptation [is] fundamentally about knowledge and power" which, as demonstrated, vary between and within groups of men and women. Adaptation options are diverse. For example, research has documented the relationship between gender and adaptations involving the re-allocation of income, food and education (Kakota et al., 2011; McKinley et al., 2016; Randell & Gray, 2016), and decisions around marriage (Ahmed et al., 2019; Gray & Wise, 2016) and family size (Brauner-Otto & Axinn, 2017). Most scholarship, however, focuses on gender in the context of adaptations involving livelihood change or migration (and, in some cases, both).

Livelihood Change Livelihoods encompass the strategies people use to secure water, food, shelter, clothing and other necessities (Ellis, 2000). Livelihood-based adaptations involve adjustments in both natural resource use (e.g. fishing intensification), and non-resource based activities (e.g. small businesses) (Yang et al., 2019). *Gender and livelihood interact* in complex ways to produce adaptive preferences and decisions (Carr & Thompson, 2014). Naturally, if men and women engage in different livelihood activities (e.g. fishing vs. forestry) the adaptive strategies available to them also differ. Yet, adaptation preferences and decisions also sometimes vary between men and women engaged in the same livelihood (Codjoe et al., 2011).

Whether a particular strategy can be pursued depends partly on *adaptive capacity*, or the ability to react to stressors (Brooks & Adger, 2005). Women are frequently disadvantaged relative to men in this regard (Lambrou & Piana, 2006; Masika, 2002; Terry, 2009) due to differences in decision-making, assets, and access to information and work opportunities (Sect. 21.2). In the Sahel, drought-linked male out-migration has led to shifts in livestock herds from cattle to small ruminants; partly due to gendered rules governing land access, the remaining women are unable to manage cattle herds (Turner, 1999). Several studies indicate male-headed households are more likely to adapt through sustainable technologies, because they have better access to information, land and credit (Jin et al., 2015; Silvestri et al., 2012).

Beyond adaptive capacity, research shows adaptive decisions are mediated through gendered *norms and social processes*. Restrictions on women's ability to travel to markets, or control income, may influence their perception of the value of farm-level adaptations and their ability to benefit from their implementation (Carr & Thompson, 2014). In Malawi, women suffering from time poverty have adapted to drought by trading sex for assistance with their on-farm labor demands (Kakota et al., 2011). Research also investigates the influence of *sociocognitive factors* such as motivation, self-efficacy and perceived adaptive capacity on adaptation choices (Cvetkovic et al., 2018; Frank et al., 2011). Finally, research attends to the impact of gendered *perceptions, knowledge, and activities* on adaptation (Cutter, 1993; Schipper, 2006). For instance, in Micronesia, women's knowledge of hydrology enabled them to locate the site for a new communal well, thereby reducing their vulnerability to drought (Anderson, 2002).

Migration Adaptation can also take ex situ forms when people move, on a permanent or temporary basis, away from their original homes and activity spaces (Bardsley & Hugo, 2010). In the context of climate change, governments have proposed adaptions involving the planned relocation of entire communities (Lopez-Carr & Marter-Kenyon, 2015). Within population-environment studies, however, research primarily focuses on adaptations involving individual or household-level migration. Although environmental displacement can result from a failure to adapt (Adamo, 2008) migration is more often a proactive diversification strategy taken in response to the experience or anticipation of changing conditions (Bardsley & Hugo, 2010).

Gender shapes environmental migration decisions, experiences, and outcomes (Chant, 1992; Chindarkar, 2012) in non-uniform ways (Demetriades & Esplen, 2008; Hunter & David, 2011). Some studies find women's roles as the *primary users* of threatened natural resources increases their likelihood of out-migration (Bohra-Mishra & Massey, 2010; Massey et al., 2007). However, migration requires *access to resources*, which can be gendered, patterned by access and control over various forms of capital (Masika, 2002). In Kenya, male labor out-migration follows *positive* environmental change (Gray & Wise, 2016). Low adaptive capacity may mean the most vulnerable people (often women) are unable to migrate (Black et al., 2011). Sedentarization, a form of constrained migration, is a coping strategy women in Mali use in response to drought and male out-migration (Djoudi & Brockhaus, 2011).

Migration is an individual behavior mediated by household-level decision-making (de Sherbinin et al., 2008), thus *sociocultural and cognitive factors* also define gendered environmental migration patterns (Chindarkar, 2012; Homewood, 1997). In sub-Saharan Africa, environmental pressure typically restricts women's mobility while increasing migration prevalence for men (Carr, 2005; Henry et al.,

2004). However, earlier marriage of daughters (resulting in out-migration) is used by households in Uganda as a way of coping with increased temperatures (Gray & Wise, 2016). Gender is one of many social identities that interact to affect migration processes; for example, higher out-migration of younger and unmarried women has been observed in Latin America (Radcliffe, 1991). Gender also interacts with sociocognitive factors to influence mobility and evacuation during fast-onset hazards. In Bangladesh, restrictions on women's mobility increased their mortality during cyclones: sometimes because they were not comfortable making evacuation decisions in the absence of their husbands, or because their clothing made it difficult to flee (Haider et al., 1991). By contrast, American women are more likely to evacuate during hurricanes due to differences in risk perception, peer networks, and caregiving responsibility (Bateman & Edwards, 2002). Likewise, men in Nicaragua and Honduras had higher mortality rates during Hurricane Mitch, partly due to norms around 'heroic' masculinity that reduced men's compliance with evacuation warnings and increased their participation in risky search and rescue missions (Delaney & Shrader, 2000; Nelson, 2002).

Environmental migration *experiences and outcomes* are gendered too. Gendered labor opportunities, for example, produce variation in migration destinations, seasonality, and permanence. In the Ecuadorian Amazon, while local resource scarcity drives men to rural frontiers, women are more likely to migrate to urban areas due to the availability of domestic jobs (Barbieri & Carr, 2005). Research documents differences in security and safety during the migration process (Chindarkar, 2012), with women experiencing a greater risk of gender-based violence (Brown, 2008). Environmental migration outcomes are discussed in more detail in Chaps. 6, 7, 8 and 9. However, it is worth noting that male out-migration affects women left behind and vice versa, both positively and negatively, for instance by leading to changes in labor availability for cultivation and other resource management activities (Buechler, 2009).

Adaptation strategies are rarely undertaken in isolation. Gender-environment research also investigates the linkages between in situ and ex situ adaptation, and the interactions between men and women's choices. The work of Djoudi and Brockhaus (2011) in Mali is a good example. Drought has led to the emergence of forest resources in a former lake, which women have adapted to by moving from agriculture into charcoal production. Yet the success of this adaptation is hindered by male out-migration, which has left women responsible for work previously managed by men. As a result, women are more likely than men to see migration as a failure to adapt.

Gendered Outcomes of Environmental Shocks and Stressors

Environmental shocks and stressors affect demographic and livelihood outcomes, often in gender-differentiated ways. We briefly discuss the literature on three of these outcomes, food insecurity, education, and poverty, with the caveat that literature on other types of outcomes where impacts are mediated by gender are addressed elsewhere in this volume, including on adverse pregnancy outcomes, migration and morbidity/mortality.

Food Insecurity Environmental shocks and stressors are one of several drivers that can precipitate food insecurity, as natural hazards may induce crop loss or reduce income earned from other sources that is used to buy food. Scholars have explored responses to these events at the individual and household levels, finding that women and girls are often at a pronounced disadvantage. One *intrahousehold response* is buffering, whereby limited food resources are disproportionately directed to certain (generally male) household members (Ezeama et al., 2015; Haddad et al., 1997). For instance, researchers working in Ethiopia note that while adolescent boys and girls are as likely to be in food insecure households, girls are significantly more likely to personally report feeling food insecure (Hadley et al., 2008). In Bangladesh, cyclones, saltwater inundation, and changes in fisheries adversely affect household food security, but women are more likely than men to report eating less during periods of food insecurity (Alston, 2015).

There are also differences *between* households in vulnerability to food insecurity. Female-headed households are often more likely to experience food insecurity because of the differences in access women have to various key resources (Sect. 21.2.2). In Malawi, researchers note that because of gendered livelihood activity patterns, women and men are often exposed to different types of environmental change, with female-headed households relying on a smaller number of livelihood activities and having less information and resources to adapt. This contributes to members of female-headed households being substantially more likely to reduce or skip meals as a way of coping with environmental stress (Kakota et al., 2011). Similar effects were also documented in Ethiopia following 2007–08 drought-related food price increases (Kumar & Quisumbing, 2013). Moreover, in addition to having fewer forms of capital, discrimination or other forms of unobservable bias appear to play a role in influencing gendered food security outcomes, as has been illustrated in Kenya, where there are gaps between female-headed and male-headed households in food insecurity, even after accounting for differences between the two on observable indicators (Kassie et al., 2014).

Livelihood diversity mitigates some of the gendered gaps in food insecurity. For instance, in urban South Africa, where women engaged in agriculture often have access to alternative methods of income generation, the gender food security gap is narrower between female- and male-headed households than in the country's rural areas (Tibesigwa & Visser, 2016).

Education Following environmental shocks, households may no longer be able to meet the costs (in terms of school fees as well as lost work opportunity) associated with sending children to school. Girls, whose education is typically devalued, are generally more likely to lose access to educational opportunities than boys. For instance, lower than normal rainfall in East Africa is associated with lower educational attainment among girls relative to boys (Björkman-Nyqvist, 2013; Randell & Gray, 2016). In Nepal, female infants exposed to an earthquake were significantly less likely to complete future schooling than girls in other parts of the country, while few effects were found for boys (Paudel & Ryu, 2018). However, disasters can also raise the opportunity costs of schooling for older boys, who are more likely than girls to work in rebuilding and construction activities. This can lead to reduced school enrollment for boys without a corresponding decline for similarly-aged girls (Mottaleb et al., 2015).

Poverty Environmental shocks and stressors can disproportionately threaten women's ability to maintain their livelihoods. For instance, female-headed households in South Africa experience greater decreases in income during rainfall shocks than dual-adult households, which is in part a result of greater poverty and less social capital among female heads (Flatø et al., 2017). However, gendered vulnerability appears to be somewhat conditioned on the type of female headship. For instance, in Vietnam, female-headed households whose heads were single or widowed experienced smaller losses from shocks than female-headed households whose partners had migrated, as the latter group of households were likely reliant on uncertain and irregular flows of remittances for support (Klasen et al., 2015).

While disasters may disproportionately affect women, they may also temporarily loosen gender roles, providing greater freedom for women. In rural Bangladesh, where traditional gender roles keep women close to the home, women had greater freedom following a cyclone to conduct livelihood activities given the urgency of meeting post-disaster household needs (Alam & Collins, 2010). Even in the high-income context of the United States, there is evidence that women experience more substantial economic impacts from disasters. Following Hurricane Katrina, women were less likely than men to recover previous employment (Zottarelli, 2008), while female business owners were more likely to seek assistance from the U.S. Small Business Administration, indicating that women-owned businesses had smaller safety nets (e.g. cash reserves or insurance) and/or were more impacted by the storm (Josephson & Marshall, 2016).

Methodological Considerations and Critiques

Gender-environment researchers draw on many of the methods found across population-environment subfields. General methodological considerations in population-environment research are discussed in greater detail in Chaps. 3, 4 and 5 of this book. Here, we briefly explore three key intersections between gender and population-environment research methods that emerged from our survey of the literature.

Gender and the Household Many population-environment studies rely on the household as their unit of analysis. While this scalar focus has merit, there are limitations that must be taken into account (McDowell, 1992; Quisumbing, 2003). Household-level analyses can obscure intra-household gender differences in environmental knowledge, vulnerability, or resource management (Andersen et al., 2014; Valdes, 1992). Furthermore, censuses and surveys frequently only ask questions of the head of household. Because of sociocultural biases, household heads are nearly always male (Arias & Palloni, 1996); they may not know – or accurately report – a female family member's activities and experiences. Thus, women are often not adequately distinguished from others living with them (Quisumbing, 2003). Girls and boys are often missing, too, despite their contributions to environmentally-relevant behavior (Johnson et al., 1995; Punch & Sugden, 2013). Finally, and relatedly, studies of environmental behavior and agricultural productivity frequently compare female and male-headed households (Doss, 2018). Households with a man living in them are male-headed by default, so female-headed

households almost always have fewer adult members (Buvinic & Gupta, 1997). Therefore, studies making this comparison are actually investigating the relationship between household structure and the environment, not differences between men and women (Doss, 2018). As a result, this approach frequently results in erroneous conclusions about gender, population, and environment.

Gender and the Research Process Feminist scholarship draws attention to issues of positionality and intersubjectivity in population-environment research, or the ways in which relationships between socioenvironmental contexts, researchers, and research subjects influence data, analysis, and the interpretation of meaning (Coghlan & Brydon-Miller, 2014; McDowell, 1992). For example, the varied social identities, including gender, of enumerators and respondents may influence the quality of results (Pini, 2005) and the content of discussions (Goss & Leinbach, 1996). Due to cultural norms, male researchers may have difficulty accessing female research subjects and vice versa (Oakley & Momsen, 2005). Researchers' experience of gender and the environment in their own lives, in combination with their class, age, and background, introduce bias in the questions they choose to study, the assumptions they bring to their work, how they are viewed by subjects, and how results are interpreted and disseminated (Nightingale, 2003). Intersubjectivity, positionality, and power in population-environment research cannot be entirely overcome (Jarviluoma et al., 2003), but paying attention to them is important for ensuring quality research. The Women's Empowerment in Agriculture Index survey, for example, instructs enumerators to record whether onlookers are present during data collection (IFPRI, 2016).

Gender and Environmental Data Finally, gender-sensitive population-environment research must consider the environmental data being integrated. Choices around which environmental variables to measure, and how to operationalize environmental change, can be unintentionally gendered. For instance, in many agricultural societies, men control the cultivation of commercial crops while women manage subsistence varieties (Carr, 2008). "Women's" crops are often particularly sensitive to environmental change (Carr & Thompson, 2014), yet researchers have traditionally prioritized 'male' crops (Greenberg, 2003), resulting in a gap in our understanding of women's (and household-level) vulnerability. Environmental knowledge and preferences vary between men and women in the same household, as do their perceptions of each other's knowledge and activities (Sect. 21.4); thus, when this information is derived only from male heads (as is common), women's perspectives and experiences are occluded (Chambers & Momsen, 2007). Researchers must also consider whether the spatial and temporal foci and resolutions of their analyses are appropriate to the issues that matter to women or men. Resource use is often highly spatialized and seasonal in nature, with gendered activities and activity spaces changing throughout the year (Gururani, 2002b).

Gaps, Challenges, and Future Directions

There has been substantial growth in the gender and environment literature during the past two decades, which has resulted in a number of new avenues of inquiry. Unsurprisingly, as researchers investigate these linkages further, gaps and challenges have emerged that will shape the next generation of scholarship. This section briefly highlights these challenges, with the caveat that this does not purport to be an exhaustive list, nor do we have the space to fully elucidate on all of these concerns. Nevertheless, we hope this discussion provides broad guidance for researchers seeking to advance dynamic understandings of gender and environment and sparks additional discussions on this important topic going forward.

Increasingly, scholars who explore gender and the environment are applying an intersectional lens to their research questions (Hawkins et al.,

2011). A growing literature outlines these perspectives, often centered on concerns of climate change and the need to develop understandings of vulnerability and adaptation capacity that more accurately reflect the social settings that women and men inhabit (Kaijser & Kronsell, 2014; Sultana, 2014).

However, to date, discussions of gender in the population and environment literature have too often centered on traditional gendered divisions of labor in agrarian regions, failing to reflect the increasingly dynamic roles and responsibilities that women and men are taking on as the natural environment and human societies change simultaneously. Women are engaging in the formal labor market and in natural resource sectors traditionally dominated by men, yet these developments are often overlooked by binary gender analyses that assess "women" and "men" as homogenous categories (Carr & Thompson, 2014). Changes in women's livelihood patterns necessarily have implications for men and boys, whose experiences are also changing under environmental stress, yet are often overlooked in scholarship centered on vulnerability and resilience to environmental change (Demetriades & Esplen, 2008).

Moreover, existing notions of gender often center on misleading dichotomies, wherein men are painted as resilient and adaptive, while women are seen as vulnerable and inflexible. At the same time, men, who are generally in positions of authority and engaged in natural resource harvesting, are often portrayed as irresponsible or profligate, while women are viewed as good environmental stewards (Arora-Jonsson, 2011). These perspectives reflect the essentialist discourses of ecofeminism (Sect. 21.2.1) and rely on a-priori assumptions about the meaning of gender and its relationship to environmental vulnerabilities (Thompson-Hall et al., 2016). By obscuring context and reinforcing narrow stereotypes, such notions likely do more harm than good (Bee, 2016); there is a growing body of work that does not "fit" neatly into these narratives, including research exploring how climate change may provide women with greater autonomy and stimulate resilience capacity in low-income settings (Pearse, 2017), or the particular challenges experienced by men and boys following natural hazards (Enarson & Pease, 2016).

A related problem is the way in which much of the empirical gender and environment research is conducted. Our field struggles with the same tensions between generalizability versus particularity that exists throughout the social sciences, which largely parallels differences in methodologies between quantitative and qualitative scholarship. This is especially challenging given that vulnerability and adaptation capacity are closely shaped by local contexts (Adger et al., 2005). While quantitative studies may facilitate generalizability, all too often this literature dichotomizes gender, centering analyses on "women" and "men", while failing to address social and power structures that help shape gender roles and relations. Even smaller-scale case studies incorporating qualitative methods often inadequately explore these ideas. This is not a new critique (see Djoudi et al., 2016; Thompson-Hall et al., 2016), but one that bears repeating given that vulnerability and adaptive capacity are functions of multiple variables, some of which are challenging to measure or quantify, creating an additional obstacle for scholars. Such dichotomization is also increasingly obsolete given the growing attention to individuals who do not identify with gender binaries, and who often simultaneously experience other forms of social marginalization which in turn shapes their experiences of environmental change (Gaard, 2015).

Additionally, there are gender dimensions to two perennial challenges in the distribution of population-environment literature. First, the bulk of population and environment literature addresses themes related to natural resources or agriculture. Because of the close relationship between natural resources and livelihoods in low- and middle-income settings, most of the literature centering on gender and natural resources focuses on these settings (Call & Sellers, 2019), even though expectations and norms concerning gender often differ in high-income societies.

Albeit with notable exceptions (e.g., Alston et al., 2018; Beaumier & Ford, 2010), relatively little is known about gender and natural resources or agriculture in high-income settings, despite the important economic, cultural, and recreational implications of such understandings. Second, the population and environment literature centers on rural areas even as cities hold a growing share of the world's population. Ajibade et al. (2013), who explore flooding vulnerability and resilience among women in Lagos, provide a model for future studies to more effectively capture the growing role of urban landscapes in mediating gender and environment experiences.

The themes above serve as a starting point for thinking about how to capture gender-environment experiences in future research. Gaps include studies lacking (1) intersectional perspectives, (2) discussions of masculinity and male vulnerability, (3) diversity in research settings and contexts, (4) understandings of gender that reach beyond binaries, and (5) explorations of change over time in gender-environment dynamics. At the same time, literature that explores localized phenomena while capturing mechanisms generalizable to other settings is especially welcomed. Such research is particularly important given the growing gap between the need for contextualized understandings (especially with regard to understanding local climate change vulnerability and adaptation) and the capacity of researchers to provide it. Collectively, these needs present many challenges, not least of which is that many of the key tools of demography, including household surveys and censuses, are designed to provide standardized information on carefully-defined indicators across multiple settings. Intersectional perspectives in particular push us towards more particular and complex understandings of gender and environment phenomena, suggesting that new methods or tools may need to be developed or further refined in order to capture these experiences (Scott, 2010). Without pushing the boundaries of existing scholarship, we cannot effectively address the gendered dynamics of global environmental change.

References

Abebaw, D. (2007). Household determinants of fuelwood choice in urban Ethiopia: A case study of Jimma Town. *The Journal of Developing Areas, 41*(1), 117–126.

Adamo, S. (2008). *Addressing environmentally induced population displacements: A delicate task.* Background Paper for the Environment Research Network Cyberseminar on Environmentally Induced Population Displacements. PERN.

Adger, W. N., Arnell, N. W., & Tompkins, E. L. (2005). Successful adaptation to climate change across scales. *Adaptation to Climate Change: Perspectives Across Scales, 15*(2), 77–86. https://doi.org/10.1016/j.gloenvcha.2004.12.005

Agarwal, B. (1992). The gender and environment debate: Lessons from India. *Feminist Studies, 18*(1), 119–158.

Agarwal, B. (1997). Environmental action, gender equity and women's participation. *Development and Change, 28*(1), 1–44.

Agarwal, B. (2000). Conceptualising environmental collective action: Why gender matters. *Cambridge Journal of Economics, 24*, 283–310.

Agarwal, B. (2009). Gender and forest conservation: The impact of women's participation in community forest governance. *Ecological Economics, 68*(11), 2785–2799. https://doi.org/10.1016/j.ecolecon.2009.04.025

Ahmed, K. J., Haq, S. M. A., & Bartiaux, F. (2019). The nexus between extreme weather events, sexual violence, and early marriage: A study of vulnerable populations in Bangladesh. *Population and Environment, 40*, 303–324.

Ahmed, M. R., & Laarman, J. G. (2000). Gender equity in social forestry programs in Bangladesh. *Human Ecology, 38*(3), 433–450.

Ajibade, I., McBean, G., & Bezner-Kerr, R. (2013). Urban flooding in Lagos, Nigeria: Patterns of vulnerability and resilience among women. *Global Environmental Change-Human and Policy Dimensions, 23*(6), 1714–1725. https://doi.org/10.1016/j.gloenvcha.2013.08.009

Alam, E., & Collins, A. E. (2010). Cyclone disaster vulnerability and response experiences in coastal Bangladesh. *Disasters, 34*(4), 931–954. https://doi.org/10.1111/j.1467-7717.2010.01176.x

Alston, M. (2015). *Women and climate change in Bangladesh.* Routledge.

Alston, M., Clarke, J., & Whittenbury, K. (2018). Contemporary feminist analysis of Australian farm women in the context of climate changes. *Social Sciences, 7*(2), 16.

Andersen, L. E., Verner, D., & Wiebelt, M. (2014). *Gender and climate change in Latin America: An analysis of vulnerability, adaptation and resilience based on household surveys.* Development Research Working Paper Series No. 08/2014, Working Paper, November. La Paz: Institute for Advanced Development Studies (INESAD).

Anderson, C. (2002). Gender matters: Implications for climate variability and climate change and disaster management in the Pacific islands. *Intercoast Network, 41*, 24–25.

Arias, E., & Palloni, A. (1996). *Prevalence and Patterns of Female-headed Households in Latin America.* CDE Working Paper No. 96–14, CDE Working Paper No. 96–14. Madison, WI.

Arora-Jonsson, S. (2011). Virtue and vulnerability: Discourses on women, gender and climate change. *Global Environmental Change, 21*(2), 744–751.

Badmos, B., Sawyerr, H., Salako, G., et al. (2018). Gender variation on the perception of climate change impact on human health in Moba Local Government Area of Ekiti State, Nigeria. *Journal of Health and Environmental Research, 4*(1), 1–9.

Barbieri, A. F., & Carr, D. L. (2005). Gender-specific out-migration, deforestation and urbanization in the Ecuadorian Amazon. *Global and Planetary Change, 47*(2–4), 99–110.

Bardsley, D. K., & Hugo, G. J. (2010). Migration and climate change: Examining thresholds of change to guide effective adaptation decision-making. *Population and Environment, 32*(2–3), 238–262.

Bateman, J., & Edwards, B. (2002). Gender and evacuation: A closer look at why women are more likely to evacuate for hurricanes. *Natural Hazards Review, 3*, 107–117.

Beaumier, M. C., & Ford, J. D. (2010). Food insecurity among Inuit women exacerbated by socio-economic stresses and climate change. *Canadian Journal of Public Health-Revue Canadienne De Sante Publique, 101*(3), 196–201. Available at: ://-WOS:000280778600003.

Bee, B. A. (2016). Power, perception, and adaptation: Exploring gender and social– Environmental risk perception in northern Guanajuato, Mexico. *Geoforum, 69*, 71–80.

Bezner-Kerr, R. (2014). Lost and found crops: Agrobiodiversity, indigenous knowledge, and a feminist political ecology of sorghum and finger millet in Northern Malawi. *Annals of the Association of American Geographers, 104*(3), 577–593.

Björkman-Nyqvist, M. (2013). Income shocks and gender gaps in education: Evidence from Uganda. *Journal of Development Economics, 105*, 237–253.

Black, R., Bennett, S. R. G., Thomas, S. M., et al. (2011). Migration as adaptation. *Nature, 478*, 447–449.

Blocker, T., & Eckberg, D. (1997). Gender and environmentalism. *Social Science Quarterly, 78*, 841–858.

Bohra-Mishra, P., & Massey, D. (2010). *Environmental degradation and out-migration: New evidence from Nepal.* Princeton University.

Boissiere, M., Locatelli, B., Padmanaba, M., et al. (2013). Local perceptions of climate variability and change in tropical forests of Papua, Indonesia. *Ecology and Society, 18*(4), 13.

Brandth, B., & Haugen, M. S. (2016). Rural Masculinity. In M. Shucksmith & D. L. Brown (Eds.), *Routledge international handbook of rural studies.* Routledge.

Brauner-Otto, S. R., & Axinn, W. G. (2017). Natural resource collection and desired family size: A longitudinal test of environment-population theories. *Population and Environment, 38*(4), 381–406.

Briggs, J., Hamed, N., & Yacoub, H. (2003). Changing women's roles, changing environmental knowledges: Evidence from Upper Egypt. *The Geographical Journal, 169*(4), 313–325.

Brooks, N., & Adger, W. N. (2005). Assessing and enhancing adaptive capacity. In *Adaptation policy frameworks for climate change: Developing strategies, policies and measures* (pp. 167–180). Cambridge University Press.

Brown, O. (2008). *Climate change and forced migration: Observations, projections and implications.* Background Paper for the 2007 Human Development Report. Geneva: UNDP.

Buechler, S. (2009). Gender, water, and climate change in Sonora, Mexico: Implications for policies and programmes on agricultural income-generation. *Gender and Development, 17,* 51–66.

Bunce, M., Rosendo, S., & Brown, K. (2010). Perceptions of climate change, multiple stressors and livelihoods on marginal African coasts. *Environment, Development and Sustainability, 12,* 407–440.

Buvinic, M., & Gupta, G. R. (1997). Female-headed households and female-maintained families: Are they worth targeting to reduce poverty in developing countries? *Economic Development and Cultural Change, 45*(2), 259–280.

Call, M., & Sellers, S. (2019). How does gendered vulnerability inform livelihood responses to climate change? *Environmental Research Letters, 14*(8). https://doi.org/10.1088/1748-9326/ab2f57

Camou-Guerrero, A., Reyes-García, V., Martinez-Ramos, M., et al. (2008). Knowledge and use value of plant species in a Rarámuri community: A gender perspective for conservation. *Human Ecology, 36*(2), 259–272.

Campos, M., McCall, M. K., & Gonzalez-Puente, M. (2014). Land-users' perceptions and adaptations to climate change in Mexico and Spain: Commonalities across cultural and geographical contexts. *Regional Environmental Change, 14,* 811–823.

Carlsson-Kanyama, A., Ekström, M. P., & Shanahan, H. (2003). Food and life cycle energy inputs: Consequences of diet and ways to increase efficiency. *Ecological Economics, 44*(2–3), 293–307. https://doi.org/10.1016/S0921-8009(02)00261-6

Carney, J. (1996). Rice milling, gender and slave labour in colonial South Carolina. *Past and Present, 153,* 108–134.

Carr, E. R. (2005). Placing the environment in migration: Environment, economy, and power in Ghana's central region. *Environment and Planning A: Economy and Space, 37*(5), 925–946.

Carr, E. R. (2008). Men's crops and women's crops: The importance of gender to the understanding of agricultural and development outcomes in Ghana's Central Region. *World Development, 36*(5), 900–915. https://doi.org/10.1016/j.worlddev.2007.05.009

Carr, E. R., & Thompson, M. C. (2014). Gender and climate change adaptation in agrarian settings: Current thinking, new directions, and research frontiers. *Geography Compass, 8,* 182–197.

Castree, N., & Braun, B. (Eds.). (2001). *Social nature: Theory, practice and politics.* Blackwell Publishing.

Chambers, K. J., & Momsen, J. H. (2007). From the kitchen and the field: Gender and maize diversity in the Bajío region of Mexico. *Singapore Journal of Tropical Geography, 28*(1), 39–56. https://doi.org/10.1111/j.1467-9493.2006.00275.x

Chant, S. (1992). *Gender and migration in developing countries.* Belhaven Press.

Chindarkar, N. (2012). Gender and climate change-induced migration: Proposing a framework for analysis. *Environmental Research Letters, 7,* 1–7.

Codjoe, S. N. A., Atidoh, L. K., & Burkett, V. (2011). Gender and occupational perspectives on adaptation to climate extremes in the Afram Plains of Ghana. *Climatic Change, 110*(1–2), 431–454.

Coghlan, D., & Brydon-Miller, M. (2014). *The SAGE encyclopedia of action research.* SAGE.

Coleman, E. A., & Mwangi, E. (2013). Women's participation in forest management: A cross-country analysis. *Global Environmental Change, 23*(1), 193–205. https://doi.org/10.1016/j.gloenvcha.2012.10.005

Cullen, A., & Anderson, C. (2017). Perception of climate risk among rural farmers in Vietnam: Consistency within households and with the empirical record. *Risk Analysis, 37*(3), 531–545.

Cullen, A., Anderson, C., Biscaye, P., et al. (2018). Variability in cross-domain risk perception among smallholder farmers in Mali by gender and other demographic and attitudinal characteristics. *Risk Analysis, 38*(7), 1361–1377.

Cutter, S. L. (1993). *Living with risk.* Edward Arnold.

Cutter, S. L., Tiefenbacher, J., & Solecki, W. D. (2006). En-gendered fears: Femininity and technological risk perception. In S. L. Cutter (Ed.), *Hazards, vulnerability and environmental justice* (pp. 177–192). Routledge.

Cvetkovic, V. M., Roder, G., Ocal, A., et al. (2018). The role of gender in preparedness and response behaviors towards flood risk in Serbia. *International Journal of Environmental Research and Public Health, 15*(12).

Daoud, M. (2016). *Living on the edge: Gender relations, climate change and livelihoods in the villages of Maryut and Nubia, Egypt.* Doctoral Thesis. University of East Anglia, Norwich, UK. Available at: https://ueaeprints.uea.ac.uk/id/eprint/59683/1/Mona_Daoud_-PhD_thesis_-5sept16.pdf. Accessed 15 Feb 2020.

Davidson, D. J., & Freudenburg, W. R. (1996). Gender and environmental risk concerns: A review and analysis of available research. *Environment and Behavior, 28*(3), 302–339. https://doi.org/10.1177/0013916596283003

de Sherbinin, A., VanWey, L. K., McSweeney, K., et al. (2008). Rural household demographics, livelihoods and the environment. *Global Environmental Change, 18,* 38–53.

Deere, C. D., Alvarado, G., & Twyman, J. (2012). Gender inequality in asset ownership in Latin America: Female owners vs household heads. *Development and Change, 43*(2), 505–530.

Delaney PL and Shrader E (2000) *Gender and post-disaster reconstruction: The case of hurricane Mitch in Honduras and Nicaragua.* The World Bank.

Demetriades, J., & Esplen, E. (2008). The gender dimensions of poverty and climate change adaptation. *IDS Bulletin, 39*(4), 24–31.

Djoudi, H., & Brockhaus, M. (2011). Is adaptation to climate change gender neutral? Lessons from communities dependent on livestock and forests in northern Mali. *International Forestry Review, 13*(2), 123–135.

Djoudi, H., Locatelli, B., Vaast, C., et al. (2016). Beyond dichotomies: Gender and intersecting inequalities in climate change studies. *Ambio, 45*(S3), 248–262.

Doss, C. (2014). If women hold up half the sky, how much of the world's food do they produce? In A. Quisumbing, R. Meinzen-Dick, T. L. Raney, et al. (Eds.), *Gender in agriculture.* Springer.

Doss, C. (2018). Women and agricultural productivity: Reframing the issues. *Development Policy Review, 36*, 35–50.

Doss, C., Kovarik, C., Peterman, A., et al. (2015). Gender inequalities in ownership and control of land in Africa: Myth and reality. *Agricultural Economics, 46*.

Edwards, R., Smith, K., Zhang, J., et al. (2004). Implications of changes in household stoves and fuel use in China. *Energy Policy, 32*, 395–411.

Eitzinger, A., Binder, C. R., & Meyer, M. A. (2018). Risk perception and decision-making: Do farmers consider risks from climate change? *Climatic Change, 151*, 507–524.

Ellis, F. (2000). *Rural livelihoods and diversity in developing countries.* Oxford University Press.

Enarson, E., & Pease, B. (Eds.). (2016). *Men, masculinities and disaster* (1st ed.). Routledge.

Ezeama, N. N., Ibeh, C., Adinma, E., et al. (2015). Coping with household food insecurity: Perspectives of mothers in Anambra State, Nigeria. *Journal of Food Security, 3*(6), 145–154.

FAO. (2008). *The state of food and agriculture (SOFA) 2008.* Food and Agriculture Organization of the United Nations.

Farsi, M., Filippini, M., & Pachauri, S. (2007). Fuel choices in urban Indian households. *Environment and Development Economics, 12*(6), 757–774. https://doi.org/10.1017/S1355770X07003932

Flaherty, M. S., & Flipchuck, V. R. (1993). Forest management in northern Thailand: A rural Thai perspective. *Geoforum, 24*, 263–276.

Flatø, M., Muttarak, R., & Pelser, A. (2017). Women, weather, and woes: The triangular dynamics of female-headed households, economic vulnerability, and climate variability in South Africa. *World Development, 90*, 41–62.

Fletschner, D., & Kenney, L. (2014). Rural women's access to financial services: Credit, savings, and insurance. In A. Quisumbing, R. Meinzen-Dick, T. L. Raney, et al. (Eds.), *Gender in agriculture.* Springer.

Forsyth, T. (2003). *Critical political ecology: The politics of environmental science* (1st ed.). Routledge.

Frank, E., Eakin, H., & Lopez-Carr, D. (2011). Social identity, perception and motivation in adaptation to climate risk in the coffee sector of Chiapas, Mexico. *Global Environmental Change, 21*, 66–76.

Gaard, G. (2015). Ecofeminism and climate change. *Womens Studies International Forum, 49*, 20–33. https://doi.org/10.1016/j.wsif.2015.02.004

Goh, A. H. X. (2012). *A literature review of the gender-differentiated impacts of climate change on women's and men's assets and well-being in developing countries.* CAPRi Working Paper No. 106, September. Washington, DC.: International Food Policy Research Institute.

Goldsmith, R., Feygina, I., & Jost, J. (2012). The gender gap in environmental attitudes: A system justification perspective. In M. Alston & K. Whittenbury (Eds.), *Research, action and policy: Addressing the gendered impacts of climate change* (p. 282). Springer.

Goss, J. D., & Leinbach, T. R. (1996). Focus groups as alternative research practice: Experience with transmigrants in Indonesia. *Area, 28*(2), 115–123.

Gray, C., & Wise, E. (2016). Country-specific effects of climate variability on human migration. *Climatic Change, 135*(3), 555–568.

Greenberg, L. (2003). Women in the garden and the kitchen: The role of cuisine in the conservation of traditional house lot crops among Yucatec Mayan immigrants. In P. Howard (Ed.), *Women and plants: Gender relations in biodiversity management and conservation* (pp. 51–65). Zed Books.

Gueye, E. F. (2000). The role of family poultry in poverty alleviation, food security and the promotion of gender equality in rural Africa. *Outlook on Agriculture, 29*(2), 129–136.

Gupta, G., & Köhlin, G. (2006). Preferences for domestic fuel: Analysis with socio-economic factors and rankings in Kolkata, India. *Ecological Economics, 57*(1), 107–121. https://doi.org/10.1016/j.ecolecon.2005.03.010

Gururani, S. (2002a). Construction of third world women's knowledge in the development discourse. *International Social Science Journal, 54*(173), 313–323.

Gururani, S. (2002b). Forests of pleasure and pain: Gendered practices of labor and livelihood in the forests of Kumaon Himalayas, India. *Gender, Place and Culture, 9*, 229–243.

Haddad, L. J., Hoddinott, J., & Alderman, H. (1997). *Intrahousehold resource allocation in developing countries: models, methods, and policy.* Johns Hopkins University Press.

Hadley, C., Lindstrom, D., Tessema, F., et al. (2008). Gender bias in the food insecurity experience of Ethiopian adolescents. *Social Science & Medicine, 66*(2), 427–438. https://doi.org/10.1016/j.socscimed.2007.08.025

Haider, R., Rahman, A. A., & Huq, S. (Eds.). (1991). *Cyclone '91: An environmental and perceptional study.* Bangladesh Centre for Advanced Studies.

Haq, S. M. A., & Ahmed, K. J. (2016). Does the perception of climate change vary with the socio-demographic dimensions? A study on vulnerable populations in Bangladesh. *Natural Hazards.*

Harman, M. A. (2013). *Using qualitative geographic information systems to explore gendered dimensions for conservation agriculture production systems in the Philippines: A mixed methods approach.* Virginia Tech, Blacksburg, VA. Available at: https://vtechworks.lib.vt.edu/handle/10919/50811

Harper, S., Zeller, D., Hauzer, M., et al. (2013). Women and fisheries: Contribution to food security and local economies. *Marine Policy, 39*, 56–63. https://doi.org/10.1016/j.marpol.2012.10.018

Harris, L. M. (2015). Hegemonic waters and rethinking natures otherwise. In *Practicing feminist political ecologies: Moving beyond the 'green economy* (pp. 157–181). Zed Books.

Hauzer, M., Dearden, P., & Murray, G. (2013). The fisherwomen of Ngazidja Island, Comoros: Fisheries livelihoods, impacts, and implications for management. *Fisheries Research, 140*, 28–35. https://doi.org/10.1016/j.fishres.2012.12.001

Hawkins, R., Ojeda, D., Asher, K., et al. (2011). A discussion. *Environment and Planning D: Society and Space, 29*(2), 237–253. https://doi.org/10.1068/d16810

Heltberg, R. (2005). Factors determining household fuel choice in Guatemala. *Environment and Development Economics, 10*(3), 337–361. https://doi.org/10.1017/S1355770X04001858

Henry, S., Schoumaker, B., & Beauchemin, C. (2004). The impact of rainfall on the first out-migration: A multilevel event-history analysis in Burkina Faso. *Population and Environment, 25*(5), 423–460.

Homewood, K. (1997). *Land use, household viability and migration in the Sahel.* Final report to INCO-DC, University College London, INERA (IRBET)/CNRST Burkina Faso, IDR/UPB Burkina Faso, Universite National du Benin, Universiteit van Amsterdam.

Howard, J. (2009). Climate change mitigation and adaptation in developed nations: A critical perspective on the adaptation turn in urban climate planning. In S. Davoudi, J. Crawford, & A. Mehmood (Eds.), *Planning for climate change: Strategies for mitigation and adaptation for spatial planners* (1st ed.). Routledge.

Hunter, L. M., & David, E. (2011). Climate change and migration: Considering gender dimensions. In E. Piguet, P. de Guchteneire, & A. Pecoud (Eds.), *Climate change and migration* (pp. 306–330). UNESCO Publishing and Cambridge University Press.

IFPRI. (2016). Updated WEAI questionnaire. Available at: https://www.ifpri.org/sites/default/files/Basic%20Page/updated_weai_module_2016.pdf. Accessed 8 Aug 2019.

Israel, D. (2002). Fuel choice in developing countries: Evidence from Bolivia. *Economic Development and Cultural Change, 50*(4), 865–890. https://doi.org/10.1086/342846

Jackson, C. (1993). Doing what comes naturally? Women and environment in development. *World Development, 21*(12), 1947–1963.

Jackson, C. (1998). Gender, irrigation, and environment: Arguing for agency. *Agriculture and Human Values, 15*, 313–324.

Jarviluoma, H., Moisala, P., & Vilkko, A. (2003). Gender and fieldwork. In H. Jarviluoma, P. Moisala, & A. Vilkko (Eds.), *Gender and qualitative methods* (pp. 27–46). SAGE.

Jin, J., Wang, X., & Gao, Y. (2015). Gender differences in farmers' responses to climate change adaptation in Yongqiao District, China. *Science of the Total Environment, 538*, 942–948.

Johnson, V., Hill, J., & Ivan-Smith, E. (1995). *Listening to smaller voices: Children in an environment of change.* ActionAid.

Josephson, A., & Marshall, M. I. (2016). The demand for post-Katrina disaster aid: SBA disaster loans and small businesses in Mississippi. *Journal of Contingencies and Crisis Management, 24*(4), 264–274.

Jost, C., Kyazze, F., Naab, J., et al. (2016). Understanding gender dimensions of agriculture and climate change in smallholder farming communities. *Climate and Development, 8*(2), 133–144. https://doi.org/10.1080/17565529.2015.1050978

Kaijser, A., & Kronsell, A. (2014). Climate change through the lens of intersectionality. *Environmental Politics, 23*(3), 417–433. https://doi.org/10.1080/09644016.2013.835203

Kakota, T., Nyariki, D., Mkwambisi, D., et al. (2011). Gender vulnerability to climate variability and household food insecurity. *Climate and Development, 3*, 298–309.

Karamba, R. W., & Winters, P. C. (2015). Gender and agricultural productivity: Implications of the farm input subsidy program in Malawi. *Agricultural Economics, 46*(3), 357–374. https://doi.org/10.1111/agec.12169

Kasperson, R. E., & Kasperson, J. X. (1996). The social amplification and attenuation of risk. *Annals of the American Academy of Political and Social Science, 545*, 95–105.

Kassie, M., Ndiritu, S., & Stage, J. (2014). What determines gender inequality in household food security in Kenya? Application of exogenous switching treatment regression. *World Development, 56*, 153–171.

Klasen, S., Lechtenfeld, T., & Povel, F. (2015). A feminization of vulnerability? Female headship, poverty, and vulnerability in Thailand and Vietnam. *World Development, 71*, 36–53.

Kleiber, D., Harris, L. M., & Vincent, A. C. J. (2015). Gender and small-scale fisheries: A case for counting women and beyond. *Fish and Fisheries, 16*(4), 547–562. https://doi.org/10.1111/faf.12075

Kleiber, D., Harris, L. M., & Vincent, A. C. J. (2018). Gender and marine protected areas: A case study of Danajon Bank, Philippines. *Maritime Studies, 17*(2), 163–175. https://doi.org/10.1007/s40152-018-0107-7

Kristjanson, P., Bernier, Q., Bryan, E., et al. (2015). *Implications of Gender-Focused Research in Senegal for Farmer's Adaptation to Climate Change.* 2, Project Note, October. Washington, DC: IFPRI.

Kumar, N., & Quisumbing, A. R. (2013). Gendered impacts of the 2007-2008 food price crisis: Evidence using panel data from rural Ethiopia. *Food Policy, 38*, 11–22. https://doi.org/10.1016/j.foodpol.2012.10.002

Lambrou, Y., & Piana, G. (2006). *Gender: The missing component of the response to climate change.* FAO.

Lazrus, H. (2015). Risk perception and climate adaptation in Tuvalu: A combined cultural theory and traditional knowledge approach. *Human Organization, 74*(1), 52–61.

Leach, M. (2007). Earth mother myths and other ecofeminist fables: How a strategic notion rose and fell. *Development and Change, 38*(1), 67–85.

Leach, M., Joekes, S., & Green, C. (1995). Editorial: Gender relations and environmental change. *IDS Bulletin, 26*(1), 1–8.

Leichenko, R. (2011). Climate change and urban resilience. *Current Opinion in Environmental Sustainability, 3*(3), 164–168.

Link, C. F., Axinn, W. G., & Ghimire, D. J. (2012). Household energy consumption: Community context and the fuelwood transition. *Social Science Research, 41*(3), 598–611. https://doi.org/10.1016/j.ssresearch.2011.12.007

Lopez-Carr, D., & Marter-Kenyon, J. (2015). Human adaptation: Manage climate-induced resettlement. *Nature, 517*(7534), 265–267.

Lorber, J. (1994). Night to his day: The social construction of gender. In *Paradoxes of gender* (pp. 13–15). Yale University Press. 32–36.

Lujala, P., Lein, H., & Rod, J. K. (2014). Climate change, natural hazards, and risk perception: The role of proximity and personal experience. *Local Environment: The International Journal of Justice and Sustainability, 20*(4), 489–509.

Magnan, N., Spelman, D., Gulati, K., et al. (2013). *Gender dimensions of social networks and technology adoption: Evidence from a field experiment in Uttar Pradesh, India.* GAAP Note, November. Washington, DC: International Food Policy Research Institute.

Masika, R. (2002). Editorial. *Gender and Development, 10*, 2–9.

Massey, D., Axinn, W., & Ghimire, D. (2007). *Environmental change and out-migration: Evidence from Nepal.* Population Studies Center, University of Michigan Institute for Social Research.

May. (2014). "Speaking into the void"? Intersectionality critiques and epistemic backlash. *Hypatia, 29*(1).

McCright, A. (2010). The effects of gender on climate change knowledge and concern in the American public. *Population and Environment, 32*, 66–87.

McDowell, L. (1992). Doing gender: Feminism, feminists and research methods in human geography. *Transactions of the Institute of British Geographers, 17*(4), 399–416.

McElroy, M. (1990). The empirical content of Nash-bargained household behavior. *Journal of Human Resources, 25*(4), 559–583.

McGuckin, N., & Murakami, E. (1999). Examining trip-chaining behavior: Comparison of travel by men and women. *Transportation Research Record: Journal of the Transportation Research Board, 1693*(1), 79–85. https://doi.org/10.3141/1693-12

McKinley, J., Adaro, C., Pede, V., et al. (2016). *Gender differences in climate change perception and adaptation strategies: A case study on three provinces in Vietnam's Mekong River Delta.* Info Note, July. CGIAR.

McStay, J. R., & Dunlap, R. E. (1983). Male–female differences in concern for environmental quality. *International Journal of Women's Studies, 6*(4).

Meinzen-Dick, R., Kovarik, C., & Quisumbing, A. (2014). Gender and sustainability. *Annual Review of Environment and Resources, 39*, 29–55.

Merchant, C. (1992). *Radical ecology: The search for a liveable world.* Routledge.

Mies, M., & Shiva, V. (1993). *Ecofeminism.* Fernwood Publishing.

Momsen, J. H. (1993). Gender and environmental perception in the Eastern Caribbean. In D. Lockhart, D. Drakakis-Smith, & J. Schrembi (Eds.), *The development process in small island states* (pp. 57–70). Routledge.

Momsen, J. H. (2000). Gender differences in environmental concern and perception. *Journal of Geography, 99*(2), 47–56.

Mottaleb, K., Mohanty, S., & Mishra, A. (2015). Intra-household resource allocation under negative income shock: A natural experiment. *World Development, 66*, 557–571.

Mudege, N. N., Mdege, N., Abidin, P. E., et al. (2017). The role of gender norms in access to agricultural training in Chikwawa and Phalombe, Malawi. *Gender, Place and Culture, 24*(12), 1689–1710.

Narasimha Rao, M., & Reddy, B. S. (2007). Variations in energy use by Indian households: An analysis of micro level data. *Energy, 32*(2), 143–153. https://doi.org/10.1016/j.energy.2006.03.012

Nelson, V. (2002). Uncertain predictions, invisible impacts, and the need to mainstream gender in climate change adaptations. *Gender and Development, 10*(2), 51–90.

Nightingale, A. (2001). Participating or just sitting in? The dynamics of gender and caste in community forestry. *Journal of Forestry and Livelihoods, 2*, 17–24.

Nightingale, A. (2006). *The nature of gender: Work, gender and environment.* Online Pap. Ser. GEO-030, Univ. Edinburgh Sch. GeoSci. Inst. Geogr.

Nightingale, A. J. (2003). A feminist in the forest: Situated knowledges and mixing methods in natural resource management. *ACME: An International E-Journal for Critical Geographies, 2*(1), 77–89.

Oakley, E., & Momsen, J. H. (2005). Gender and agro-biodiversity: A case study from Bangladesh. *The Geographical Journal, 171*(3), 195–208.

OECD. (2017). *WEO-2017 special report: Energy access outlook.* World Energy Outlook. International Energy Agency.

Orlove, B., & Caton, S. (2010). Water sustainability: Anthropological approaches and prospects. *Annual Review of Anthropology, 39*, 401–415.

Ouedraogo, B. (2006). Household energy preferences for cooking in urban Ouagadougou, Burkina Faso. *Energy Policy, 34*(18), 3787–3795. https://doi.org/10.1016/j.enpol.2005.09.006

Parpart, J. L., Connelly, M. P., & Barriteau, V. E. (Eds.). (2000). *Theoretical perspectives on gender and development*. International Development Research Centre.

Patt, A., & Schroter, D. (2008). Perceptions of climate risk in Mozambique: Implications for the success of adaptation strategies. *Global Environmental Change, 18*, 458–467.

Paudel, J., & Ryu, H. (2018). Natural disasters and human capital: The case of Nepal's earthquake. *World Development, 111*, 1–12.

Pearce, T., Wright, H., Notaina, R., et al. (2011). Transmission of environmental knowledge and land skills among Inuit men in Ulukhaktok, Northwest Territories, Canada. *Human Ecology, 39*, 271–288.

Pearse, R. (2017). Gender and climate change. *Wiley Interdisciplinary Reviews: Climate Change, 8*(2), e451.

Peet, R., & Watts, M. (Eds.). (2004). *Liberation ecologies: Environment, development and social movements* (2nd ed.). Routledge.

Pini, B. (2005). Interviewing men: Gender and the collection and interpretation of qualitative data. *Journal of Sociology, 41*(2), 201–216.

Punch, S., & Sugden, F. (2013). Work, education and out-migration among children and youth in upland Asia: Changing patterns of labour and ecological knowledge in an era of globalisation. *Local Environment: The International Journal of Justice and Sustainability, 18*(3), 255–270.

Quinn, C. H., Huby, M., Kiwasila, H., et al. (2003). Local perceptions of risk to livelihoods in semi-arid Tanzania. *Journal of Environmental Management, 68*, 111–119.

Quisumbing, A. (2003). *Household decisions, gender and development: A synthesis of recent research*. IFPRI.

Radcliffe, S. (1991). The role of gender in peasant migration: Conceptual issues from the Peruvian Andes. *Revolutionary and Radical Political Economy, 23*(3–4), 129–147.

Ragasa, C., Berhane, G., Tadesse, F., et al. (2013). Gender differences in access to extension services and agricultural productivity. *Journal of Agricultural Education and Extension, 19*(5), 437–468.

Randell, H., & Gray, C. (2016). Climate variability and educational attainment: Evidence from rural Ethiopia. *Global Environmental Change, 41*, 111–123.

Rao, B. (1991). *Dominant constructions of women and nature in the social science literature*. CES/CNS Pamphlet 2: University of California Santa Cruz.

Räty, R., & Carlsson-Kanyama, A. (2010). Energy consumption by gender in some European countries. *Energy Policy, 38*(1), 646–649. https://doi.org/10.1016/j.enpol.2009.08.010

Ravera, F., Martin-Lopez, B., Pascual, U., et al. (2016). The diversity of gendered adaptation strategies to climate change of Indian farmers: A feminist intersectional approach. *Ambio, 45*(S3), S335–S351.

Resurreccion, B. P. (2017). Gender and environment from 'women, environment and development' to feminist political ecology. In *Routledge handbook of gender and environment, Part I* (pp. 71–85). Routledge.

Resurreccion, B. P., & Elmhirst, R. (Eds.). (2008). *Gender and natural resource management: Livelihoods, mobility and interventions*. Earthscan.

Rico, M. N. (1998). *Gender, the environment and the sustainability of development*. Serie Mujer Y Desarrollo, October. Santiago, Chile: United Nations.

Rocheleau, D., Thomas-Slayter, B., & Wangari, E. (Eds.). (1996). *Feminist political ecology: Global issues and local experience*. Routledge.

Rodda, A. (1993). *Women and the environment*. Zed Books.

Rohe, J., Schlüter, A., & Ferse, S. C. A. (2018). A gender lens on women's harvesting activities and interactions with local marine governance in a South Pacific fishing community. *Maritime Studies, 17*(2), 155–162. https://doi.org/10.1007/s40152-018-0106-8

Sachs, C. E. (1996). *Gendered fields: Rural women, agriculture, and environment* (1st ed.). Routledge.

Sapkota, T. B., Aryal, J. P., Khatri-Chhetri, A., et al. (2018). Identifying high-yield low-emission pathways for the cereal production in South Asia. *Mitigation and Adaptation Strategies for Global Change, 23*(4), 621–641. https://doi.org/10.1007/s11027-017-9752-1

Schipper, E. L. (2006). *Climate risk, perceptions and development in El Salvador*. Tyndall Center for Climate Change Research & School of Environmental Sciences.

Schroeder, R. A. (1999). *Shady practices: Agroforestry and gender politics in the Gambia*. University of California Press.

Scott, J. (2010). Quantitative methods and gender inequalities. *International Journal of Social Research Methodology, 13*(3), 223–236.

Sen, G., & Grown, C. (1987). *Development crises and alternative visions: Third world Women's perspectives* (1st ed.). Routledge.

Shi, X. (2019). Inequality of opportunity in energy consumption in China. *Energy Policy, 124*, 371–382. https://doi.org/10.1016/j.enpol.2018.09.029

Shiva, V. (1989). *Staying alive. women, ecology and development*. Zed Books.

Silvestri, S., Bryan, E., Ringler, C., et al. (2012). Climate change perception and adaptation of agro-pastoral communities in Kenya. *Regional Environmental Change, 12*, 791–802.

Singh, C., Osbahr, H., & Dorward, P. (2018). The implications of rural perceptions of water scarcity on differential adaptation behaviour in Rajasthan, India. *Regional Environmental Change, 18*, 2417–2432.

Smit, B., & Wandel, J. (2006). Adaptation, adaptive capacity and vulnerability. *Global Environmental Change, 16*, 282–292.

Snapp, S. S., Grabowski, P., Chikowo, R., et al. (2018). Maize yield and profitability tradeoffs with social, human and environmental performance: Is sustainable intensification feasible? *Agricultural Systems, 162*, 77–88. https://doi.org/10.1016/j.agsy.2018.01.012

Sontheimer, S. (1991). *Women and the environment*. Monthly Review Press.

Sultana, F. (2014). Gendering climate change: Geographical insights. *The Professional Geographer, 66*(3), 372–381. https://doi.org/10.1080/00330124.2013.821730

Sultana, P., & Thompson, P. (2008). Gender and local floodplain management institutions: A case study from Bangladesh. *Journal of International Development, 20*, 53–68. https://doi.org/10.1002/jid.1427

Sun, Y., Mwangi, E., & Meinzen-Dick, R. (2011). Is gender an important factor influencing user groups' property rights and forestry governance? Empirical analysis from East Africa and Latin America. *International Forestry Review, 13*(2), 205–219.

Sunderland, T., Achdiawan, R., Angelsen, A., et al. (2014). Challenging perceptions about men, women, and forest product use: A global comparative study. *World Development, 64*, S56–S66. https://doi.org/10.1016/j.worlddev.2014.03.003

Terry, G. (2009). No climate justice without gender justice: An overview of the issues. *Gender and Development, 17*(1), 5–18.

Theriault, V., Smale, M., & Haider, H. (2017). How does gender affect sustainable intensification of cereal production in the West African Sahel? Evidence from Burkina Faso. *World Development, 92*, 177–191. https://doi.org/10.1016/j.worlddev.2016.12.003

Thompson-Hall, M., Carr, E. R., & Pascual, U. (2016). Enhancing and expanding intersectional research for climate change adaptation in agrarian settings. *Ambio, 45*(Suppl. 3), S373–S382.

Tibesigwa, B., & Visser, M. (2016). Assessing gender inequality in food security among small-holder farm households in urban and rural South Africa. *World Development, 88*, 33–49.

Trawick, P. (2003). *The struggle for water in Peru: Comedy and tragedy in the Andean commons*. Stanford University Press.

Turner, B. L. (1999). Merging local and regional analysis of land-use change: The case of livestock in the Sahel. *Annals of the Association of American Geographers, 89*(2), 191–219.

Valdes, X. (1992). *Mujer, trabajo y medio ambiente. Los nudos de la modernización agrícola*. CEDEM.

van Dijk, H., & Bose, P. (2016). Dryland landscapes: Forest management, gender and social diversity in Asia and Africa. In *Dryland forests. Management and social diversity in Africa and Asia* (pp. 3–21). Springer.

Villamor, G., Desrianti, F., Akiefnawati, R., et al. (2014). Gender influences decision to change land use practices in the tropical forest margins of Jambi, Indonesia. *Mitigation and Adaptation Strategies for Global Change, 19*(6), 733–755.

West, C., & Zimmerman, D. (1987). Doing gender. *Gender & Society, 1*, 125–151.

Wutich, A. (2012). Gender, water scarcity, and sustainability tradeoffs. In M. L. Cruz-Torres & P. McElwee (Eds.), *Gender and sustainability: Lessons from Asia and Latin America* (pp. 97–120). University of Arizona Press.

Yang, F., & Xi, Y. (2001). Naxi women: Protection and management of forests in Lijiang, China. *Gender Technology and Development, 5*(2), 199–222.

Yang, J., Owusu, V., Andriesse, E., et al. (2019). In-situ adaptation and coastal vulnerabilities in Ghana and Tanzania. *The Journal of Environment and Development, 28*(3), 282–308.

Zelezny, L., Chua, P.-P., & Aldrich, C. (2000). Elaborating on gender differences in environmentalism. *Journal of Social Issues, 56*, 443–457.

Zottarelli, L. (2008). Post-hurricane Katrina employment recovery: The interaction of race and place. *Social Science Quarterly, 89*(3), 592–607.

Zwarteveen, M. Z., & Neupane, N. (1996). *Free-riders or victims: Women's nonparticipation in irrigation management in Nepal's Chhattis Mauja Irrigation Scheme*. 52731, IWMI Research Reports. International Water Management Institute.

Socio-demographic Inequalities in Environmental Exposures

22

James R. Elliott and Kevin T. Smiley

Abstract

Contemporary social scientific research emphasizes that populations commonly interact with their environments not as monolithic entities but, instead, through stratified social systems that result in unequal socio-demographic access and exposure to different types of transformed nature. This piece reviews the study of those systemic inequities through a survey of major areas of ongoing research, ending with potential directions for future investigation. A running theme throughout is how the twin forces of urbanization and industrialization not only continue to change the natural world of which we are all part but also conjoin with ongoing forces of racial oppression and economic exploitation to refract that shared fate into highly unequal exposures to hazards of growing scale, scope and intensity.

J. R. Elliott (✉)
Department of Sociology, Kraft Hall, Rice University, Houston, TX, USA
e-mail: james.r.elliott@rice.edu; jre5@rice.edu

K. T. Smiley
Department of Sociology, 126 Stubbs Hall, Louisiana State University, Baton Rouge, LA, USA
e-mail: ksmiley@lsu.edu

This chapter reviews major themes and directions in research on social inequalities in environmental exposures. We begin with the historical forces of urbanization and industrialization that have long transformed the social as well as physical environments in which people live, especially after capitalism yoked them to the production of commodities for market exchange. We then follow these forces into increasingly toxic and unequal exposures highlighted by the environmental justice movement before reviewing efforts to expand and enrich their empirical investigation, drawing heavily from the U.S. experience. We then consider the rising and unequal impacts of natural hazards in the age of climate change before concluding with a couple of directions we think important for future research. Running throughout are several fundamental points.

First, societies especially affluent ones have been altering the environment – intentionally or otherwise – for a long time. That work started on a local scale thousands of years ago with the advent of agriculture, followed by the rise of towns and cities. It has since become increasingly planetary, as resources are globally extracted, wastes are shipped from place to place, and sea levels rise. Second, these environmental changes have real consequences because they expose people not just to new social orders but also to new material

© Springer Nature Switzerland AG 2022
L. M. Hunter et al. (eds.), *International Handbook of Population and Environment*, International Handbooks of Population 10, https://doi.org/10.1007/978-3-030-76433-3_22

threats, whether it is the plague, toxic pollution, increased flooding, or other hazards. Third, these consequences tend to be unequally distributed socially as well as spatially. This is especially true in the United States, where divisions of race, class, and citizenship have long produced deep and lasting housing segregation, which exposes different groups to different local environments. Finally, the fact that these unequal exposures favor socially privileged groups is both unjust and ultimately temporary because the hazards now being produced do not degrade so much as accumulate and spread with time. We review this dynamic in the section on relict pollutants buried in the land by unregulated industries of the past, but it is also evident in the ongoing accumulation of greenhouse gases in the atmosphere.

These observations inform our understanding of environmental exposures and thus the rest of this chapter. Typically, to be exposed to something is to suggest that it is separate from us. This is how many Americans, including many environmental scientists, in particular view the environment and by extension nature: as something distinct from the social. That perspective has been dismantled within the social and behavioral sciences. Rather than seeing infectious diseases or hazardous industrial wastes or powerful floods as being "outside" society, researchers now emphasize the mutual constitution and ongoing interaction of society and environment and thus exposures of each to one another.

The Urbanization and Industrialization of New Environmental Exposures

For much of history, the chief environmental threat facing humans was that the planet provided ample predators but limited food, water, and shelter. So our ancestors adapted by moving with the seasons to secure different resources in different places at different times to survive. As they did, clans and tribes remained small in number and navigated more than altered the natural world around them. To be sure, members suffered and died from exposure to heat, cold, rain, drought, and other earthly forces, but for the most part they did not create new environmental forms to which they were then adversely exposed. That began to change with the rise of agriculture and the first towns and cities it enabled nearly ten millennia ago.

In these new environments, humans began to live in closer proximity not just to one another but also to the animals they were now domesticating to sustain their settled ways. This combination of denser living, settled residence, and comingling with other species would eventually expose entire towns and villages to some of the world's most devastating infectious diseases – cholera, small pox, and the plague among them in the past and COVID-19 more recently and globally. Exposures to these and other illnesses were further exacerbated by squalid and unsanitary conditions in which many but especially poorer residents found themselves, as well by smoke inhalation from cooking and heating fires that still contribute to adult and especially childhood mortality today. As these early settlements matured into Medieval cities thousands of years later, they provided new advances in governance, medicine, and other technological breakthroughs, what Edward Glaeser (2011) contemporaneously terms the "triumph of the city." But, they also continued to expose residents to epidemics that kept their mortality rates well above those of surrounding countrysides. Indeed, historical records indicate that medieval cities were only able to survive because of the constant influx of migrants from their hinterlands, who replaced those who died (Woods, 2003). In towns and cities, it seems, people were not only building new social orders but also facilitating new and dangerous environmental exposures.

Scholars now refer to these exposures collectively as the "urban penalty," which by the mid-1800s and early 1900s became increasingly attributed to overcrowding, poor sanitation, and lack of clean water relative to what could be found in more rural areas. These conditions ushered in new epidemics of influenza, typhus, cholera, and tuberculosis that killed millions of urban residents in the United States and other wealthier countries (Leon, 2008). During this time not only do population health records become more reliable on these points, early social scientists begin to take note of their links to new modes of production and staggering social inequalities that were arising alongside unprecedented volumes of industrial wastes that Europeans from Dickens to de Tocqueville would take time to note. Friedrich Engels offered perhaps the richest descriptions of the exceptionally poor quality of urban environments despoiled by the industrialization of work in his classic statement on *The Condition of the Working Class in England in 1844* (1968). Here, he identifies Manchester's "Little Ireland" as "the most disgusting spot of all":

> From [the river Medwell's] entry into Manchester to its confluence with the Irwell, this coal-Black, stagnant, stinking river is lined on both sides by a broad belt of factories and workers' dwellings . . . .The cottages are very small, old and dirty, while the streets are uneven, partly unpaved, not properly drained and full of ruts. Heaps of refuse, offal and sickening filth are everywhere interspersed with pools of stagnant liquid. The atmosphere is polluted by the stench and is darkened by the thick smoke of a dozen factory chimneys . . . .The inhabitants live in dilapidated cottages, the windows of which are broken and patched with oilskin. The doors and the door posts are broken and rotten. The creatures who inhabit these dwellings and even their dark wet cellar, and who live confined amidst all this filth and foul air – which cannot be dissipated because of the surrounding lofty buildings – must surely have sunk to the lowest level of humanity. (p. 71)

Thus, as capitalism was harnessing the power of industrialization to the production of commodities for capital accumulation, local environments were not the only thing to suffer. So too did a growing working class who had little choice but to live and labor amidst the new and harmful environments they were helping to create but did not control.

On the job, this lack of control meant exposures to a wide range of new hazards, including cramped working quarters with poor ventilation and trauma from machinery as well as toxic heavy metals, dust, and solvents. At home, it meant exposures to incessant noise, poor sanitation, and unsafe housing that contributed to infectious diseases, including cholera, which ushered in modern studies of epidemiology such

as with John Snow's famous work on environmental exposures and Cholera in 1850s London (Cameron & Jones, 1983). But it was not just radical social critics and medical experts who noticed these developments. A newly emergent if racially divided field of urban sociology did, too. In his 1898 study of Philadelphia, W.E.B. DuBois connected the poor housing and health conditions of African Americans to the neighborhood environments into which they were segregated (see also Taylor, 2014). In the 1920s and 1930s, the Chicago School of urban ecology expressed a clear, if implicit, understanding that more privileged groups deliberately avoided the city's increasingly hazardous industrial facilities (Harris & Ullman, 1945; Hoyt [1933] 2000; Park & Burgess [1925] 1967).

Evidence of the latter lies just beneath the surface of the concentric zone model of urbanization. In that classic morphology, Park and Burgess ([1925] 1967) explain that the (white) gentry pushed the problem of noxious industry away from the urban core where they lived and worked, thereby relieving themselves of exposure to it throughout the day. Non-gentry, could also live farther away from the polluting factories where most of their male constituents worked, but that was expensive in both housing and transportation costs. So many working-class families, including those of color, ended up living close to factories where members of their families worked (Hoyt [1933] 2000). In these neighborhoods, a walking commute was common, as was an implicit tradeoff of pollution for wages and lower housing costs.

Harris and Ullman (1945, p. 15–17) described these dynamics explicitly in their classic discussion of why urban industrial zones formed where they did, away from the central business district but still far from affluent suburbs:

> [T]he noise of boiler works, the odors of stockyards, the waste disposal problems of smelters and iron and steel mills, the fire hazards of petroleum refineries, and the space and transportation needs which interrupt streets and accessibility – all these favor the growth of heavy industry away from the main center of the large city. . . . In general, high-class districts are likely to be on well-drained, high land and away from nuisances such as noise, odors,

smoke, and railroad lines. Low-class districts are likely to arise near factories and railroad districts, wherever located in the city.

From this perspective the modern metropolis is not an inert space to be filled. Instead, it is actively produced by actors and flows embedded within unequal social relations. French sociologist Henri Lefebvre ([1974] 1991) developed this point explicitly in his classic work *The Production of Space*. Here, he begins with an assumption now shared by many contemporary social theorists, namely that urbanization reflects the sociocultural context in which it occurs. In societies such as the United States, this context includes political tensions of race and class but also and just as fundamentally, the unequal production of use and exchange values under capitalism. As this unequal production unfolds unequally for different groups, Lefebvre argued, two ideal-typical spaces tend emerge at opposite ends of the sociospatial spectrum.

At one end of the spectrum are "dominated spaces" that tend to be "exploited for the purpose of and by means of production" for purposes of market exchange (1991, p. 353). These areas are where negative externalities such as industrial pollution and other forms of environmental degradation cluster. At the other end of the spectrum are "appropriated spaces" that tend to be "exploited for the purpose of and by means of the consumption of space." These areas are where positive externalities such as clean air, cultural amenities, and green spaces tend to cluster. According to Lefebvre, a root cause of separation between these two types of spaces lies in how more privileged members of society develop, control, and restrict access to their areas of consumption while pushing areas of hazardous production elsewhere – a process that Harvey Molotch (1976, p. 328) has called aristocratic conservation; Thomas Rudel (2013) terms defensive environmentalism; and others have linked to LULUs (locally unwanted land uses) and NIMBY (not in my backyard) movements (e.g. Bullard, 1990) that have accompanied the increasingly toxic and unequal intertwining of society and hazardous industry.

The Failure of New Regulations to Curb Unjust Environmental Exposures

As these developments unfolded into the twentieth century, American cities allowed manufacturers to release their hazardous wastes largely unprocessed into the local environment, including into the air and water that promised to carry them elsewhere (Tarr, 1996). For decades, this "out of sight, out of mind" strategy provoked conflict with downstream and downwind communities, who held little sway in the politics and policies of their larger, urban neighbors. But after World War II the problem of hazardous industrial waste became difficult for even elites to ignore as highly visible smog and massive fish kills sparked widespread public concern and community action in cities across the country. In response and after protracted struggle with industry, the federal government began to pass new environmental guidelines and regulations to monitor and protect local residents from exposures to increasingly polluted air, water, and land, in that order.

First came the Air Pollution Control Act of 1955, then the Federal Water Pollution Control Act of 1972, and then the Resource Conservation and Recovery Act (RCRA) of 1976. The latter aimed to further coordinate and rationalize waste management policy, in part by distinguishing industrially produced "hazardous" waste from other, more benign forms of municipal solid waste. Hazardous waste was defined as that which was dangerous to public health. In addition to making this clarification, RCRA also took a number of direct, actionable steps. It created new "best available technology" standards for treating, storing, and disposing of hazardous waste, and it required larger hazardous waste producers to file "cradle-to-grave" manifests. These efforts marked a significant advance over earlier, largely unregulated waste disposal practices, but they still contained no provision for identifying or remedying unequal exposures of different social groups.

By the late 1970s and early 1980s, however, it was clear that the nation's growing raft of old and new environmental regulations was not protecting all groups similarly from the harms of industrial pollution, prompting a new social movement centered on environmental justice (Taylor, 1997). This movement identified environmental racism, in particular, as a pernicious and lingering form of inequality that disproportionately exposed communities of color to the harmful effects of ongoing industrial pollution (Mohai et al., 2009). As this social movement gained political traction it became the subject of new research aimed at better documenting prevailing patterns and uncovering their underlying processes. This research began with relatively straightforward accountings brought to national prominence by the case of Warren County, North Carolina, where many observers say the American environmental justice movement began.

In 1978, government officials chose this specific county – one of the most socially vulnerable in state, with a quarter of its population living in poverty and with one of the highest proportions of Black residents of any in the state – as the site for a new hazardous landfill, which was to hold oil containing polychlorinated biphenyls (PCBs), a class of chemicals so toxic that Congress banned their production the following year. As construction of the site proceeded, civil rights organizations such as the United Church of Christ and the Southern Christian Leadership Conference sent organizers to assist in local protests (Newkirk II, 2016). In 1987, as part of that larger ongoing effort, the United Church of Christ (1987), through its Commission for Racial Justice, prepared a report, "Toxic Wastes and Race in the United States," which provided some of the first evidence that race correlated more strongly with the placement of hazardous waste facilities than any other demographic factor, including income. The report also indicated that three of the country's five largest commercial hazardous-waste landfills, comprising 40% of the nation's entire commercial-landfill capacity, were located in predominantly Black or Hispanic communities.

Thus began not only the national environmental justice movement but also a growing body of academic research on social inequalities in environmental exposures to harmful industrial pollution. Through these efforts two hypotheses eventually pushed to the fore (Mohai & Saha, 2015a). The first hypothesis – sometimes termed the racial discrimination hypothesis – posits that locally unwanted land uses (LULUs) are purposefully sited in neighborhoods already disproportionately comprised of racial minorities. This targeting occurs because such areas tend to have less social and political power to resist, thereby offering a "path of least resistance" for polluters (Ard & Fairbrother, 2017; Pastor et al., 2001; see also Bullard, 1983; Saha & Mohai, 2005). Although events in Warren County clearly support this hypothesis, efforts to generalize it to other places gave rise to a second hypothesis sometimes dubbed the minority move-in hypothesis. It posits that areas with poorer environmental quality (such as those near hazardous waste sites or polluting industrial facilities) attract minority residents because housing costs tend to be lower there (Been, 1994; Been & Gupta, 1997).

Efforts to adjudicate these two hypotheses sometimes forget that the class inequalities of different racial groups highlighted by the second hypothesis have historical roots in racist policies, practices, and customs that have suppressed the wealth of black and Hispanic Americans. These class inequalities intertwine with housing choices that remain locally unequal for Americans of color due not only to economic differences and ongoing discrimination but also to variations in local housing search processes that continue to reproduce the segregated neighborhoods of American cities (Krysan & Crowder, 2017; Pulido, 2000). In this context, the real question is not *if* racism is operating, but *how*. Answering this question pushed researchers beyond point-in-time snapshots of who lived where to pursue longitudinal research on their mutual unfolding over time (see Been & Gupta, 1997; Oakes et al., 1996). The most conclusive work to date in this vein comes from Paul Mohai and Robin Saha's (Mohai & Saha, 2015b) national analysis of the location of hazardous waste facilities. The authors find that those types of facilities have indeed been disproportionately sited in minority neighborhoods or in places in which whites were already moving out in decades prior.

As these and related findings mounted, however, they also encouraged researchers to shift and expand how they conceptualize and thus measure environmental exposure. The earliest and most prominent approach was the *spatial coincidence* method. It focuses on whether a given geographic unit – most often a census tract or other neighborhood analog – of a particular demographic composition hosts an environmentally degraded site or heavy polluter. This is the method used by the United Church of Christ (1987) in their groundbreaking study as well as by pioneering scholars such as Robert Bullard (1983) and researchers at the Governmental Accountability Office (1983) – (see also Anderton et al., 1994; Bryant & Mohai, 1992; Downey, 1998, 2005). Recently, however, investigators have begun to critique this approach for its over-reliance on predetermined boundaries of the geographic units constructed for purposes other than the study of environmental inequalities (Downey, 2006a; Mohai & Saha, 2006; Mohai & Saha, 2007).

Consider, for example, a street running north and south that serves as a dividing line between two census tracts, with a large petrochemical facility located on its west side. The conventional spatial coincidence method would assign that facility's exposure to the west side only. However, the facility is still likely to expose residents on the east side, perhaps even more so if prevailing winds and waterways flow in that direction. To address this issue, Liam Downey (2003, 2006a, b) and Mohai and Saha (2006, 2007), among others (see Maantay, 2007), advanced a *distance decay* model of air pollution that conceptualizes environmental exposure in radii, or concentric circles, from a given facility. This approach – made possible by new advances in geographic information systems (GIS) software – pushes past concerns about spatial coincidence within geographic boundaries premade for other purposes to offer more valid measurement of local environmental exposures.

Even with these analytic refinements, however, there remains the question of exposure to what and how much, given that not all hazardous facilities are equally dangerous to humans. This issue is especially pertinent with air pollution because of vast disparities in the volume and toxicity of reported releases in government databases such as the U.S. Environmental Protection Agency's (EPA) Toxic Release Inventory (Collins et al., 2016). One approach that researchers have used to quantify this variation is to measure the total amount of hazardous releases reported to air (in pounds), regardless of what those releases are (e.g., Crowder & Downey, 2010; Pais et al., 2013). While this approach is better than treating all reporting facilities the same, it ignores the fact that some chemicals are more toxic, or harmful to human health, than others. One pound of emitted arsenic, for example, is currently considered to be 2000 times more harmful to humans than one pound of emitted benzene.

To better account for this variability in toxicity, federal officials commissioned health experts to convert pounds of hazardous releases to air reported in the agency's Toxic Release Inventory to relative toxicity scores indexed to the estimated health risks of specific chemicals released. The result is the Risk-Screening Environmental Indicators – Geographic Microdata (RSEI-GM), which measures air toxicities in relatively small 810-meter by 810-meter grid cells, which researchers can readily scale up or down to create their own geographic units. These features make the new database a valuable resource for research, which continues to document racial disparities not just in local exposures to hazardous facilities and total amounts of pollution but also to air toxicity (Ash et al., 2013; Ash & Fetter, 2004; Collins et al., 2016; Downey, 2007; Downey & Hawkins, 2008; Downey et al., 2008; Elliott & Smiley, 2019; Grant et al., 2010; Prechel & Zheng, 2012; Sicotte & Swanson, 2007; Smiley, 2019; 2020).

Despite these advances, however, the RSEI-GM is still limited to data contained in the EPA's Toxic Release Inventory, which collects information only from industrial facilities that voluntarily report their hazardous releases, employ at least ten full-time employees, operate in certain industrial sectors (e.g., manufacturing or mining), and manufacture more than 25,000 pounds of hazardous chemicals or use more than 10,000 pounds in a given year. To correct for these limitations, researchers now also monitor air quality directly with stationary monitors rather than relying on circumscribed facility reporting, which allows them to account for other types of air pollution, too, such as vehicular emissions, particulate matter (PM), and nitrogen dioxide (NO_2), which can then be converted into estimated levels of cancer risk as well as daily respiratory hazards. With these advances, too, research continues to document racial as well as class inequalities in harmful exposures. Unequal pollution burdens, in other words, not only remain alive and well in the United States, they are remarkably robust, whether measured in total industrial air pollution (Ard, 2015, 2016; Abel & White, 2012; Ash & Fetter, 2004; Downey & Hawkins, 2008; Smiley, 2019, 2020), cancer risks from an array of pollutants (Alvarez & Norton-Smith, 2018; Chakraborty et al., 2014; Collins et al., 2015; Liévanos, 2015), or exposure to PM and NO_2 emissions (Grineski et al., 2007; Kravitz-Wirtz et al., 2016).

Advancing Contextual and Demographic Understandings of Unequal Exposures

Along with the increasingly sophisticated measurement of local environmental exposures, there have also been efforts to enrich understanding of the different spatial and temporal contexts in which these exposures occur as well as who demographically is affected. One path emphasizes the geographically nested scales in which local environmental exposures take place, noting for example how disproportionately exposed neighborhoods are themselves located within cities, counties, and regions whose histories, economies, and politics differ in ways that can affect local exposures (Boyce et al., 2016; Downey, 2007; Downey et al., 2008; Konisky & Reenock, 2018; Lopez, 2002; Morello-Frosch & Jesdale, 2006; Pulido, 2000; Smiley, 2019; Zwickl et al., 2014).

Across regions, for example, research finds that racial inequalities remain, even in areas

where releases of industrial pollution are decreasing (Ard, 2015). Across states, research also shows wide inequalities in overall exposure including larger disparities than even those found locally by income (Boyce et al., 2016). And across metropolitan areas, research shows that local inequalities in hazardous exposure tend to be higher in areas with higher racial inequalities generally (Ash et al., 2013), in areas that are more residentially segregated by race (Ard, 2016), and in areas with more bonding social capital organizations such as religious organizations like conservative Protestant groups (Smiley, 2019, 2020). Local exposures, in other words, vary not only by the racial and class composition of respective neighborhoods but also by the racial, class, and social dynamics of their encompassing towns, cities, and regions.

Another path emphasizes how individuals move during the course of their average day, creating "activity spaces" and thus environmental exposures that extend beyond their residential neighborhoods to include other areas such as where they work (Park & Kwan, 2017; Kwan, 2018; Sharp et al., 2015). While these multidimensional exposures may seem obvious, their empirical investigation has remained limited because data on where people reside are more readily available than data on where people work, which creates blind spots for most quantitative analyses. Yet at the same time, qualitative studies have documented how workers of color in urban areas such as Chicago (Pellow, 2002), Gary, Indiana (Hurley, 1995), and Silicon Valley (Pellow and Park, 2002) experience more toxic pollution than whites at their places of employment (see also Auyero & Swistun, 2008; Taylor, 1997) – findings which parallel research in occupational health, which has long drawn attention to racial disparities in health risks at sites of work (Clougherty et al., 2010; Landsbergis, 2010; Lipscomb et al., 2006).

To get at these dynamics quantitatively, Elliott and Smiley (2019) have used commuting data from the Census Transportation Planning Products (CTPP). This dataset is compiled by the U.S. Census Bureau and the American Association of State Highway and Transportation Officials using original source data from recent American Community Surveys. Conceptually, commuting underscores the fact that nearly all areas have many different users, or stakeholders, over a given 24-hour period (Ellis et al., 2004). As Pellow (2002) explains, this multiple, interlocking form of environmental inequality matters because it moves away from a simple "perpetrator-victim" assessment of environmental injustice to one that acknowledges a multitude of actors who can have contradictory and crosscutting interests and allegiances. By acknowledging this complexity, researchers can better conceptualize underlying processes of inequality not just as intentional acts (say, of facility siting) but also as more encompassing socio-spatial systems that define and divide towns more generally (Pulido, 2000). Conceptualizing commuting dynamics in this way can improve understanding not only of unequally polluted places but also unequally polluted *spaces* forged and sustained through daily flows of socially unequal groups.

Here, we note that prior research on environmental injustice has hinted at the importance of these types of inequalities (Chakraborty et al., 2014; Morello-Frosch & Jesdale, 2006; Pellow & Park, 2002), but it has not developed the idea explicitly. For example, the U.S. Environmental Protection Agency now incorporates commuting data into the creation of its National Air Toxics Assessment (NATA) database but mainly for the sake of automobile emissions accounting. It does not consider the role that commuting plays in the spatial reproduction of unequal exposures to industrial pollution across places throughout the day (Environmental Protection Agency, 2018). To examine these dynamics directly, Elliott and Smiley (2019) studied commuting and pollution exposure at home and work in the ten-county Houston Metropolitan Statistical Area. Their results indicate that Blacks and Latinos not only tend to reside in more toxic areas than whites, they also tend to *work* in more toxic areas than whites, even if, as Ash and Boyce (2018) show, these groups are less likely to work in jobs in the toxic facilities themselves (and therefore are not attaining the "silver lining" of economic benefits

alongside toxic exposures). This finding not only indicates elevated health risks; it also implies that residential segregation and related housing inequalities are not the only social drivers of racially unequal exposures to toxic industrial pollution.

These findings have important implications for how researchers and policymakers conceptualize and study inequalities in exposure to industrial pollution. As discussed above, prior research challenged cross-sectional investigations by asking who moved in first, the hazardous facility or the residents. While this line of investigation remains useful, findings based on local commuting patterns underscore the point that people move not just by changing residences but also by commuting daily from home to work, that is, through their daily mobilities. As a result of these routine movements, the most and least polluted neighborhoods in any given metropolitan area often are not separate and unequal. They are connected and unequal. Findings show, for example, that the more a neighborhood of color is linked to whiter neighborhoods through the in-commutes of workers from elsewhere in the metropolitan area, the more polluted that neighborhood of color tends to be (Elliott & Smiley, 2019).

Another path of recent research extends this empirical focus on geographic movement over time from daily commutes to individual lifetimes. Here, the work of Kyle Crowder and colleagues has been particularly insightful. By matching longitudinal data on individuals from the Panel Study of Income Dynamics with neighborhood-level data on hazardous pollution from the EPA's Toxic Release Inventory, Crowder and Downey (2010) find that Black and Latino householders not only tend to reside in highly polluted areas relative to whites, they also tend to move into them as well (see also Pais et al., 2013). In other words, the high levels of local pollution experienced by Black and Latino households appear to be maintained by both lower odds of escaping highly polluted neighborhoods in which they reside and a tendency to relocate to destinations with higher levels of local pollution when they do. As Crowder and

Downey (2010: 1144) state, this pattern, which persists even after controlling for individual-level income, education, and other sociodemographic characteristics "is consistent with the argument that discriminatory real estate practices restrict residential options for members of at least some minority groups and that these restrictions are especially virulent in limiting opportunities for Black householders."

Yet another path of recent research has begun to extend analyses to a wider array of exposed groups. To start, researchers pushed beyond disparities between whites and minorities generally, and Blacks specifically, to incorporate other racial and ethnic groups, including first Latinos, then Asians. Inequalities among Asians, for example, show that neighborhoods with more Asian residents are linked to higher levels of air pollution, especially for Chinese Americans, Korean Americans, and in neighborhoods with a higher percentage of Asians speaking languages other than English (Grineski et al., 2017). In addition, research on Native Americans shows that they too face unequal environmental burdens, often linked to state-sanctioned policies and environmental degradation by the military (Hooks & Smith, 2004; see also Downey et al., 2008; Vickery & Hunter, 2016).

There is also research now showing that in the United States immigrants are located in areas with greater environmental risks (Hunter, 2000; Liévanos, 2015), with the same also true for areas with higher percentages of people with disabilities (Chakraborty, 2019).

Collectively, these recent and ongoing paths of research continue to deepen and expand our understanding of unequal exposures to environmental hazards, extending more recently into explicitly intersectional analyses of how various identities overlap and intertwine to influence those exposures (Malin & Ryder, 2018; Pellow, 2018). This work remains important. So, too, does a recognition that most of it continues to focus almost entirely on exposures to known hazards reported in government databases, ignoring vast quantities of undocumented, relict pollutants left behind by past industrial activities.

Uncovering Exposures to Relict Industrial Wastes

Before the U.S. Congress instituted the Toxic Release Inventory in the 1980s, industrial facilities could release their hazardous wastes into on-site lands without documenting these releases. These pollutants still linger, hidden in the ground without official record. Researchers Scott Frickel and James Elliott (Frickel & Elliott, 2018) care about these sites of relict waste because in the place of the factories and workshops that produced them, we now find restaurants, coffee shops, playgrounds, retail clothing stores, artist studios, and other seemingly benign land uses that arouse little if any scrutiny. This re-use occurs because the historical processes that once generated older, now-relict sites of industrial waste are ongoing. In time, even today's newer factories will go out of business, and their industrialized lots will be sold or rented and put to other uses. And, because a large percentage of these newer factories are small, they too will escape current federal reporting requirements and eventually go missing too as they contribute to the relict on-site wastes of tomorrow. Thus, despite what the term "relict" might imply, hidden industrial wastes buried in the ground are not a problem of history so much as a problem that extends in complicated ways from the past, through the present, into the future – in no small part because the increasingly toxic wastes that industries began to produce and release by the 1930s can take thousands of years to degrade.

To illuminate these concerns, Frickel and Elliott (2018) assembled data from more than 60 years of state manufacturing directories in four cities to identify sites where hazardous industry operated in the past and likely deposited wastes into the ground (see also Marlow et al., 2019). The result is an inventory of tens of thousands of sites of concern, less than 10% of which have ever been formally inspected for environmental safety, often because they changed to non-industrial uses years ago. Thus, despite their risks, they do not look, feel, or smell like risky places today. Analyses also show that many of these redeveloped industrial sites are *not* located in predominantly low-income or minority neighborhoods, and the percentage of those that are has been declining over recent decades, as new generations of white residents move back into gentrifying urban neighborhoods such as Philadelphia's Northern Liberties, Portland's Pearl, and New Orleans's warehouse district. Indeed, even Manchester's "Little Ireland" slum of the mid 1800s – once described by Engels as "the lowest stage of humanity" – is now home to some of the most expensive apartments in the city, including now renovated and up-market housing of Chorlton Mills and Liberty Heights.

At first the latter finding might seem to contradict evidence that environmental burdens are unequally borne by people of color and that during recent decades this situation has gotten worse, not better. But in fact the two dynamics are compatible, one nested temporally within the other. Exposures to hidden deposits of land-based industrial pollution is expanding not only because it is ongoing but because many former industrial areas are now being renovated into glitzy spaces, replete with gastropubs, dog parks, and crossfit facilities that denote the churning of different social groups in and out of different neighborhoods over time. These dynamics do not mean that minorities and low-income groups face less risk than earlier studies have indicated; instead, they imply that whites and middle-income groups face more risk than they and scholars have previously realized. In these ways, the unjust exposure of marginalized groups to industrial hazards, especially known ones, is the contemporaneous process, the now. The lasting environmental consequence of that unjust exposure occurring over and over again in neighborhood after neighborhood is the cumulative process, the later. In this way, environmental injustices of today contribute to *and* hide increasingly systemic and often hidden environmental risks of tomorrow.

With these insights, we can better see the different domains of knowledge that different researchers are accessing. More conventional studies of environmental inequality (reviewed above) typically begin with what is already known. They start with information contained in regulatory databases that catalogue reported toxic releases and associated risks at known and soon-to-be-opened sites. By contrast, research on relict pollutants begins from what is *not known*. It mines historical sources for data on former hazardous sites and processes that came and went long before regulatory attention began to develop in the late twentieth century. These hidden sites of contamination present serious concerns, especially those that have operated and then disappeared since the Second World War, when industrial production and use of new hazardous chemicals rose exponentially, as did their discard into local ecosystems. This point was made dreadfully clear in the United States with infamous case of Love Canal.

In 1978, unsuspecting residents of the Love Canal neighborhood of Niagara Falls, New York discovered 22,000 tons of hazardous chemical and municipal wastes buried beneath their subdivision, which included a school that many of the area's children attended. Both had been built atop a vast site formerly occupied by the city landfill and then by Hooker Chemical Company, which dumped its toxic pollutants on-site, covered them with topsoil, and then sold the property back to the unsuspecting city for $1 (Levine, 1982; Fletcher, 2003). Around the same time, other large, unregulated landfills also began to enter public knowledge in Riverside County (California), Hardeman County (Tennessee), Bullitt County (Kentucky), and elsewhere, making it clear that Love Canal was not an isolated case. It was symptomatic of waste disposal practices that had gone on for quite some time and which now pose real, direct threats to local communities and the ecosystems they inhabit (Andrews, 1999).

Estimates of the number of such relict, or brownfield, sites has risen steadily since those initial cases, from approximately 27,000 in 1987 (Colten, 1990) to 450,000 by the mid-1990s (Simons, 1998), to a million today (U.S. Government Accounting Office, 2004:1). Yet these increasing estimates still tell us little about how many relict sites have yet to be identified (Coffin, 2003). The reality is that five decades after initial public attention and government action surrounding Love Canal, the existence of hazardous wastes left behind on relict industrial sites remains a mystery. In the meantime, relict chemicals have become so pervasive that scientists no longer consider them to be "out there," but instead and increasingly in our bodies. PCBs, for example, are now so ubiquitous that medical experts must assume an average "background level" of 0.18 parts per million for each person against which to assess individual lab tests (Spears, 2014).

The Unequal Impacts of "Natural" Hazards in the Age of Climate Change

Of course, relict industrial wastes do not just accumulate in the soil and in our bodies. They also collect in the atmosphere in the form of greenhouse gases – a fact that is now driving concerns about global climate change. Warming bodies of water, for example, catalyze hurricanes; melting ice sheets contribute to sea level rise; and hotter temperatures bring more record heat days. Because these developments are human-driven, they problematize the "natural" in "natural disasters" and once again foreground a running theme of this essay: that environmental hazards are deeply entwined with social systems, which are highly unequal. If two concepts animate this understanding in the area of natural hazards, they are vulnerability and resilience. Both terms are contested, so some initial orientation is useful.

By *vulnerability*, most social scientists refer to the potential for worse case outcomes during and after natural hazards occur (Cutter et al., 2003; Tierney, 2019). By *resilience*, they refer to the capacities of individuals and communities to plan for and cope with related impacts (Aldrich & Meyer, 2015; Cutter et al., 2008; Tierney, 2014). Of chief interest with respect to unequal exposures is vulnerability, which a growing body of research emphasizes is unequally distributed not just by race, class, and ethnicity but also by age, immigrant status, housing tenure, and the local built environment. As a result of these inequalities, the short- and long-term effects of natu-

ral hazards can vary greatly for different groups in the same disaster zone (Chakraborty et al., 2019; Cutter et al., 2003; Cutter & Finch, 2008; Fothergill & Peek, 2004; Howell & Elliott, 2018; Loughran et al., 2019; Schultz & Elliott, 2013; Tierney, 2019; Walker & Burningham, 2011).

To illuminate these dynamics, most research to date has focused on extreme cases, as researchers have sought to dispel myths about disaster panic (Quarantelli, 1954; Quarantelli & Dynes, 1977; Solnit, 2009), highlight unjust recoveries (Fothergill & Peek, 2015; Gotham & Greenberg, 2014), or investigate long-term impacts on individual and community well-being (Aldrich, 2019; Cope et al., 2013; Erikson, 1976; Gill et al., 2012; Ritchie & Gill, 2007). Less attention, by contrast, has been directed toward understanding how the growing costs of natural disasters are playing out among different social groups experiencing varying levels of impact. That gap is beginning to fill, however, with the construction of innovative datasets that assemble information across a wide range of local areas and events over time. A leading example is the Spatial Hazards Events and Losses Database for the United States (SHELDUS). Originally assembled at the University of South Carolina and rooted in Susan Cutter's work on disaster vulnerability (e.g. Cutter, 1996), SHELDUS combines multiple datasets on natural hazard impacts to provide event-specific estimates of deaths and direct property damages for all counties in the United States dating back to 1960. These data reveal that nearly all counties have experienced significant disaster impacts over recent years, whether it is from fires in the West, from flooding and tornados in the Midwest, or from storms and hurricanes along the Gulf and Atlantic coasts (Howell & Elliott, 2018; Tierney, 2019).

In addition to establishing the ubiquity and rising costs of natural hazard impacts, these datasets have helped to illuminate the long-term consequences of natural hazards for different social groups. One of those consequences is residential instability. Demographers have long been interested in how large-scale disasters can drive the movement of people; in fact, the U.S. Census Bureau first began collecting data on internal migration in order to track residential movements away from the dust bowl droughts of the 1930s (McLeman et al., 2014). In more recent research, Elliott and Howell (2017) merged county-level SHELDUS data to nationally representative, longitudinal data on residential mobility from the Panel Study of Income Dynamics for 1999 through 2011. Overall, they found a positive relationship between county-level natural hazard damages and residential instability, measured as the number of moves undertaken during the 12-year period. They also found that this relationship is stronger for Black and Latino residents compared to their White counterparts. Similarly, less educated, renting, and non-married residents are more likely to experience greater residential instability after more costly natural hazards than their more educated, home owning, and married counterparts (see also Elliott, 2015; as well as related research by Curtis et al., 2015; Fussell et al., 2017). The rising costs of natural disasters, in other words, are having unequal effects on the housing outcomes of Americans across the country – a point echoed in a growing literature on unequal disaster recoveries (duPont et al., 2015; Elliott & Clement, 2017; Gotham & Greenberg, 2014; Karim & Noy, 2016; Kroll-Smith, 2018; Noy & Vu, 2010; Pais & Elliott, 2006; Pais & Elliott, 2008; Pelling, 2003; Preston, 2013; Smiley et al., 2018; Tierney, 2014).

Steve Kroll-Smith's (2018) recent comparative study of the 1906 San Francisco earthquake and Hurricane Katrina in 2005 offers insights into how these inequalities unfold. Broadly, he argues that the United States' long-standing approach to disaster recovery is best viewed as a three act play that loops back on itself. During the first act, before disaster, there is the making of local place, space, and social inequalities. As cities San Francisco and New Orleans offer stark illustrations because they not only concentrate vast sums of people and wealth but also divide them locally. Through capitalism and racism, Kroll-Smith (2018: 36) explains, "The sharp scissors of money and color cut each urban space into warrens of social kinds whose disparate relations to capital accumulation and the privileges of skin shade [bring] a modicum of both cognitive and social order to the cacophony of urban life."

In the next act, when catastrophe strikes, local residents shift into a kinder, gentler mode. Amidst the immediate destruction and uncertainty caused by the disaster, categorical divides recede. Strangers help strangers. People of stature call for rebirth. And, local media celebrate the heroism of first responders and the spirit of cooperation. Through these actions, disaster-hit places not only assert their relevance to outsiders but also become therapeutic communities to those within them, as a more just and caring society emerges to help those in need. Soon after, though, comes act three, which sternly and steadily reconstructs not just the damaged buildings, roads, and other infrastructures damaged by the disaster but also the lost and highly unequal social order. Kroll-Smith uncovers how this recovery of inequality unfolds in a variety of unintentional and uncoordinated ways to restore categorical divides threatened by local disasters.

Early on, he explains that media coverage and law enforcement begin distinguishing "looters" from survivors based more on the color of people's bodies than on what they are actually doing, such as securing food, water, or other provisions. As Kroll-Smith explains, "If disaster acts to disorder the market hierarchy that organizes urban life, 'the looter' is the scapegoat, the victim sacrificed to launch the rescue of order" (p.83). As the rescued order in restored, inequality is further propelled by opportunistic developers looking to accumulate land and wealth through the dispossession of marginalized victims. And, it is further fueled by how long-term relief is imagined and administered, with fraud and entitlement increasingly framed as the true threats to recovery. In the shadow of these twin specters, government agencies and private insurers categorically divide deserving from undeserving victims. Kroll-Smith argues that this division is based less on need than on how far down the social ladder respective parties have fallen, which of course depends on how far up the ladder they already were. This tendency minimizes the plight of marginalized residents – Asians in Chinatown for example, or low-income African Americans in public housing. It also extends upward to differentiate working and middle class homeowners, as well as the neighborhoods they anchor. A case in point is the Road Home Program in New Orleans. By tying the amount of housing restoration grants to pre-disaster property values, this relief effort underfunded the reconstruction of Black communities, where a long history of oppression and reduced public investment suppressed market values below those of similar properties in predominantly white neighborhoods.

By comparing two iconic disasters separated by nearly a century, Kroll-Smith reveals the common and highly unequal recoveries that Americans have long experienced in the wake of natural disasters. Yet, for all its value this approach is still unable to disentangle the influence of natural hazards from the influence of other national factors and trends or to examine the cumulative effect of habitual disasters on different social groups over time. In an recent effort to fill that gap, Howell and Elliott (2018) advocate for an alternative "population centered" approach to the study of natural hazard impacts – one that follows households over time as they experience different levels of cumulative disaster damages in the counties where they live. To conduct these analyses, they link a representative sample of Americans from the Panel Study of Income Dynamics to local information on natural hazard damages, FEMA assistance, and social demographics to evaluate how damages from local impacts affect residents' wealth accumulation over a 15-year timespan. From previous research, they expected socially marginalized residents to suffer the most from local hazard damages, but they also expected all residents who lived through them to suffer financially. To their surprise, they were wrong.

Instead of financially suffering, socially advantaged residents actually seem to accumulate more wealth than similar counterparts who did not live through natural hazards or lived through ones that were less costly. And, the more social advantage one has in terms of race, education, and homeowner status, the more this is so. Thus, after disasters strike, Whites tend to accumulate more wealth than people of Color, homeowners more

than renters, and the well-educated more than the less-educated. Furthermore, the more damage that occurs in an area over time, the more these inequalities grow. As a result, not only do rising natural hazard costs and federal disaster aid seem to be contributing to increased wealth inequality across traditional social divides, they also seem to be benefitting more socially privileged groups, even if they are unaware of it. Why might this be the case?

Howell and Elliott do not identify why exactly privileged Americans benefit financially from natural disasters while more marginalized groups suffer. Yet, one overriding factor is that disaster recoveries focus primarily on restoring property, not people. To rebuild damaged homes, buildings, and infrastructure, government assistance and insurance payouts allocate funds depending on the value of the properties damaged. Thus, the higher the property's value, the more money often flows in to restore it. However, property values are not merely determined by the structures themselves. Instead, the value of property is tightly tied to neighborhood racial composition. This means that a two-bedroom, two-bathroom house located in a Whiter neighborhood is likely to receive more recovery capital than the same two-bedroom, two-bathroom house in a neighborhood of color. This additional recovery aid helps families restore their homes and make renovations, and as a result, those homes increase in value, increasing their owners' wealth.

Yet, it is not just the immediate recipients of insurance payouts and recovery aid that benefit from the influx of recovery capital. Communities that have sufficient capital to rebuild also experience corporate investments, infrastructure advancements, and associated increases in property values. These collective resources create positive ripple effects for all residents of more socially advantaged neighborhoods, all the while requiring no explicit conspiracy or negative intent. The market logic upon which long-term recovery assistance rests is sufficient, with government programs offering no exception. Because they typically operate through rather than counter to the market, the more FEMA aid a county receives for the same level of disaster damage, the more wealth inequity grows between more and less privileged households and communities (Howell & Elliott, 2018).

These findings suggest that the rising costs of disasters are inseparable from rising levels of inequality. They also hint at unexpected benefits of climate change to those positioned further up the social ladder. Because these benefits are counterintuitive and go largely unrecognized and because they are grounded in seemingly status-neutral private property rights and market logics, they are unlikely to change without deliberate action. In the meantime and unwittingly, disaster recovery has turned into a kind of opportunity hoarding in which collective resources are gathered via taxes and insurance and then disproportionately funneled to more socially advantaged populations, who are unlikely to see current approaches as a problem in need of reform.

Conclusion

The foregoing review traced the increasing scale, scope, and harm of new environmental threats produced by the march of affluent societies. With each new advance, we see how socially marginalized groups disproportionately bear the brunt of those threats, regardless of how associated hazards are conceptualized, measured, or analyzed at different points in time. We conclude with two broad directions for future research.

One direction is to chip away at the increasingly artificial divide between exposures to natural hazards and industrial hazards. Hurricane Harvey offers a telling example. In August 2017 historic rains from the storm flooded more than a hundred thousand homes in the greater Houston area. They also triggered more than a hundred industrial spills, releasing a half billion gallons of hazardous chemicals and wastewater into the local environment, including well known carcinogens such as dioxin, ethlyne, and PCBs (Bajak & Olsen, 2018; see also Bernier et al., 2018; Evans, 2017; Flitter & Valdmanis, 2017; Thomas et al., 2018; Valdmanis & Gardner, 2017). These releases are not exceptions. Instead,

they underscore the fact that when natural hazards strike, they now often do so in highly developed areas with significant industrial infrastructures and relict wastes, creating the potential for what researchers call a natural-technological, or "natech," disaster (Freudenburg, 1993, 1997; Gill & Steven Picou, 1998; Kroll-Smith & Couch, 1991; Nascimento & Alencar, 2016). In the case of Hurricane Katrina, for example, levee failures not only displaced an entire city's population but also contributed to the second largest oil spill in North America to that date, releasing eight million gallons of petroleum while also spreading dangerous amounts of arsenic, diesel fuel, and other toxic chemicals throughout the greater New Orleans area (Frickel & Moore, 2006; Picou & Marshall, 2006, 2007). In the wake of those releases, the city was dubbed the "Big Uneasy," with litigation and loss of trust in government agencies distressing residents and slowing recovery. Similar experiences have been documented after the 2011 earthquake, tsunami, and meltdown of the Daiichi nuclear power plant in Fukushima Japan (Aldrich, 2019; Miller, 2016).

In addition to the immediate health and environmental impacts of such events, prior sociological research highlights their long-term corrosive potential, including social fragmentation, loss of trust in social institutions, and negative mental health effects resulting from the disruption of individual and collective life (Adeola & Steven Picou, 2017; Erikson, 1976; Freudenburg, 1997; Picou et al., 2004). As valuable as these insights remain, however, the studies generating them often share two common methodological shortcomings with other studies of disasters. One is that they typically begin investigation *after* a natech event has occurred. As a result, we gain little insight into how residents experience nearby industrial risks before such events occur as well as how those experiences shift thereafter. This analytical gap is important because natech disasters and other harmful environmental exposures do not happen spontaneously. They typically unfold in settings where related hazards and surrounding communities have long histories together, giving residents ample time to develop collective understandings of related injustices as well as local attachments to place (Auyero & Swistun, 2009; Erikson, 1994; Pellow, 2002; Spears, 2014).

Research on such events has also tended to occur in relatively isolated fashion, with climate scientists for example studying rising disaster risks, engineers studying the structural integrity of vulnerable infrastructures, planners studying local risk management policies, and social scientists studying related social vulnerabilities. These efforts should continue, but they must also converge to offer solutions to increasingly complex and pressing threats posed by the intersection of our changing natural, industrial and social worlds. This work will not be easy. Since the inception of the social sciences, there has been a strong sense that they differ from the natural sciences, with the latter focused more on facts and the former on values and meanings. This understanding sets up an epistemological tension that can never be fully resolved. So, instead, the aim should be to harness that tension by deliberately working back and forth among different fields. In those efforts the goal is not a unified science but rather collaborative efforts to innovate, integrate and extend the impact of research beyond what one field alone can accomplish, with deliberate attention to the social inequalities involved.

A second direction for future research involves moving beyond a focus on unequal exposures to environmental threats to also consider unequal access to environmental amenities, both natural and produced. One example is the small but growing body of work on parks and other public greenspaces that nine in ten Americans agree constitute important local services (National Recreation and Park Association, 2016). Today, a good deal of investment in parks comes from public-private partnerships that leverage significant private and philanthropic investment to develop new, repurposed greenspaces largely for the urban elite. Leading examples include New York City's High Line, Chicago's Millennium Park, and Houston's Discovery Green, but similar efforts are evident throughout the United States (Loughran, 2014; Smiley et al., 2016). Notably, this latest wave of park development has not replaced older public park systems. Instead, it has inserted itself in front of that system, eclipsing older, smaller parks that have been left to decay, under-funded in many neighborhoods of color (Elliott et al., 2019).

Improved access and upkeep of these existing park infrastructures are associated with increased opportunities for physical activity (McCormack et al., 2010), enhanced psychological well-being (Tinsley et al., 2002), improved property values (Wu & Dong, 2014), and a strengthened sense of community (Jennings et al., 2012). Research also credits parks with reducing health disparities in obesity and cardiovascular disease associated with low-income and racial minority communities (Jennings et al., 2017), especially when residents of those communities perceive nearby parks to be safe (Echeverria et al., 2014). To help develop those capacities, researchers should work with policy makers, communities, and philanthropists to illuminate and address inequities in environmental goods as well as bads (see Boone et al., 2009; Bruton & Floyd, 2014; Dahmann et al., 2010; Jennings et al., 2012; Cutts et al., 2009; Sister et al., 2010; Wolch et al., 2014; Smiley et al., 2016b; Suminski et al., 2012; Vaughan et al., 2013). Like inter-disciplinary research on the increasingly precarious intersections of our natural, industrial and social worlds, this work will not be easy. But, it is important if we hope not only to document but also remedy the long and varied history of social inequalities in environmental exposures.

References

Abel, T. D., & White, J. (2012). Skewed riskscapes and gentrified inequities: Environmental disparities in Seattle, Washington. *American Journal of Public Health, 101*(S1), S246–S254.

Adeola, F. O., & Steven Picou, J. (2017). Hurricane Katrina-linked environmental injustice: Race, class, and place differentials in attitudes. *Disasters, 41*(2), 228–257.

Aldrich, D. P. (2019). *Black wave: How networks and governance shaped Japan's 3/11 disasters*. University of Chicago Press.

Aldrich, D. P., & Meyer, M. A. (2015). Social capital and community resilience. *American Behavioral Scientist, 59*(2), 254–269.

Alvarez, C. H., & Norton-Smith, K. G. (2018). Latino destinations and environmental inequality: Estimated cancer risk from air toxics in Latino traditional and new destinations. *Socius, 4*, 1–11.

Andrews, R. N. (1999). *Managing the environment, managing ourselves: A history of American environmental policy*. Yale University Press.

Anderton, D. L., Anderson, A. B., Oakes, J. M., & Fraser, M. R. (1994). Environmental equity: The demographics of dumping. *Demography, 31*(2), 229–248.

Ard, K. (2015). Trends in exposure to industrial air toxins for different racial and socioeconomic groups: A spatial and temporal examination of environmental inequality in the U.S. from 1995 to 2004. *Social Science Research, 53*, 375–390.

Ard, K. (2016). By all measures: An examination of the relationship between segregation and health risk from air pollution. *Population and Environment, 38*(1), 1–20.

Ard, K., & Fairbrother, M. (2017). Pollution prophylaxis? Social capital and environmental inequality. *Social Science Quarterly, 98*(2), 584–607.

Ash, M., & Boyce, J. K. (2018). Racial disparities in pollution exposure and employment at US industrial facilities. *Proceedings of the National Academies of Sciences, 115*(42), 10636–10641.

Ash, M., & Fetter, T. R. (2004). Who lives on the wrong side of the environmental tracks? Evidence from the EPA's risk-screening environmental indicators model. *Social Science Quarterly, 85*(2), 441–462.

Ash, M., Boyce, J. K., Chang, G., & Scharber, H. (2013). Is environmental justice good for white folks? Industrial air toxics and exposure in urban America. *Social Science Quarterly, 94*(3), 616–636.

Auyero, J., & Swistun, D. (2008). The social production of toxic uncertainty. *American Sociological Review, 73*(3), 357–379.

Auyero, J., & Swistun, D. A. (2009). *Flammable: Environmental suffering in an argentine shantytown*. Oxford University Press.

Bajak, F., & Olsen, L. (2018). Silent spills: Part 1: In Houston and beyond, Harvey's spills leave a toxic legacy. *Houston Chronicle*. Retrieved April 2, 2018. https://www.houstonchronicle.com/news/houston-texas/houston/article/In-Houston-and-beyond-Harvey-s-spills-leave-a-12771237.php

Been, V. (1994). Locally undesirable land uses in minority neighborhoods: Disproportionate siting or market dynamics? *The Yale Law Journal, 103*(6), 1383–1422.

Been, V., & Gupta, F. (1997). Coming to the nuisance or going to the barrios – A longitudinal analysis of environmental justice claims. *Ecology Law Quarterly, 24*, 1–56.

Bernier, C., Kameshwar, S., Elliott, J. R., Padgett, J. E., & Bedient, P. B. (2018). Evaluation of mitigation strategies to protect petrochemical infrastructure and nearby communities during storm surge. *Natural Hazards Review, 19*(4), 1–18.

Boone, C. G., Buckley, G. L., Morgan Grove, J., & Sister, C. (2009). Parks and people: An environmental justice inquiry in Baltimore, Maryland. *Annals of the Association of American Geographers, 99*(4), 767–787.

Boyce, J. K., Zwickl, K., & Ash, M. (2016). Measuring environmental inequality. *Ecological Economics, 124*, 114–123.

Bruton, C. M., & Floyd, M. F. (2014). Disparities in built and natural features of urban parks: Comparisons by neighborhood level race/ethnicity and income. *Journal of Urban Health, 91*(5), 894–907.

Bryant, B., & Mohai, P. (1992). *Race and the incidence of environmental hazards*. Westview Press.

Bullard, R. D. (1983). Solid waste sites and the black Houston community. *Sociological Inquiry, 53*(2–3), 273–288.

Bullard, R. D. (1990). *Dumping in dixie: Race, class, and environmental quality*. Westview Press.

Cameron, D., & Jones, I. G. (1983). John snow, the broad street pump and modern epidemiology. *International Journal of Epidemiology, 12*(4), 393–396.

Chakraborty, J. (2019). Proximity to extremely hazardous substances for people with disabilities: A case study in Houston, Texas. *Disability and Health Journal, 12*(1), 121–125.

Chakraborty, J., Collins, T. W., Grineski, S. E., Montgomery, M. C., & Hernandez, M. (2014). Comparing disproportionate exposure to acute and chronic pollution risks: A case study in Houston, Texas. *Risk Analysis, 34*(11), 2005–2020.

Chakraborty, J., Collins, T. W., & Grineski, S. E. (2019). Exploring the environmental justice implications of Hurricane Harvey flooding in greater Houston, Texas. *American Journal of Public Health, 109*, 244–250.

Clougherty, J. E., Rossi, C. A., Lawrence, J., Long, M. S., Diaz, E. A., Lim, R. H., … Godleski, J. J. (2010). Chronic social stress and susceptibility to concentrated ambient fine particles in rats. *Environmental Health Perspectives, 118*(6), 769–775.

Coffin, S. L. (2003). Closing the brownfield information gap: Some practical methods for identifying brownfields. *Environmental Practice, 5*, 34–39.

Collins, T. W., Grineski, S. E., Chakraborty, J., Montgomery, M. C., & Hernandez, M. (2015). Downscaling environmental justice analysis: Determiannts of household-level hazardous air pollutant exposure in greater Houston. *Annals of the Association of American Geographers, 105*(4), 684–703.

Collins, M. B., Munoz, I., & JaJa, J. (2016). Linking 'toxic outliers' to environmental justice communities. *Environmental Research Letters, 11*, 1–9.

Colten, C. E. (1990). Historical hazards: The geography of relict industrial wastes. *The Professional Geographer, 42*, 143–156.

Cope, M. R., Slack, T., Blanchard, T. C., & Lee, M. R. (2013). Does time heal all wounds? Community attachment, natural resource employment, and health impacts in the wake of the BP Deepwater Horizon disaster. *Social Science Research, 42*, 872–881.

Crowder, K., & Downey, L. (2010). Interneighborhood migration, race, and environmental hazards: Modeling microlevel processes of environmental inequality. *American Journal of Sociology, 115*(4), 1110–1149.

Curtis, K. J., Fussell, E., & DeWaard, J. (2015). Recovery migration after Hurricanes Katrina and Rita: Spatial concentration and intensification in the migration system. *Demography, 52*, 1269–1293.

Cutter, S. L. (1996). Vulnerability to environmental hazards. *Progress in Human Geography, 20*(4), 529–539.

Cutter, S. L., & Finch, C. (2008). Temporal and spatial changes in social vulnerability to natural hazards. *Proceedings of the National Academy of Sciences, 105*(7), 2301–2306.

Cutter, S. L., Boruff, B. J., & Lynn Shirley, W. (2003). Social vulnerability to environmental hazards. *Social Science Quarterly, 84*(2), 242–261.

Cutter, S. L., Barnes, L., Berry, M., Burton, C., Evans, E., Tate, E., & Webb, J. (2008). A place-based model for understanding community resilience to natural disasters. *Global Environmental Change, 18*, 598–606.

Cutts, B. B., Darby, K. J., Boone, C. G., & Brewis, A. (2009). City structure, obesity, and environmental justice: An integrated analysis of physical and social barriers to walkable streets and park access. *Social Science and Medicine, 69*, 314–1322.

Dahmann, N., Wolch, J., Joassart-Marcelli, P., Reynolds, K., & Jerrett, M. (2010). The active city? Disparities in provision of urban public recreation resources. *Health & Place, 16*, 431–445.

Downey, L. (1998). Environmental injustice: Is race or income a better predictor? *Social Science Quarterly, 79*(4), 766–778.

Downey, L. (2003). Spatial measurement, geography, and urban racial inequality. *Social Forces, 81*(3), 937–952.

Downey, L. (2005). The unintended significance of race: Environmental racial inequality in Detroit. *Social Forces, 83*(03), 971–1008.

Downey, L. (2006a). Using geographic information systems to reconceptualize spatial relationships and ecological context. *American Journal of Sociology, 112*(2), 567–612.

Downey, L. (2006b). Environmental racial inequality in Detroit. *Social Forces, 85*(2), 771–796.

Downey, L. (2007). US metropolitan-area variation in environmental inequality outcomes. *Urban Studies, 44*(5/6), 953–977.

Downey, L., & Hawkins, B. (2008). Race, income, and environmental inequality in the United States. *Sociological Perspectives, 51*(4), 759–781.

Downey, L., Dubois, S., Hawkins, B., & Walker, M. (2008). Environmental inequality in metropolitan America. *Organization & Environment, 21*(3), 270–294.

DuBois, W. E. B. (1898). *The Philadelphia negro: A social study*. University of Pennsylvania.

duPont, I. V., William, I. N., Okuyama, Y., & Sawada, Y. (2015). The long-run socio-economic consequences of a large disaster: The 1995 earthquake in Kobe. *PLoSOne, 10*(10), 1–17.

Echeverria, S. E., Kang, A. L., Isasi, C. R., Johnson-Dias, J., & Pacquiao, D. (2014). A community survey on neighborhood violence, park use, and physical activity among urban youth. *Journal of Physical Health and Activity, 11*(1), 186–194.

Ellis, M., Wright, R., & Parks, V. (2004). Work together, live apart? Geographies of racial and ethnic segregation at home and at work. *Annals of the Association of American Geographers, 94*(3), 620–637.

Elliott, J. R. (2015). Natural hazards and residential mobility: General patterns and racially unequal outcomes in the United States. *Social Forces, 93*(4), 1723–1747.

Elliott, J. R., & Clement, M. T. (2017). Natural hazards and local development: The successive nature of landscape transformation in the United States. *Social Forces, 96*(2), 851–876.

Elliott, J. R., & Howell, J. (2017). Beyond disasters: A longitudinal analysis of natural hazards' unequal impacts on residential instability. *Social Forces, 95*(3), 1181–1207.

Elliott, J. R., & Smiley, K. T. (2019). Place, space, and racially unequal exposures to pollution at home and work. *Social Currents, 6*(1), 32–50.

Elliott, J. R., Korvoer-Glenn, E., & Bolger, D. (2019). The successive nature of city parks: Making and remaking unequal across over time. *City & Community, 18*(1), 109–127.

Engels, F. (1968). *The condition of the working class in England in 1844*. Stanford University Press.

Erikson, K. (1994). *A new species of trouble*. WW Norton & Company.

Environmental Protection Agency. (2018). *Technical support document EPA's 2014 national air toxics assessment*. EPA Office of Air Quality Planning and Standards.

Erikson, K. T. (1976). *Everything in its path: Destruction of Community in the Buffalo Creek Flood*. Simon & Schuster.

Evans, M. (2017). After oil refinery is damaged by Harvey, Benzene is detected in Houston Area. New York City, NY: *Wall Street Journal*. Retrieved January 25, 2018. https://www.wsj.com/articles/after-oil-refinery-is-damaged-by-harvey-benzene-is-detected-in-the-air-in-houston-area-1504638772

Fletcher, T. H. (2003). *From Love Canal to environmental justice: The politics of hazardous waste on the Canada-US border*. Broadview.

Flitter, E., & Valdmanis, R. (2017). Oil and chemical spills from Hurricane Harvey big, but dwarfed by Katrina. *Reuters*. Retrieved January 25, 2018. https://www.reuters.com/article/us-storm-harvey-spills/oil-and-chemical-spills-from-hurricane-harvey-big-but-dwarfed-by-katrina-idUSKCN1BQ1E8

Fothergill, A., & Peek, L. A. (2004). Poverty and disasters in the United States: A review of recent sociological findings. *Natural Hazards, 32*, 89–110.

Freudenburg, W. R. (1993). Risk and recreancy: Weber, the division of labor, and the rationality of risk perceptions. *Social Forces, 71*(4), 909–932.

Freudenburg, W. R. (1997). Contamination, corrosion and the social order: An overview. *Current Sociology, 45*(3), 19–39.

Frickel, S., & Elliott, J. R. (2018). *Sites unseen: Uncovering hidden hazards in American cities*. Russell Sage Foundation.

Frickel, S., & Moore, K. (Eds.). (2006). *The new political sociology of science: Institutions, networks and power*. The University of Wisconsin Press.

Fussell, E., Curran, S. R., Dunbar, M. D., Babb, M. A., Thompson, L., & Meijer-Irons, J. (2017). Weather-related hazards and population change: A study of hurricanes and tropical storms in the United States, 1980–2012. *Annals of the American Academy of Political and Social Science, 669*, 146–167.

Fothergill, A., & Peek, L. (2015). *Children of Katrina*. University of Texas Press.

Gill, D. A., & Steven Picou, J. (1998). Technological disaster and chronic community stress. *Society & Natural Resources, 11*, 795–815.

Gill, D. A., Steven Picou, J., & Ritchie, L. A. (2012). The *Exxon Valdez* and BP oil spills: A comparison of initial social and psychological impacts. *American Behavioral Scientist, 56*(1), 3–23.

Glaeser, E. (2011). *Triumph of the city: How our greatest invention makes us richer, smarter, greener, healthier, and happier*. Penguin Books.

Gotham, K. F., & Greenberg, M. (2014). *Crisis cities: Disaster and redevelopment in New York and New Orleans*. Oxford University Press.

Grant, D., Trautner, M. N., Downey, L., & Thiebaud, L. (2010). Bringing the polluters back in: Environmental inequality and the organization of chemical production. *American Sociological Review, 75*(4),479–504.

Grineski, S., Bolin, B., & Boone, C. (2007). Criteria air pollution and marginalized populations: Environmental inequity in metropolitan Phoenix, Arizona. *Social Science Quarterly, 88*(2), 535–554.

Grineski, S. E., Collins, T. W., & Morales, D. X. (2017). Asian Americans and disproportionate exposure to carcinogenic hazardous air pollutants: A national study. *Social Science & Medicine, 185*, 71–80.

Harris, C. D., & Ullman, E. L. (1945). The nature of cities. *The Annals of the American Academy of Political and Social Science, 242*, 7–17.

Hooks, G., & Smith, C. L. (2004). The treadmill of destruction: National sacrifice areas and native Americans. *American Sociological Review, 69*(4), 558–575.

Howell, J., & Elliott, J. R. (2018). Damages done: The longitudinal impacts of natural hazards on wealth inequality in the United States. *Social Problems*. https://doi.org/10.1093/socpro/spy016/5074453

Hoyt, H. ([1933] 2000). One hundred years of land values in Chicago: The relationship of the growth of Chicago to the rise of its land values, 1830–1933. : BeardBooks.

Hunter, L. M. (2000). The spatial association between U.S. immigrant residential concentration and environmental hazards. *International Migration Review, 34*(2), 460–488.

Hurley, A. (1995). *Environmental inequalities: Class, race, and industrial pollution in Gary, Indiana, 1945-1980*. The University of North Carolina Press.

Jennings, V., Gaither, C. J., & Gragg, R. S. (2012). Promoting environmental justice through urban green space access: A synopsis. *Environmental Justice, 5*(1), 1–7.

Jennings, V., Floyd, M. F., Shanahan, D., Coutts, C., & Sinykin, A. (2017). Emerging issues in urban ecology: Implications for research, social justice, human health, and Well-being. *Population and Environment, 39*(1), 69–86.

Karim, A., & Noy, I. (2016). Poverty and natural disasters: A meta-regression analysis. *Review of Economics and Institutions, 7*(2), 1–26.

Konisky, D. M., & Reenock, C. (2018). Regulatory enforcement, riskscapes, and environmental justice. *Policy Studies Journal, 46*(1), 7–36.

Kravitz-Wirtz, N., Crowder, K., Hajat, A., & Sass, V. (2016). The long-term dynamics of racial/ethnic inequality in neighborhood air pollution exposure, 1990–2009. *Du Bois Review, 13*(2), 237–259.

Kroll-Smith, S. (2018). *Recovering inequality: Hurricane Katrina, the San Francisco earthquake of 1906, and the aftermath of disaster*. University of Texas Press.

Kroll-Smith, J. S., & Couch, S. R. (1991). As if exposure to toxins were not enough: The social and cultural system as a secondary stressor. *Environmental Health Perspectives, 95*, 61–66.

Krysan, M., & Crowder, K. (2017). *Cycle of segregation: Social processes and residential stratification*. Russell Sage Foundation.

Kwan, M.-P. (2018). The limits of the neighborhood effect: Contextual uncertainties in geographic, environmental health, and social science research. *Annals of the Association of American Geographers, 108*(6), 1482–1490.

Landsbergis, P. A. (2010). Assessing the contribution of working conditions to socioeconomic disparities in health: A commentary. *American Journal of Industrial Medicine, 53*(2), 95–103.

Lefebvre, H. ([1974] 1991). *The production of space* (D. Nicholson-Smith, Trans.). : Wiley-Blackwell.

Leon, D. A. (2008). Cities, urbanization and health. *International Journal of Epidemiology, 37*(1), 4–8.

Levine, A. G. (1982). *Love canal: Science, politics and people*. D.C. Heath and Company.

Liévanos, R. S. (2015). Race, deprivation, and immigrant isolation: The spatial demography of air-toxic clusters in the continental United States. *Social Science Research, 54*, 50–67.

Lopez, R. (2002). Segregation and black/white differences in exposure to air toxics in 1990. *Environmental Health Perspectives, 110*(Supplement 2), 289–295.

Lipscomb, H. J., Loomis, D., McDonald, M. A., Argue, R. A., & Wing, S. (2006). A conceptual model of work and health disparities in the United States. *International Journal of Health Services, 36*(1), 25–50.

Loughran, K. (2014). Parks for profit: The high line, growth machines, and the uneven development of urban public spaces. *City & Community, 13*, 49–68.

Loughran, K., Elliott, J. R., & Wright Kennedy, S. (2019). Urban ecology in the time of climate change: Houston, flooding, and the case of Federal Buyouts. *Social Currents, 6*(2), 121–140.

Maantay, J. (2007). Asthma and air pollution in the Bronx: Methodological and data considerations in using GIS for environmental justice and health research. *Health & Place, 13*, 32–56.

Malin, S. A., & Ryder, S. R. (2018). Developing deeply intersectional environmental justice scholarship. *Environmental Sociology, 4*(1), 1–7.

Marlow, T., Frickel, S., & Elliott, J. R. (2019). Do legacy industrial sites produce legacy effects? Environmental inequality formation in Rhode Island's industrial core. *Sociological Forum*. In press.

McCormack, G. R., Toohey, A. M., & Hignell, A. D. (2010). Characteristics of urban parks associated with park use and physical activity. *Health and Place, 16*, 712–726.

McLeman, R. A., Dupre, J., Ford, L. B., Ford, J., Gajewski, K., & Marchildon, G. (2014). What we learned from the dust bowl: Lessons in science, policy and adaptation. *Population and Environment, 35*, 417–440.

Miller, D. S. (2016). Public trust in the aftermath of natural and na-technological disasters: Hurricane Katrina and the Fukushima Daiichi nuclear incident. *International Journal of Sociology and Social Policy, 36*(5/6), 410–431.

Mohai, P., & Saha, R. (2006). Reassessing racial and socioeconomic disparities in environmental justice research. *Demography, 43*(2), 383–399.

Mohai, P., & Saha, R. (2007). Racial inequality in the distribution of hazardous waste: A national-level reassessment. *Social Problems, 54*(3), 343–370.

Mohai, P., & Saha, R. (2015a). Which came first, people or pollution? A review of theory and evidence from longitudinal environmental justice studies. *Environmental Research Letters, 10*, 1–9.

Mohai, P., & Saha, R. (2015b). Which came first, people or pollution? Assessing the disparate siting and post-siting demographic change hypotheses of environmental injustice. *Environmental Research Letters*, 1–17.

Mohai, P., Pellow, D., & Timmons Roberts, J. (2009). Environmental justice. *Annual Review of Environment and Resources, 34*, 405–430.

Molotch, H. (1976). The city as a growth machine: Toward a political economy of place. *American Journal of Sociology, 82*(2), 309–332.

Morello-Frosch, R., & Jesdale, B. M. (2006). Separate and unequal: Residential segregation and estimated cancer risks associated with ambient air toxics in U.S. metropolitan areas. *Environmental Health Perspectives, 114*(3), 386–393.

Nascimento, K. R. D. S., & Alencar, M. H. (2016). Management of risks in natural disasters: A systematic review of the literature on NATECH events. *Journal of Loss Prevention in the Process Industries, 44*, 347–359.

National Recreation and Park Association. (2016). 'NRPA Americans' engagement with parks survey. Retrieved November 2, 2016 at http://www.nrpa.org/uploadedFiles/nrpa.org/Publications_and_Research/Research/Papers/Engagement-Survey-Report.pdf

Newkirk, V. R. II. (2016). Fighting environmental racism in North Carolina. *The New Yorker*, January 2016. Accessed online July 25, 2019 at https://www.newyorker.com/news/news-desk/fighting-environmental-racism-in-north-carolina

Noy, I., & Vu, T. B. (2010). The economics of natural disasters in a developing country: The case of Vietnam. *Journal of Asian Economics, 21*(4), 345–354.

Oakes, J. M., Anderton, D. L., & Anderson, A. B. (1996). A longitudinal analysis of environmental equity in communities with hazardous waste facilities. *Social Science Research, 25*, 125–148.

Pais, J. F., & Elliott, J. R. (2006). Race, class, and Hurricane Katrina: Social differences in human responses to disaster. *Social Science Research, 35*, 395–321.

Pais, J., & Elliott, J. R. (2008). Places as recovery machines: Vulnerability and neighborhood change after major hurricanes. *Social Forces, 86*(4), 1415–1453.

Pais, J., Crowder, K., & Downey, L. (2013). Unequal trajectories: Racial and class differences in residential exposure to industrial hazard. *Social Forces, 92*(3), 1189–1215.

Park, R. E. & Burgess, E. W. ([1925] 1967). *The city: Suggestions for investigation of human behavior in the urban environment* (M. Janowitz, Ed.). : University of Chicago Press.

Park, Y. M., & Kwan, M.-P. (2017). Multi-contextual segregation and environmental justice research: Toward fine-scale spatiotemporal approaches. *International Journal of Environmental Research and Public Health, 14*(2105), 1–19.

Pastor, J., Manuel, J. S., & Hipp, J. (2001). Which came first? Toxic facilities, minority move-in, and environmental justice. *Journal of Urban Affairs, 23*(1), 1–21.

Pelling, M. (2003). *The vulnerability of cities: Natural disasters and social resilience*. Routledge.

Pellow, D. N., & Park, L. S.-H. (2002). *The Silicon Valley of dreams: Environmental injustice, immigrant workers, and the high-tech global economy*. New York University Press.

Pellow, D. N. (2002). *Garbage wars: The struggle for environmental justice in Chicago*. MIT Press.

Pellow, D. N. (2018). *What is critical environmental justice?* Polity Press.

Picou, J. S., Marshall, B. K., & Gill, D. A. (2004). Disaster, litigation, and the corrosive community. *Social Forces, 82*(4), 1493–1522.

Picou, J. S., & Marshall, B. K. (2006, April). Katrina as a natech disaster. Paper presented at the annual meeting of the Southern Sociological Society. New Orleans, LA.

Picou, J. S., & Marshall, B. K. (2007). Katrina as a paradigm-shift: Reflections on disaster research in the Twenty-First Century. In D. L. Brunsma, D. Overfelt, & J. S. Picou (Eds.), *The sociology of Katrina: Perspectives on a modern catastrophe* (pp. 1–20). Rowman & Littlefield Publishers, Inc.

Prechel, H., & Zheng, L. (2012). Corporate characteristics, political embeddedness and environmental pollution by large U.S. corporations. *Social Forces, 90*(3), 947–970.

Preston, B. L. (2013). Local path dependence of U.S. socioeconomic exposure to climate extremes and the vulnerability commitment. *Global Environmental Change, 23*, 719–732.

Pulido, L. (2000). Rethinking environmental racism: White privilege and urban development in Southern California. *Annals fo the Association of American Geographers, 90*(1), 12–40.

Quarantelli, E. L. (1954). The nature and conditions of panic. *American Journal of Sociology, 60*(3), 267–275.

Quarantelli, E. L., & Dynes, R. R. (1977). Response to social crisis and disaster. *Annual Review of Sociology, 3*, 23–49.

Ritchie, L. A., & Gill, D. A. (2007). Social capital theory as an integrating theoretical framework in technological disaster research. *Sociological Spectrum, 27*(1), 103–129.

Rudel, T. K. (2013). *Defensive environmentalists and the dynamics of global reform.* Cambridge University Press.

Saha, R., & Mohai, P. (2005). Historical context and hazardous waste facility siting: Understanding temporal patterns in Michigan. *Social Problems, 52*(4), 618–648.

Schultz, J., & Elliott, J. R. (2013). Natural disasters and local demographic change in the United States. *Population Environment, 34*, 293–312.

Sharp, G., Denney, J. T., & Kimbro, R. T. (2015). Multiple contexts of exposure: Activity spaces, residential neighborhoods, and self-rated health. *Social Science & Medicine, 146*, 204–213.

Sicotte, D., & Swanson, S. (2007). Whose risk in Philadelphia? Proximity to unequally hazardous industrial facilities. *Social Science Quarterly, 88*(2), 515–534.

Simons, R. A. (1998). How many urban brownfields are out there? An economic base contraction analysis of 31 U.S. cities. *Public Works Management & Policy, 2*(3), 267–273.

Sister, C., Wolch, J., & Wilson, J. (2010). Got green? Addressing environmental justice in park provision. *GeoJournal, 75*, 229–248.

Smiley, K. T. (2019). A polluting creed: Religion and environmental inequality in the United States. *Sociological Perspectives, 62*(6), 980–1000.

Smiley, K. T. (2020). Social capital and industrial air pollution in metropolitan America. *The Sociological Quarterly.* https://doi.org/10.1080/00380253.2019.1711252

Smiley, K. T., Rushing, W., & Scott, M. (2016a). Behind a bicycling boom: Governance, cultural change and place character in Memphis, Tennessee. *Urban Studies, 53*(1), 193–209.

Smiley, K. T., Sharma, T., Steinberg, A., Hodges-Copple, S., Jacobson, E., & Matveeva, L. (2016b). More inclusive parks planning: Park quality and preferences for park access and amenities. *Environmental Justice, 9*(1), 1–7.

Smiley, K. T., Howell, J., & Elliott, J. R. (2018). Disasters, local organizations, and poverty in the USA, 1998 to 2015. *Population and Environment, 40*(2), 115–135.

Solnit, R. (2009).*A paradise built in hell: The extraordinary communicates that arise in disasters.* Penguin Books.

Spears, E. G. (2014). *Baptized in PCBs: Race, pollution, and justice in an all-American town.* University of North Carolina.

Suminski, R. R., Connolly, E. K., May, L. E., Wasserman, J., Olivera, N., & Lee, R. E. (2012). Park quality in racial/ethnic minority neighborhoods. *Environmental Justice, 5*(6), 271–278.

Tarr, J. A. (1996). *The search for the ultimate sink: Urban pollution in historical perspective.* University of Akron Press.

Taylor, D. E. (1997). American environmentalism: The role of race, class and gender in shaping activism 1820–1995. *Race, Gender & Class, 5*(1), 16–62.

Taylor, D. E. (2014). *Toxic communities: Environmental racism, industrial pollution, and residential mobility.* New York University Press.

Thomas, K. A., Elliott, J. R., & Chavez, S. (2018). Community perceptions of industrial risks before and after a toxic flood: The case of Houston and Hurricane Harvey. *Sociological Spectrum, 38*(6), 371–386.

Tierney, K. (2014). *The social roots of risk: Producing disasters, producing resilience.* Stanford University Press.

Tierney, K. (2019). *Disasters: A sociological approach.* Polity Press.

Tinsley, H., et al. (2002). Park usage, social milieu, and psychosocial benefits of park use. *Leisure Sciences, 24*, 199–218.

U.S. Government Accounting Office. 2004. *Brownfield redevelopment: Stakeholders report that EPA's program helps to develop sites, but additional measures could complement agency action.* Report to Congressional Requesters. GAO-05-94. Washington, DC. http://www.gao.gov/products/GAO-05-94. Retrieved June 2, 2015.

U.S. Governmental Accountability Office. (1983). *Siting of hazardous waste landfills and their correlation with racial and economic status of surrounding communities.* U.S. Governmental Accountability Office.

United Church of Christ. (1987). *Toxic waste and race in the United States: A national report on the racial and socio-economic characteristics of communities with hazardous waste sites.* United Church of Christ Commission for Racial Justice.

Valdmanis, R., & Gardner, T. (2017). Update 2-Harvey floods or damages 161 Texas Superfund sites – EPA. *Reuters.* Retrieved January 25, 2018 at https://www.reuters.com/article/storm-harvey-superfund/update-2-harvey-floods-or-damages-13-texas-superfund-sites-epa-idUSL2N1LJ0LL

Vaughan, K. B., Kaczynski, A. T., Wilhelm Stanis, S. A., Besenyi, G. M., Bergstrom, R., & Heinrich, K. M. (2013). Exploring the distribution of park availability, features, and quality across Kansas City, Missouri by income and race/ethnicity: An environmental justice investigation. *Annals of Behavioral Medicine, 45*(Supplement 1), S28–S38.

Vickery, J., & Hunter, L. M. (2016). Native Americans: Where in environmental justice research? *Society & Natural Resources, 29*(1), 36–52.

Walker, G., & Burningham, K. (2011). Flood risk, vulnerability, and environmental justice: Evidence and evaluation of inequality in a UK context. *Critical Social Policy, 31*(2), 216–240.

Wolch, J. R., Bryne, J., & Newell, J. P. (2014). Urban green space, public health, and environmental justice: The challenge of making cities 'just green enough.'. *Landscape and Urban Planning, 125*, 234–244.

Woods, R. (2003). Urban-rural mortality differentials: An unresolved debate. *Population and Development Review, 29*, 29–46.

Wu, W., & Dong, G. (2014). Valuing the 'green' amenities in a spatial context. *Journal of Regional Science, 54*(4), 569–585.

Zwickl, K., Ash, M., & Boyce, J. K. (2014). Regional variation in environmental inequality: Industrial air toxics exposure in U.S. cities. *Ecological Economics, 107*, 494–509.

Part VII

Conclusion & Reflections

Reflections on the Past, Present, and Future of Population-Environment Research

23

Barbara Entwisle

Abstract

This essay briefly describes the current state of population-environment research, highlighting some key developments that have occurred over the past 25 years and drawing on Handbook chapters to illustrate the state of the field now. With this as foundation, the essay then identifies and discusses promising directions for future research: (1) develop a more holistic account of population-environment interrelations by combining population size, composition, and components of change; (2) build bridges between the literature on climate change and the literature on pollution and other hazards; (3) incorporate social, economic, and cultural domains into a multidimensional conceptualization of the natural and social environment; (4) utilize a comparative perspective to better understand how and why population-environment interrelations vary from place to place; (5) further develop the tools needed to accomplish all of these

B. Entwisle (✉)
Department of Sociology and Carolina Population Center, University of North Carolina at Chapel Hill, Chapel Hill, NC, USA
e-mail: entwisle@unc.edu

tasks. Major advances over the past 25 years notwithstanding, there is still room for fresh ideas and innovation.

A *Handbook of Population and Environment* was unthinkable 25 years ago when the National Institute for Child Health and Human Development (NICHD) awarded its first grants in an "emerging area of research" called "Population and the Environment." As described in the Request for Applications (RFA), the goal was to establish a broad foundation of research on "the effect of population change on the environment, ... the effect of environmental change on factors such as fertility, mortality, migration, and distribution that determine population change, and ... the reciprocal influences of population and environmental change" in a "variety of physical settings worldwide" (https://grants.nih.gov/grants/guide/rfa-files/RFA-HD-95-002.html). Existing research at that time consisted mainly of cross-national studies (Bilsborrow, 1987; National Research Council, 1993), which while promising, were not designed to shed light on mechanisms. The intent of the RFA was to develop a series of case studies in which the generality of population-environment

© Springer Nature Switzerland AG 2022
L. M. Hunter et al. (eds.), *International Handbook of Population and Environment*, International Handbooks of Population 10, https://doi.org/10.1007/978-3-030-76433-3_23

interrelationships could be assessed, mechanisms explored, and the likely relevance of institutions and other contextual factors examined.

The *Handbook of Population and Environment* is testimony to the value and success of NICHD's investment a quarter of a century ago, as well as the vision, innovation, and hard work of population scientists before and since. There is now a full- fledged portfolio of data resources for the study of population and environment along with the tools for integrating these resources in a GIS. Building on early work in the Ecuadorean Amazon (Murphy et al., 1997), Chitwan Valley, Nepal (Axinn et al., 1991; Axinn & Fricke, 1996), Nang Rong, Thailand (Entwisle et al., 1996, 1998), and Wolong, China (Liu et al., 1999), among others, it is now routine to collect geographic information in surveys and demographic surveillance systems. Indeed, the Demographic and Health Surveys (DHS) Program collects locational coordinates in virtually all of its surveys. Cluster locations, displaced in such a way as to preserve confidentiality, have been made available for 60 countries so far, from Afghanistan to Zimbabwe (https://dhsprogram.com/data/available-datasets. cfm). As a parallel development, integrated coverages of land cover/land use, night lights, road networks, weather patterns, and the like have become broadly available, for multiple time points (e.g. Blumenstock et al., 2015; Bruederle & Hodler, 2018; Los, 2015). Further, in addition to data resources and tools, investment was made in training an interdisciplinary cadre of scientists poised to address critical questions in the study of population and environment (e.g., the National Science Foundation-funded Integrated Graduate Education and Research Training (IGERT) Program in Population and Environment at the University of North Carolina at Chapel Hill). Many are authors of chapters in this volume.

Thanks to these developments, there is now a considerable research record on population-environment relationships available to review. The *Handbook* does an excellent job covering this large and growing literature, laying the foundation for continued research in the future.

To build toward this future, I will highlight five "to do" items that remain, and that if addressed, would continue the forward momentum.

To begin, perhaps partly of necessity but also reflective of the scope and organization of the literature, *Handbook* chapters address parts of complex interdependencies, but not how they might fit together. For example, there are separate chapters on each of the components of population change: "Health and Mortality Consequences of Natural Disasters," "Land Use Change and Health," "Air Pollution, Health, and Mortality," "Heat, Mortality, and Health," "Environment and Fertility," "Building a Population-Relevant Research Agenda on Environmental Migration in Africa," and "Environmental Migration in Latin America." Of course, as every author would quickly acknowledge, population size, structure, and change are interrelated, whether at the national, regional, community, neighborhood, family or household level. More needs to be done to integrate the components of population size, composition, and change into a holistic account of population-environment interrelations.

Second, in contrast to the explicit identification of components of population size, structure and change in the RFA on "Population and the Environment," the purview of "environment" was not as well defined. The initial focus recommended in the RFA was "land use, flora and fauna, soil and water quality," all clearly aspects of the natural environment although a limited set from the standpoint of where the literature stands now. What was not foreseen 25 years ago was the emergence of a strong interest in hazards and natural disasters, especially in relation to climate change (although see Pebley, 1998). The literature on demographic responses to floods, droughts, heat waves, catastrophic storms, and sea-level rise is large and growing (e.g., Hauer, 2017; Klineberg, 2015; Raker, 2020; VanLandingham, 2017). Pollution is also a hazard, but interestingly, research on its interrelations with population size, structure, and change largely constitutes a separate stream in the population-environment literature (e.g., see the chapter by Brantley Liddle and Gregory Casey). One place the two streams of literature come together is in the study of water scarcity, which is related to climate change as well as the politics of its use (e.g., McDonald et al., 2011), and of water quality, related to agricultural practices, the use of antibiotics, infrastructure, and sanitation (e.g., Hopewell & Graham, 2014). As argued in the chapter by Stéphanie Dos Santos, Bénédicte Gastineau and Valérie Golaz, a blended approach looks promising.

Third, still needed is a multidimensional conceptualization of environment. While the natural environment is itself multidimensional, it is time to fully incorporate social, economic, and cultural domains as well (Entwisle, 2007). Doing so may be important for causal inference. As a simplified example, illness and mortality could be due to poverty, or to outdoor air pollution, but most likely both. Looking at one to the exclusion of the other, as is so often done in the literature, risks attributing effects of poverty to pollution, or pollution to poverty. Research incorporating both is just now beginning to emerge (e.g., Humphrey et al., 2019; Kravitz-Wirtz et al., 2018). As a related development, the concepts of locational and social vulnerability are increasingly intertwined in the hazards literature (Raker, 2020). As an additional point, it is important to understand the potential conditioning effects of the social, economic, or cultural environment. This role is illustrated in research on trapped versus displaced populations following disasters (Logan et al., 2016), where the relationship between household resources and migration can range from negative to positive. Sometimes, it is the least well off who *must* move; other times, it is the most well off who *can* move. Comparative analysis within and across countries has the potential to shed light on these differences as well as on population-environment relationships generally.

Fourth, a comparative perspective would also help bridge a tendency to treat questions about environmental effects and population impacts in lower and middle income countries (LMIC's) in ways that are separate and different from the way they are addressed in wealthy industrialized countries. At the risk of oversimplifying, with the exception of the important work of Elizabeth Fussell, Matthew Hauer, Mark VanLandingham,

and their colleagues, much of the research to date on the demographic consequences of climate change and the shocks and strains caused by it is set in the rural areas of LMICs (e.g., Call et al., 2019; Entwisle et al., 2020; Riosmena et al., 2018; Sellers & Gray, 2019). Frequently, this research is organized within a sustainable livelihoods framework especially appropriate for those settings. Again with exceptions, as Pebley (1998) noted two decades ago, research on the demographic effects of pollution is generally set in industrialized contexts. Much of it is organized around a susceptibility and exposure framework very familiar to epidemiologists (e.g., see chapter by Elliott and Smiley in this volume). It is past time to unite these research streams. A good place to focus is in the study of urban areas in LMICs, where hazards related to climate change (e.g., flooding and also water scarcity) coincide with hazards related to environmental pollution (water quality). The chapter contributed by Mark Montgomery and his colleagues provides a valuable starting point for future research along these lines.

Fifth and finally, compared to the blunt instruments available 25 years ago, there have been impressive developments in data and tools. Even so, the challenge of "putting people into place" is still not completely solved. For example, while addresses link houses (or other place of residence) to households (a social group), household members may cross multiple political and administrative boundaries in the course of their routine activities, whether it is to gather wood for fuel or to travel back and forth to school or work (Jagger & Heydrich, 2016; Browning et al., 2017; Elliott & Smiley, 2019). Whether one is interested in environmental exposure or environmental impact, the environmental unit of interest will extend beyond a specific address. Further complicating matters, it is rare that a socially relevant unit such as neighborhood or county maps directly to an environmentally meaningful unit such as a river basin. The boundaries do not generally translate. As a related point, in most instances, the territory represented by a population sample rarely generalizes to an environment, nor do the people associated with a sample of environmental units typically generalize to a population of interest. That is, unless special steps are taken to jointly reflect both in the definition of the target population (e.g., Moran et al., 2005).

To conclude, the RFA on "Population and the Environment" issued a quarter century ago serves as a useful benchmark for reflecting on the past, present, and future of population-environment research. Comparing then and now demonstrates the tremendous progress that has been made in addressing the environmental determinants of population size, structure, and change as well as the impact of these on such outcomes as deforestation and land use/land cover change, pollution and other environmental hazards, and water scarcity, to name a few. Demographers and others have assembled an impressive collection of data, tools, and cases and produced a growing set of findings informed by and informing a variety of theoretical and organizing frameworks. This *Handbook* is a worthy effort to survey these developments. All of this progress notwithstanding, more work remains to be done. This reflection has given particular attention to the need to further integrate the components of population size, structure, and change in the study of population-environment relationships; to incorporate multiple dimensions of environment, including the social, economic, and cultural environment; and to take full advantage of multiple contexts within and across environments for insights about the institutional setting as well as conclusions that can or cannot be made about the generality of results. There is still plenty of room for fresh perspectives and innovation.

References

Axinn, W. G., & Fricke, T. (1996). Community context, women's natal kin ties, and demand for children: Macro-micro linkages in social demography. *Rural Sociology, 61*, 249–271.

Axinn, W. G., Fricke, T. E., & Thornton, A. (1991). The microdemographic community-study approach: Improving survey data by integrating the ethnographic method. *Sociological Methods & Research, 20*, 187–217.

Bilsborrow, R. E. (1987). Population pressures and agricultural development in developing countries: A conceptual framework and recent evidence. *World Development, 15*, 183–203.

Blumenstock, J., Cadamuro, G., & On, R. (2015). Predicting poverty and wealth from mobile phone metadata. *Science, 350*, 1073–1076.

Browning, C. R., Calder, C. A., Ford, J. L., Boettner, B., Smith, A. L., & Haynie, D. (2017). Understanding racial differences in exposure to violent areas: Integrating survey, smartphone, and administrative data resources. *The Annals of the American Academy of Political and Social Science, 669*, 41–62.

Bruederle, A., & Hodler, R. (2018). Nighttime lights as a proxy for human development at the local level. *PLoS One, 13*, e0202231.

Call, M., Gray, C., & Jagger, P. (2019). Smallholder responses to climate anomalies in rural Uganda. *World Development, 115*, 132–144.

Elliott, J. R., & Smiley, K. T. (2019). Place, space, and racially unequal exposures to pollution at home and work. *Social Currents, 6*, 32–50.

Entwisle, B. (2007). Putting people into place. *Demography, 44*, 687–703.

Entwisle, B., Rindfuss, R. R., Guilkey, D. K., Chamratrithriong, A., Curran, S. R., & Sawangdee, Y. (1996). Community and contraceptive choice in rural Thailand: A case study of Nang Rong. *Demography, 33*, 1–11.

Entwisle, B., Walsh, S. J., Rindfuss, R. R., & Chamratrithirong, A. (1998). Landuse/Landcover (LULC) and population dynamics, Nang Rong, Thailand. In D. Liverman, E. F. Moran, R. R. Rindfuss, & P. C. Stern (Eds.), *People and pixels: Using remotely sensed data in social science research* (pp. 121–144). National Academy Press.

Entwisle, B., Verdery, A. M., & Williams, N. (2020). Climate change and migration: New insights from a model of out-migration and return migration. Forthcoming in the *American Journal of Sociology*.

Hauer, M. E. (2017). Sea level rise induced migration could reshape the United States population landscape. *Nature Climate Change, 7*, 321–325.

Hopewell, M. R., & Graham, J. P. (2014). Trends in access to water supply and sanitation in 31 major sub-Saharan African cities: An analysis of DHS data from 2000 to 2012. *BMC Public Health, 14*, 208.

Humphrey, J. L., Reid, C. E., Kinnee, E. J., Kubzansky, L. D., Robinson, L. F., & Cloughtery, J. E. (2019). Putting co-exposures on equal footing: An ecological analysis of same-scale measures of air pollution and social factors on cardiovascular disease in New York City. *International Journal of Environmental Research and Public Health, 16*, 4621.

Jagger, P., & Perez-Heydrich, C. (2016). Land use and household energy dynamics in Malawi. *Environmental Research Letters, 11*, 125004.

Klineberg, E. (2015). *Heat wave: A social autopsy of disaster in Chicago* (2nd ed.). University of Chicago Press.

Kravitz-Wirtz, N., Teixeira, S., Hajat, A., Woo, B., Crowder, K., & Takeuchi, D. (2018). Early-life air pollution exposure, neighborhood poverty, and childhood asthma in the United States, 1990-2014. *InternationalJournal of Environmental Research and Public Health, 15*, 1114.

Liu, J., Ouyang, Z., Tan, Y., Yang, J., & Zhang, H. (1999). Changes in human population structure: Implications for biodiversity conservation. *Population and Environment, 21*, 45–58.

Logan, J. R., Issar, S., & Zengwang, X. (2016). Trapped in place? Segmented resilience to hurricanes in the Gulf Coast, 1970–2005. *Demography, 53*, 1511–1534.

Los, S. O. (2015). Testing gridded land precipitation data and precipitation and runoff reanalyses (1982–2010) between 45 S and 45 N with normalised difference vegetation index data. *Hydrology and Earth System Sciences, 19*, 1713–1725.

McDonald, R. I., Green, P., Balk, D., Fekete, B. M., Revenga, C., Todd, M., & Montgomery, M. (2011). Urban growth, climate change, and freshwater availability. *PNAS, 108*(15), 6312–6317.

Moran, E. F., Brondizio, E. S., & VanWey, L. (2005). Population and environment in Amazonia: Landscape, household dynamics. In B. Entwisle & P. C. Stern (Eds.), *Population, land use, and environment* (pp. 106–134). The National Academies Press.

Murphy, L., Bilsborrow, R., & Pichón, F. (1997). Poverty and prosperity among migrant settlers in the Amazon rainforest frontier of Ecuador. *Journal of Development Studies, 34*, 35–65.

National Research Council. (1993). *Population and land use in developing countries: Report of a workshop*. The National Academies Press.

Pebley, A. R. (1998). Demography and the environment. *Demography, 35*, 377–389.

Raker, E. J. (2020). Natural hazards, disasters, and demographic change: The case of severe tornadoes in the United States, 1980–2010. *Demography, 57*, 653–674.

Riosmena, F., Nawrotzki, R., & Hunter, L. (2018). Climate migration at the height and end of the Great Mexican Emigration Era. *Population and Development Review, 44*, 455.

Sellers, S., & Gray, C. (2019). Climate shocks constrain human fertility in Indonesia. *World Development, 117*, 357–369.

VanLandingham, M. J. (2017). *Weathering Katrina: Culture and recovery among Vietnamese Americans*. Russell Sage.

Environmental Migration Scholarship and Policy: Recent Progress, Future Challenges

24

Robert McLeman

Abstract

This chapter reviews the evolution of environmental migration research over recent decades and summarizes key conceptual developments. It identifies priority areas for current and future research, with a particular focus on research that supports policymaking. Environmental migration processes are much more complex than once thought, and there is more still to learn about drivers and connections with broader socio-economic processes. Particular attention should be given to linkages between climate adaptation and labour migration flows, and identification of thresholds at which adaptation switches from in situ options to migration. The gender dimensions of environmental migration are especially understudied. Existing international agreements provide reliable policy guidance – particularly the Global Compact on Safe and Orderly Migration – but have yet to be widely implemented, providing impetus for continued scholarly efforts to identify how migration

patterns are likely to evolve in a climate-disrupted future.

The first scholarly article I ever read on the topic of environmental migration was by Swedish meteorologist Bo Döös (1994). It warned that global population growth was rapidly outstripping global food production,[1] and identified a number of scenarios under which future involuntary migrations and population displacements would occur, primarily in developing countries. It was published at a time of rapid expansion in the volume and quality of environmental migration research, growth that continues to this day. Between the infamous and forgettable 'environmental determinism' research of the early twentieth century[2] and the 1994

R. McLeman (✉)
Wilfrid Laurier University, Waterloo, ON, Canada
e-mail: rmcleman@wlu.ca

---

[1] In doing so, the author echoed similar, earlier neo-Malthusian warnings made by Paul Ehrlich, the Club of Rome, the WorldWatch Institute, and others of an impending global food crisis.

[2] See McLeman and Gemenne (2018) for a detailed review.

© Springer Nature Switzerland AG 2022
L. M. Hunter et al. (eds.), *International Handbook of Population and Environment*, International Handbooks of Population 10, https://doi.org/10.1007/978-3-030-76433-3_24

Döös article, only a smattering of environmental migration research was published (see Hunter, 2005), with little of it focussed on environmental challenges that preoccupy us today, such as deforestation, land degradation, and climate change.

When Döös published his 1994 study, the ink was barely dry on United Nations conventions on biodiversity protection, climate change, and desertification, and scholars and policymakers alike were trying to figure out the human impacts of these newly recognized global environmental challenges. The resulting environmental migration scholarship was framed using security language left over from the Cold War, and featured normative warnings of millions of "environmental refugees" to come (El-Hinnawi, 1985; Westing, 1992; Homer-Dixon, 1994; Myers & Kent, 1995; Richmond, 1995; Dabelko, 1999). Since then, environmental migration has expanded into a broad sub-disciplinary field that engages researchers from multiple disciplines, using a range of theoretical, conceptual, and empirical approaches. The data and tools available to Döös and his contemporaries were crude; with a few clicks today's researcher can download complete climate and population datasets, and enter these into Geographical Information Systems software to generate high resolution maps and models of environmental migration past, present and future. Along the way, the scholarly paradigm has shifted from warnings of environmental refugees to a focus on socio-economic adaptation to environmental challenges (McLeman & Smit, 2006; Black, 2011; Adger et al., 2015). The policy environment has also evolved. The 2016 Paris Agreement instructed policymakers to seek guidance on involuntary displacements due to climate change, and the 2018 United Nations 'Global Compact for Safe, Orderly and Regular Migration' sets out objectives for orderly management of environmental migration.

Scholars now realize that environmental migration processes are much more complex and variable than the old environmental refugee paradigm would have had us believe. An environmental event or condition that stimulates a particular type of migration in location A might have an entirely different effect in location B, depending on the cultural, demographic, economic, political and social contexts of each location (McLeman, 2014). As populations adjust and adapt over time, environmental events or conditions that once stimulated a particular migration response in a given region may no longer do so. But as we look to the future, there is valid concern that rapid modification of the climate and degradation of lands and forests may amplify existing migration flows and generate new ones, potentially displacing people in numbers never before seen (Rigaud et al., 2018).

Although we know quite a lot about the environmental influence on migration, there is still much more yet to learn, in terms of drivers, mechanisms, connections with other socio-economic processes, and broader implications. A first need is for greater work on what is described by the Intergovernmental Panel on Climate Change as 'detection and attribution'. We have an abundance of research showing that migration patterns can potentially be influenced by or associated with changes in environmental conditions, but environmental factors alone rarely determine migration decisions or outcomes; these are invariably filtered through socio-economic, cultural and other non-environmental processes (Black et al., 2011). A question of great importance to planners and policymakers is the degree or weight of the environmental influence relative to other, non-environmental drivers. Such information is of considerable benefit in deciding how and where to channel scarce institutional resources, so that the most important drivers of involuntary migration and displacement receive priority attention. Few studies to date have made such attempts.

A related concern is that of the 'streetlight effect'. Many of the case studies from which we draw information about environmental migration processes are selected precisely because we know environmental migration happened there and then. We too often select our cases on the basis of the outcome and not the stimulus.[3] We have as much to learn about migration processes from places where an environmental event did not cause people to move as we do from places where it did.

Governments are keenly interested in forecasts of future population movements under various standardized climate change and socio-economic pathway scenarios. An example of this type of work is the World Bank's 'Groundswell' project which projected future climate-related internal migration for selected developing regions (Rigaud et al., 2018). A small number of other country- and regional-level forecasts have been made, but these typically have not employed the Representative Concentration Pathways (RCP) and Shared Socio-economic Pathways (SSP) scenarios encouraged by the IPCC (Moss et al., 2010; Riahi et al., 2017). Although there is no requirement that environmental migration scholars use these scenarios when considering the future, doing so makes the research more likely to be taken up and disseminated through the IPCC reporting process, and therefore potentially more likely to be noticed by decision makers and others outside the field. Similarly, policymakers are keenly interested in knowing more about thresholds – for example, how much drought or crop loss can farmers handle before migration becomes the best remaining adaptation option? How many residents can a given community lose before their departure starts affecting the adaptive capacity of those left behind? There has been some conceptual research in this direction (Meze-Hausken, 2008; Bardsley & Hugo, 2010; McLeman, 2017) and a few examples of empirical research (e.g. Nawrotzki et al., 2017), but more concerted efforts would be welcome.

There is a need for studying the gender dimensions of environmental migration, not as a separate area of inquiry, but as an integrated component of larger research projects. Scholars – myself included – are all too quick to submit that our research findings, be they quantitative or qualitatively derived, apply generally across the entire population we have studied, without really tackling differentiation within the population along lines of gender, age, social class. We know in practice that the costs and benefits, risks and reward of migration are never equally distributed, but we often do not go far enough in assessing this empirically.

My final observation concerns the current trend in high-income countries toward tighter border controls and greater restrictions on migration from low-income countries, which I have written about elsewhere in greater detail (McLeman, 2019). In a rapidly changing climate in which the earliest, most severe impacts will be experienced in low-income countries, the ability for people of working age to move out of highly exposed areas to seek employment and wages that can be remitted home will be an important means of adaptation. The previously mentioned Global Compact on Migration provides a well-designed framework for integrating migration and development policies that benefit both sending and receiving countries, while ensuring the agency and safety of migrants and their families. Unfortunately, the Compact is a non-binding guidance document. My own hope for the future is that the high quality environmental migration scholarship being done today will persuade western policymakers of the importance orderly, well managed international migration.

References

Adger, W. N., Arnell, N. W., Black, R., Dercon, S., Geddes, A., & Thomas, D. (2015). Focus on environmental risks and migration: Causes and consequences. *Environmental Research Letters, 10*(6), 060201.

Bardsley, D. K., & Hugo, G. J. (2010). Migration and climate change: Examining thresholds of change to guide effective adaptation decision-making. *Population and Environment, 32*(2–3), 238–262.

Black, R. (2011). Climate change: Migration as adaptation. *Nature, 478*, 447–449. https://doi.org/10.1038/478477a

Black, R., Adger, W. N., Arnell, N. W., Dercon, S., Geddes, A., & Thomas, D. (2011). The effect of environmental change on human migration. *Global Environmental Change, 21*(S1), S3–S11.

---

[3]This observation was made to me a decade ago by the accomplished scholar Myron Gutmann, and I have repeated it often.

Dabelko, G. D. (1999). The environmental factor. *The Wilson Quarterly, 23*(4), 14–19.

Döös, B. R. (1994). Environmental degradation, global food production, and risk for large-scale migrations. *Ambio, 23*(2), 124–130.

El-Hinnawi, E. (1985). *Environmental refugees*. United Nations Environmental Program.

Homer-Dixon, T. (1994). Environmental scarcities and violent conflict: Evidence from cases. *International Security, 19*(1), 5–40.

Hunter, L. M. (2005). Migration and environmental hazards. *Population and Environment, 26*(4), 273–302.

McLeman, R. (2014). *Climate and human migration: Past experiences, future challenges*. Cambridge University Press.

McLeman, R. (2017). Thresholds in climate migration. *Population and Environment, 39*(4), 319–338.

McLeman, R. (2019). International migration and climate adaptation in an era of hardening borders. *Nature Climate Change, 9*(12), 911–918.

McLeman, R., & Gemenne, F. (2018). Environmental migration research: Evolution and current state of the science. In R. McLeman & F. Gemenne (Eds.), *The Routledge handbook of environmental migration and displacement*. Routledge.

McLeman, R., & Smit, B. (2006). Migration as an adaptation to climate change. *Climatic Change, 76*(1–2), 31–53.

Meze-Hausken, E. (2008). On the (im-)possibilities of defining human climate thresholds. *Climatic Change, 89*(3–4), 299–324.

Moss, R. H., Edmonds, J. A., Hibbard, K. A., Manning, M. R., Rose, S. K., VanVuuren, D. P., Carter, T. R., Emori, S., Kainuma, M., Kram, T., Meehl, G. A., Mitchell, J. F. B., Nakicenovic, N., Riahi, K., Smith, S. J., Stouffer, R. J., Thomson, A. M., Weyant, J. P., & Wilbank, T. J. (2010). The next generation of scenarios for climate change research and assessment. *Nature, 463*, 747–756.

Myers, N., & Kent, J. (1995). *Environmental exodus: An emergent crisis in the global arena* (214 p). Climate Institute.

Nawrotzki, R. J., DeWaard, J., Bakhtsiyarava, M., & Ha, J. T. (2017). Climate shocks and rural-urban migration in Mexico: Exploring nonlinearities and thresholds. *Climatic Change, 140*(2), 243–258.

Riahi, K., et al. (2017). The shared socioeconomic pathways and their energy, land use, and greenhouse gas emissions implications: An overview. *Global Environmental Change, 42*, 153–168.

Richmond, A. H. (1995). The environment and refugees: Theoretical and policy issues. *Population Bulletin of the United Nations, 39*, 1–17.

Rigaud, K. K., de Sherbinin, A., Jones, B., Bergmann, J., Clement, V., Ober, K., Schewe, J., Adamo, S., McCusker, B., Heuser, S., & Midgley, A. (2018). *Groundswell: Preparing for internal climate migration*. World Bank.

Westing, A. H. (1992). Environmental refugees: A growing category of displaced persons. *Environmental Conservation, 19*(3), 201–207.

Printed in the United States
by Baker & Taylor Publisher Services